Advances in
Carbohydrate Chemistry and Biochemistry

Volume 39

Advances in Carbohydrate Chemistry and Biochemistry

Editors

R. STUART TIPSON

DEREK HORTON

Board of Advisors

LAURENS ANDERSON
STEPHEN J. ANGYAL
GUY G. S. DUTTON
ALLAN B. FOSTER
DEXTER FRENCH

BENGT LINDBERG
HANS PAULSEN
NATHAN SHARON
MAURICE STACEY
ROY L. WHISTLER

Volume 39

1981

ACADEMIC PRESS

A Subsidiary of Harcourt Brace Jovanovich, Publishers

New York   London   Toronto   Sydney   San Francisco

COPYRIGHT © 1981, BY ACADEMIC PRESS, INC.
ALL RIGHTS RESERVED.
NO PART OF THIS PUBLICATION MAY BE REPRODUCED OR
TRANSMITTED IN ANY FORM OR BY ANY MEANS, ELECTRONIC
OR MECHANICAL, INCLUDING PHOTOCOPY, RECORDING, OR ANY
INFORMATION STORAGE AND RETRIEVAL SYSTEM, WITHOUT
PERMISSION IN WRITING FROM THE PUBLISHER.

ACADEMIC PRESS, INC.
111 Fifth Avenue, New York, New York 10003

United Kingdom Edition published by
ACADEMIC PRESS, INC. (LONDON) LTD.
24/28 Oval Road, London NW1 7DX

LIBRARY OF CONGRESS CATALOG CARD NUMBER: 45–11351

ISBN 0–12–007239–4

PRINTED IN THE UNITED STATES OF AMERICA

81 82 83 84   9 8 7 6 5 4 3 2 1

CONTENTS

LIST OF CONTRIBUTORS vii
PREFACE ix

Karl Paul Gerhardt Link (1901–1978)

CLINTON E. BALLOU

Text 1

The Selective Removal of Protecting Groups in Carbohydrate Chemistry

ALAN H. HAINES

    I. Introduction 13
   II. Deacetalation 14
  III. Deacylation 28
  IV. Dealkylation 44
   V. Deborination and Deboronation 53
 VI. Denitration 55
VII. Dephosphonylation 58
VIII. De(trialkylsily)ation 62

The Reactivity of Cyclic Acetals of Aldoses and Aldosides

JACQUES GELAS

  I. Introduction 71
 II. Summary of Developments Concerning the Synthesis of Cyclic Acetals . . 73
III. Reactivity 81
IV. Conclusion 155

Synthesis and Polymerization of Anhydro Sugars

CONRAD SCHUERCH

  I. Introduction 157
 II. Synthesis of Anhydro-aldoses and -ketoses 160
III. Polymerization of Anhydroaldoses 173
IV. Summary and Prospects 211

The Chemistry of Maltose

RIAZ KHAN

 I. Introduction 214
II. Structure and Synthesis 214

III.	Glycosides	217
IV.	Ethers	219
V.	Esters	223
VI.	Nucleophilic-displacement Reactions	229
VII.	Anhydro Derivatives	235
VIII.	Cyclic Acetal Derivatives	237
IX.	Aminodeoxy Derivatives	239
X.	Miscellaneous Compounds	242
XI.	Physical Methods	249
XII.	Tables of Properties of Maltose Derivatives	263

Chemistry and Biochemistry of D- and L-Fucose

Harold M. Flowers

I.	Introduction	280
II.	Chemistry	283
III.	Fucose in Glycans	301
IV.	Immunological Aspects of Complex Fucans	311
V.	Metabolism	316
VI.	Clinical	329
VII.	Concluding Remarks	335
VIII.	Tables of Some Properties of the Fucoses and Their Derivatives	337

The Utilization of Disaccharides and Some Other Sugars by Yeasts

James A. Barnett

I.	Introduction	347
II.	Glycoside Structure and Hydrolysis	349
III.	The Utilization of Glycosides Hydrolyzed Outside the Plasmalemma	353
IV.	The Utilization of Glycosides Hydrolyzed Inside the Plasmalemma	378
V.	The Requirement of Oxygen for Utilizing Glycosides and D-Galactose	397
VI.	Addendum	401

Affinity Chromatography of Macromolecular Substances on Adsorbents Bearing Carbohydrate Ligands

John H. Pazur

I.	Introduction	405
II.	General Considerations	408
III.	Activation of Supports	412
IV.	Derivatives of Carbohydrates	419
V.	Coupling Reactions	430
VI.	Applications	437

Author Index . . . 449
Subject Index . . . 479

LIST OF CONTRIBUTORS

Numbers in parentheses indicate the pages on which the authors' contributions begin.

CLINTON E. BALLOU, *Department of Biochemistry, University of California, Berkeley, California 94720* (1)

JAMES A. BARNETT, *School of Biological Sciences, University of East Anglia, Norwich NR4 7TJ, England* (347)

HAROLD M. FLOWERS, *Department of Biophysics, The Weizmann Institute of Science, Rehovot, Israel* (279)

JACQUES GELAS, *École Nationale Supérieure de Chimie, Université de Clermont-Ferrand, Ensemble Scientifique des Cézeaux, BP 45, 63170 Aubière, France* (71)

ALAN H. HAINES, *School of Chemical Sciences, University of East Anglia, Norwich NR4 7TJ, England* (13)

RIAZ KHAN, *Tate & Lyle Limited, Group Research & Development, Philip Lyle Memorial Research Laboratory, P. O. Box 68, Reading, Berkshire RG6 2BX, England* (213)

JOHN H. PAZUR, *Paul M. Althouse Laboratory, The Pennsylvania State University, University Park, Pennsylvania 16802* (405)

CONRAD SCHUERCH, *Department of Chemistry, College of Environmental Science and Forestry, State University of New York, Syracuse, New York 13210* (157)

PREFACE

A subject of continuing interest to carbohydrate chemists, namely, the selective removal of protecting groups, is here discussed by A. H. Haines (Norwich) in an article that updates and extends those of the same author in Vol. 33 and of J. M. Sugihara in Vol. 8, which were devoted more to the selective introduction of protecting groups. Of such protecting groups, the cyclic acetals, particularly 1,3-dioxolanes and 1,3-dioxanes, have played a major role since 1895, when Emil Fischer first described such derivatives of glycoses; and articles by A. N. de Belder, in Vols. 20 and 34 (on the cyclic acetals of the aldoses and aldosides), by R. F. Brady, Jr., in Vol. 26 (on those of the ketoses), and by S. A. Barker and E. J. Bourne in Vol. 7 (on the acetals of tetritols, pentitols, and hexitols) have discussed their synthesis and utility as intermediates. However, in this volume, J. Gelas (Clermont-Ferrand) describes an entirely different aspect of their chemistry in which the cyclic acetal groupings themselves serve as functional groups in synthetic transformations. A subject of both academic and industrial importance is the synthesis and polymerization of anhydro sugars, and C. Schuerch (Syracuse) contributes an extremely valuable account of the breakthroughs achieved by German and Latvian chemists, and in his own laboratory, in the application of cationic polymerization in this field. He also brings articles on the synthesis of anhydro sugars, by S. Peat (Vol. 2), R. J. Dimler (Vol. 7), N. R. Williams (Vol. 25), J. Defaye (Vol. 25), and M. Černý and J. Staněk, Jr. (Vol. 34), up to date. An article by R. Khan (Reading) constitutes an all-encompassing treatment of the chemistry of maltose, a disaccharide whose industrial utility as a raw material has yet to be realized. Both enantiomers of fucose are widespread in nature, and developments in their chemistry and their metabolism and biochemistry are treated by H. M. Flowers (Rehovot). In Vol. 32, J. A. Barnett (Norwich) surveyed the utilization of the monosaccharide components of the common glycosides by yeasts, and he now extends this to an examination of their utilization of disaccharides. Some attention had been devoted by J. F. Kennedy (Vol. 29) to affinity-chromatography matrices derived from polysaccharides, and the subject is discussed extensively in this volume by J. H. Pazur (University Park). A highly interesting obituary article by C. E. Ballou (Berkeley) describes the career of K. P. (G.) Link and his contributions to carbohydrate chemistry.

The Subject Index was compiled by Dr. L. T. Capell.

Kensington, Maryland R. STUART TIPSON
Columbus, Ohio DEREK HORTON
March, 1981

Karl Paul Link

1901–1978

KARL PAUL GERHARDT LINK

1901–1978

Karl Paul Link (partly named after the German poet Paul Gerhardt) was born in the Lutheran parsonage of La Porte, Indiana, on January 31, 1901, and he died peacefully at his home on Willow Lane in Madison, Wisconsin, on November 21, 1978. He was the eighth of ten children born to the Reverend and Mrs. George Link. When Karl was 11, his father died, and the children banded together to provide for the family. This shared responsibility was accepted by Karl, even through his school years, and he continued to send money home while he supported himself at the University of Wisconsin, where he received the Bachelor of Science degree in 1922 and the Doctorate in 1925, both in Agricultural Chemistry.

Karl grew up in a stimulating family environment with five brothers and four sisters, all of whom entered professional or business careers, or gave their time freely to service activities. While he was still in high school, Karl's aptitude for chemistry revealed itself—when the chemistry teacher was called away for service in the First World War, Karl was asked to teach his fellow students until a new teacher could be found. His choice to study Agricultural Chemistry in the University appears to have been determined in part by finances, because Link once said that, if he had had just a bit more money, he would have entered Medical School.

In retrospect, it seems probable that the medical profession gained far more from Link's successes as an agricultural chemist than it could have from his practice of medicine. The discovery of Dicumarol and its adoption as an important clinical anticoagulant, and the synthesis of Warfarin and its development as a widely used rodenticide—these accomplishments, which will not be described in this article, alone were more than sufficient to make Karl Link an important name in science. For his work on Dicumarol, Link received the Cameron Award from the University of Edinburgh in 1952, the Lasker Award in 1955 and 1960, the John Scott Award in 1959, and the Kovalenko Medal of the National Academy of Sciences in 1967. He was a Harvey Lecturer in 1944. In addition, of course, he was a carbohydrate chemist of note, and to that part of his career the remainder of this article is devoted.

Link's contribution to carbohydrate chemistry can be reviewed as two distinct, although interconnected, periods in his career that spanned 50 years. His interest in carbohydrates clearly began while he was in graduate school at Wisconsin, and it originated from studies on the effects of the methods of desiccation on the carbohydrates of plant tissue, the subject of his doctoral dissertation carried out under the supervision of W. E. Tottingham. This led naturally to a concern with plant polysaccharides that centered, eventually, on the hexuronic acids and led him and his coworkers to pioneer in the preparation and study of a large number of these important sugars and their derivatives. This work was published from about 1930 to 1940, after which time, Link's efforts became more and more concerned with the hemorrhagic, sweet-clover disease of cattle. This is a disease of cattle caused by their eating improperly stored, sweet-clover hay, in which the inadvertent growth of mold leads to the formation of formaldehyde that reacts spontaneously with nontoxic 4-hydroxycoumarin, a normal constituent of the clover, to produce 4,4'-methylenebis(4-hydroxycoumarin), a potent blood-anticoagulant. Such spoiled hay, consumed by the cattle in small amounts over a period of several days or a few weeks, causes a diminution in the prothrombin level of the serum with a loss in the clottability of the blood, resulting in internal hemorrhage and, often, death of the animal. The second period of carbohydrate research ran from about 1945 to 1953, and it grew out of the observation that D-glucosides of 4-hydroxycoumarins are alkali-sensitive. Several publications from this time dealt with systematic efforts to relate glycosidic structure to reactivity under alkaline conditions. These various studies on carbohydrates will be reviewed in detail in the following paragraphs.

Link received his Ph.D. degree from the University of Wisconsin in 1925, and he left for two years in Europe as a Fellow of the International Education Board. During this stay, he worked with Sir James C. Irvine at the University of St. Andrews in Scotland, with Paul Karrer at the University of Zürich in Switzerland, and with Fritz Pregl at the University of Graz in Austria. These postdoctoral experiences had a strong influence on him, and he returned to America impressed with the excellence of the practice of carbohydrate chemistry in Britain, and the rigor of the Germanic approach to organic chemistry, and with a conviction of the value of the microanalytical methods being developed by Pregl and others. He was so assured of the latter that he brought back with him the instruments and equipment for a complete analytical laboratory. This micro lab developed over the years into a kind of "inner sanctum" around which the fortunes of the laboratory centered. When Link gave up doing the analyses himself and turned the assignment over to a senior graduate student, not just anyone was allowed the privilege of doing the C and H determinations for the lab. The most important qualification was to have "good hands"; not only should they be clean and steady, they should be naturally "dry," because the Pregl procedures required a precise and reproducible handling of the combustion tubes and balances. The wiping of the tubes with chamois was a ritual as fixed as that of a Japanese tea ceremony, and only the person with good hands could do a reliable job.

In the early period of carbohydrate research, the names of four graduate students who worked with Link stand out—these are Carl Niemann, Sam Morell, Harold Campbell, and Stanford Moore. Before describing the research collaborations of these four students with Link, the stage should be set. The Biochemistry Department at Wisconsin was in the College of Agriculture, and the research emphasis was on agricultural problems. Link was appointed an Assistant Professor in 1927 on his return from Europe, and he undertook "an investigation of corn seedling blight . . . to correlate chemical composition with disease resistance." As a preliminary study, Link analyzed Zea mays xylan and cellulose, and concluded that they were similar to those polysaccharides from other plants. Although he had a brief interest in the occurrence of protocatechuic acid in onions and its possible relation to disease resistance, he soon settled on a study of polysaccharide structure and composition. The need for a good method of estimating uronic acids became apparent, and Link, familiar with the report by Lefevre and Tollens (1892) that heating D-glucuronic acid with concentrated hydrochloric acid yielded 2-furaldehyde and a quantitative amount of carbon dioxide, investigated the method systematically. He soon found that, by using a higher temperature for a shorter period of time, he could get good quantitation. The method was further improved (A. D. Dickson, H. Otterson, and Link, 1930), and converted to the 10-milligram scale by applying the microanalytical methods that Link had learned from Pregl.

At this time, Link was also investigating methods for the preparation of glycuronic acids, and he developed (with Dickson, 1930) an excellent procedure for the isolation of D-galacturonic acid in 20% yield by hydrolysis of lemon pectic acid obtained from the California Fruit Growers Exchange. General interest in hexuronic acids was growing rapidly, in part owing to their isolation from pneumococcal polysaccharides (Heidelberger and Goebel, 1927), and their use in the characterization and synthesis of vitamin C (L-ascorbic acid). Link (1932) also described a preparation of methyl D-galactosiduronic acid, and E. Schoeffel and Link (1933) reported isolation of crystalline forms of both anomers of D-mannuronic acid.

Carl Niemann was an undergraduate student at Wisconsin when he became associated with the Link lab. Niemann's first paper with Link, published in the *Journal of the American Chemical Society* in 1930, while he was still an undergraduate student, concerned the degradative effect on D-glucuronic and D-galacturonic acids of the dilute acid normally employed to hydrolyze polysaccharides in order to determine their sugar compositions. His later research as a graduate student involved the synthesis of a number of hexuronic acids by reduction of the saccharic (aldaric) acids with sodium amalgam. In 1891, Fischer and Piloty had used this method to prepare D-glucuronic acid, and they had noted, but not isolated, an intermediate in the reduction of mucic acid (galactaric acid) to DL-galactonic acid that would correspond to DL-galacturonic acid. Niemann and Link (1932) repeated this reduction under controlled conditions, and isolated the DL-galacturonic acid as a crystalline product. The same procedure was then applied for the preparation of D-mannuronic acid (Niemann and Link, 1933), L-mannuronic acid (Niemann, R. J. McCubbin, and Link, 1934), DL-alluronic acid (Niemann, S. Karjala, and Link, 1934), and L-glucuronic acid (Niemann and Link, 1934). Niemann and Link (1934) also extended the work of Ohle and Berend (1925), who had prepared D-galacturonic acid by oxidation of 1,2:3,4-di-O-isopropylidene-α-D-galactopyranose with permanganate; the yield was improved and crystalline D-galacturonic acid was obtained. The same method was then used for preparing L-galacturonic acid (Niemann and Link, 1934). During these investigations, it was noted that preparations of D-glucuronic and D-mannuronic acid tended to crystallize preferentially as the free acids. This difference was attributed to differences in "stereochemical configurations," but no suggestion was made to rationalize the differences in more detail. This reticence to "push arrows" was a Link characteristic that stayed with him through all his scientific career, with only minor lapses. Niemann was involved with several other studies on glycuronic acids, including a confirmation of the structure of an aldobiouronic acid obtained from flaxseed mucilage that was shown to contain D-galacturonic acid (Niemann and Link, 1934), the resolution of DL-galacturonic acid as its brucine salts (Niemann and Link, 1934), the preparation of (p-bromophenyl)hydrazide derivatives of D-galacturonic and D-mannuronic acids (Niemann, Schoeffel, and Link, 1933), and the characterization of starch isolated from the woody tissue of the apple tree (Niemann, R. H. Roberts, and Link, 1935). This starch was found to be very similar to that obtained from other plant tissues.

Sam Morell, also an undergraduate student at the University of Wisconsin, began his graduate studies with Link in 1930. While Niemann was preparing hexuronic acids, Morell set out to prepare derivatives of D-galacturonic acid and study their properties. First extending Link's earlier work, Morell and Link (1933) published an improved preparation of methyl α-D-galactosiduronic acid, studied the kinetics of its acid hydrolysis (Morell and Link, 1934) to confirm the presumed pyranoid ring-form, and analyzed the partial methanolyzate of pectin (Morell, L. Baur, and Link, 1934) in order to demonstrate that it is a linear polymer, and to disprove the then-accepted idea (Ehrlich, 1932) that it was a cyclic tetrasaccharide. Crystalline methyl tetra-O-acetyl-D-galacturonates were prepared (Morell and Link, 1935), and these were converted into the peracetylated bromide and, thence, into methyl β-D-galactosiduronic acid in a process that appears to have been the first application of the Koenigs–Knorr synthesis to a hexuronic acid (Morell, Baur, and Link, 1935). This procedure was later used by Link and H. M. Sell (1938) to prepare D-galactosiduronic acids of cholesterol, sitosterol, and ergosterol, and by C. F. Huebner, R. S. Overman, and Link (1944) to prepare pregnanediol-3 β-D-glucosiduronic acid.

Harold Campbell entered the Link lab in 1935, and he took over the study of D-galacturonic acid derivatives from Morell. With Link (1937), he described the synthesis of D-galacturonic acid diethyl dithioacetal and, from it, methyl tetra-O-acetyl-*aldehydo*-D-galacturonate. Long before a synthesis of D-galactose dimethyl acetal had been published (Campbell and Link, 1938), Campbell had begun to work on the isolation of the hemorrhagic agent of sweet-clover disease. The latter studies eventually led to the preparation, in crystalline form, of the causative agent (Campbell and Link, 1941), and opened the way to its characterization as a coumarin derivative. R. S. Dimler and Link (1940) later described the isolation of both forms of the ethyl hemiacetal of methyl tetra-O-acetyl-*aldehydo*-D-galacturonate, and a study of their mutarotation.

The fourth student in this group, Stanford Moore, began his doctoral thesis research with Link in 1935. His work involved the development of a new, and still useful, procedure for oxidizing aldoses to the aldonic acids by potassium hypoiodite in methanol solution; the resulting aldonic acids were then converted into nicely crystalline benzimidazole derivatives that proved more suitable for the characterization of carbohydrates than the classical osazones. This method, developed by Moore and Link (1940), was extended to the characterization of hexuronic acids, as their bis(benzimidazole)s, after oxidation to the glycaric acid (R. Lohmar, Dimler, Moore, and Link, 1942), of lactic acid (Moore, Dimler, and Link, 1941; Dimler and Link, 1942), and of ribose, fucose, and digitoxose (Dimler and Link, 1943). During these studies, it was observed that heating pentose-derived benzimidazoles at 180° led to formation of 1,4-anhydrides, and it was established by synthesis of reference compounds that this occurred without inversion (Huebner and Link, 1950).

Work on the sweet-clover disease had been started by Link about 1933, and the isolation of the crystalline factor was reported, as already mentioned, in 1941. Carbohydrate chemistry was put aside for several years, but it gained a new life with an important observation by Huebner, Karjala, Sullivan, and Link (1944), who had prepared acetylated β-D-glucopyranosides of 4-hydroxycoumarins to be tested as analogs of the blood anticoagulant Dicumarol, which had by this time been shown to be 4,4′-methylenebis(4-hydroxycoumarin) (M. A. Stahmann, Huebner, and Link, 1941). The derivatives were prepared by the Koenigs–Knorr method, and the intermediates required deacetylation, which, in this instance, proved impossible without alkaline cleavage of the D-glucosidic bond. It was a striking observation that the cleavage occurred by alcoholysis if the reaction was performed in methanolic sodium methoxide, and that the products were the 4-hydroxycoumarin and methyl α-D-glucopyranoside; this implied that the reaction proceeded by a rearside attack by methoxide anion at C-1 of the D-glucose, with elimination of the aglycon anion, resulting in inversion of the anomeric configuration. This mechanism was substantiated when it was shown (L. Spero, C. E. Ballou, and Link, 1949) that β-D-glycopyranosides of 4-hydroxy-3-phenylcoumarin always yield methyl α-D-glycopyranosides, whereas α-D-glycopyranosides of 4-hydroxy-3-phenylcoumarin yield methyl β-D-glycopyranosides.

A general study of alkali-sensitive glycosides was then undertaken by Link. Whereas the 4-hydroxycoumarin glycopyranosides underwent alcoholysis by cleavage of the sugar–oxygen bond, theobromine β-D-glucopyranoside tetraacetate was degraded with the formation of theobromine methyl ether (6-methoxy-3,7-dimethyl-2-purinol) and free D-glucose (Ballou and Link, 1949). J. A. Snyder and Link (1953) observed a similar reaction on treatment of 4-(β-D-glucopyranosyloxy)-2,6-pyrimidinedione with methoxide ion in methanol. In other studies (Ballou and Link, 1950), it was found that enolic D-glucopyranosides are alkali-sensitive if sufficiently activated (acetoacetaldehyde β-D-glycopyranoside tetraacetate), but that deactivation of the carbonyl function could sufficiently stabilize the glycosidic bond (ethyl acetoacetate β-D-glycopyranoside tetraacetate) to permit catalytic deacetylation without cleavage of the glycosidic bond. The alkaline degradation by β-elimination of glycopyranosides of β-hydroxycarbonyl compounds had been reported (Kuhn and Löw, 1941), and this reaction was also investigated by Ballou and Link (1950). A clear correlation was observed between the electronegativity of the carbonyl function and the rate of alkaline degradation of the glycoside.

In 1950, M. Seidman and Link prepared o-nitrophenyl β-D-galactopyranoside as a service to J. Lederberg, then in the Genetics Department at the University of Wisconsin, who needed a chromogenic substrate for studies that he was initiating on the β-D-galactosidase of *Escherichia coli*. This experience brought Link's attention to the previously reported alkali-sensitivity of nitrophenyl glycosides and to the "anomalously" positive optical rotations shown by o-nitrophenyl β-D-glycopyranoside acetates. Snyder and Link (1952) studied the kinetics of degradation of such compounds, and found that the alkaline degradation is much faster than acid hydrolysis. They obtained data that were consistent with the mechanism of alkaline degradation, which is considered to involve loss of a phenolic anion, with intermediate formation of the 1,2-epoxide of the sugar.

Pigman (1944) had observed that o-nitrophenyl β-D-glucopyranoside tetraacetate has a large, positive, optical rotation in chloroform, whereas the p-nitrophenyl derivative gives a typical, negative value. When he found that the deacetylated o-nitrophenyl β-D-glucopyranoside also has a negative rotation (in water), he attributed the "abnormal" value for the tetraacetate to a steric interaction of the o-nitrophenyl group with acetyl groups on the sugar ring. Snyder and Link (1953) studied these compounds in more detail, and concluded that steric hindrance is present, even in the deacetylated compounds; they made the interesting observation that the deacetylated o-nitrophenyl glycopyranosides show a high temperature-coefficient of rotation that is not found for the p-nitrophenyl derivatives, suggesting that movement of the o-nitrophenyl group is restricted, even in the deacetylated compounds.

"Anomalous" optical rotations were also observed by Ballou and Link (1951) for the methyl acetoacetate β-D-glycopyranosides. Those having the trans orientation about the enolic glycosidic bond have high positive rotations, whereas the cis derivatives have rotations normal for β anomers. This was rationalized as being the result of restricted rotation of the aglycon, owing to an internal hydrogen-bond between the methyl ester group and a hydroxyl group on the sugar ring that is possible only in the trans derivative. This postulate has since been tested by a crystal-structure determination made by Ruble and Jeffrey (1973).

The unusual properties of the methyl acetoacetate *trans*-β-D-galactopyranoside were also apparent in its susceptibility to enzymic hydrolysis (Ballou, Snyder, and Link, 1952). Whereas the cis derivative was readily hydrolyzed and showed a K_m value of 0.5 mM with the β-D-galactosidase of *Escherichia coli*, the trans form was not attacked. Surprisingly, the trans compound inhibited the enzymic reaction, and had a K_i value of 75 μM, which indicated that it is bound to the enzyme better than the cis form. Thus, the optical rotation and the resistance to enzymic hydrolysis suggest that the compound retains an "abnormal" conformation in solution that is, perhaps, analogous to that determined for the crystals.

Several other papers related to carbohydrate chemistry were published by Link. Some work dealt with the synthesis of 2-(aminoacyl)amino-2-deoxy-D-glucose by the coupling of 1,3,4,6-tetra-O-acetyl-2-amino-2-deoxy-D-glucopyranose with N-substituted, amino acid chlorides, which gave the expected products (D. G. Doherty, E. A. Popenoe, and Link, 1953), and another, with the "anomalous" formation of urea derivatives of 2-amino-2-deoxy-D-glucose when the same reaction was attempted with N-substituted, amino acid azides (Popenoe, Doherty, and Link, 1953). The latter reaction was rationalized as the result of a Curtius rearrangement of the azide to afford a urea derivative. The synthesis of L-arabinose dimethyl acetal from the diethyl dithioacetal was reported by Huebner, R. A. Pankratz, and Link (1950), who prepared the 2,3:4,5-di-O-benzylidene dimethyl acetal intermediate, and then used hydrogenolysis to remove the benzylidene groups. Ballou, S. Roseman, and Link (1951) employed hydrogenolysis of benzyl α- and β-D-glycopyranoside acetates in ethyl ether to prepare the α- and β-D-glycopyranose acetates of a number of hexoses and pentoses. This permitted a direct comparison of the anomeric configurations of the glycoside and aldose, and it is also a useful, general method for preparing such compounds.

Work on carbohydrates almost ceased in the laboratory after 1954, when Roseman, Huebner, Pankratz, and Link published studies on the metabolism of 4-hydroxycoumarin in an effort to learn something about the fate of Dicumarol in animals. They showed that the β-D-glucosiduronic acid of this compound is secreted in the urine of dogs after it had been injected intravenously. The identity of the D-glucosiduronic acid was confirmed by its chemical synthesis from 4-hydroxycoumarin and methyl tetra-O-acetyl-α-D-glucopyranosyluronate chloride.

How are Karl Link's contributions to the field of carbohydrate chemistry to be judged, what was his professional standing, and how did he influence developments in the field? Fundamentally, he was interested in natural products from the viewpoint of an organic chemist, but he never considered himself to be a "modern" organic chemist, and he appears to have made little effort to apply the tools of physical organic chemistry to carbohydrates. Link did not often go to scientific meetings to present the latest word on his carbohydrate research, although he did occasionally send students to the meetings of the American Chemical Society. His major contact with other professionals was at the Starch Round Table organized by the Corn Industries Research Foundation. These unusual meetings of academic and industrial scientists, a forerunner of what the Gordon Conferences now strive to be, gave Link an ideal stage, and he made the most of it. On my first attendance at the Round Table, the chance to associate with famous carbohydrate chemists from around the world made an impression on me, but I found that Karl Link was clearly one of the most highly respected members of the establishment. On his 70th birthday, he was honored, along with his friends Zev Hassid and Horace Isbell, at the 6th International Symposium on Carbohydrate Chemistry (1972) as one of three outstanding, senior American carbohydrate chemists.

Karl Link was recognized as an authority on pectin [D-galacturonan, poly(D-galactosiduronic acid)], and, in a review, he wrote of the book "The Pectic Substances," by Z. I. Kertesz (1951), the following paragraph that reveals his feelings for the subject and gives a hint of his characteristic crustiness. After noting the summary mention of pectin in Baldwin's "Dynamic Aspects of Biochemistry," 2nd edition (1952), and the lack of any mention of D-galacturonic acid, he commented: "After the ABC's of the aliphatics, aromatics, etc., our chemists are fed a classroom diet rich in hormones, vitamins, carcinogenic agents, antibiotics and enzymes. Pectin—thank heavens—is acquired in a natural form through the vegetables, fruits and fruit juices they eat. Who knows, perhaps the physical health of our students is maintained in part through 'an apple a day keeps the doctor away.' Apples are rich in pectin."

Perhaps the clearest influence that Link exerted was in his teaching of carbohydrate chemistry, or, in modern terminology, it might better be called "carbohydrate chemistry and society." His formal teaching was factual and descriptive, as might be expected, but it was never dull or uninteresting. With the chalk board full of detailed reactions describing the interconversions of the hexoses, written in a precise and careful hand well before the class began, he then had time to sketch the "personality" of science as he related the important facts concerning the follies of mankind.

Of the 55 doctoral candidates whom Link supervised, over 30 worked partly or wholly on carbohydrates. Life in the laboratory was centered around the regular, Friday lab cleanup and research conference. To perform poorly in either activity was something to be avoided. In research conferences, Link aimed to help rather than embarrass. The help could come in unexpected ways. I recall an occasion on which I was frustrated in trying to find a solvent in which benzyl 2,3,4,6-tetra-O-acetyl-β-D-glucoside could be hydrogenolyzed with H_2–Pd without subsequent mutarotation of the free 2,3,4,6-tetra-O-acetyl-β-D-glucose. After I related all of my unsuccessful efforts, Link said, "Try ethyl ether." At the time, this seemed like an invitation to disaster, to mix such a flammable solvent with palladium–hydrogen catalyst, which reacts violently on exposure to oxygen. However, the suggestion proved to be a good one and the problem was solved.

Link married Elizabeth Feldman in 1930. They had 3 sons and, as of this writing, 5 grandchildren. Early in their life together, the Links obtained a large plot of land in The Highlands near Madison, where they built a lovely home with a sweeping lawn and garden that sloped off toward Lake Mendota in the far distance. This provided an ideal place for entertaining indigent graduate students, and the Links were generous with their hospitality. The only hazard was that, after a strenuous game of softball on a particularly hot Wisconsin summer day, one could consume too much cold beer. Even this, however, could be remedied by a sleep under a nearby shade tree with one's face in the cool grass.

Karl loved the home that he and Elizabeth built on Willow Lane. He was usually up early in the morning, monitoring the movements of wild rabbits, and pheasants and other birds. After he retired from the University in 1971, Link particularly liked to conduct experiments with different mixtures of bird feed to assess their qualities in carrying the birds through the often harsh Wisconsin winters. During these later years, he also enjoyed the visits from former students and other friends, and he would converse for hours and even days if the visitor had the stamina to continue. It was an exhilarating and renewing experience for Karl and for his visitors.

Link was impressive in both appearance and demeanor. He projected a size that exceeded his real physical dimensions. His casual dress conveyed the air of a relaxed country squire completely at ease in a formal and distrustful world. A well-worn black herringbone-tweed jacket, flannel shirt, large knotted tie, and unpressed grey trousers were usually accompanied by a wide-brimmed hat, and brown work-shoes suited for tramping the countryside as well as the halls of academia. His somewhat swarthy features and long flowing hair added to the image, intended or not, of a character actor, and he once confided that he might have had a successful career on the stage had he chosen acting instead of science. In any event, one sensed that he was always "on stage" and was aware of an obligation to pique or stimulate the audience. There are as many "Link stories" as he had acquaintances, and these anecdotes I leave for others to recount. It is enough to say that, when one was with Karl Link, there was the sense of being in a charged environment with high potential for pleasure and intellectual exchange.

Karl once related the family story that, when he was two, he became seriously ill with pneumonia and was not expected to survive. When he did, his mother declared, "The Lord has spared Karl so that he can do something great for humanity." Although spared, Link was encumbered with fragile health, in spite of his robust appearance, such that, in middle life, he was stricken with tuberculosis and spent many months in a sanitorium in the winter of 1945 and most of 1946, and again in 1958-1959 for about 6 months. Even during those discouraging periods, however, Link's character showed through. It is not unexpected that, at such a time, a scientist would want to learn about his disease, but what Link concluded from his readings was not so expected. He put his thoughts in writing in an article entitled "As seen from 214," which was a criticism of the practices at Lake View Sanitorium and a suggestion that, with the help of a proper environment, the tubercular patient would usually cure himself. Link quoted from an article in *The Lancet* (1932) by British physician A. Morland who said, "The patient who shows a proper combination of courage and caution, adaptability to a new environment, and the capacity to form fresh interests when the old ones are inadvisable, has a good prospect of recovery even when the disease is of moderate severity." This was the advice Link took, and he followed it for over 30 years as he fought off the ever present danger of a relapse. His death, the result of circulatory disease, was painless.

Writing this article stimulated me to look back at my own doctoral dissertation, prepared under the tutelage of K. P. L., and I note the following acknowledgment: "The author is sincerely thankful for the opportunity to work in this excellent laboratory with such fine people. The former is directly due to the continued efforts of Professor Karl Paul Link in behalf of his students, the latter to his discerning selection of men. These will remain unforgettable years." If I were to restate my feelings in this regard, I would change only one phrase, to read "his discerning selection of men and women." One of Link's students, Dorothy Lun Wu, and I were married in the home of Karl and Elizabeth in December of 1949.

It is not a happy moment when writing on the passing of a close friend, colleague, and mentor. In this instance, the sadness is tempered by the recognition that Karl Link had a long, creative, productive, and eminently successful life. He left his mark on science, and he left it on the people who knew him and on the many who had only heard his name. The farmers of Wisconsin probably recognized his true worth better than any others, and it was their judgment he valued most. But the medical profession was generous in its praise, and his students and the general public will not forget.

CLINTON E. BALLOU

THE SELECTIVE REMOVAL OF PROTECTING GROUPS IN CARBOHYDRATE CHEMISTRY

By Alan H. Haines

School of Chemical Sciences, University of East Anglia, Norwich, England

I. Introduction	13
II. Deacetalation	14
1. Hydrolysis	14
2. Acetolysis	24
3. Acetal Isomerization	26
III. Deacylation	28
1. Selective Deacylation of Derivatives Containing the Same Type of Acyl Group	28
2. Selective Deacylation of Derivatives Containing Different Types of Acyl Groups	34
3. Selective Deacylation in Nucleoside Derivatives	36
IV. Dealkylation	44
V. Deboronation and Deboronation	53
VI. Denitration	55
VII. Dephosphonylation	58
VIII. De(trialkylsilyl)ation	62

I. Introduction

The subject of relative reactivities of hydroxyl groups in carbohydrates has been discussed previously in this Series.[1,2] In these articles, emphasis was placed on the selective *introduction* of substituents into carbohydrates. A much less exploited approach for the preparation of partially substituted carbohydrates involves the selective *removal* of hydroxyl-protecting groups from carbohydrate derivatives. The purpose of the present article is to draw attention to this relatively neglected aspect of synthesis in the belief that the greater use of de-

---

(1) A. H. Haines, *Adv. Carbohydr. Chem. Biochem.*, 33 (1976) 11–109.
(2) J. M. Sugihara, *Adv. Carbohydr. Chem.*, 8 (1953) 1–44.

protection procedures will significantly increase the armory of the worker in synthetic carbohydrate chemistry.

This article is divided into Sections, based on the type of derivative that is subjected to deprotection. The literature was surveyed up to December 1979.

II. DEACETALATION

In this Section, hydrolysis, acetolysis, and isomerization of acetals are considered. Selective deprotection of acetals may also be achieved through halogenation, hydrogenolysis, ozonolysis, and photolysis, but these topics, are covered in an accompanying article in this Series, on the reactivity of cyclic acetals of aldoses and aldosides,[3] and will not be discussed here.

1. Hydrolysis

Of all the selective, deprotection procedures that are available to carbohydrate chemists, the partial hydrolysis of polyacetals is probably the most familiar. Articles by de Belder[4,5] and Brady[6] contained examples of this type of reaction for aldose and ketose derivatives, respectively, and an article by Barker and Bourne[7] gave useful information from the early literature on the graded, acid hydrolysis of acetal derivatives of polyols. A discussion of the stereochemistry of cyclic acetals of carbohydrates was included in an article by Mills [7a] and in one by Ferrier and Overend,[7b] and a survey of the formation and migration of carbohydrate cyclic acetals was made by Clode.[7c]

In so far as the relative rates of hydrolysis and acetolysis, or the migratory tendencies of acetal groups, reflect their relative, thermodynamic stabilities, the selective removal, and the isomerization, of such groups may often be understood in terms of their relative free-energies. An indication of the difference in free energy between two acetal groups may often be obtained by a consideration of their constitution and conformation. However, a reaction rate depends on the difference in free energy between reactants and the transition state for that reaction. Therefore, a rationalization of the relative reaction-rates of two acetal groups towards hydrolysis, acetolysis, or isomerization often re-

---

(3) J. Gelas, *Adv. Carbohydr. Chem. Biochem.*, 39 (1981) 71–156.
(4) A. N. de Belder, *Adv. Carbohydr. Chem.*, 20 (1965) 219–302.
(5) A. N. de Belder, *Adv. Carbohydr. Chem. Biochem.*, 34 (1977) 179–241.
(6) R. F. Brady, Jr., *Adv. Carbohydr. Chem. Biochem.*, 26 (1971) 197–278.
(7) S. A. Barker and E. J. Bourne, *Adv. Carbohydr. Chem.*, 7 (1952) 137–207.
(7a) J. A. Mills, *Adv. Carbohydr. Chem.*, 10 (1955) 1–53.
(7b) R. J. Ferrier and W. G. Overend, *Q. Rev. Chem. Soc.*, (1959) 13.
(7c) D. M. Clode, *Chem. Rev.*, 79 (1979) 491–513.

quires, additionally, a consideration of the relative stabilities of the transition states arising from the two acetal groups in these reactions, because the stabilities of such transition states can differ significantly.[7c]

Many examples of the selective hydrolysis of aldose diacetals are known that indicate that acetal groups incorporating the oxygen atom at the anomeric center are hydrolyzed less readily than those attached at other positions, but this is not invariably the case. Such selective hydrolyses can be understood in terms of the special, polar environment at the anomeric center[4,5] and the energetically favorable, bicyclic ring-system present in those compounds containing a 1,3-dioxolane ring cis-fused to a furanose (oxolane) ring.[7a,7d] Among the 1,2:5,6-di-O-isopropylidene acetals of the aldohexoses, those of D-glucose,[8] L-idose,[9] and D-allose[10] may be partially hydrolyzed to their respective 1,2-O-isopropylidene acetals, but it has been alleged that those of D-gulose[11,12] and D-galactose[12] cannot. The fact that 3-(S-methyldithiocarbonates) of the 1,2:5,6-diacetals of D-gulose and D-galactose, also, could not be selectively hydrolyzed,[12] and yet 3-deoxy-1,2:5,6-di-O-isopropylidene-α-D-*ribo*-hexofuranose gave[13] the 1,2-acetal in high yield on partial hydrolysis, led to questioning[12] of the suggestion[14] that electronegative substituents at O-3 of 1,2-O-isopropylidenealdoses stabilize the acetal group against hydrolysis. Furthermore, it was suggested[12] that the similar rates of hydrolysis of the 1,2- and 5,6-acetal groups in the D-*galacto* and D-*gulo* derivatives is a result of the capability of the initially formed, 5,6-diol grouping, which in both cases lies on the endo face of the fused, bicyclic system, of assisting delivery of a proton to O-1. However, doubt is thrown on this rationalization by the fact that both 3-O-benzoyl-1,2:5,6-di-O-isopropylidene-α-D-galactofuranose[15] and 3-O-benzyl-1,2:5,6-di-O-isopropylidene-α-D-gulofuranose[11] are readily hydrolyzed, to give 1,2-O-isopropylidene derivatives, and also by other work[16] which suggested that partial hydrolysis of 1,2:5,6-di-O-isopropylidene-α-D-galactofuranose is, indeed, possible.

(7d) B. Capon, *Chem. Rev.*, 69 (1969) 407–498.
(8) O. T. Schmidt, *Methods Carbohydr. Chem.*, 2 (1963) 318–325.
(9) R. Schaffer, *Natl. Bur. Stand. U.S. Tech. Note*, 427 (1967) 60–63.
(10) O. Theander, *Acta Chem. Scand.*, 18 (1964) 2209–2216.
(11) H. Kuzuhara, H. Terayama, H. Ohrui, and S. Emoto, *Carbohydr. Res.*, 20 (1971) 165–169.
(12) C. Copeland and R. V. Stick, *Aust. J. Chem.*, 30 (1977) 1269–1273.
(13) E. J. Hedgley, W. G. Overend, and R. A. C. Rennie, *J. Chem. Soc.*, (1963) 4701–4711.
(14) J. Prokop and D. H. Murray, *J. Pharm. Sci.*, 54 (1965) 359–365.
(15) C. L. Brewer, S. David, and A. Veyrières, *Carbohydr. Res.*, 36 (1974) 188–190.
(16) S. Morgenlie, *Acta Chem. Scand.*, 27 (1973) 3609–3610.

Kinetic studies[17] showed that the rate of hydrolysis of the 5,6-acetal group in 1,2:5,6-di-O-isopropylidenealdofuranoses is not influenced by the proximity of the hydroxyl group at C-3.

Among other diacetals of the aldoses, 1,2:3,5-diacetals of D-xylose[18-21] and L-xylose,[22,23] and 1,2:3,4-diacetals of L-arabinose,[24] D-ribose,[25] and D-glucoseptanose[26] can be selectively hydrolyzed to 1,2-acetals, but, in the case of the 1,2:3,4-diacetals, the yields of the corresponding 1,2-monoacetals are often low, indicating little difference in the rates of hydrolysis of the two 1,3-dioxolane rings. The good selectivity found in the hydrolysis of 1,2:3,5-diacetals of D- and L-xylose may reflect the energetically favorable nature of the bicyclic system formed by *cis*-fusion of the furanose (oxolane) and 1,3-dioxolane rings. However, it is pertinent to consider this selectivity, and other selective hydrolyses described later, in the light of the report by Angyal and Beveridge,[26a] that secondary, compared to primary, hydroxyl groups have a greater tendency to participate in the formation of cyclic acetals. The 3,5-acetal groups in the 1,2:3,5-diacetals of D- and L-xylose are derived by reaction of a primary (HO-5) and a secondary (HO-3) hydroxyl group with the carbonyl component, and might, therefore, be expected to be less stable than the 3,4-acetal groups in the 1,2:3,4-diacetals, which are derived by reaction at two secondary hydroxyl groups (HO-3 and HO-4). However, as the 3,5- and 3,4-cyclic acetal rings are six- and five-membered, respectively, it is also probable that differing steric interactions in the two types of ring could contribute significantly to their different stability towards acid hydrolysis.

Studies on the rates of hydrolysis of di-O-isopropylidene derivatives of ketoses indicated that the relative order of stability of 2,2-dimethyl-1,3-dioxolane rings is in the order: *cis*-fused to a furanoid ring > *spiro*-fused to a pyranoid ring > *cis*-fused to a pyranoid ring > *spiro*-fused to a furanoid ring.[6] Thus, partial hydrolysis of 1,2:3,4-di-O-isopropylidene-β-D-*erythro*-2-pentulofuranose (**1**) to the 3,4-acetal (**2**)

(17) P. M. Collins, *Tetrahedron*, 21 (1965) 1809–1815.
(18) O. Svanberg and K. Sjöberg, *Ber.*, 56 (1923) 863–869.
(19) R. S. Tipson, B. F. West, and R. F. Brady, Jr., *Carbohydr. Res.*, 10 (1969) 181–183.
(20) R. J. Ferrier and L. R. Hatton, *Carbohydr. Res.*, 5 (1967) 132–139.
(21) K. Heyns and J. Lenz, *Chem. Ber.*, 94 (1961) 348–352.
(22) H. Müller and T. Reichstein, *Helv. Chim. Acta*, 21 (1938) 251–262.
(23) S. von Schuching and G. H. Frye, *J. Org. Chem.*, 30 (1965) 1288–1291.
(24) B. E. Stacey and B. Tierney, *Carbohydr. Res.*, 49 (1976) 129–140.
(25) N. A. Hughes and P. R. H. Speakman, *Carbohydr. Res.*, 1 (1965) 171–175.
(26) J. D. Stevens, *Aust. J. Chem.*, 28 (1975) 525–557.
(26a) S. J. Angyal and R. J. Beveridge, *Carbohydr. Res.*, 65 (1978) 229–234.

is possible,[27] and this contrasts with the selective hydrolysis[28] of 1,2:4,5-di-O-isopropylidene-β-D-fructopyranose (3) to the 1,2-acetal (4). In the case of diacetal 1, the strain present in the *spiro*-fused system is, presumably, sufficient to override the stability normally shown by acetals attached at the anomeric center.

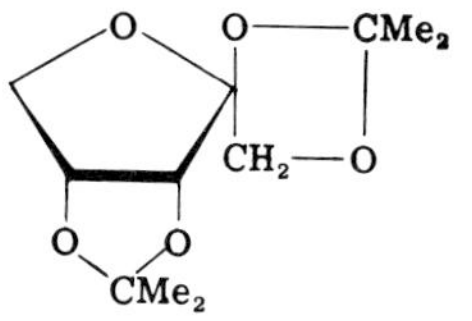

A further interesting example of the effect of structure on the stability of acetal rings towards hydrolysis is provided by 1,2:3,4-di-O-isopropylidene-5-thio-α-D-ribopyranose, for which no conditions for partial hydrolysis could be found,[29] in contrast to the oxygen analog.[25] However, selective cleavage of the 3,4-acetal ring was possible in 1,2:3,4-di-O-isopropylidene-5-thio-α-D-xylopyranose, this difference in behavior of *ribo* and *xylo* 5-thioaldoses presumably being, at least partly, due to the different types of fusion (*cis* and *trans*, respectively) of the 3,4-acetal ring to the thiopyranose ring in the two compounds.

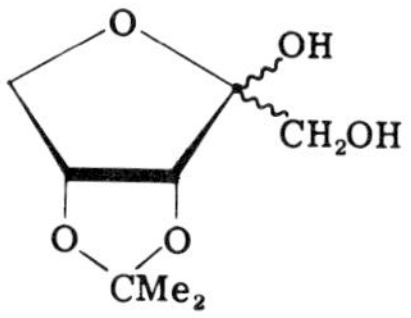

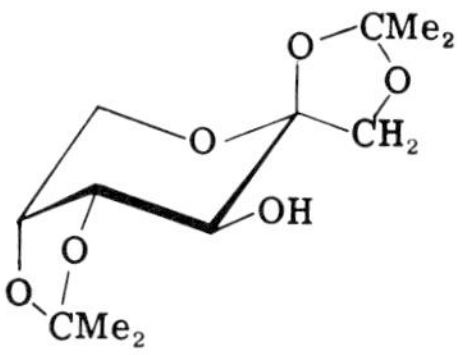

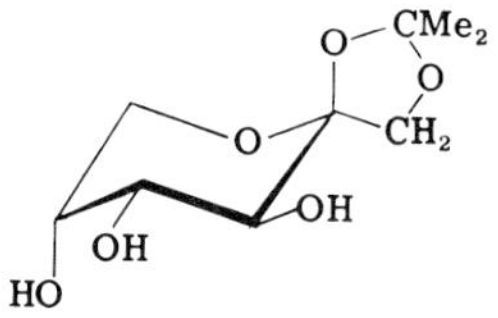

The selective deacetalation of 2,1':4,6-di-O-isopropylidenesucrose tetraacetate (5) may be achieved[30] by hydrolysis in aqueous acetic

(27) R. S. Tipson and R. F. Brady, Jr., *Carbohydr. Res.*, 10 (1969) 549–563.
(28) J. C. Irvine and C. S. Garrett, *J. Chem. Soc.*, 97 (1910) 1277.
(29) N. A. Hughes and C. J. Wood, *Carbohydr. Res.*, 49 (1976) 225–232.
(30) R. Khan and H. Lindseth, *Carbohydr. Res.*, 71 (1979) 327–330.

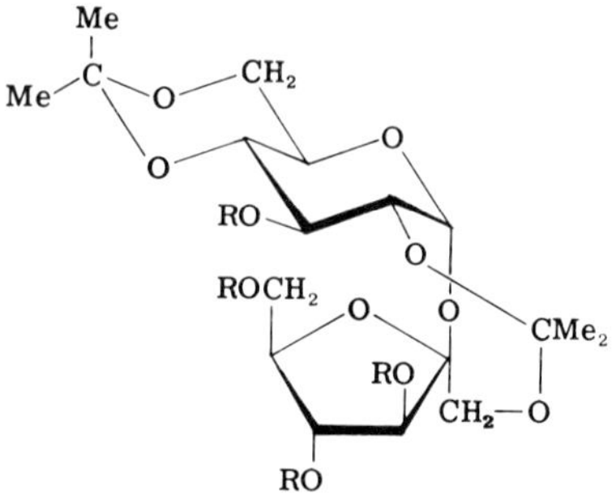

5 R = Ac

acid, which gives, in addition to **5** (11%) 3,3',4',6'-tetra-O-acetyl-4,6-O-isopropylidene- and 3,3',4',6'-tetra-O-acetyl-2,1'-O-isopropylidene-sucrose (4.7 and 23%, respectively), together with 3,3',4',6'-tetra-O-acetylsucrose (28%). Therefore, it appears that the eight-membered, 2,1'-cyclic acetal ring is more stable towards hydrolysis than the six-membered, 4,6-ring under these conditions.

For acetals in which the acetal rings do not involve the anomeric center, it appears that, in general, a 1,3-dioxolane ring *cis*-fused to a furanose[31-35] (oxolane) or a pyranose[36] (oxane) ring is more stable towards hydrolysis than a 1,3-dioxolane ring formed within the same molecule from a side-chain, vicinal-diol grouping which comprises a primary and a secondary hydroxyl group. An explanation of this relative stability that is based on the observation by Angyal and Beveridge[26a] that acetal formation by a primary hydroxyl group is disfavored over that by a secondary group is particularly attractive in these cases. The acetal rings being compared in each compound are of the same size (five-membered) and, therefore, free-energy differences due to constitutional differences in the acetal groups are lessened.

The correct choice of the type of acetal group can be of some importance when routes to partially protected sugars are needed. For example, although the two benzylidene groups in methyl 2,3:4,6-di-O-benzylidene-α-D-mannopyranoside are removed[37] at similar rates during acid hydrolysis, methyl 2,3:4,6-di-O-isopropylidene-α-D-mannopyran-

(31) H. Ohrui and S. Emoto, *Tetrahedron Lett.*, (1975) 2765–2766.
(32) W. E. Dick, Jr., and D. Weisleder, *Carbohydr. Res.*, 39 (1975) 87–96.
(33) G. Mackenzie and G. Shaw, *J. Chem. Soc. Chem. Commun.*, (1977) 753–754.
(34) P. Koell, *Tetrahedron Lett.*, (1978) 51–52.
(35) L. M. Lerner, *J. Org. Chem.*, 43 (1978) 2469–2473.
(36) P. Szabó, *Carbohydr. Res.*, 59 (1977) 580–583.
(37) G. J. Robertson, *J. Chem. Soc.*, (1934) 330–332.

oside undergoes[38] selective acetal-cleavage, to afford the 2,3-acetal in 75% yield. The greater lability of the six-membered acetal-ring in the di-O-isopropylidene derivative is most probably of steric origin; one of the methyl groups attached to the 1,3-dioxane ring must occupy an axial position if this ring is in a chair conformation, leading to energetically unfavorable, syn-diaxial interactions.[7a,38a]

The rate of hydrolysis of a benzylidene acetal derived from a vicinal-diol grouping of a carbohydrate may depend markedly on the configuration at C-2 in the 1,3-dioxolane ring. In 1946, it was noted[39] that the diastereoisomers of methyl 2-O-benzoyl-3,4-O-benzylidene-β-L-arabinopyranoside differ in their stability towards 0.05 M hydrochloric acid, and later work,[40] using ^{1}H-n.m.r. spectroscopy, showed that the isomer having the (S) configuration at the benzylidene acetal center (endo phenyl group) is more acid-labile than the (R) isomer (exo phenyl group). Selective hydrolysis of an equimolar mixture of the diastereoisomeric benzoates led to isolation of the (R) isomer in 35–40% yields; after benzoylation of the benzylidenation reaction-mixture, the (S) isomer could be obtained directly by crystallization. The selective hydrolysis of the (S) isomer cannot be rationalized satisfactorily in terms of a difference in free energy between the (R) and (S) isomers, as this difference is likely to be small; in chloroform-d containing p-toluenesulfonic acid, they exist[40] in equilibrium in a 1:1 mixture. The different rates of hydrolysis of the two diastereoisomers are most readily rationalized in terms of a difference in the energies of the transition states for hydrolysis of the two acetals.[7c,40a]

The selective removal of one of two acetal groups that are symmetrically disposed within a diacetal may be achieved. 1,3:2,4-Di-O-benzylidene derivatives of erythritol and L-threitol have been converted[41] into their 1,3-acetals, and selective methanolysis of 2,3,2′,3′-tetra-O-benzoyl-4,6:4′,6′-di-O-benzylidene-α,α-trehalose gave[42] the monobenzylidene derivative in 47% yield, thereby offering a route to unsymmetrically substituted α,α-trehalose derivatives.

It is known[43] that trans-fused, 4,6-benzylidene acetals of hexopyran-

(38) M. E. Evans and F. W. Parrish, Carbohydr. Res., 54 (1977) 105–114.
(38a) H. C. Brown, J. H. Brewster, and H. Shechter, J. Am. Chem. Soc., 76 (1954) 467–474; see footnote 46.
(39) M. A. Oldham and J. Honeyman, J. Chem. Soc., (1946) 986–989.
(40) D. M. Clode, Can. J. Chem., 55 (1977) 4066–4070.
(40a) The author acknowledges helpful discussion with Drs. N. Baggett and D. M. Clode on this point.
(41) A. B. Foster, A. H. Haines, and J. Lehmann, J. Chem. Soc., (1961) 5011–5014.
(42) A. C. Richardson and E. Tarelli, J. Chem. Soc., (1971) 3733–3735.
(43) B. Capon, W. G. Overend, and M. Sobell, Tetrahedron, 16 (1961) 106–112.

osides are hydrolyzed faster than the corresponding *cis*-fused acetals, and related results have been obtained[41] with the 1,3:2,4-di-*O*-benzylidenetetritols. 2,6-Anhydro-1,3:5,7-di-*O*-benzylidene-D-*glycero*-L-*manno*-heptitol[44] (**6**) contains a *cis*-fused (5,7) and a *trans*-fused (1,3) acetal ring, and it is interesting that, on graded, acid hydrolysis, it affords the 5,7-monoacetal in high yield (82–85%). The acid-lability of a *p*-anisylidene acetal is greater than that of a corresponding benzylidene acetal as a result of the methoxyl substituent in the former type of compound.[45] The *p*-anisylidene group in 2,6-anhydro-1,3-*O*-*p*-anisylidene-5,7-*O*-benzylidene-D-*glycero*-L-*manno*-heptitol (**7**) could be quantitatively removed,[44] without loss of the benzylidene group, by graded, acid hydrolysis.

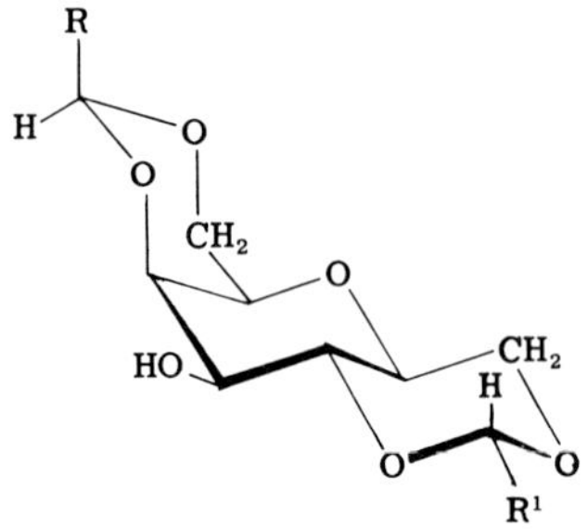

6 R = R¹ = Ph
7 R = Ph, R¹ = C₆H₄OMe-*p*

Considerable data on the graded, acid hydrolysis of di- and tri-acetals of acyclic sugar derivatives, especially the alditols, are contained in an article by Barker and Bourne.[7] These authors indicated the types of acetal rings by the Greek letters α, β, and γ, which describe the relative positions of the hydroxyl groups involved in acetal formation, and which represent five-, six-, and seven-membered acetal-rings, respectively. In those cases where neither of the hydroxyl groups is primary, the letters C and T were used to describe the relative configurations (*cis* or *trans*, respectively) of the two hydroxyl groups as they appear in the Fischer projection formulas. However, the use of *cis* and *trans* in this context is misleading and *erythro* and *threo*, respectively, have

(44) J. Lehmann and B. Schwesinger, *Carbohydr. Res.*, 72 (1979) 267–271.
(45) M. Smith, D. H. Rammler, I. H. Goldberg, and H. G. Khorana, *J. Am. Chem. Soc.*, 84 (1962) 430–440.

since been proposed[46,46a] as alternative descriptors of relative configuration. To aid an understanding of the relative stabilities of the various types of acetals, and thus gain an insight into their relative lability towards acid hydrolysis, the following points should be noted.

(a) Reactions of aldehydes with alditols show a preference for the formation of six-membered, cyclic acetals (1,3-dioxanes), containing β-erythro- or β rings, with the former type being favored over the latter.[7] A six-membered, β-erythro ring can adopt a stable chair conformation in which the substituent groups at C-4 and C-6 in the 1,3-dioxane of the ring are favorably disposed in equatorial positions.[7a,46a] It has been noted[26a] that the order of preference for cyclic acetal formation, namely, β-erythro > β, is in accordance with the greater tendency of secondary hydroxyl groups to be involved in cyclic acetal formation, compared to primary hydroxyl groups. Less favored than β-erythro or β-rings are α-, α-threo, β-threo, or γ-threo rings; the order of preference amongst this last sequence depends on the nature of the aldehyde.[7] However, it should be noted that, in a six-membered β-threo ring, in a chair conformation, one of the substituent groups, either at C-4 or C-6 of the 1,3-dioxane ring, must occupy an energetically unfavorable, axial position, whilst the other, at C-6 or C-4, respectively, is in an equatorial position.[7a,46a] Also noteworthy is the expectation, based on the results of Angyal and Beveridge,[26a] that an α-threo ring should be more stable than a terminal α-ring.

(b) Reactions of ketones with alditols show a marked preference[7,46a] for the formation of five-membered, cyclic acetals (1,3-dioxolanes). Where both hydroxyl groups involved in acetal formation are secondary, a *threo* rather than an *erythro* configuration is favored, as, in the resulting α-*threo* ring, the remainders of the carbon chain are in a *trans* relationship, thereby minimizing steric interactions between them.[7a,46a] It is to be expected that an α-*threo* ring is favored over an α-ring, based on the preferential involvement[26a] of secondary, compared to primary, hydroxyl groups in acetal formation.

With due consideration of the explanations just presented for the observed, relative stabilities of cyclic acetals derived from polyols, in terms of their constitution and conformation, nearly all of the following observations on the selective hydrolysis of cyclic acetals of alditols and dialkyl dithioacetals may be readily understood.

For benzylidene, ethylidene, and methylene acetals, α- and γ-*threo*

---

(46) K. W. Buck, A. B. Foster, B. H. Rees, J. M. Webber, and J. Lehmann, *Carbohydr. Res.*, 1 (1965) 329–331.

(46a) L. Hough and A. C. Richardson, in S. Coffey (Ed.), *Rodd's Chemistry of the Carbon Compounds*," Vol. 1F, Elsevier, Amsterdam, 2nd edn., 1967, pp. 1–66.

rings are hydrolyzed more readily than β rings, which, in turn, are less stable than β-*erythro* rings. As an example, acid hydrolysis of 1,3:2,4:5,6-tri-*O*-benzylidene-D-glucitol affords the 1,3:2,4-di-*O*-benzylidene derivative, which, in turn, is degraded to 2,4-*O*-benzylidene-D-glucitol.[47] The relative lability of γ-*threo* acetal rings is substantiated by the partial, acid hydrolysis of 1,3:2,5:4,6-tri-*O*-ethylidene-D-mannitol, which affords[48] a mixture of the 1,3:4,6-di- and 1,3-mono-acetals. Similarly, partial hydrolysis of 1,3:2,5:4,6-tri-*O*-methylene-D-mannitol gives[49,50] mainly the 1,3:4,6-diacetal, but it is interesting that, whereas the 2,5-methylene ring in the triacetal is the least stable towards acid hydrolysis, it is the most stable towards acetolysis (see later). The same sequence of stability among acetal rings is found in acetal derivatives of the dithioacetals of sugars. Thus, 2,4:3,5-di-*O*-benzylidene derivatives of D-xylose diethyl dithioacetal[51] and D-ribose ethylene dithioacetal[52] may be selectively hydrolyzed to 2,4-benzylidene acetals. Partial hydrolysis of 2,3:4,5-di-*O*-benzylidene-D-[53] or -L-arabinose[54] diethyl dithioacetal to the 2,3-acetal indicates that, for benzylidene acetals, an α-*threo* acetal ring is more stable than an α ring.

Less is known concerning the partial, acid hydrolysis of alditol polyacetals derived from ketones, compared to those derived from aldehydes. The acid hydrolysis of 1,2:3,4:5,6-tri-*O*-isopropylidene-D-mannitol to 3,4-*O*-isopropylidene-D-mannitol[55,56] and of 1,2:3,4-di-*O*-isopropylidene-L-rhamnitol to 3,4-*O*-isopropylidene-L-rhamnitol[57] indicates an order of isopropylidene acetal stability of α-*threo* > α, and this order is supported by the partial hydrolysis of 2,3:4,5-di-*O*-isopropylidene derivatives of dialkyl dithioacetals of D-arabinose[58] and D-xylose[59] to 2,3-acetals.

Although there appear to be no partial hydrolysis data to support the

(47) S. J. Angyal and J. V. Lawler, *J. Am. Chem. Soc.*, 66 (1944) 837–838.
(48) E. J. Bourne, G. T. Bruce, and L. F. Wiggins, *J. Chem. Soc.*, (1951) 2708–2713.
(49) H. G. Fletcher, Jr., and H. W. Diehl, *J. Am. Chem. Soc.*, 74 (1952) 3797–3799.
(50) T. G. Bonner, E. J. Bourne, and D. Lewis, *J. Chem. Soc.*, C, (1967) 2321–2326.
(51) E. J. C. Curtis and J. K. N. Jones, *Can. J. Chem.*, 38 (1960) 1305–1315.
(52) H. Zinner and E. Wittenburg, *Chem. Ber.*, 94 (1961) 1298–1303.
(53) H. Zinner, B. Richard, M. Blessmann, and M. Schlutt, *Carbohydr. Res.*, 2 (1966) 197–203.
(54) C. F. Huebner, R. A. Pankratz, and K. P. Link, *J. Am. Chem. Soc.*, 72 (1950) 4811–4812.
(55) L. F. Wiggins, *J. Chem. Soc.*, (1946) 13–14.
(56) N. Baggett, K. W. Buck, A. B. Foster, R. Jefferis, B. H. Rees, and J. M. Webber, *J. Chem. Soc.*, (1965) 3382–3388.
(57) M. A. Bukhari, A. B. Foster, J. Lehmann, and J. M. Webber, *J. Chem. Soc.*, (1963) 2287–2290.
(58) H. Zinner, G. Rembarz, and H. P. Klöcking, *Chem. Ber.*, 90 (1957) 2688–2696.
(59) D. G. Lance and J. K. N. Jones, *Can. J. Chem.*, 45 (1967) 1533–1538.

stability sequence α-*threo* > α-*erythro* for isopropylidene acetals, the greater thermodynamic stability of the trans- (α-*threo*) over the cis- (α-*erythro*) substituted dioxolane ring was recognized,[7a,46a] and is supported by an example of acetal isomerization (see Section II,3).

Hydrolytic studies on 1,2:4,5-di-*O*-isopropylidene-D-mannitol showed[56] that the two acetal rings are cleaved at similar rates, suggesting for the relative stabilities of the two rings, α ≈ α-*erythro*.

An interesting distinction between two isopropylidene α-acetal rings is provided by the acid hydrolysis of 5-*O*-benzoyl-4-*C*-(benzoyloxymethyl)-1,2:3,4-di-*O*-isopropylidene-L-ribitol (**8**), in which the

$$\text{Me}_2\text{C} \begin{matrix} \diagup \text{OCH}_2 \\ \diagdown \text{OCH} \end{matrix}$$
$$\text{Me}_2\text{C} \begin{matrix} \diagup \text{OCH} \\ \diagdown \text{OC}-\text{CH}_2\text{OBz} \\ | \\ \text{CH}_2\text{OBz} \end{matrix}$$

8

more highly substituted ring (that incorporating O-3 and O-4) appears to be the more stable.[60]

Evidence for the stability sequence, for isopropylidene acetals, of β-*erythro* > α comes from the graded, acid hydrolysis of 3,5:6,7-di-*O*-isopropylidene-D-*glycero*-D-*gulo*-heptitol and 2,4:5,6-di-*O*-isopropylidene-D-glucitol, which respectively give,[61] as the preponderant product, the 3,5- and 2,4-monoacetal. This greater stability of a six- over a five-membered acetal ring in acetals derived from a ketone is unusual. The same, relative stability was also indicated by the hydrolysis of 3,5:6,7-di-*O*-isopropylidene-D-*glycero*-D-*gulo*-heptono-1,4-lactone,[61,62] in which the β-*erythro* acetal ring is part of a ring system of the *cis*-hydrindanone type. Interestingly, when an isopropylidene β-acetal is part of a ring system of the *cis*-decalin type, it may be cleaved in preference to an α-ring. Thus, 1,3:5,6-di-*O*-isopropylidene-2,4-*O*-methylene-D-glucitol gives[46] 5,6-*O*-isopropylidene-2,4-*O*-methylene-D-glucitol on acid hydrolysis, supporting the stability order α > β. It is noteworthy that the previously mentioned, selective hydrolysis of the six-membered, isopropylidene acetal ring in methyl 2,3:4,6-di-*O*-isopro-

(60) A. Gonzalez, M. Orzaez, E. Martinez, V. Custardoy, and R. Mestres, *An. Quím.*, 70 (1974) 1073–1076; *Chem. Abstr.*, 83 (1975) 164,447r.
(61) J. S. Brimacombe, A. B. Foster, and L. C. N. Tucker, *Carbohydr. Res.*, 3 (1966) 76–80.
(62) J. S. Brimacombe and L. C. N. Tucker, *Carbohydr. Res.*, 1 (1965) 332–333.

pylidene-α-D-mannopyranoside is an example[38] of the favored removal of a β ring in the presence of an α-erythro ring.

Acid hydrolysis of 2,4:3,5-di-O-isopropylidene-D-xylose diethyl dithioacetal to give[63] the 2,4-acetal afforded a rare opportunity to establish that, in isopropylidene acetals, as with acetals derived from aldehydes, β-erythro rings are more stable than β rings.

An indirect method[64] for the deacetalation of 1,2:3,5-di-O-methylene-α-D-xylofuranose avoids problems that were encountered in the attempted, partial hydrolysis of this compound with acid. Oxidation of the diacetal with potassium permanganate in dilute, aqueous phosphoric acid gave 3,5-O-methylene-D-xylofuranose 1,2-carbonate in 36% yield (based on unrecovered starting-material) and base-catalyzed hydrolysis of the cyclic ester gave 3,5-O-methylene-D-xylofuranose. A similar sequence of reactions has been applied[65] to 6-O-acetyl-1,2:3,5-di-O-methylene-α-D-glucofuranose. It is noteworthy that the D-xylofuranose derivative is protected at O-3 and O-5 by this procedure, whereas acid hydrolysis of the corresponding 1,2:3,5-di-O-isopropylidene acetal affords[18,19] a derivative protected at O-1 and O-2.

2. Acetolysis

An article[66] in this Series summarized many important aspects of the acetolysis of acetals. For methylene acetals, the principal findings were that those spanning (i) a primary and a secondary position are cleaved, to produce a primary acetate and an acetoxymethyl ether at the secondary position, and (ii) secondary positions are the most difficult to cleave. These observations are consistent with an acetolysis mechanism which involves favored attack of an acetylium ion on an acetal oxygen atom attached to a less hindered, primary position. Accumulated evidence suggests that α and β rings are cleaved before β-erythro rings, β rings before γ-threo rings, and β-threo rings before β-erythro rings. The results may readily be rationalized by consideration of the relative stabilities of the acetal rings, and recognition of the sterically demanding nature of the acetylium ion. The following results of acetolysis with acetic acid–acetic anhydride–sulfuric acid illustrate some of these points: 1,3:2,4:5,6-tri-O-methylene-D-talitol affords[67] 3,5-di-O-(acetoxymethyl)-1,6-di-O-acetyl-2,4-O-methylene-

(63) T. van Es, Carbohydr. Res., 32 (1974) 370–374.
(64) O. T. Schmidt and G. Nieswandt, Chem. Ber., 82 (1949) 1–7.
(65) O. T. Schmidt, A. Distelmaier, and H. Reinhard, Chem. Ber., 86 (1953) 741–749.
(66) R. D. Guthrie and J. F. McCarthy, Adv. Carbohydr. Chem., 22 (1967) 11–23.
(67) R. M. Hann, W. T. Haskins, and C. S. Hudson, J. Am. Chem. Soc., 69 (1947) 624–629.

D-talitol, 1,3:2,5:4,6-tri-O-methylene-D-mannitol gives[68] 3,4-di-O-(acetoxymethyl)-1,6-di-O-acetyl-2,5-O-methylene-D-mannitol, and 2,4:3,5-di-O-methylene-D-glucitol yields[69] 3-O-(acetoxymethyl)-1,5,6-tri-O-acetyl-2,4-O-methylene-D-glucitol. Because treatment with base readily removes the O-acetyl and O-(acetoxymethyl) groups from these products, to liberate the parent hydroxy compound, acetolysis offers a useful means for selectively removing acetal groups, to yield compounds in which two secondary hydroxyl groups are protected by an acetal group. It should be noted that the selective removal of an O-(acetoxymethyl) group from an acetolysis product is possible by controlled, acid hydrolysis,[70] and further investigation of the synthetic utility of this procedure may be warranted.

The acetolysis reagent trifluoroacetic anhydride–acetic acid[70,71] shows a specificity similar to that of the acetic anhydride–acetic acid –sulfuric acid reagent, but offers the important advantage that, owing to the greater hydrolytic lability of the O-(trifluoroacetoxymethyl) group compared to the O-acetyl group, it is possible to obtain acetate–hydroxy derivatives under mild conditions. Through use of this reagent, 5-O-acetyl-1,6-di-O-benzoyl-2,4-O-methylene-D-glucitol has been prepared[70,72] in high yield by cleavage of the β-threo acetal ring in 1,6-di-O-benzoyl-2,4:3,5-di-O-methylene-D-glucitol. The highly regioselective cleavage of the 3,5-acetal (β-threo) ring in this compound has been rationalized[70] in terms of favored attack, at O-5, by the sterically demanding acetylium ion. Making the reasonable assumption that the favored conformation of the diacetal is the "O"-inside form,[7a] attack along an equatorial direction at O-5 is not subject to significant, steric hindrance. However, attack along an equatorial direction at O-3 leads to considerable, nonbonded interaction with the benzoyloxymethyl group joined to C-2, and attack along an axial direction at either O-3 or O-5 is also sterically disfavored.

By acetolysis of α- or β-methylene acetals of polyols with trifluoroacetic anhydride–carboxylic acid mixtures, derivatives may be prepared[71] that are O-acylated at the primary hydroxyl group, and that have free hydroxyl groups at certain secondary positions.

In general, only methylene acetals have been selectively deprotected in this way; on acetolysis, benzylidene and isopropylidene acetals are readily converted into the corresponding peracetates of the

(68) A. T. Ness, R. M. Hann, and C. S. Hudson, *J. Am. Chem. Soc.*, 65 (1943) 2215–2222.
(69) R. M. Hann, J. K. Wolfe, and C. S. Hudson, *J. Am. Chem. Soc.*, 66 (1944) 1898–1901.
(70) E. J. Bourne, J. Burdon, and J. C. Tatlow, *J. Chem. Soc.*, (1958) 1274–1279.
(71) E. J. Bourne, J. Burdon, and J. C. Tatlow, *J. Chem. Soc.*, (1959) 1864–1870.
(72) T. G. Bonner, *Methods Carbohydr. Chem.*, 2 (1963) 314–317.

parent alcohols, and little work has been reported on the acetolysis of ethylidene acetals.

3. Acetal Isomerization

Solely on the basis of product structure, it appears that, generally, acid-catalyzed isomerization of carbohydrate acetals occurs with favored cleavage of only one of the two acetal, carbon-to-oxygen bonds. The isomerization of methyl 3,4-O-(R)-benzylidene-β-D-ribopyranoside into the 2,3-(R)-acetal[73] and of 1,2:5,6-di-O-isopropylidene-α-D-allofuranose into the 2,3:5,6-diacetal[74,75] serve as examples. It appears reasonable to formulate isomerizations occurring under anhydrous conditions in terms of an intramolecular mechanism.[7c] It has been suggested that the conversion, in N,N-dimethylformamide containing p-toluenesulfonic acid, of 2,4-O-benzylidene-D-erythrose into *endo*-2,3-O-benzylidene-D-erythrofuranose,[76] and of 1,4-anhydro-3,5-O-benzylidene-D-mannitol into the corresponding 2,3-acetal,[77] occurs in this way.

The isomerization of 2,4-O-ethylidene-D-erythrose into 2,3-O-ethylidene-D-erythrofuranose occurs[78] in 0.05 M sulfuric acid, and acetal migration has been reported[79] to accompany the hydrolysis of 1,6-anhydro-3,4-O-isopropylidene-β-D-talopyranose in 80% acetic acid, from which the 2,3-acetal was isolated. It would be of interest to know whether these isomerizations occurring under hydrolysis conditions are truly of an intramolecular type. Investigations by proton-n.m.r. spectroscopy suggested[56] that acetal migration does not occur to a significant extent with isopropylidene derivatives of acyclic polyols under certain hydrolyzing conditions (3:1 methanol–water containing 1.6% of p-toluenesulfonic acid). Significantly, butanal reacts[80] with D-glucitol in M hydrochloric acid to form initially a 2,3-O-butylidene derivative, which is gradually replaced by the 2,4-acetal, but there is no evidence for intramolecular conversion. However, in anhydrous N,N-dimethylformamide in the presence of hydrogen chloride, the

(73) D. M. Clode, *Can. J. Chem.*, 55 (1977) 4071–4077.
(74) J. M. Ballard and B. E. Stacey, *Carbohydr. Res.*, 12 (1970) 37–41.
(75) M. Haga, M. Takano, and S. Tejima, *Carbohydr. Res.*, 14 (1970) 237–244.
(76) N. Baggett, K. W. Buck, A. B. Foster, B. H. Rees, and J. M. Webber, *J. Chem. Soc., C*, (1966) 212–215.
(77) F. S. Al-Jeboury, N. Baggett, A. B. Foster, and J. M. Webber, *Chem. Commun.*, (1965) 222–224.
(78) C. E. Ballou, *J. Am. Chem. Soc.*, 82 (1960) 2585–2588.
(79) N. A. Hughes, *Carbohydr. Res.*, 7 (1968) 474–479.
(80) T. G. Bonner, E. J. Bourne, P. J. V. Cleare, and D. Lewis, *J. Chem. Soc., B*, (1968) 822–827.

acetal rearrangement appears to be intramolecular. Information supporting the foregoing observations on the butylidene acetals has come from a study of 2,3-O-ethylidene-[81] and 2,3-O-benzylidene-D-glucitol,[81] and related results have been obtained in the condensation of butanal with 2-deoxy-D-*arabino*-hexitol.[82] For the last, the 1,3-acetal is the kinetic product, and the 3,4-acetal, the thermodynamic product, and there was no evidence for intramolecular, acetal migration in aqueous acid.

Acetals comprising a 1,3-dioxane ring fused to another cyclic system may often be rearranged to isomeric 1,3-dioxolane structures; for example, 3,5-O-benzylidene-1,2-O-isopropylidene-α-D-glucofuranose to the 5,6-benzylidene acetal,[83] 1,4-anhydro-3,5-O-benzylidene-D-mannitol to the 2,3-acetal,[77,84] methyl 4,6-O-benzylidene-2-deoxy-2-iodo-α-D-altropyranoside to the 3,4-(R)- and 3,4-(S)-acetals,[85] and methyl 4,6-O-isopropylidene-α-D-altropyranoside to the 3,4-acetal.[86] Interestingly, a number of examples of isomerization occurring in the opposite direction, that is, 1,3-dioxolane to 1,3-dioxane, have been found for acetals of acyclic polyols. The *"cis"*-2,3-O-benzylidene-erythritol rearranges to 1,3-O-benzylidene-DL-erythritol,[76] and 2,3-O-butylidene-,[80] 2,3-O-ethylidene-,[81] and 2,3-O-benzylidene[81]-D-glucitol, and 2,3-O-butylidene-1-deoxy-D-glucitol[82] are all converted into 2,4-acetals in acidic media.

A consideration of the foregoing examples of acetal isomerization indicates that they may, in some cases, be put to synthetic use. A study of the acid-catalyzed interconversion of 1,2-O- and 1,3-O-ethylidene-DL-glycerol indicated[87] that the product composition is strongly temperature-dependent, and that the formation of 1,3-dioxanes is more favored at low temperatures. Use of this fact has been made[87] in preparing 1,3-O-isopropylidene-DL-glycerol, albeit in only 2% yield, by storage of 1,2-O-isopropylidene-DL-glycerol containing an acidic catalyst below 0° for several days.

During a preparation of a *C*-nucleoside analog, Buchanan and coworkers[87a] discovered that treatment of 3(5)-(1,2:4,5-di-O-isopropyli-

(81) T. G. Bonner, E. J. Bourne, P. J. V. Cleare, and D. Lewis, *J. Chem. Soc., B*, (1968) 827–830.
(82) T. G. Bonner, E. J. Bourne, P. J. V. Cleare, R. F. J. Cole, and D. Lewis, *J. Chem. Soc., B*, (1971) 957–962.
(83) P. A. Levene and A. L. Raymond, *Ber.*, 66 (1933) 384–386.
(84) R. E. Reeves, *J. Am. Chem. Soc.*, 71 (1949) 2868–2870.
(85) T. D. Inch, *Carbohydr. Res.*, 21 (1972) 37–43.
(86) M. E. Evans, *Carbohydr. Res.*, 30 (1973) 215–217.
(87) G. Aksnes, P. Albriktsen, and P. Juvvik, *Acta Chem Scand.*, 19 (1965) 920–930.
(87a) J. G. Buchanan, M. E. Chacón-Fuertes, and R. H. Wightman, *J. Chem. Soc. Perkin Trans. 1*, (1979) 244–248.

dene-D-*manno*-pentitol-1-yl)pyrazole with acetone and concentrated sulfuric acid caused its isomerization to 3(5)-(2,3:4,5-di-O-isopropylidene-D-*manno*-pentitol-1-yl)pyrazole, which suggests a stability sequence for isopropylidene acetals of α-*threo* > α-*erythro* (see Section II,1).

III. DEACYLATION

The selective removal of an acyl group at the anomeric center will not be considered here, unless it is accompanied by selective deacylation at another position. Also, the selective removal of O-acyl groups in the presence of N-acyl groups, which is generally achieved without difficulty, is omitted. The material considered here is divided into Subsections based, essentially, on the type of substrate.

1. Selective Deacylation of Derivatives Containing the Same Type of Acyl Group

Selective hydrolysis of β-D-glucopyranose pentaacetate with potassium hydroxide in acetone–water yielded[88,89] 6-O-acetyl-α-D-glucose in 20–22% yield. Maltose octaacetate also underwent partial de-esterification with the reagent, to afford[89] an α,β mixture of a diacetate, and sucrose octaacetate gave a variety of products from which a crystalline diacetate and triacetate were isolated. Methyl α- and β-D-glucopyranoside tetraacetate[89] and o-nitrophenyl β-D-galactopyranoside tetraacetate[90] were converted into the corresponding 6-acetates in low yields by the same means, and an alternative procedure using a basic ion-exchange resin has been reported.[91] These apparently selective hydrolyses of secondary over primary ester groups are surprising in view of the greater steric hindrance to be expected in the former type, but it is possible that, under the experimental conditions, intramolecular, acyl migration[1] occurs, and that acyl groups are liberated, largely, from the primary position (to which secondary acyl groups then migrate). However, the remarkable conversion[92] of 1,2,6-tri-O-acetyl-3,5-di-O-methyl-α-D-glucofuranose into the corresponding 6-O-acetyl derivative in 88% yield, in methanol containing a catalytic amount of sodium methoxide, cannot be similarly rationalized, because of the

(88) Y. Z. Frohwein and J. Leibowitz, *Nature (London)*, 186 (1960) 153.
(89) Y. Z. Frohwein and J. Leibowitz, *Bull. Res. Counc. Isr., Sect. A11*, (1963) 330, *Chem. Abstr.*, 59 (1963) 15,363.
(90) Y. Z. Frohwein, *Nature (London)*, 196 (1962) 775–776.
(91) Y. Z. Frohwein, *Isr. J. Chem.*, 5 (1967) 141P.
(92) J. Kuszmann, P. Sohár, and L. Kiss, *Carbohydr. Res.*, 63 (1978) 115–125.

stereochemistry, which is unfavorable for intramolecular, acyl migration.

It is noteworthy that the favored route to methyl 6-O-benzoyl-α-D-galactopyranoside involves[93] partial deprotection of the fully benzoylated glycoside.

Although trifluoroacetates are very susceptible to hydrolysis and methanolysis,[94] preferential removal of one trifluoroacetyl group, is often possible by controlled alcoholysis. Thus, a concentrated solution of methyl 4,6-O-benzylidene-2,3-di-O-(trifluoroacetyl)-α-D-glucopyranoside in methanol or in methanol–carbon tetrachloride gradually deposited a mono(trifluoroacetate),[95] which, on the basis of further transformations, was identified[96] as the 3-ester. Also, 1,3:2,4-di-O-ethylidene-D-glucitol 5,6-bis(trifluoroacetate) was converted[97] into the 6-trifluoroacetate in 52% yield by partial alcoholysis in 3-methyl-1-butanol (isopentyl alcohol) solution, and the controlled alcoholysis of methyl 4-O-methyl-2,3-di-O-(trifluoroacetyl)-α-L-rhamnopyranoside in methanol–carbon tetrachloride gave[98] a mono(trifluoroacetate), regarded as being the 3-ester.

Spontaneous, partial deacetylation of benzyl 3,4,6-tri-O-acetyl-2-amino-2-deoxy-β-D-glucopyranoside was reported[99] to occur in ethanol at the oxygen atom adjacent to the amino group. Related observations have been made during studies on the partial deacylation of other amino sugar derivatives. Adsorption of methyl 3-acetamido-2,4-di-O-acetyl-3,6-dideoxy-α-D-altropyranoside onto a column of alkaline alumina, followed by elution of the column with benzene–methanol, afforded[100,101] methyl 3-acetamido-4-O-acetyl-3,6-dideoxy-α-D-altropyranoside in 70% yield. Similar treatment of methyl 3-acetamido-2,4-di-O-acetyl-3,6-dideoxy-α-L-glucopyranoside gave[102] a 35% yield of a

(93) D. H. Hollenberg, K. A. Watanabe, and J. J. Fox, Carbohydr. Res., 28 (1973) 135–139.
(94) E. J. Bourne, C. E. M. Tatlow, and J. C. Tatlow, J. Chem. Soc., (1950) 1367–1369.
(95) E. J. Bourne, M. Stacey, C. E. M. Tatlow, and J. C. Tatlow, J. Chem. Soc., (1951) 826–833.
(96) E. J. Bourne, A. J. Huggard, and J. C. Tatlow, J. Chem. Soc., (1953) 735–741.
(97) E. J. Bourne, C. E. M. Tatlow, J. C. Tatlow, and R. Worrall, J. Chem. Soc., (1958) 3945–3950.
(98) K. Butler, P. F. Lloyd, and M. Stacey, J. Chem. Soc., (1955) 1531–1536.
(99) L. Ötvös and A. Bórbás, Acta Univ. Szeged., Acta Phys. Chem. [N. S.], 3 (1957) 151–157; Chem. Abstr., 53 (1959) 2103.
(100) J. Jarý, J. Kovář, and K. Čapek, Collect. Czech. Chem. Commun., 30 (1965) 1144–1150.
(101) K. Čapek, J. Šteffková, and J. Jarý, Collect. Czech. Chem. Commun., 32 (1967) 2491–2497.
(102) K. Čapek, J. Šteffková, and J. Jarý, Collect. Czech. Chem. Commun., 31 (1966) 1854–1863.

mixture of methyl 3-acetamido-4-O-acetyl-3,6-dideoxy- and (most probably) methyl 3-acetamido-2-O-acetyl-3,6-dideoxy-α-L-glucopyranoside in the ratio of 7:1. Peracetylated derivatives of methyl 3-acetamido-3,6-dideoxy-α-D-gulopyranoside and methyl 3-acetamido-3,6-dideoxy-α-L-allopyranoside also underwent preferential acyl cleavage at O-2, to afford[103] the corresponding 4-O-acetyl derivatives in yields of 33 and 74%, respectively.

Reaction of penta-O-acetyl-α- or -β-D-glucopyranose with 3 molar equivalents of piperidine gives[104] 1-(3,4,6-tri-O-acetyl-D-glucopyranosyl)piperidine, which is a useful intermediate for the preparation of certain 2-O-substituted derivatives of D-glucose. 2-O-Methyl-,[104] 2-O-carbanilino-,[104] and 2-O-benzyl-D-glucose[105] have been synthesized by this route. This reaction has been extended to the selective O-deacylation of β-cellobiose octaacetate[106] (**9**), with subsequent formation of the 2-methyl ether of cellobiose. Interestingly, reaction of **9** with piperidine for a short time gave[107] 2,3,6,2',3',4',6'-hepta-O-acetyl-α-cellobiose (**10**) in 81% yield, and this compound, on further treatment with piperidine, gave 1-(3,6,2',3',4',6'-hexa-O-acetyl-β-cellobiosyl)piperidine (**11**), which was obtained[106] in the longer-term reaction of **9**

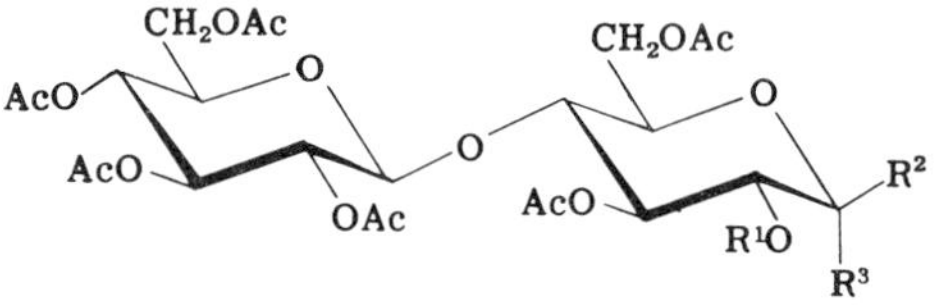

9 R^1 = Ac, R^2 = OAc, R^3 = H
10 R^1 = Ac, R^2 = H, R^3 = OH
11 R^1 = H, R^2 = HNC$_5$H$_{11}$, R^3 = H

with piperidine. From experiments in the D-glucose series, it had been noted earlier[104] that removal of the 2-O-acetyl group must accompany or precede the formation of the glycosylamine linkage.

Ammonolysis of perbenzoylated derivatives of mono- and di-saccharides provides interesting examples of selective deacylation. Treatment of the benzoylated nitriles of D-galactonic, D-gluconic, and

(103) K. Čapek, J. Šteffková, and J. Jarý, *Collect. Czech. Chem. Commun.*, 35 (1970) 107–115.
(104) J. E. Hodge and C. E. Rist, *J. Am. Chem. Soc.*, 74 (1952) 1498–1500.
(105) A. Klemer, G. Drolshagen, and H. Lukowski, *Chem. Ber.*, 96 (1963) 634–635.
(106) B. Lindberg, O. Theander, and M. S. Feather, *Acta Chem. Scand.*, 20 (1966) 206–210.
(107) R. M. Rowell and M. S. Feather, *Carbohydr. Res.*, 4 (1967) 486–491.

D-mannonic acid in ethanol saturated with ammonia gave[108] the corresponding 1,1-bis(benzamido)-1-deoxy-mono-O-benzoylpentitol derivative, most probably the 5-O-benzoyl derivative, as, on similar treatment, tetra-O-benzoyl-L-rhamnonic acid nitrile gave 1,1-bis-(benzamido)-1,5-dideoxy-L-arabinitol. From the ammonolysis of 1,2,3,4,6-penta-O-benzoyl-D-glucose and 1,2,3,4,6-penta-O-benzoyl-D-galactose was obtained[109] the corresponding 1,1-bis(benzamido)-1-deoxy-D-hexitol 6-benzoate. The formation of 6-O-benzoylated disaccharides has been observed in the ammonolysis of β-cellobiose octabenzoate,[110] β-maltose octabenzoate,[111] β-maltose 1,2,6,2′,3′,4′,6′-heptabenzoate,[112] and β-lactose 1,2,6,2′,3′,4′,6′-heptabenzoate[113]; particularly favorable yields were obtained in the case of the maltose derivatives. It is noteworthy that the benzoyl group in 6-O-benzoyl-maltose is very resistant to ammonolysis, complete removal requiring[111] 15 days. It has been suggested,[111] based on a consideration of models, that this stability might be attributable to an interaction of the carbonyl carbon atom with the oxygen atoms of the pyranose rings (O-5 and O-5′) and with that of the glycosidic linkage [1′-O-4], as shown in formula **12**. Presumably, if any one of these stabilizing influences is

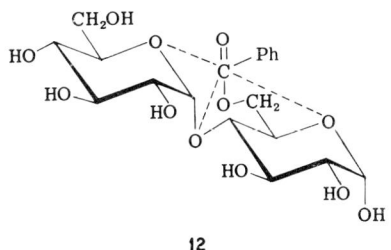

12

absent, ammonolysis should proceed more rapidly, and supporting evidence for this idea is the higher rate of hydrolysis of 6-O-benzoyl-maltitol (2 days), 6-O-benzoyl-D-glucose (2 hours), and 6-O-benzoyl-D-glucitol (2 hours). 6-O-Benzoylcellobiose, which was obtained[110]

(108) E. Restelli de Labriola and V. Deulofeu, *J. Org. Chem.*, 12 (1947) 726–730.
(109) E. G. Gros, A. Lezerovich, E. F. Recondo, V. Deulofeu, and J. O. Deferrari, *An. Asoc. Quím. Argent.*, 50 (1962) 185–197; *Chem. Abstr.*, 59 (1963) 5244.
(110) J. O. Deferrari, I. M. E. Thiel, and R. A. Cadenas, *J. Org. Chem.*, 30 (1965) 3053–3055.
(111) I. M. E. Thiel, J. O. Deferrari, and R. A. Cadenas, *J. Org. Chem.*, 31 (1966) 3704–3707.
(112) I. M. E. Thiel, J. O. Deferrari, and R. A. Cadenas, *Ann.*, 723 (1969) 192–197.
(113) I. M. Vazquez, I. M. E. Thiel, and J. O. Deferrari, *Carbohydr. Res.*, 26 (1973) 351–356.

in much lower yield than 6-O-benzoylmaltose from the parent octabenzoate, required 7 days for complete ammonolysis; interestingly, 6-O-benzoylcellobiitol underwent[111] complete ammonolysis in only 6 hours, which is further evidence for the influence of molecular geometry on ester stability in this range of compounds.

Ammonolysis of β-maltose 1,2,6,2',3',4',6'-heptabenzoate[112] and β-lactose 1,2,6,2',3',4',6'-heptabenzoate,[113] which gave no nitrogenous derivatives at C-1, in contrast to corresponding reactions[111,114] of the disaccharide octabenzoates, afforded 6-O-benzoylmaltose and 6-O-benzoyllactose in yields of 40 and 21.5%, respectively, together with the parent disaccharide. The lower yield obtained for the lactose mono-ester has been rationalized[113] on the basis that the (1→4)-β-D-glycosidic linkage of lactose allows interaction of the carbonyl carbon atom of the benzoyl group on O-6 with only two oxygen atoms [O-1'(4) and O-5'], and therefore the benzoyl group is less stable to hydrolysis than that in the corresponding maltose derivative 12. The importance of a cooperative interaction in stabilizing the lactose 6-ester towards hydrolysis is indicated by the fact that 6-O-benzoyllactose required 4 days in methanolic ammonia for removal of the benzoyl group, whereas 6-O-benzoyl-lactitol required only 4 hours.

Further examples of selective deacylation in disaccharide derivatives were provided by the reaction of 2,3,6,2',3',4',6'-hepta-O-acetylmaltosyl bromide and 2,3,6,2',3',4',6'-hepta-O-acetyllactosyl bromide with pyridine[115]; the former yielded 3,6,2',3',4',6'-hexa-O-acetylmaltosylpyridinium bromide, whereas the latter gave 3,3',4'-tri-O-acetyl- and 3,6,3',4',6'-penta-O-acetyl-lactosylpyridinium bromide.

Attempts have been made to solve the problem of selective substitution in sucrose octaacetate. Adsorption of the latter onto alumina for 46 hours gave a complex mixture from which, after extensive chromatography, were isolated the 2,3,4,6,1',3',4'-heptaacetate in 9% yield,[116] the 2,3,6,1',3',4',6'-heptaacetate in 2.7% yield,[117] and the 2,3,4,6,1',3',6'-heptaacetate in 6% yield.[117] Although selective deacylation of the octaacetate apparently occurs at secondary positions, as well as at O-6', it is likely that heptaacetates containing 4- and 4'-OH groups

---

(114) J. O. Deferrari, I. M. E. Thiel, and R. A. Cadenas, *Carbohydr. Res.*, 29 (1973) 141–146.

(115) A. Piskorsha-Chlebowska, *Rocz. Chem.*, 47 (1973) 49–57; *Chem. Abstr.*, 79 (1973) 32,217y.

(116) J. M. Ballard, L. Hough, and A. C. Richardson, *Carbohydr. Res.*, 24 (1972) 152–153.

(117) J. M. Ballard, L. Hough, and A. C. Richardson, *Carbohydr. Res.*, 34 (1974) 184–188.

arise from the 6- and 6'-OH derivatives, respectively, as a result of acyl migration.

Partial, de-esterification of 2,1':4,6-di-O-isopropylidenesucrose tetraacetate with methanolic ammonia at −10° gave[118] an inseparable mixture of the 3,4',6'- and 3,3',6'-triacetates (32%) and also the 3,6'-diacetate (30%); from a similar reaction at 5°, the 3',6'-diacetate (35%) and the 3-acetate (45%) were isolated.

Selective deprotection at the primary ester groups in cyclohexaamylose octadecabenzoate was the crucial step that allowed selective modification of all the primary hydroxyl groups in this cyclo-oligosaccharide.[118a] Lehn and coworkers provided[118a] an illuminating analysis of the problem of discriminating between one set of six primary hydroxyl groups and one set of 12 secondary hydroxyl groups, and pointed out the advantages of an "indirect," or reverse, method based on nonselective reaction with all of the hydroxyl groups followed by selective deprotection of one set, over the "direct" approach involving selective reaction with only one set. In their work, direct p-toluenesulfonylation or mesitylenesulfonylation did not lead to the desired, selectively substituted products in a pure state. Although the hexakis(6-azido-6-deoxy)cyclohexaamylose was prepared in ∼25% overall yield through reaction of the cyclohexaamylose with triphenylphosphine, carbon tetrabromide, and lithium azide, the same compound was produced in ∼55% yield by a route consisting of more steps, in which the key reaction was selective deacylation of the octadecabenzoate at the primary positions in 62% (isolated) yield, using potassium isopropoxide in 2-propanol. Predictably, both the rate and selectivity in the deacylation step were dependent on the steric bulk of the alkoxide used.

Interestingly, application of the selective deprotection procedure to cycloheptaamylose did not give satisfactory results, a large number of products being produced. However, in this case, the "direct" approach afforded the heptakis(6-azido-6-deoxy)cycloheptaamylose [as its hepta(2,3-diacetate)] in 57% yield, that is, over twice the yield for the corresponding reaction on cyclohexaamylose. The different outcome of these two reactions when applied to the two closely related cyclo-oligosaccharides illustrates the subtle factors that may influence relative reactivity in complex molecules.

A study[119] of the enzymic deacylation of esterified mono- and di-saccharides indicated that acetyl groups are hydrolyzed 12 times as rap-

(118) R. Khan, M. R. Jenner, and H. Lindseth, *Carbohydr. Res.*, 65 (1978) 99–108.
(118a) J. Boger, R. J. Corcoran, and J. M. Lehn, *Helv. Chim. Acta*, 61 (1978) 2190–2218.
(119) A. L. Fink and G. W. Hay, *Can. J. Biochem.*, 47 (1969) 353–359.

idly as propanoyl groups, and 25 times as rapidly as benzoyl groups, and that the qualitative order of the relative rates of enzymic deacetylation is C-1 > C-6 > C-4 > C-3 > C-2 for peracetylated D-glucose, and C-1 > C-6', C-6, C-4' > C-3', C-3 > C-2', C-2 for peracetylated maltose and cellobiose.

Selective deacetylation has played a useful part in the preparation of phosphatidyl-*muco*-inositol[120] and streptamine.[121]

2. Selective Deacylation of Derivatives Containing Different Types of Acyl Groups

Hydrolysis of esters of 2,4,6-trimethylbenzoic acid is difficult under basic conditions, because of the steric hindrance provided by the 2- and 6-methyl groups to nucleophilic attack at the carbonyl carbon atom.[122,123] It is understandable, therefore, that 2,3,4,6-tetra-*O*-acetyl-β-D-glucopyranosyl 2,4,6-trimethylbenzoate may be readily deacetylated on treatment with sodium methoxide in methanol,[124] or with methanolic ammonia,[125] to afford β-D-glucopyranosyl 2,4,6-trimethylbenzoate. However, similar deacylations can be considerably more complicated when a hydroxyl group liberated at C-2 is *cis*-disposed to the 1-*O*-acyl group. Thus, 2,3,4,6-tetra-*O*-acetyl-α-D-glucopyranosyl 2,4,6-trimethylbenzoate yielded 2-*O*-(2,4,6-trimethylbenzoyl)-D-glucose[126] as a result of deacetylation and subsequent acyl migration.[127] Analogous results were obtained on alkaline deacetylation of 2,3,4,6-tetra-*O*-acetyl-1-*O*-(tri-*O*-acetylgalloyl)-α- and -β-D-glucopyranose and of 1-*O*-(*p*-acetoxybenzoyl)-2,3,4,6-tetra-*O*-acetyl-α- and -β-D-glucopyranose.[128,129] In these cases, resistance of the galloyl and *p*-hydroxybenzoyl groups to removal under conditions of hydrolysis that remove acetyl groups is, presumably, a result of mesomeric interaction of the carbonyl group with a *p*-disposed oxygen atom. Interestingly, the cinnamoyl group may also be selectively removed in the presence of a

(120) V. P. Shevchenko, T. Yu. Lazurkina, Yu. G. Molotkovskii, and L. D. Bergel'son, *Bioorg. Khim.*, 2 (1976) 923–926; *Chem. Abstr.*, 86 (1977) 16,856v.
(121) S. Ogawa, T. Abe, H. Sano, K. Kotera, and T. Suami, *Bull. Chem. Soc. Jpn.*, 40 (1967) 2405–2409.
(122) H. L. Goering, T. Rubin, and M. S. Newman, *J. Am. Chem. Soc.*, 76 (1954) 787–791.
(123) M. L. Bender and R. S. Dewey, *J. Am. Chem. Soc.*, 78 (1956) 317–319.
(124) F. Micheel and G. Baum, *Chem. Ber.*, 88 (1955) 2020–2025.
(125) H. B. Wood, Jr., and H. G. Fletcher, Jr., *J. Am. Chem. Soc.*, 78 (1956) 207–210.
(126) H. B. Wood, Jr., and H. G. Fletcher, Jr., *J. Am. Chem. Soc.*, 78 (1956) 2849–2851.
(127) C. Pedersen and H. G. Fletcher, Jr., *J. Am. Chem. Soc.*, 82 (1960) 3215–3217.
(128) O. T. Schmidt and H. Reuss, *Ann.*, 649 (1961) 137–148.
(129) O. T. Schmidt and H. Schmadel, *Ann.*, 649 (1961) 157–167.

galloyl group, hydrolysis of 2-O-cinnamoyl-β-D-glucopyranosyl gallate giving [130] the 1-gallate in 72% yield.

It has long been known[131] that treatment of β-D-glucopyranose pentaacetate with phosphorus pentachloride gives 3,4,6-tri-O-acetyl-2-(trichloroacetyl)-β-D-glucopyranosyl chloride, from which the 2-O-acyl group may be selectively cleaved by reaction with an ethereal solution of ammonia. This procedure has been applied to β-maltose octaacetate[132,133] and to β-maltotriose hendecaacetate,[134] in order to synthesize the corresponding 2-hydroxy derivatives.

The chloroacetyl group has found considerable use in the synthesis of partially acylated sugars, its particular advantage being that it can be selectively removed under mild, neutral conditions on treatment with thiourea,[135,136] and problems of acyl migration are thereby avoided. 1,2,4,6-Tetra-O-acetyl-3-O-(chloroacetyl)-β-D-glucopyranose gave[137] the corresponding 1,2,4,6-tetraacetate in 43% yield, and β-D-glucopyranosyl benzoate was obtained[137] from the corresponding 2,3,4,6-tetrakis(chloroacetate) in 14% yield, but some difficulties were noted[138] in the removal of the 6-O-(chloroacetyl) group in D-glucose derivatives. 2,3-Di-O-acetyl-5-O-(chloroacetyl)-D-xylofuranose could be partially de-esterified to give[139] 2,3-di-O-acetyl-D-xylopyranose, and, similarly, 1,2,3-tri-O-acetyl-β-D-xylopyranose and 1,2,3,4-tetra-O-acetyl-β-D-glucopyranose were obtained[140] from the corresponding chloroacetyl derivatives; in the latter instance, reaction with thiourea in acetonitrile–water appeared advantageous as compared to reaction[138] in methanol. The chloroacetyl group has been used as a temporary blocking group in the synthesis of 1,3,4,6-tetra-O-acetyl-β-D-galactopyranose,[141] 1,3,4,6-tetra-O-acetyl-β-D-glucopyranose,[141] and benzyl 2-acetamido-3,4-di-O-acetyl-2-deoxy-β-D-glucopyranoside.[142]

The selective removal of a formyl group in the presence of a ben-

(130) W. Mayer, G. Schultz, S. Wrede, and G. Schilling, Ann., (1975) 946–952.
(131) P. Brigl, Z. Physiol. Chem., 116 (1921) 1.
(132) B. H. Koeppen, Carbohydr. Res., 13 (1970) 193–198.
(133) K. Takeo, Carbohydr. Res., 48 (1976) 290–293.
(134) K. Takeo, K. Mine, and T. Kuge, Carbohydr. Res., 48 (1976) 197–208.
(135) M. Masaki, T. Kitahara, H. Kurita, and M. Ohta, J. Am. Chem. Soc., 90 (1968) 4508–4509.
(136) A. Fontana and E. Scoffone, Gazz. Chim. Ital., 98 (1968) 1261–1269.
(137) M. Bertolini and C. P. J. Glaudemans, Carbohydr. Res., 15 (1970) 263–270.
(138) D. Y. Gagnaire and P. J. A. Vottero, Carbohydr. Res., 28 (1973) 165–170.
(139) J. P. Utille and P. J. A. Vottero, Carbohydr. Res., 52 (1976) 241–245.
(140) N. Héran, J. P. Utille, and P. J. A. Vottero, Carbohydr. Res., 53 (1977) 268–275.
(141) V. M. Chari, M. Jordan, and H. Wagner, Acta Chim. Acad. Sci. Hung., 93 (1977) 99–102; Chem. Abstr., 88 (1978) 51,103k.
(142) K. L. Matta and J. J. Barlow, Carbohydr. Res., 51 (1976) 215–222.

zoyl group has been achieved[143] by careful treatment of 5-O-benzoyl-2-deoxy-2-fluoro-3-O-formyl-D-arabinofuranose with sodium methoxide in methanol. In contrast, attempts to hydrolyze selectively the formyl group in 1,2,4,6-tetra-O-acetyl-3-O-formyl-β-D-glucopyranose were unsuccessful,[144] which is surprising in view of rate differences observed in the hydrolysis of formic and acetic ester groups in nucleosides (see Section III,3).

The trifluoroacetyl group, which is extremely labile towards hydrolysis, may be readily removed under mild conditions from carbohydrate mixed-esters. Thus, treatment of an acetone solution of methyl 3-O-benzoyl-4,6-O-benzylidene-2-O-(trifluoroacetyl)-α-D-glucopyranoside with dilute sodium hydroxide solution, or with water only, or simple storage of the mixed ester in ethanol, gave the 3-benzoate.[96]

Enzymic hydrolysis can often provide the means for solving a problem involving the partial hydrolysis of esters. A synthesis[145] of 1-O-abscisoyl-β-D-glucopyranose proved difficult at the last step, but the use of a crude enzyme-preparation obtained from ripe seeds of *Helianthus annus* allowed deacetylation to be achieved at O-2, 3, 4, and 6 of the D-glucosyl group without rupture of the ester bond to the anomeric center.

3. Selective Deacylation in Nucleoside Derivatives

The selective deacylation of nucleoside derivatives has played an important part in the development of synthetic methods for the preparation of oligonucleotides of defined structure by providing partially protected monomer units capable of being phosphorylated and then linked together in a controlled way. Todd and coworkers[146] obtained partially protected 2'-deoxyadenosine and 2'-deoxyguanosine by treatment of the 3',5'-di-O-acetyl derivatives with ammonia in alcoholic solution, followed by separation of the resultant mixtures by countercurrent distribution. Similarly, 5'- and 3'-O-acetylthymidine were prepared[147] from 3', 5'-di-O-acetylthymidine, and 5'-O-acetyluridine[148] from 2',3',5'-tri-O-acetyluridine. 5'-O-Acetyladenosine was obtained[149] through partial hydrolysis of the corresponding triacetate, without recourse to countercurrent distribution.

(143) U. Reichman, K. A. Watanabe, and J. J. Fox, *Carbohydr. Res.*, 42 (1975) 233–240.
(144) S. J. Angyal and K. James, *Carbohydr. Res.*, 12 (1970) 147–149.
(145) H. Lehmann, O. Miersch, and H. R. Schütte, *Z. Chem.*, 15 (1975) 443.
(146) D. H. Hayes, A. M. Michelson, and A. R. Todd, *J. Chem. Soc.*, (1955) 808–815.
(147) A. M. Michelson and A. R. Todd, *J. Chem. Soc.*, (1955) 2632–2638.
(148) D. M. Brown, A. R. Todd, and S. Varadarajan, *J. Chem. Soc.*, (1956) 2388–2393.
(149) A. M. Michelson, L. Szabó, and A. R. Todd, *J. Chem. Soc.*, (1956) 1546–1549.

In related reactions, Rammler and Khorana reported[150] that methanolysis of N^4-benzoyl-2',3',5'-tri-O-benzoylcytidine in methanol–oxolane containing sodium methoxide gave N^4-benzoyl-5'-O-benzoyl-, N^4-benzoyl-3',5'-di-O-benzoyl-, and N^4-benzoyl-cytidine. No N^4-benzoyl-2',5'-di-O-benzoyl derivative was found, presumably as a consequence of benzoyl migration[1] from O-2' to O-3'.

Smrt and coworkers compared[151] the behavior of O-formyl and O-acetyl derivatives of certain nucleosides towards methanolysis and ascertained that the formyl group is removed at a considerably higher rate than the acetyl group in boiling methanol. In addition, methanolysis of the 2',3',5'-tri-O-formyl derivatives of uridine, 6-azauridine, and 6-azacytidine indicated that both of the secondary formic ester groups in each of these compounds were hydrolyzed more quickly than the primary formic ester group. Based on these observations, a synthesis of 2',3'-di-O-acetyluridine was developed.[151,152]

The greater hydrolytic lability of O-formyl than of O-benzoyl groups has been used in the preparation of ribonucleoside 2'-acetal 5'-esters.[153] Acid-catalyzed reaction between 5,6-dihydro-4-methoxy-$2H$-pyran and N^2-benzoyl-5'-O-benzoyl-3'-O-formylguanosine, followed by selective deformylation of the product with dilute methanolic ammonia, gave N^2-benzoyl-5'-O-benzoyl-2'-O-(4-methoxytetrahydropyran-4-yl)guanosine.

An interesting approach[149] to the synthesis of partially protected nucleosides involved fusion for 12 hours at 130° of 5'-O-acetyladenosine and 2',3',5'-tri-O-acetyladenosine, followed by separation, by countercurrent distribution, of the so-formed mixture of 3',5'-di-O-acetyl-, 5'-O-acetyl-, and 2',3',5'-tri-O-acetyladenosine. It is noteworthy that no 2',5'-diester was isolated. A variation[154] of this procedure was the fusion, at ~200°, of a mixture of 2',3',5'-tri-O-acetyladenosine and adenosine, which yielded the 2',3',5'-triacetate, the 2',3'-diacetate, and the 3',5'-diacetate, with the last preponderating.

It has been suggested[155] that a base-catalyzed, acyl-transfer reaction occurs in certain reactions involving 3',5'-di-O-aroyl derivatives of

---

(150) D. H. Rammler and H. G. Khorana, *J. Am. Chem. Soc.*, 84 (1962) 3112–3122.
(151) J. Žemlička, J. Beránek, and J. Smrt, *Collect. Czech. Chem. Commun.*, 27 (1962) 2784–2795.
(152) F. Šorm, J. Žemlička, J. Beránek, and J. Smrt, Czech. Pat. 109,963 (1964); *Chem. Abstr.*, 61 (1964) 1932.
(153) J. H. van Boom, G. R. Owen, J. Preston, T. Ravindranathan, and C. B. Reese, *J. Chem. Soc., C*, (1971) 3230–3237.
(154) L. Szabó, *Bull. Soc. Chim. Fr.*, (1966) 3159.
(155) T. Sasaki, K. Minamoto, and T. Okugawa, *J. Org. Chem.*, 43 (1978) 350–353.

TABLE I

Half-times of Reaction[157] for Deacylation of
5'-O-Acyluridines with Ammonia at 22°

Acyl group	$t_{1/2}$ (min)	
	Reagent I[a]	Reagent II[b]
MeCO	191	59
MeOCH$_2$CO—	10.4	2.5
PhOCH$_2$CO—	3.9	<1[c]
HCO—	0.4	(0.22)[d]
ClCH$_2$CO—	0.28	(0.17)[d]

[a] Aqueous 155 mM ammonia was used in at least tenfold excess. [b] This reagent was prepared by diluting methanol, presaturated with ammonia at 0°, with an equal volume of methanol. [c] Reaction too fast to measure. [d] Figures in parentheses represent times by which complete solvolysis of the substrate had occurred.

1-β-D-arabinofuranosyluracil, leading to the formation of 2',3',5'-tri-O-aroyl and 3'-O-aroyl derivatives.

The observation[156] that treatment of 3',5'-di-O-acetyladenosine with morpholine afforded 5'-O-acetyladenosine in ~95% yield, whereas, under the same conditions, 3',5'-di-O-acetyl-2'-deoxyadenosine and 2',5'-di-O-acetyl-3'-deoxyadenosine were unchanged, emphasized the importance of the vicinal hydroxyl group in this selective hydrolysis. In an attempt to gain a fuller understanding of selective de-esterification of nucleoside derivatives, the rates of deacylation of 5'-O-acetyl-, 5'-O-(methoxyacetyl)-, 5'-O-(phenoxyacetyl)-, 5'-O-formyl-, and 5'-O-(chloroacetyl)-uridines were measured[157] in aqueous and in methanolic ammonia. The half-times of reaction ($t_{1/2}$) with the two reagents are recorded in Table I; the relative rates of solvolysis of the acyl groups in dilute aqueous ammonia are as follows: MeCO (1), MeOCH$_2$CO (17), PhOCH$_2$CO (49), HCO (480), and ClCH$_2$CO (680). On the basis of these data, a general procedure was developed for the preparation of 3'-O-acyl-2'-O-(4-methoxytetrahydropyran-4-yl)ribonucleosides by selective deacylation of suitable 3',5'-di-O-acyl derivatives. The selective removal of the 5'- over the 3'-ester group is further favored by the primary nature of O-5', and also, presumably, by the presence of the bulky (4-methoxytetrahydropyran-4-yl) group on O-2'.

(156) B. E. Griffin and C. B. Reese, *Proc. Natl. Acad. Sci. USA*, 51 (1964) 440–444.
(157) C. B. Reese, J. C. M. Stewart, J. H. van Boom, H. P. M. de Leeuw, J. Nagel, and J. F. M. de Rooij, *J. Chem. Soc. Perkin Trans. 1*, (1975) 934–942.

In contrast, selective removal of a 3'-O-acetyl group was achieved[158] on treatment of 3'-O-acetyl-5'-O-pivaloyl-2'-O-(tetrahydropyran-2-yl)uridine with methanolic ammonia. Pivalic (trimethylacetic) esters are comparatively stable to this reagent, presumably as a result of the bulky trimethylmethyl groups hindering the approach of nucleophiles to the carbonyl carbon atom, but they may conveniently be hydrolyzed by using aqueous, methanolic tetraethylammonium hydroxide.[158,159] It is interesting that, for the hydrolysis of 5'-O-pivaloylthymidine in M sodium hydroxide solution at 0°, $t_{1/2}$ was ~3 minutes, and hydrolysis was complete within ~30 minutes.[160] Under these conditions 5'-O-acetylthymidine was completely hydrolyzed to thymidine within 30 seconds. At 23°, the half-life for the hydrolysis of the 2,4,6-trimethylbenzoate group in 5'-O-(2,4,6-trimethylbenzoyl)thymidine 3'-phosphate was ~80 hours.[160]

Protection of a hydroxyl group at C-5' in a nucleoside has been performed[161] by using crotonic (2-butenoic) and related esters. A remarkably selective removal of the 5'-O-acyl group was observed on treatment of 3'-O-acetyl-5'-O-(4-methoxycrotonoyl)thymidine with 0.05 M hydrazine hydrate in methanol–pyridine for 2 hours at 20°.

In connection with the synthesis of C-glycosyl compounds, Moffatt and coworkers[162] developed the selective debenzoylation of 2,3,5-tri-O-benzoyl-β-D-ribofuranosyl cyanide, which had been reported earlier[163]; reaction in methanolic ammonia–chloroform at 0° gave the crystalline 5'-O-benzoyl derivative in 83% yield, without recourse to chromatography.

A notable, selective deacylation at O-2' in N^6,N^6-dibenzoyl-2',3',5'-tri-O-benzoyladenosine was achieved[164] by treatment of the ester with hydrazine hydrate in acetic acid–pyridine for 7 days at room temperature; chromatography of the crude product, followed by crystallization, gave 3',5'-di-O-benzoyladenosine in 70% yield. N^2-Benzoyl-2',3',5'-tri-O-benzoylguanosine and 2',3',5'-tri-O-benzoylinosine gave lower yields (48 and 52%, respectively) of the corresponding 2'-hydroxy compounds. 2',3',5'-Tri-O-benzoyluridine gave a 2:1 mixture of 3',5'- and 2',5'-di-O-benzoyluridine in 65% yield, whereas N^4-ben-

(158) B. E. Griffin, M. Jarman, and C. B. Reese, *Tetrahedron*, 24 (1968) 639–662.
(159) B. E. Griffin and C. B. Reese, *Tetrahedron Lett.*, (1964) 2925–2931.
(160) G. Weimann and H. G. Khorana, *J. Am. Chem. Soc.*, 84 (1962) 4329–4341.
(161) R. Arentzen and C. B. Reese, *J. Chem. Soc. Chem. Commun.*, (1977) 270–272.
(162) H. P. Albrecht, D. B. Repke, and J. G. Moffatt, *J. Org. Chem.*, 38 (1973) 1836–1840.
(163) J. A. Montgomery and K. Hewson, *J. Heterocycl. Chem.*, 7 (1970) 443–445.
(164) Y. Ishido, N. Nakazaki, and N. Sakairi, *J. Chem. Soc. Chem. Commun.*, (1976) 832–833.

zoyl-2',3',5'-tri-O-benzoylcytidine gave a mixture of the di-O-benzoyl derivatives in 21% yield. It is interesting that the same reagent leads to specific N-debenzoylation of poly-N,O-benzoyl 2'-deoxynucleosides[165]; that is, in such cases, amide bonds are cleaved in preference to ester bonds. A similar selectivity is found in the reaction of some phenols and alcohols with poly-N,O-benzoyl derivatives of adenosine and cytidine.[166]

Other workers[167] found difficulty in monitoring the hydrazine-mediated debenzoylation of N^2-benzoyl-2',3',5'-tri-O-benzoylguanosine, and failed to obtain a crystalline product. However, deacetylation of 2',3',5'-tri-O-acetyl-N^2-benzoylguanosine was conveniently monitored by ^{1}H-n.m.r. spectroscopy, and the 3',5'-di-O-acetyl-N^2-benzoyl derivative was obtained crystalline in 61% yield from a crude product indicated by t.l.c. to be a mixture of diacetates (~90%), monoacetates (~10%), and a trace of the starting material. Similar hydrazinolysis of 2',3',5'-tri-O-acetyladenosine gave the 3',5'-di-O-acetyl derivative in 59% yield. It was noted[167] that, although the original report[164] on this procedure of deacylation by hydrazinolysis claimed the reaction to be regioselective, base-catalyzed acyl migration was operative in the case of the acetates. The diacetates resulting directly from the acetyl cleavage were the 2',5'- and 3',5'-diacetates in the ratio of ~1:5, and the latter ester could be obtained from the mixture by crystallization.

Regioselective deacylation at O-2' has been reported[168] on treatment of fully acylated purine and pyrimidine ribonucleosides with hydroxylaminium acetate in pyridine, and, by using an excess of the reagent, 5'-O-acylribonucleosides were obtained in some cases. A later publication[168a] from the same researchers pointed out that they had incorrectly interpreted their results on the supposed, regioselective hydrazinolysis[164] and hydroxylaminolysis[168] of fully acylated purine and pyrimidine nucleosides; although partial deacylation of fully acylated ribonucleosides does appear to occur preferentially at O-2', the highly favored formation of the 3',5'-di-O-acyl derivatives was due to a re-equilibration of the 2',5'- and 3',5'-diesters on the sil-

(165) R. L. Letsinger, P. S. Miller, and G. W. Grams, *Tetrahedron Lett.*, (1968) 2621–2624.
(166) Y. Ishido, N. Nakazaki, and N. Sakairi, *J. Chem. Soc. Perkin Trans. 1*, (1977) 657–660.
(167) R. J. Gregoire and T. Neilson, *Can. J. Chem.*, 56 (1978) 487–490.
(168) Y. Ishido and N. Sakairi, *Nucleic Acids Res., Spec. Publ.*, 3 (1977) 13–15; *Chem. Abstr.*, 88 (1978) 136,890q.
(168a) Y. Ishido, N. Sakairi, and I. Hirao, *Nucleic Acids Res., Spec. Publ.*, 5 (1978) 263–265.

ica gel used for column chromatography, to afford a mixture in which the 3′,5′-isomer preponderated. The effectiveness of the re-equilibration seemed to depend on the type of silica gel. Other workers who examined[167] the hydrazinolysis procedure had suggested, implicitly, that base-catalyzed, acyl migration might be the cause of the apparent regiospecificity.

Regioselective deacylation has also been observed[169] in the acid-catalyzed transesterification of some 3′,5′-di-O-acetyl-2′-deoxy-2′-halo-uridines in methanol to give the corresponding 3′-O-acetyl derivatives in yields of 36–66%. The presence of the 2-halogen substituent seems essential for the selectivity; in the case of 2′,3′,5′-tri-O-acetyluridine, no significant differences in the reaction rates of the acetyl groups towards transesterification were apparent.

Subtle variations in the structures of acyl groups attached to nucleosides have been made in order to allow such groups to be removed under mild or special conditions that do not affect other acyl groups. The (2,2,2-tribromoethoxy)carbonyl group is susceptible to non-hydrolytic removal by using a zinc–copper couple in acetic acid; thus, 3′-O-acetyl-5′-O-(2,2,2-tribromoethoxycarbonyl)- and 5′-O-acetyl-3′-O-(2,2,2-tribromoethoxycarbonyl)-thymidine gave[170] 3′-O-acetyl- and 5′-O-acetyl-thymidine in a yield of 75 and 60%, respectively.

The use of 3-benzoylpropanoyl[171,172] and benzoylformyl[172] groups in nucleoside chemistry has been described. The former group is readily removed by treatment with hydrazine hydrate in pyridine buffered with acetic acid, under conditions that, essentially, do not affect O-acetyl groups[171]; the ketone function serves as a trigger for the cleavage by reacting preferentially with the hydrazine to give a hydrazone in which the nucleophilic amino group is favorably disposed to attack the neighboring, ester-carbonyl group. The benzoylformyl group is selectively removed by hydrolysis in aqueous pyridine at room temperature. As an example of the use of the latter group, 5′-O-(3-benzoylpropanoyl)thymidine was prepared[172] in 31% yield by esterifying 3′-O-(benzoylformyl)thymidine with 3-benzoylpropanoic acid, and then selectively hydrolyzing the diester so obtained.

Unfortunately, under conditions used for removal of the 3-benzoylpropanoyl group, N-protecting benzoyl groups on 2′-deoxycytidine and 2′-deoxyadenosine are unstable[165,172]; the presence of such N-pro-

(169) M. Merész, P. Soháh, and J. Kuszmann, *Tetrahedron*, 33 (1977) 2131–2133.
(170) A. F. Cook, *J. Org. Chem.*, 33 (1968) 3589–3593.
(171) R. L. Letsinger, M. H. Caruthers, P. S. Miller, and K. K. Ogilvie, *J. Am. Chem. Soc.*, 89 (1967) 7146–7147.
(172) R. L. Letsinger and P. S. Miller, *J. Am. Chem. Soc.*, 91 (1969) 3356–3359.

tecting groups in these and related nucleosides is necessary during phosphorylation reactions. This problem was overcome[173] by recognition of the fact that nucleophilic attack on a ketone function is much slower when the α-substituent is an aryl rather than an alkyl group. Use of the 4-oxopentanoyl (levulinoyl) protecting group, which was found to be ~100 times as labile as the 3-benzoylpropanoyl group under the conditions of the hydrazinolysis, allowed selective removal of this acyl group from an ester, in N,O-protected nucleosides, without the concurrent removal of a protecting N-aroyl group.

The 4-oxopentanoyl group has found useful application[173a] as a protecting group for the 5'-hydroxyl group in a series of nucleotide derivatives used in the synthesis of a tetradeca(ribonucleotide), the particularly important property of this protecting group being the ease and selectivity with which it is removed by hydrazinolysis under essentially neutral conditions. The selective removal of 4-oxopentanoyl groups under such conditions in the presence of acetyl groups has been exploited[173b] in the synthesis of a trisaccharide derivative; in this synthesis, acetyl groups are used as "persistent," and 4-oxopentanoyl groups as "temporary" protecting groups. Thus, reaction of 1,2:3,4-di-O-isopropylidene-α-D-galactopyranose with 3,4,6-tri-O-acetyl-2-O-(4-oxopentanoyl)-α-D-galactopyranosyl bromide gave 1,2:3,4-di-O-isopropylidene-6-O-[3,4,6-tri-O-acetyl-2-O-(4-oxopentanoyl)-β-D-galactopyranosyl]-α-D-galactopyranose, from which the 4-oxopentanoyl group was removed selectively, in < 10 minutes, by treatment, in pyridine solution, with a solution of hydrazine hydrate in pyridine–acetic acid. Glycosylation of the disaccharide derivative, at the hydroxyl group so liberated, with 3,4,6-tri-O-acetyl-2-O-(4-oxopentanoyl)-α-D-galactopyranosyl bromide then gave the trisaccharide derivative O-[3,4,6-tri-O-acetyl-2-O-(4-oxopentanoyl)-β-D-galactopyranosyl]-(1→2)-O-(3,4,6-tri-O-acetyl-β-D-galactopyranosyl)-(1→6)-(1,2:3,4-di-O-isopropylidene-α-D-galactopyranose), from which the 4-oxopentanoyl group could again be removed selectively. Repetition of the glycosylation–deprotection sequence would, of course, allow the preparation of higher oligosaccharides, and it is clear that this approach to the synthesis of complex oligosaccharides is of considerable promise and, potentially, of great versatility.

In Section III,2, mention was made of the use of the chloroacetyl

(173) J. H. van Boom and P. M. J. Burgers, *Tetrahedron Lett.*, (1976) 4875–4878.
(173a) J. H. van Boom and P. M. J. Burgers, *Recl. Trav. Chim. Pays-Bas*, 97 (1978) 73–80.
(173b) H. J. Koeners, J. Verhoeven, and J. H. van Boom, *Tetrahedron Lett.*, (1980) 381–382.

group in the synthesis of partially acylated sugars, and investigations into its use in nucleoside chemistry have been conducted.[174] 3'-O-Acetyl-5'-O-(chloroacetyl)thymidine was readily deacylated at O-5' on treatment with thiourea in refluxing ethanol, to give the 3'-acetate in 81% yield. Although similar treatment of N^6-benzoyl-3',5'-di-O-(chloroacetyl)adenosine led to cleavage of the C-1'–N-9 bond, with the formation of N^6-benzoyladenine, O-deacylation could be achieved with 2-aminoethanethiol hydrochloride or ethylenediamine dihydrochloride in pyridine–methanol–triethylamine, or with o-phenylenediamine in pyridine–ethanol. This means of O-deprotection of an N,O-protected purine nucleoside is particularly important in view of the acid-lability of the purine–glycosyl bond often found for derivatives in which the purine bears an N-acyl group. This instability can limit the use of acid-labile, O-protecting groups in such compounds. O-Deacylation of N^4-anisoyl-3'-O-(chloroacetyl)-2'-deoxy-5'-O-(p-methoxytrityl)cytidine with thiourea in ethanol–pyridine proceeded satisfactorily, without loss of the other protecting groups.

An O-(2,4-dinitrophenylsulfenyl) group may be readily removed on treatment of the sulfenic ester with benzenethiol in pyridine,[175] or with hydrogen sulfide in pyridine,[176] and reports of its use in nucleoside chemistry include the transformation of 3'-O-(2,4-dinitrophenylsulfonyl)thymidine into 5'-O-acetylthymidine,[175] and the preparation of N^6-benzoyl-3'-O-(3-benzoylpropanoyl)-2'-deoxyadenosine[176] from the corresponding 5'-(2,4-dinitrobenzenesulfenate). The 2,4-dinitrophenylsulfenyl group may also be removed under neutral conditions by treatment with Raney nickel,[177] and it would appear that, by this method, O- or N-(2,4-dinitrophenylsulfenyl) groups in nucleosides may be removed in the respective presence of N- or O-acetyl groups.

Surprisingly few investigations appear to have been made into selective deacylation with the aid of enzymes. α-Chymotrypsin is very sluggish in hydrolyzing acetates of simple nucleosides and nucleotides, but dihydrocinnamic (3-phenylpropanoic) esters appear to be more satisfactory as substrates, and, with such derivatives of nucleosides, it seems that selective deacylation may be achieved enzymically.[178] Thus, enzymic hydrolysis of 2'-deoxy-3',5'-di-O-(dihydrocinnamoyl)uridine gave the 3'-ester as the only intermediate to the

(174) A. F. Cook and D. T. Maichuk, *J. Org. Chem.*, 35 (1970) 1940–1943.
(175) R. L. Letsinger, J. Fontaine, V. Mahadevan, D. A. Schexnayder, and R. E. Leone, *J. Org. Chem.*, 29 (1964) 2615–2618.
(176) G. W. Grams and R. L. Letsinger, *J. Org. Chem.*, 33 (1968) 2589–2590.
(177) F. Eckstein, *Tetrahedron Lett.*, (1965) 531–535.
(178) H. S. Sachdev and N. A. Starkovsky, *Tetrahedron Lett.*, (1969) 733–736.

formation of 2′-deoxyuridine; for comparison, it may be noted that deacylation of the diester with methanolic ammonia gave a 1:1 mixture of the two monoesters as intermediates to 2′-deoxyuridine.

The dihydrocinnamoyl, 2-(o-phenylenedioxy)acetyl, and D-(+)-dihydrocoumariloyl groups have been compared[179] as enzymically removable, O-protecting groups, and a stepwise synthesis of a tetranucleotide was achieved by using the dihydrocinnamoyl group as an HO-3′ protecting group, and α-chymotrypsin for the deprotection of intermediates. No loss of any base-protecting groups was observed under the conditions of enzymic hydrolysis.

IV. DEALKYLATION

In this Section is considered the selective cleavage of $O-CR_3$ bonds in sugar derivatives, where R may be hydrogen, alkyl, or aryl. Cleavage of triphenylmethyl (trityl) groups or benzyl groups in the presence of methyl groups will not be discussed. Such operations are usually straightforward, as a result of the relative stability of methyl ethers, and the ready cleavage of benzyl ethers by hydrogenolysis,[180] by sodium–alcohol,[180] or sodium–ammonia[181,182] reduction, or bromination–hydrolysis,[183] and of trityl ethers by hydrogenolysis[184] and by treatment with an acid.[184]

There are surprisingly few methods for the selective removal of one or more O-alkyl groups from a carbohydrate derivative that is multi-O-substituted with the same type of alkyl group. Boron trichloride[185,186] and boron tribromide[186] were found to cause demethylation of methyl ethers of sugars, the latter reagent being more effective in achieving this result in a single treatment, but, in many cases, partially demethylated products were formed in minor proportions. The composition of the mono-O-methylated product obtained[186] on treatment of 2,3,4,6-tetra-O-methyl-D-glucose with boron trichloride indicated that selective demethylation had not occurred.

Demethylation of methylated sugars by means of hydrogen peroxide in the presence of ferrous ions (the Fenton reagent) has been re-

(179) A. Taunton-Rigby, *J. Org. Chem.*, 38 (1973) 977–985.
(180) C. M. McCloskey, *Adv. Carbohydr. Chem.*, 12 (1957) 137–156.
(181) E. J. Reist, V. J. Bartuska, and L. Goodman, *J. Org. Chem.*, 29 (1964) 3725–3726.
(182) U. G. Nayak and R. L. Whistler, *J. Org. Chem.*, 34 (1969) 97–100.
(183) J. N. BeMiller, R. E. Wing, and C. Y. Meyers, *J. Org. Chem.*, 33 (1968) 4292–4294.
(184) B. Helferich, *Adv. Carbohydr. Chem.*, 3 (1948) 79–111.
(185) S. Allen, T. G. Bonner, E. J. Bourne, and N. M. Saville, *Chem. Ind. (London)*, (1958) 630.
(186) T. G. Bonner, E. J. Bourne, and S. McNally, *J. Chem. Soc.*, (1960) 2929–2934.

ported,[187] but the reaction appears to show little specificity, and further oxidation of the sugar occurs. Thus, treatment of 3,4-di-O-methyl-D-mannitol afforded 3,4-di-O-methyl-D-mannose, 3- and 4-O-methyl-D-mannose, 3-O-methyl-D-mannitol, D-mannitol, and D-mannose. Methyl 2,3,4,6-tetra-O-methyl-α-D-glucopyranoside gave, amongst other products, D-glucose and the possible mono-, di-, tri-, and tetra-O-methyl-D-glucoses after acid hydrolysis of the reaction mixture. There was an indication that the methoxyl groups on C-2 and C-4 are the most strongly attacked, followed by those on C-6, and that those on C-3 are the least strongly attacked.

Demethylation of methylated sugars may be brought about by the ethylamine–lithium reagent.[188] Two aminoglycosteroids were demethylated in the sugar moiety by this reagent without cleavage of the glycoside bond, and the methyl group at O-3 in methyl 2,6-dideoxy-3-O-methyl-α-D-*ribo*-hexopyranoside was also cleanly removed by the reagent. It would be of interest to ascertain whether selective demethylation of a poly-O-methyl derivative of a sugar is possible.

Benzyl ethers are very stable to alkaline reagents, and sufficiently stable to aqueous acids and to hydrogen bromide in acetic acid to permit the removal of trityl groups in their presence.[180] Acetolyzing agents may induce debenzylation if the conditions are sufficiently drastic.[189] Acetolysis of 1,6-anhydro-2,3,4-tri-O-benzyl-β-D-glucose for 3 minutes with sulfuric acid at low concentration (0.22%) gave[190] 1,6-di-O-acetyl-2,3,4-tri-O-benzyl-D-glucopyranose, but reaction for 18 hours at a higher concentration (2%) of acid afforded[191] α-D-glucopyranose pentaacetate. A study[191] of hexitol derivatives indicated that useful differences in reactivity towards acetolysis may exist between benzyl groups and other hydroxyl-protecting groups; primary benzyl ethers are cleaved more readily than methylene acetal groups linking secondary positions, although methylene acetal groups linking primary and secondary positions are cleaved more readily than secondary benzyl ethers. There is limited evidence,[191] based on the acetolysis of 1,4:3,6-dianhydro-2,5-di-O-benzyl-D-mannitol, that O-benzyl groups may be cleaved more rapidly than oxolane rings.

Selective acetolysis has been achieved in useful yields with some

(187) B. Fraser-Reid, J. K. N. Jones, and M. B. Perry, *Can. J. Chem.*, 39 (1961) 555–563.
(188) Q. Khuong-Huu, C. Monneret, L. Kaboré, and R. Goutarel, *Tetrahedron Lett.*, (1971) 1935–1938.
(189) H. Burton and P. G. F. Praill, *J. Chem. Soc.*, (1951) 522–529.
(190) G. Zemplén, Z. Csürös, and S. J. Angyal, *Ber.*, 70 (1937) 1848–1856.
(191) R. Allerton and H. G. Fletcher, Jr., *J. Am. Chem. Soc.*, 76 (1954) 1757–1760.

benzyl ethers of *myo*-inositol.[192] Treatment of DL-1,4,5,6-tetra-*O*-benzyl-3-*O*-methyl-*myo*-inositol (**13**) with acetic anhydride–perchloric acid for 2 hours at 100° gave DL-2,4,5,6-tetra-*O*-acetyl-1-*O*-benzyl-3-*O*-methyl-*myo*-inositol (**14**) in 61% yield. DL-1,4,5,6-Tetra-*O*-benzyl-*myo*-inositol (**15**) and 1,3,4,5,6-penta-*O*-benzyl-*myo*-inositol (**16**) gave, at 0°, DL-1,2,4,5,6-penta-*O*-acetyl-3-*O*-benzyl- and 2,4,5,6-tetra-*O*-acetyl-1,3-di-*O*-benzyl-*myo*-inositol, (**17**) and (**18**), respectively, in yields of 83 and 60%. The acetolysis of hexa-*O*-benzyl-*myo*-inositol (**19**) was less selective. From the reaction conducted for 100 minutes at 0° was isolated 4,5,6-tri-*O*-acetyl-1,2,3-tri-*O*-benzyl-*myo*-inositol (**20**) in low yield (~10%), whereas reaction for 48 hours at 0° gave a range of products, from which **17**, **18**, **20**, and 1,3,4,5,6-penta-*O*-acetyl-2-*O*-benzyl-*myo*-inositol (**21**) were isolated. Compounds **17** and **21** were obtained

13 $R^1 = R^4 = R^5 = R^6 = CH_2Ph$, $R^2 = H$, $R^3 = Me$
14 $R^1 = CH_2Ph$, $R^2 = R^4 = R^5 = R^6 = Ac$, $R^3 = Me$
15 $R^1 = R^4 = R^5 = R^6 = CH_2Ph$, $R^2 = R^3 = H$
16 $R^1 = R^3 = R^4 = R^5 = R^6 = CH_2Ph$, $R^2 = H$
17 $R^1 = R^2 = R^4 = R^5 = R^6 = Ac$, $R^3 = CH_2Ph$
18 $R^1 = R^3 = CH_2Ph$, $R^2 = R^4 = R^5 = R^6 = Ac$
19 $R^1 = R^2 = R^3 = R^4 = R^5 = R^6 = CH_2Ph$
20 $R^1 = R^2 = R^3 = CH_2Ph$, $R^4 = R^5 = R^6 = Ac$
21 $R^1 = R^3 = R^4 = R^5 = R^6 = Ac$, $R^2 = CH_2Ph$

For racemic compounds, only one enantiomer is depicted.

in the ratio of ~2:1, suggesting that the rates of acetolysis at O-1 and O-2 are approximately equal. The results indicated that these reactions are subject to considerable steric influences, and that the rates of acetolysis at O-1 and O-2 in *myo*-inositol derivatives are both much less than those at O-4 and O-5. It appears that addition of an acetylium ion to the ether oxygen atom is subject to hindrance by neighboring groups, that an acetoxyl groups exerts greater hindrance than a benzyloxy group, and that *cis*-vicinal cause greater hindrance than *trans*-vicinal groups. These observations are in agreement with the result of acetolysis of 3,4-6-tri-*O*-benzyl-1,2-*O*-(1 methoxyethylidene)-β-D-mannopyranose, which, on treatment at room temperature with an acetolysis mixture containing 2% of sulfuric acid, gave[193] 1,2,4,6-tetra-*O*-ace-

(192) S. J. Angyal, M. H. Randall, and M. E. Tate, *J. Chem. Soc.*, C, (1967) 919–922.
(193) M. M. Ponpipom, *Carbohydr. Res.*, 59 (1977) 311–317.

tyl-3-O-benzyl-α-D-mannopyranose in 59% yield. Treatment of 0° for a shorter time gave 1,2,6-tri-O-acetyl-3,4-di-O-benzyl-α-D-mannopyranose in 71% yield, and the ease of cleavage of the O-benzyl groups therefore appeared to be 6 > 4 > 3.

Selective acetolysis of primary benzyl ethers has been reported[193a] as a convenient alternative to the conventional reaction-sequence of tritylation, etherification, detritylation, in order to obtain carbohydrate derivatives having free primary hydroxyl groups and ether-protected secondary hydroxyl groups. Careful monitoring of the reaction by ^{1}H-n.m.r. spectroscopy (in preference to t.l.c.) gives satisfactory control; the rate of this type of reaction appears to vary markedly with slight variations in the acidity. By using this procedure, primary benzyl ethers have been cleaved in the presence of secondary benzyl and allyl substituents. Thus 2,3,4,6-tetra-O-benzyl-D-glucopyranose gave 2,3,4-tri-O-benzyl-D-glucopyranose, and methyl 2-O-allyl-3,4,6-tri-O-benzyl-α-D-glucopyranose gave 2-O-allyl-3,4-di-O-benzyl-D-glucopyranose in good yield, after acetolysis, and deacetylation of the acetylated intermediate.

Trityl ethers are readily cleaved under acidic conditions,[184] and, by introduction of a p-methoxyl group into one or more of the phenyl groups in the trityl group, the lability of the triarylmethyl group towards acidic cleavage may be increased.[45] Thus, the times needed for complete hydrolysis in 80% acetic acid of 5'-O-trityluridine, 5'-O-(p-anisyldiphenylmethyl)uridine [5'-O-(mono-p-methoxytrityl)uridine], 5'-O-(di-p-anisylphenylmethyl)uridine, and 5'-O-(tri-p-anisylmethyl)uridine at 25° were 48 hours, 2 hours, 15 minutes, and <1 minute, respectively. The introduction of each p-methoxyl group increased the rate of hydrolysis under these conditions by a factor of ~10. Although this control of acid-lability has found wide application in nucleotide chemistry, its further investigation in the chemistry of the saccharides appears warranted.

Interesting differences in reactivity between O-trityl groups contained within the same molecule have been reported. On treatment[194] with hydrogen chloride in benzene, 2,3-di-O-benzoyl-1,5-di-O-trityl-D-xylofuranose underwent selective cleavage at O-5, and a similar selectivity has been observed[195] on hydrogenolysis (over palladium catalyst) of 1,3(2),5-tri-O-trityl-D-ribofuranose (or its acetate), and of 1,5-di-O-trityl-D-ribofuranose. Interestingly, treatment[195] of 1,3(2),5-tri-O-

---

(193a) R. Eby, S. J. Sondheimer, and C. Schuerch, *Carbohydr. Res.*, 73 (1979) 273–276.
(194) K. Zeile and W. Kruckenberg, *Ber.*, 75 (1942) 1127–1140.
(195) H. Bredereck and W. Greiner, *Chem. Ber.*, 86 (1953) 717–722.

trityl-2(3)-O-acetyl-D-ribofuranose with acetyl bromide in acetic anhydride afforded, in 70% yield, the product of selective cleavage at O-5, whereas a similar reaction in chloroform gave a product supposedly a mixture of tetra-O-acetyl-α- and -β-D-ribofuranose.

Partial detritylation has been achieved[196] in the case of 3′,5′- and 2′,5′-di-O-trityluridine. With the former derivative, on heating with 80% acetic acid for 1 hour at 50°, partial detritylation occurred to about the same extent at O-5′ and O-3′, but, by contrast, the 2′,5′ derivative, which required a temperature of 80° for reaction to occur, gave 2′-O-trityluridine and uridine, with only traces of 5′-O-trityluridine.

Selective detritylation of 2′,5′-di-O-trityluridine on reaction with methyltriphenoxyphosphonium iodide has been reported.[197] In benzene solution at 50°, 1-(3-deoxy-3-iodo-5-O-trityl-β-D-xylofuranosyl)uracil was isolated in 41% yield after processing, even though the crude reaction-mixture appeared to contain the expected di-O-trityl compound. Interestingly, when the reaction was conducted at room temperature in N,N-dimethylformamide containing pyridine, selective loss of the 5′-O-trityl group occurred, although 1-(3-deoxy-3-iodo-2,5-di-O-trityl-β-D-xylofuranosyl)uracil was also isolated. The reason for the differing selectivities in the two solvents is not yet apparent. In model experiments on the reaction of 2′,5′-di-O-trityluridine with about one molar equivalent of trifluoroacetic acid in benzene, a mixture of monotrityluridine and uridine was produced, but no clearcut preference for removal of a specific trityl group was noted. No detritylation occurred under these conditions in N,N-dimethylformamide.

A subtle approach[198] to the stepwise removal of one member of a family of protecting groups is illustrated by the reductive cleavage of the (1-naphthyldiphenylmethyl) group from 3′-O-(p-anisyldiphenylmethyl)-5′-O-(1-naphthyldiphenylmethyl)thymidine on treatment with the anthracene radical-anion in oxolane. 3′-O-(p-Anisyldiphenylmethyl)thymidine was isolated in 87% yield, and the yield of thymidine was only 3%. It is noteworthy that the relative reactivity of 5′-O-(1-naphthyldiphenylmethyl) and 5′-O-(p-anisyldiphenylmethyl) derivatives of thymidine towards radical anions is the reverse of that towards acid.

The tritylone group (**22**), which has been used[199] for the protection of alcohols, resembles the trityl group, but it is more stable to acid. It

---

(196) G. Kowollik, K. Gaertner, and P. Langen, *Tetrahedron Lett.*, (1972) 3345–3346.
(197) J. P. H. Verheyden and J. G. Moffatt, *J. Org. Chem.*, 35 (1970) 2868–2877.
(198) R. L. Letsinger and J. L. Finnan, *J. Am. Chem. Soc.*, 97 (1975) 7197–7198.
(199) W. E. Barnett, L. L. Needham, and R. W. Powell, *Tetrahedron*, 28 (1972) 419–424.

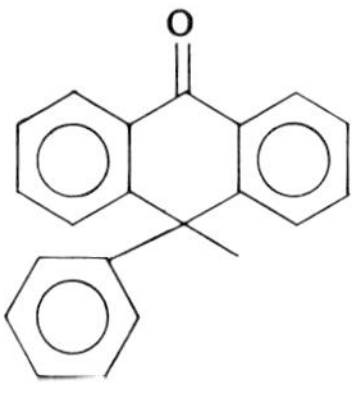

Me
|
R¹OCH(CH₂)₃OR²

23 R¹ = CH₂C₆H₄CN-p, R² = **22**
24 R¹ = CH₂C₆H₄CN-p, R² = H
25 R¹ = H, R² = H

22

has been shown[200] that alcohols protected with this group may be deprotected at room temperature under neutral conditions by cathodic cleavage. Controlled, potential electrolysis at −1.4 V (vs. Ag/AgCl) of diether **23** allowed selective deprotection of the primary hydroxyl group, to afford monoether **24**, which could itself be converted into 1,4-pentanediol (**25**) on electrolysis at −2.1 V. Such controlled deprotection by electrolytic means could find useful applications in carbohydrate chemistry.

Allyl ethers are readily prepared, and are stable to aqueous alkali and to moderately acidic conditions. The allyl group has gained considerable importance as a temporary, protecting group for hydroxyl groups in carbohydrates, largely as a result of the studies of Gigg and coworkers,[201] following reports[202,203] that allyl ethers undergo isomerization to cis-l-propenyl ethers under basic conditions, as shown in equation 1.

$$ROCH_2CH=CH_2 \xrightarrow{\text{base}} \underset{RO}{\overset{H}{\diagdown}}C=C\underset{Me}{\overset{H}{\diagup}} \quad (1)$$

As the latter compounds are vinyl ethers, which are readily hydrolyzed under mildly acidic conditions, the base–acid sequence of reactions applied to an allyl ether brings about dealkylation. The acid hydrolysis in the second step can cause problems if other acid-labile protecting groups are present; for example, 1,2:5,6-di-O-isopropylidene-3-O-(l-propenyl)-α-D-glucofuranose gave[204,205] 1,2-O-isopropylidene-α-D-glucofuranose.

(200) C. van der Stouwe and H. J. Schaefer, *Tetrahedron Lett.*, (1979) 2643–2646.
(201) For a useful summary of the extensive work of this group, see R. Gigg, *ACS Symp. Ser.*, 39 (1977) 253–278.
(202) T. J. Prosser, *J. Am. Chem. Soc.*, 83 (1961) 1701–1704.
(203) C. C. Price and W. H. Snyder, *J. Am. Chem. Soc.*, 83 (1961) 1773.
(204) J. Cunningham, R. Gigg, and C. D. Warren, *Tetrahedron Lett.*, (1964) 1191–1196.
(205) J. Gigg and R. Gigg, *J. Chem. Soc., C*, (1966) 82–86.

Alternative methods for removing 1-propenyl groups under non-acidic conditions have been reported; these include reaction with alkaline potassium permanganate solution,[204,205] ozonolysis followed by alkaline hydrolysis,[204,205] treatment with a mercuric chloride–mercuric oxide reagent,[206,207] and treatment with palladium on activated charcoal in a hydroxylic solvent containing a trace of an acid.[207a] The last procedure is a combined isomerization–hydrolysis, but the isomerization step may be independently achieved by omission of the acid.[207a,207b]

The isomerization of allyl ethers to 1-propenyl ethers, which is usually performed with potassium *tert*-butoxide in dimethyl sulfoxide, can also be carried out under milder conditions using tris(triphenylphosphine)rhodium chloride,[208] and by an ene reaction with diethyl azodicarboxylate,[209,210] which affords a vinyl ether adduct. Removal of an O-allyl group may be achieved by oxidation with selenium dioxide in acetic acid,[211] and by treatment with N-bromosuccinimide, followed by an aqueous base.[201,212]

The facile hydrolysis of 1-propenyl glycosides, which may be prepared from the corresponding allyl glycosides, has been used to advantage in the preparation of some benzylated monosaccharides, in order to overcome the problem of the benzyl ether cleavage that may occur[180] under acidic conditions. Thus, synthesis[205] of the 2,3,4,6-tetrabenzyl ethers of D-glucose and D-galactose is considerably improved by starting with the allyl instead of the methyl glycopyranoside, as hydrolysis of the benzylated glycoside occurs under milder conditions with the former.

In a preparation[213] of 2,3,4-tri-O-benzyl-D-galactopyranose, allyl protecting-groups were removed from O-1 and O-6 of allyl 6-O-allyl-2,3,4-tri-O-benzyl-α-D-galactopyranoside, and the allyl group has been used[214] for the temporary protection of O-2, O-3, O-4, or O-6 in benzyl

---

(206) R. Gigg and C. D. Warren, *Tetrahedron Lett.*, (1967) 1683–1684.
(207) R. Gigg and C. D. Warren, *J. Chem. Soc., C,* (1968) 1903–1911.
(207a) R. Boss and R. Scheffold, *Angew. Chem. Int. Ed. Engl.*, 15 (1976) 558–559.
(207b) M. A. Nashed, *Carbohydr. Res.*, 71 (1979) 299–304.
(208) E. J. Corey and J. W. Suggs, *J. Org. Chem.*, 38 (1973) 3224.
(209) T.-L. Ho and C. M. Wong, *Synth. Commun.*, 4 (1974) 109–111.
(210) E. J. Corey and J. W. Suggs, *Tetrahedron Lett.*, (1975) 3775–3778.
(211) K. Kariyone and H. Yazawa, *Tetrahedron Lett.*, (1970) 2885–2888.
(212) P. A. Gent, R. Gigg, and A. A. E. Penglis, *J. Chem. Soc. Perkin Trans. 1*, (1976) 1395–1404.
(213) R. Gigg and C. D. Warren, *J. Chem. Soc.*, (1965) 2205–2210.
(214) P. A. Gent and R. Gigg, *J. Chem. Soc. Perkin Trans. 1*, (1974) 1446–1455.

α-D-galactopyranoside during the preparation of the four tribenzyl ethers of this glycoside.

An intriguing, preferential rearrangement[207] of the 2-O-allyl group in methyl 2,3-di-O-allyl-4,6-O-benzylidene-α-D-glucopyranoside allowed methylation to be achieved at O-2 or O-3, depending on the reaction sequence used after isomerization.

γ-Substituted allyl ethers are readily cleaved by the action of potassium *tert*-butoxide in dimethyl sulfoxide[207,215] and the O-(2-butenyl) group, which is removed more quickly than the allyl group is isomerized,[216] extends the synthetic utility of such protective groups. As an illustration,[216] the 2-butenyl group of allyl 2,4,6-tri-O-benzyl-3-O-(2-butenyl)-α-D-galactopyranoside (**26**) was completely eliminated by the action of potassium *tert*-butoxide in dimethyl sulfoxide at room temperature, to give the allyl glycoside **27** with only slight isomeriza-

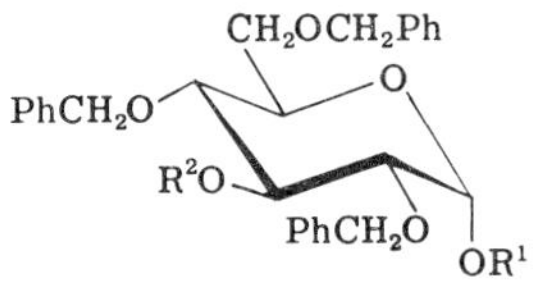

26 $R^1 = CH_2CH=CH_2$,
     $R^2 = CH_2CH=CHMe$
27 $R^1 = CH_2CH=CH_2$, $R^2 = H$
28 $R^1 = CH=CHMe$, $R^2 = H$

tion to the 1-propenyl glycoside **28**. Further treatment of **27** with the same, basic reagent at a higher temperature gave **28**. It is noteworthy that preparation of **26** by benzylation of allyl 2-O-benzyl-3-O-(2-butenyl)-α-D-galactopyranoside with the benzyl chloride–sodium hydride reagent occurred with no evidence of decomposition of the 2-butenyl ether. Further examples of the selective removal of the 2-butenyl groups from compounds containing, initially, an O-(2-butenyl) and an O-allyl group have been reported in the synthesis of benzyl and allyl ethers of D-glucose,[217] of derivatives of D-galactose,[212] and in the synthesis[218] of "seminolipid."

Although an O-(2-butenyl) group may be selectively cleaved in the

(215) J. Cunningham and R. Gigg, *J. Chem. Soc.*, (1965) 2968–2975.
(216) P. A. Gent, R. Gigg, and R. Conant, *J. Chem. Soc. Perkin Trans. 1*, (1972) 1535–1542.
(217) P. A. Gent and R. Gigg, *Carbohydr. Res.*, 49 (1976) 325–333.
(218) R. Gigg, *J. Chem. Soc. Perkin Trans. 1*, (1979) 712–718.

presence of an O-allyl ether group by treatment with a base at room temperature, some concomitant isomerization of the allyl group to the l-propenyl group has, nevertheless, been observed.[216,217] Both the O-(1-methylallyl)[216] and the O-(2-methylallyl)[207,219] group are rearranged at considerably lower rates than an O-allyl group on treatment with potassium *tert*-butoxide in dimethyl sulfoxide, and the 2-methylallyl group has been used[219] for the protection of a hydroxyl group in the presence of an O-(2-butenyl) group, as the latter can be readily removed without simultaneous rearrangement of the O-(2-methylallyl) group.

A further increase in the versatility of the allyl and substituted-allyl groups as selective, protecting groups resulted from the observation[220,221] that an O-(2-butenyl) group is isomerized much more slowly than an allyl group into the corresponding vinyl ether by tris(triphenylphosphine)rhodium chloride (the 2-methyallyl group is isomerized at a slightly lower rate than the allyl group). Thus, on treatment[220,221] with the rhodium complex, allyl 2,3,4-tri-O-benzyl-6-O-(2-butenyl)-α-D-galactopyranoside (**29**) gave the l-propenyl glycoside **30** as the major product, and the hydrolytic removal of the l-propenyl group by mercuric chloride gave 2,3,4-tri-O-benzyl-6-O-(2-butenyl)-D-galactopyranose (**31**).

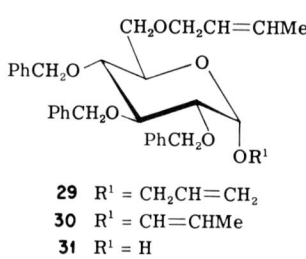

29 R^1 = $CH_2CH=CH_2$
30 R^1 = $CH=CHMe$
31 R^1 = H

The allyl group in 2-O-allyl-3,4,6-tri-O-benzyl-D-galactopyranosyl chloride has been used as a non-participating, protecting group at O-2 in a synthesis[222] of a 1,2-*cis*-glycoside with benzyl 2,3,4-tri-O-benzyl-α-D-galactopyranoside. Selective removal of the allyl group in the product afforded an α-linked disaccharide that had a free hydroxyl group on C-2, but which was otherwise fully protected with benzyl

(219) P. A. Gent, R. Gigg, and R. Conant, *J. Chem. Soc. Perkin Trans. 1*, (1973) 1858–1863.
(220) P. A. Gent and R. Gigg, *J. Chem. Soc. Chem. Commun.*, (1974) 277–278.
(221) P. A. Gent and R. Gigg, *J. Chem. Soc. Perkin Trans. 1*, (1974) 1835–1839.
(222) P. A. Gent and R. Gigg, *J. Chem. Soc. Perkin Trans. 1*, (1975) 361–363.

groups. Such compounds offer a possibility for the preparation of certain trisaccharides, and of the corresponding 2-keto derivatives, which can be transformed into other important derivatives.

In the synthesis[223] of a trisaccharide suggested to be the antigenic determinant Le^c, the allyl group played an important part in protection of O-3 in the intermediate disaccharide benzyl 2-acetamido-6-O-acetyl 3-O-allyl-2-deoxy-4-O-(2,3,4,6-tetra-O-acetyl-β-D-galactopyranosyl)-α-D-glucopyranoside. O-Deallylation with tris(triphenylphosphine)rhodium chloride, and condensation of the resulting alcohol with 2,3,4-tri-O-benzyl-α-L-fucosyl bromide gave, after deprotection, O-α-L-fucopyranosyl-(1→3)-O-β-D-galactopyranosyl-(1→4)-2-acetamido-2-deoxy-D-glucopyranose.

V. Deborination and Deboronation

Boron derivatives of carbohydrates, particularly borinates (**32**) and cyclic boronates (**33**), find considerable use as partially protected, syn-

$$R^1_2BOR^2 \qquad R^1B\diamond R^2$$

32 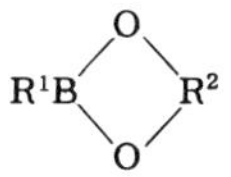   **33**

thetic intermediates.[224,225] Although there have been relatively few reports on the selective deprotection of these compounds, there are indications that such reactions offer considerable scope for the controlled modification of carbohydrate structures.

The seven-membered ring of methyl α-D-glucopyranoside 2,3-(diphenylcyclodiboronate) 4,6-phenylboronate (**34**) is selectively

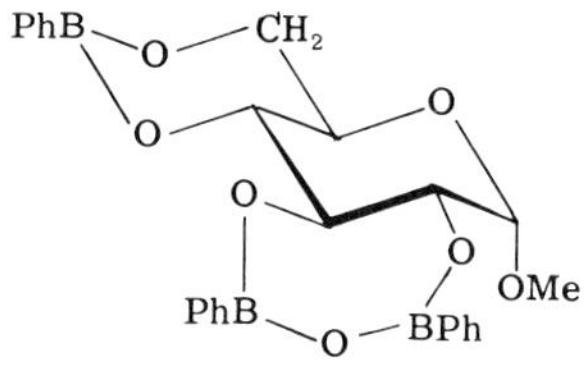

34

---

(223) J.-C. Jacquinet and P. Sinaÿ, *J. Chem. Soc. Perkin Trans. 1*, (1979) 314–318.
(224) R. J. Ferrier, *Adv. Carbohydr. Chem. Biochem.*, 35 (1978) 31–80.
(225) R. Köster and W. V. Dahlhoff, *ACS Symp. Ser.*, 39 (1977) 1–21.

cleaved when the compound is stored in wet benzene, to yield the corresponding 4,6-phenylboronate in good yield.[226]

Several examples of selective cleavage among oxygen–boron bonds are found in the work of Köster and coworkers[225] on diethylborinates and ethylboronates of carbohydrates. A diethylborinate group can usually be cleaved in the presence of a cyclic ethylboronate group by careful treatment with methanol[227,228] or 2,4-pentanedione,[227–231] and the preferential methanolysis of a nine-membered,[228] and a six-membered,[230] in the presence of a five-membered cyclic boronate group is possible. Selectivity towards methanolysis has also been observed between five-membered, cyclic boronate groups.[231] The following reactions provide examples of these selective cleavages.

On treatment with methanol at room temperature for 0.75 hour, DL-glycerol 1-(diethylborinate) 2,3-ethylboronate (**35**) gave[227] the DL-2,3-ethylboronate **36**, and this could be used for preparing DL-1-O-acyl derivatives of glycerol in good yields. Similarly, 2-deoxy-DL-*glycero*-tetritol 4-diethylborinate 1,3-ethylboronate (**37**) and 2,3,4-trideoxy-DL-*glycero*-hexitol 1-diethylborinate 5,6-ethylboronate (**39**) could be converted[227] into the DL-1,3-ethylboronate **38** and the DL-5,6-ethylboronate **40**, respectively. Reaction[230] of DL-xylitol 4-diethylborinate 1,2:3,5-bis(ethylboronate) (**41**) with 2,4-pentanedione, and with methanol, at room temperature gave DL-xylitol 1,2:3,5-bis(ethylboronate) (**42**) and DL-xylitol 1,2-ethylboronate (**43**), respectively, in high yields. The free hydroxyl groups in **42** and **43** could be acylated, and DL-2-O-acyl- and DL-1,2,3-tri-O-acyl derivatives of xylitol thereby prepared.

In the hexitol series, D-mannitol 3,4-ethylboronate has been obtained[231] by partial deboronation of D-mannitol 1,2:3,4:5,6-tris(ethylboronate) with methanol, and by deborination of D-mannitol 1,2:5,6-

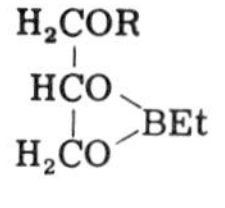

```
    H₂COR              OCH₂              CH₂OR
     |               EtB   CH₂            (CH₂)₃
    HCO   BEt           \  |                OCH
     |                   OCH              EtB  |
    H₂CO                 CH₂OR               OCH₂

  35 R = BEt₂         37 R = BEt₂         39 R = BEt₂
  36 R = H            38 R = H            40 R = H
```

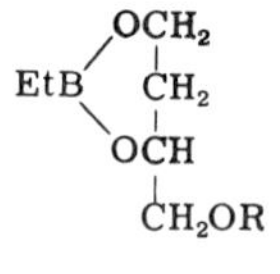

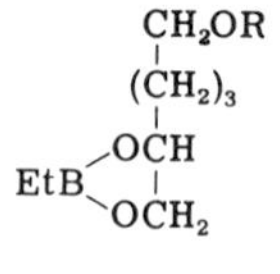

(226) R. J. Ferrier, *J. Chem. Soc.*, (1961) 2325–2330.
(227) W. V. Dahlhoff and R. Köster, *Ann.*, (1975) 1914–1925.
(228) W. V. Dahlhoff and R. Köster, *J. Org. Chem.*, 41 (1976) 2316–2320.
(229) R. Köster, K.-L. Amen, and W. V. Dahlhoff, *Ann.*, (1975) 752–788.
(230) W. V. Dahlhoff and R. Köster, *Ann.*, (1975) 1926–1933.
(231) W. V. Dahlhoff, W. Schussler, and R. Köster, *Ann.*, (1976) 387–394.

```
    H₂CO                  H₂CO
       \BEt                  \BEt
    HCO/                  HCO/
     |                     |
    /OCH                  HOCH
   /  |                    |
EtB   HCOR                HCOH
   \  |                    |
    \OCH₂                 H₂COH

  41 R = BEt₂                43
  42 R = H
```

Only one enantiomer is depicted.

tetrakis(diethylborinate) 3,4-ethylboronate with 2,4-pentanedione. Selective deboronation of galactitol 1,6:2,3:4,5-tris(ethylboronate) with methanol, and deborination of galactitol 1,6-bis(diethylborinate) 2,3:4,5-bis(ethylboronate) with methanol or 2,4-pentanedione, gave[228] galactitol 2,3:4,5-bis(ethylboronate).

Selective deprotection has been performed on related boron derivatives of glycosides.[225] On treatment with 2,4-pentanedione at room temperature, methyl α-D-glucopyranoside 2,3-bis(diethylborinate) 4,6-ethylboronate and methyl β-D-xyloyranoside 3-diethylborinate 2,4-ethylboronate were converted into the corresponding 4,6-ethylboronate and 2,4-ethylboronate, respectively. Acylation of these boronates, followed by deboronation with methanol, led to convenient preparations of 2,3-di-O-acyl derivatives of the D-glucoside and 3-O-acyl derivatives of the D-xyloside.

Interestingly, attempted selective-deboronation of L-rhamnopyranose 1,2:3,5-bis(ethylboronate) with methanol did not yield[225] the 1,2- or 3,5-ethylboronate, but gave, instead, the 2,3-ethylboronate as a result of an intramolecular transesterification. Acetylation of this compound, followed by deboronation with 1,2-ethanediol, gave 1,5-di-O-acetyl-L-rhamnofuranose, which could not be prepared through the corresponding 2,3-O-isopropylidene intermediate.

VI. DENITRATION

Sugar nitrates formed the subject of a Chapter in an earlier Volume in this Series,[232] and that article contained several examples of selective denitration. As nitro groups do not appear to undergo oxygen-to-oxygen migration to any great extent,[233] the process of partial denitra-

(232) J. Honeyman and J. W. W. Morgan, Adv. Carbohydr. Chem., 12 (1957) 117–135.
(233) However, for an apparent O-3 to O-2 migration of a nitro group during the alkaline methanolysis of methyl 4,6-O-benzylidene-α-D-glucopyranoside 3-nitrate, see J. Honeyman and J. W. W. Morgan, J. Chem. Soc., (1955) 3660–3674.

tion of a polynitrate is not usually complicated by this type of rearrangement, which can occur during partial deacylation.[1]

Several reagents have been used for the selective deprotection of carbohydrate nitrates. Sodium iodide in acetone causes replacement of a nitrate group at a primary center by an iodo group, but may transform a nitrate group at a secondary carbon atom into the parent alcohol group, with retention of configuration. Thus, methyl 3-O-acetyl-2-O-methyl-β-D-glucopyranoside 4,6-dinitrate[234] and the corresponding 2,3-dimethyl ether[235] afforded 2,3-di-O-substituted methyl 6-deoxy-6-iodo-β-D-glucopyranoside derivatives.

Selectivity in deprotection with sodium iodide in acetone is often observed between nitrate groups attached to secondary carbon atoms. For example, methyl 4,6-O-ethylidene-β-D-glucopyranoside 2,3-dinitrate,[234,236,237] methyl 4,6-O-benzylidene-,[233] methyl 4,6-O-propylidene-,[237] and methyl 4,6-O-ethylidene-α-D-glucopyranoside 2,3-dinitrate,[233] and methyl 4,6-O-ethylidene-α-D-mannopyranoside 2,3-dinitrate[238] all afforded the corresponding 3-nitrate in reasonable yields. Reaction[239] of methyl β-D-glucopyranoside 2,3,4,6-tetranitrate with the reagent gave a mixture of products that, after treatment with silver nitrate to replace the 6-iodo groups by nitrate groups, yielded the 2,6-dinitrate (as its diacetate) and a smaller amount of 3,6-dinitrate of the parent glycoside, suggesting that the nitro group on O-4 is particularly labile under these conditions.

Selective denitration has also been effected by performing the reaction with sodium iodide in 2-pentanone,[240] methanol,[233] pyridine,[233] and acetic anhydride.[237]

Selective denitration of methyl 4,6-O-benzylidene- and 4,6-O-alkylidene-D-hexopyranoside 2,3-dinitrates to yield 3-nitrates has also been achieved by using sodium nitrite in ethanol as the reagent.[233,238,241,242] Interestingly, when the nitrate group is on a primary carbon atom, reaction with this reagent takes place to give the primary alcohol,[232,241,243] but the conversion may, on occasion, proceed only with

(234) J. Dewar and G. Fort, *J. Chem. Soc.*, (1944) 492–496.
(235) J. W. H. Oldham and J. K. Rutherford, *J. Am. Chem. Soc.*, 54 (1932) 366–378.
(236) D. J. Bell and R. L. M. Synge, *J. Chem. Soc.*, (1938) 833–836.
(237) E. G. Ansell and J. Honeyman, *J. Chem. Soc.*, (1952) 2778–2789.
(238) J. Honeyman and T. C. Stening, *J. Chem. Soc.*, (1957) 2278–2280.
(239) J. Dewar, G. Fort, and N. McArthur, *J. Chem. Soc.*, (1944) 499–501.
(240) D. O'Meara and D. M. Shepherd, *J. Chem. Soc.*, (1955) 4232–4235.
(241) J. Honeyman and T. C. Stening, *J. Chem. Soc.*, (1958) 537–546.
(242) K. S. Ennor, J. Honeyman, C. J. G. Shaw, and T. C. Stening, *J. Chem. Soc.*, (1958) 2921–2925.
(243) K. S. Ennor, J. Honeyman, and T. C. Stening, *Chem. Ind. (London)*, (1956) 1308–1309.

extreme difficulty. For example, 1,2:3,4-di-O-isopropylidene-α-D-galactopyranose 6-nitrate is largely unchanged after treatment for several days.[241,243] On reaction with the reagent for 5.5 hours, methyl β-D-glucopyranoside 2,3,4,6-tetranitrate gave[241] the 2,3,6-trinitrate as the major product. The considerable selectivity shown for denitration at O-4 led to a useful route to the 4-methyl ether of the glycoside. A further example of the selective cleavage of a nitrate group at a secondary position with sodium nitrite in ethanol is found in the reaction of 1,3:2,4-di-O-ethylidene-D-glucitol 5,6-dinitrate, which gave[244] the 6-nitrate, albeit in a yield of only 25%.

Although the alkaline hydrolysis or alcoholysis of alkyl nitrates appears to be a complex process,[232] some selectivity in the removal of nitro groups from carbohydrate nitrates has been noted under these conditions. Treatment[233] of methyl 4,6-O-benzylidene-α-D-glucopyranoside 2,3-dinitrate with sodium methoxide in chloroform–methanol for 3 days at 0°, or for 5 days at room temperature, gave the corresponding 3-nitrate in ~25% yield, together with some unchanged starting-material, and, under the latter conditions, the 2-nitrate in 6% yield.

Nitrogen-containing bases have been used to bring about partial deprotection of carbohydrate nitrates. Considerable selectivity was achieved in the reaction[245] of 6-O-acetyl-1,2-O-isopropylidene-α-D-glucofuranose 3,5-dinitrate in benzene solution with a 30% solution of dimethylamine in ethanol, from which 1,2-O-isopropylidene-α-D-glucofuranose 5-nitrate was isolated in 47% yield.

The action of hydroxylamine in pyridine on methyl β-D-glucopyranoside tetranitrate gave,[246] in order of decreasing yields, the 2,3,6-trinitrate, the 3,6-dinitrate, and an unidentified trinitrate. On similar treatment, cellulose trinitrate underwent smooth reaction,[247] to give a partially denitrated product (degree of substitution 1.7) which, on the basis of methylation analysis,[248] was shown to have nitrate groups almost exclusively on C-3 and C-6 of the D-glucosyl residues. The highly selective denitration at C-2 in this case is difficult to rationalize.

Perhaps the most fascinating examples of selective denitration are those observed in the hexitol series. For example, the action of pyridine on D-mannitol hexanitrate[249–251] yields D-mannitol 1,2,3,5,6-pen-

(244) K. S. Ennor and J. Honeyman, *J. Chem. Soc.*, (1958) 2586–2594.
(245) D. J. Bell, *J. Chem. Soc.*, (1936) 1553–1554.
(246) L. D. Hayward and C. B. Purves, *Can. J. Chem.*, 32 (1954) 19–30.
(247) G. H. Segall and C. B. Purves, *Can. J. Chem.*, 30 (1952) 860–871.
(248) E. L. Falconer and C. B. Purves, *J. Am. Chem. Soc.*, 79 (1957) 5308–5310.
(249) J. H. Wigner, *Ber.*, 36 (1903) 794–800.
(250) C. R. Marshall and J. H. Wigner, *Br. Med. J.*, 2 (1902) 1231–1233.
(251) L. D. Hayward, *J. Am. Chem. Soc.*, 73 (1951) 1974–1975.

tanitrate[251] in high yield. Later workers[252,253] showed that partial denitration of the D-mannitol derivative could also be achieved, equally effectively, by the use of ammonium carbonate in acetone–water. Galactitol hexanitrate also undergoes[249,254] regioselective deprotection in pyridine, to give a racemic mixture of D- and L-galactitol 1,2,3,5,6-pentanitrate in good yield.

The high regioselectivity observed on partial de-esterification of the foregoing hexitol hexanitrates is intriguing, but largely unexplained. A closer study[255] of the reaction of D-mannitol hexanitrate revealed that approximately two moles of pyridine suffered ring-cleavage while 0.25 mole of hexanitrate was completely denitrated and 0.75 mole of pentanitrate was formed. Gas evolved from the reaction consisted of nitric oxide, nitrous oxide, and nitrogen, and the proportion and exact composition of the mixture were sensitive to traces of moisture in the pyridine. Pyridinum nitrate was also formed.

D-Glucitol hexanitrate[256] and the pentanitrates of xylitol,[257] ribitol,[257] and L-arabinitol[257] have been partially denitrated by the action of pyridine, but the structures of the reaction products were not ascertained. The dinitrates of the 1,4:3,6-dianhydro derivatives of D-mannitol, D-glucitol, and L-iditol react only slowly in anhydrous pyridine,[258] and only traces of mononitrates could be detected; in aqueous pyridine, mononitrates accounted for a maximum of 10% of the decomposed dinitrates.

Although hydrazine hydrate in ethanol is effective in achieving denitration of carbohydrate nitrates,[243,244] the reagent showed little selectivity in its reactions[244] with the 2,3-dinitrates of methyl 4,6-O-benzylidene-α- and methyl 4,6-O-ethylidene-α- and -β-D-glucopyranoside.

VII. DEPHOSPHONYLATION

The hydrolysis of a monoalkyl phosphate may also be classed as a dephosphonylation reaction, that is, one involving removal of the O-phosphono [$(HO)_2P(O)$-] group.[259]

(252) D. E. Elrick, N. S. Marans, and R. F. Preckel, *J. Am. Chem. Soc.*, 76 (1954) 1373–1374.
(253) M. L. Wolfrom, E. P. Swan, K. S. Ennor, and A. Chaney, *J. Am. Chem. Soc.*, 81 (1959) 5701–5705.
(254) G. G. McKeown and L. D. Hayward, *Can. J. Chem.*, 33 (1955) 1392–1398.
(255) J. R. Brown and L. D. Hayward, *Can. J. Chem.*, 33 (1955) 1735–1745.
(256) T. Urbański and S. Kwiatkowska, *Rocz. Chem.*, 25 (1951) 312–314.
(257) G. Wright and L. D. Hayward, *Can. J. Chem.*, 38 (1960) 316–319.
(258) M. Jackson and L. D. Hayward, *Can. J. Chem.*, 38 (1960) 496–502.
(259) See *Chem. Abstr. Index Guide*, 76 (1972) subsection H, paragraph 294, p. 131 I, Column 3.

A considerable amount of our knowledge on the phosphoric esters of carbohydrates and related compounds stems from the extensive studies that have been made on nucleotides and the cyclitol phosphates. The chemistry of the mononucleotides was treated in an earlier article in this Series,[260] and useful accounts of cyclitol phosphates are available.[261] Although cleavage of preformed cyclic phosphates is considered in this Section, no attempt has been made to cover the cleavage of α,β-cyclic phosphates formed in the course of phosphono migration, or to cover regioselective phosphate cleavage brought about enzymically. The use of enzymes in nucleic acid research has been described in detail.[262]

Relatively little work has been reported on the sequential removal of phosphate groups by chemical means from compounds containing several such groups. Partial hydrolysis of myo-inositol hexaphosphate (phytic acid) by acid[263] or alkali[264] gives myo-inositol 2-phosphate.[265,266] Presumably, the axially disposed phosphate group at C-2 of the six-membered ring suffers greater steric hindrance towards hydrolysis than the other phosphate groups, which are all attached equatorially.

Although the 2′,3′-cyclic phosphates of adenosine, cytidine, guanosine, and uridine appear to be stable in aqueous solution between, approximately, pH 4 and 9, they are hydrolyzed under more-strenuous conditions (for example, 0.1 M hydrochloric acid for 1 hour at room temperature or 0.5 M sodium hydroxide for 12 hours at 37°) to give a mixture of 2′- and 3′-phosphates[267] in approximately equal amounts. An appreciation of this aspect of the chemistry of cyclic phosphates was of considerable importance in elucidating the mechanism of the chemical degradation of ribonucleic acid.[268]

Both base-catalyzed[269,270] and acid-catalyzed[271] esterification of primary alcohols by nucleoside 2′,3′-cyclic phosphates gave mixtures of

(260) T. Ueda and J. J. Fox, *Adv. Carbohydr. Chem. Biochem.*, 22 (1967) 307–419.
(261) T. Posternak, *The Cyclitols*, Holden–Day, San Francisco, 1965, pp. 221–243; D. J. Cosgrave, *Inositol Phosphates: Their Chemistry, Biochemistry and Physiology*, Elsevier, Amsterdam, 1980.
(262) M. Privat de Garilhe, *Enzymes in Nucleic Acid Research*, Holden–Day, San Francisco, 1967.
(263) J. Courtois and M. Masson, *Bull. Soc. Chim. Biol.*, 32 (1950) 314–325.
(264) A. Desjobert, *Bull. Soc. Chim. Biol.*, 36 (1954) 1293–1299.
(265) D. M. Brown and G. E. Hall, *J. Chem. Soc.*, (1959) 357–359.
(266) F. L. Pizer and C. E. Ballou, *J. Am. Chem. Soc.*, 81 (1959) 915–921.
(267) D. M. Brown, D. I. Magrath, and A. R. Todd, *J. Chem. Soc.*, (1952) 2708–2714.
(268) D. M. Brown and A. R. Todd, *J. Chem. Soc.*, (1952) 52–58.
(269) C. A. Dekker and H. G. Khorana, *J. Am. Chem. Soc.*, 76 (1954) 3522–3527.
(270) G. R. Barker, M. D. Montague, R. J. Moss, and M. A. Parsons, *J. Chem. Soc.*, (1957) 3786–3793.
(271) G. M. Tener and H. G. Khorana, *J. Am. Chem. Soc.*, 77 (1955) 5349–5351.

nucleoside 2'- and 3'-(alkyl phosphates), and it appears that the steric bulk of the attacking nucleophile can control the ratio of the two isomers so obtained.[269–271] For alkoxide attack on the cyclic phosphate, the larger the attacking group, the higher the proportion of 3'-ester.[270] Riboflavine 5'-(alkyl phosphate) has been prepared[272] by acid-catalyzed esterification of an alcohol with riboflavine 4',5'-cyclic phosphate.

The favored formation of a nucleoside 3'-(alkyl phosphate) was observed on treating the diphenyl phosphate anhydride of uridine 2',3'-cyclic phosphate with benzyl alcohol.[273] In a somewhat related reaction, on treatment with dicyclohexylcarbodiimide in aqueous pyridine, adenosine 2'(3')-phosphate gave, initially, the 2',3'-cyclic phosphate, which, on further reaction with the diimide, gave a mixture of the N-phosphonourea nucleoside **44** and its 2'-isomer, in unequal amounts.[269] This type of reaction does not seem to occur with

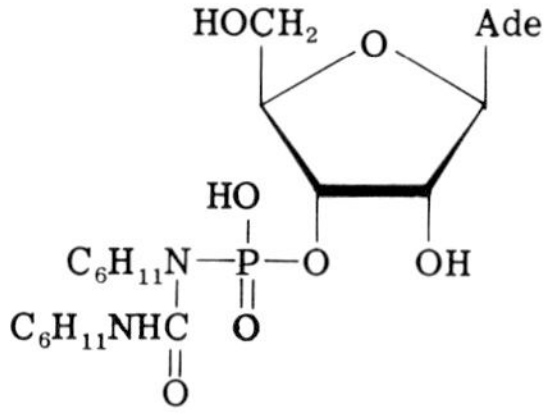

44

six- or seven-membered-ring, cyclic phosphates,[274] and the greater lability of cyclic phosphates containing a five-membered ring towards hydrolysis has been noted.[274]

Hydrolysis of *myo*-inositol 1(3),2-cyclic phosphate under acidic or alkaline conditions affords a mixture of *myo*-inositol 1(3)-phosphate and *myo*-inositol 2-phosphate, with the former preponderating.[275] As the cyclic phosphate may be prepared[275] from the 2-phosphate, which is itself readily obtained from *myo*-inositol hexaphosphate by enzymic cleavage,[266,276] and as the 1(3)-phosphate may be isolated[277] from the hydrolysis mixture, the reaction sequence provides a convenient way

---

(272) T. Tanaka, *Yakugaku Zasshi*, 78 (1958) 627–631; *Chem. Abstr.*, 52 (1958) 18,460g.
(273) A. M. Michelson, *J. Chem. Soc.*, (1959) 1371–1394.
(274) H. G. Khorana, G. M. Tener, R. S. Wright, and J. G. Moffatt, *J. Am. Chem. Soc.*, 79 (1957) 430–436.
(275) T. Posternak, *Helv. Chim. Acta*, 41 (1958) 1891–1898.
(276) M. H. McCormick and H. E. Carter, *Biochem. Prep.*, 2 (1952) 65–68.
(277) T. Posternak, *Helv. Chim. Acta*, 42 (1959) 390–393.

in which to transpose a phosphono group from O-2 to O-1(3) in the cyclitol.

The difference in hydrolytic stability between five- and six-membered, cyclic phosphates[274] is well illustrated by the far greater stability of adenosine 3′,5′-cyclic phosphate compared to its 2′,3′-isomer.[278] It is noteworthy that, in the alkaline hydrolysis of the 3′,5′-cyclic phosphates of adenosine,[278,279] cytidine,[279] guanosine,[279] uridine,[279] thymidine,[280] and 2′-deoxycytidine,[281] the 3′-phosphate is formed in greater proportion than the 5′-phosphate, the ratio in most cases being ~5:1. Acid-catalyzed hydrolysis of the 3′,5′-cyclic phosphates of cytidine[279] and uridine[279] gave the 2′(3′)- and 5′-phosphates, with the former preponderating.

Other examples of the ring opening of a cyclic phosphate to form a preponderant proportion of a secondary phosphate are found in the alkaline hydrolysis of methyl α-D-glucopyranoside 4,6-phosphate[282,283] and methyl α-D-galactopyranoside 4,6-phosphate[281,283]; in each case, the 4- and 6-phosphates were formed in the ratio of ~4 or 5:1. The alkaline hydrolysis of phenyl β-D-glucopyranoside 4,6-(phenyl phosphate) and related derivatives[284,285] is somewhat more complex, as a result of intramolecular transesterification of two of the initial products, the 4- and 6-(phenyl hydrogenphosphates). Studies[285a] on the ring opening of D-glucose cyclic 4,6-(ethyl phosphate) in dilute sodium methoxide–methanol solutions suggested that D-glucose 4-(ethyl methyl phosphate)s are the kinetically favored products, and that ring opening occurs with predominant inversion of configuration at the phosphorus atom; on storage in sodium methoxide–methanol, these 4-(ethyl methyl phosphate)s afforded D-glucose 6-(ethyl methyl phosphate)s, with retention of configuration at the phosphorus atom.

It has been pointed out[279] that the greater stability of a secondary phosphoric ester linkage relative to a primary linkage is analogous to the greater stability of secondary alkyl esters of carboxylic acids, rela-

---

(278) D. Lipkin, W. H. Cook, and R. Markham, *J. Am. Chem. Soc.*, 81 (1959) 6198–6203.
(279) M. Smith, G. I. Drummond, and H. G. Khorana, *J. Am. Chem. Soc.*, 83 (1961) 698–706.
(280) G. M. Tener, H. G. Khorana, R. Markham, and E. H. Pol, *J. Am. Chem. Soc.*, 80 (1958) 6223–6230.
(281) H. G. Khorana, A. F. Turner, and J. P. Vizsolyi, *J. Am. Chem. Soc.*, 83 (1961) 686–698.
(282) P. Szabó and L. Szabó, *C. R. Acad. Sci. Ser. A.*, 247 (1958) 1748–1750.
(283) P. Szabó and L. Szabó, *J. Chem. Soc.*, (1960) 3758–3762.
(284) P. Rivaille and L. Szabó, *C. R. Acad. Sci. Ser. A*, 254 (1962) 3705–3707.
(285) P. Rivaille and L. Szabó, *Bull. Soc. Chim. Fr.*, (1963) 712–716.
(285a) J. M. Harrison and T. D. Inch, *J. Chem. Soc. Perkin Trans. 1*, (1979) 2855–2862.

tive to primary alkyl esters.[286] However, examples of cyclic phosphates undergoing hydrolysis to afford, as the major product, a compound containing the phosphate group attached to a primary carbon atom are known. Thus, on treatment with aqueous sodium hydroxide, followed by deacetalation, 1,2-O-isopropylidene-α-D-xylofuranose 3,5-cyclic phosphate gave[287] D-xylose 5- and 3-phosphate in the ratio of 5:2, and pantothenic acid 2,4-cyclic phosphate yielded,[288] exclusively, the primary 4-phosphate on reaction with barium hydroxide solution. In the former instance, it was suggested[279] that the 3-isomer might be disfavored because of steric interaction between the phosphate group and the methylene group on C-4; in the latter example, polar or chelating effects associated with the amide group in pantothenic acid could exert a controlling influence.[279] Clearly, in all such cases, it is necessary to show that the reaction products are formed under thermodynamic control, before deductions based on product ratios can be made regarding the relative stability of primary and secondary phosphates.

VIII. DE(TRIALKYLSILYL)ATION

The selective removal of trialkylsilyl groups in O-(trialkylsilyl)ated carbohydrate derivatives seems to offer considerable scope in synthesis. McInnes appears to have been the first to have realized[289] that trialkysilyl derivatives of primary, secondary, and tertiary alcohols might have different rates of alcoholysis. Kinetic measurements on methyl 2,3,4,6-tetra-O-(trimethylsilyl)-α-D-glucopyranoside (**45**) showed[289] that selective methanolysis of the trimethylsilyl group at the primary position (O-6) was catalyzed by acid or base, and that, in the presence of potassium methoxide, the rate constant for the methanolysis of trimethylsilyl groups at the primary position was ~25 times that for removal of the corresponding groups at the secondary positions. Methyl 2,3,4-tri-O-(trimethylsilyl)-α-D-glucopyranoside (**46**)

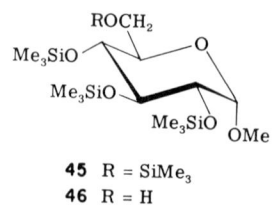

45 R = SiMe$_3$
46 R = H

(286) M. S. Newman, *Steric Effects in Organic Chemistry*, Wiley, New York, 1956, p. 219.
(287) J. G. Moffatt and H. G. Khorana, *J. Am. Chem. Soc.*, 79 (1957) 1194–1200.
(288) J. Baddiley and E. M. Thain, *J. Chem. Soc.*, (1951) 3421–3424.
(289) A. G. McInnes, *Can. J. Chem.*, 43 (1965) 1998–2003.

was isolated[290] crystalline in ~80% yield by methanolysis of the tetrakis(trimethylsilyl) ether **45** at 0° in the presence of potassium carbonate, and **46** proved to be a useful intermediate for the preparation of 6-substituted derivatives of the glycoside.

The rate of alcoholysis of the tetrakis(trimethylsilyl) ether **45** was found to be strongly dependent on the structure of the alcohol used.[289] Relative, second-order rate-constants using methoxide, ethoxide, isopropoxide, and *tert*-butoxide as catalysts dissolved in the corresponding alcohol were in the ratio $10^4 : 1.7 \times 10^3 : 40 : 1$.

Following the work of McInnes,[289,290] methanolysis in the presence of potassium carbonate was used to prepare methyl 6-*O*-acetyl- and methyl 6-*O*-octadecanoyl-β-D-glucopyranoside,[291] *o*-nitrophenyl β-D-galactopyranoside 6-phosphate,[292] methyl 5-thio-α-D-glucopyranoside 6-phosphate,[293] and 6,6′-di-*O*-hexadecanoyl-α,α-trehalose,[294] from the appropriate, per-*O*-(trialkylsilyl)ated precursors.

Methanolysis of the tetrakis(trimethylsilyl) ether **45** in the presence of mineral acids appeared[289] to be less selective than the alkoxide-catalyzed reaction, but catalysis by acetic acid afforded the tris(trimethylsilyl) ether **46** in yields comparable to those obtained by methoxide-ion catalysis, but at a lower rate for a given concentration of the catalyst. An improved method for the selective removal of *O*-(trimethylsilyl) groups at primary positions in carbohydrate derivatives involves[295] a proton-catalyzed reaction with water in pyridine, with acetic acid as the proton source, which liberates a free primary hydroxyl group in high yield. On the other hand, controlled reaction with acetic anhydride in pyridine in the presence of acetic acid affords, directly, a derivative that is acylated at the primary hydroxyl group. Such procedures have been used in the synthesis of 7-amino-7-deoxyheptonic acid derivatives[296] and 4-deoxy-D-*xylo*-hexose,[297] and in the selective 6-*O*-acetylation of amylose.[298]

As a result of investigations directed towards obtaining new, selective, hydroxyl-protecting groups having differing stabilities towards solvolytic conditions and yet favorable stability towards synthetic re-

(290) D. T. Hurst and A. G. McInnes, *Can. J. Chem.*, 43 (1965) 2004–2011.
(291) A. P. Tulloch and A. Hill, *Can. J. Chem.*, 46 (1968) 2485–2493.
(292) W. Hengstenberg and M. L. Morse, *Carbohydr. Res.*, 7 (1968) 180–183.
(293) R. L. Whistler and J. H. Stark, *Carbohydr. Res.*, 13 (1970) 15–21.
(294) R. Toubiana, B. C. Das, J. Defaye, B. Mompon, and M. J. Toubiana, *Carbohydr. Res.*, 44 (1975) 308–312; R. Toubiana, M. J. Toubiana, B. C. Das, and E. Lederer, Ger. Offen. 2,633,690; *Chem. Abstr.*, 86 (1977) 155,893t.
(295) E. F. Fuchs and J. Lehmann, *Chem. Ber.*, 107 (1974) 721–724.
(296) E. F. Fuchs and J. Lehmann, *Chem. Ber.*, 108 (1975) 2254–2260.
(297) M. Brockhaus, E. F. Fuchs, and J. Lehmann, *Chem. Ber.*, 111 (1978) 811–813.
(298) D. Horton and J. Lehmann, *Carbohydr. Res.*, 61 (1978) 553–556.

agents, a wide range of silyl ethers has been prepared. Triphenylsilyl,[299] tricyclohexylsilyl,[299] isopropyldimethylsilyl,[300] tert-butyldimethylsilyl,[301] tert-butyldiphenylsilyl,[302] diisopropylmethylsilyl,[303,304] triisopropylsilyl,[303–305] isopropyl(tetramethylene)silyl,[303,304] tert-butyl(tetramethylene)silyl,[303,304] and tert-butyl(methyl)octadecylsilyl[304] ethers have been described, and, except for the first three of this series, the preparation of such ethers from nucleosides has been investigated,[303–309] with a view to providing suitably protected intermediates for oligonucleotide synthesis. The feasibility of this approach for the preparation of oligodeoxyribonucleotides[304,310] and of oligoribonucleotides[308,309] has been demonstrated. The tert-butyldimethylsilyl group seems particularly suited[308,309] as an O-protecting group in this area of nucleotide chemistry, and tetrabutylammonium fluoride is a remarkably useful reagent for the removal of trialkylsilyl groups, even from nucleotide derivatives containing phosphoric triester groups.[304,310] Importantly, the p-anisyldiphenylmethyl (mono-p-methoxytrityl) group, which finds considerable use as an O-protecting group in nucleotide synthesis, may be selectively removed in the presence of an O-(tert-butyldimethylsilyl) group at a secondary position by treatment with 80% acetic acid.[304,308,309]

The dependence of the acid-lability of trialkylsilyl and related ethers on the type of substitution at silicon is illustrated by the formation[302] of 1-O-(tert-butyldiphenylsilyl)-2,3:4,5-di-O-methylene-D-mannitol (**47**) in 96% yield on treatment of 6-O-(tert-butyldimethylsilyl)-1-O-(tert-butyldiphenylsilyl)-2,3:4,5-di-O-methylene-D-mannitol (**48**) with 80% acetic acid.

(299) S. A. Barker, J. S. Brimacombe, M. R. Harnden, and J. A. Jarvis, *J. Chem. Soc.*, (1963) 3403–3406.
(300) E. J. Corey and R. K. Varma, *J. Am. Chem. Soc.*, 93 (1971) 7319–7320.
(301) E. J. Corey and A. Venkateswarlu, *J. Am. Chem. Soc.*, 94 (1972) 6190–6191.
(302) S. Hanessian and P. Lavallée, *Can. J. Chem.*, 53 (1975) 2975–2977.
(303) K. K. Ogilvie, E. A. Thompson, M. A. Quilliam, and J. B. Westmore, *Tetrahedron Lett.*, (1974) 2865–2868.
(304) K. K. Ogilvie, S. L. Beaucage, D. W. Entwistle, E. A. Thompson, M. A. Quilliam, and J. B. Westmore, *J. Carbohydr. Nucleos. Nucleot.*, 3 (1976) 197–227.
(305) K. K. Ogilvie, K. L. Sadana, E. A. Thompson, M. A. Quilliam, and J. B. Westmore, *Tetrahedron Lett.*, (1974) 2861–2863.
(306) K. K. Ogilvie and D. J. Iwacha, *Tetrahedron Lett.*, (1973) 317–319.
(307) K. K. Ogilvie, *Can. J. Chem.*, 51 (1973) 3799–3807.
(308) K. K. Ogilvie, N. Y. Theriault, and K. L. Sadana, *J. Am. Chem. Soc.*, 99 (1977) 7741–7743.
(309) K. K. Ogilvie, S. L. Beaucage, A. L. Schifman, N. Y. Theriault, and K. L. Sadana, *Can. J. Chem.*, 56 (1978) 2768–2780.
(310) K. K. Ogilvie, S. L. Beaucage, and D. W. Entwistle, *Tetrahedron Lett.*, (1976) 1255–1256.

```
         H₂COSiPh₂(CMe₃)
         |
      ╱OCH
   H₂C    |
      ╲OCH
         |
         HCO╲
         |    CH₂
         HCO╱
         |
         H₂COR
```
47 R = H
48 R = Si(CMe₃)Me₂

Ogilvie and coworkers[303,304] measured the stabilities, towards aqueous acid, of a range of 5'- and 3'-(trialkysilyl) ethers of thymidine (**49a–e** and **50a–e**, respectively), and their results are given in Table II.

49a R = Si(CHMe₂)₂Me
49b R = Si(CHMe₂)(─(CH₂)₄─)
49c R = Si(CMe₃)Me₂
49d R = Si(CHMe₂)₃
49e R = Si(CMe₃)(─(CH₂)₄─)

50a R = Si(CHMe₂)₂Me
50b R = Si(CHMe₂)(─(CH₂)₄─)
50c R = Si(CMe₃)Me₂
50d R = Si(CHMe₂)₃
50e R = Si(CMe₃)(─(CH₂)₄─)

TABLE II

Acid Hydrolysis[303,304] of Some Trialkylsilyl Ethers of Thymidine

Compound	Conditions[a]	Time for complete hydrolysis[b] (minutes)
49a	A	90
50a	A	360
49b	A	120
50b	A	360
49c	A	300
50c	A	1500
49d	B	15
50d	B	80
49e	B	35
50e	B	120

[a] Key: A, 80% acetic acid at room temperature; B, 0.01M hydrochloric acid in 1:1 (v/v) water–ethanol at 100°. [b] "Complete hydrolysis" refers to >98% hydrolysis.

The stability sequence for the O-trialkysilyl groups appears to be tert-butyl(tetramethylene)silyl > triisopropylsilyl > tert-butyldimethylsilyl > isopropyl(tetramethylene)silyl > diisopropylmethylsilyl. In all cases, the 3'-isomer (secondary position) is considerably more stable than the 5'-isomer (primary position), and this difference in reactivity permits[303,304] the synthesis of a 3'-(trialkylsilyl) ether of the nucleoside from the corresponding 3',5'-bis(trialkylsilyl) ether, in yields of ~50%.

For nucleoside derivatives containing two different types of trialkylsilyl group, considerable selectivity in hydrolysis may be achieved. For example, on hydrolysis with 80% acetic acid for 18 hours at 20°, 5'-O-[tert-butyl(tetramethylene)silyl]-3'-O-[isopropyl(tetramethylene)silyl]thymidine gave[304] the 5'-ether in 95% yield, whereas treatment of 3'-O-[tert-butyl(tetramethylene)silyl]-5'-O-[isopropyl(tetramethylene)silyl]thymidine for 3 minutes at 100° with 80% acetic acid gave[304] the 3'-ether, quantitatively. Because of their wide range of stabilities towards 80% acetic acid, these trialkylsilyl groups provide an extremely useful system for the protection of hydroxyl groups. Although all of these trialkylsilyl derivatives show[304] considerable stability towards ethanolic ammonium hydroxide, deprotection is readily achieved[303,304] on treatment with tetrabutylammonium fluoride in oxolane at room temperature.

The study of such protecting groups has been extended to their employment in the synthesis of oligoribonucleotides,[308,309] and use has been made of the greater lability of a given trialkylsilyl group towards acid when at O-5' (the primary position) than at O-2' (a secondary position) of a nucleoside in order to convert 2',5'-di-O-(tert-butyldimethylsilyl) derivatives of adenosine,[308,309] cytidine,[308] guanosine,[308] and uridine[308,309] into their corresponding 2'-ethers in good yields by selective hydrolysis with 80% acetic acid. Similar selective hydrolysis at a primary position can also be achieved in a dinucleotide derivative,[308,309] and the selective removal of a 5'-O-(p-anisyldiphenylmethyl) (mono-p-methoxytrityl) group from a 5'-O-(p-anisyldiphenylmethyl)oligonucleotide derivative containing one or more O-(tert-butyl)dimethylsilyl) groups at secondary positions can be achieved[304,308,309] in good yield.

It should be noted that there are some limitations to the use of the tert-butyldimethylsilyl group (and, presumably, other trialkylsilyl groups) for the selective protection of hydroxyl groups. Isomerization between O-2' and O-3' occurs with 2'- and 3'-O-(tert-butyldimethylsilyl) derivatives of nucleosides when they are in prolonged contact with silica gel,[309] or when stored in aqueous pyridine, aqueous triethylamine, N,N-dimethylformamide, N,N-dimethylformamide containing imidazole, 95% ethanol, or dimethyl sulfoxide.[309] Köhler and Pfleiderer[311] made a detailed study of this type of isomerization in cytidine derivatives, and concluded that the rate of isomerization is solvent-dependent, base-catalyzed, and markedly influenced by the purity of the solvent; in pure pyridine, the isomerization is extremely slow. Methanol is a particularly favorable solvent for the equilibration, and, in the presence of sodium methoxide, the equilibrium is attained in a few seconds. In contrast, addition of acid to a methanol solution of the nucleoside derivative drastically lowers the rate of isomerization. Other workers have reported[312] that heating a solution of 2-O-(tert-butyldimethylsilyl)glycerol in pyridine leads to slow interconversion with the corresponding O-1 derivative. Migration of a tert-butyldimethylsilyl group was also noted[313] when methyl 3,6-di-O-(tert-butyldimethylsilyl)-β-D-glucopyranoside was treated with triphenylphosphine–diethyldiazodicarboxylate in benzene, methyl 4,6-di-O-(tert-butyldimethylsilyl)- and methyl 2,6-di-O-(tert-butyldimethylsilyl)-β-D-glucopyranoside being formed in the ratio of ~8:1.

A further limitation on the use of the tert-butyldimethylsilyl group arises from the observation that cleavage of such a group from an acylated trialkylsilyl ether of a carbohydrate, notably by treatment with tetrabutylammonium fluoride in oxolane, may be accompanied by acyl migration.[312,314–316] However, it appears[316] that deprotection with dilute acid may circumvent this problem. Cleavage of the tert-butyldimethylsilyl groups from 2,3,4,1',3',4'-hexa-O-benzoyl-6,6'-di-O-(tert-butyldimethylsilyl)sucrose and from the corresponding acetyl derivative was achieved,[316] without acyl migration, by use of 0.2% methanolic sulfuric acid. In contrast, other workers,[312,314] who attempted O-de(trialkylsilyl)ation of 1,3-di-O-acyl-2-O-(tert-butyldimethylsilyl)glycerols with hydrogen fluoride in aqueous ethanol, with anhydrous hydrogen fluoride–pyridine in oxolane, and with hydrochloric acid in chloroform–methanol, found that O-de(trialkylsilyl)ation under acidic conditions was only partly effective, and that concomitant acyl migration and deacylation took place.

(311) W. Köhler and W. Pfleiderer, Ann., (1979) 1885–1871.
(312) G. H. Dodd, B. T. Golding and P. V. Ioannou, J. Chem. Soc. Perkin Trans. 1, (1976) 2273–2277.
(313) H. H. Brandstetter and E. Zbiral, Helv. Chim. Acta, 61 (1978) 1832–1841.
(314) G. H. Dodd, B. T. Golding, and P. V. Ioannou, J. Chem. Soc. Chem. Commun., (1975) 249–250.
(315) F. Franke and R. D. Guthrie, Aust. J. Chem., 30 (1977) 639–647.
(316) F. Franke and R. D. Guthrie, Aust. J. Chem., 31 (1978) 1285–1290.

Regioselective ring-cleavage of cyclic derivatives of carbohydrates is a useful method for achieving their selective deprotection. 3',5'-O-(Tetraisopropyldisiloxane-1,3-diyl) derivatives of nucleosides (**51**, B = uracil-1-yl, cytosin-1-yl, N^4-benzoylcytosin-1-yl, guanin-9-yl, or adenin-9-yl) have been prepared,[317] and selectively cleaved in 95% yield at the silicon–O-3' bond on treatment with 0.2 M sodium hy-

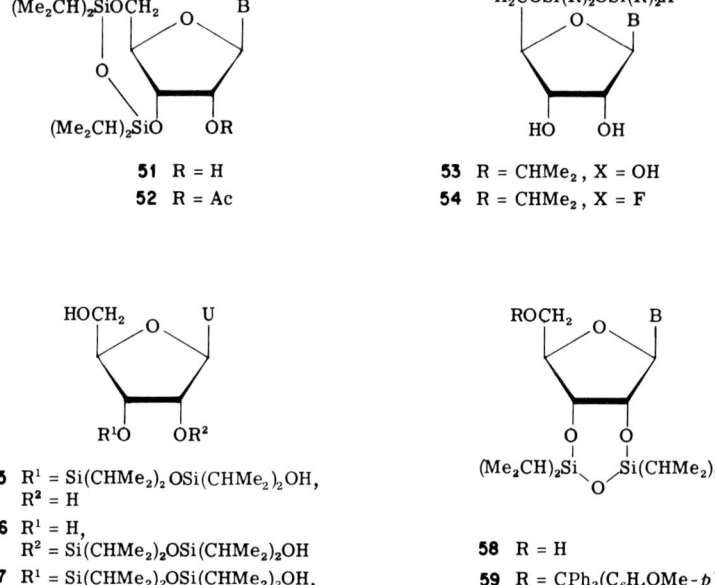

51 R = H
52 R = Ac

53 R = CHMe$_2$, X = OH
54 R = CHMe$_2$, X = F

55 R^1 = Si(CHMe$_2$)$_2$OSi(CHMe$_2$)$_2$OH,
    R^2 = H
56 R^1 = H,
    R^2 = Si(CHMe$_2$)$_2$OSi(CHMe$_2$)$_2$OH
57 R^1 = Si(CHMe$_2$)$_2$OSi(CHMe$_2$)$_2$OH,
    R^2 = Ac

58 R = H
59 R = CPh$_2$(C$_6$H$_4$OMe-p)

droxide in 4:1 1,4-dioxane–water at room temperature, to give 5'-O-substituted nucleosides (**53**). Tetrabutylammonium fluoride in oxolane is too reactive to bring about selective cleavage in this ring system, but tributylammonium fluoride, which is much less reactive in this type of reaction, converts the uridine derivative **51** (B = uracil-1-yl) in 0.5 hour at room temperature into **54** (B = uracil-1-yl) in ~60% yield in this solvent. When treated with 0.2 M hydrochloric acid in 4:1 1,4-dioxane–water for 2.5 hours at room temperature, **51** (B = uracil-1-yl) gave a mixture of **53** (B = uracil-1-yl), **55**, and **56** in the ratios of 2:5:3, indicating that O-3' to O-2' migration of the substituted-silyl groups occurs under these conditions. 2'-O-Acetyl-3',5'-O-(tetraisopropyldisiloxane-1,3-diyl)uridine (**52**, B = uracil-1-yl) underwent silicon–O-5' bond-cleavage, almost exclusively, under the acidic conditions, to yield the 3'-ether **57**, but O-3' to O-2' isomerization took place during removal of the acetyl group under alkaline conditions.

The preparation of 2',3'-O-(tetraisopropyldisiloxane-1,3-diyl)uridine (**58**, B = uracil-1-yl) through its 5'-O-(p-anisyldiphenylmethyl) derivative (**59**, B = uracil-1-yl) has been described,[317] but regioselective ring-cleavage at the Si–O-2' or –O-3' bond in such compounds to produce nucleoside derivatives having, respectively, either O-3' or O-2' protected, may be difficult to achieve unless the problem of migration of such protecting groups between O-2' and O-3' can be overcome.

It has been suggested[317] that the position of ring cleavage in such cyclic, (tetraisopropyldisiloxane-1,3-diyl) derivatives on partial hydrolysis could possibly be rationalized in terms of the acidity of the carbon-bound hydroxyl group that is formed on hydrolysis of a silicon–oxygen bond.

Other workers have noted[318] that, on treatment with mesitylenesulfonic acid in N,N-dimethylformamide, 3',5'-O-(tetraisopropyldisiloxane-1,3-diyl) derivatives (**51**) of nucleosides are isomerized into the corresponding 2',3'-O-(tetraisopropyldisiloxane-1,3-diyl) derivatives (**58**). Yields in these isomerizations were increased when pyridine hydrochloride was used as the acidic catalyst, and it is noteworthy that the rearrangement did not proceed effectively in acetonitrile, chloroform, 1,4-dioxane, and oxolane. In this rearrangement, the position of the (tetraisopropyldisiloxane-1,3-diyl) group changes, from one in which it spans an oxygen atom at a primary and an oxygen atom at a secondary position, to one in which it spans two oxygen atoms that are both at secondary positions. In order to investigate the generality of this type of process, methyl 4,6-O-(tetraisopropyldisiloxane-1,3-diyl)-α-D-glucopyranoside (**60**), prepared in 60% yield by reaction of methyl α-D-glucopyranoside with 1,3-dichloro-1,1,3,3-tetraisopropyldisiloxane (**61**), was treated with mesitylenesulfonic acid in N,N-dimethylformamide. After 6 hours at room temperature, complete isomerization of **60** had occurred, and methyl 3,4-O-(tetraisopropyldisiloxane-1,3-diyl)-α-D-glucopyranoside (**62**) was isolated in 60% yield.

(317) W. T. Markiewicz, *J. Chem. Res.* (S), (1979) 24–25; *J. Chem. Res.* (M), (1979) 181–197.
(318) C. H. M. Verdegaal, P. L. Jansse, J. F. M. de Rooij, and J. H. van Boom, *Tetrahedron Lett.*, (1980) 1571–1574.

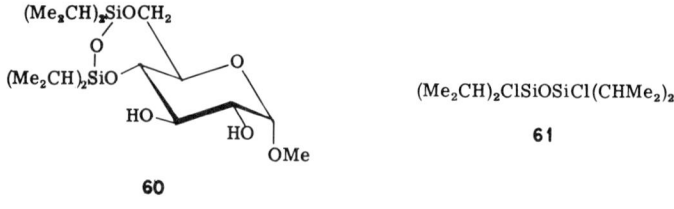

It is clear that an important new principle of selective protection in compounds containing both primary and secondary hydroxyl groups is embodied in the foregoing examples involving the use of the (tetraisopropyldisiloxane-1,3-diyl) group. Initial reaction of the bifunctional reagent **61** appears to be controlled by selective reaction at a primary position, and the position of ring closure will be governed by the ability of the newly introduced group to span the two positions. Assuming that acid-catalyzed ring-opening of the cyclic derivative occurs at the primary position, and that the bond from silicon to oxygen at the secondary position is not affected under these conditions, the position of any new ring-closure will be governed by the ability of the disiloxane-1,3-diyl group to span the distance between the new site and the original, secondary position. Thus, the location of the primary hydroxyl group within the molecule may influence the structure of the final product, even though this hydroxyl group is not incorporated into the silicon-containing ring in the rearranged compound.

THE REACTIVITY OF CYCLIC ACETALS OF ALDOSES AND ALDOSIDES*

By Jacques Gelas

École Nationale Supérieure de Chimie, Université de Clermont-Ferrand, Ensemble Scientifique des Cézeaux, BP 45, 63170 Aubière, France

I. Introduction 71
II. Summary of Developments Concerning the Synthesis of Cyclic Acetals 73
    1. Synthesis of Acetals of Free Sugars 73
    2. Synthesis of *trans*-Fused 1,3-Dioxolanes 77
    3. Miscellaneous Methods 78
III. Reactivity 81
    1. Oxidation 81
    2. Photolysis 86
    3. Halogenation 95
    4. Hydrogenolysis 121
    5. Action of Strong Bases 138
    6. Miscellaneous Reactions 148
IV. Conclusion 155

I. Introduction

Since 1895, when Emil Fischer[1] described the reaction of aldehydes and ketones with glycoses, an impressive part of the chemistry of carbohydrates has dealt with acetals, and especially cyclic acetals (mainly 1,3-dioxolanes and 1,3-dioxanes). There are probably relatively few studies on the synthetic chemistry of monosaccharides that do not describe at least one acetal of a carbohydrate, be it for routine protection, or for use in an original synthesis. At least, in this Series, three articles have appeared on the cyclic acetals of the aldoses and aldosides[2,3] and of the ketoses[4], one article dealt with acetals of tetri-

---

* Dedicated to the memory of Professor René Rambaud, former Director of the École Nationale Supérieure de Chimie de Clermont-Ferrand.

(1) E. Fischer, *Ber.*, 28 (1895) 1145–1167.
(2) A. N. de Belder, *Adv. Carbohydr. Chem.*, 20 (1965) 219–302.
(3) A. N. de Belder, *Adv. Carbohydr. Chem. Biochem.*, 34 (1977) 179–241.
(4) R. F. Brady, Jr., *Adv. Carbohydr. Chem. Biochem.*, 26 (1971) 197–278.

tols, pentitols, and hexitols,[5] and a discussion of the stereochemistry of cyclic derivatives of carbohydrates[6] was partly devoted to cyclic acetals.

There is no doubt that the objective of many of the studies that have been reported has concerned the synthesis of cyclic acetals as temporary protecting groups. Qualification as a good protecting group requires: (1) general accessibility of cheap reagents; (2) ease of procedures, which must lead to the protected substrate quickly and with a good yield; (3) inertness of the protecting group to a large variety of reagents, to permit structural modifications; and (4) ease of methods for deprotection in high yield. The popularity of cyclic acetals has evidently been due to the reasonable fulfilment of all of these conditions. Although this could seem paradoxical regarding point (3), during the past few years, the role of acetals, especially in carbohydrate chemistry, has been changing: routes for opening cyclic acetals to afford synthetically useful derivatives have been introduced, and some of them have already proved to be major tools for structural modifications of carbohydrates. This area is growing regularly, and now constitutes an expanded utility of derivatives that had hitherto been largely confined to the fundamental, but limited, role of protecting groups. The aim of the present article is to convince the reader that cyclic acetals can also be functional groups.

This article is limited to the reactions involving cyclic acetals of aldoses and aldosides. However, for comprehensiveness, or for their potential usefulness in carbohydrate chemistry, some properties of cyclic acetals not directly relevant to aldoses and aldosides are included. It may be recalled that the chemistry of dithioacetals had already been reviewed in this Series.[7]

It is intended that the present Chapter shall be complementary to Haines's article in this Volume[8] that deals with the selective deprotection of protected sugars. The reader interested in all aspects of the hydrolysis, the alcoholysis, the acetolysis, and the isomerization of cyclic acetals of sugars should refer to Haines's article. A review has also been published concerning the formation and migration of cyclic acetals of carbohydrates.[8a]

It seems useful to begin with a summary of recent advances concerning the synthesis of cyclic acetals as a supplement to de Belder's discussion[3] of the literature up to 1975.

(5) S. A. Barker and E. J. Bourne, *Adv. Carbohydr. Chem.*, 7 (1952) 137–207.
(6) J. A. Mills, *Adv. Carbohydr. Chem.*, 10 (1955) 1–53.
(7) D. Horton and J. D. Wander, *Adv. Carbohydr. Chem. Biochem.*, 32 (1976) 15–123.
(8) A. H. Haines, *Adv. Carbohydr. Chem. Biochem.*, 39 (1981) 13–70.
(8a) D. M. Clode, *Chem. Rev.*, 79 (1979) 491–513.

II. Summary of Developments Concerning the Synthesis of Cyclic Acetals

1. Synthesis of Acetals of Free Sugars

The condensation of a sugar with an aldehyde, a ketone, or a dialkoxyalkane in the presence of a desiccant, with (or without) an acid catalyst, has continued to be routinely used for synthesizing cyclic acetals. The investigation of the kinetic control of the acetonation of free sugars by the action of enol ethers has been completed.[8b] Whereas D-mannose is well known to give 2,3:5,6-di-O-isopropylidene-D-mannofuranose by reaction of the free sugar with either acetone[2,3] or 2,2-dimethoxypropane,[9] the major compound (yield ~85%) obtained from the reaction with 2-methoxypropene in N,N-dimethylformamide, at 0°, in the presence of p-toluenesulfonic acid is[10] 4,6-O-isopropylidene-D-mannopyranose (**1**). An excess of reagent gave a high yield of the diisopropylidene acetal **2**. Similar reactions were conducted

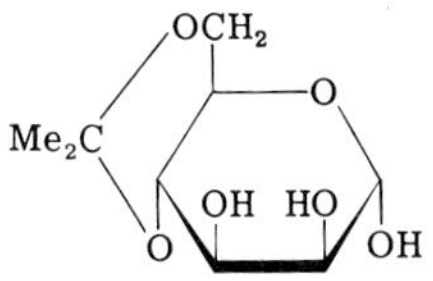

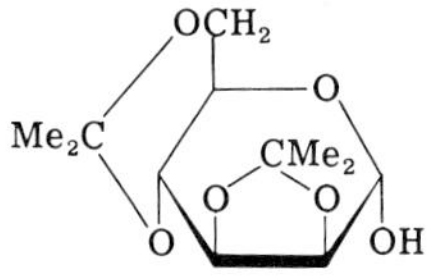

with D-allose and D-talose[11]: from each, the 4,6-O-isopropylidene-aldohexopyranose (**3** and **4**, respectively) was obtained in high yield (conventional procedures give the di-O-isopropylidene-aldohexo-

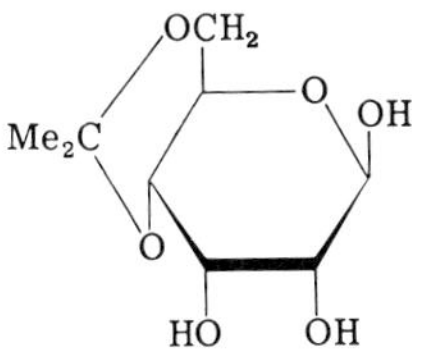

furanose derivatives[2,3]). In the case of D-galactose, concurrent 3,4-monoacetonation giving compound **5** could not be avoided, but the major compound isolated (yield ~65%) was 4,6-O-isopropylidene-D-galactopyranose (**6**).

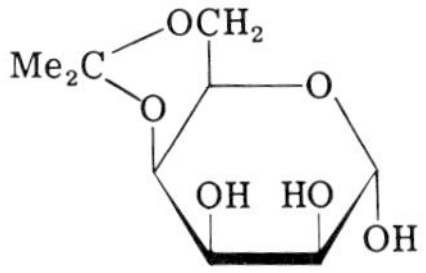

(8b) J. Gelas and D. Horton, to be published.
(9) A. Hasegawa and H. G. Fletcher, Jr., *Carbohydr. Res.*, 29 (1973) 209–222.
(10) J. Gelas and D. Horton, *Carbohydr. Res.*, 67 (1978) 371–387.
(11) J. Gelas and D. Horton, *Carbohydr. Res.*, 71 (1979) 103–121.

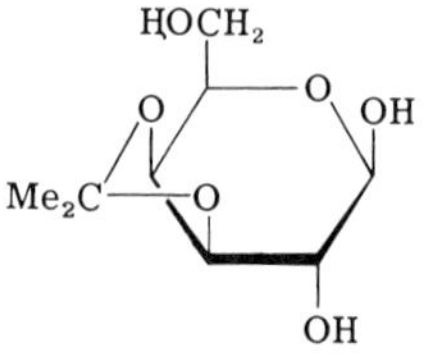

                    5 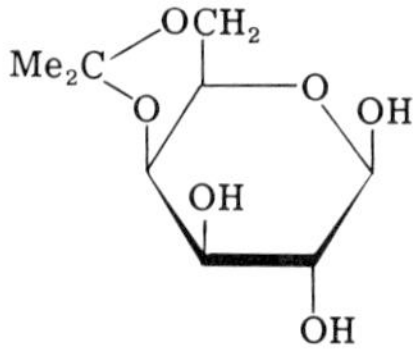   6

These results, and those obtained previously for D-glucose[12] and D-ribose and D-arabinose,[13] now permit definition of the general scope of the reaction of free sugars with alkyl isopropenyl ethers: (1) alkyl isopropenyl ethers effect acetonation under kinetic control; (2) primary hydroxyl groups are the favored sites for initial attack of the reagent[14]; (3) sugars not having a primary hydroxyl group in the favored tautomer in solution react without tautomerization, to give 1,3-dioxolanes; (4) the anomeric center does not take part in the reaction; (5) stoichiometric control of the reaction is possible, permitting access to either mono- or di-acetals; (6) the method is sufficiently mild for use in the synthesis of acetals of oligo- or poly-saccharides[16], and (7) strained rings or medium-sized rings are accessible by this method.

The reaction of D-lyxose (7) with 2-methoxypropene gave[17] a high yield of 2,3-O-isopropylidene-D-lyxofuranose (9), previously obtained[18] in low yield by the conventional procedure [with acetone in the presence of copper(II) sulfate and sulfuric acid]. This reaction probably proceeds through the pyranoid tautomer 8, a kinetic product of the reaction that tautomerizes rapidly to the thermodynamic prod-

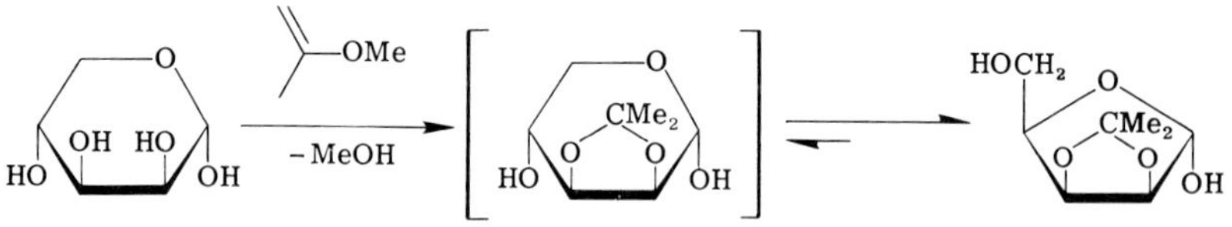

      7   8   9

(12) M. L. Wolfrom, A. B. Diwadkar, J. Gelas, and D. Horton, *Carbohydr. Res.*, 35 (1974) 87–96.
(13) J. Gelas and D. Horton, *Carbohydr. Res.*, 45 (1975) 181–195.
(14) As a consequence, 4,6-O-isopropylidene derivatives are obtained from D-aldohexoses; this has now been demonstrated not only for D-allose, D-galactose, D-glucose, D-mannose, and D-talose, but also for D-gulose (J. Gelas and D. Horton, unpublished results; see also, Ref. 15); it may reasonably be assumed that it will also be found for D-altrose and D-idose.
(15) C. Copeland and R. V. Stick, *Aust. J. Chem.*, 31 (1978) 1371–1374.

uct 9. This conformational effect, which disfavors a five-membered ring fused to a six-membered ring-system at C-2,3 (a bicyclo[4.3.0] system) as against a five-membered ring fused to a five-membered ring system (a bicyclo[3.3.0] system), had previously been found in the L-rhamnose series,[19] and was confirmed in the D-mannose series: selective hydrolysis of the 4,6-O-isopropylidene group was found to occur when the acetylated derivative 10 was treated with an aqueous solution of acetic acid at 0°, and the 2,3-O-isopropylidene derivative 11 was thus obtained in high yield[10]; when the hydrolysis was conducted with compound 2, in which the anomeric hydroxyl group is free, the pyranoid sugar 12 was not detected, but the useful, synthetic intermediate 13 was obtained directly, in high yield.[20] The formation

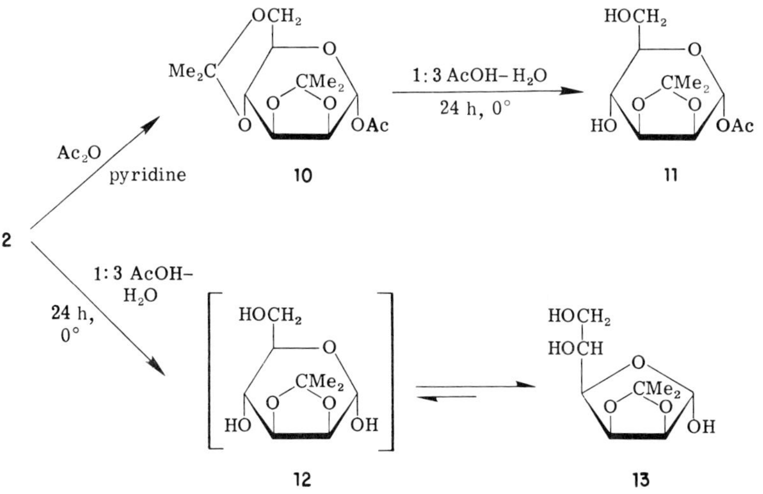

of 2,3-O-isopropylidene-D-ribofuranose as a minor product of the reaction of 2-methoxypropene with D-ribose[13] can now be explained by the same conformational effect, which probably operates when D-ribopyranose is concurrently acetonated at O-2,3 and O-3,4, the major product isolated being 3,4-O-isopropylidene-D-ribopyranose; for D-lyxose, the process leading to a *trans*-acetal at O-3,4 is too much

---

(16) E. Fanton, J. Gelas, and D. Horton, *J. Chem. Soc. Chem. Commun.*, (1980) 21–22.
(17) J. Gelas and D. Horton, to be published.
(18) R. Schaffer, *J. Res. Natl. Bur. Stand.*, Sect. A, 65 (1961) 507–512.
(19) S. J. Angyal, V. A. Pickles, and R. Ahluwalia, *Carbohydr. Res.*, 3 (1967) 300–307.
(20) E. Fanton and J. Gelas, to be published.

disfavored to be observed. These investigations have been extended to ketoses.[16]

A known method of acetalation by means of an enol acetate (2-acetoxypropene),[21] previously reported for some monosaccharides,[22] has been reinvestigated.[23] As it gave thermodynamically formed acetals in low to moderate yield (for instance, 30 and 54% yield, respectively, for the preparation of 2,3-O-isopropylidene-D-ribofuranose and 1,2:5,6-di-O-isopropylidene-α-D-glucofuranose), it appears to have no special advantages over the methods already available.

Although the synthesis of acetals by means of dialkoxyalkanes, especially 2,2-dimethoxypropane or 1,1-dimethoxycyclohexane, is a well known method, and continues to stimulate interest (see Section II, 3), less classic is the use of cyclohexanone ethylene acetal. The reaction using this reagent in the presence of sulfuric acid has been described[24] for some monosaccharides and, notably, some free sugars. In the case of D-mannose, 2,3:4,6-di-O-cyclohexylidene-α-D-mannofuranose was obtained either alone (yield 46%), or as a contaminant (yield 19%) of 2,3-O-cyclohexylidene-α-D-mannofuranose (yield 24%), depending on the reaction conditions. When the reaction was performed in the presence of methanol, the corresponding methyl D-mannofuranosides were obtained. More original and interesting was the application of this reaction to D-xylose (**14**), which gave 2,3:4,5-di-O-cyclohexylidene-D-xylose ethylene acetal (**15**, yield 55%), an analog of 2,3:4,5-di-

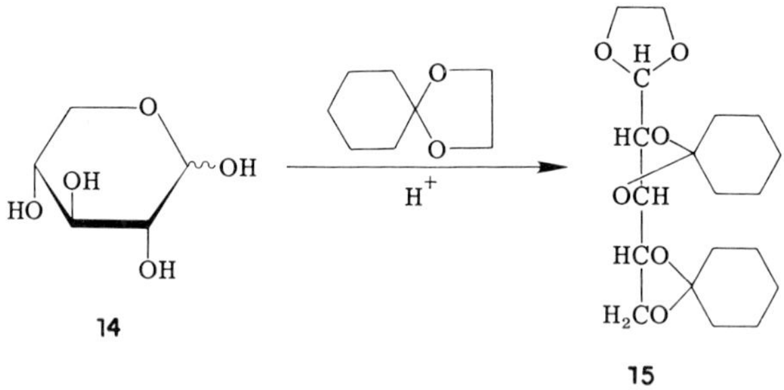

O-isopropylidene-D-xylose diethyl dithioacetal.[7] The same reaction was observed for 2-deoxy-D-*arabino*-hexopyranose and 4,6-dichloro-4,6-dideoxy-D-galactopyranose, giving acetals **16** and **17**, respectively.

(21) W. J. Croxall, F. J. Glavis, and H. T. Neher, *J. Am. Chem. Soc.*, 70 (1948) 2805–2807.
(22) T. Sato and R. Ishido, Jpn. Pat. 9024 (1965); *Chem. Abstr.*, 63 (1965) 7094.
(23) Y. Araki, Y. Hijioka, Y. Ishido, and T. Sato, *Carbohydr. Res.*, 64 (1978) 309–314.
(24) H. Paulsen, H. Salzburg, and H. Redlich, *Chem. Ber.*, 109 (1976) 3598–3605.

```
    O   H   O              O   H   O
     \ / \ /                \ / \ /
      C                      C
      |                      |
     CH₂                    HCO
      |                      |    \
     OCH                    OCH    \
      |    \                 |      |
     HCO    \               ClCH    |
      |     |                |      |
     HCO    |               HCOH   /
      |    /                 |    /
     H₂CO/                  H₂CCl
      16                     17
```

2. Synthesis of *trans*-Fused 1,3-Dioxolanes

Interesting progress has also been made on that part of the studies concerning the synthesis of acetals that deals with the strained ring formed from diequatorial, vicinal diols.

It is known that a very low yield of methyl 2,3:4,6-di-*O*-isopropylidene-α-D-glucopyranoside (18) may be obtained by the reaction of methyl α-D-glucopyranoside with 2,2-dimethoxypropane.[25] When 2-methoxypropene was used, an almost quantitative yield of the diacetal 18 was obtained.[26] It is noteworthy that, with 5-thioaldose analogs, examples of unusual acetal formation seem to be encountered more frequently. For instance, 1,2:3,4-di-*O*-isopropylidene derivatives of 5-thio-D-ribose and 5-thio-D-glucose have been prepared by reaction of the sugars with acetone containing 2,2-dimethoxypropane,[27,28] and the *trans*-fused 1,3-dioxolane, namely, 2,3:4,6-di-*O*-isopropylidene-5-thio-α-D-glucopyranose (19) was isolated, and its structure confirmed by X-ray crystallography,[29] which showed that the $^4C_1$ conformation is less distorted than that of the isomeric 1,2:4,6-diacetal.

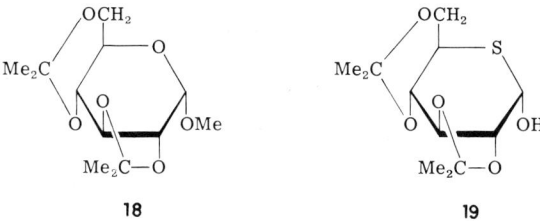

(25) M. E. Evans, F. W. Parrish, and L. Long, Jr., *Carbohydr. Res.*, 3 (1967) 453–462.
(26) J. L. Debost, J. Gelas, and D. Horton, to be published.
(27) N. A. Hughes and C. J. Wood, *J. Chem. Soc. Chem. Commun.*, (1975) 294–295.
(28) N. A. Hughes and C. J. Wood, *Carbohydr. Res.*, 49 (1976) 225–232.
(29) W. Clegg, N. A. Hughes, and N. Al-Masoudi, *J. Chem. Soc. Chem. Commun.*, (1979) 320–321.

Very interesting progress has been made in attempts to overcome the difficulty encountered for a *trans*, vicinal diol to undergo acetalation in an acid-catalyzed system. It concerns the irreversible reaction of a *gem*-dihalide under basic conditions, and using phase-transfer catalysis. Following a similar method previously introduced into carbohydrate chemistry for the *cis*-2,3-O-methylenation of methyl 4,6-O-benzylidene-α-D-mannopyranoside,[30] *trans*-2,3-O-methylenation of methyl 4,6-O-benzylidene-D-hexopyranosides (**20–23**) was performed by efficient stirring of the glycosides at 60–65° with dibromomethane under phase-transfer catalysis conditions,[31] affording the products in yields of 58 to 65%.

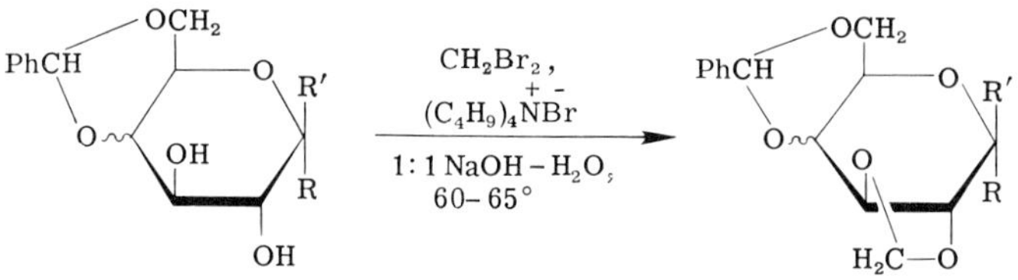

20 (C-4-O *eq*) R = OMe, R' = H (α-D-*gluco*)
21 (C-4-O *eq*) R = H, R' = OMe (β-D-*gluco*)
22 (C-4-O *ax*) R = OMe, R' = H (α-D-*galacto*)
23 (C-4-O *ax*) R = H, R' = OMe (β-D-*galacto*)

3. Miscellaneous Methods

Transacetalation has continued to be a useful method. We have already mentioned (see Section II,1) the behavior of cyclohexanone ethylene acetal with some free sugars.[24] Synthesis of methyl 4,6-O-benzylidene-α-D-gluco- and -galacto-pyranosides, of methyl 2,3-O-benzylidene-α-L-rhamnopyranoside, and of methyl 2,3:4,6-di-O-benzylidene-α-D-mannopyranoside in high yield by heating the glycosides with α,α-diethoxytoluene in 1,4-dioxane containing an acidic resin during 4 h at 50° has been reported.[32] This study merely confirmed results previously described for the synthesis of most of these compounds; for instance, from methyl α-D-glucopyranoside[33] and from

(30) P. Di Cesare and B. Gross, *Carbohydr. Res.*, 48 (1976) 271–276.
(31) K. S. Kim and W. A. Szarek, *Synthesis*, (1978) 48–50.
(32) A. Yu. Romanovich, A. F. Sviridov, and S. V. Yarotskii, *Izv. Akad. Nauk SSSR, Ser. Khim.*, (1977) 2160–2161.
(33) M. E. Evans, *Carbohydr. Res.*, 21 (1972) 473–475.

methyl α-D-mannopyranoside.[34] Preparation of some cyclic acetals of methyl α-D-glucopyranoside, namely, the 4,6-O-anisylidene,-syringylidene, and -veratrylidene derivatives from the dimethyl acetals of the corresponding aromatic aldehydes has also been described.[35] Usually, stoichiometric control of the acetalation of the carbohydrate, to give either mono- or di-acetals when competition is possible, is not easy to obtain; some progress has been made as regards monomolar acetalation of methyl α-D-mannopyranoside and -mannofuranoside.[36] For instance, methyl α-D-mannopyranoside in N,N-dimethylformamide gave the 4,6-O-isopropylidene derivative when treated at 25° with one molar proportion of 2,2-dimethoxypropane in the presence of p-toluenesulfonic acid; at 60°, in the presence of sulfuric acid, the 2,3-O-isopropylidene derivative was the major product.

Hydrogenolysis of cyclic orthoesters with diborane (a method using lithium aluminum hydride–aluminum chloride had been described for 1,2-orthoesters[37]) has been shown to be a route to cyclic acetals[38]; thus, a mixture of the *endo* and *exo* isomers of methyl 3,4-O-(ethoxyethylidene)-β-L-arabinopyranoside led to methyl *endo*-3,4-O-ethylidene-β-L-arabinopyranoside (yield 66%).

A potentially very useful method of protection by use of insoluble polystyrene resins has appeared.[39] A polymer containing vinylbenzaldehyde units was used for preparing 4,6-acetals of methyl α-D-glucopyranoside in good yield.

Synthesis of the cyclic acetal **25** in 63% yield by reaction of the unsaturated bromo-nitrilo sugar **24** (*E* configuration) with ethylene glycol in the presence of sodium hydroxide has been reported.[40] An extension of this work was given for sugars having a vicinal diol

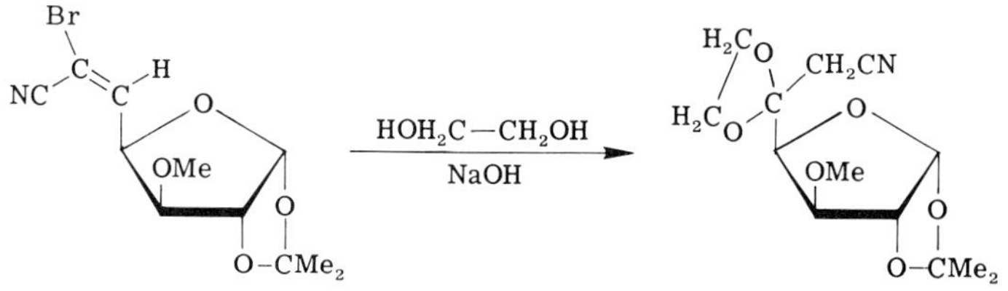

24   →   **25**

HOH$_2$C—CH$_2$OH, NaOH

(34) D. Horton and W. Weckerle, *Carbohydr. Res.*, 44 (1975) 227–240.
(35) D. Joniak, B. Košíková, and L. Kosáková, *Chem. Zvesti*, 31 (1977) 106–108.
(36) M. E. Evans and F. W. Parrish, *Carbohydr. Res.*, 54 (1977) 105–114.
(37) S. S. Bhattacharjee and P. A. J. Gorin, *Carbohydr. Res.*, 12 (1970) 57–68.
(38) J. G. Buchanan and A. R. Edgar, *Carbohydr. Res.*, 49 (1975) 289–304.
(39) J. M. J. Fréchet and G. Pellé, *J. Chem. Soc. Chem. Commun.*, (1975) 225–226.
(40) J. M. J. Tronchet and O. R. Martin, *Helv. Chim. Acta*, 59 (1976) 945–948.

grouping replacing ethylene glycol.[40a] It may be recalled that cyclic acetals may also be prepared by irradiation of a mixture of 1,3-dioxolane and 5,6-dideoxy-1,2-O-isopropylidene-α-D-*xylo*-hex-5-enofuranose (**26**) in acetone[41], the adduct **27** corresponding to an anti-Markov-

nikov reaction was isolated (yield 45%). Also studied was 3-deoxy-1,2:5,6-di-O-isopropylidene-3-C-methylene-α-D-*ribo*-hexofuranose. The reaction probably proceeds through abstraction of a hydrogen atom from C-2 of the 1,3-dioxolane, and trapping of the free radical by the alkene. This method constitutes a useful route for a one-carbon chain-extension or for chain branching. For other, various examples in the carbohydrate series of this well known photoaddition of 1,3-dioxolanes to alkenes,[42] see, for instance, Refs. 43–45.

The search for new catalysts that would be effective for acetalations still stimulates interest. Better yields, and increased rates, have been claimed for the reaction of monosaccharides with acetone in the presence of ferric chloride.[46] (Ferric chloride was already known to catalyze acetalation in other series.[47]) Not yet applied in the carbohydrate field, but potentially useful for sugars, is the separate use of two different catalysts. The first one recommended is pyridinium *p*-toluenesulfonate as a mild and efficient catalyst for the tetrahydropyranylation of alcohols.[48] The main interest in this catalyst lies in the excellent yields

(40a) O. R. Martin, Thèse, Université de Genève, No. 1962, June 1980.
(41) J. S. Jewell and W. A. Szarek, *Tetrahedron Lett.*, (1969) 43–46.
(42) I. Rosenthal and D. Elad, *J. Org. Chem.*, 33 (1968) 805–811, and preceding papers.
(43) K. Matsuura, S. Maeda, Y. Araki, Y. Ishido, and T. Murai, *Tetrahedron Lett.*, (1970) 2869–2872.
(44) K. Matsuura, K. Nishiyama, K. Yamada, K. Araki, and Y. Ishido, *Bull. Chem. Soc. Jpn.*, 46 (1973) 2538–2542.
(45) B. Fraser-Reid, D. R. Hicks, D. L. Walker, D. E. Iley, M. B. Yunker, S. Y. K. Tam, R. C. Anderson, and J. Saunders, *Tetrahedron Lett.*, (1975) 297–300.
(46) P. P. Singh, M. M. Gharia, F. Dasgupta, and H. C. Srivastava, *Tetrahedron Lett.*, (1977) 439–440.
(47) J. Bornstein, S. F. Bedell, P. E. Drummond, and C. L. Kosloski, *J. Am. Chem. Soc.*, 78 (1956) 83–86.
(48) N. Miyashita, A. Yoshikoshi, and P. A. Grieco, *J. Org. Chem.*, 42 (1977) 3772–3774.

reported (95–100% for alcohols of the steroid and terpene series), the short reaction-time, and the weak acidity. Use of this catalyst would probably be a satisfactory method to be applied to highly sensitive carbohydrates. This catalyst is structurally similar to pyridinium chloride, which had been reported to be a catalyst for acetalation of non-carbohydrates.[49,50] The other catalysts introduced as an alternative for the acetalation of aldehydes are the rare-earth chlorides,[51] which are mild and efficient (excellent yields, generally after a reaction time of 10 min); but appropriate choice of the rare-earth chloride seems critical for the best efficiency.

III. REACTIVITY

1. Oxidation

a. Ozonolysis.—In 1971, Deslongchamps and Moreau[52] discovered that ozone could react readily with acetals to give a high yield of the corresponding esters. This reaction, demonstrated for acetals derived from different aldehydes (1-heptanal, cyclohexanecarboxaldehyde, and benzaldehyde) and various alcohols (methanol, ethanol, ethylene glycol, 1,3-propanediol, and 2,2-dimethyl-1,3-propanediol), was shown also to be effective for the ring opening of tetrahydropyranyl ethers. These results were extended to sugars, and it was shown that methyl β-D-glucopyranosides are oxidized, whereas the corresponding α anomers do not react; this was explained by the assumption that ozonolysis is operative only if the hydrogen atom of the acetal grouping could be *trans*-antiparallel to the lone-pair orbitals of both oxygen atoms of this function. Finally, typical cyclic acetals (O-ethylidene and O-benzylidene derivatives) were also cleaved to esters. Thus, a mixture of methyl 2,3-O-acetyl-4,6-O-ethylidene-α-D-glucopyranoside, acetic anhydride, and sodium acetate gave methyl α-D-glucopyranoside peracetate after treatment with ozone during 1 h at room temperature, followed by heating to reflux. The influence of the conformation of the acetal grouping upon the reactivity of the substrate was discussed in detail for tetrahydropyran-2-yl ethers, 1,3-dioxanes, and glycosides.[53,54]

(49) J. Gelas, *Tetrahedron Lett.*, (1971) 509–512.
(50) J. Egyed, P. Demerseman, and R. Royer, *Bull. Soc. Chim. Fr.*, (1972) 2287–2288, and references cited therein.
(51) J. L. Luche and A. L. Gemal, *J. Chem. Soc. Chem. Commun.*, (1978) 976–977.
(52) P. Deslongchamps and C. Moreau, *Can. J. Chem.*, 49 (1971) 2465–2467.
(53) P. Deslongchamps, C. Moreau, D. Fréhel, and P. Atlani, *Can. J. Chem.*, 50 (1972) 3402–3404.
(54) P. Deslongchamps, P. Atlani, D. Fréhel, A. Malaval, and C. Moreau, *Can. J. Chem.*, 52 (1974) 3651–3664.

The oxidation of benzylidene acetals was further investigated[55] for two different types of cyclic acetal. Ozonolysis of 1,3-dioxolanes, benzylidene acetals *cis*-fused to a rigid, six-membered ring, was examined for examples chosen from the decalindiol and the cholestanediol series; formation of a cyclic, orthoacid (hemiorthoester) intermediate, leading finally to the *axial* benzoate (only slightly contaminated by the isomeric *equatorial* benzoate) was observed. Directly relevant to the present article were the results concerning the second type of acetal, namely, 4,6-O-benzylidene derivatives of sugars. Ozonolysis of 2,3-di-O-acetyl, -methyl, and -p-tolylsulfonyl derivatives of methyl 4,6-O-benzylidene-α-D-glucopyranoside gave a mixture of the 6- and the 4-benzoates. The regiospecificity of the cleavage seemed to depend on the structure of the starting material; the possible influence of the stereochemistry at the acetal carbon atom (phenyl group *axial* or *equatorial*) was examined; it is noteworthy that oxidation of benzylidene acetals was much more rapid than the cleavage of methyl β-D-glucopyranoside. Thus, for methyl α-D-glucopyranoside, as well as for the β anomer, ozonolysis of benzylidene acetals (and this was confirmed for O-ethylidene derivatives) can be used in order to provide partially esterified sugars.

b. Oxidation by Potassium Permanganate.—Two examples of oxidation of methylene acetals of aldoses by the action of potassium permanganate in neutral solution, to give 1,2-carbonates, have been reported by Schmidt and coworkers.[56,57] Thus, 1,2:3,5-di-O-methylene-α-D-xylofuranose[56] and 6-O-acetyl-1,2:3,5-di-O-methylene-α-D-glucofuranose[57] (compounds **28** and **29**, respectively) were selectively converted into the corresponding *mono*-1,2-carbonates, **30** and **31**.

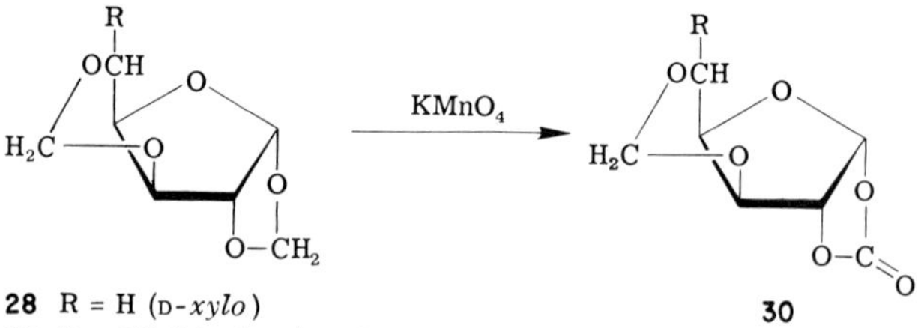

28 R = H (D-*xylo*)
29 R = CH$_2$OAc (D-*gluco*)

30
31

c. Oxidation by Chromium Trioxide in Acetic Acid.—Although it does not appear that acetals of aldoses and aldosides have been tested

(55) P. Deslongchamps, C. Moreau, D. Fréhel, and R. Chênevert, *Can. J. Chem.*, 53 (1975) 1204–1211.
(56) O. T. Schmidt, *Angew. Chem., Teil A*, 60 (1948) 252.
(57) O. T. Schmidt, A. Distelmaier, and H. Reinhard, *Chem. Ber.*, 86 (1953) 741–749.

with this reagent, it is worth reporting here the useful method introduced by Angyal and coworkers[58-62] as a route to 3-hexuloses. Peracetylated, cyclic methylene (or benzylidene) acetals of alditols were oxidized at room temperature by chromium trioxide in acetic acid to formyl (or benzoyl) derivatives of aldloses: one of the alcohol groups, originally from the acetal, had been retained, the other being oxidized to a ketone. The scope of the reaction was presented through examples described for (a) peracetylated methyl D-hexopyranosides and lactose[59] (in these cases, the acetal function was the glycosidic group, and, for instance, methyl 2,3,4,6-tetra-O-acetyl-β-D-glucopyranoside gave a 76% yield of methyl 2,3,4,6-tetra-O-acetyl-D-*xylo*-5-hexulosonate, which could be a very useful intermediate for the synthesis of L-idose or L-sorbose); (b) peracetylated methylene and benzylidene acetals of alditols[60] (symmetrical acetals) gave only one product, and, for instance, 1,2,5,6-tetra-O-acetyl-4-O-formyl-*keto*-D-*arabino*-3-hexulose (**32**) was obtained in excellent yield from 3,4-O-methylene-D-mannitol peracetate; on the other hand, unsymmetrical acetals gave two prod-

```
H₂COAc            H₂COAc
 |                 |
AcOCH             AcOCH
 |                 |
 OCH   CrO₃-AcOH   C=O
  \CH₂  ──────→    |
 HCO               HCOCHO
 |                 |
 HCOAc             HCOAc
 |                 |
 H₂COAc           H₂COAc

                    32
```

ucts corresponding to the two cleavages possible, and, for instance, oxidation of 2,4-O-benzylidene-D-glucitol peracetate (**33**) gave a 69% yield of 4-O-benzoyl-*keto*-D-fructose peracetate (**34**), but evidence was given that the mother liquors contained its regio-isomer (**35**); and

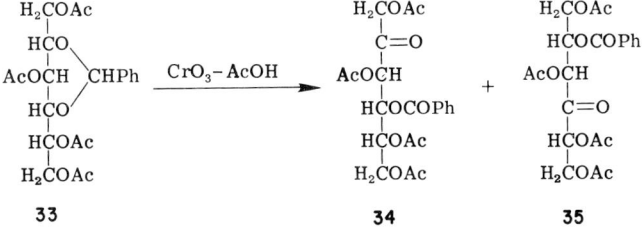

(58) S. J. Angyal and K. James, *Chem. Commun.*, (1970) 320–321.
(59) S. J. Angyal and K. James, *Aust. J. Chem.*, 23 (1970) 1209–1221.
(60) S. J. Angyal and K. James, *Aust. J. Chem.*, 24 (1971) 1219–1227.
(61) S. J. Angyal and M. E. Evans, *Aust. J. Chem.*, 25 (1972) 1495–1512.
(62) S. J. Angyal and M. E. Evans, *Aust. J. Chem.*, 25 (1972) 1513–1520.

(c) peracetylated 3,4-O-ethylidene-alditols,[61] which appeared to be more suitable than the benzylidene acetals because of the easier removal of the acetate than of the benzoate groups. Finally, it was found that addition of acetic anhydride to the mixed reagent (chromium trioxide–acetic acid) induced a new type of oxidation of methylene acetals of alditols,[62] which thus gave carbonates, isolated as their isopropylidene acetals (after acid hydrolysis of the acetates and action of 2,2-dimethoxypropane). Thus, 1,2,5,6-tetra-O-acetyl-3,4-O-methylene-D-mannitol was converted into carbonate 36 in an overall yield of 48%.

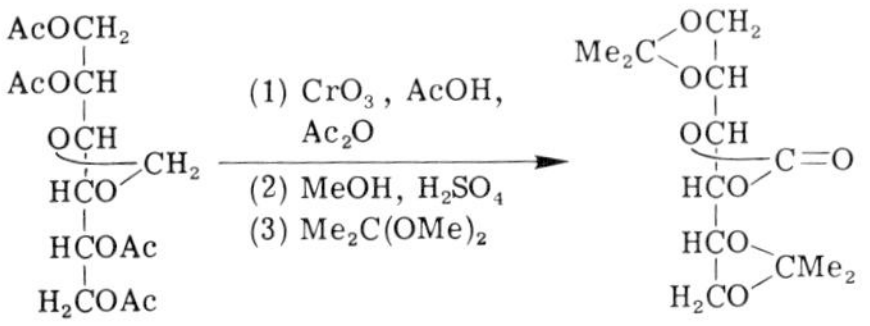

An interpretation of the oxidation of acetals by chromium trioxide is summarized in Scheme 1: the reagent could initially remove a proton from the acetalic carbon atom, to give a dioxolan-(or dioxan-)2-ylium ion, rapidly hydrolyzed to an ester (formyl, acetyl, or benzoyl) of an α-alcohol, which would be further oxidized to a ketone [see Scheme 1, path (a)]; in the case of methylene acetals, in the presence of acetic anhydride (which acts as a water scavenger), the intermediate is further oxidized, probably through a chromic ester.

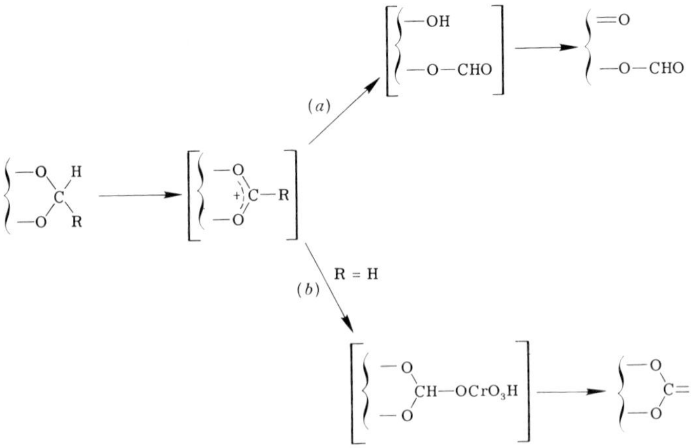

Scheme 1

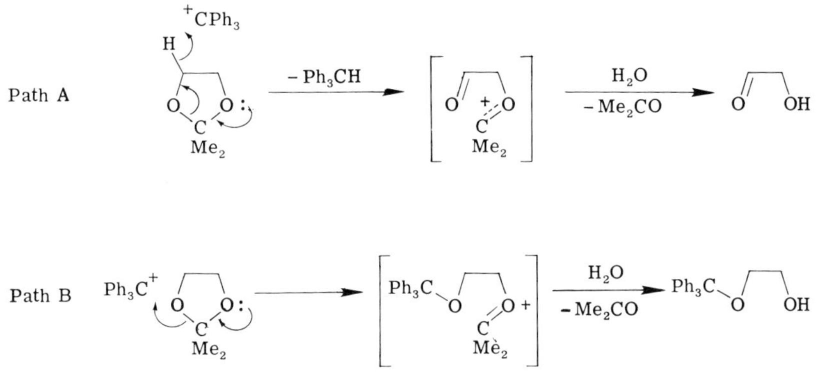

Scheme 2

d. Oxidation by Triphenylmethyl Fluoroborate.

—An alternative method for the synthesis of 3-hexuloses was later proposed by Barton and coworkers,[63,64] who gave three examples of conversion of isopropylidene acetals of monosaccharides into compounds assumed, without proof, to be α-ketols, by treatment with triphenylmethyl fluoroborate in dichloromethane. This procedure would have been a route to the selective oxidation of a diol. The mechanism proposed[64] involved hydride abstraction by the acceptor (triphenylmethyl cation) from C-4 (or C-5) of the 1,3-dioxolane (see Scheme 2, path A). Thus, after treatment with triphenylmethyl fluoroborate, followed by benzoylation, 3,5,6-tri-*O*-benzoyl-1,2-*O*-isopropylidene-α-D-glucofuranose (**37**) supposedly gave ketone **38** (overall yield 20%). Two other alleged examples were described for this reaction as applied to acetals of alditols, namely, 1,2,5,6-tetra-*O*-benzoyl-3,4-*O*-isopropylidene-D-mannitol (the perbenzoylated analog of the compound that

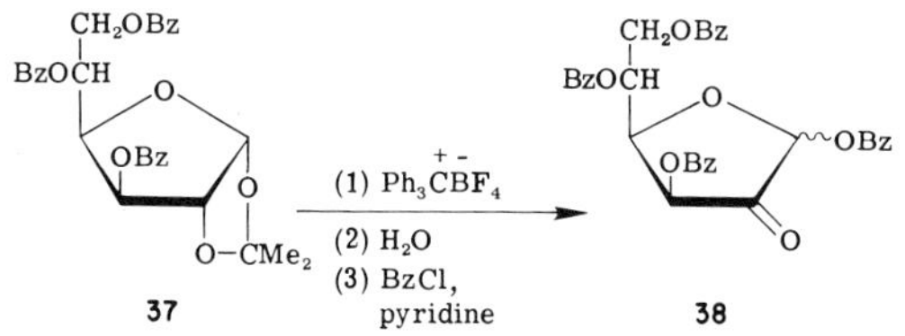

---

(63) D. H. R. Barton, P. D. Magnus, G. Smith, and D. Zurr, *Chem. Commun.*, (1971) 861–863.

(64) D. H. R. Barton, P. D. Magnus, G. Smith, G. Streckert, and D. Zurr, *J. Chem. Soc. Perkin Trans. 1*, (1972) 542–552.

was oxidized by Angyal and coworkers, see Section III,1,c) and 1,2,3,4-tetra-O-benzoyl-5,6-O-isopropylidene-D-glucitol.

In fact, for all three of these examples, it was later demonstrated, in a reappraisal of the method,[64a] that triphenylmethane was not produced, and, therefore, that dehydrogenation had not occurred. Finally, only glycols were obtained (isolated as the corresponding perbenzoylated derivatives) by hydrolysis of the isopropylidene group and, presumably, the cleavage of the triphenylmethyl ether (Scheme 2, path B). Hydride abstraction seems to be operative only for acetals formed from glycols having a benzylic hydrogen atom.

2. Photolysis

The possibility of using light-induced cleavage of cyclic acetals as a mild and selective procedure for the removal of an acetal protecting-group was pioneered a long time ago by Tănăsescu and his coworkers.[65-72] It was demonstrated that nitroso derivatives could be obtained when o-nitrobenzylidene acetals of such polyols as pentaerythritol,[65] ethylene glycol,[66] or erythritol[66] were exposed to summer sunlight. Thus, 2-(o-nitrophenyl)-1,3-dioxolane (**39**) gave the nitroso derivative **40**, which probably had the structure of 2-hydroxyethyl o-nitrosobenzoate (**41**).

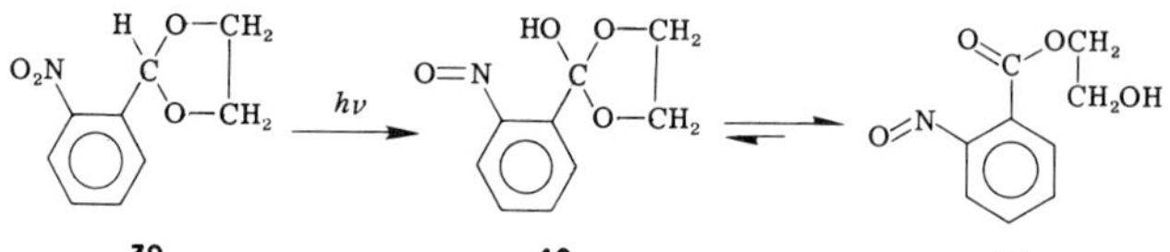

This reaction was extended to O-(o-nitrobenzylidene) derivatives of glycerol,[67] D-mannitol,[68] and D-glucitol.[69] Some regioselectivity seems to have been observed, as two acetal groups [of the three for the tri-O-(o-nitrobenzylidene) derivatives of D-mannitol and D-glucitol] were found to have reacted. The identity and the structure of the photo-

(64a) D. H. R. Barton, P. D. Magnus, J. A. Garbarino, and R. N. Young, *J. Chem. Soc. Perkin Trans. 1*, (1974) 2101–2107.
(65) I. Tănăsescu, *Bul. Soc. Stiinte Cluj*, 2 (1924) 111–120.
(66) I. Tănăsescu and H. Tănăsescu, *Bul. Soc. Stiinte Cluj*, 2 (1925) 369–382.
(67) I. Tănăsescu and E. Macovski, *Bull. Soc. Chim. Fr.*, 45 (1929) 1022–1030.
(68) I. Tănăsescu and E. Macovski, *Bull. Soc. Chim. Fr.*, 47 (1930) 86–93.
(69) I. Tănăsescu and E. Macovski, *Bull. Soc. Chim. Fr.*, 47 (1930) 457–463.
(70) I. Tănăsescu and E. Macovski, *Bull. Soc. Chim. Fr.*, 51 (1932) 1371–1377.
(71) I. Tănăsescu and E. Macovski, *Bull. Soc. Chim. Fr.*, 51 (1932) 1556–1565.
(72) I. Tănăsescu and E. Macovski, *Bull. Soc. Chim. Fr.*, 53 (1932) 1097–1102.

products were discussed in the case of nitrobenzylidene acetals of ethylene glycol[70] and glycerol.[71] Further examples were studied for nitrobenzylidene acetals of alditols (D-glucitol[72]), aldoses (D-galactose, D-mannose, and D-rhamnose,[73] or D-glucose[74]), and oligosaccharides (lactose and sucrose[73]). Other observations were also published concerning acetals of aldoses and aldosides,[75,76] alditols,[77] and 1,2-cyclohexanediol.[78] Some developments have since been published regarding the D-glucitol series.[79] Unfortunately, this very useful work was performed without recourse to modern methods of analytical and structural identification. The supposed structure of many acetals among the starting materials was also questionable; for instance, either as regards the positions of the cyclic acetal grouping on the sugar chain, or the tautomeric form of the protected sugar. Furthermore, the nitroso photoproducts are very unstable species, and their structure, or their stereochemistry, was difficult to determine. Thus, there was a need to reinvestigate the scope of the photolysis of cyclic acetals of sugars.

In 1963, Elad and Youssefyeh[80] described the photochemical conversion, in the presence of acetone, of a number of 2-alkyl- or 2-aryl-substituted 1,3-dioxolanes (see Scheme 3, $n=0$) or 1,3-dioxanes (Scheme 3, $n=1$) into carboxylic esters. They suggested that the reaction proceeds by initial, hydrogen abstraction from the acetal carbon

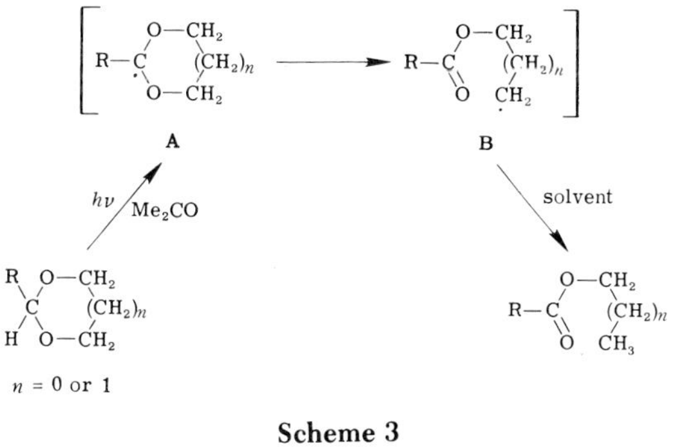

Scheme 3

(73) I. Tănăsescu and E. Craciunescu, *Bull. Soc. Chim. Fr.*, 3 (1936) 581–598.
(74) I. Tănăsescu and M. Ionescu, *Bull. Soc. Chim. Fr.*, 3 (1936) 1511–1517.
(75) I. Tănăsescu and E. Craciunescu, *Bull. Soc. Chim. Fr.*, 3 (1936) 1517–1527.
(76) I. Tănăsescu and M. Ionescu, *Bull. Soc. Chim. Fr.*, 7 (1940) 84–90.
(77) I. Tănăsescu and I. Iliescu, *Bull. Soc. Chim. Fr.*, 5 (1938) 1446–1457.
(78) I. Tănăsescu and M. Ionescu, *Bull. Soc. Chim. Fr.*, 7 (1940) 77–83.
(79) I. Tănăsescu and A. Otea, *Stud. Univ. Babes-Bolyai, Ser. Chem.*, 17 (1972) 113–119.
(80) D. Elad and R. D. Youssefyeh, *Tetrahedron Lett.*, (1963) 2189–2191.

atom, generating the radical species **A** and **B**, which lead to the ester, by reaction with the solvent (*tert*-butanol), in yields of 14 to 52%. A few years later, this reaction was applied to sugars, in an attempt to find a route to deoxy sugars.[81] In fact, when a solution of methyl 2,3-*O*-benzylidene-β-D-ribofuranoside (**42**) in acetone and *tert*-butanol was

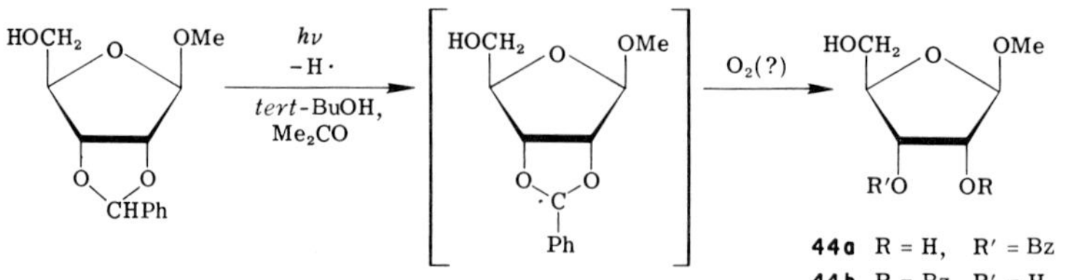

irradiated at room temperature, the formation of methyl 2(or 3)-*O*-benzoyl-β-D-ribofuranoside (**44**) resulted (yield 58%; it was not specified whether a single compound was obtained, or a mixture of the two possible regio-isomers, **44a** and **44b**). It was suspected that the presence of oxygen during the reaction was responsible for the formation of an α-hydroxybenzoate from the intermediate **43**. When 3,5-*O*-benzylidene-1,2-*O*-isopropylidene-α-D-xylofuranose (**45**) was irradiated under the same conditions, a mixture of the 5- and 3-monobenzoate (**46a** and

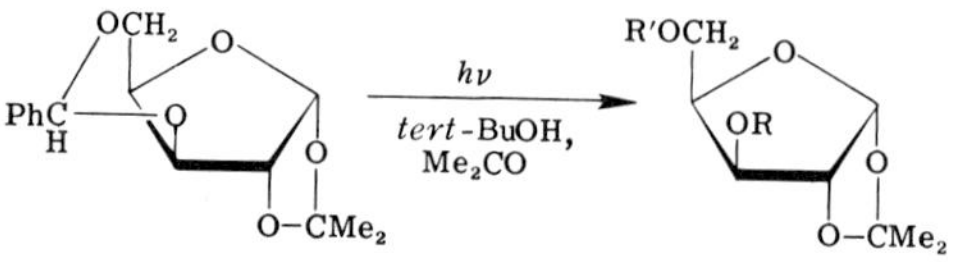

46b) resulted.[81] The selective removal of 5,6- or 3,5-*O*-isopropylidene groups by u.v. irradiation of 1,2:5,6-di-*O*-isopropylidene-α-D-glucofuranose and 1,2:3,5-di-*O*-isopropylidene-α-D-xylofuranose respectively was claimed[82] to be a new method for the synthesis of 1,2-*O*-isopropylidene derivatives of monosaccharides (yield in the range of 30 to 70%, depending on the polarity of the solvent). Under the same conditions, 1,2-*O*-isopropylidene-3,5,6-tri-*O*-methyl-α-D-glucofuranose was unreactive.

A thorough re-examination of the important, pioneering work of Tănăsescu and coworkers[65-72] was performed, with modern tech-

(81) K. Matsuura, S. Maeda, Y. Araki, and Y. Ishido, *Bull. Chem. Soc. Jpn.*, 44 (1971) 292.
(82) I. V. Balanina, G. M. Zarubinskii, and S. N. Danilov, *Zh. Obshch. Khim.*, 42 (1972) 1876.

niques, by Collins and coworkers.[83-85] Methanolic solutions of 2,3-, 3,4-, and 4,6-O-(o-nitrobenzylidene)aldopyranosides were separately subjected to irradiation (λ >290 nm). In each experiment, a yellow–green powder, difficult to purify, was obtained, and assumed to be the hydroxynitrosobenzoate obtained in the earlier studies.[65-72] To overcome the problem of isolation, and in order to avoid further reaction or degradation of the photoproducts (well known to dimerize), the crude products of the irradiation were oxidized directly with trifluoroperoxyacetic acid. Thus, methyl 4-O-acetyl-6-deoxy-2,3-O-(o-nitrobenzylidene)-α-L-mannopyranoside (**47**) gave the 4-O-acetyl-2-O-(o-nitrobenzoyl) derivative **48**. Irradiation of a mixture of the *endo* and *exo* iso-

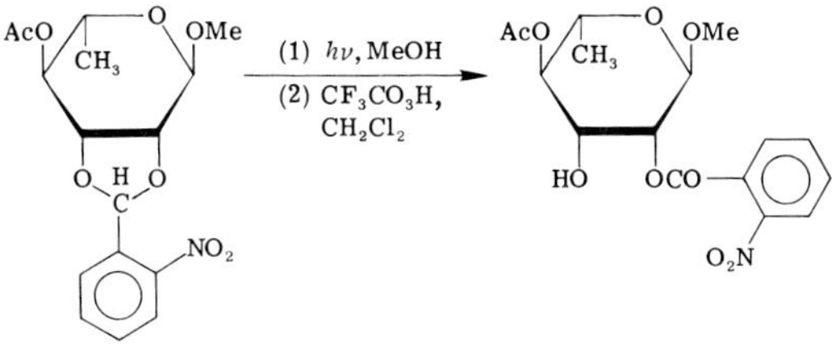

mers of compound **47**, and separate irradiation of both of the pure isomers, gave similar results. Compound **48** was obtained crystalline in a yield of 63%, but the mother liquor was shown to contain essentially the same derivative (overall yield > 90%) only slightly contaminated by its 3-o-nitrobenzoate regio-isomer (ratio in the crude mixture ~ 19:1).[84] This result was confirmed by using the same sequence starting from methyl 4,6-di-O-methyl-2,3-O-(o-nitrobenzylidene)-β-D-allopyranoside (**49**); a crystalline mixture was obtained (93% yield) that contained mainly the 3-o-nitrobenzoate **50**, with a small proportion

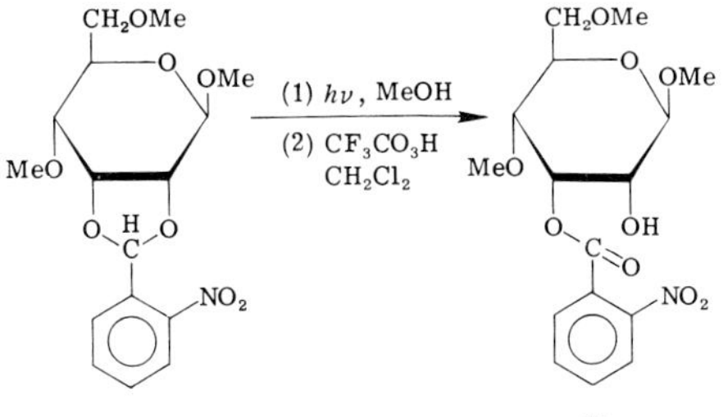

(83) P. M. Collins and N. N. Oparaeche, *J. Chem. Soc. Chem. Commun.*, (1972) 532–533.
(84) P. M. Collins and N. N. Oparaeche, *J. Chem. Soc. Perkin Trans. 1*, (1975) 1695–1700.
(85) P. M. Collins, N. N. Oparaeche, and V. R. N. Munasinghe, *J. Chem. Soc. Perkin Trans. 1*, (1975) 1700–1706.

(ratio ~ 19:1) of its 2-*o*-nitrobenzoate regioisomer. On the other hand, two examples of photorearrangements of 3,4-cyclic acetals of aldosides were given for the α-L-fuco- (compound **51**) and the β-L-arabinopyranoside (compound **53**) series; for both, a high regioselectivity (more than 95%) was observed in favor of the 4-*O*-(*o*-nitrobenzoyl) derivatives (**52** and **54**, respectively), isolated in excellent yields. The

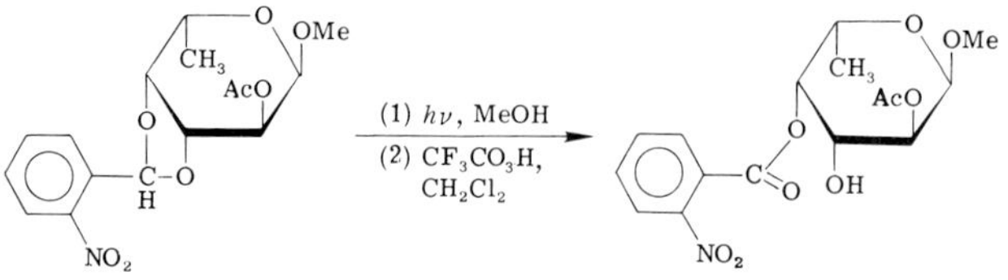

mechanism proposed for this reaction is summarized in Scheme 4, the intervention of other intermediates also being postulated. It is noteworthy that an orthoacid was probably involved; the regiospecificity of the reaction was therefore explained on the same basis as the interpretation of the known (see, especially, the references cited in Ref. 84 and in Sections 3, *b* and 3, *c*) regiospecificity of the hydrolysis of orthoesters (favored formation of the axial ester). This was further exemplified by the photorearrangement of 1,6-anhydro-2-*O*-(methylsulfonyl)-3,4-*O*-(*o*-nitrobenzylidene)-β-D-galactopyranose, which gave the corresponding, axial 3-(*o*-nitrobenzoate) in 89% yield. The same photorearrangement–oxidation sequence was applied to some methyl 4,6-*O*-(*o*-nitrobenzylidene)aldopyranosides.[85] Mixtures of 4- and 6-(*o*-nitrobenzoate)s were obtained in good yield from methyl 2,3-di-*O*-acetyl-D-gluco-, -galacto-, and -mannopyranoside in both the α and the

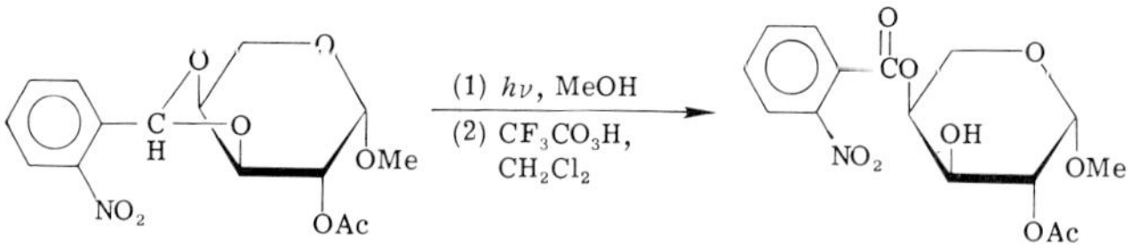

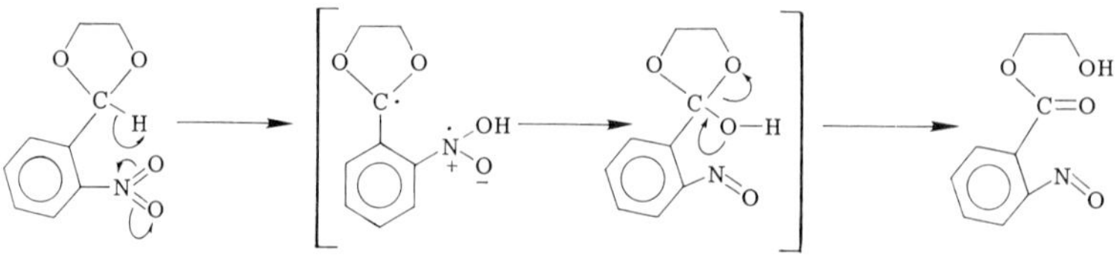

Scheme 4

β series. For instance, the D-glucoside **55** and the D-mannoside **58** respectively gave the 6-*O*-(*o*-nitrobenzoyl) derivatives **56** (isolated in 48% yield) and **59** (isolated in 53% yield). The minor regioisomers, **57** and **60**, were also purified and identified. The regioselectivity was

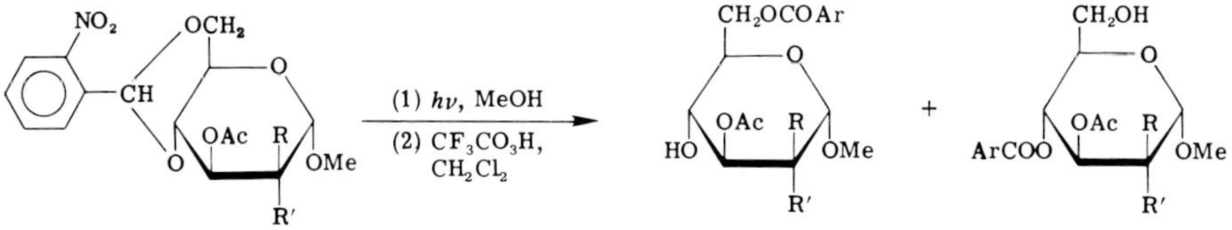

55 R = H, R' = OAc (D-*gluco*)
58 R = OAc, R' = H (D-*manno*)

56 ratio 66:34
59 ratio 70:30

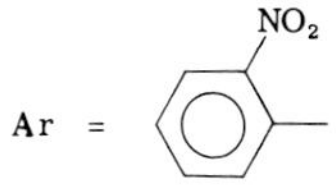

found to be reversed on starting with the D-*galacto* analog, which gave the corresponding 6- and 4-*O*-(*o*-nitrobenzoyl) derivatives in the ratio of 3:7. Possible explanations for this result were discussed.[85] Finally, the regioselectivity of the reaction was tested upon a diacetal, namely, methyl 2,3:4,6-di-*O*-(*o*-nitrobenzylidene)-α-D-mannopyranoside (**61**);

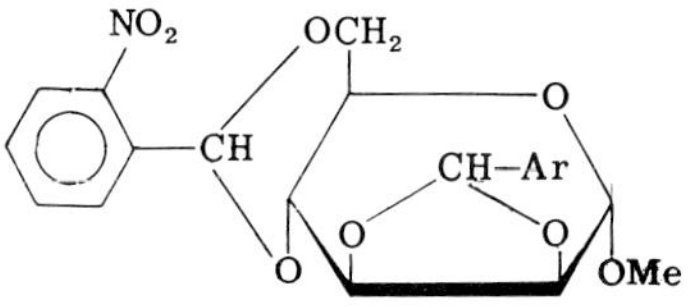

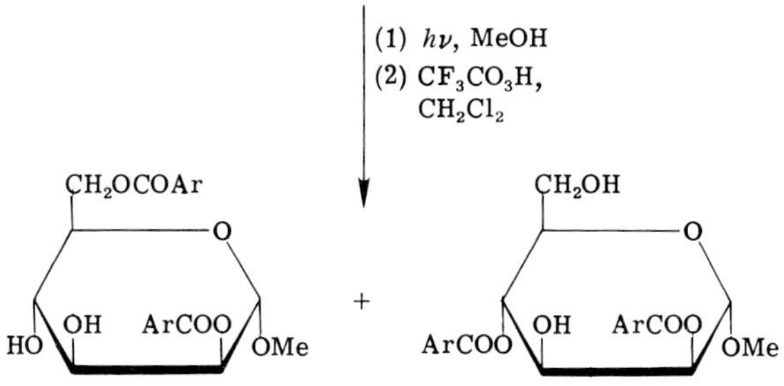

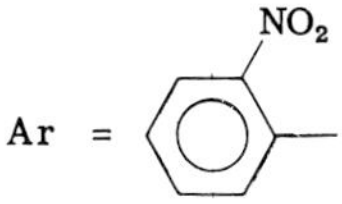

only the two o-nitrobenzoates **62** and **63** were isolated in significant proportions (respectively, in 45 and 32% yield). This result was in good accord with the regioselectivity observed both for 2,3-acetals[84] and 4,6-acetals.[85] The mechanism postulated was similar to that proposed for explaining the rearrangement of 1,3-dioxolanes (see Scheme 4). It should be emphasized that the nitro group of the O-(o-nitrobenzylidene) group was involved in the reaction, as O-(o-chloro- and O-(o-methoxy-benzylidene) derivatives were found unreactive. The initial, intramolecular, hydrogen abstraction (see Scheme 4) seemed to be demonstrated by the fact that O-(p-nitrobenzylidene) derivatives were found to be photostable.

As benzylidene acetals are more commonly used in carbohydrate chemistry than their o-nitrobenzylidene relatives, and as hydrogen abstraction from the benzylidene grouping is a known process occurring in different kinds of reactions (see Sections III, 3 and III, 6), it was of interest to reinvestigate the behavior of benzylidene acetals under irradiation in the presence of a photosensitizer, a process that had only briefly been reported for two examples.[81] When a benzene solution of 2-phenyl-1,3-dioxane was irradiated in the presence of a ketone under oxygen-free nitrogen, a pinacol, a cross-adduct, and a dimer were obtained.[86] When air was bubbled through the solution, the benzoate expected from ring cleavage was obtained (compare Ref. 81). This model reaction was then applied to 1,2,3-tri-O-acetyl-4,6-O-benzylidene-β-D-glucopyranose (**64**), for which ring cleavage was only observed if oxygen was bubbled into the solution, giving compounds **65** and **66**. In the absence of oxygen, the pinacol and the cross-adduct derivatives (whose respective structures, **67a**–**67c** and **68a**–**68c**, depended on the nature of the photosensitizer) were obtained concurrently with the dimer **69**.

Quantum yields were determined, and the mechanism was discussed as a function of the temperature of the reaction, the structure of the substrate, and the effect of the dissolved oxygen. The formation of the 4- and 6-O-benzoyl derivatives **65** and **66** was later confirmed,[87] and preliminary experiments conducted on methyl 2,3:4,6-di-O-benzylidene-α-D-mannopyranoside and on methyl 2,3-di-O-benzoyl-4,6-O-benzylidene-α-D-gluco- and -galacto-pyranosides were reported.[88] Further results in the field were obtained for an O-ethylidene deriva-

(86) M. Suzuki, T. Inai, and R. Matsushima, *Bull. Chem. Soc. Jpn.*, 49 (1976) 1585–1589.
(87) M. Suzuki, R. Matsushima, T. Inai, and S. Tsujimoto, *Asahi Garasu Kogyo Gijutsu Shoreikai Kenkyu Hokoku*, 30 (1977) 235–237; *Chem. Abstr.*, 89 (1978) 180,240.
(88) W. Szeja and M. Łapkowski, *Pol. J. Chem.*, 52 (1978) 673–675.

65 66 64

67a R = R′ = Ph
67b R = Ph, R′ = Me
67c R = R′ = Me

68a
68b
68c

69

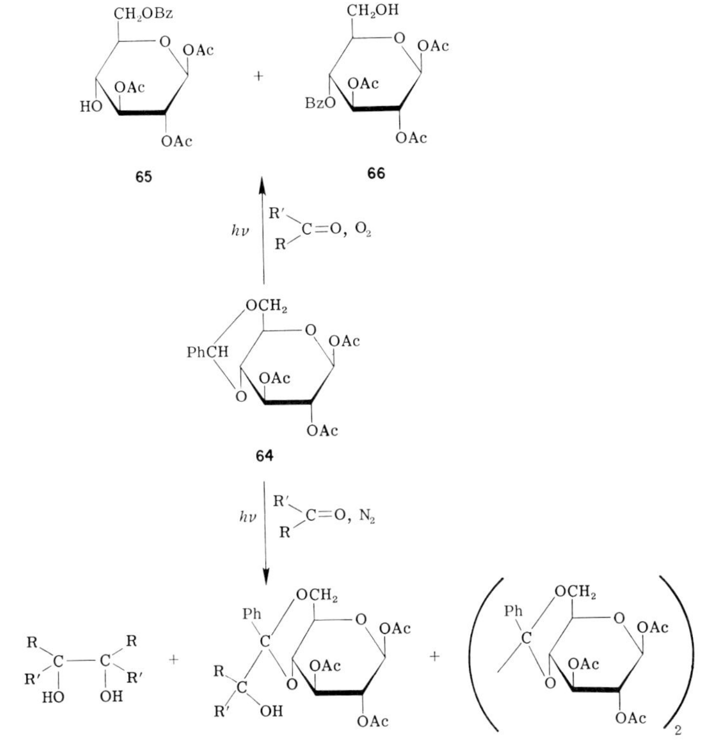

tive: irradiation of a solution of methyl 3,4-*O*-*endo*-ethylidene-β-L-arabinopyranoside (**70**) in *tert*-butanol, in the presence of acetone, led to a complex mixture.[89] The major products were the two acetates **71** and **72**, a mixture of which could be isolated in a yield of 59% (compare Ref. 84 for similar formation of compound **54** and its very minor regio-isomer). Their formation was assumed to be due to the presence of oxygen, and this supposition is consistent with observations just reported (see Refs. 81 and 86). Other compounds isolated and identified were the deoxy sugars **73** and **74** (combined yield, 7%) and the *endo*-methyl (compound **75**, yield 17%) and the *exo*-methyl (compound **76**, yield 7%) isomers of methyl 3,4-*O*-(2-hydroxy-1,2-dimethylpropylidene)-β-L-arabinopyranoside. The formation of the deoxy sugars was explained by ring opening of the acetal radical formed by initial hy-

(89) W. A. Szarek, R. J. Beveridge, and K. S. Kim, *J. Carbohydr. Nucleos. Nucleot.*, 5 (1978) 273–284.

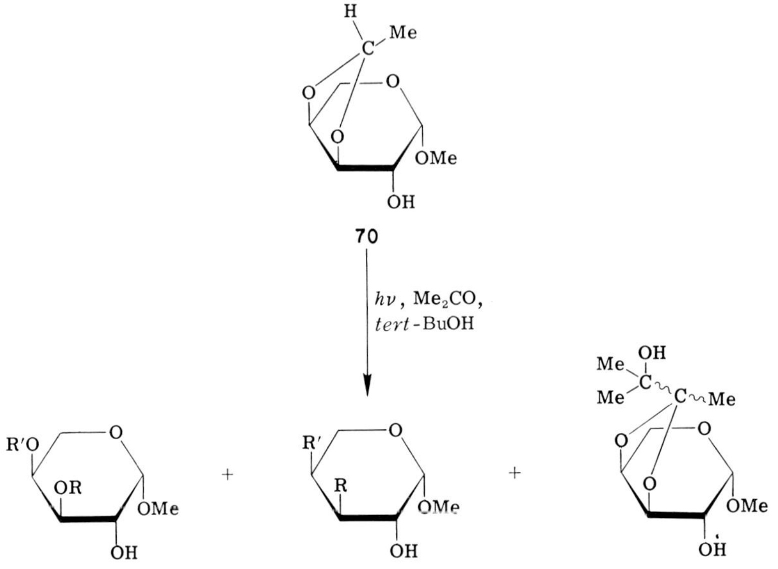

70

hν, Me₂CO, *tert*-BuOH

71 R = Ac, R' = H
72 R = H, R' = Ac

73 R = OAc, R' = H
74 R = H, R' = OAc

75 Me *endo*
76 Me *exo*

drogen-abstraction, followed by addition of a proton taken from the solvent. This acetal-radical intermediate was also very probably responsible for the formation of acetals **75** and **76** by the addition of the acetone-ketyl radical.

Descriptions of a number of interesting, photochemical reactions have been published concerning acyclic acetals and mixed acetals. Although not directly relevant to this Chapter, a few examples will be given from among the most significant and useful for synthesis in both series. By irradiation, diethyl dithioacetals of monosaccharides could be converted into acyclic thioethers (1-S-ethyl-1-thioalditols) in good yield. First demonstrated by Horton and Jewell[90] by irradiation of a methanolic solution of D-galactose diethyl dithioacetal, this reaction was later applied to diethyl dithioacetals of D-glucose, D-arabinose, D-ribose, and D-xylose,[91] the yields being improved by starting from peracetylated substrates in solution in *tert*-butanol.

As already mentioned (see Section III, 3 and Refs. 41–45), an acetal can be added to a double bond under irradiation by initial hydrogen-abstraction from the acetal carbon atom. If both the acetal grouping

(90) D. Horton and J. S. Jewell, *J. Org. Chem.*, 31 (1966) 509–513.
(91) D. Horton and J. S. Jewell, *Abstr. Pap. Am. Chem. Soc. Winter Meet.*, (1966) C11; K. Matsuura, Y. Araki, and Y. Ishido, *Bull. Chem. Soc. Jpn.*, 46 (1973) 2261–2262.

and the double bond belong to the same molecule (for instance, an acyclic sugar acetal or dithioacetal), cyclization would be expected, and this has been shown to be a possible route to cyclitol derivatives related to aminoglycosidic antibiotics[92]: irradiation of 2,3,4-tri-O-acetyl-5,6-dideoxy-D-*xylo*-hex-5-enose diethyl acetal (or diethyl dithioacetal) gave the corresponding cyclohexanetetrol derivative. For mixed acetals, Descotes and coworkers[93-99] developed an elegant method of photolysis of 2-alkoxytetrahydropyrans. The corresponding lactone was readily obtained from the nonsubstituted 2-alkoxytetrahydropyran when the alkyl group of the alkoxyl substituent was allyl or propargyl; on the other hand, ring cleavage giving an ester was observed when C-5 was substituted.[94]

This discovery was applied to model compounds wherein the alkoxyl substituent was a carbonylated chain, the number of carbon atoms of which determined the course of the reaction, to either a lactone or a dioxaspiranol.[95] The extension of these results to the sugar series was found to be useful for glycosides of monosaccharides, unsaturated sugars, and deoxy sugars,[96] especially for the synthesis of C-1-*spiro*-sugars,[97] and vinyl glycosides.[98] During the course of study of photolysis of xanthates of monosaccharides,[99] it was confirmed (compare Ref. 82) that O-isopropylidene groups are more photostable when on O-1,2 than on O-5,6; as it was also found that, in polar media, xanthates are more stable than isopropylidene groups, this result should find applications in the selective protection of sugars.

3. Halogenation

The halogenation of cyclic acetals, be it halogenation of the sidechain at C-2 of a 1,3-dioxolane or halogenation corresponding to a ring-cleavage or a ring-rearrangement, has probably been one of the most studied reactions in the field of the reactivity of the cyclic acetal grouping, and the reaction of acetals with N-bromosuccinimide will be discussed before other methods of halogenation are treated.

(92) A. A. Othman, N. A. Al-Masudi, and U. S. Al-Timari, *J. Antibiot.*, 31 (1978) 1007–1012.
(93) C. Bernasconi and G. Descotes, *C. R. Acad. Sci. Ser. C*, 280 (1975) 469–472.
(94) C. Bernasconi, L. Cottier, and G. Descotes, *Bull. Soc. Chim. Fr.*, (1977) 101–106.
(95) C. Bernasconi, L. Cottier, and G. Descotes, *Bull. Soc. Chim. Fr.*, (1977) 107–112.
(96) G. Bernasconi, L. Cottier, G. Descotes, and G. Rémy, *Bull. Soc. Chim. Fr.*, (1979) 332–336.
(97) G. Rémy, L. Cottier, and G. Descotes, *Tetrahedron Lett.*, (1979) 1847–1850.
(98) L. Cottier, G. Rémy, and G. Descotes, *Synthesis*, (1979) 711–712.
(99) G. Descotes, A. Faure, B. Kryczka, and M. N. Bouchu, *Bull. Acad. Pol. Sci.*, 27 (1979) 173–179.

a. **Action of N-Bromosuccinimide.**—The fact that acetals could be cleaved by N-bromosuccinimide was discovered about thirty years ago, for noncarbohydrate, acyclic derivatives, by Marvell and Joncich,[100] who demonstrated that ethyl benzoate is produced on refluxing a solution of benzaldehyde diethyl acetal in carbon tetrachloride with N-bromosuccinimide under irradiation by light. This observation was later extended to acyclic acetals of α-ketoaldehydes[101]; it was assumed that an unstable intermediate (a *gem*-bromoacetal) was produced, and that it decomposed immediately into an ester and alkyl bromide. Halogenation of cyclic acetals was then described by Rieche and coworkers,[102] who showed that benzaldehyde ethylene acetal is brominated, and cleaved, to give $PhCO_2CH_2CH_2Br$; they also later demonstrated[103] that the benzylidene acetal of *cis*-1,2-cyclohexanediol gives *trans*-2-bromocyclohexyl benzoate. These results should be compared to those obtained concurrently[104,105] for the chlorination of cyclic acetals of formaldehyde, which gave formic esters of chlorinated alcohols. They must also be compared to those describing the formation of ethyl benzoate by peroxide decomposition of benzaldehyde ethylene acetal to ethyl benzoate (see Section III, 6,*a*). The great potential utility of this reaction was recognized by Prugh and McCarthy,[106] who obtained various bromoesters from 1,3-dioxolanes unsubstituted at C-2, or substituted by a methyl, a propyl, or a phenyl group. This very useful tool for organic synthesis rapidly became one of the most elegant methods for the structural modification of sugars when it was introduced into carbohydrate chemistry by Hanessian,[107] and, very shortly thereafter, independently by Hullar and coworkers.[108] The first results described in the field, mainly by Hanessian and coworkers, have already been partly reviewed elsewhere,[109] and have been discussed in this Series[110,111] in articles devoted to deoxyhalogeno sugars and cyclic acyloxonium ions.

A study conducted on model compounds (1,3-dioxolanes variously

---

(100) E. N. Marvell and M. J. Joncich, *J. Am. Chem. Soc.*, 73 (1951) 973–975.
(101) J. B. Wright, *J. Am. Chem. Soc.*, 77 (1955) 4883–4884.
(102) A. Rieche, E. Schmitz, and E. Beyer, *Chem. Ber.*, 91 (1958) 1935–1941.
(103) A. Rieche, E. Schmitz, W. Shade, and E. Beyer, *Chem. Ber.*, 94 (1961) 2926–2932.
(104) H. Baganz and L. Domaschke, *Chem. Ber.*, 91 (1958) 653–656.
(105) L. A. Cort and R. G. Pearson, *J. Chem. Soc.*, (1960) 1682–1687.
(106) J. D. Prugh and W. C. McCarthy, *Tetrahedron Lett.*, (1966) 1351–1356.
(107) S. Hanessian, *Carbohydr. Res.*, 2 (1966) 86–88.
(108) D. L. Failla, T. L. Hullar, and S. B. Siskin, *Chem. Commun.*, (1966) 716–717.
(109) S. Hanessian, *Adv. Chem. Ser.*, 74 (1968) 159–201.
(110) W. A. Szarek, *Adv. Carbohydr. Chem. Biochem.*, 28 (1973) 225–306.
(111) H. Paulsen, *Adv. Carbohydr. Chem. Biochem.*, 26 (1971) 127–195.

substituted at C-2, C-4, and C-5) demonstrated the following points [112,113]: (1) the presence of a hydrogen atom on the acetal carbon atom is indispensable to ring cleavage; 2,2-disubstituted 1,3-dioxolanes underwent halogenation of the side chain on C-2, probably through the intermediacy of an enol ether[113,114]; (2) O-methylene and O-ethylidene acetals react to give, respectively, formyl and acetyl α-bromo esters; (3) the presence of a substituent at C-4 (methyl or chloromethyl) induces favored cleavage of the C-2–O-5 bond; (4) O-ethylidene derivatives of meso-2,3-butanediol and (±)-threo-2,3-butanediol, respectively, give α-bromoacetates corresponding to an inversion of the configuration of the carbon atom attacked by the bromine ion (S_N2-like reaction). The last two points are particularly in agreement with the conclusions of a study of a photochemically initiated reaction of 1,3-dioxolanes with trichlorofluoromethane, leading to chloroesters.[115]

The first example of application in the carbohydrate series was the demonstration by Hanessian[107] that treatment of methyl 4,6-O-

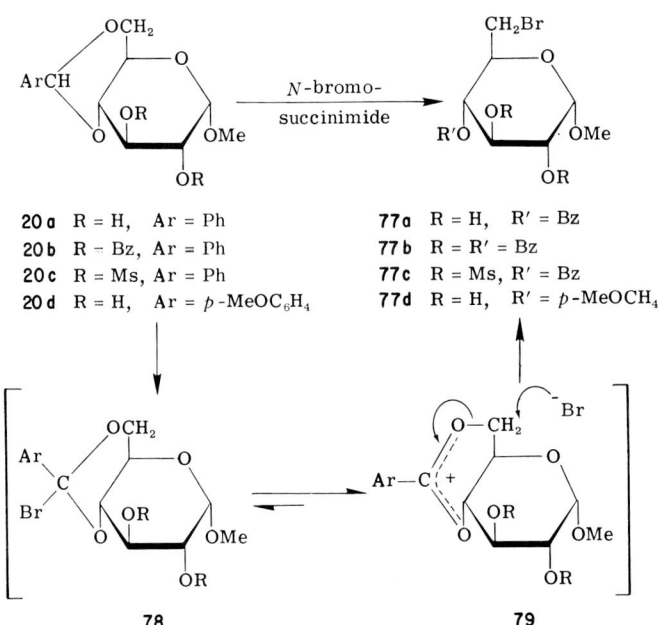

(112) J. Gelas and S. Michaud, C. R. Acad. Sci. Ser. C, 270 (1970) 1614–1616.
(113) J. Gelas and S. Michaud, Bull. Soc. Chim. Fr., (1972) 2445–2459.
(114) M. Gaudry and A. Marquet, Bull. Soc. Chim. Fr., (1969) 4169–4178, and references cited therein.
(115) J. W. Hartgerink, L. C. J. Van der Laan, J. B. F. N. Engberts, and T. J. de Boer, Tetrahedron, 27 (1971) 4323–4334.

benzylidene-α-D-glucopyranoside (**20a**) with N-bromosuccinimide in refluxing carbon tetrachloride, in the presence of barium carbonate, during ~2.5 h gives a good yield of methyl 4-O-benzoyl-6-bromo-6-deoxy-α-D-glucopyranoside (**77a**). The presence of a free-radical initiator was the only difference in reactions conducted by Hullar and coworkers[108] upon three substituted methyl 4,6-O-benzylidenehexopyranosides. The same results were obtained[116] for various 4-O-aroyl derivatives (**20a–20d**).

The most probable mechanism involves, first, the attack of N-bromosuccinimide (or bromide), giving the *gem*-bromoacetal **78** by removal of the hydrogen atom from the acetal carbon atom. This step could be either a free-radical process, or an ionic one; however, the general behavior of N-bromosuccinimide, and the influence of irradiation by light[116] and of free-radical initiators, at least on the rate of the reaction (which seemed to be established[108,116]), were in favor of a free-radical process. The *gem*-bromoacetal **78** (or orthoester halide), a very unstable species,[117] probably rearranges to the benzoxonium ion **79**, which is then attacked by bromine ion, to give the 4-O-aroyl-6-bromo derivatives **77**. The intervention of a benzoxonium ion, not a free radical, seems to be supported by the observation that the action of N-bromosuccinimide on 4,5-dimethyl-2-phenyl-1,3-dioxolane in the presence of water gives 2-(3-hydroxybutyl)benzoate; this hydroxy ester is probably formed through the attack of a benzoxonium ion on water, to give an orthoacid which rearranges without inversion of configuration[122] (for some aspects of the latter reaction in carbohydrate chemistry, see Section III, 3,*b*).

Further evidence for the intervention of 1,2-dioxolan-2-ylium ions has been given. First, a rather stable, tribromide aroyl ion was isolated from the reaction of the *p*-nitrobenzylidene acetal of norbornanediol

---

(116) S. Hanessian and N. R. Plessas, *J. Org. Chem.*, 34 (1969) 1035–1044.
(117) All attempts to isolate 2-halogeno-1,3-dioxolanes or -1,3-dioxanes were unsuccessful. For instance, 2-chloro-1,3-dioxolane was detected only by photochlorination of 1,3-dioxolane (Ref. 118) at low temperature and 2-chloro-2-methyl-1,3-dioxolane was prepared by treatment of 2-methyl-1,3-dioxolane-2-carboxylic acid with phosphorus pentachloride at −60°, and shown[119] to rearrange to 2-chloroethyl acetate on warming to 0°. For other examples of such problems, see especially, references 120 and 121.
(118) J. Jonas, T. P. Forrest, M. Kratochvíl, and H. Gross, *J. Org. Chem.*, 33 (1968) 2126–2127.
(119) M. S. Newman and C. H. Chen, *J. Am. Chem. Soc.*, 94 (1972) 2149–2150.
(120) S. Hünig, *Angew. Chem. Int. Ed. Engl.*, 3 (1964) 548–560.
(121) H. Gross, J. Freiberg, and B. Costisella, *Chem. Ber.*, 101 (1968) 1250–1256.
(122) D. A. Seeley and J. McElwee, *J. Org. Chem.*, 38 (1973) 1691–1693.

with bromine or N-bromosuccinimide.[123] Independently, dioxolanylium trihalides were isolated from the reaction of 4,4,5,5-tetramethyl-2-phenyl-1,3-dioxolane with halogens or interhalogens.[124] (For related studies involving the characterization and the relative stability of dioxocarbonium ions corresponding to acyclic acetals, see, for instance, Refs. 125 and 126, and references cited therein.)

The results obtained in the methyl D-glucopyranoside series were thoroughly extended to other benzylidene acetals. From methyl 4,6-O-benzylidene-α-D-galactopyranoside (**22**), the expected 6-bromo derivative **80** was obtained in high yield (90%),[116] and methyl 2,3-di-O-

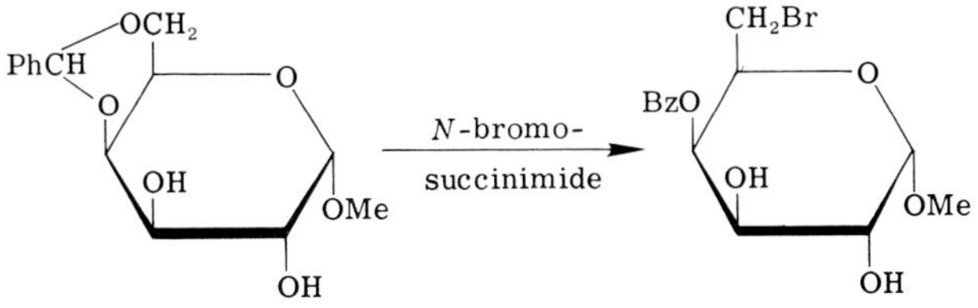

acetyl-4,6-O-benzylidene-α-D-mannopyranoside (**81**) gave the 6-bromo derivative **82** (yield 71%).[127] Examples were given for the 3-branched 2-deoxy sugar series: methyl 4-O-benzoyl-6-bromo-2,6-dideoxy-3-C-methyl-3-O-methyl-α-D-*xylo*-hexopyranoside (yield 91%)[128,129] and its α-L-*arabino*-hexopyranoside analog (yield 86%)[130]

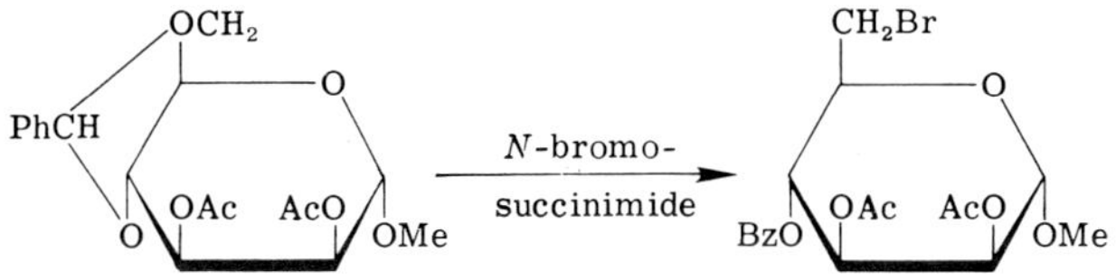

(123) A. Babouz, J. Coste, H. Christol, and F. Plénat, *Tetrahedron Lett.*, (1979) 11–14.
(124) A. Goosen and C. W. McCleland, *J. Chem. Soc. Chem. Commun.*, (1979) 751–752.
(125) J. W. Scheeren, *Tetrahedron Lett.*, (1968) 5613–5614.
(126) C. H. V. Dusseau, S. E. Schaafsma, and T. J. de Boer, *Recl. Trav. Chim. Pays-Bas*, 89 (1970) 535–544.
(127) D. Horton and A. E. Luetzow, *Carbohydr. Res.*, 7 (1968) 101–105.
(128) G. B. Howarth, W. A. Szarek, and J. K. N. Jones, *Chem. Commun.*, (1968) 62–63.
(129) G. B. Howarth, W. A. Szarek, and J. K. N. Jones, *Carbohydr. Res.*, 7 (1968) 284–290.
(130) E. H. Williams, W. A. Szarek, and J. K. N. Jones, *Can. J. Chem.*, 47 (1969) 4467–4471.

were prepared from the corresponding 4,6-benzylidene acetals. Although the regiospecificity of the reaction seemed to be rather general, two products were isolated when methyl 4,6-O-benzylidene-β-D-galactopyranoside (**23**) was treated with N-bromosuccinimide[116]: one of them was the expected, 6-bromo derivative (**83**) and the second was postulated to be methyl 6-O-benzoyl-3-bromo-3-deoxy-β-D-gulopyranoside (**84**), probably resulting from a rearrangement of the benzoxonium ion **85** to **86**, followed by a benzoyl migration from O-4

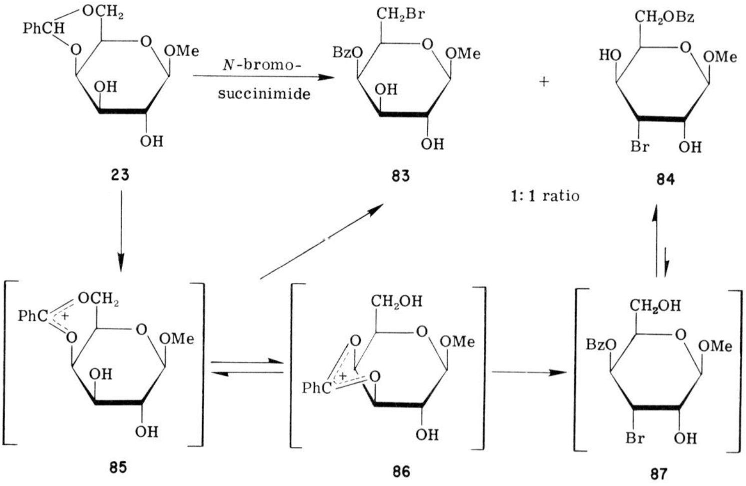

(compound **87**, not isolated) to O-6 (compound **84**). When the hydroxyl groups at C-2 and C-3 of compound **23** were protected in order to avoid this rearrangement, only the 6-bromo derivative expected was produced[116]; thus, methyl 2,3,4-tri-O-benzoyl-6-bromo-6-deoxy-β-D-galactopyranoside and its 2,3-di-O-methylsulfonyl analog were obtained from the corresponding 4,6-benzylidene acetals (respective yield, 65 and 84%). This finding has been confirmed,[131] starting from the 2,3-di-O-methyl analog of compound **23**.

The evident interest attending this method of bromination was reinforced by the observation that the reaction may be conducted in the presence of a wide range of other groups that are unaffected under these conditions (for instance, O-mesyl, -tosyl, -acetyl, -benzoyl, and -methyl groups, and α-epoxides).[116]

The procedure was extended to O-benzylidene derivatives of some disaccharides,[116,132] 2-acetamido-2-deoxyhexopyranosides, 3-azido-3-

(131) T. Fujikawa, *Carbohydr. Res.*, 38 (1974) 325–327.
(132) S. Hanessian and N. R. Plessas, *J. Org. Chem.*, 34 (1969) 1045–1053.

deoxyhexopyranosides, and 2,3-dideoxyhexopyranosides.[132] Further observations were reported for benzylidene acetals of two vicinal, secondary hydroxyl groups of sugars. As an example in the furanoid series,[133] methyl 2,3-O-benzylidene-5-O-methyl-β-D-ribofuranoside (**88**) gave a mixture (yield 85%) of the two α-bromobenzoates **90** and **91** corresponding to the possible pathways for the attack of the interme-

diate **89**. In the pyranoid series,[133] methyl 2-O-benzoyl-3,4-O-benzylidene-β-D-arabinopyranoside (**92**) led to a mixture of two bromo deriv-

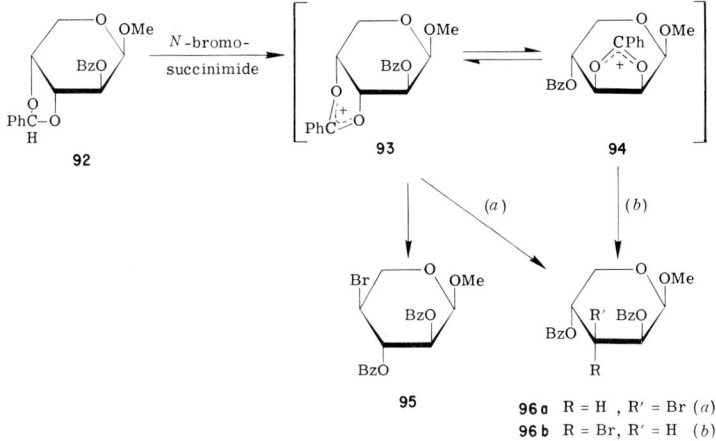

(133) S. Hanessian and N. R. Plessas, *J. Org. Chem.*, 34 (1969) 1053–1058.

atives, to which structures **95** and **96a** (D-*lyxo* configuration) or **96b** (D-*arabino* configuration) were assigned. The presence of compound **96b** would prove the reality of the rearrangement of the 3,4-benzoxonium ion **93** to its 2,3-isomer **94**. Other examples involving 3,4-*O*-benzylidene pyranosides gave further information concerning the regioselectivity of the reaction. Methyl 3,4-*O*-benzylidene-2,6-dichloro-2,6-dideoxy-α-D-altropyranoside (**97**) and its 2-azido analog **98** gave the corresponding 3-bromo derivatives **99** and **100** as major products of the reaction.[134]

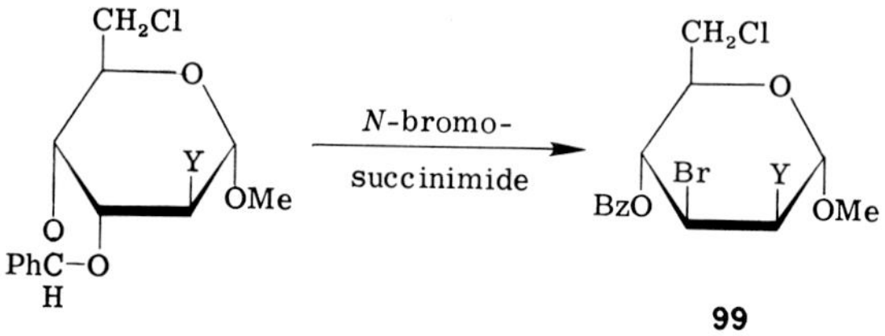

97 Y = Cl
98 Y = N$_3$

99
100

A similar orientation was found for the bromination of methyl 3,4-*O*-benzylidene-6-deoxy-β-D-fucopyranoside (**101**), which gave[135] methyl 4-*O*-benzoyl-3-bromo-3,6-dideoxy-D-gulopyranoside (**102**). It appears

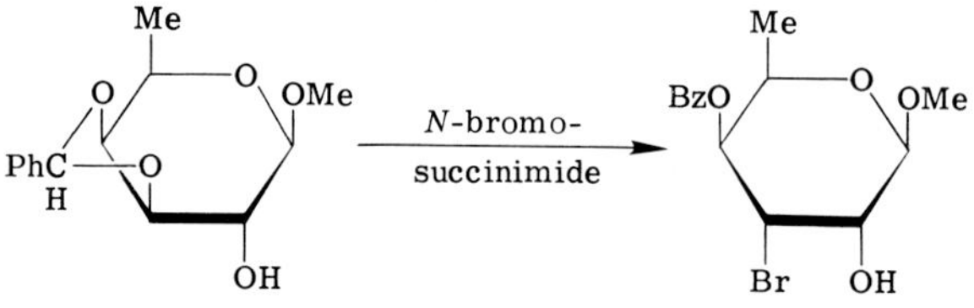

101

102

that a *trans*-diaxial ring-opening, which is favored for 1,3-dioxolan-2-ylium ions[136-138] (see Ref. 139 and references cited therein), obtains for 3,4-*O*-isopropylidene pyranosides (attack of bromide ion upon intermediate **86**, or **101a**), except when *syn*-axial interactions prevent

(134) S. Hanessian and N. R. Plessas, *J. Org. Chem.*, 34 (1969) 2163–2170.
(135) K. Eklind, P. J. Garegg, and B. Gotthammar, *Acta Chem. Scand. Ser. B*, 29 (1975) 633–634.
(136) J. F. King and A. D. Albutt, *Chem. Commun.*, (1966) 14–15.
(137) J. F. King and A. D. Albutt, *Can. J. Chem.*, 47 (1969) 1445–1459.
(138) J. F. King and A. D. Albutt, *Can. J. Chem.*, 48 (1970) 1754–1769.
(139) C. U. Pittman, Jr., S. P. McManus, and J. W. Larsen, *Chem. Rev.*, 72 (1972) 357–438.

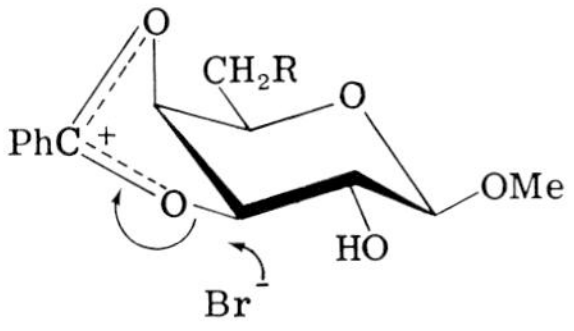

86 R = OH
101a R = H

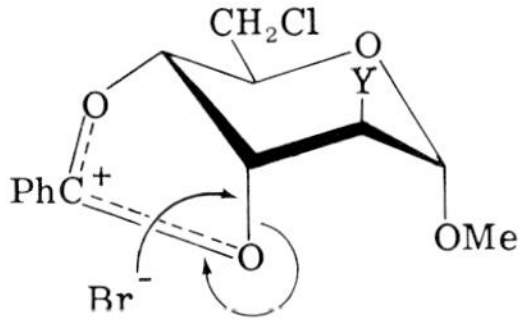

97a Y = Cl
98a Y = N$_3$

the approach of bromide ion (attack of bromide ion upon intermediate **97a** or **98a**). It would be of value to compare the ring opening of intermediates **86** and **101a** (β anomers) with that of their α anomers.

The reaction of an another, bicyclic ring-system with N-bromosuccinimide was investigated: it concerned an O-benzylidene grouping formed from two secondary hydroxyl groups, one of which is not attached to the sugar ring. Thus, 3,5-O-benzylidene-1,2-O-isopropylidene-α-D-glucofuranose (**103**) gave a mixture of two unexpected compounds, namely, 3,6-anhydro-5-O-benzoyl-1,2-O-isopropylidene-α-D-glucofuranose (**106**) and the 6-bromo derivative **107**, in the ratio of 2:3 after chromatographic purification.[133] This result clearly demonstrated the rearrangement of the benzoxonium ion **104**, formed initially, to the ion **105**, because of the favorable position of the 6-hydroxyl group for an intramolecular attack by the positive charge.

The reaction of benzylidene acetals of sugars with N-bromosuccinimide has since known great popularity, and, especially for 4,6-O-benzylidene derivatives, is now a well established, laboratory procedure.[140] It has been extended to 2′,3′-O-benzylidene derivatives of nucleosides.[141,142]

Further developments in the field were introduced by the study of dibenzylidene acetals. It was claimed[143] that methyl 2,3:4,6-di-O-benzylidene-α-D-mannopyranoside gave, by treatment with N-bromosuccinimide, a 2,6-dibromo derivative (yield 88%), contaminated by traces of an unidentified compound. This dibromo derivative was shown, by n.m.r. spectroscopy, to be actually a mixture of two compounds for which the respective configurations D-*manno* and D-*gluco* were proposed on the basis of the reduction of the mixture to the corresponding 2-deoxy sugar only. No selective ring-cleavage of only one O-benzylidene group could be observed. This work was later reinves-

(140) S. Hanessian, *Methods Carbohydr. Chem.*, 6 (1972) 183–189.
(141) M. M. Ponpipom and S. Hanessian, *Can. J. Chem.*, 50 (1972) 246–252.
(142) M. M. Ponpipom and S. Hanessian, *Can. J. Chem.*, 50 (1972) 253–258.
(143) M. Haga, M. Chonan, and S. Tejima, *Carbohydr. Res.*, 16 (1971) 486–491.

104 103 105 106 107

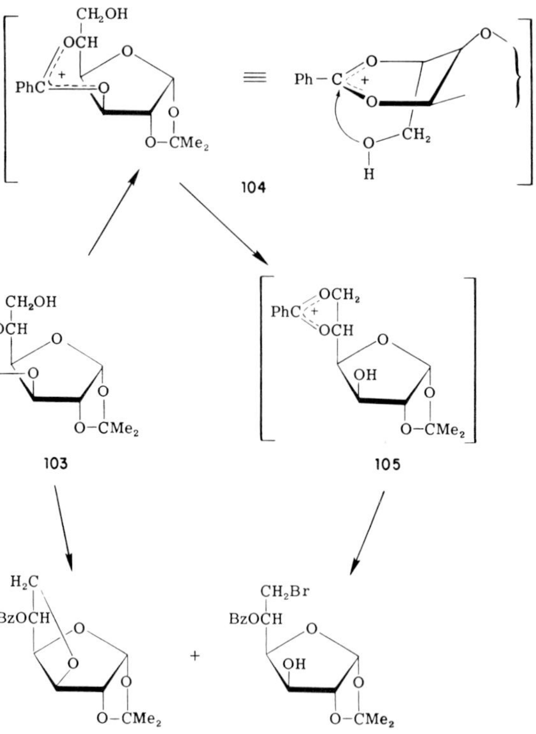

tigated,[144] and it was demonstrated that the reduction, by lithium aluminum hydride, of the dibromo derivative obtained from the methyl

108 109 110

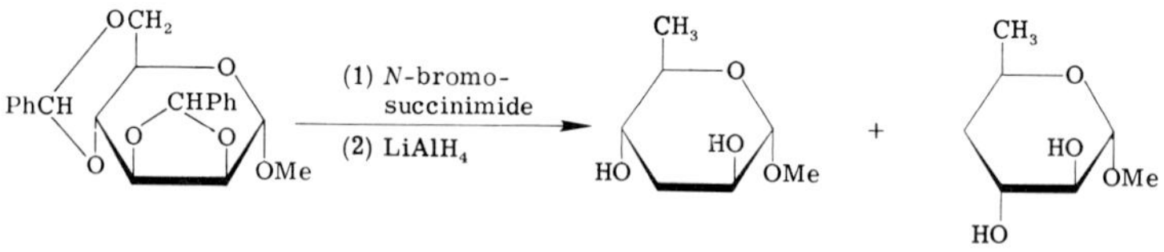

α-D-mannoside acetal **108** gave, in fact, two distinct deoxy sugars, **109** and **110** (respective overall yield for the purified compounds 54 and 27%). The formation of the 3,6-dideoxy sugar **109** was interpreted on the basis of the expected reaction for both the 4,6- and the 2,3-*O*-benzylidene group, the attack of the bromine ion being favored at C-3 of

(144) C. Monneret, J.-C. Florent, N. Gladieux, and Q. Khuong-Huu, *Carbohydr. Res.*, 50 (1976) 35–44.

the corresponding benzoxonium ion **112**. On the other hand, the rearrangement of **112** to the 3,4-benzoxonium ion **113** opened the way to the 3,6-dideoxy sugar **110**, by way of bromide **116**. This interpretation

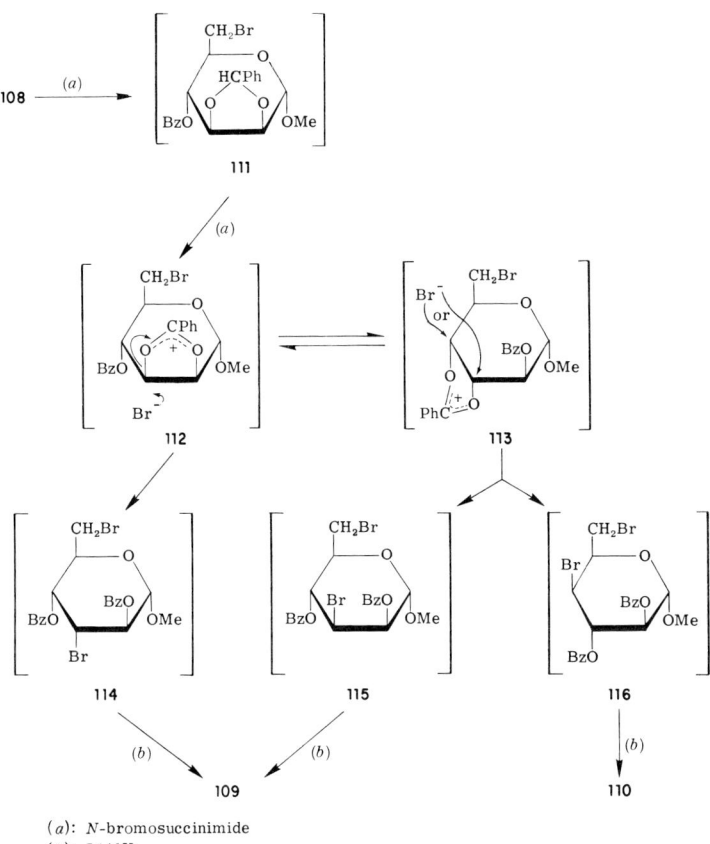

(*a*): *N*-bromosuccinimide
(*b*): LiAlH$_4$

was supported by the observation of the regiospecific cleavage of compound **117**, which was treated with *N*-bromosuccinimide to give the 3-bromo compound **118** (yield 87%) as the only detectable product.[144]

The same, regiospecific cleavage was observed[145] for the 6-azido analog of compound **117**. Meanwhile, another, independent reexamination of the ring opening of the methyl α-D-mannoside diacetal **108** was published,[146] and this confirmed, at all points, the scope of the reaction given in Ref. 144; the 3,6-dibromo-D-*altro* (**114**), 3,6-dibromo-D-*manno* (**115**), and 4,6-dibromo-D-*ido* (**116**) derivatives were isolated from the reaction mixture, and identified. A very small proportion of a fourth product was identified as being methyl 2,4-di-*O*-benzoyl-6-bromo-6-deoxy-α-D-mannopyranoside. Definitive identification of compounds **114**, **115**, and **116** was also achieved after their isolation, by h.p.l.c., from the crude mixture.[147] In the L-rhamnose series, it was shown that compound **119** gives[144] a good yield (70%) of the 3-bromo derivative **120**. The 4-hydroxy analog **121** was also shown to give

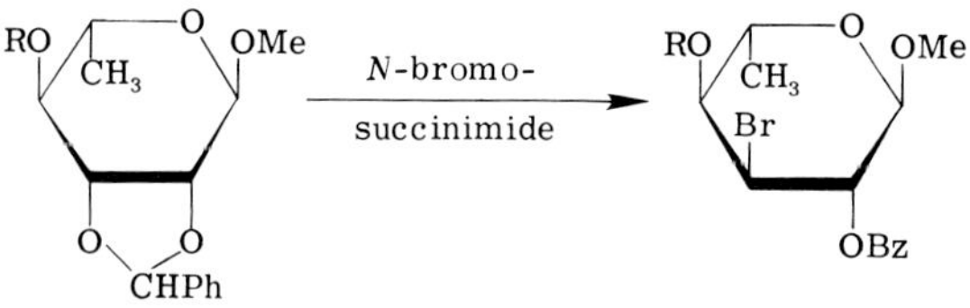

119 R = Me
121 R = H

120
122

methyl 2-*O*-benzoyl-3-bromo-3,6-dideoxy-α-L-altropyranoside[148] (**122**) in a yield of 81%. The same reaction was conducted for methyl 2,3:4,6-di-*O*-benzylidene-α-D-allopyranoside, which gave[146] a mixture (yield 60%) of the 2,6-dibromo-D-*altro* and the 3,6-dibromo-D-*gluco* derivatives in the ratio of 10:17. An attempt was made to rationalize the regioselectivity of the bromination of 2,3-*O*-benzylidene pyranosides.[147] It had been observed that, even when *syn*-axial interactions were developed for the attack of bromine ion on C-3 (ions **117a** and **119a**), no formation of derivatives brominated at C-2 could be detected; this was attributed to either a strong distortion of the (theoretical) chair conformation, or the known reluctance, in the carbohydrate series, for reaction at C-2. In favor of the latter hypothesis, the ring opening of methyl 2,3:4,6-di-*O*-benzylidene-α-D-allopyranoside to the 2,6-dibromo-D-*altro* derivative through *trans*-diaxial stereochemistry

---

(145) C. Monneret, J.-C. Florent, and Y. Chapleur, *C. R. Acad. Sci. Ser. C*, 285 (1977) 587–589.
(146) J. Thiem and J. Elvers, *Carbohydr. Res.*, 60 (1978) 63–73.
(147) J.-C. Florent and C. Monneret, *Carbohydr. Res.*, 85 (1980) 243–257.
(148) J.-C. Florent, C. Monneret, and Q. Khuong-Huu, *Carbohydr. Res.*, 56 (1977) 301–314.

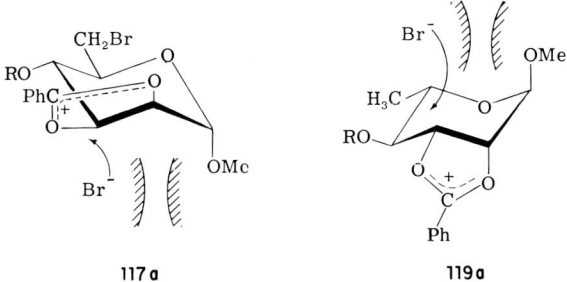

117a   119a

showed poor regioselectivity,[146] despite the absence of a strong *syn*-axial interaction during the attack of bromine ion on C-2.

Reactions involving other ring-systems were further investigated, especially 1,6-anhydro-β-D-mannopyranose, and furanoid sugars. Thus, 1,6-anhydro-2,3-*O*-benzylidene-β-D-mannopyranose was treated with *N*-bromosuccinimide in order to avoid the rearrangement demonstrated for compound **108** (see earlier); a regiospecific reaction led to the corresponding 3-bromo-D-*altro* derivative.[147] On the other hand, methyl 2,3-*O*-benzylidene-5-*O*-acetyl-α-L-rhamnofuranoside (**123**) gave, regiospecifically,[147,149] the 3-bromo-L-*altro* derivative **124** (yield 95%).

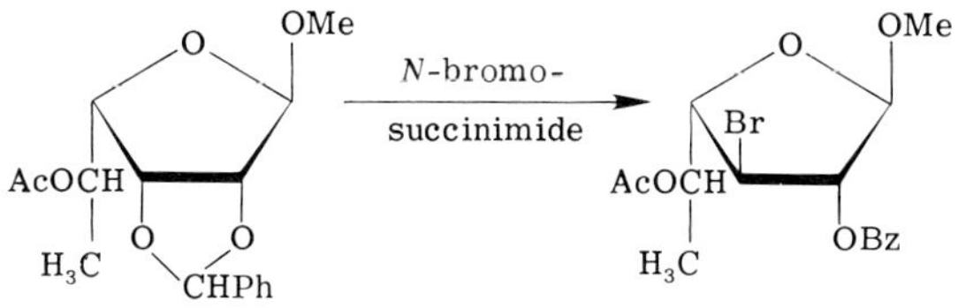

b. Action of Triphenylmethyl Fluoroborate Followed by Addition of Halide Ions.

—The reaction involving the removal of a proton (if any) from the acetal carbon atom of a 1,3-dioxolane by strong hydride-acceptors, notably triphenylmethyl fluoroborate, to generate dioxolan-ylium ions has been known for about twenty years,[150,151] and is well documented,[139] especially in carbohydrate chemistry.[111] It may be noted that triphenylmethyl fluoroborate is able to remove a proton on

---

(149) J.-C. Florent and C. Monneret, personal communication.
(150) H. Meerwein, V. Hederich, H. Morschel, and K. Wunderlich, *Justus Liebigs Ann. Chem.*, 635 (1960) 1–21.
(151) C. B. Anderson, E. C. Friedrich, and S. Winstein, *Tetrahedron Lett.*, (1963) 2037–2044.

C-3 (or C-4) from a 1,3-dioxolane; actually, this reaction (see Section III, 1,d) was observed for 2,2-disubstituted 1,3-dioxolanes, and it may be assumed that, if there is any competition between removal of a proton from C-2 and from C-3 (or C-4), the rate of the first reaction is far greater than that of the second. The possibility of dioxolanylium ions reacting with nucleophilic reagents was first demonstrated by Hanessian and Staub,[152] and the scope of these reactions was independently investigated by Pedersen and coworkers.[153–157] Hanessian and Staub[152] treated methyl 5-O-benzoyl-2,3-O-benzylidene-β-D-ribofuranoside (**125**) with triphenylmethyl fluoroborate in acetonitrile, and obtained, after hydrolysis, a 1:1 mixture of the two benzoates **127** and **128** (yield 83%), probably through the benzoxonium ion **126**. The same reaction was repeated,[152] starting from the 5-O-methyl analog of compound **125**. This procedure was applied to 5,6-O-benzylidene-1,2-O-isopropylidene-3-O-(methylsulfonyl)-α-D-glucofuranose (**129**) which gave,

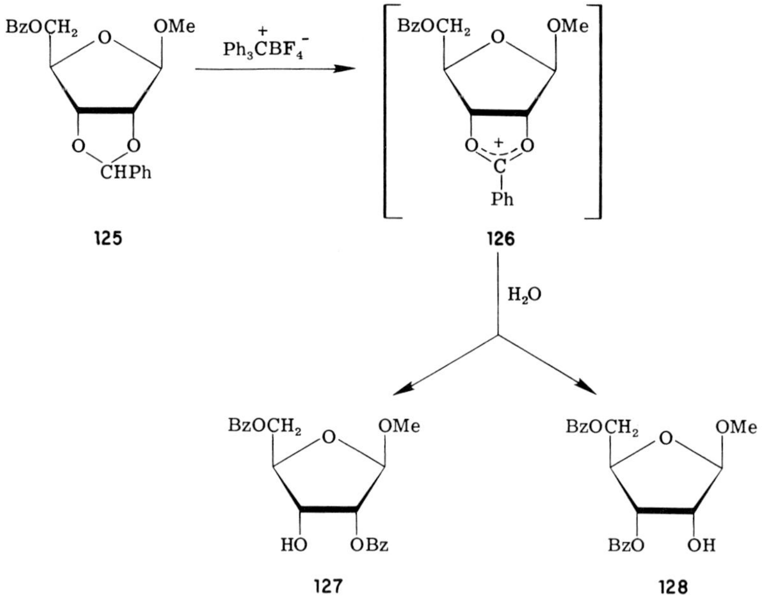

(152) S. Hanessian and A. P. A. Staub, *Tetrahedron Lett.*, (1973) 3551–3554.
(153) S. Jacobsen and C. Pedersen, *Acta Chem. Scand. Ser. B*, 28 (1974) 866–872.
(154) S. Jacobsen and C. Pedersen, *Acta Chem. Scand. Ser. B*, 28 (1974) 1024–1028.
(155) S. Jacobsen, B. Nielsen, and C. Pedersen, *Acta Chem. Scand. Ser. B*, 31 (1977) 359–364.
(156) S. Jacobsen and C. Pedersen, *Acta Chem. Scand. Ser. B*, 31 (1977) 365–368.
(157) L. Hoffmeyer, S. Jacobsen, O. Mols, and C. Pedersen, *Acta Chem. Scand. Ser. B*, 33 (1979) 175–186.

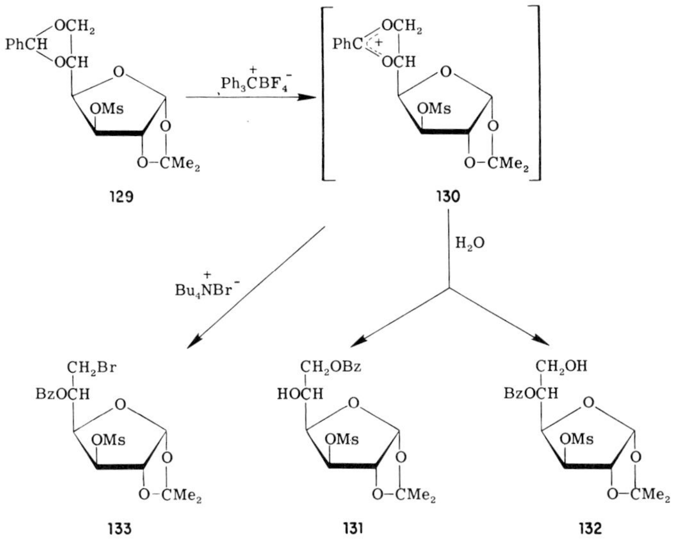

after hydrolysis, a 1:1 mixture of the expected benzoates **131** and **132**. To benefit from the intermediacy of the benzoxonium ion **130**, tetrabutylammonium bromide was added immediately after the formation of **130**, affording the 5-O-benzoyl-6-bromo-6-deoxy derivative **133** (yield 80%). Under these conditions, compound **129** behaved exactly as it does with N-bromosuccinimide (compare with the formation of compound **107** from the ion **105**). Similar results were obtained in the generation of benzoxonium ions with chlorotriphenylmethane and stannic chloride, or tropylium fluoroborate, or 2,4,6-trichlorobenzenediazonium fluoroborate, or triphenylmethyl hexachloroantimoniate, but without any advantages over triphenylmethyl fluoroborate.[152]

Jacobsen and Pedersen[153] confirmed the generation of benzoxonium ion **126** by treatment of compound **125** with triphenylmethyl fluoroborate in acetonitrile, but they treated it with tetraethylammonium bromide, obtaining a 3:2 mixture of the α and β anomers of methyl 2,5-di-O-benzoyl-3-bromo-3-deoxy-D-xylofuranoside (**134**) in a yield of 74%. This mixture could be anomerized to give the pure β anomer (overall yield 60%) by treatment with hydrogen bromide followed by the action of methanol.

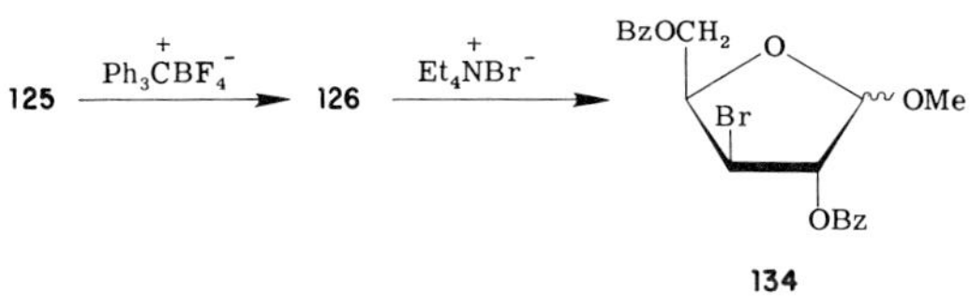

The first example in the pyranoid series was afforded[153] by methyl 2-O-benzoyl-3,4-O-benzylidene-β-D-arabinopyranoside (92), which gave the corresponding benzoxonium ion (93); this ion was shown to be in equilibrium[158] with its regio- and stereo-isomer 94, the latter

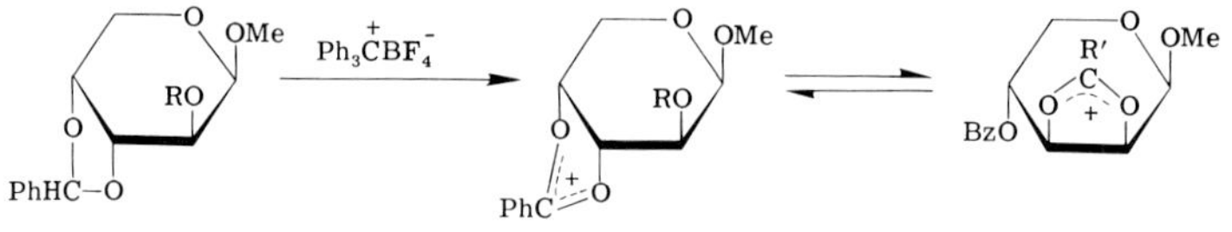

92	R = Bz	93	R = Bz	94	R′ = Ph
92a	R = Me	93a	R = Me		
92b	R = p-MeOC$_6$H$_4$CO	93b	R = p-MeOC$_6$H$_4$CO	94b	R′ = p-MeOC$_6$H$_4$

being largely disfavored and almost absent, as was demonstrated by comparison of the n.m.r. spectrum of ion 93, only slightly contaminated by ion 94, with that of the benzoxonium ion 93a, for which no further rearrangement could occur, and which had been produced from the 2-O-methyl analog 92a. Additional evidence was afforded by shifting the equilibrium in the opposite direction; replacement of the benzoyl group on O-2 by a p-methoxybenzoyl group (compound 92b) largely stabilized the p-methoxybenzoxonium ion 94b in respect to its isomer 93b, now disfavored (9:1 ratio, estimated from the n.m.r. spectra). As in the preceding work,[152] two types of treatment of these benzoxonium ions were investigated: (1) their hydrolysis, and (2) their reaction with tetraethylammonium bromide.

(1) For instance, hydrolysis of benzoxonium ions 93 and 93a gave, respectively, the 4-O-benzoyl derivatives 135 and 135a (respective yield, 67 and 46%). The preservation of the configuration of the start-

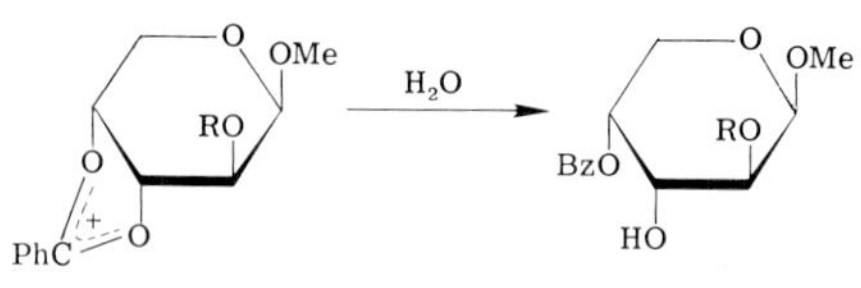

| 93 | R = Bz | 135 |
| 93a | R = Me | 135a |

ing material demonstrated the *cis* ring-opening of the ion leading to the compound having an axial O-acyl group (for a discussion, see, especially, Ref. 139, and references cited therein). On the other hand, treatment of methyl 4-O-benzoyl-2,3-O-benzylidene-α-D-lyxopyranoside (136) with triphenylmethyl fluoroborate, followed by hydrolysis,

(158) For discussions and papers related to dioxolanylium ions and their rearrangements, see, for instance, Refs. 111 and 139.

gave a mixture of the 4-O-benzoyl derivative **139** (yield 42%) contaminated by its 3-O-benzoyl regio-isomer **140** (yield 8%), the latter proba-

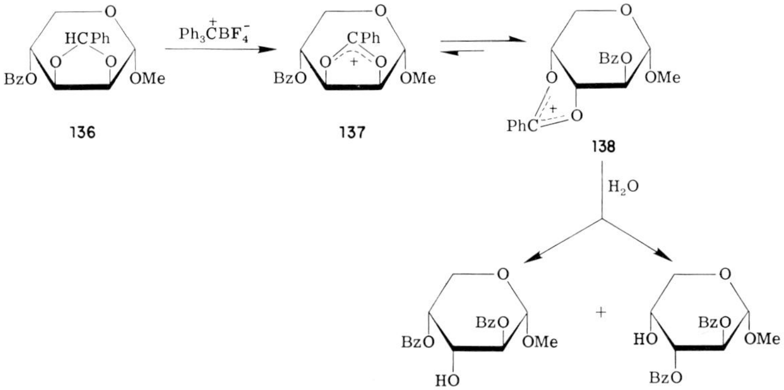

bly being produced from the former by an acyl migration. No compound that could have arisen from the 2,3-benzoxonium ion **137** could be detected, and this was assumed to be due to a rearrangement favoring ion **138**.

(2) The reaction of benzoxonium ions with tetraethylammonium bromide showed, within a few minutes, disappearance of the starting material, probably corresponding to the formation of a bromoorthoester. The latter was slowly (within a few hours) converted into bromodeoxy derivatives through a *trans* ring-opening. Thus, ion **93** gave a 1:1 mixture of methyl 2,3-di-O-benzoyl-4-bromo-4-deoxy-α-L-xylopyranoside (**95**) in a yield of 33% with methyl 2,4-di-O-benzyl-3-bromo-3-deoxy-β-D-arabinopyranoside (**96b**) in a yield of 31%. The reaction was interpreted[153] as involving the attack of bromine ion at C-4 of ion **93** concurrently (but more rapidly, as ion **94** is disfavored in its equilibrium with ion **93**) with the attack at C-3 of ion **94**. These results may be compared with those obtained for the same starting materials (or related derivatives), but involving the reaction of benzylidene acetals with N-bromosuccinimide (see the preceding Section 3,a). For the 2-O-methyl derivative **92a**, the reaction gave[153] the 2-O-methyl analog **141** (yield 41%), only slightly contaminated with methyl 4-O-benzoyl-3-bromo-3-deoxy-2-O-methyl-β-D-lyxopyranoside

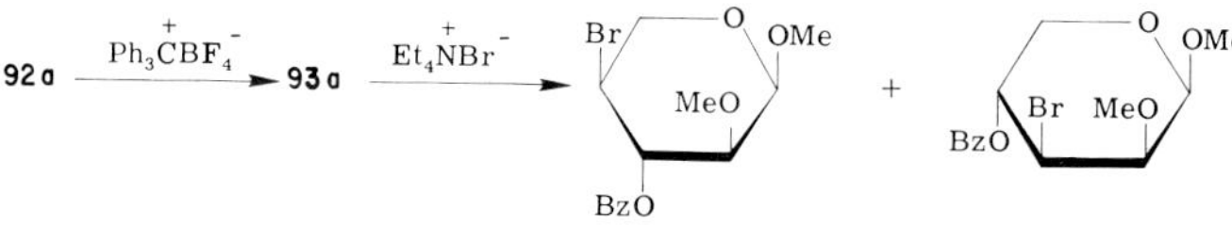

(**142**) in a yield of 5%, assumed to result from concurrent attack of bromine ion at C-3 of the ion **93a**. Methyl 2,4-di-O-benzoyl-3-bromo-3-deoxy-α-D-lyxopyranoside (**143**) (yield 49%) was obtained from compound **136**, and was accompanied by a small proportion of the bromo

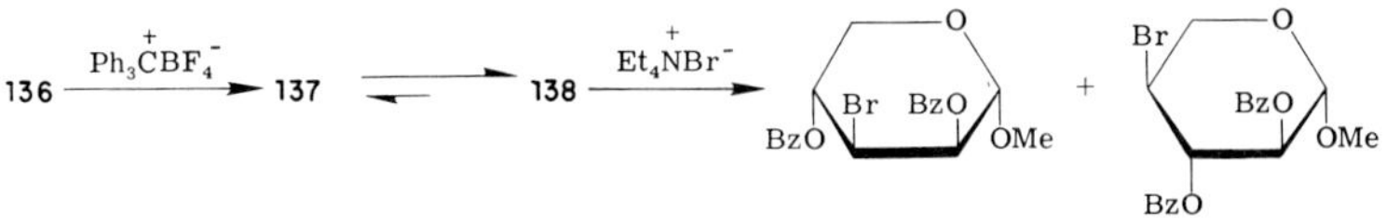

$$136 \xrightarrow{Ph_3\overset{+}{C}BF_4^-} 137 \rightleftharpoons 138 \xrightarrow{Et_4\overset{+}{N}Br^-}$$ **143** + **144**

derivative **144** (yield 6%).[153] The course of the reaction was interpreted by steric hindrance to the attack of bromide ion at C-3 of the ion **137**, because of a *syn*-axial interaction with the O-methyl group at the anomeric center (compare with related reactions involving N-bromosuccinimide that were reported in Section 3a).

In contrast to 3,4-O-benzylidene derivatives, 4,6-benzylidene acetals were found unreactive, and they gave no reaction with triphenylmethyl fluoroborate.[154] In fact, it seems that it is impossible to generate a benzoxonium ion from a six-membered ring, as not only 4,6-O-benzylidene derivatives (in the methyl α-D-gluco-, methyl α-D-manno-, and methyl α-D-galacto-pyranoside series) but also 3,5-O-benzylidene derivatives (in the α-D-glucopyranose series) were found to be unreactive. Furthermore, there also appeared to be a lack of reactivity of 1,2-O-benzylidene groups, at least for the examples given.[154] Thus, regiospecificity was observed in the reaction of di-O-benzylidene derivatives. From methyl 2,3:4,6-di-O-benzylidene-α-D-mannopyranoside (**108**), only the ion **145** was obtained, and, after addition of bromide ion, it gave the 3-bromo-3-deoxy-D-altroside **146** (yield 50%).[154] (Hy-

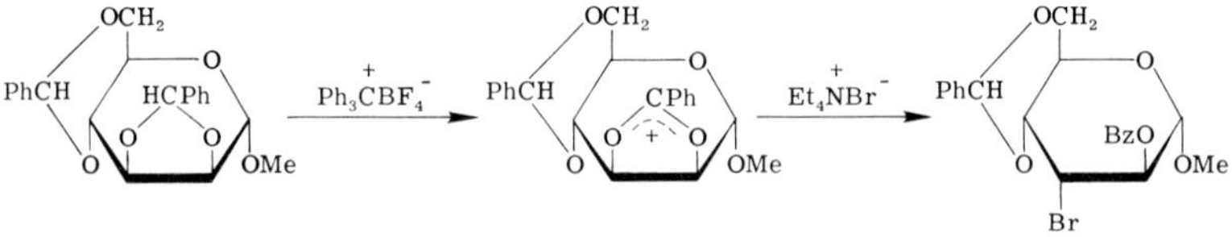

108 $\xrightarrow{Ph_3\overset{+}{C}BF_4^-}$ **145** $\xrightarrow{Et_4\overset{+}{N}Br^-}$ **146**

drolysis of ion **145** also gave the expected D-*manno* hydroxybenzoate having an axial benzoyl group on O-2.) Regiospecificity was also demonstrated for 1,2:3,4-di-O-benzylidene-β-D-arabinopyranose (**147**), which gave only the ion **148**, leading finally to 3-O-benzoyl-1,2-O-benzylidene-4-bromo-4-deoxy-α-L-xylopyranose (**149**) in a yield of 51% for the *endo-exo* mixture of diastereoisomers.[154]

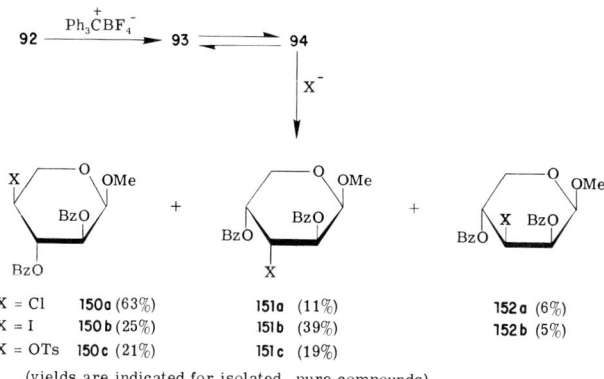

Addition of nucleophiles other than bromide ion to benzoxonium ions has been studied.[155] A *trans* ring-opening was demonstrated for various reagents, such as tetraethylammonium chloride, potassium iodide, potassium thiocyanate, tetraethylammonium *p*-toluenesulfonate, and sodium trifluoroacetate. For instance, with chloride, iodide, and *p*-toluenesulfonate ions, the equilibrated ions **93** and **94**, obtained from compound **92**, gave a mixture of two major compounds (**150a, b,** or **c,** and **151a, b,** or **c**), only slightly contaminated by a third (**152a** or **b**) for the first two examples. On the other hand, a *cis* ring-opening giving hydroxybenzoates was obtained by reaction of the same ions **93**⇌**94** with fluoride, acetate, azide, amide, cyanide, or methoxide ions, followed by the classical, aqueous processing.[156] The transitory formation of an orthoacid was invoked to interpret this observation. A ^{19}F-n.m.r. spectrum indicative of the presence of an orthoacid fluoride supported this assumption.

Other examples of formation of esters from 1,3-dioxolanylium ions were reported for the D-ribofuranose series[156] (see Section III,6,*a*), and formation of halogeno derivatives has been described for 1,6-anhydro-β-D-glycopyranoses.[157]

c. Action of Hydrogen Bromide in Acetic Acid.—Whereas acetolysis of cyclic acetals is rather well documented,[8] acetolysis conducted in the presence of hydrogen bromide seems to be a rather seldom used procedure for halogenation. Bock and Pedersen[159] treated 1-(R)-1,3,4,5,6-penta-O-acetyl-1,2-O-isopropylidene-*aldehydo*-D-glucose aldehydrol (**153**), obtained by acetolysis of 3,5,6-tri-O-acetyl-1,2-O-isopropylidene-α-D-glucofuranose, with a 30% solution of hydrogen bromide in glacial acetic acid; after classical processing, three fractions were obtained by preparative t.l.c. of the mixture. The first contained the 6-bromo derivative **154** (2:5 Z:E mixture; yield 6%); from the second was obtained (E)-4,5,6-tri-O-acetyl-2,3-dideoxy-*aldehydo*-D-*erythro*-hex-2-enose (**155**) in a yield of 45%; the third gave a 2,3,5,6-tetra-O-acetyl-D-glucofuranose (**156**) in a yield of 9%. A mechanism involving the formation of a cation on C-1 was discussed.

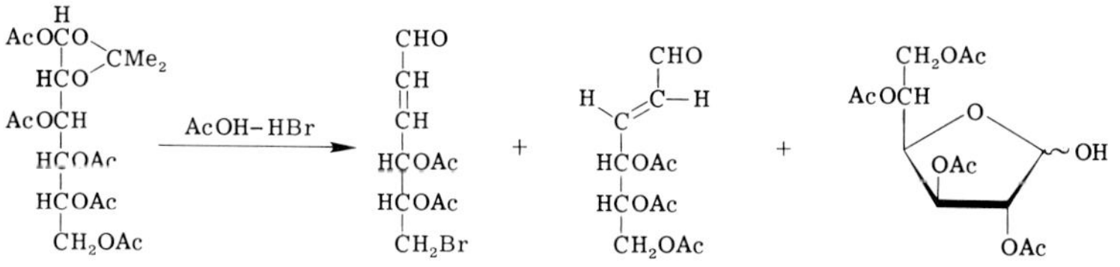

In the alditol series, examples of acetobrominolysis of O-isopropylidene groups of derivatives of D-mannitol and D-iditol were given by McCasland and coworkers.[160,161] Thus, the D-mannitol trithiocarbonate derivative **157** gave the 1,6-dibromo derivative **158** (yield 55%).[160] The same bromation with ring-opening was described for the D-iditol trithiocarbonate analog (yield 77%).

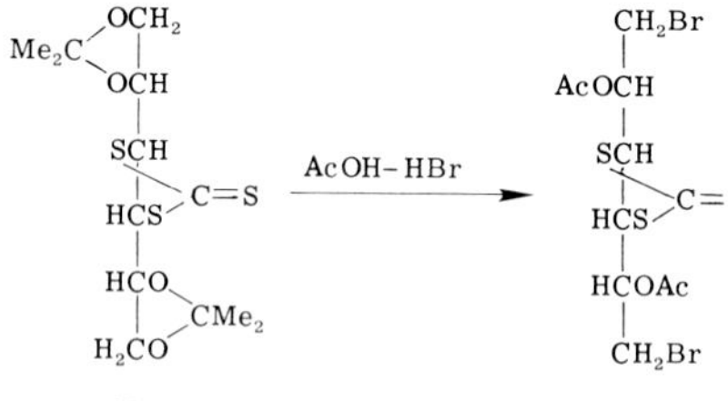

In one example, it was shown that O-isopropylidene derivatives of aldoses may react like those of alditols. When 2,3:5,6-di-O-isopropyli-

(159) K. Bock and C. Pedersen, *Acta Chem. Scand. Ser. B*, 28 (1974) 853–856.
(160) G. E. McCasland, A. B. Zanlungo, and L. J. Durham, *J. Org. Chem.*, 39 (1974) 1462–1466.
(161) G. E. McCasland, A. B. Zanlungo, and L. J. Durham, *J. Org. Chem.*, 41 (1976) 1125–1128.

dene-α-D-mannofuranose (**159**) was treated with acetic acid–hydrogen bromide at room temperature, only 5-O-acetyl-6-bromo-6-deoxy-2,3-O-isopropylidene-α-D-mannofuranosyl bromide (**160**) was obtained (in almost quantitative yield).[162] It was proposed that the reaction proceeds by removal of the 5,6-O-isopropylidene group, acetylation at O-5 (or O-6), and formation of a 5,6-acetoxonium ion which is finally attacked by bromine ion. As support for this interpretation, it was indicated that 2,3-O-isopropylidene-D-mannofuranose gives compound **160** in high yield under the same conditions.

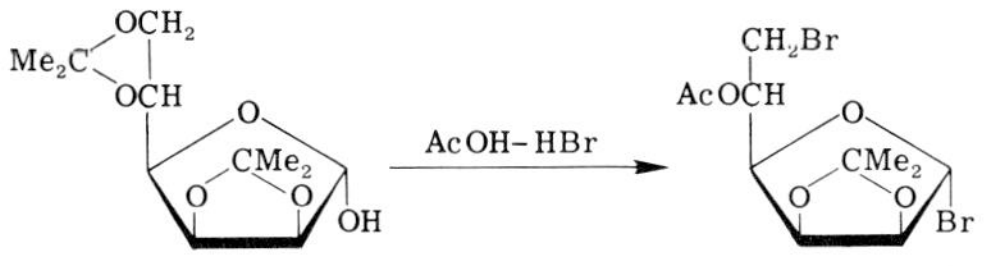

A few other examples may be found in related series. For instance, the methylenoxy bridge of 1,6-anhydro-β-D-glucopyranose can be cleaved, with subsequent bromination at both C-6 and the anomeric center.[163]

d. Action of Miscellaneous Reagents for the Halogenation of Cyclic Acetals.—(i) *Migration of Cyclic Acetals due to the Action of Various Halogenation Reagents Usually Employed for Isolated Hydroxyl Groups.* These reagents include halophosphites, phosphorus pentachloride, (chloromethylene)dimethyliminium chloride, and dichlorocarbene. Although the following reactions do not exactly provide an illustration of the reactivity of cyclic acetals, they are, however, not merely migrations that should have been reviewed elsewhere.[8]

Kochetkov and Usov[164] showed that the 5,6-acetal group tends to migrate during the reaction of 1,2:5,6-di-O-isopropylidene-α-D-glucofuranose (**161**) with triphenylphosphite methiodide, leading to the 6-iodo derivative **162a** (instead of the previously reported,[165] alleged 3-halo derivative, which would have been expected). The 6-bromo analog **162b** was similarly obtained[164] from the reaction of compound **161** with triphenylphosphite dibromide. This kind of reagent could, however, be used for the iodination of the 6-hydroxyl group of 2,3:4,5-di-O-isopropylidene-D-galactose diethyl dithioacetal without acetal mi-

(162) K. Bock and C. Pedersen, *Acta Chem. Scand. Ser. B*, 31 (1977) 248–250.
(163) M. Černý, V. Přikrylová, and J. Pacák, *Collect. Czech. Chem. Commun.*, 37 (1972) 2978–2984.
(164) N. K. Kochetkov and A. I. Usov, *Tetrahedron*, 19 (1963) 973–983.
(165) J. B. Lee and M. M. El Sawi, *Chem. Ind. (London)*, (1960) 839.

161

162a X = I
162b X = Br

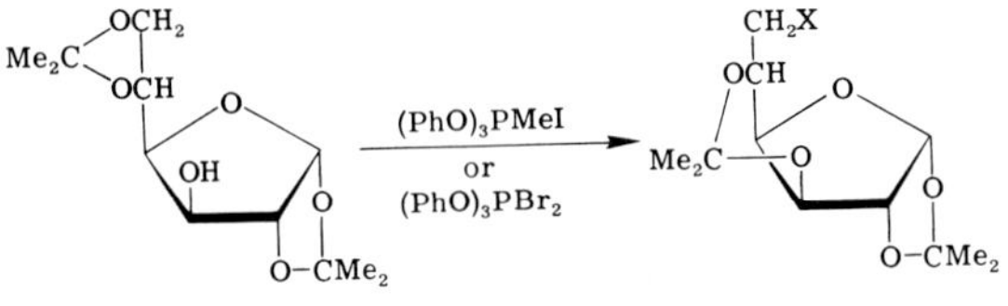

gration.[164] This reaction is very similar to the chlorination of 1,2:5,6-di-*O*-isopropylidene-α-D-glucose (**161**) with phosphorus pentachloride, which, it was erroneously supposed, gives the corresponding 3-chloro derivative[166]; but it was later demonstrated[167] to lead, actually, to the rearranged, 6-chloro compound **162c**. A mechanism proposed[168] to account for this acetal migration involved coordination of the oxygen atom on C-6 with phosphorus pentachloride.[169] The 6-chloro deriva-

163 164 165 162c

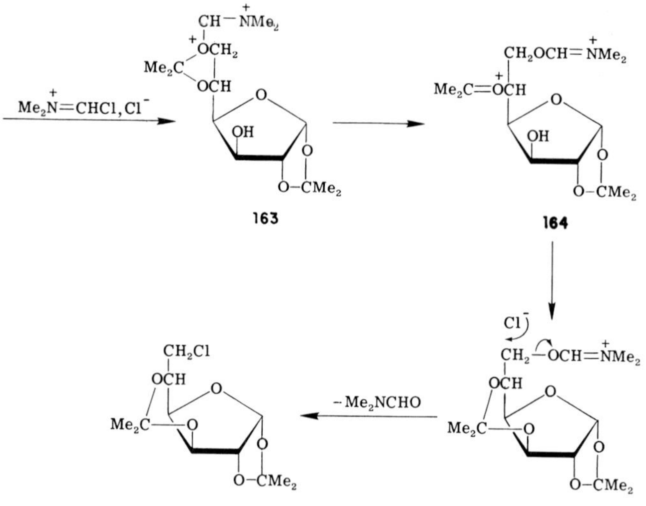

(166) J. B. Allison and R. M. Hixon, *J. Am. Chem. Soc.*, 48 (1926) 406–410.
(167) D. C. C. Smith, *J. Chem. Soc.*, (1956) 1244–1247.
(168) J. Baddiley, J. G. Buchanan, and F. E. Hardy, *J. Chem. Soc.*, (1961) 2180–2186.
(169) The course of the reaction of simple 1,3-dioxolanes or 1,3-dioxanes either with phosphorus pentachloride (see notably Refs. 170 and 171, and references cited therein), or with chlorodiethylphosphine (see Ref. 172) has been investigated.
(170) S. V. Fridland, S. K. Chirkunova, V. A. Kataeva, and G. K. Kamai, *Zh. Obshch. Khim.*, 41 (1971) 554–556.
(171) S. V. Fridland, S. K. Chirkunova, and Yu. K. Malkov, *Zh. Obshch. Khim.*, 43 (1973) 279–280.
(172) A. I. Razumov, A. I. Majkova, and V. V. Moskva, *Izv. Vyssh. Uchebn. Zaved., Khim. Khim. Tekhnol.*, 16 (1973) 1600–1602.

tive **162c** was obtained by the reaction of the diacetal **161** with triphenylphosphine in carbon tetrachloride.[173] An interpretation of this reaction was given in order to explain the same rearrangement observed when (chloromethylene)dimethyliminium chloride was used. This reagent, prepared by the action of phosgene on *N,N*-dimethylformamide, was employed by Hanessian and Plessas[134,174] to chlorinate compound **161**, to give the derivative **162c**, probably through the intermediates **163–165**.

Di Cesare and Gross[175] introduced a procedure using phase-transfer catalysis to induce the action of dichlorocarbene on various protected sugars,[176] and obtained the chloro derivative **162c** from compound **161**. Another, similar migration was observed,[178] and confirmed,[134,174] for the chlorination of methyl 2,3-anhydro-4,6-*O*-benzylidene-α-D-allopyranoside (**166**) with (chloromethylene)dimethyliminium chloride, which gave the rearranged chloro derivative **167**.

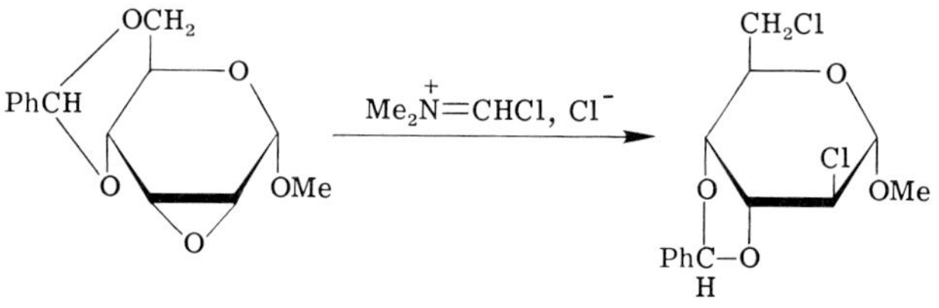

(ii) **Hydrogen Fluoride**. Anhydrous hydrogen fluoride at −70° has been used to transform 3,5,6-tri-*O*-acetyl-1,2-*O*-isopropylidene-α-D-glucofuranose (**168**) into a mixture of the anomers of the fluoride **168** (yield 91%) which was resolved by preparative t.l.c.[179] When the reaction was conducted for 15 min at 0°, a dimeric material resulted, in addition to compound **169**. Finally, during 24 h at room temperature, a 2,3:5,6-diacetoxonium ion (D-*manno* configuration) was obtained.

(173) C. R. Haylock, L. D. Melton, K. N. Slessor, and A. S. Tracey, *Carbohydr. Res.*, 16 (1971) 375–382.
(174) S. Hanessian and N. R. Plessas, *Chem. Commun.*, (1967) 1152–1155.
(175) P. Di Cesare and B. Gross, *Carbohydr. Res.*, 58 (1977) c1–c3.
(176) For reactions involving sugars protected with a cyclic acetal grouping having at least one hydrogen atom on the acetal carbon atom with dihalocarbenes, attention should be drawn to a study (Ref. 177) using phase-transfer catalysis to insert a dihalocarbene on the C-2–H bond, leading to 2-(dihalomethyl)-1,3-dioxolanes or -dioxanes.
(177) K. Steinbeck, *Chem. Ber.*, 112 (1979) 2402–2412.
(178) A. Klemer and K. Homberg, *Chem. Ber.*, 94 (1961) 2747–2754.
(179) K. Bock and C. Pedersen, *Acta Chem. Scand. Ser. B*, 26 (1972) 2360–2366.

CH₂OAc   CH₂OAc
 | |
AcOCH   O   AcOCH   O
   ___/   HF   ___/
   /   \   ─────────────→   /   \ ~ F
  OAc   O   10 min, at −70° OAc
       / \
      O—CMe₂   OH

     168   **169**

(iii) **Bromine, *N*-Chlorosuccinimide, Trichlorotetrahydrotriazine-trione, and *N,N*-Dibromobenzenesulfonamide.** Some reagents can behave like *N*-bromosuccinimide towards 1,3-dioxolanes or 1,3-dioxanes having at least one hydrogen atom at C-2. Thus, bromine was found[116] to be effective, either in the presence, or absence, of peroxides, to give bromoesters from 4,6-*O*-benzylidene derivatives. It may be noted that the complex 1,4-dioxane dibromide was also shown to react with 1,3-dioxacycloalkanes, to give bromoesters.[105,180] Use of trimethyl phosphate, (MeO)₃PO, at 0–10° with bromine or chlorine under anhydrous conditions was recommended as a simple, high-yielding method for halogenative cleavage of cyclic acetals of noncarbohydrates.[180a]

Chloroesters may be simply prepared by using *N*-chlorosuccinimide instead of *N*-bromosuccinimide,[181,182] Use of 1,3,5-trichloro-1,3,5-triazine-2,4,6-trione, a known reagent for halogenation of cyclic ethers,[183] has been introduced as a mild procedure for ring opening of acetals to chloroesters.[182] The scope of the reaction was studied on model compounds. The reaction of 1,3,5-trichlorotriazine-2,4,6-trione **171** with the 1,3-dioxolane **170** gave a mixture (crude yield 90%) in which the expected chloroester **172a** was largely preponderant. Two minor processes of chlorination were also demonstrated. The first corresponded to the halogenation of the 2-alkyl group, which gave compound **173a**, a precursor of the chloroester **172b**. Dichlorination was also observed, giving compound **172c**, probably through the precursor **173b'**(not isolated). The second, very minor, process was chlorination at C-4, giving the halogenodioxolane **174** (not isolated), a precursor of the ester **172d**.

(180) D. L. Rakhamankulov, V. S. Martem'yanov, S. S. Zlotskii, Z. L. Ayupova, and T. S. Artamonova, *Zh. Obshch. Khim.*, 45 (1975) 2739–2741.
(180a) S. D. Venkataramu, J. H. Cleveland, and D. E. Pearson, *J. Org. Chem.*, 44 (1979) 3082–3084.
(181) J. Gelas and S. Michaud, unpublished results.
(182) J. Gelas and D. Pétrequin, *Carbohydr. Res.*, 36 (1974) 227–230.
(183) W. P. Duncan, G. D. Strate, and B. G. Adcock, *Org. Prep. Proced. Int.*, 3 (1971) 149–154.

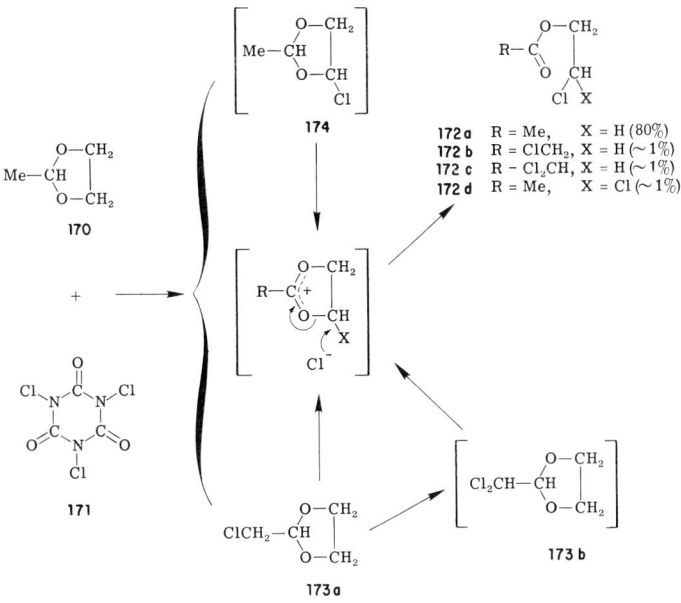

When 2,2-disubstituted 1,3-dioxolanes were employed, mono- or poly-chlorination of the side chain was observed on the carbon atom α to the acetal grouping. Incidentally, it should be noted that N-bromosuccinimide, N-chlorosuccinimide, and trichlorotetrahydrotriazinetrione were found to be effective for the preparation of bromoacetates from O-ethylidene derivatives; this might be useful when O-benzylidene derivatives are not readily available, or when a problem arises due to the fact that O-benzoyl groups are, in general, more difficult to remove than O-acetyl groups.

A procedure for bromination of 2-phenyl-1,3-dioxolanes with N,N-dibromobenzenesulfonamide ($C_6H_5SO_2NBr_2$) has been described,[184] but it has not yet been applied in the carbohydrate field. The behavior of this reagent is essentially identical to that of N-bromosuccinimide, but some differences were noted, and a suggested mechanism was discussed.[184]

(iv) **Dibromomethyl Methyl Ether.** Use of this reagent was introduced[185] as a mild procedure for the synthesis of glycosyl bromides. Its reaction on the O-isopropylidene groups of some aldoside acetals

(184) Y. Kamiya and S. Takemura, *Chem. Pharm. Bull.*, 22 (1974) 201–206.
(185) R. Bognár, I. Farkas-Szabó, I. Farkas, and H. Gross, *Carbohydr. Res.*, 5 (1967) 241–243.

has been investigated.[186] The reaction of dibromomethyl methyl ether in the presence of zinc bromide with methyl 4-O-benzoyl-2,3-O-isopropylidene-α-L-rhamnopyranoside (**175**) for several hours at room temperature gave 4-O-benzoyl-2-bromo-2,6-dideoxy-3-O-formyl-α-L-glucopyranosyl bromide (**177**). The mechanism postulated involves the successive formation of a glycosyl bromide, cleavage of the isopropylidene group, formylation, and formation of the formoxonium ion

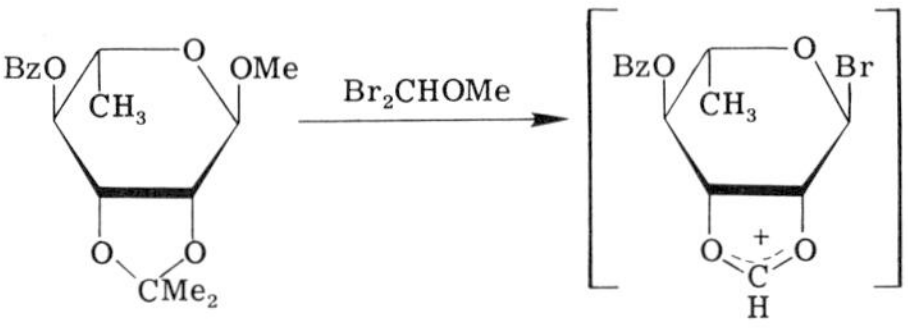

176. Compound **177** was purified through the corresponding methyl β-L-glucopyranoside by treatment with methanol, ether, and silver carbonate (overall yield 42%). The same reaction was applied[186] to methyl 5-O-benzoyl-2,3-O-isopropylidene-α-D-lyxofuranoside (**178**), which gave the bromide **180** (crude yield 98%). Evidently, compounds for which rearrangement of the intermediate dioxolanylium ion is prevented behave differently. For instance, methyl 4,6-di-O-benzoyl-2,3-O-isopropylidene-α-D-mannopyranoside (**181**) gave the 6-bromo-α-D-idopyranosyl bromide **185**, isolated as the corresponding methyl β-D-idoside (after treatment with methanol in the presence of silver carbonate) in an overall yield of 57%.[186] The reaction probably proceeds through a sequence of rearrangements of the 2,3-formoxonium ion **182**, to the more stable 3,4-benzoxonium ion **183**, and finally, to the more abundant 4,6-benzoxonium ion **184** (compare with Ref. 111, and references cited therein). Results concerning the reaction of dibromethyl methyl ether with methyl 5-O-benzoyl-2,3-O-isopropylidene-β-D-ribofuranoside were also given[186]; a mixture was obtained from which methyl 2-bromo-2-deoxy-3-O-formyl-5-O-benzoyl-β-D-xylofuranoside was isolated in a yield of 45%.

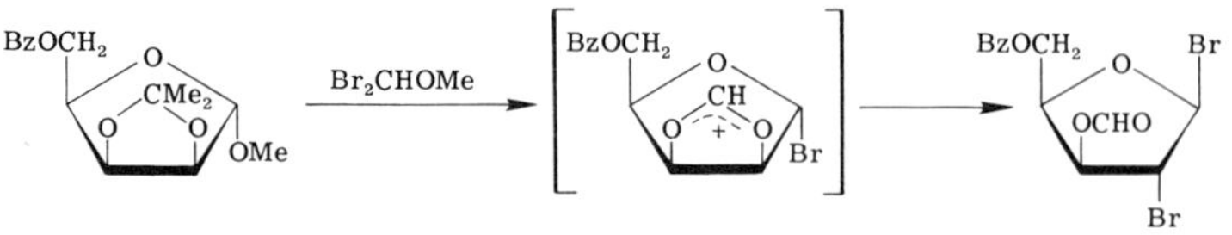

(186) K. Bock, C. Pedersen, and J. Thiem, *Carbohydr. Res.*, 73 (1979) 85–91.

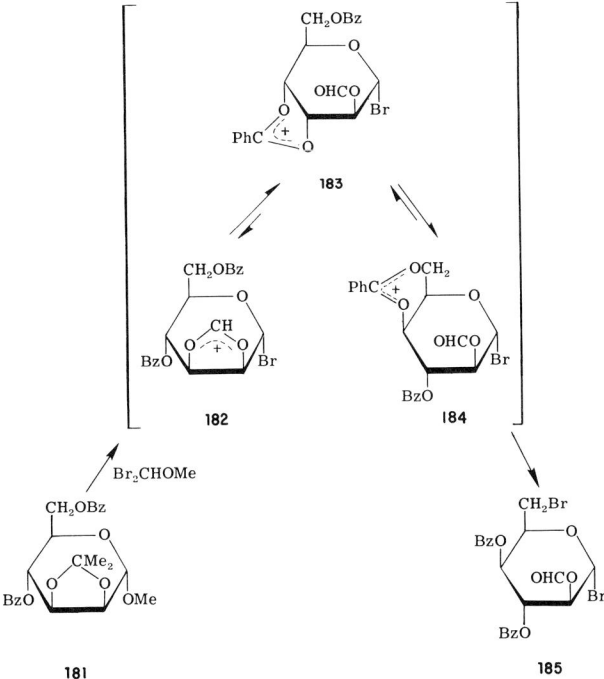

4. Hydrogenolysis

It has been known for a long time that acetal groups in alkaline media are resistant to reducing agents. In the aliphatic series, very drastic conditions were needed (8h, 190°, 200 atm. hydrogen pressure) in order to achieve[187] the catalytic reduction of an acyclic acetal to the corresponding ether. Under the usual conditions, acetal groups are not reduced by lithium aluminum hydride or sodium borohydride.[188] Thus, they are classical protecting groups in reducing reactions involving these reagents. The benzylidene acetal grouping constitutes a particular functional group in this respect, as it has been well known, for a long time,[189-191] that such groups are readily removed, totally, from sugars by hydrogenolysis in the presence of palladium-on-charcoal. This type of reactivity does not need to be emphasized here.

(187) U. Schmidt and P. Grafen, *Chem. Ber.*, 92 (1959) 1177–1184.
(188) N. G. Gaylord, *Reduction with Complex Metal Hydrides*, Interscience, New York, 1956.
(189) S. Peat and L. F. Wiggins, *J. Chem. Soc.*, (1938) 1088–1097.
(190) G. R. Barker and J. W. Spoors, *J. Chem. Soc.*, (1956) 1192–1195, 2656–2658.
(191) H. B. Wood, Jr., H. W. Diehl, and H. G. Fletcher, Jr., *J. Am. Chem. Soc.*, 79 (1957) 1986–1988.

In 1953, the reduction was reported[192] of diosgenine (a spiro-linked, cyclic acetal in the steroid series) to dihydrodiosgenine (an ether) by the addition of lithium aluminum hydride to a solution of the starting material in anhydrous ether saturated with hydrogen chloride (or bromide) gas. Eliel and Rerick[193] gave evidence for the formation of a *"mixed reagent"* or a *"mixed hydride"* (lithium aluminum hydride–aluminum chloride) responsible for the reduction. A long series of reports then followed that described the scope of the reaction of the mixed hydride (AlH_2Cl probably being the reducing species when $AlCl_3$ and $LiAlH_4$ were used in the ratio of 1:1; see, for example, Ref. 194) with cyclic acetals and thioacetals [1,3-dioxolanes, 1,3-dioxanes, 1,3-oxathiolanes and O-tetrahydropyran(or furan)-2-yl derivatives], these reports mainly emanating from Eliel[195–198] and R. K. Brown[199–207] and their coworkers. As these results were useful to further investigations of this reaction in the carbohydrate series, as regards the rate, and the regio- and stereo-selectivity of the hydrogenolysis, a brief summary of some of the most important features will now be given.

a. Rate of Reductive Cleavage.—A comparative study, as between 1,3-dioxolanes and 1,3-dioxanes, showed[206] that, for 2-O-alkyl or -aryl monosubstituted acetals, 1,3-dioxolanes are hydrogenolyzed markedly more rapidly than the corresponding 1,3-dioxanes; in contrast, six-membered acetals are reduced slightly more rapidly than the five-membered acetal when C-2 is disubstituted, and formyl acetals are hydrogenolyzed with greater difficulty.[200] Electron-donating substi-

---

(192) H. M. Doukas and T. D. Fontaine, *J. Am. Chem. Soc.*, 75 (1953) 5355–5356.
(193) E. L. Eliel and M. N. Rerick, *J. Org. Chem.*, 23 (1958) 1088.
(194) E. C. Ashby and J. Prather, *J. Am. Chem. Soc.*, 88 (1966) 729–733.
(195) E. L. Eliel and V. G. Badding, *J. Am. Chem. Soc.*, 81 (1959) 6087.
(196) E. L. Eliel, V. G. Badding, and M. N. Rerick, *J. Am. Chem. Soc.*, 84 (1962) 2371–2377.
(197) E. L. Eliel, L. A. Pilato, and V. G. Badding, *J. Am. Chem. Soc.*, 84 (1962) 2377–2384.
(198) E. L. Eliel, B. E. Nowak, R. A. Daignault, and V. G. Badding, *J. Org. Chem.*, 30 (1965) 2441–2447.
(199) B. E. Leggetter and R. K. Brown, *Can. J. Chem.*, 41 (1963) 2671–2682.
(200) B. E. Leggetter and R. K. Brown, *Can. J. Chem.*, 42 (1964) 990–1004.
(201) B. E. Leggetter, U. E. Diner, and R. K. Brown, *Can. J. Chem.*, 42 (1964) 2113–2118.
(202) B. E. Leggetter and R. K. Brown, *Can. J. Chem.*, 42 (1964) 1005–1008.
(203) B. E. Leggetter and R. K. Brown, *Can. J. Chem.*, 43 (1965) 1030–1035.
(204) W. W. Zajac, Jr., B. Rhee, and R. K. Brown, *Can. J. Chem.*, 44 (1966) 1547–1550.
(205) U. E. Diner, H. A. Davis, and R. K. Brown, *Can. J. Chem.*, 45 (1967) 207–211.
(206) U. E. Diner and R. K. Brown, *Can. J. Chem.*, 45 (1967) 1297–1299.
(207) U. E. Diner and R. K. Brown, *Can. J. Chem.*, 45 (1967) 2547–2558.

tuents (methyl or phenyl) at C-2 increase the rate of reduction of 1,3-dioxolanes (the same effect was less pronounced when the substituents were on O-4 or O-5). On the other hand, such electron-withdrawing substituents as chloromethyl and dichloromethyl retard the hydrogenolysis. The effectiveness of heteroatoms in decreasing the ease of hydrogenolysis when they replace a hydrogen atom of the methyl group on C-2 was determined[208] to be $H < S < O < Br < NR_2$.

b. Regioslectivity.—Two reaction paths have been invoked to explain the regioselectivity of the reduction,[202] as shown in Scheme 5.[209] Electron-donor substituents at C-4 (Scheme 5, R = Me, for instance)

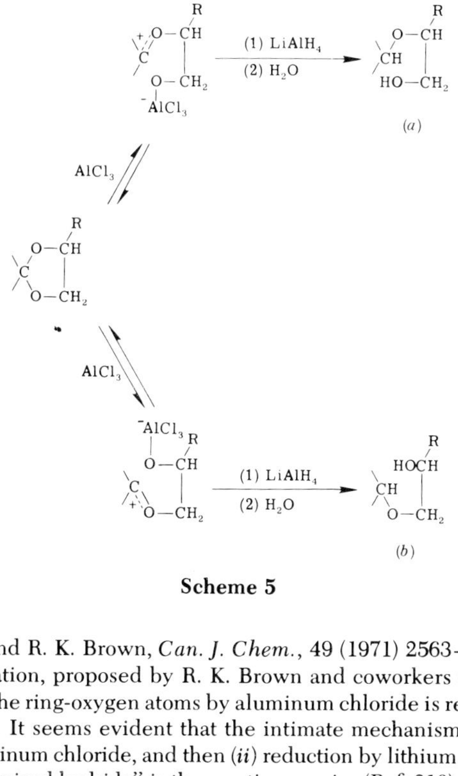

Scheme 5

(208) H. A. Davis and R. K. Brown, *Can. J. Chem.*, 49 (1971) 2563–2577.
(209) The complexation, proposed by R. K. Brown and coworkers (Refs. 199–207), of either one of the ring-oxygen atoms by aluminum chloride is reproduced here as a simplification. It seems evident that the intimate mechanism does not imply (*i*) attack by aluminum chloride, and then (*ii*) reduction by lithium aluminum hydride; actually, the "mixed hydride" is the reactive species (Ref. 210), and its identity depends on the ratio between the Lewis acid and the hydride (see, for instance, Refs. 211 and 212, and references cited therein, for a discussion of the nature of mixed hydrides).
(210) H. A. Davis and R. K. Brown, *Can. J. Chem.*, 49 (1971) 2166–2168.
(211) E. L. Eliel, *Rec. Chem. Prog.*, 23 (1961) 129–145.
(212) M. N. Rerick, in R. L. Augustine (Ed.). *Reduction*, Dekker, New York, 1968, pp. 2–94.

favor the cleavage of the C-2–O-1 bond (pathway *a*). The reverse situation was found for electron-withdrawing groups (Scheme 1, R =CMe$_2$Cl, for instance) which favor the rupture of the C-2–O-3 bond (pathway *b*). This effect was simply rationalized on the basis of the stabilizing or destabilizing influence exerted by the substituent on the intermediate, oxocarbonium ions. "Anomalous" ring-cleavage could, however, be observed when strong steric effects were involved: for instance, the reverse of the preceding behavior was found for dioxolanes disubstituted at both C-2 and C-4, for which, scission of the C-2–C-3 bond occurs[202] because of the steric destabilization of the intermediate involved in pathway *a*.

c. Stereoselectivity.—For 2,4-dialkyl-substituted-1,3-dioxolanes, *cis* isomers were found[203] to be hydrogenolyzed faster (~6 to 10 times) than their corresponding *trans* isomers. The regioselectivity of the cleavage was found to be different for each diastereoisomer: for *cis* isomers the ratio between the C 2–O-1 and C-2–O-3 scissions was 15:1, whereas that for the *trans* isomers was ~1:2. There is as yet no clear explanation for such a difference.

One typical example, a model that presents some analogy with carbohydrates, is the hydrogenolysis of (racemic) 1,2(or 2,3)-*O*-isopropylidene-DL-glycerol (**186**) or -3(or 1)-*O*-alkylglycerol (**189**), which showed[200] favored cleavage of the C-2–O-3 bond to give, respectively,

$$\underset{\underset{\text{CH}_2\text{OR}}{|}}{\text{Me}_2\text{C}\overset{\text{OCH}_2}{\underset{\text{OCH}}{\diagdown|}}} \quad \xrightarrow[(2)\ \text{H}_2\text{O}]{(1)\ \text{LiAlH}_4-\text{AlCl}_3} \quad \underset{\underset{\text{CH}_2\text{OR}}{|}}{\overset{\text{Me}_2\text{CHOCH}_2}{\underset{\text{HOCH}}{|}}} \quad + \quad \underset{\underset{\text{CH}_2\text{OR}}{|}}{\overset{\text{HOCH}_2}{\underset{\text{Me}_2\text{CHOCH}}{|}}}$$

186 R = H   **187** (81%)   **188** (19%)
189 R = CHMe$_2$ **190** (86%)   **191** (14%)

compounds **187** and **190** as the main, and **188** and **191** as the minor, constituents of the mixture. This finding was later confirmed[213] for some analogs having various alkoxyl groups on C-4 and alkyl groups on C-2, clearly showing the influence, previously postulated,[200] of the more or less powerful influence of the electronic character of the substituent on C-4 (42 to 98% of the secondary alcohol, as against the primary alcohol).

The first, direct application of mixed hydrides in the carbohydrate series was described by Bhattacharjee and Gorin.[214] Using a 1:1 ratio of lithium aluminum hydride to aluminum chloride, and a one-molar equivalent of this reagent, they obtained, after 40 hours, a 64% yield of

(213) F. G. Ponomarev and N. N. Chernousova, *Zh. Org. Khim.*, 7 (1971) 890–892.
(214) S. S. Bhattacharjee and P. A. J. Gorin, *Can. J. Chem.*, 47 (1969) 1195–1206.

the 6-hydroxy derivative **193** from methyl 4,6-*O*-benzylidene-2,3-di-*O*-benzyl-β-D-glucopyranoside (**192**). The high regioselectivity was

192 **193**

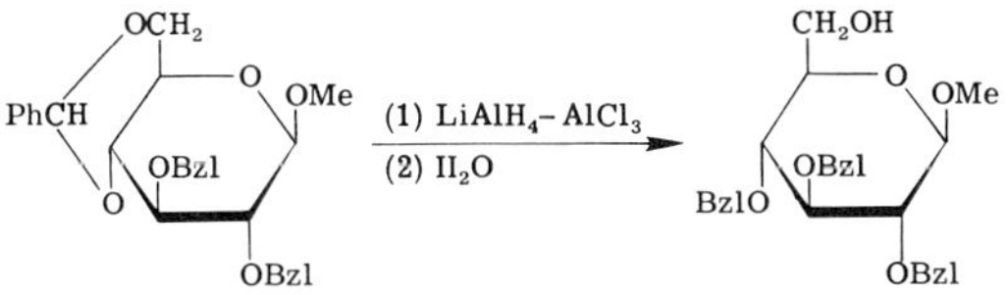

confirmed by the results of hydrogenolysis of methyl 4,6-*O*-benzylidene-α-D-gluco- (**20a**) and -mannopyranoside (**196**). However, hydrogenolysis of methyl 4,6-*O*-benzylidene-α-D-galactopyranoside (**22**) gave potentially the 6-*O*-benzyl derivative **199**, the yields being[214] in

20a R = H, R′ = OH
196 R = OH, R′ = H

194 (3:2 ratio) **195**
197 (4:1 ratio) **198**

22 **199**

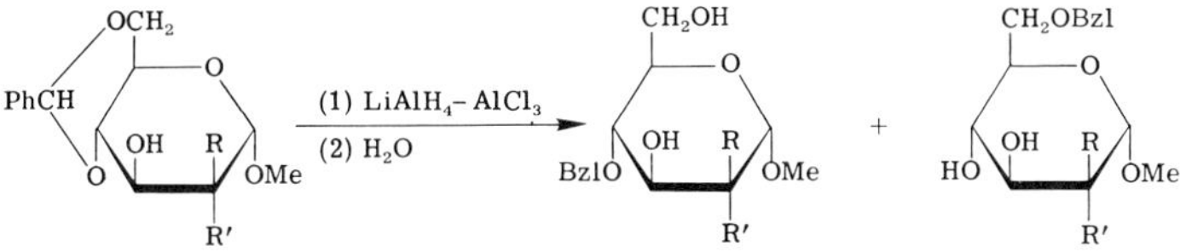

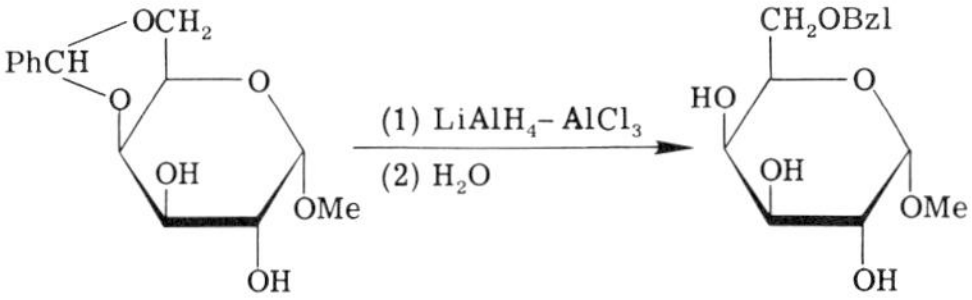

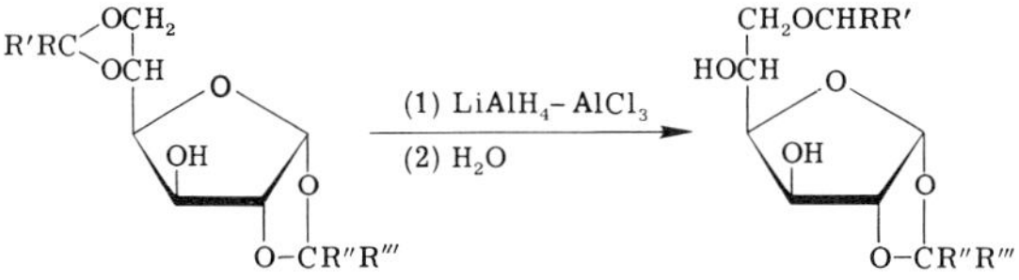

the range of 27 to 38%. Diacetal derivatives were found particularly interesting substrates, as they could be used to determine regioselectivity between the cleavage of two different rings. In the D-glucofuranose series (compounds **161**, **161a**, and **161b**), the 5,6-*O*-isopropyl-

161 R = R′ = R″ = R‴ = Me **200**
161a R = Ph, R′ = H, R″ = R‴ = Me **200a**
161b RR′ = R″R‴ = $(CH_2)_5$ **200b**

idene group was cleaved in preference to the 1,2-*O*-isopropylidene or the 1,2-*O*-cyclohexylidene group, giving **200**, **200a**, or **200b**. Depending on the reaction time, the yields lay between 49 and 60%. Competition between the cleavage of five- and six-membered-ring acetals was investigated,[214] particularly through the following two examples. The six-membered ring of 3,5-*O*-benzylidene-1,2-*O*-isopropylidene-α-D-glucofuranose (**103**) was preferentially cleaved, with good regioselectivity, to the 3-benzyl ether **201** (yield 42%). On the other

103 201

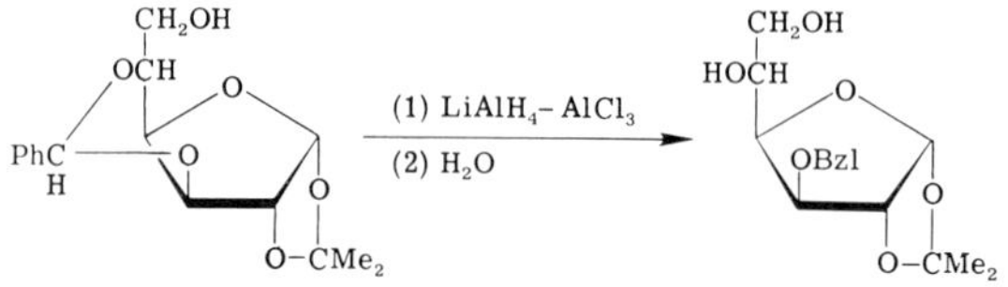

hand, with methyl 2,3:5,6-di-*O*-benzylidene-α-D-mannopyranoside (**108**), a compound previously alleged to give one major compound, the 3-*O*-benzyl derivative **202** was obtained, with formation of only traces of the concurrent regio-isomer **203**, and without noticeable cleavage of the 4,6-*O*-benzylidene group (two molar equivalents of the reducing agent and a reaction time of 35 hours).[214] Actually, when the reaction time was 6 days (2.1 molar equivalents of reducing agent), the reaction was shown[215] to give a mixture from which three fractions, in approxi-

202 203 (trace)

108

204 205 206

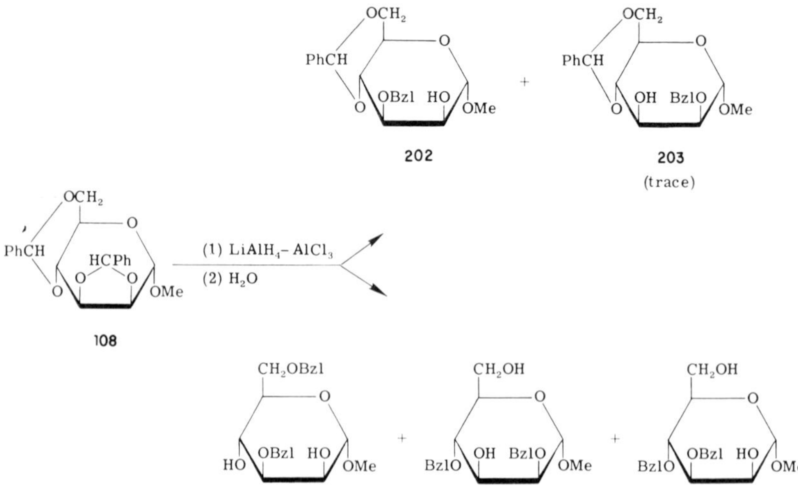

(215) S. S. Bhattacharjee and P. A. J. Gorin, *Can. J. Chem.*, 47 (1969) 1207–1215.

mately equal proportions, could be obtained by column chromatography: the first gave the monobenzyl derivative **202**, the second contained the two dibenzyl derivatives **202** and **205**, and a third dibenzyl derivative (**206**) was found in the last. Except for the particular situation of 1,2-acetals (known, for instance, to be resistant to hydrolysis as compared to 5,6-acetals), it is rather clear that six-membered acetals are more resistant to hydrogenolysis than five-membered acetals (compare with Refs. 201 and 206). Methyl 2,3:5,6-di-O-benzylidene-α-D-mannofuranoside (**207**) was hydrogenolyzed to a major compound considered[214] to be a methyl di-O-benzyl-α-D-mannofuranoside, and then characterized[215] as the 2,6-di-O-benzyl derivative **208** (yield in

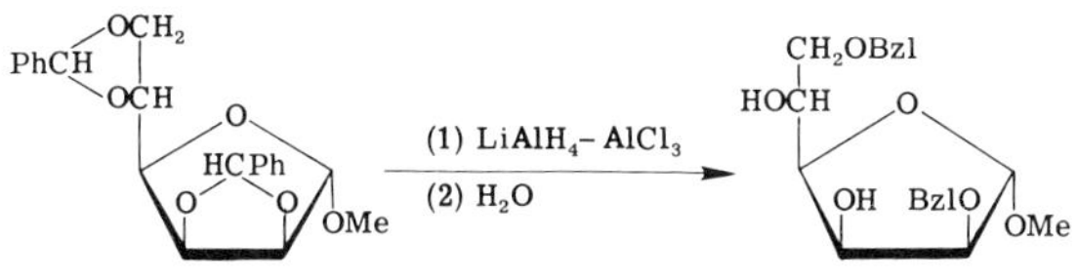

the range of 30 to 50%). This kind of cleavage of a 2,3-O-benzylidene group was also observed for the hydrogenolysis of methyl 2,3-O-benzylidene-5,6-di-O-methyl-α-D-mannofuranoside, which was mainly converted into methyl 2-O-benzyl-5,6-di-O-methyl-α-D-mannofuranoside[215]; examples in the 1,6-anhydro-β-D-mannopyranose series were also given. It should be noted that O-methyl derivatives may be obtained when O-methylene acetals are reduced.[214]

These studies were extended to orthoesters, which gave acetals, the latter eventually being cleaved, with an excess of the reagent, to alkoxy derivatives.[216]

Summarizing, it may be concluded that the ease of hydrogenolysis is (1) cyclic orthoester > isopropylidene acetal, cyclohexylidene acetal > benzylidene acetal > ethylidene acetal > methylene acetal, and (2) 5,6-O-linked and 3,5-O-linked > 1,2-O-linked acetals.

Application of hydrogenolysis of benzylidene acetals to 2-amino-2-deoxy sugar derivatives was found to be possible.[217] The reductive cleavage of benzyl 4,6-O-benzylidene-2-(benzyloxycarbonyl)amino-2-deoxy-α-D-glucopyranoside (**209**) gave, in a yield of 50%, a mixture of the 4-O-benzyl (**210**) and the 6-O-benzyl (**211**) derivatives. If the regioselectivity of this reaction, although notably inferior, was in ac-

(216) S. S. Bhattacharjee and P. A. J. Gorin, *Carbohydr. Res.*, 12 (1970) 57–68.
(217) P. A. J. Gorin and A. J. Finlayson, *Carbohydr. Res.*, 18 (1971) 269–279.

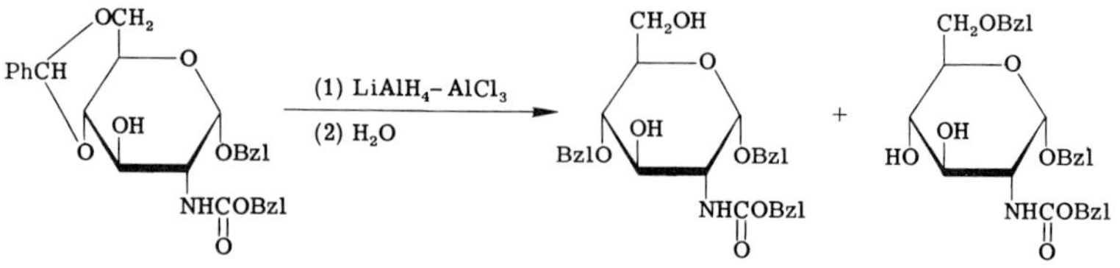

209 → **210** + **211**

cordance with that observed for the analog **20a**, the observation[218] of the favored formation of the 4-O-benzyl derivative **213** over **214** in the α-D-*galacto* series (from compound **212**) contrasted with the formation of the 6-O-benzyl derivative **199** from the analog **22**.

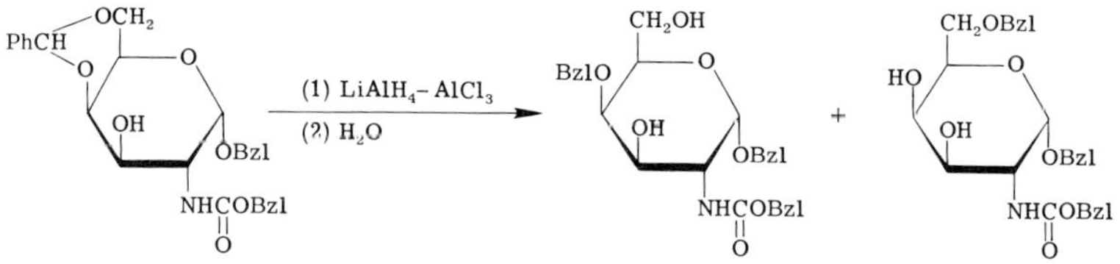

212 → **213** + **214**

(ratio ~ 2.5:1)

A more-precise determination of the factors that orient the cleavage of a 4,6-O-benzylidene group was undertaken by Lipták and coworkers.[219,220] First, confirmation of the observation made by Bhattacharjee and Gorin[214] of the high regioselectivity of the cleavage for the methyl β-D-glucopyranoside derivative **192** was provided[219] for the benzyl analog (benzyl 2,3-di-O-benzyl-4,6-O-benzylidene-β-D-glucopyranoside, **215a**), which led only to the 4-O-benzyl derivative **216a** in a yield of 90%. This reaction was successfully repeated for the α anomer **215b**, to give phenyl 2,3,4-tri-O-benzyl-α-D-glucopyranoside (**216b**) in a yield of 68%.[220] The fact that methyl 2,3-di-O-benzyl-4,6-O-benzylidene-α-D-glucopyranoside (**215c**), methyl 3-O-benzyl-4,6-O-benzylidene-α-D-glucopyranoside (**215d**), and benzyl 2,3-di-O-benzyl-4,6-O-benzylidene-α-D-mannopyranoside (**215e**) gave only the 4-O-benzyl derivatives (**216c**, **216d**, and **216e**) in high yield (compare with the cleavage of compounds **20a** and **196**, which both gave mixtures[214]), and the fact that this regiospecificity was independent of the anomeric configuration, led to the conclusion that the bulkiness of the substituent on O-3 determines the direction of the cleavage, and is the main,

(218) P. A. J. Gorin, *Carbohydr. Res.*, 18 (1971) 281–289.
(219) P. Nánási and A. Lipták, *Mag. Kem. Foly.*, 80 (1974) 217–225.
(220) A. Lipták, I. Jodál, and P. Nánási, *Carbohydr. Res.*, 44 (1975) 1–11.

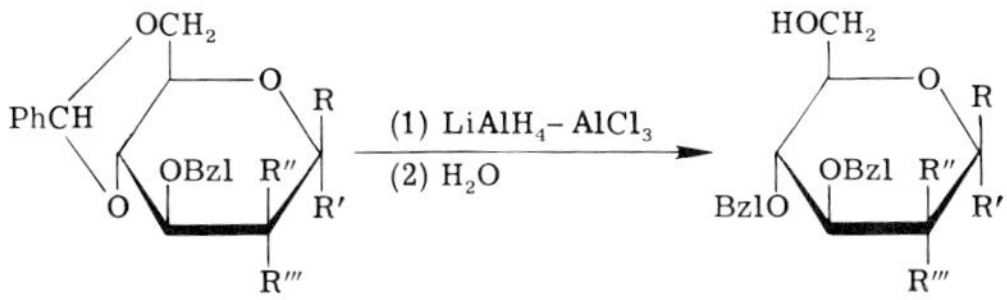

215a R = OBzl, R' = H, R" = H, R''' = OBzl (β-D-*gluco*)		**216a**
215b R = H, R' = OPh, R" = H, R''' = OBzl (α-D-*gluco*)		**216b**
215c R = H, R' = OMe, R" = H, R''' = OBzl (α-D-*gluco*)		**216c**
215d R = H, R' = OMe, R" = H, R''' = OH (α-D-*gluco*)		**216d**
215e R = H, R' = OBzl, R" = OBzl, R''' = H (α-D-*manno*)		**216e**

structural factor involved with O-6, because O-4 was too hindered by the bulk of the benzyl group on O-3. For non-carbohydrate series, factors involved in reduction with mixed hydrides have been discussed; Eliel[211] emphasized that, considering the kinetic control of the reaction, it could be assumed that the mixed hydride was a reagent more sensitive to "steric-approach control" than lithium aluminum hydride alone. It was therefore considered that approach of the reducing agent from the crowded side was more impeded when the mixed hydride was used. Other information concerning the mechanism of the hydrogenolysis of acyclic acetals has also been reported.[221,222]

These results were extended[220] to the D-galactopyranoside series, and it was demonstrated, for compounds **217a–217e**, that the cleavage, although not regiospecific, is highly selective and clearly favors the formation of the 4-O-benzyl derivatives (**218**) over the 6-benzyl ethers (**219**). The ratio between the 4- and the 6-O-benzyl derivatives

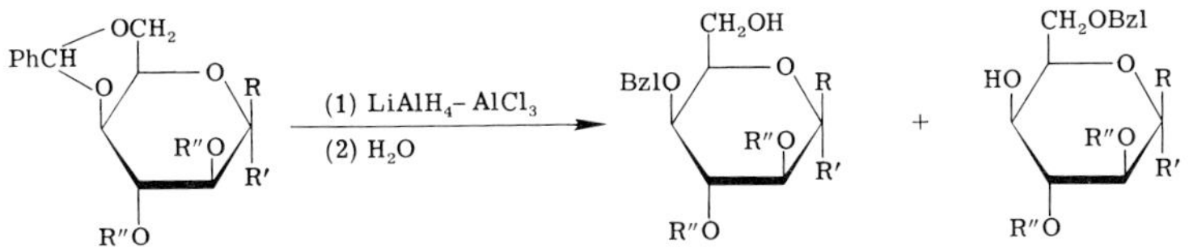

	218	**219**
217a R = H, R' = OBzl, R" = Bzl (α-D-*galacto*)	82%	<10%
217b R = H, R' = OPh, R" = Bzl (α-D-*galacto*)	88%	<10%
217c R = OBzl, R' = H, R" = Bzl (β-D-*galacto*)	92%	4%
217d R = OPh, R' = H, R" = Bzl (β-D-*galacto*)	68%	12%
217e R = OBzl, R' = H, R" = Me (β-D-*galacto*)	26%	7%

yields are indicated for isolated, pure compounds

(221) W. W. Zajac, Jr., and K. J. Byrne, *J. Org. Chem.*, 35 (1970) 3375–3377.
(222) W. W. Zajac, Jr., and K. J. Byrne, *J. Org. Chem.*, 37 (1972) 521–524.

thus obtained was altered from ~9:1 to ~4:1 when methyl groups replaced the benzyl groups on O-2 and O-3, reflecting an increase of the accessibility of O-4 to the mixed hydride.

This method of protection of O-4 with a benzyl group was applied to certain disaccharides: the 4′,6′-O-benzylidene derivatives of benzyl β-cellobioside, -maltoside, and -allolactoside were regiospecifically cleaved.[223] In the case of benzyl penta-O-benzyl-4′,6′-O-benzylidene-β-lactoside, hydrogenolysis of which gave an ~5:1 ratio between the 6- and 4-O-benzyl isomers, a possible role of the substituent on C-1 of the D-galactosyl group in directing the acetal ring-cleavage was postulated. Hydrogenolysis of benzyl 3-O-benzyl-4,6-O-benzylidene-2-O-(2,3,4,6-tetra-O-benzyl-β-D-galactopyranosyl)-β-D-galactopyranoside was also shown[224] to give a 79% yield of the corresponding 3,4-di-O-benzyl derivative (6-OH and 4-OBzl for the benzyl D-galactoside moiety).

The regioselectivity of the cleavage of methyl 4,6-O-benzylidene-α-D-glucopyranoside (**20a**), to give the 4- and 6-benzyl ethers (**194** and **195**, respectively) in the ratio of 3:2 (Ref. 214) has been confirmed[225]. Moreover, the regiospecificity of the hydrogenolysis of methyl 4,6-O-(4-methoxybenzylidene)-2,3-di-O-methyl-α-D-glucopyranoside (affording the 4-aryloxy derivative in a yield of 92%, whereas the 6-ether was not detected) was explained[225] by the presence of a methoxyl group that, it was supposed, completely hindered the access of the reducing agent to the electron pair at O-4. This regiospecificity had previously been reported[214] for the α-glucoside **192** (see earlier) and was later observed[220] for some 3-O-benzyl- and 2,3-di-O-benzyl-4,6-O-benzylidenehexopyranosides.

The usefulness in synthetic work of the hydrogenolysis of 4,6-O-benzylidenehexopyranosides was further exemplified by the reductive cleavage of derivatives protected at O-2 and O-3 by allyl groups (known[226] to be readily removable). Thus, benzyl 4-O-benzyl-β-D-glucopyranoside was obtained[227] by reduction of benzyl 2,3-di-O-allyl-4-O-benzyl-β-D-glucopyranoside, followed by removal of the allyl groups on O-2 and O-3 (yield, 44% for the three steps from benzyl 4,6-O-benzylidene-β-D-glucopyranoside). Likewise, benzyl 4-O-benzyl-α-D-mannopyranoside (yield 41%) and benzyl 4-O-benzyl-β-D-galac-

(223) A. Lipták, I. Jodál, and P. Nánási, *Carbohydr. Res.*, 52 (1976) 17–22.
(224) A. Lipták and P. Nánási, *Tetrahedron Lett.*, (1977) 921–924.
(225) D. Joniak, B. Košiková, and L. Kosáková, *Collect. Czech. Chem. Commun.*, 43 (1978) 769–773.
(226) R. Gigg and C. D. Warren, *J. Chem. Soc., C*, (1968) 1903–1911.
(227) A. Lipták, P. Fügedi, and P. Nánási, *Carbohydr. Res.*, 68 (1979) 151–154.

topyranoside (yield, only ~15%, because reductive cleavage of the cis-fused, acetal ring was not regiospecific, and also gave an ~20% yield of the 6-O-benzyl regio-isomer) were prepared.[227]

Interesting progress in the field was made after the observation of a strong regioselectivity (and even regiospecificity) of the hydrogenolysis of *exo*- and *endo*-3,4-O-benzylidene acetals. For 3,4-O-benzylidene derivatives in the D-arabino-, D-galacto-, and D-fuco-pyranoside series, it was found[228,229] that the hydrogenolysis is highly regioselective, and that this selectivity (compounds **220**, **221**, and **226**), or even this specificity (compounds **226** and **227**), depends on the stereochemistry of the starting material: for the *exo* isomers, the reducing agent attacks the *axial* oxygen atom (O-4), finally affording an *axial*-4-hydroxyl–*equatorial*-3-O-benzyl derivative. The situation was exactly

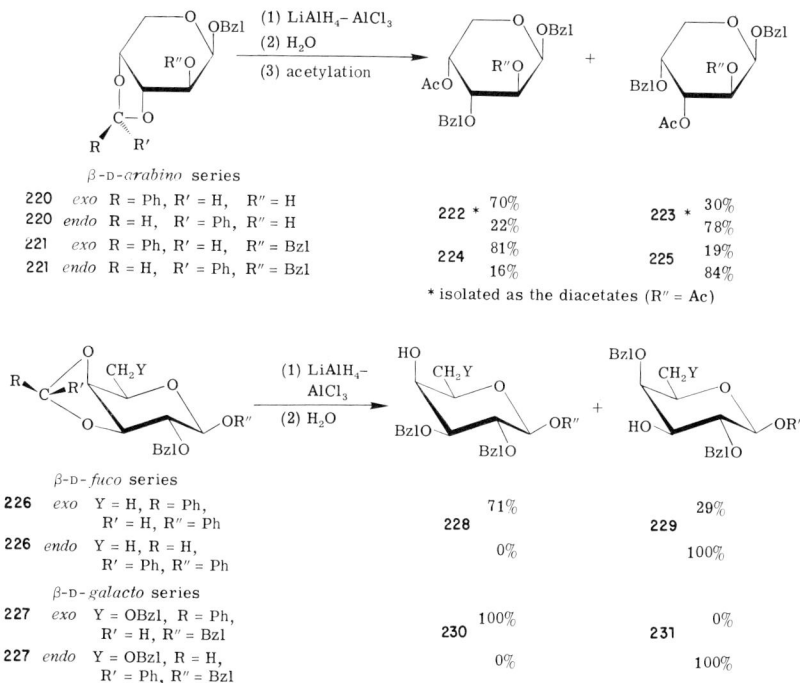

reversed for O-benzyl derivatives obtained from *endo*-benzylidene acetals. Similar results were independently obtained[230] for the hydro-

(228) A. Lipták, *Tetrahedron Lett.*, (1976) 3551–3554.
(229) A. Lipták, *Carbohydr. Res.*, 63 (1978) 69–75.
(230) P. Rollin and P. Sinaÿ, *C. R. Acad. Sci. Ser. C*, 284 (1977) 65–68.

genolysis of benzyl 6-O-allyl-2-O-benzyl-*exo*-3,4-O-benzylidene-α-D-galactopyranoside (**232**) and its *endo* isomer: pure 4-O-benzyl (**233**) and pure 3-O-benzyl (**234**) regio-isomers were successfully obtained (separately) in good yield (85 and 87%, respectively) from *exo*- and from *endo*-O-benzylidene acetals.

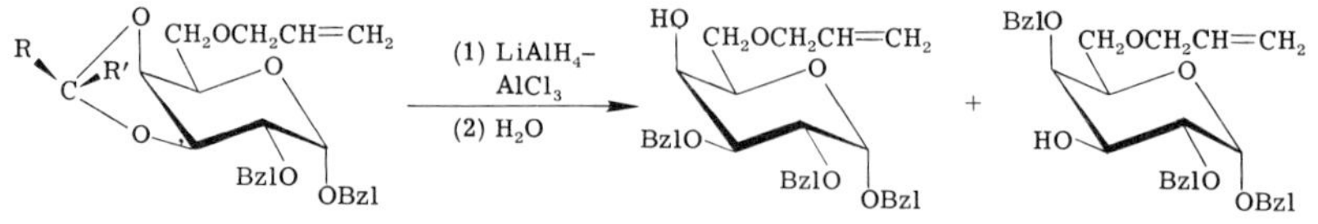

α-D-*galacto* series
232 *exo* R = Ph, R' = H
232 *endo* R = H, R' = Ph

233 100%
     0%

234 0%
     100%

The direction of the cleavage of 2,3-benzylidene acetals of hexopyranosides was examined first through examples chosen from benzyl L-rhamnopyranoside series.[231,232] For compounds **235** and **236**, it clearly appeared that the regioselectivity of the hydrogenolytic ring-cleavage was determined by the configuration of the acetal carbon atom, and followed exactly the same rule formulated for 3,4-O-benzylidene derivatives (see earlier), giving **238** and **239**. The experiment

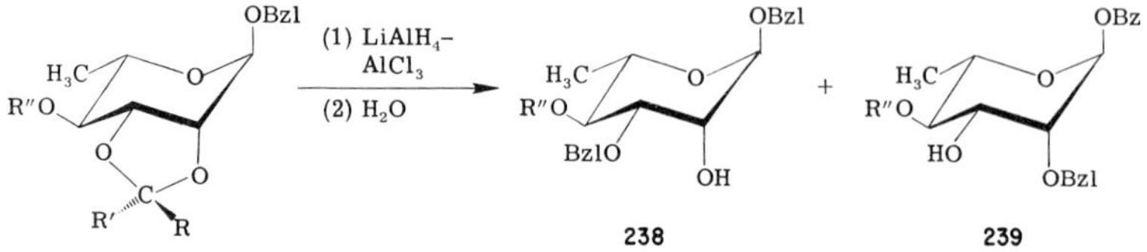

		238	**239**
235 *exo*	R = Ph, R' = H, R" = H	98%	2%
236 *exo*	R" = Bzl	94%	6%
237 *exo*	R" = CH$_2$CH=CH$_2$	85%	15%
235 *endo*	R = H, R = Ph, R" = H	2%	98%
236 *endo*	R" = Bzl	18%	82%

previously described[214] for compound **108** was repeated[231,232] on the benzyl analog of known configuration at the acetal carbon atom: benzyl *exo*-2,3:4,6-di-O-benzylidene-α-D-mannopyranoside gave (yield 76%) benzyl 3-O-benzyl-4,6-O-benzylidene-α-D-mannopyranoside (a benzyl analog of compound **202**) with only traces of the isomeric, 2-O-benzyl derivative. The highly selective cleavage of a 1,3-dioxolane (at C-2,3) compared to a 1,3-dioxane (at C-4,6) is potentially useful in synthesis, and the work of Bhattacharjee and Gorin[214] was therefore con-

(231) A. Lipták, P. Fügedi, and P. Nánási, *Carbohydr. Res.*, 51 (1976) c19–c21.
(232) A. Lipták, P. Fügedi, and P. Nánási, *Carbohydr. Res.*, 65 (1978) 209–217.

firmed. Finally, methyl *exo*- and *endo*-2,3:4,6-di-*O*-benzylidene-α-D-mannopyranoside (**108**) were separately hydrogenolyzed, and respectively gave,[233] as expected, methyl 3-*O*-benzyl-4,6-*O*-benzylidene-α-D-mannopyranoside (**202**) (yield 67%; 96:4 ratio of **202** to **203**) and its regioisomer, the 2-*O*-benzyl derivative **203** (yield 46%; 22:79 ratio of **202** to **203**). The usefulness of the hydrogenolysis of benzylidene acetals for the selective deprotection of sugars involved in glycosidic synthesis has been shown for a disaccharide, namely, benzyl *exo*-2,3-*O*-benzylidene-4-*O*-(2,3,4,6-tetra-*O*-benzyl-α-D-galactopyranosyl)-α-L-rhamnopyranoside.[233a] Selective ring-opening gave (yield 64%) the corresponding free alcohol group on C-2 of the rhamnoside residue (*axial* OH-2, *equatorial* OBzl-3).

Although not directly related to this article, a brief summary of the hydrogenolysis of bicyclic acetals, a synthetically useful reaction, is given.

Hydrogenolysis of the acetal group of anhydro sugars has not been extensively studied. Cleavage of the C-1–O-6 bond was observed during the reaction of 1,6-anhydro-β-D-glucopyranose (levoglucosan) with hydrogen over copper chromite.[234] Ring cleavage of model compounds for the 1,6-anhydropyranose series gave confirmation of this mode of scission. In the 6,8-dioxabicyclo[3.2.1]octane series, it was found for the unsubstituted bicyclic[235] (**240**), as well as for the dimethyl derivative[236] (**241**) and several alkyl-substituted derivatives,[237] that no product corresponding to the rupture of the bond in the six-membered ring (C-1–O-5 bond in a 1,6-anhydropyranose) could be detected; instead, **242** and **243** were respectively formed. This unique mode of

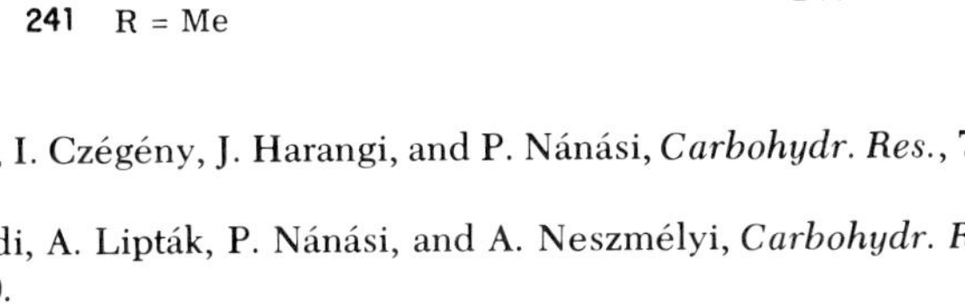

240 R = H
241 R = Me

(1) LiAlH$_4$– AlCl$_3$
(2) H$_2$O

242
243

(233) A. Lipták, I. Czégény, J. Harangi, and P. Nánási, *Carbohydr. Res.*, 73 (1979) 327–331.
(233a) P. Fügedi, A. Lipták, P. Nánási, and A. Neszmélyi, *Carbohydr. Res.*, 80 (1980) 233–239.
(234) P. A. J. Gorin, *J. Org. Chem.*, 24 (1959) 49–53.
(235) P. Clasper and R. K. Brown, *J. Org. Chem.*, 37 (1972) 3346–3347.
(236) Y. Naya and M. Kotake, *Tetrahedron Lett.*, (1967) 2459–2460.
(237) J. Colonge, J. Buendia, and H. Guignard, *Bull. Soc. Chim. Fr.*, (1969) 956–962.

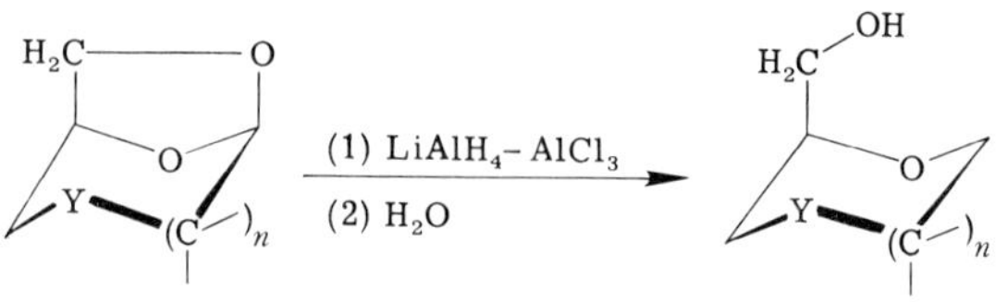

Model **A** Y = O, n = 1
      B Y = NR, n = 1
      C Y = S, n = 1
      D Y = O, n = 2

Scheme 6

cleavage was also observed in homologous series, where a carbon atom had been replaced by an oxygen, a sulfur, or a nitrogen atom. Thus, racemic, alkyl-substituted 3,6,8-trioxabicyclo[3.2.1]octanes (see Scheme 6, model **A**, alkyl-substituted) were hydrogenolyzed as a route to functionalized 1,4-dioxanes.[238] Similarly, sulfur[239] and nitrogen[240] analogs (Scheme 6, model **B** and model **C**, alkyl-substituted) could be cleaved, to give functionalized, 1,4-oxathianes and tetrahydro-1,4-oxazines. By analogy, seven-membered-ring compounds (1,4-dioxepanes) were obtained[239] from 3,7,9-trioxabicyclo[4.2.1]nonanes (Scheme 6, model **D**, alkyl-substituted), or from isomeric trioxabicyclononanes. In these series, epimerization of the primary product of the reductive cleavage was observed, and this was probably due to the sensitivity of the corresponding heterocycles to Lewis acids. Regarding this aspect of the reaction, it is interesting that no noticeable equilibration was observed when diisobutylaluminum hydride was used instead of the mixed hydride.[240] The regiospecificity of the bond cleavage in these series is probably due to thermodynamic factors, and may be explained by a release of the ring strain following cleavage of the five-membered ring, more than to a (dubious) greater accessibility of one of the oxygen atoms of the acetal group to the mixed hydride. Attention is drawn to the excessive generalization of this concept, as both of the bond cleavages possible have been observed when different atoms are involved. Thus, catalytic hydrogenation of aza analogs of 1,6-anhydropyranoses (compounds **244** and **246**) gave,[241] respectively, the six- and the seven-membered ring-systems (compounds **245** and **247**).

(238) J. Gelas and S. Veyssières-Rambaud, *Carbohydr. Res.*, 37 (1974) 293–301.
(239) J. Gelas and S. Veyssières-Rambaud, *Carbohydr. Res.*, 37 (1974) 303–308.
(240) P. Calinaud and J. Gelas, to be published; P. Calinaud, Thèse d'Etat, Clermont-Ferrand, France, Sept, 1977.
(241) H. Paulsen and K. Todt, *Chem. Ber.*, 100 (1967) 512–520.

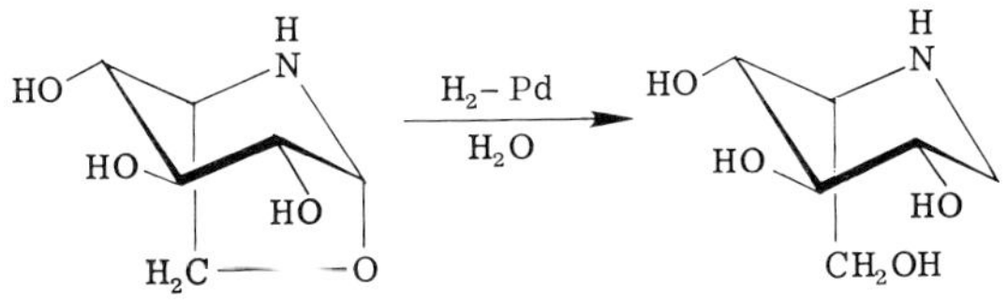

[Structures 244, 245, 246, 247 shown with H₂-Pd/H₂O reductions]

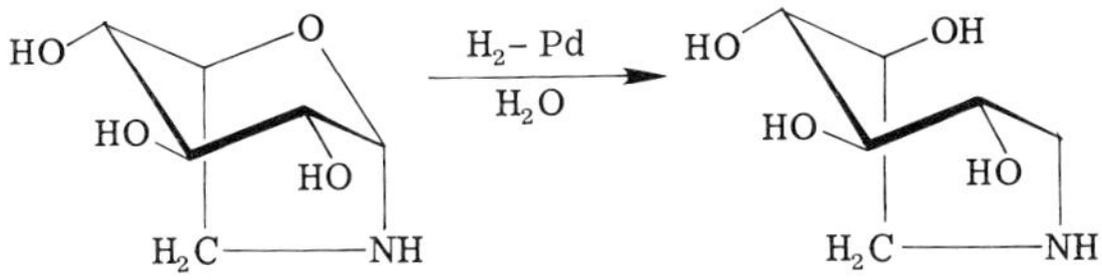

c. Some Other Reducing Systems for the Hydrogenolysis of Acetals.—Nearly all of the reactions thus far reported in this article were performed with lithium aluminum hydride–aluminum chloride as the mixed hydride (the Brown–Eliel method), but it may be useful to summarize, briefly, other reductive methods that are effective for acetals, although only a few of them have as yet been tested in the carbohydrate series. First, it is not necessary to insist on the catalytic hydrogenolysis of O-benzylidene groups as a routine deprotection method. Some Lewis acids can be as effective as aluminum chloride, and have been employed in the reduction of molecules that could serve as models for carbohydrates.The reaction of lithium aluminum hydride on an ethereal solution of the substrate saturated with hydrogen chloride (or bromide) gas has already been mentioned, as well as hydrogenation under pressure at 200° over copper chromite[234]: 1,2-O-isopropylidene-α-D-glucofuranose led[242] to tetra- and tri-deoxyhexitols, with inversion of the configuration of C-5, whereas 1,2:5,6-di-O-isopropylidene-α-D-glucofuranose gave[234] the 6-isopropyl ethers of 1,2-O-isopropylidene-α-D-gluco- and -L-ido-furanose. Use of boron trifluoride[198,243] alone, or with hydridoboron fluoride[198] or titanium chloride[244] added to lithium aluminum hydride, has been reported. Other methods have been proposed in which lithium aluminum hydride is

---

(242) P. A. J. Gorin and A. S. Perlin, *Can. J. Chem.*, 36 (1958) 661–666.
(243) A. R. Abdun-Nur and C. H. Issidorides, *J. Org. Chem.*, 27 (1962) 67–70.
(244) H. Hishikawa and T. Mukaiyama, *Bull. Chem. Soc. Jpn.*, 51 (1978) 2059–2063.

replaced by another reducing agent: for example, catalytic hydrogenation in the presence of hydrogen chloride,[245] lithium in oxolane,[246] lithium hydride and aluminum chloride in the presence of a catalytic amount of lithium aluminum hydride,[196] triethylsilane with zinc chloride,[247,248] sodium borohydride with either boron trifluoride or aluminum chloride,[249,250] decaborane,[251] diisobutylaluminum hydride,[240,251] borane in oxolane,[252] alane (AlH_3),[208] and alkoxyalanes and alkoxychloroalanes.[253] A particular kind of reaction has been described, namely, the reaction of sodium cyanoborohydride, methanol, and gaseous hydrogen chloride with 2-methyl-2-phenyl-1,3-dioxolane, to give 1-methoxy-1-phenylethane.[254]

d. Hydrogenolysis of Unsaturated Acetals.—A somewhat particular situation to be discussed concerns the case of the presence of an alkenyl group attached to C-2 of 1,3-dioxolanes or 1,3-dioxanes. Because the ease of hydrogenolysis of the 1,3-dioxolane ring is greatly increased by electron-donating substituents at C-2 (see earlier), it was assumed[208] that 2-alkylvinyl-1,3-dioxolanes would probably be cleaved without addition of a Lewis acid to the reducing agent. It has thus been observed[208] that, although 2-vinyl-1,3-dioxolane (**248**) is reduced to the expected ether (**249**) by reaction with the mixed-hydride reagent (yield 80%), on the other hand, lithium aluminum hydride alone in diethyl ether at room temperature (or in 1,2-dimethoxyethane at 80°, or dipropyl ether at 90°) was able to cleave **248** to the α-propenyl-β-hydroxyalkyl ether **250** (yield 80%). Similar results were ob-

$$H_2C=CH-CH\begin{matrix}O-CH_2\\|\\O-CH_2\end{matrix} \xrightarrow{\text{(a) or (b)}} H_2C=CH-H_2C\begin{matrix}O-CH_2\\|\\HOCH_2\end{matrix} + H_3C-CH=CH\begin{matrix}O-CH_2\\|\\HOCH_2\end{matrix}$$

248   **249**   **250**

(a): (1) $LiAlH_4-AlCl_3$
     (2) H_2O 100%   0%

(b) (1) $LiAlH_4$,
    $MeOCH_2CH_2OMe$, 80°
    (2) H_2O 0   100%

(245) W. L. Howard and J. H. Brown, Jr., *J. Org. Chem.*, 26 (1961) 1026–1028.
(246) E. M. Kaiser, C. G. Edmonds, S. D. Grubb, J. W. Smith, and D. Tramp, *J. Org. Chem.*, 36 (1971) 330–335.
(247) E. Frainnet, R. Calas, and A. Bazouin, *Bull. Soc. Chim. Fr.*, (1960) 1480.
(248) E. Frainnet and C. Esclamadou, *C. R. Acad. Sci. Ser. C*, 254 (1962) 1814–1816.
(249) N. Janaki, K. D. Pathak, and B. C. Subba Rao, *Curr. Sci.*, 32 (1963) 404.
(250) N. Janaki, K. D. Pathak, and B. C. Subba Rao, *Indian J. Chem.*, 3 (1965) 123–125.
(251) L. I. Zakharkin, V. I. Stanko, and Yu. A. Charpovskii, *Izv. Akad. Nauk SSSR, Otd. Khim. Nauk*, (1962) 1118–1119.
(252) B. Fleming and H. I. Bolker, *Can. J. Chem.*, 52 (1974) 888–893.
(253) W. W. Zajac, Jr., and K. J. Byrne, *J. Org. Chem.*, 38 (1973) 384–387.
(254) D. A. Horne and A. Jordan, *Tetrahedron Lett.*, (1978) 1357–1358.

tained with various other 2-alkylvinyl- and 2-arylvinyl-1,3-dioxolanes. Remarkable differences were observed in the reactivity of the substrate, depending not only on the reducing agent but also on the solvent,[255] and a rationale was suggested to account for all of these results.

This reaction is of especial interest to carbohydrate chemists, as considerable interest in O-allyl and O-propenyl groups has been evinced as a protection of hydroxyl groups (see, for instance, Ref. 258, and references cited therein). In a preliminary study, the utility of the hydrogenolysis of unsaturated acetals was investigated by Gigg.[259] The reaction of the acrolein acetal **251** with the mixed hydride was shown to give, in a yield of 20%, the 4-allyl (**252**) and 4-(1-propenyl) (**253**) ethers with high regioselectivity (presence of only traces of the 6-alkenyl ethers) in the ratio of 3:2. Formation of a high yield of compound **253** (from which the temporary protecting group at O-4 could

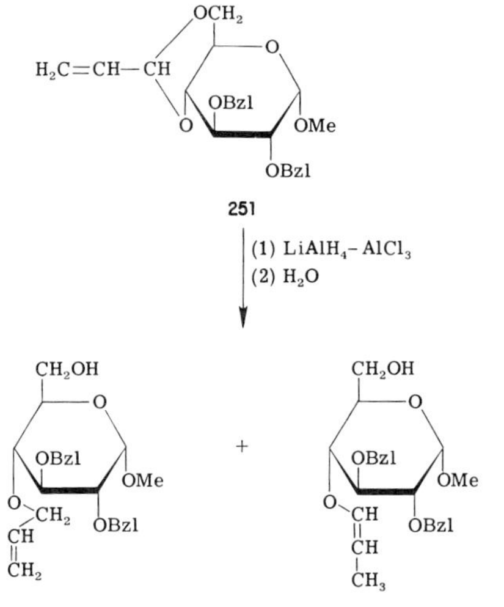

(255) It may be noted that most of the reactions conducted with mixed hydrides employed diethyl ether as the solvent. Some other solvents may be not suitable for reductive cleavages, as they can themselves be hydrogenolyzed by the reducing agent. For instance, oxolane is cleaved both by lithium aluminum hydride–aluminum chloride (Ref. 256), and by tri-*tert*-butoxyaluminohydride in the presence of triethylborane (Ref. 257).
(256) W. J. Bailey and F. Marktscheffel, *J. Org. Chem.*, 25 (1960) 1797–1800.
(257) H. C. Brown, S. Krishnamurthy, and R. A. Coleman, *J. Am. Chem. Soc.*, 94 (1972) 1750–1751.
(258) R. Gigg, *A.C.S. Symp. Ser.*, 39 (1977) 253–278.
(259) R. Gigg, personal communication; see, R. Gigg, *A.C.S. Symp. Ser.*, 77 (1978) 44–66.

be readily removed) would reasonably be expected on using the conditions described for the formation of compound **250**.

5. Action of Strong Bases

a. Action of Butyllithium.—The reaction of 2-aryl-1,3-dioxolanes with alkyllithium reagents to give alkylaryl ketones and ethylene (see Scheme 7) was discovered for noncarbohydrates.[260–263] Phenyllithium

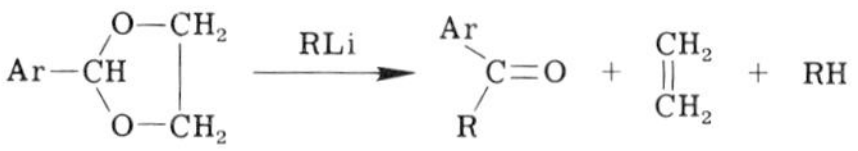

$$Ar-CH \underset{O-CH_2}{\overset{O-CH_2}{\diagup}} \xrightarrow{RLi} \underset{R}{\overset{Ar}{\diagup}}C=O + \underset{CH_2}{\overset{CH_2}{\|}} + RH$$

Scheme 7

was first reported to react with 2-phenyl-1,3-dioxolane,[260,262] but evidence for its potential usefulness in synthesis was given later,[262] and a possible mechanism for the ring opening was suggested. As 1,3-dioxolane is commonly used to protect ketones during reactions with organometallic reagents, attention was directed[264] to a limitation on the protection by acetals when highly reactive, organolithium reagents are used. It was found that 1,1-ethylenedioxacyclopentane (**254**) or 1,1-ethylenedioxacyclohexane (**255**) react with three equivalents of isopropyllithium or *tert*-butyllithium in pentane during 17 h at room temperature to give, respectively, the alcohols **256** and **257**. A mechanism was suggested that involved concerted fragmentation of the ace-

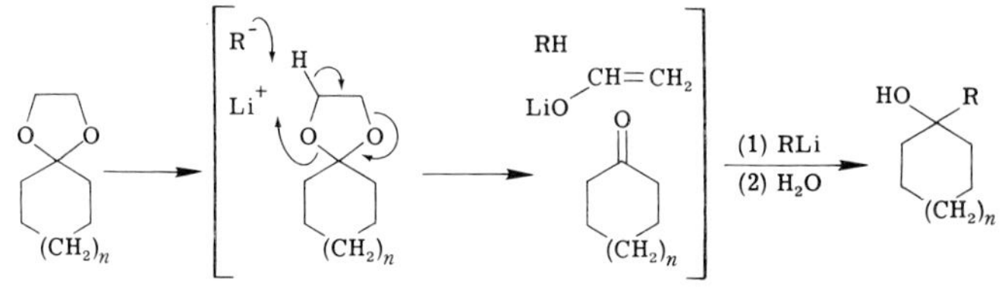

254 $n = 0$, R = Me$_2$CH
255 $n = 1$, R = Me$_3$C

256
257

tal, leading to the lithium enolate of acetaldehyde. It was also observed that 1,1-propylenedioxacyclohexane (the 1,3-dioxane homolog of compound **254**) does not react with *tert*-butyllithium under the same conditions.

(260) L. J. Nehmsmann, *Diss. Abstr.*, 23 (1962) 1929.
(261) P. S. Wharton, G. A. Hiegel, and S. Ramaswami, *J. Org. Chem.*, 29 (1964) 2441–2442.
(262) K. D. Berlin, B. S. Rathore, and M. Peterson, *J. Org. Chem.*, 30 (1965) 226–228.
(263) T. L. V. Ulbricht, *J. Chem. Soc.*, (1965) 6649–6650.
(264) C. H. Heathcock, J. E. Ellis, and R. A. Badger, *J. Heterocycl. Chem.*, 6 (1969) 139–140.

The scope of the mechanism of this type of reaction was later given by Whitman and coworkers,[265] who studied the action of butyllithium notably on benzylidene derivatives of *trans*-1,2-cyclooctanediol or hydrobenzoin, and on 2-phenyl-1,3-dioxane or 2-phenyl-1,3-dioxolane. It was found that (*1*) the anion which derives from benzaldehyde diethyl acetal gives mainly an internal *E2* elimination; (*2*) the anion derived from 2-phenyl-1,3-dioxane gives a Wittig-like rearrangement; (*3*) the anion from 1,3-dioxolanes gives a cyclo-elimination, affording either an alkene and the benzoate ion (proton abstraction from C-2) or benzaldehyde and an enolate ion (proton abstraction from C-4 or C-5). For interpretation of the following reactions reported on carbohydrates, it should be noted that the eventual, favored removal of a proton from C-4 or C-5 of a 1,3-dioxolane may be due to the higher stability of the carbanion thus formed, compared to the concurrent carbanium adjacent to two oxygen atoms which would be produced by removal of the proton on C-2 (see Ref. 265, and references cited therein). Formation of *trans*-cyclooctene derivatives from 1,3-dioxolanes was further discussed.[266,267] Other examples of this type of 1,3-anionic cycloreversion, in, for instance, the 1,3-oxathiolane series were reported (see Ref. 268, and references cited therein).

A very useful, synthetic application of the reaction of cyclic acetals with butyllithium was first described in carbohydrate chemistry by Klemer and Rodemeyer[269]: methyl 2,3:4,6-di-*O*-benzylidene-α-D-mannopyranoside (**108**) reacted with two molar equivalents of butyllithium in oxolane at $-30°$, to give a 70% yield of the hexos-3-ulose **258**, the side-products being a trace of the 2,3-unsaturated sugar **259** and a 1-phenyl-1-pentanol (**260**). An interpretation of this reaction was given (see Scheme 8) by assuming that the strong base removed a proton from one of the two possible positions α to an oxygen atom of the acetal. The regioselectivity depended on which proton was easier to be abstracted. The ring fragmentation then occurred, to give the hexosulose derivative [see Scheme 8, pathway (*a*)]. A concurrent route was the abstraction of the benzylic proton from the acetal group, giving an anion that led to the unsaturated sugar [pathway (*b*)]. Some difficulties were experienced in increasing the scale of this reaction, and because

---

(265) J. N. Hines, M. J. Peagram, G. H. Whitham, and M. Wright, *Chem. Commun.*, (1968) 1593–1594.
(266) G. H. Whitham and M. Wright, *J. Chem. Soc., C*, (1971) 886–891.
(267) J. N. Hines, M. J. Peagram, E. J. Thomas, and G. H. Whitham, *J. Chem. Soc. Perkin Trans. 1*, (1973) 2332–2337.
(268) G. Bianchi, C. De Micheli, and R. Gandolfi, *Angew. Chem. Int. Ed. Engl.*, 18 (1979) 721–738.
(269) A. Klemer and G. Rodemeyer, *Chem. Ber.*, 107 (1974) 2612–2614.

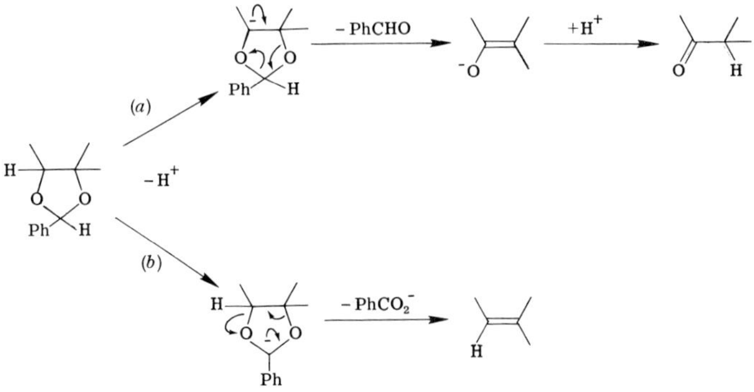

Scheme 8

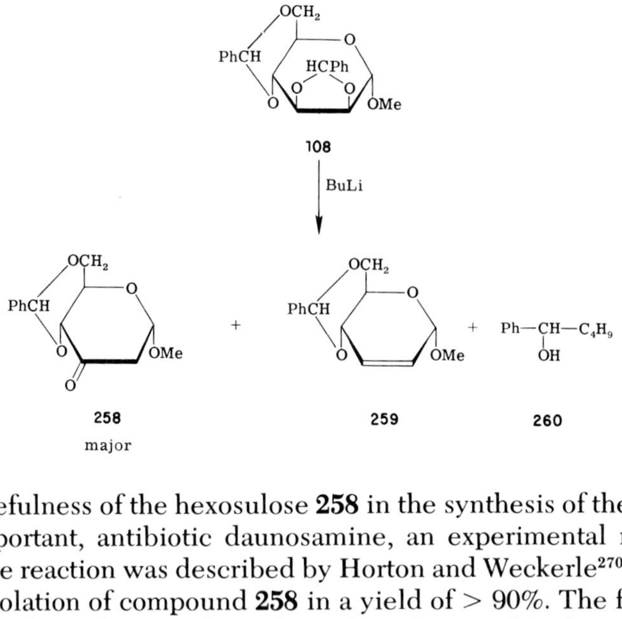

of the usefulness of the hexosulose **258** in the synthesis of the biologically important, antibiotic daunosamine, an experimental modification of the reaction was described by Horton and Weckerle[270] that permitted isolation of compound **258** in a yield of > 90%. The formation of side products due mainly to chain degradation of the hexosulose to the derivative **261** (in the presence of a strong base during processing) was thus avoided.

The reaction of butyllithium with methyl 4,6-O-benzylidene-2,3-di-O-methyl-α-D-hexopyranosides was investigated.[271] A rationale was proposed, to account for the different rates of elimination of methanol (leading to the 2-O-methyl-2,3-unsaturated derivative) depending on

(270) D. Horton and W. Weckerle, *Carbohydr. Res.*, 44 (1975) 227–240.
(271) A. Klemer and G. Rodemeyer, *Chem. Ber.*, 108 (1975) 1896–1901.

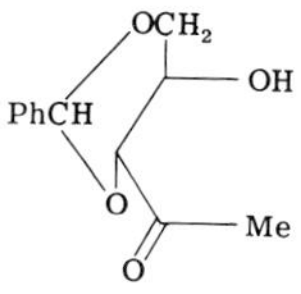

261

the stereochemistry of the 2- and the 3-O-methyl groups (compounds having α-D-*gluco*, α-D-*altro*, and α-D-*allo* configurations reacted with respective yields of 30, 45, and 55%, but the α-D-*manno* derivative did not react).

The reaction involving 2,3-O-benzylidene sugars was then extended to a study of the behavior of 1,2-O-isopropylidene derivatives.[272] Furanoid derivatives **262** and **263** gave, respectively, (+)-(S)-(E)-5-butyl-1-methoxy-3-nonen-2,5-diol (**264**) (yield 23%) and (+)-2(R),3(S)-(E)-6-butyl-1,2-dimethoxy-4-decen-3,6-diol (**265**) (yield 21%). On the other hand, acetal derivatives of the D-*arabino* (**266**) and

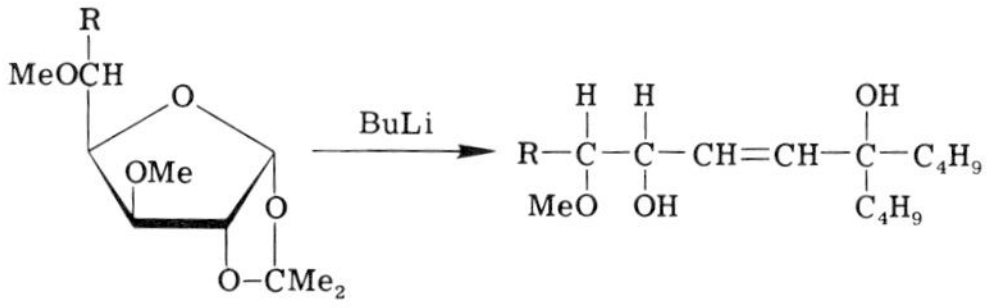

262 R = H
263 R = CH$_2$OMe

264 (E)
265 (E)

the D-*galacto* (**268**) configurations led, respectively, to the acyclic, ethylenic alcohols **267** [(E) and (Z) stereoisomers isolated in ~2 and 16% yields] and **269** [(E) and (Z) stereoisomers isolated in ~3 and 15%

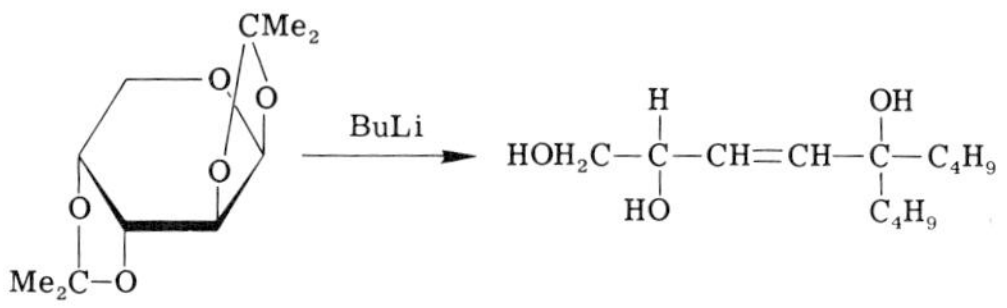

266

267 (E) + (Z)

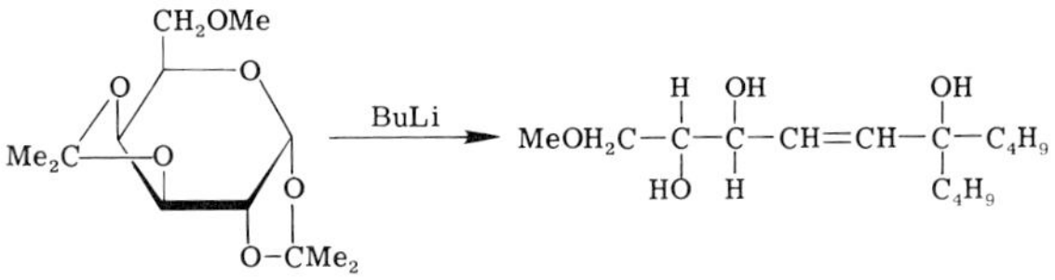

268

269 (E) + (Z)

yields]. An interpretation of the fragmentation was proposed and discussed.[272]

Further observations were published concerning 3,4- and 4,5-O-isopropylidene derivatives of D-arabinose and D-fructose.[273] Thus, 1,2:3,4-di-O-isopropylidene-β-D-arabinose (266) gave a mixture of (−)-(S)-(Z)-5-butyl-3-nonen-1,2,5-triol (267), its (E) stereoisomer, and 4-deoxy-1,2-O-isopropylidene-β-D-*threo*-pent-4-*exo*-enopyranose, isolated as pure compounds in rather low yields (respectively, 9, 1, and 16%). On the other hand, methyl 3,4-O-isopropylidene-2-O-methyl-β-D-arabinopyranoside (270) gave methyl 2-O-methyl-β-D-*threo*-pent-4-

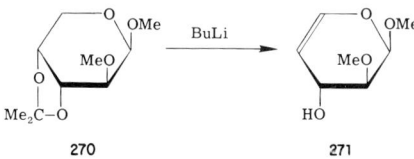

enopyranoside (271). Consideration of the configurational and conformational criteria for these starting materials, and for those reported in the D-fructose series, was given in order to account for these reactions.[273]

As part of a program designed to give new, practical routes to daunosamine,[270] and related sugars (D-ristosamine, for instance[274]), the reaction of butyllithium with methyl 2,3-O-benzylidene-α-L-rhamnopyranoside (121) was investigated.[275] When the reaction was conducted at −30°, starting from the 4-hydroxy derivative 121, a mixture of three products was obtained, and this was separated (low yields): two compounds were identified as 1,5-anhydro-3-C-butyl-1,2,6-trideoxy-L-*ribo*-hex-1-enitol and its L-*arabino* analog (compounds 274; no individual attribution was given for the derivatives isolated), and the third was methyl 2,3,6-trideoxy-α-L-*erythro*-hex-2-enopyranoside (275). When the reaction was repeated at room temperature with a large excess of butyllithium, only compounds 274a (yield 17%) and 274b (yield 44%) were isolated. The failure to obtain the expected 2-deoxy-3-ketone derivative from compound 121 was ascribed to the presence of the 4-hydroxyl group, which allowed the formation of the 4-oxy anion (see Scheme 9), leading to the enolate of the desired 2-

(272) G. Rodemeyer and A. Klemer, *Chem. Ber.*, 109 (1976) 1708–1723.
(273) A. Klemer, G. Rodemeyer, and F. J. Linnenbaum, *Chem. Ber.*, 109 (1976) 2849–2861.
(274) D. Horton and W. Weckerle, *Carbohydr. Res.*, 46 (1976) 227–235.
(275) D. M. Clode, D. Horton, and W. Weckerle, *Carbohydr. Res.*, 49 (1976) 305–314.

REACTIVITY OF CYCLIC ACETALS

Scheme 9

deoxy-3-ketone; subsequent elimination of methoxide ion from the anomeric center gave the 1-en-3-one, which suffered attack by butyllithium, finally giving the alcohols **274** (as the reaction involved abstraction of two hydrogen atoms, it is evident that the experimental conditions had to be more vigorous).[275] This explanation seems to be validated by the observation of the formation of the hexos-3-ulose **276**

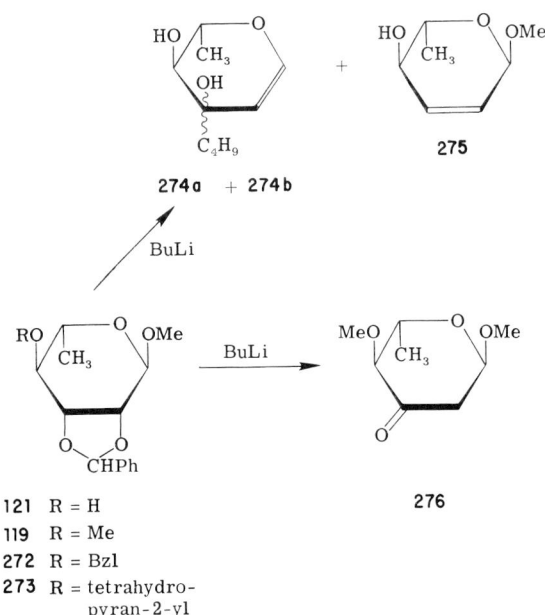

121 R = H
119 R = Me
272 R = Bzl
273 R = tetrahydro-
      pyran-2-yl

(yield 40%) from the reaction of the 4-O-methyl derivative **119**. Obviously, the 4-O-benzyl derivative **272** was not a satisfactory starting-material, as, due to concurrent attack on the benzylic methylene group by butyllithium, it gave a complex mixture.[275] Because an O-methyl group would be rather difficult to remove, it was found that the tetrahydropyran-2-yl group (as in compound **273**) is a satisfactory protecting-group for a hydroxyl group during reactions with butyllithium[276]; thus, the 4-O-(tetrahydropyran-2-yl) analog of compound **276** was obtained in a yield of 40%.

Information concerning the dependence of the regiospecificity of the reaction on the configuration of the starting material was provided by the reaction of methyl 1,3:4,6-di-O-benzylidene-β-D-allopyranoside **277** with butyllithium[277]; the major product (contaminated by only a trace of the unsaturated derivative **279**, the β anomer of compound **259**) was the 3-deoxyhexos-2-ulose **278** (β anomer, the regio-

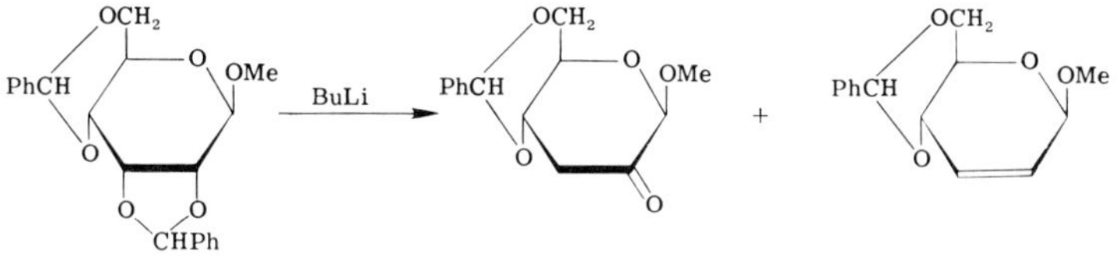

isomer of compound **258**), obtained in excellent yield (86%). It was suggested that the direction of the reaction was determined by the configuration of the carbon atom bearing the hydrogen atom that could be abstracted. The initial step (see Scheme 10) was the attack of the base of the *axial* hydrogen atom, leading finally to the product having the ketone group at this position.

Study of this reaction has not been limited to cyclic acetals of aldoses and aldosides, and synthetic applications in the ketose series have been developed,[273,276,278,279] particularly for isopropylidene acetals of D-fructose. As an illustration of these reactions, two examples are given here. The action of butyllithium on 2,3:4,5-di-O-isopropylidene-1-O-methyl-β-D-fructopyranose (**280**) gave[273] a 30% yield of the 5-enopyranose **282**. Abstraction by the base of the *axial* hydrogen atom from the 6-methylene group, giving the anion **281**, was invoked

(276) A. Klemer and D. Balkau, *J. Chem. Res.*, (1978) 3823–3830.
(277) D. Horton and W. Weckerle, *A.C.S. Symp. Ser.*, 39 (1977) 22–35.
(278) A. Klemer and D. Balkau, *Justus Liebigs Ann. Chem.*, (1977) 181–183.
(279) A. Klemer and D. Balkau, *Chem. Ber.*, 111 (1978) 1514–1520.

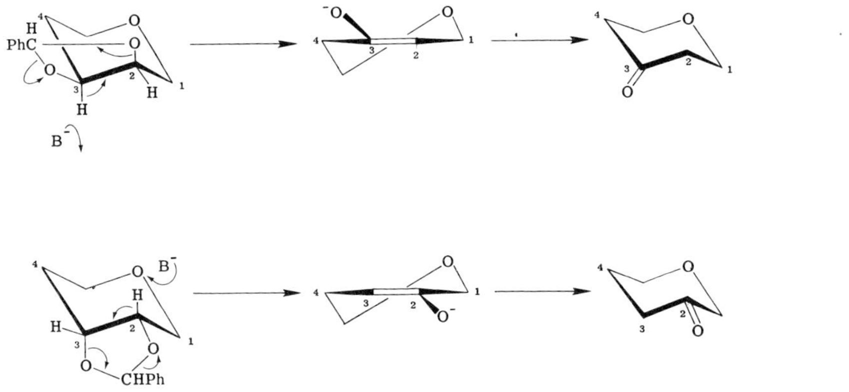

REACTIVITY OF CYCLIC ACETALS 145

Scheme 10

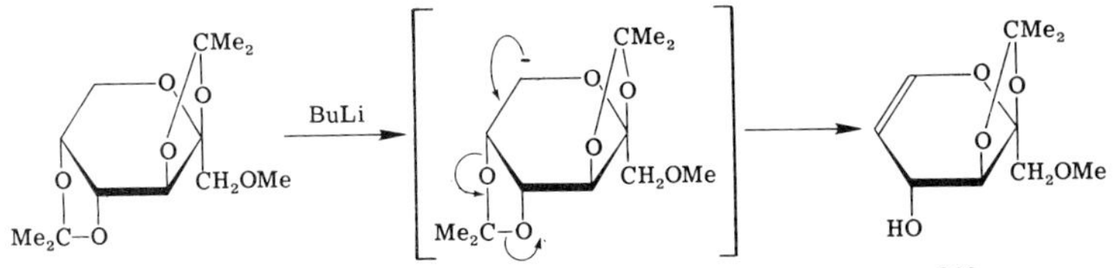

as an interpretation (for a discussion, see also, Ref. 278). On the other hand, when 1,2:4,5-di-*O*-isopropylidene-3-*O*-methyl-β-D-fructopyranose (**283**) reacted with butyllithium, 1,2:4,5-di-*O*-isopropylidene-3-*O*-methyl-D-*arabino*-hex-1-enitol (**285**) was obtained (yield 15%),[279] probably through the intermediacy of the anion **284**.

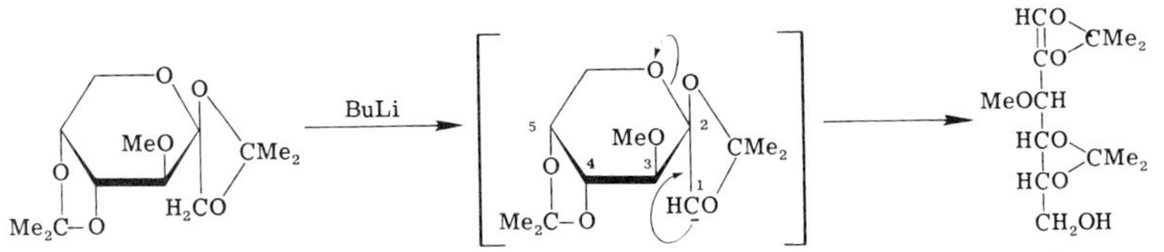

b. Action of Other Bases.— Strong bases other than butyllithium are able to induce similar reactions of cyclic acetals. For instance, Klemer and Balkau[276] tested the possibilities afforded by lithium diisopropylamide in diethyl ether at room temperature: thus, the 1-*O*-(tetrahydropyran-2-yl) analog of compound **282** (namely, 5-deoxy-2,3-*O*-isopropylidene-1-*O*-(tetrahydropyran-2-yl)-β-D-*threo*-hex-5-enulopyranose)

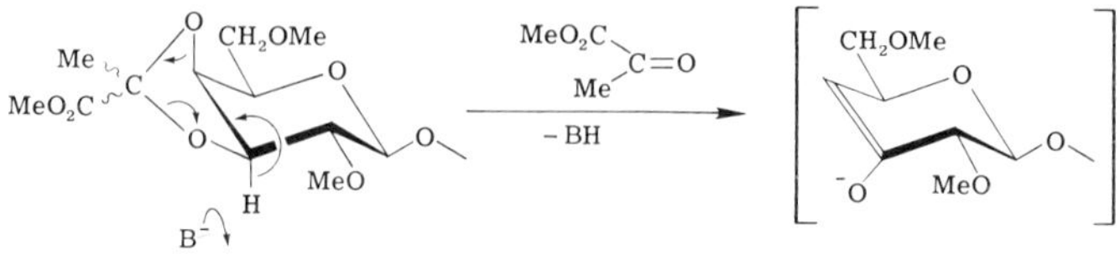

Scheme 11

was obtained (yield, 38%) from 2,3:4,5-di-O-isopropylidene-1-O-(tetrahydropyran-2-yl)-β-D-fructopyranose.[276]

Some examples of reactions having similarities to those involving attack on an O-isopropylidene sugar by butyllithium will now be briefly reported. Generally speaking, they are related to base degradation leading to the elimination of the acetal group. For instance, the 3,4-O-(1-carboxyethylidene)-D-galactopyranosyl group of the capsular polysaccharide from *Klebsiella* type 33 has been degraded with strong bases; a mechanism similar to those reported for the reaction of butyllithium, and involving an enolate (see Scheme 11), was proposed.[280]

A practical synthesis of ketene dithioacetals of sugars was discovered by Horton and Wander.[281,282] This reaction does not need to be emphasized here, as it has been discussed in this Series (see Ref. 7, and references cited therein). We only recall as an example that the

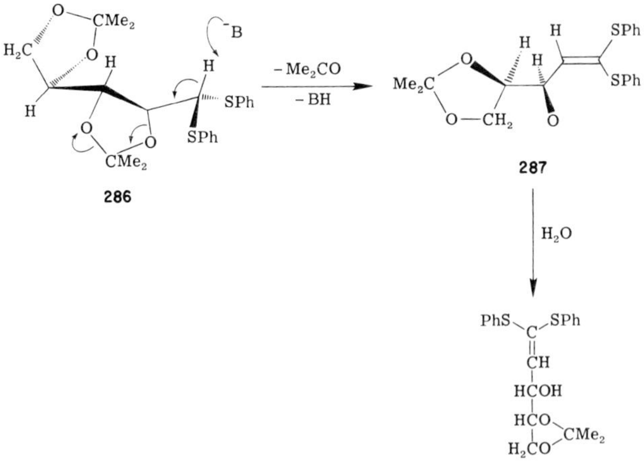

(280) B. Lindberg, F. Lindh, J. Lönngren, and W. Nimmich, *Carbohydr. Res.*, 70 (1979) 135–144.
(281) D. Horton and J. D. Wander, *Carbohydr. Res.*, 13 (1970) 33–47.
(282) B. Berrang, D. Horton, and J. D. Wander, *J. Org. Chem.*, 38 (1973) 187–192.

treatment of 2,3:4,5-di-O-isopropylidene-D-arabinose diphenyl dithioacetal (**286**) by sodium methylsulfinyl carbanion (other bases, such as potassium *tert*-butoxide, or butyllithium, were also found to be effective) gave 1,2-dideoxy-4,5-O-isopropylidene-D-*erythro*-pent-1-enitol (**288**; yield 68%), probably through the oxyanion **287** formed by an E-2-like process.[281] This synthetic method was later extended to diethyl dithioacetals, thereby providing a route to 2-deoxyaldoses.[283,284]

An interesting application of acetal elimination from sugars that is promoted by base is the formation of furanoid and pyranoid glycals from isopropylidene acetals of furanosyl and pyranosyl halides.[285–287] Treatment of 2,3:5,6-di-O-isopropylidene-α-D-mannofuranosyl bromide (**289**) with a base gave the furanoid glycal **290**. The yield was

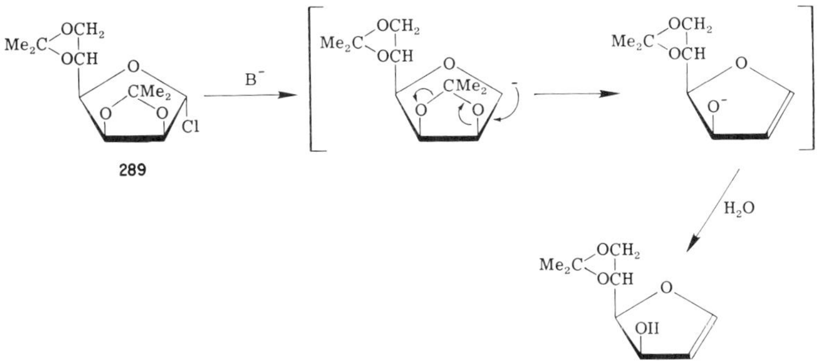

11% when the base was sodium in oxolane,[285] and 59% when sodium naphthalide was employed.[286] The same reaction was described starting from 2,3:5,6-di-O-isopropylidene-β-D-allofuranosyl bromide. A mechanism involving the formation of a carbanion species at the anomeric center, followed by elimination of acetone, was presumed.

Other examples of furanoid and pyranoid glycals were independently prepared[287] by using lithium in liquid ammonia, a procedure claimed to be more efficient and general. For instance, 2,3-O-isopropylidene-4-O-methyl-α-L-rhamnopyranosyl chloride (**291**) and 6-

(283) M. Y. H. Wong and G. R. Gray, *Tetrahedron Lett.*, (1977) 1617–1620.
(284) M. Y. H. Wong and G. R. Gray, *J. Am. Chem. Soc.*, 100 (1978) 3548–3553.
(285) S. J. Eitelman and A. Jordaan, *J. Chem. Soc. Chem. Commun.*, (1977) 552–553.
(286) S. J. Eitelman, R. H. Hall, and A. Jordaan, *J. Chem. Soc. Perkin Trans. 1*, (1978) 595–600.
(287) R. E. Ireland, C. S. Wilcox, and S. Thaisrivongs, *J. Org. Chem.*, 43 (1978) 786–787.

deoxy-2,3-O-isopropylidene-4-O-methyl-β-L-gulopyranosyl chloride (**293**) gave[287] the corresponding glycals, **292** and **294**, in yields of 90%.

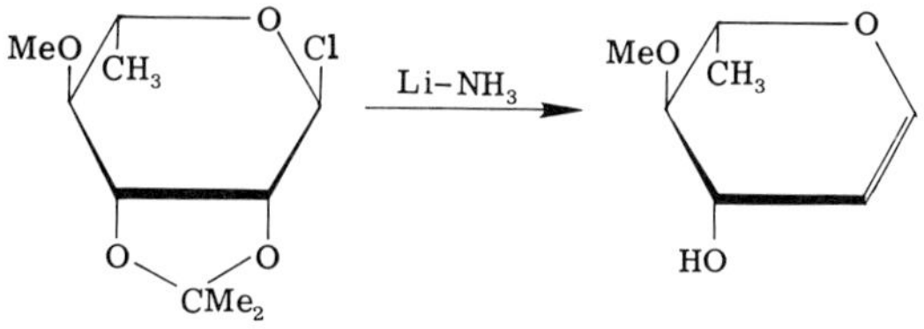

Finally, it may be concluded that a sugar bearing a cyclic acetal group may react in the presence of a strong base by generating an anion through at least three different ways, summarized in Scheme 12: (*a*) removal of the hydrogen atom, if any, from the acetalic carbon atom, giving the possibility of the final formation of an unsaturated derivative of type **A** (for instance, compound **259** or compound **279**); (*b*) removal of one hydrogen atom from one of the two carbon atoms in the α position to the oxygen atoms of the 1,3-dioxolane, giving the possibility of the final formation of a deoxyglycosulose derivative of type **B** or type **B′** (for instance, compound **258** or compound **278**); and (*c*) removal of the cyclic acetal group, promoted by the generation of a vicinal carbanion, giving the possibility of the final formation of glycals of type **C** or type **D** (for instance, compound **282** and compounds **290**, **292**, and **294**).

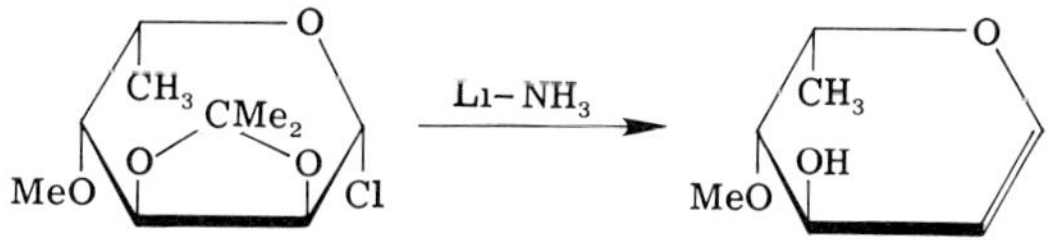

6. Miscellaneous Reactions

The reactivity of functional groups presenting some similarities to cyclic acetals, such as cyclic orthoesters, and stannylidene and amide acetals, falls outside the scope of the present article, but attention has to be drawn to the fact that they may be considered to be special types of "acetals" having distinct properties, and a growing interest is focused

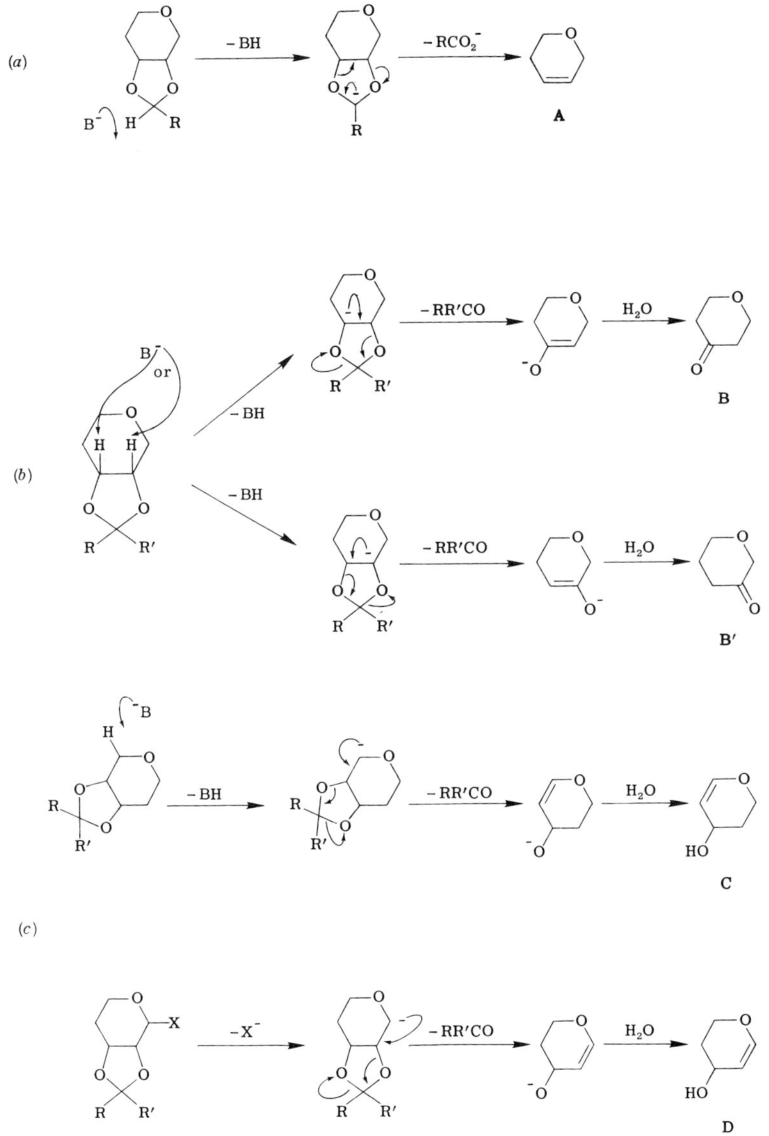

Scheme 12[288]

(288) For the sake of convenience, only pyranoid rings have been depicted, without indication of a particular configuration. Evidently, reactions starting from furanoid structures have been described,[286,287] or are potentially feasible (see for instance, p. 34 of Ref. 277).

on them. Notably, a review appeared concerning the chemistry of formamide acetals, but it was only partly devoted to carbohydrate derivatives.[289]

Reactions of cyclic acetals other than those already described in this Chapter are possible, and some of them have been described for non-carbohydrates. A few reactions of interest in the sugar field will now be briefly reported.

a. Formation of Esters.—(i) Peroxide-induced Rearrangement. The formation of esters by the photochemical rearrangement of cyclic acetals may be expected (see Section III, 2), but it is known that esters may also be obtained by decomposition of *tert*-butyl peroxide in acyclic acetals (1,1-dialkoxybutanes) in the liquid phase (at 120–140°).[290] This result was extended to the rearrangement of tetrahydro-2-methoxypyran to methyl pentanoate induced by alkoxyl radicals,[291] and to the formation of benzoic esters from cyclic acetals prepared from benzaldehyde and 1,2- or 1,3-diols.[292] A systematic study was made concerning various derivatives of 1,3-dioxanes.[293–296] To the best of our knowledge, only one report constituted a direct extension of this reaction to sugars. Thus, treatment of methyl 2,3-di-O-acetyl-4,6-O-ben-

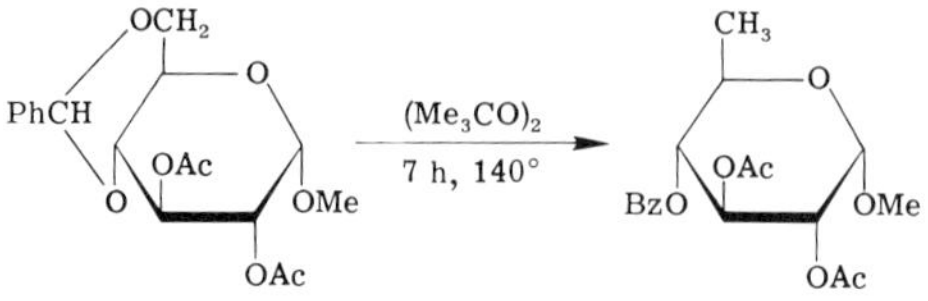

zylidene-α-D-glucopyranoside (**295**) with di-*tert*-butyl peroxide for 7 h at 140° gave[297] methyl 2,3-di-O-acetyl-4-O-benzoyl-6-deoxy-α-D-glucopyranoside (**296**) in a yield of 41%. The 2,3-di-O-dibenzoyl analog of

(289) R. F. Abdulla and R. S. Brinkmeyer, *Tetrahedron*, 35 (1979) 1679–1735.
(290) L. P. Kuhn and C. Wellman, *J. Org. Chem.*, 22 (1957) 774–776.
(291) E. S. Huyser, *J. Org. Chem.*, 25 (1960) 1820–1822.
(292) E. S. Huyser and Z. Garcia, *J. Org. Chem.*, 27 (1962) 2716–2719.
(293) D. L. Rakhmankulov, V. I. Isagulyants, and S. S. Zlotskii, *Neftekhimiya*, 13 (1973) 254–258.
(294) D. L. Rakhmankulov, V. I. Isagulyants, and S. S. Zlotskii, *Zh. Prikl. Khim. (Moscow)*, 46 (1973) 477–479.
(295) D. L. Rakhmankulov, S. S. Zlotskii, and V. N. Uzikova, *Zh. Org. Khim.*, 10 (1974) 2625–2626.
(296) D. L. Rakhmankulov, S. S. Zlotskii, V. N. Uzikova, and Ya. M. Paushkin, *Dokl. Akad. Nauk SSSR*, 218 (1974) 156–157.
(297) L. M. Jeppesen, I. Lundt, and C. Pedersen, *Acta Chem. Scand. Ser. B*, 27 (1973) 3579–3585.

compound **295** gave the same rearrangement, but in a lower yield (27%). Unacylated derivatives were found to be unreactive, because of their low solubility in di-*tert*-butyl peroxide. On the other hand, the D-*galacto* derivative **297** yielded, under similar conditions, a mixture of the two rearranged compounds possible, namely, **298** (yield 16%)

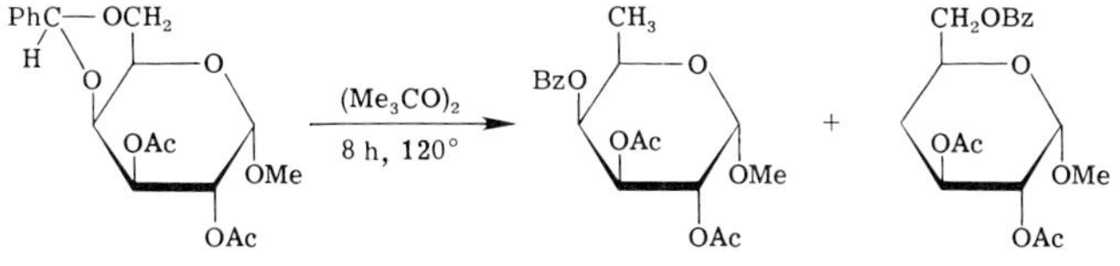

and **299** (yield 25%).[297] Other examples of the formation of mixtures of the two rearranged compounds expected were described from 3,4-O-benzylidene-D-arabinopyranoside and 2,3-O-benzylidene-D-ribofuranoside derivatives. In contrast, a good regioselectivity was observed for the benzylidene derivatives **300** and **301**, which both gave mainly the ester **302** (respective yield, 45 and 32%). The reaction

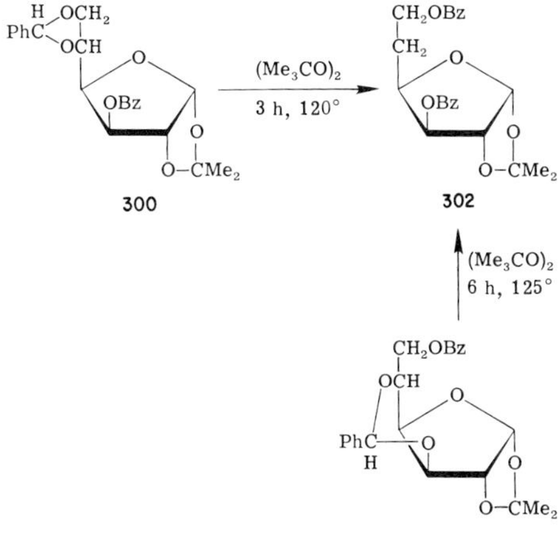

probably proceeds through initial abstraction of the acetalic hydrogen atom, giving a cyclic free-radical that rearranges, and then recombines with a hydrogen atom.

(ii) **Reaction of Nucleophiles with Benzoxonium Ions.** The work done by Jacobsen and Pedersen on the hydrolysis of benzoxonium ions, to give α-hydroxybenzoates, has already been discussed (see Section III,3,*b*). Further observations were reported[156] for the reaction of methyl 5-O-benzoyl-2,3-O-benzylidene-β-D-ribofuranoside (**125**) with triphenylmethyl fluoroborate. It has been shown that the corresponding benzoxonium ion (**126**) gave, on treatment with anhydrous sodium acetate, and hydrolysis of the product, a mixture of methyl 2,5- and 3,5-di-O-benzoyl-β-D-ribofuranoside (**127** and **128**), probably through the formation of an orthoacetate. This *cis* ring-opening was unexpected, as the opposite result was obtained[151] for the acetoxonium ion derived from *cis*-1,2-cyclohexanediol; this was attributed to the experimental conditions, rather than to a difference between the two reactions.[156] On the other hand, the same experiment conducted with tetraethylammonium *p*-toluenesulfonate gave[156] methyl 2,3-di-O-benzoyl-5-O-*p*-tolylsulfonyl-β-D-xylofuranoside (**303**) and methyl 2,5-di-O-benzoyl-3-O-*p*-tolylsulfonyl-β-D-xylofuranoside (**304**) (yield

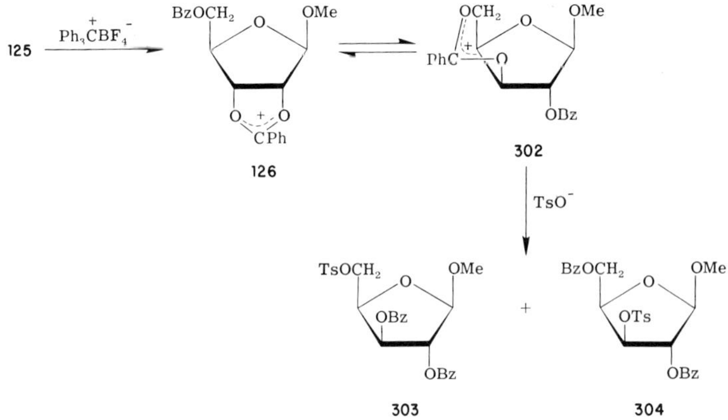

36%, for the ~1:1 mixture), besides some unreacted material and a small proportion of the benzoates **128** and **129**. The course of this reaction was rationalized by assuming formation of an equilibrium between the 2,3-benzononium ion (**126**, D-*ribo*) and its 3,5-isomer (**302**, D-*xylo*). The hypothesis of the intervention of a small proportion of ion **302**, reacting very rapidly, was supported by the stabilization of this type of ion starting from the 5-O-(*p*-methoxybenzoyl) analog of compound **125**; it was then possible to detect (by n.m.r. spectroscopy) 16% of the analog of ion **302**.

b. Polymerization of Acetals.—If such unsaturated sugar acetals as 4,6-O-acrylidene-D-glucopyranosides[298] can be polymerized by using cationic initiator, the acetal group cannot be considered to be respon-

---

(298) Z. Jedliński and J. Maślinska-Solich, *J. Polym. Sci. Part C*, (1968) 3611–3618.

sible for the reaction. On the other hand, it is now well known, particularly since Schuerch's work (see Refs. 299 and 300, and references cited therein), that, due to their acetal group, acetals of 1,6-anhydrohexopyranoses, and related derivatives, may be effectively polymerized. A review has appeared concerning this topic.[301]

c. Formation of Nucleosides from 1,2-O-(Diphenylmethylidene)aldoses.—An interesting ring-cleavage of a 1,2-acetal by the action of the trimethylsilyloxy derivative of a pyrimidine has been described,[302] and it constitutes a novel, facile mode of coupling for preparing nucleosides. Thus, an almost quantitative yield of 3′,5′-di-O-benzoyluridine (**306**) was obtained by treatment of 3,5-di-O-benzoyl-1,2-O-(diphenylmethylidene)-α-D-ribofuranose (**305**) with 2,4-bis(trimethylsilyloxy)pyrimidine and stannic chloride.

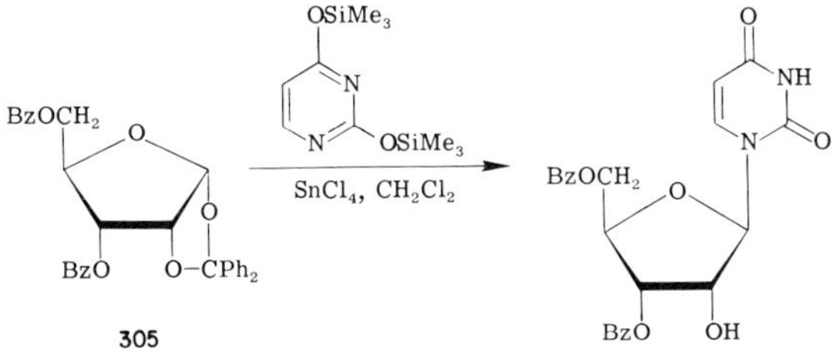

d. Cleavage of Cyclic Acetals with Grignard Reagents.—Cyclic acetals may be considered good protecting groups for either a vicinal diol or a carbonyl function during reactions using Grignard reagents under classical conditions, especially if ether is the solvent. When the reaction is deliberately conducted under drastic conditions (for instance, in refluxing benzene or toluene), ring opening may be observed. This behavior may be compared with that already described for the reaction of acetals with lithium aluminum hydride alone ("protecting group" behavior) and with mixed hydrides ("functional group" behavior).

The observation of ring cleavage of 1,3-dioxolanes and 1,3-dioxanes by Grignard reagents was first reported for isosafrole treated with

(299) C. Schuerch, *Adv. Polym. Sci.*, 10 (1972) 173–194; *Acc. Chem. Res.*, 6 (1973) 184–191.
(300) C. Schuerch, *Adv. Carbohydr. Chem. Biochem.*, 39 (1981) 157–212.
(301) H. Sumitomo and M. Okada, *Adv. Polym. Sci.*, 28 (1978) 47–82.
(302) G. Ritzmann, R. S. Klein, H. Ohrui, and J. J. Fox, *Tetrahedron Lett.*, (1974) 1519–1521.

methylmagnesium iodide.[303] The synthetic utility of the reaction was shown by the preparation of 2-(1-methylcyclohexyloxy)ethanol from the action of methylmagnesium iodide on cyclohexanone ethylene acetal in benzene at 75° (yield 91%),[304] and by examination of the behavior of steroidal acetals.[305] Some systematic studies have, for instance, been conducted by Blomberg and coworkers[306–309] and Shostakovskii and coworkers[310]; for the scope of the reaction, see, for example, the references cited in Ref. 307. Bicyclic acetals structurally related to 1,6-anhydropyranoses were found to react with alkyl(or aryl)magnesium halides in boiling benzene under reflux by opening of the five-membered ring.[240]

In the carbohydrate series, this reaction has received attention by Kawana and coworkers.[311,312] However, it is evident that ring opening may occur to some extent whenever a cyclic acetal is refluxed with a Grignard reagent in a solvent having a boiling point higher than that of ether. For instance, a small proportion of the compound corresponding to 5,6-dioxolane ring-opening (giving, selectively, the OH-5, OCPhMe$_2$-6 derivative) of 1,2:5,6-di-O-isopropylidene-α-D-ribo-hexofuranos-3-ulose was detected after the normal reaction of addition of one molar proportion of phenylmagnesium bromide at the carbonyl group, the reaction being conducted in oxolane.[313]

The same regiospecificity was observed[311] for cleavage of the 5,6-dioxolane ring of methyl 5,6-O-cyclohexylidene-3-deoxy-2-C-methyl-β-D-ribo-hexofuranoside; scission of the aglycon was not observed, but anomerization of the original furanoside to a mixture of the α- and β-glycosides resulted. A more-detailed study showed[312] that the

(303) N. Hirao, *Nippon Kagaku Zasshi*, 52 (1931) 153–155; *Chem. Abstr.*, 25 (1931) 5156.
(304) R. A. Mallory, S. Rovinski, and I. Scheer, *Proc. Chem. Soc.*, (1964) 416; *Chem. Abstr.*, 62 (1964) 9027–9028.
(305) R. A. Mallory, S. Rovinski, F. Kohen, and I. Scheer, *J. Org. Chem.*, 32 (1967) 1417–1422.
(306) C. Blomberg, A. D. Vreugdenhil, and T. Homsma, *Recl. Trav. Chim. Pays-Bas*, 82 (1963) 355–360.
(307) G. Westera, C. Blomberg, and F. Bickelhaupt, *J. Organomet. Chem.*, 82 (1974) 291–299.
(308) G. Westera, C. Blomberg, and F. Bickelhaupt, *J. Organomet. Chem.*, 144 (1978) 285–290.
(309) G. Westera, C. Blomberg, and F. Bickelhaupt, *J. Organomet. Chem.*, 144 (1978) 291–301.
(310) M. F. Shostakovskii, A. S. Atavin, and B. A. Trofimov, *Zh. Obshch. Khim.*, 34 (1964) 2088–2089.
(311) M. Kawana and S. Emoto, *Tetrahedron Lett.*, (1978) 1561–1562.
(312) M. Kawana and S. Emoto, *Bull. Chem. Soc. Jpn.*, 53 (1980) 230–235.
(313) J. C. Fischer and D. Horton, *Carbohydr. Res.*, 59 (1977) 477–503.

course of the reaction depends on the nature of the Grignard reagent. With methylmagnesium iodide (4 molar equivalents) in benzene–ether during 20 h under reflux (~85°), 5,6-O-cyclohexylidene derivatives in the D-*gluco* (**307**) and D-*allo* (**308**) series were regiospecifi-

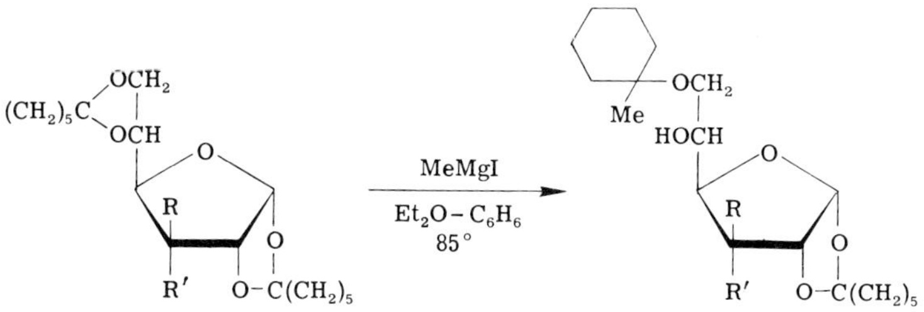

307 R = OH, R' = H
308 R = H, R' = OH

309
310

cally cleaved to the corresponding 6-(1-methylcyclohexyl) ethers (**309**, in a yield of 58%, and **310**).

Although a similar reaction was observed with ethylmagnesium iodide, the use of such a sterically bulky reagent as isopropyl(or *tert*-butyl)magnesium iodide gave the 6-cyclohexenyl and the 6-cyclohexyl ethers, concurrently with the 6-(1-alkylcyclohexyl) ether. A rationale was proposed involving a cyclohexyl cation that could be attacked by the reagent in any of three ways: (*a*) direct attack on the cation; (*b*) abstraction of a proton α to the cation; and (*c*) reduction of the cation. The reaction was also tested with an acetal of an aldopentose, namely, 1,2,:3,5-di-O-cyclohexylidene-α-D-xylofuranose (**311**),

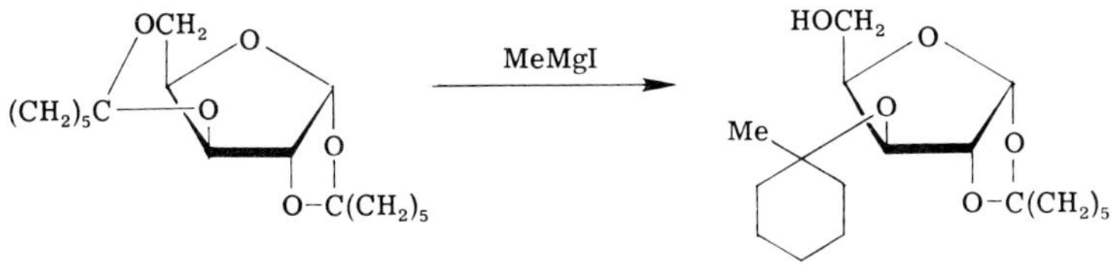

311

312

which gave a 58% yield of ether **312**, corresponding to cleavage of the O-5–cyclohexyl carbon atom bond.

IV. CONCLUSION

It is rather obvious that cyclic acetals may now be considered to be not only classical protecting-groups but also functional groups. The work done on both aspects has led to important improvements in, or

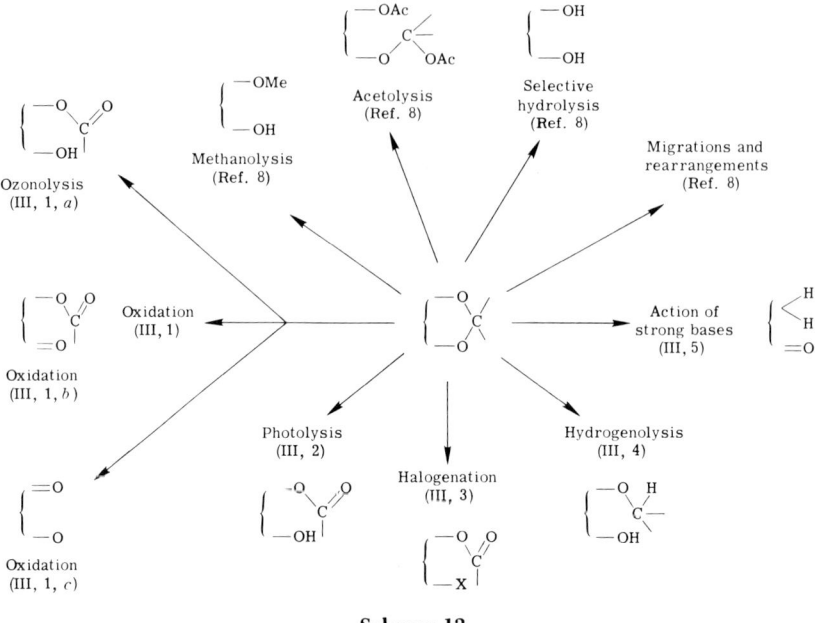

Scheme 13

simplifications of, the synthesis or structural modification of carbohydrates. In the near future, further developments concerning the utilization of new reagents, or procedures, and, probably, the discovery of new aspects of the reactivity of cyclic acetals may be expected. Scheme 13 summarizes most of the reactions that have been applied in carbohydrate chemistry, and that have been discussed either in the present Chapter or in the related one in this Volume.[8]

SYNTHESIS AND POLYMERIZATION OF ANHYDRO SUGARS[*]

By Conrad Schuerch

Department of Chemistry,
College of Environmental Science and Forestry,
State University of New York, Syracuse, New York 13210

I. Introduction 157
II. Synthesis of Anhydro-aldoses and -ketoses 160
   1. General 160
   2. 1,6-Anhydroaldopyranoses and 1,6-Anhydroaldofuranoses
      (6,8-Dioxabicyclo[3.2.1]octanes and 2,8-Dioxabicyclo[3.2.1]octanes) 161
   3. 1,4-Anhydroaldopyranoses or 1,5-Anhydroaldofuranoses
      (2,7-Dioxabicyclo[2.2.1]heptanes) 164
   4. 1,3-Anhydroaldopyranoses (2,6-Dioxabicyclo[3.1.1]heptanes) 169
   5. 1,2-Anhydroaldopyranoses (2,7-Dioxabicyclo[4.1.0]heptanes) 171
   6. Miscellaneous 172
III. Polymerization of Anhydroaldoses 173
   1. General Considerations: Mechanism 173
   2. General Considerations: Copolymerization 176
   3. Polymerization of 1,6-Anhydroaldo-pyranoses and -furanoses 179
   4. 1,6-Anhydroaldopyranans: Characterization and Applications 197
   5. Polymerization of 1,4-Anhydroaldopyranoses 204
   6. Polymerization of 1,3-Anhydroaldopyranoses 207
   7. Polymerization of 1,2-Anhydroaldopyranoses 207
   8. Miscellaneous Polymerizations 209
IV. Summary and Prospects 211

I. Introduction

Anhydro sugars have been of recurring interest for nearly a century. They were first obtained by alkaline[1] and pyrolytic degradation[2] of natural products and have since found varied uses. The rigid, bicyclic structures of some classes have made them useful in spectral investi-

---

[*] This work was largely supported by a research grant (5 RO1 GM 06168) of the National Institute of General Medical Sciences, National Institutes of Health, U. S. Public Health Service, which is gratefully acknowledged.
(1) C. Tanret, *C. R. Acad. Sci.*, 119 (1894) 158–161.
(2) A. Pictet and V. Sarasin, *Helv. Chim. Acta*, 1 (1918) 87–96.

gations and in studies of the relation between conformation and reaction rates and equilibria.[3,4] In syntheses, anhydro rings have served as both protecting groups[5] and as reaction sites.[6] One of the more recent synthetic applications of this class of compounds has been to serve as monomers in the preparation of stereoregular polysaccharides.[7-10] For the last of these applications, the "glycosans" (anhydroglycoses), or internal, bicyclic acetals, are of the most interest, as their polymerization leads to the synthesis of hydrolyzable polymers, resembling model, natural polysaccharides, whereas the polymerization of other anhydro sugars leads to nonhydrolyzable products[11-14] for which no special use or interest has as yet developed.

Anhydro sugars have been discussed in a number of earlier Chapters in this Series,[15-19] two of which dealt exclusively with the bicyclic acetals, the 1,6-anhydroglyco-pyranoses and -furanoses. A brief, systematic treatment of anhydro sugars[20] has also appeared, but a large body of knowledge exists only with respect to 1,6-anhydro sugars.[19] The main emphases of the present article are on the preparation of anhydro sugar derivatives that are appropriate starting-materials for the synthesis of stereoregular polysaccharides, methods and mechanisms of polymerization, properties and characterization of the resultant polymers, and their developing applications.

The first investigation of the polymerization of an anhydro sugar was reported by Pictet[21] shortly after his preparation of 1,6-anhydro-β-

(3) R. E. Reeves, *Adv. Carbohydr. Chem.*, 6 (1951) 107–134.
(4) S. J. Angyal, in E. L. Eliel, N. Allinger, S. J. Angyal, and C. A. Morrison (Eds.), *Conformational Analysis*, Interscience, New York, 1966.
(5) Y. Rabinsohn, A. J. Acher, and D. Shapiro, *J. Org. Chem.*, 38 (1973) 202–204.
(6) W. J. Hickinbottom, *J. Chem. Soc.*, (1928) 3140–3147.
(7) C. Schuerch, *Adv. Polym. Sci.*, 10 (1972) 173–194.
(8) C. Schuerch, *Acc. Chem. Res.*, 6 (1973) 184–191.
(9) C. Schuerch, *Encycl. Polym. Sci. Techn., Suppl.*, 1 (1976) 510–521.
(10) C. Schuerch, J. Zachoval, and B. Veruovic, *Chem. Listy*, 66 (1972) 1124–1148.
(11) R. S. Nevin, K. Sarkanen, and C. Schuerch, *J. Am. Chem. Soc.*, 84 (1962) 78–82.
(12) T. Uryu, K. Kitano, H. Tachikawa, K. Ito, and K. Matsuzaki, *Makromol. Chem.*, 179 (1978) 1773–1782.
(13) T. Uryu, K. Kitano, and K. Matsuzaki, *Makromol. Chem.*, 180 (1979) 1135–1144.
(14) T. Uryu, Y. Koyama, and K. Matsuzaki, *J. Polym. Sci., Polym. Lett.*, 17 (1979) 673–678.
(15) S. Peat, *Adv. Carbohydr. Chem.*, 2 (1946) 37–77.
(16) R. J. Dimler, *Adv. Carbohydr. Chem.*, 7 (1952) 37–52.
(17) N. R. Williams, *Adv. Carbohydr. Chem. Biochem.*, 25 (1970) 109–179.
(18) J. Defaye, *Adv. Carbohydr. Chem. Biochem.*, 25 (1970) 181–228.
(19) M. Černý and J. Staněk, Jr., *Adv. Carbohydr. Chem. Biochem.*, 34 (1977) 23–177.
(20) R. D. Guthrie, in W. Pigman and D. Horton (Eds.), *The Carbohydrates: Chemistry and Biochemistry*, Vol. IA, Academic Press, New York, 1972, pp. 423–478.
(21) A. Pictet, *Helv. Chim. Acta*, 1 (1918) 226–230.

D-glucopyranose (levoglucosan) by the pyrolysis of cellulose.[2] The polymerization was explored in detail many years later, and the products obtained were generally similar to those obtained by condensation polymerization of D-glucose; their structures were found to be multibranched, with a random distribution of glycosyl linkages to oxygen atoms on pyranosyl residues, and a small percentage of furanosyl units. Because of the complexity of their structures, research on these products is no longer active, and their chemistry was well covered in an extensive article.[22]

The first, attempted synthesis of a linear polysaccharide by polymerization of a fully protected anhydro sugar, the trimethyl ether of levoglucosan, was made in 1925 and was unsuccessful.[23] A number of additional attempts to prepare polysaccharides was made over the following decades by both condensation[24] and ring-opening polymerization.[25] The most important of these were experiments conducted in Bredereck's laboratory in Germany[26] and in the Institute of Wood Chemistry of the Latvian Academy of Sciences.[27] The two groups independently demonstrated the cationic polymerization of fully substituted levoglucosan derivatives in solution. Bredereck and Hutten[26] used carbocation perchlorates to initiate polymerization, and benzyl ethers as removable protecting groups. The parent polysaccharides obtained by debenzylation proved to be not stereoregular, and the work was reported in the open literature only in a review.[26] The publication from Riga described the first conversion of levoglucosan trimethyl ether into a crystalline polymer by means of boron trifluoride etherate. The use of a fluorinated Lewis acid seems, in retrospect, of especially critical importance to the future development of the field, and indeed, the achievement was hailed with enthusiasm in the *Moscow News* of October 10, 1964, and credited to a young, woman staff-member, Ruta Ya. Pernikis. The *News* reported that "this green-eyed, present-day magician in white overalls" had succeeded, where "American scientists had failed to obtain any positive results in polymerizing levoglucosan derivatives"[28] and had[29] then "started search-

---

(22) I. J. Goldstein and T. L. Hullar, *Adv. Carbohydr. Chem.*, 21 (1966) 431–512.
(23) J. C. Irvine and J. W. H. Oldham, *J. Chem. Soc.*, 127 (1925) 2903–2922.
(24) S. Haq and W. J. Whelan, *J. Chem. Soc.*, (1956) 4543–4549.
(25) S. Haq and W. J. Whelan, *Nature*, 178 (1956) 1222–1223.
(26) H. Bredereck and V. Hutten, unpublished data cited in J. Klar, *Chem. Ztg.*, 87 (1963) 731–740; V. Hutten, Ph.D. Dissertation, Technische Hochschule, Stuttgart, 1961.
(27) V. V. Korshak, O. P. Golova, V. A. Sergeev, N. M. Merlis, and R. Ya. (Schneer) Pernikis, *Vysokomol. Soedin.*, 3 (1961) 477–485.
(28) A. J. Mian, E. V. Quinn, and C. Schuerch, *J. Org. Chem.*, 27 (1962) 1895–1896.
(29) A. Bhattacharya and C. Schuerch, *J. Org. Chem.*, 26 (1961) 3101–3104.

TABLE I

Ring Indexing of Anhydroaldoses[a]

Name of carbohydrate	(2-Ring) Size	(2-Ring) Atoms	Systematic name	CAS parent compound identifier
Anhydroaldo-furanose				
1,2-	3,5	C_2O-C_4O	2,6-dioxabicyclo[3.1.0]hexane	FPVPM
1,3-	4,5	C_3O-C_4O	2,5-dioxabicyclo[2.1.1]hexane	—
1,5-	5,5	$C_3O_2-C_4O$	2,7-dioxabicyclo[2.2.1]heptane	FPVFY
1,6-	5,6	$C_4O-C_4O_2$	2,8-dioxabicyclo[3.2.1]octane	FPWSQ
Anhydroaldo-pyranose				
1,2-	3,6	C_2O-C_5O	2,7-dioxabicyclo[4.1.0]heptane	FPVHR
1,3-	4,6	C_3O-C_5O	2,6-dioxabicyclo[3.1.1]heptane	JXVLQ
1,6-	5,6	$C_3O_2-C_5O$	6,8-dioxabicyclo[3.2.1]octane	FPXBP

[a] Carbohydrate names of all parent anhydro sugars that are reported in a given volume of *Chemical Abstracts* may be found under these *Ring-Index* headings in the *Index of Ring Systems*. A complete listing of these names may be found in the *Ring Analysis Index* of the CAS Parent Compound Handbook. For 1,4-anhydroglycopyranose, see 1,5-anhydroglycofuranose.

ing for a theoretical explanation why these white crystals could not be polymerized." The enthusiasm was justified. Refinements and extensions of these early studies have now given a fairly complete understanding of the polymerization of 1,6-anhydroglycopyranose derivatives, and have given impetus to the synthesis and polymerization of other classes of anhydro sugars.

The polymerizable anhydro sugars have a variety of parent sugars, substituents, and ring sizes, and the polysaccharides that may be prepared from them constitute a major extension of polymer research. In order to facilitate interaction between carbohydrate chemists and polymer scientists, it will be advantageous in the future to make use of the systematic nomenclature for bicyclic compounds used in *The Ring Index* of *Chemical Abstracts*. Literature surveys of related anhydro sugars can be greatly facilitated by use of the *Ring Index*, as the examples in Table I demonstrate.

II. Synthesis of Anhydro-aldoses and -ketoses

1. General

The two best-known syntheses of anhydro sugars are clearly those of 1,6-anhydro-β-D-glucopyranose by alkaline treatment of phenyl β-D-glucopyranoside[30] and of the Brigl anhydride by treatment of 3,4,6-tri-*O*-acetyl-β-D-glucopyranosyl chloride with ammonia.[31] These syntheses involve certain, specific complications that have, perhaps, obscured the fact that the best, general method for the synthesis of bicyclic acetals of sugars is the same as that for the synthesis of other anhydro sugars, namely, the rearside attack of an alkoxide ion on a carbon atom bearing a leaving group.[15] The main restraints on this general approach appear to be that (*1*) the reacting alcohol and the leaving group should bear a trans relationship, (*2*) in approaching the transition state, no steric restraints should develop that might direct reaction elsewhere, and (*3*) the reactivity of the leaving group and reactant alcohol, or alkoxide, should permit reaction under conditions that do not rupture the resultant, anhydro ring. (Many examples of configurational inversion in the literature are very reasonably ascribed to the transient formation of anhydro sugars.[30,32–34] In some cases, they undoubtedly have a measurable lifetime, but, in other cases, they should, perhaps, be considered to be in an activated state having so small an energy barrier to final product that they can scarcely be considered to have an independent existence.) In individual cases, another synthetic method may be preferable, but the use of internal, nucleophilic substitution of the S_N2 type should always be considered.

2. 1,6-Anhydroaldopyranoses and 1,6 Anhydroaldofuranoses (6,8-Dioxabicylo[3.2.1]octanes and 2,8-Dioxabicyclo[3.2.1]octanes)

The synthesis of 1,6-anhydrohexoses has been the most fully investigated of the internal anhydro sugars (glycosans) and has been described in this Series.[19] It will, therefore, be necessary only to summarize the more practical approaches for synthesis of a quantity of a monomer and to present a few significant findings since made. Only the 1,6-anhydro sugars can be prepared directly from the free aldohexoses. Angyal and Beveridge[35] emphasized that direct synthesis may be achieved by methods largely under either thermodynamic or kinetic control. When an aldohexose is heated in aqueous acid, an equilibrium is established between the free sugar and its 1,6-anhydride. The yield of the anhydrohexopyranose varies within wide limits (0.2–86%), depending on its configuration. When the configura-

(30) G. H. Coleman, *Methods Carbohydr. Chem.*, 2 (1963) 397–399.
(31) R. U. Lemieux and J. Howard, *Methods Carbohydr. Chem.*, 2 (1963) 400–402.
(32) D. C. C. Smith, *J. Chem. Soc.*, (1957) 2690–2697.
(33) J. K. N. Jones and W. H. Nicholson, *J. Chem. Soc.*, (1955) 3050–3053.
(34) C. P. J. Glaudemans and H. G. Fletcher, Jr., *J. Org. Chem.*, 29 (1964) 3286–3290.
(35) S. J. Angyal and R. J. Beveridge, *Aust. J. Chem.*, 31 (1978) 1151–1155.

tion is energetically favorable for ring stabilization (for example, idose, altrose, or gulose), the anhydride is formed in good yield. However, anhydrohexopyranoses that have conformational instabilities (for example, glucose, mannose, or galactose) are obtained only in minute proportion by this method. Small proportions of the 1,6-anhydrohexofuranoses are also formed under these conditions. Angyal and Beveridge also showed that, on acid-catalyzed equilibration in aqueous solution, internal acetals are more readily formed with secondary than with primary hydroxyl groups.[36] Therefore, substantially higher proportions of the 1,6-anhydroaldopyranose are formed from aldoheptoses than from the homomorphous aldohexose, at least in those cases in which the additional hydroxymethyl group is exo in the anhydride and therefore does not introduce any additional steric interaction. Because much of the driving force for the polymerization of anhydro sugars comes from the release of strain in the structure of the anhydro sugar, the most readily polymerized monomers will, however, not be obtainable by equilibrium dehydration.

For the synthesis of 1,6-anhydro sugars from free sugars, at least two methods of interest are known that are under kinetic control, and, in both, 1,6-anhydroglyco-pyranoses and -furanoses are formed in comparable quantities. The older method is by pyrolysis of aldoses, disaccharides, and polysaccharides.[19] Historically, the method was used for the preparation of 1,6-anhydro-β-D-gluco-, -galacto-, -and mannopyranoses, and the presence of 1,6-anhydrofuranoses was only recognized later. A variety of by-products and a char are usually formed; lengthy purifications and slow crystallizations are not unusual. Negative catalysis, variable results, and unpleasant esthetics discourage casual use of this method, although various modifications and improvements have been developed, and, with experience, large preparative quantities of these compounds may be obtained.

Anhydro sugars have also been prepared under kinetic control by the acid-catalyzed dehydration of the aldohexoses in N,N-dimethylformamide while water is removed azeotropically.[35] In this system, the main products are, in some instances, the 1,6-anhydrohexofuranoses, and the anhydrohexofuranose may be separated from the anhydropyranose by chromatography on a column of Dowex AG-50W X2 (Ca^{2+}) ion-exchange resin, with water as the eluant. To avoid an excessive extent of oligosaccharide formation, high dilution is necessary during the reaction. The necessity for high dilution and chromatographic separation probably limits the quantity practically accessible

(36) S. J. Angyal and R. J. Beveridge, *Carbohydr. Res.*, 65 (1978) 229–234.

by this process, but reasonable yields have been obtained of 1,6-anhydro-α-D-galactofuranose (33%), 1,6-anhydro-β-D-allofuranose (32%), 1,6-anhydro-α-D-talofuranose (10%), and 1,6-anhydro-β-D-mannofuranose (9%). The yields of anhydrohexopyranose obtained simultaneously were, respectively, 3, 4, 2, and 23%. Although these kinetic methods can have application for the synthesis of some of the more common 1,6-anhydrohexoses, they cannot be considered to be viable, general methods, and syntheses from sugar derivatives provide a wider scope. These were treated in detail by Černý and Staněk.[19]

Of the syntheses available for 1,6-anhydro-β-D-glucopyranose, the alkaline treatment of phenyl tetra-O-acetyl-β-D-glucopyranoside and related glycosides is the most fully developed. The reaction also provides a satisfactory method for the synthesis of 1,6-anhydro-β-D-galactopyranose, although, for the best results, the preparation involves the synthesis of β-D-galactose pentaacetate (a low-yield step), the careful purification of the phenyl β-glycoside, and an ester-interchange elimination of ester groups prior to alkaline ring-closure. The generally accepted mechanism of the ring closure involves an attack at C-1 by O-2 alkoxide, to form a transient epoxide as an intermediate that is attacked by O-6 alkoxide to afford the alkali-stable, 1,6-anhydro ring.[30] As this reaction proceeds by double inversion, other phenyl 1,2-*trans*-β-glycosides should give the corresponding 1,6-anhydroglycopyranoses.

As phenyl β-D-mannopyranoside is not readily available, and the steric relationship on C-1 and C-2 is cis, not trans, another approach is desirable for synthesis of this monomer. A satisfactory sequence from the readily available methyl α-D-mannopyranoside is that of tritylation of O-6, benzylation of O-2, -3, and -4, detritylation, tosylation of O-6, acetolysis of the methyl glycoside, and treatment of the product with alkoxide.[37] Alkoxide removes the ester group on O-1, and the resultant, O-1 anion eliminates the tosyl group, with ring closure. The product, namely, 1,6-anhydro-2,3,4-tri-O-benzyl-β-D-mannopyranose, is produced in high yield, and, after slight purification, is suitable for polymerization. This monomer is dimorphic as a crystalline solid, and may exist in solution as an equlibrium mixture of two conformations.[38] Presumably, analogous, synthetic sequences will be satisfactory for other sugars, although but few similar ring-closures were noted by Černý and Staněk.[19]

There appears to have been little effort expended to date on the synthesis of 1,6-anhydroglycofuranoses. They have usually been pre-

(37) S. J. Sondheimer, R. Eby, and C. Schuerch, *Carbohydr. Res.*, 60 (1978) 187–192.
(38) H. Ito and T. Uryu, unpublished observations.

pared by pyrolysis, or by the acid-catalyzed, kinetically controlled, dehydration methods already discussed.[19] The more-common sugar derivatives have been prepared in this way. The less-common sugar derivatives have been prepared by various, acid-catalyzed rearrangements of triacetates of 1,6-anhydro sugars, and by reduction of 1,6-anhydrohexofuranosuloses.[19] There is no obvious reason why complete, synthetic sequences could not be devised.

3. 1,4-Anhydroaldopyranoses or 1,5-Anhydroaldofuranoses (2,7-Dioxabicyclo[2.2.1]heptanes)

The early literature on the synthesis of 1,4-anhydro sugars is clouded by controversy and by conclusions based on inadequately characterized products. Micheel and Hess[39] attempted to prepare 1,4-anhydro-2,3,6-tri-O-methyl-D-glucopyranose by treating (4-O-acetyl-2,3,6-tri-O-methyl-β-D-glucopyranosyl)trimethylammonium chloride with a base, but did not make clear whether the expected product was formed. Consequently, Freudenberg and Braun[40] questioned the result, and reported a synthesis of the anhydro sugar by reaction of 2,3,6-tri-O-methyl-α-D-glucopyranosyl chloride with sodium metal; this synthesis was later repeated and confirmed.[41] Alternative syntheses have also been reported: by treatment *in vacuo* of "2,3,6-tri-O-methyl glucose mercaptal" with silver carbonate,[41] by pyrolysis of 2,3,6-tri-O-methyl-D-glucopyranose,[41] and by pyrolysis of tri-O-methylcellulose.[42]

Micheel and Kreuzer[43] developed a superior synthesis of 1,4-anhydro-2,3,6-tri-O-benzyl-α-D-glucopyranose that is mechanistically "clean" and has two generally useful, synthetic steps. 2,3,6-Tri-O-benzyl-D-glucopyranose was allowed to react with hydrogen chloride in ether, to form the related D-glucosyl chloride. The elimination of an anomeric hydroxyl group by hydrogen chloride to afford a glycosyl chloride is a relatively unusual, but very useful, synthetic step. The use of ether as the solvent appears to decrease the acidity of the acid sufficiently that debenzylation is not a problem. The anomeric chloride and 4-hydroxyl group are *cis* to each other, and are therefore in steric relationship unsatisfactory for ring closure. Treatment with silver fluoride produced the β-fluoride, which was readily eliminated by

(39) F. Micheel and K. Hess, *Ber.*, 60 (1927) 1898–1906.
(40) K. Freudenberg and E. Braun, *Ber.*, 66 (1933) 780–781.
(41) E. Huseman and J. Klar, *Makromol. Chem.*, 53 (1962) 223–224.
(42) A. M. Pakhomov, O. P. Golova, and I. Nikolaeva, *Bull. Acad. Sci. USSR, Div. Chem. Sci.*, (1957) 535–537.
(43) F. Micheel and U. Kreuzer, *Justus Liebigs Ann. Chem.*, 722 (1969) 228–231.

rearside attack of O-4 alkoxide. The benzyl ether groups were removed by catalytic hydrogenolysis, to afford stable, crystalline 1,4-anhydro-β-D-glucopyranose. A somewhat related synthesis of 1,5-anhydro-β-D-ribofuranose by treatment of phenyl β-D-ribofuranoside with a base gave a poor yield of the desired product.[44]

The reaction of a sugar derivative having only one, free, alcoholic hydroxyl group and a *trans*-positioned leaving-group at the glycosidic center often proves to be the method of choice for the formation of 1,4-anhydro sugars and more highly strained systems. There are, however, two similar methods that have also been successfully applied to the formation of the 1,4-anhydroaldopyranoses. These are the reaction of a base with the anomeric hydroxyl group of an otherwise fully protected, pyranoid or furanoid sugar. In the case of a pyranose derivative, a leaving group, typically tosyl or mesyl, is the substituent on O-4, and, for a furanose, the leaving group is on O-5; inversion occurs at the carbon atom bearing the leaving group. Reactions of this type were introduced by Hess, and have since been used repeatedly. A number of examples demonstrate, however, that alternative pathways may be followed if the ring-closure reaction is sluggish due to steric restraints.

Hess and Heumann[45] treated 1-O-acetyl-2,3,6-tri-O-methyl-5-O-tosyl-D-glucofuranose with a base, and obtained 1,4-anhydro-2,3,6-tri-O-methyl-L-idopyranose (1,5-anhydro-2,3,6-tri-O-methyl-L-idofuranose), although they were not certain of its identity. Vis and Fletcher[46] treated 2,3-O-benzylidene-5-O-tosyl-D-ribofuranose with sodium isopropoxide, and proved the product to be 1,5-anhydro-2,3-O-benzylidene-β-D-ribofuranose. They also prepared the free anhydro sugar, and some derivatives. 1,4-Anhydro-2,3,6-tri-O-methyl-β-D-galactopyranose was formed by inversion of 2,3,6-tri-O-methyl-4-O-tosyl-D-glucopyranose with sodium isopropoxide, and a 1,4-anhydro-2,3-di-O-methyl-α-L-arabinose was similarly prepared (in a somewhat impure state) from the corresponding D-xylose derivative by Kops and Schuerch.[47] All of these syntheses proceeded with simple inversion at either carbon atom 4 or 5.

Brimacombe and Tucker[48] were able to prepare 1,4-anhydro-6-deoxy-2,3-O-isopropylidene-β-L-talopyranose by an uncomplicated ring-closure, with inversion, of 6-deoxy-2,3-O-isopropylidene-4-O-

(44) E. Vis and H. G. Fletcher, Jr., *J. Org. Chem.*, 23 (1958) 1393.
(45) K. Hess and K. E. Heumann, *Ber.*, 72 (1939) 137–148.
(46) E. Vis and H. G. Fletcher, Jr., *J. Am. Chem. Soc.*, 79 (1957) 1182–1185.
(47) J. Kops and C. Schuerch, *J. Org. Chem.*, 30 (1965) 3951–3953.
(48) J. S. Brimacombe and L. C. N. Tucker, *Carbohydr. Res.*, 5 (1967) 36–44.

mesyl-L-mannopyranose, and by inversion of 6-deoxy-2,3-O-isopropylidene-5-O-mesyl-D-allofuranose. In contrast, when 6-deoxy-2,3-O-isopropylidene-4-O-mesyl-α-L-talopyranose (1) was treated with a base, direct inversion to 1,4-anhydro-6-deoxy-2,3-O-isopropylidene-α-L-mannopyranose (3) occurred only to the extent[49] of 58%. In this reaction, the 1-alkoxide ion can cause direct inversion at C-4 only if in a boat conformation (2) in which both the C-methyl group and the 1,3-dioxolane ring assume an endo orientation with respect to the bicyclic ring-system under formation. These nonbonded interactions are sufficient in magnitude to divert the reaction, in part, to alternative pathways. In addition, methyl 6-deoxy-2,3-O-isopropylidene-α-L-talofuranoside (6) (26%) and methyl 2,3-O-isopropylidene-α-L-rhamnofuranoside (7) (12%) were formed. The two by-products were apparently produced from the acyclic form (4) of the sugar. Elimination of the mesyl group by the alkoxide of O-5 resulted in formation of epoxide 5, with inversion at C-4. Methoxide attack on C-1 followed, and the resulting O-1 alkoxide ion attacked C-4, with a second inversion at the same center, to give the methyl α-L-talofuranoside derivative (6) in 26% yield. To a lesser extent (12%) the O-1 alkoxide ion eliminated 4-mesyl, directly leading to the formation[49] of the methyl α-L-rhamnofuranoside 7 (see Scheme 1).

Quite analogous ring-closures occur when the 1-O-acetyl derivatives of the rhamnopyranose and talopyranose derivatives are treated with sodium azide in N,N-dimethylformamide. 1-O-Acetyl-6-deoxy-2,3-O-isopropylidene-4-O-mesyl-α-L-mannopyranose is converted exclusively into 1,4-anhydro-6-deoxy-2,3-O-isopropylidene-β-L-talopyranose. In this instance, the azide nucleophile attacks the 1-O-acetyl group, liberating an O-1 oxide ion which reacts with inversion of C-4. The 4-epimeric, 1-O-acetyl-6-deoxy-talose derivative gives 60% of the direct inversion product 1,4-anhydro-6-deoxy-2,3-O-isopropylidene-α-L-mannopyranose, together with other products.[50]

Similarly, various α-D-glucopyranose derivatives having a mesyloxy group at C-4 and an acetoxyl group at C-1 are also converted into 1,4-anhydro-β-D-galactopyranose derivatives on treatment with sodium azide in such aprotic solvents as N,N-dimethylformamide.[51] The use of sodium azide in N,N-dimethylformamide under forcing conditions originated in attempts at nucleophilic displacements to form azido,

(49) J. S. Brimacombe, F. Hunedy, and A. K. Al-Radhi, *Carbohydr. Res.*, 11 (1969) 331–340.
(50) J. S. Brimacombe, J. Minshall, and L. C. N. Tucker, *Chem. Commun.*, (1973) 142–143, *J. Chem. Soc., Perkin Trans. 1*, (1973) 2691–2694.
(51) C. Bullock, L. Hough, and A. C. Richardson, *Chem. Commun.*, (1971) 1276–1277.

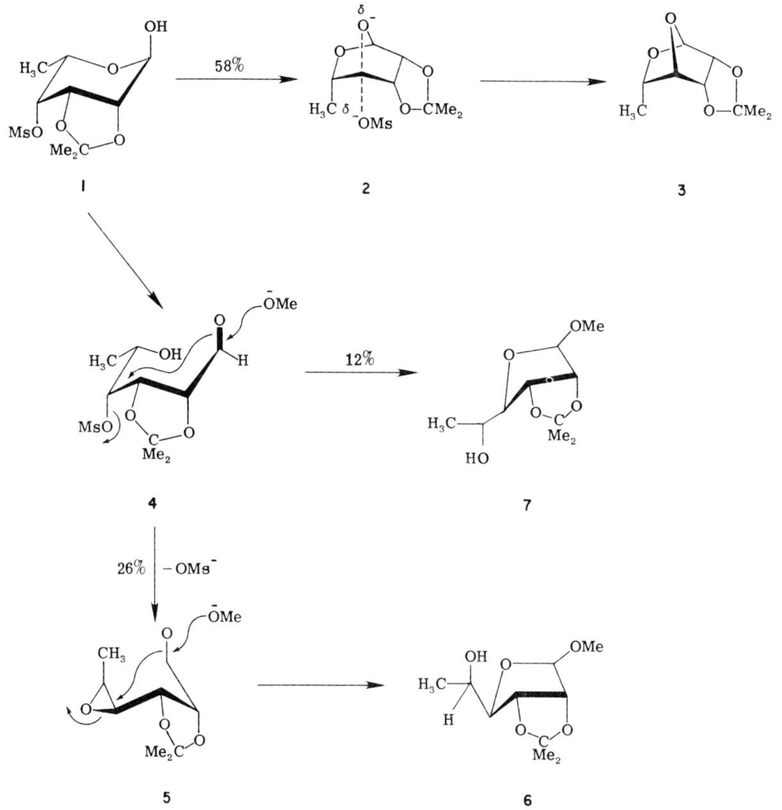

Scheme 1

and ultimately amino, sugars.[52] In some cases, these displacements proceed normally, as on glycosides. In others, complex intramolecular-reactions intervene, because the 1-acetoxyl group is not sufficiently stable under these conditions to serve as a protecting group. Sodium azide does not appear to have any special preparative utility for the formation of anhydro sugars, however, as basic conditions can produce the same ring-closure under milder conditions.

In a number of reactions on furanose derivatives, steric restraints interfere with anhydro-ring formation. 6-Deoxy-2,3-O-isopropylidene-5-O-tosyl-L-mannofuranose is readily converted by sodium methoxide into methyl 6-deoxy-2,3-O-isopropylidene-β-D-allofuranoside. In this

(52) J. F. Batey, C. Bullock, E. O'Brien, and J. M. Williams, Carbohydr., Res., 43 (1975) 43–50.

reaction, inversion of both C-4 and C-5 occurs.[53] Carbon atom 5 is most probably inverted by formation of an epoxide with C-4 of the acyclic form of the sugar. Attack at C-1 by methoxide results in an O-1 alkoxide, which then attacks C-4, with inversion, to give the observed product.[54]

The reaction of 2,3-O-isopropylidene-5-O-mesyl-D-lyxofuranose (**8**) with sodium methoxide proceeds similarly. The main product, methyl 2,3-O-isopropylidene-β-L-ribofuranoside (**10**), is formed (with its anomer) by opening of the furanose ring, elimination of the primary, 5-methanesulfonate anion (inversion not observable) by O-4 alkoxide ion to form epoxide **9**, and attack at C-1 by methoxide to form an O-1 alkoxide ion which attacks C-4 with inversion. Some attack at C-5 gives minor proportions of methyl 2,3-O-isopropylidene-β-D-lyxopyranoside and its anomer (**11**). The formation of 1,5-anhydro-2,3-O-isopropylidene-β-D-lyxofuranose (1,4-anhydro-2,3-O-isopropylidene-β-D-lyxopyranose), **12**, occurs to only a minor extent[55,56] (see Scheme 2).

A more unusual complication has been encountered that involves the base-catalyzed migration of a *p*-tolylsulfonyl group between vicinal oxygen atoms (O-4→O-5). 2,3:6,7-Di-O-isopropylidene-5-O-*p*-tolylsulfonyl-D-*glycero*-D-*gulo*-heptose reacts with sodium methoxide to form (in 16% yield) 1,4-anhydro-2,3:6,7-di-O-isopropylidene-α-D-*glycero*-D-*allo*-heptopyranose. The anhydro compound differs from the original 5-sulfonate in that the configuration of C-4 (rather than C-5) is inverted. Presumably, migration occurs and is then followed by ring closure. Other products are formed, as well.[57]

In summary, 1,4-anhydroaldopyranose derivatives can be formed by treating with a base either pyranose or furanose derivatives having a leaving group on C-4 or C-5 and a free hydroxyl group on C-1 only. These may be the methods of choice for some of the rarer sugars. However, rearrangements are always a possibility, and therefore new, alternative sequences should be evaluated by conformational analysis, and the identity of the products must be rigorously proved. When the parent sugar is available, the simpler and cleaner reaction-sequence will usually involve displacement of a leaving group at the anomeric center.

(53) P. A. Levene and J. Compton, *J. Biol. Chem.*, 116 (1936) 169–188.
(54) E. J. Reist, L. Goodman, R. R. Spencer, and B. R. Baker, *J. Am. Chem. Soc.*, 80 (1958) 3962–3966.
(55) J. S. Brimacombe and F. Hunedy, *J. Chem. Soc.*, C, (1968) 2701–2703.
(56) J. S. Brimacombe, F. Hunedy, and L. C. N. Tucker, *J. Chem. Soc.*, C, (1968) 1381–1384.
(57) J. S. Brimacombe and L. C. N. Tucker, *Chem. Commun.*, (1966) 903–904.

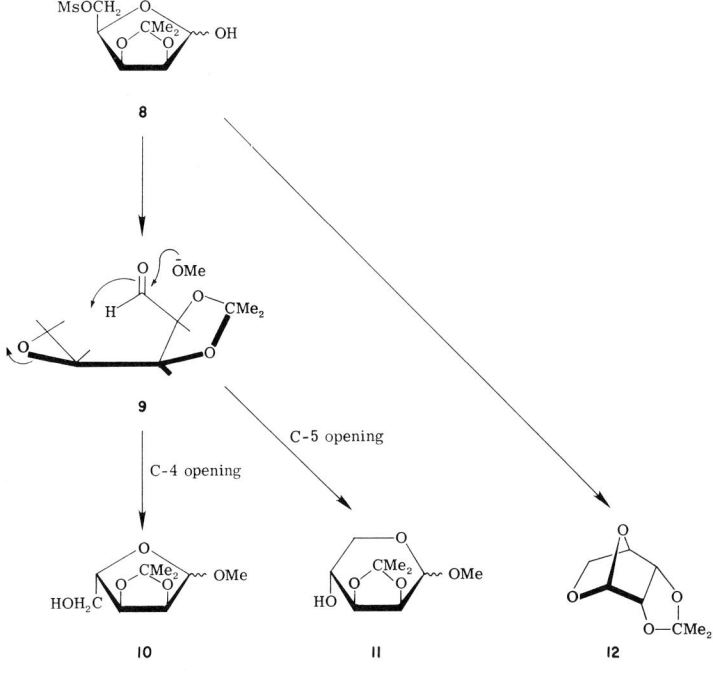

Scheme 2

4. 1,3-Anhydroaldopyranoses (2,6-Dioxabicyclo[3.1.1]heptanes)

When 1,6-anhydro-β-D-galactofuranose was first obtained from the pyrolysis distillate of D-galactose,[58] it was erroneously identified as 1,3-anhydro-β-D-galactopyranose because of its resistance to periodate oxidation, but correct identity was later established.[59] Except for these, there appears to be no report of any 1,3-anhydroglycopyranose. Nevertheless, application of the methodology just described leads without undue difficulty to members of this class of compound.

In derivatives of the common sugars, D-glucose, D-mannose, and D-galactose, the 3-hydroxyl group is trans to the thermodynamically favored position for an electronegative leaving-group on C-1. In the $^4C_1$ (D) conformation, the 3-hydroxyl group is equatorial and the C-1 leaving-group is axial. However, in the skew[60] conformation $^OS_2$, the

(58) R. M. Hann and C. S. Hudson, *J. Am. Chem. Soc.*, 63 (1941) 2241–2242.
(59) B. H. Alexander, R. J. Dimler, and C. L. Mehltretter, *J. Am. Chem. Soc.*, 73 (1951) 4658–4659.
(60) Rules for Conformational Nomenclature for Five- and Six-membered Rings in Monosaccharides and their Derivatives, *Chem. Commun.*, (1973) 505–508.

steric relationship for O-3 and the C-1 leaving-group is nearly ideal for ring closure. As the C-6 atom is essentially equatorial in this skew conformation, the energy requirements for the molecule to adopt this position must not be excessive, and, indeed, ring-closure does occur on treatment with a base.

If, however, the leaving group were situated on C-3 of the sugar and O-1 were to serve as the nucleophile, a conformation close to $^5S_1$ would be required for ring closure to afford the C-3-epimeric, anhydro sugar. In this conformation, C-6 is in an axial position, and greater resistance to ring closure might be expected, and, perhaps, alternative pathways would predominate.

The synthesis of 1,3-anhydro-2,4,6-tri-O-benzyl-β-D-glucopyranose was first accomplished by Eby and Kramer (see Ref. 61), who used, as the starting material, methyl tri-O-benzyl-α-D-glucopyranoside, prepared by partial benzylation of the parent glucoside, which proved to be a mixture of the 2,4,6- and 2,3,6-tri-O-benzyl isomers; but the desired isomer was formed in larger proportion and could be separated. Hydrolysis, followed by treatment with hydrogen chloride in ether, gave 2,4,6-tri-O-benzyl-α-D-glucopyranosyl chloride. Ring closure with potassium *tert*-butoxide was accompanied by some elimination and some glycoside formation. A preferable synthesis was developed by Ito and coworkers[61] from 1,2:5,6-di-O-isopropylidene-α-D-glucofuranose. Allyl 3-O-allyl-2,3,6-tri-O-benzyl-D-glucopyranoside was synthesized, and deallylated by conventional methods, and the glucopyranosyl chloride was again formed by treatment with hydrogen chloride in ether. Sodium hydride in oxolane caused ring closure, with formation of only a small proportion of the elimination product.

Varma[62] synthesized the corresponding 1,3-anhydro-2,4,6-tri-O-benzyl-β-D-mannopyranose from methyl 3-O-allyl-2,4,6-tri-O-benzyl-α-D-mannopyranoside similarly, with some variation in the method. Elimination was not a significant problem in this synthesis, as a *trans*-diaxial relationship for H-2 and the halogen is not possible.

For both compounds, the ^{13}C-n.m.r. spectrum was consistent with the assigned structure, the anomeric carbon resonance being observed at 5-7 p.p.m. to lower field than that of the corresponding 1,6-anhydride; this difference in chemical shift parallels that between oxetane (72.8 p.p.m.) and oxolane[63] (68.6 p.p.m.).

Ito (see Ref. 61) interpreted the ^{1}H-n.m.r. spectra of these compounds. Because of their rigid structure, only C-5 is capable of confor-

(61) H. Ito, R. Eby, S. Kramer, and C. Schuerch, *Carbohydr. Res.*, 86 (1980) 193–202.
(62) A. J. Varma, unpublished results.
(63) J. B. Stothers, *Carbon-13 NMR Spectroscopy*, Academic Press, New York, 1972.

mational motion, and its favored position has C-6 equatorial; thus, the pyranoid ring has a boat, and the 1,3-dioxane ring a chair, conformation (13). The anomeric-proton resonance of 1,3-anhydro-2,4,6-tri-O-

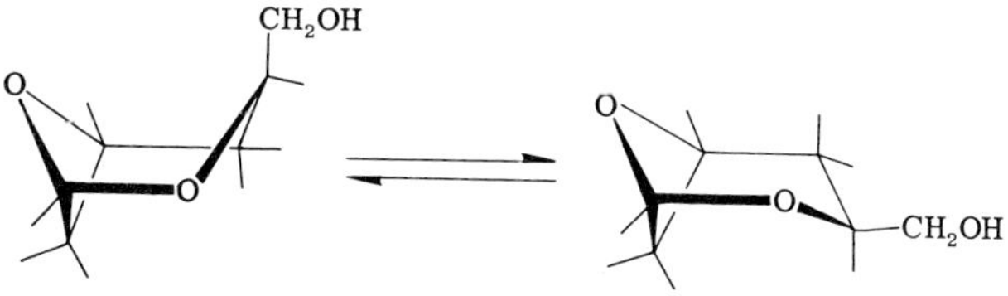

13

benzyl-β-D-glucopyranose appears at 5.48 p.p.m. as a triplet, demonstrating the presence of vicinal (1,2) and long-range coupling with coupling constants of ~4 Hz. The long-range coupling must be to H-3, as the dihedral angle between the anomeric C–H bond and the C-5–H bond is near the minimum range (~90°) of the Karplus relationship. For the 2-epimeric, D-mannose derivative, the anomeric proton appears as a doublet, with only long-range coupling ($^4J_{1,3}$ 4 Hz) evident.[62] These coupling constants are consistent with data for related bicyclo compounds.

No other 1,3-anhydroaldopyranose derivatives have been reported, and no 1,3-anhydroaldofuranoses are known.

5. 1,2-Anhydroaldopyranoses (2,7-Dioxabicyclo[4.1.0]heptanes)

Brigl's early synthesis[64] of 3,4,6-tri-O-acetyl-1,2-anhydro-α-D-glucopyranose was accomplished through a series of reactions not generally applicable to the synthesis of other 1,2-anhydroglycopyranoses. However, the structural features of the key intermediate, namely, 3,4,6-tri-O-acetyl-β-D-glucopyranosyl chloride, and Brigl's mode of ring closure are those that have already been emphasized as being the most suitable for the synthesis of anhydroaldoses. An alternative method,[65] the epoxidation of a substituted glucal, led to related compounds based on maltose, namely, 3,6-di-O-acetyl-1,2-anhydro-4-O-(2,3,4,6-tetra-O-acetyl-α-D-glucopyranosyl)-α-D-glucopyranose and its epimer, but this reaction sequence is clearly inferior.

Experience in the synthesis of 1,2-anhydro-3,4,6-tri-O-benzyl-β-D-

(64) P. Brigl. Z. Physiol. Chem., 122 (1922) 245–262.
(65) H. Arita, T. Ikenaka, and Y. Matsushima, J. Biochem. (Tokyo), 69 (1971) 401–407.

mannopyranose[66] and -α-D-glucopyranose[67] suggests that this class of anhydroaldoses should be fairly readily accessible. For both, perbenzylation of an orthoester, followed by hydrolysis, gave the corresponding 3,4,6-tri-O-benzyl-D-aldopyranose, and reaction of this with hydrogen chloride in ether yielded the α-D-aldosyl chloride. The D-mannopyranosyl chloride has OH-2 and the chlorine in the *trans*-diaxial arrangement, is highly reactive, and must be promptly treated with ammonia in order to achieve ring closure (rather than formation of oligomeric by-products). The anhydro sugar is somewhat labile, and must be carefully purified, but is a well characterized compound. The α-D-glucopyranosyl chloride undergoes dimerization on treatment with base, as the steric relationship of OH-2 and the chlorine are unfavorable for ring closure, but, if the α-chloride is converted into the β-fluoride, the steric relationship of the two groups is suitable, and ring closure occurs. In both instances, the formation of the anhydro ring may be readily monitored by n.m.r. spectroscopy, because of the upfield absorptions characteristic of three-membered rings.

6. Miscellaneous

A number of anhydro sugar derivatives have been prepared by treatment of orthoesters with mercuric bromide in nitromethane. Among them are 3-O-acetyl-1,5-anhydro-2-O-benzoyl-β-L-arabinofuranose,[68] 2,3-di-O-acetyl-1,5-anhydro-β-D-xylofuranose,[69] and 2,3,6-tri-O-acetyl-1,4-anhydro-α-D-glucopyranose.[70] These reactions might be considered as syntheses alternative to those already presented.

Anhydrides of ketoses have not been tested for polymerizability, but two monomeric anhydrides of D-fructose have been reported, namely, 2,3-anhydro-1,4,6-tri-O-nitro-D-fructofuranose[71] and 2,6-anhydro-1-O-methyl-D-fructofuranose.[72] The structures of both suggest that they have sufficient strain to be of interest. The first is the product of a

---

(66) S. J. Sondheimer, H. Yamaguchi, and C. Schuerch, *Carbohydr. Res.*, 74 (1979) 327–332.
(67) H. Yamaguchi and C. Schuerch, *Carbohydr. Res.*, 81 (1980) 192–195.
(68) A. F. Bochkov, I. V. Obruchnikov, V. N. Chernetsky, and N. K. Kochetkov, *Carbohydr. Res.*, 36 (1974) 191–195.
(69) A. F. Bochkov and A. V. Rodionov, *Izv. Akad. Nauk SSSR, Ser. Khim.*, (1975) 667–670; *Chem. Abstr.*, 83 (1975) 43,625p.
(70) A. F. Bochkov and Ya. V. Boznyi, *Zh. Org. Khim.*, 13 (1977) 750–755; A. F. Bochkov and G. E. Zarkov, *Chemistry of the O-Glycosidic Bond*, Pergamon, Oxford, 1979.
(71) M. Sarel-Imber and J. Liebowitz, *J. Org. Chem.*, 24 (1959) 141, 1897–1900.
(72) F. Micheel and E. A. Kleinheidt, *Chem. Ber.*, 98 (1965) 1668–1672.

nitration. The second is formed by the action of alkali on the ketosyl fluoride, a reaction analogous to others already discussed.

The synthesis of other anhydro sugar derivatives is usually accomplished by an internal displacement of a leaving group with inversion, as has already been emphasized for those anhydrides in which the anhydride bridge engages the anomeric position. Three such compounds that have been used in polymerizations deserve specific mention: 5,6-anhydro-1,2-O-isopropylidene-α-D-glucofuranose,[12,13] its 3-methyl ether,[11] and 3,5-anhydro-1,2-O-isopropylidene-α-D-xylofuranose,[14] all of which were synthesized by p-toluenesulfonylation of the primary hydroxyl group of the parent, isopropylidene derivative, followed by treatment with base.

III. Polymerization of Anhydroaldoses

1. General Considerations: Mechanism

There are two characteristic classes of polymerization process, those of step growth and of chain growth.[73] In the former, monomers combine to form dimers, and dimers to form tetramers; and all oligomers of intermediate size combine to form larger molecules at random. In this kind of process, typical of classical, equilibrium-condensation polymerization, products of high molecular weight are obtained at very high conversion only.

In chain-growth processes, however, a monomer is activated in an initiation step, and reacts rapidly with other monomers in sequence, forming a chain of mers having an active, growing end. Chain growth is typically rapid, and high-polymeric molecules are formed in the presence of a large excess of the monomer. Termination, or chain-transfer, reactions give completed, inactive product of high molecular weight, even at low conversion of the monomer. As a practical matter, the latter characteristic provides an enormous, synthetic advantage to chain-growth processes, and the great advantage of anhydro sugars as the monomers for polysaccharide synthesis is the fact that they polymerize by a chain-growth mechanism.

There are, however, synthetic constraints on the use of anhydro sugars as a source of polysaccharides, because chain-growth processes generally lead to the formation of homopolymers only, or, if two or more monomers are used, to the formation of random copolymers. In these processes, the polymer composition and mer sequence-

(73) R. W. Lenz, *Organic Chemistry of Synthetic High Polymers*, Interscience, New York, 1967.

distribution are determined by statistical considerations based on monomer reactivity and the proportion of comonomers in the feed. Even within these limitations, there are, in principle, great possibilities for polysaccharide synthesis yet to be realized.

The details of ring-opening, chain polymerization vary with the different heterocycles, and have been well reviewed.[74-79] Only common features will be emphasized here. Typical monomers are monocyclic ethers, and acetals having ring strain. The ring strain, and thus the polymerizability of these compounds, is predictably influenced by ring size, and, less predictably, by the number and distribution of the substituents. Substituents frequently, but not always, stabilize small rings, and shift ring-chain equilibria in favor of the ring. For bicyclic ethers and acetals, the influence of substituents is more complex; they may increase polymerizability by introducing additional, conformational strains, but reactivity may be adversely affected by eclipsed conformations in the transition state of propagation. As a result, the relative reactivities of closely similar monomers cannot always be foreseen.

In the three-membered heterocycles, the oxiranes, polymerization has been induced with basic catalysts. Usually, these polymerizations are slow, and the polymers formed are of relatively low molecular weight. Certain, specially prepared carbonates and amides of calcium constitute a limited exception to this generalization, as they allow the formation of a solid polymer from ethylene oxide.

Cationic mechanisms are much more characteristic of the polymerization of oxygen heterocycles, both ethers and acetals. A wide variety of catalysts has been used, including protonic acids, such Lewis acids as boron trifluoride, phosphorus pentafluoride, stannic chloride, antimony pentachloride, titanium tetrachloride, zinc chloride, and ferric chloride, and salts of carbocations or trialkyloxonium ions having anions derived from Lewis acids. Some complex, coordination catalysts that appear to operate by a mechanism

(74) A. M. Eastham, *Adv. Polym. Sci.*, 2 (1960) 18–50.
(75) J. Furukawa and T. Saegusa, *Polymerization of Aldehydes and Oxides*, Interscience, New York, 1963.
(76) K. C. Frisch and S. L. Reegen, *Ring-Opening Polymerization*, Dekker, New York, 1969.
(77) T. Saegusa and E. Goethals, *Ring-Opening Polymerization*, A.C.S. Symp. Ser. No. 59, American Chemical Society, Washington, D.C., 1977.
(78) J. P. Kennedy, *Int. Symp. Cationic Polymerization, 4th, Polym. Symp.*, No. 56, Interscience, New York, 1976.
(79) J. B. Rose, in P. H. Plesch (Ed.), *The Chemistry of Cationic Polymerization*, Macmillan, New York, 1963.

having both cationic and anionic features have also been found effective; these are, typically, partially hydrolyzed aluminum-, zinc-, or magnesium-alkyls containing multiple metal atoms linked by bridging oxygen atoms. In some instances, chelating agents and alcohols are added to the recipe. The effectiveness of the catalyst is a function of the ratio of water to metal-alkyl, the nature of the metal, and the structure of the monomer. These catalysts are often complex, and are not single molecular species.[80] A number of more-complex, catalyst systems have also been reported.[73]

Although the mechanisms of most of these polymerizations are cationic, and have been investigated widely, the specific details should probably be accepted with some reserve, and their descriptions for the most part considered to be illustrative, rather than definitive. The kinetics of cationic polymerization are notoriously difficult to determine with accuracy. The level of purity of solvents, catalysts, and reactants that is necessary for rigorous results, especially with oxygenated monomers, is rarely obtained. The definition of initiation, transfer, and termination processes, and the recognition of cationic side-reactions are serious problems. In many systems, the necessity for, or absence of, cocatalysis has not been established. In other systems, protonic cocatalysts may cause very great increases in propagation rate and molecular weight; however, this effect usually passes through a maximum at a specific ratio of catalyst to cocatalyst, and, beyond that point, more and more cocatalyst causes poorer and poorer results. Frequently, two kinds of product are obtained from a single polymerization, those of high molecular weight, and an oligomeric fraction. Sometimes, the oligomers are cyclic or atactic, whereas the high polymer may be linear and largely isotactic. In such systems, two reaction mechanisms must be operative. For successful results with these polymerizations, it is probably more important to understand the effect of pertinent, empirical parameters than to have completely rigorous, mechanistic models of the systems.

From the standpoint of synthesis of polysaccharides, the most significant aspect of a polymerization mechanism is whether or not it involves regio- and stereo-selective control. However, the structural and stereochemical problems with polysaccharide synthesis are generally simpler than those that obtain with a racemic, unsymmetrical monomer, such as (R, S)-propylene oxide. For example, a variable percentage of head-to-head and tail-to-tail sequences is found in

(80) E. J. Vandenberg, *J. Polym Sci. Part A1*, 7 (1969) 525–567.

poly(propylene oxide). To date, none have been found in the synthetic polysaccharides. Acid hydrolysis invariably yields the original sugar exclusively. If tail-to-tail (2→2')-linkages had been formed, a nonhydrolyzable, disaccharide fraction would be present. In addition, as no epimeric sugars are released on hydrolysis, bond breaking during polymerization must have occurred solely between the ring-oxygen atom and C-1. Furthermore, in all examples so far examined, except the 1,4-anhydroaldopyranoses, one ring is opened selectively. Therefore, the only stereochemical problem to be solved is the configuration at C-1 of the polymer. No case of configurational retention at C-1 has been observed, and polymerization occurs either by total inversion at C-1, or with some racemization. These steric relationships may be readily interpreted by assuming simultaneous bond-breaking and bond-formation during propagation with inversion, or, in racemization, some degree of prior bond-breaking. In the former, the proposed intermediate is treated as a trialkyloxonium ion, and in the latter, as a C-1 carbocation.

This mechanistic model cannot be complete, however, as it gives no explicable role to the counter-ion or other features of the system. Yet, in the polymerization of 1,6-anhydroaldopyranose derivatives, the conditions under which the monomers polymerize with steric control, and their rates, are quite different with phosphorus pentafluoride, boron trifluoride, and trialkyloxonium hexafluorophosphates, but the propagating cation is formally the same in each reaction and, in two, so is the counter-ion.

2. General Considerations: Copolymerization

Copolymers are readily prepared by conducting polymerizations of a mixture of monomers. However, to obtain a product having any reasonable, structural homogeneity, it is necessary to take the reaction mechanism into account, and to perform the experiment under conditions consistent with classical, copolymerization theory. With properly controlled experiments, it is possible to determine the relative reactivities of the monomers, and the range of compositions and mer sequence-length distributions in any copolymer produced.[81,82]

The assumptions underlying classical copolymerization theory are the following. (a) The composition of the copolymer is not affected by the initiation or termination processes, but is determined by the prop-

---

(81) T. Alfrey, Jr., J. J. Bohrer, and H. Mark, *Copolymerization*, Interscience, New York, 1952.
(82) G. E. Ham, *Copolymerization*, Interscience, New York, 1964.

agation reactions. (b) The reactivity of a growing chain is determined by the terminal mer at the active end, not by the penultimate or earlier units. (c) Disappearance of the monomers from the feed is only by virtue of their being incorporated into the copolymer, and the process is not reversible. (d) In order to eliminate the (unknown) concentration of growing ends from the rate expressions, it is customary to invoke a steady-state assumption (although other treatments of this problem have been proposed[83]).

In binary copolymerizations, therefore, four propagation reactions, having four rate constants, are considered.

$$\sim\!\!\sim\!\!\sim m_1^+ + M_1 \xrightarrow{k_{11}} \sim\!\!\sim\!\!\sim m_1^+$$

$$\sim\!\!\sim\!\!\sim m_1^+ + M_2 \xrightarrow{k_{12}} \sim\!\!\sim\!\!\sim m_2^+$$

$$\sim\!\!\sim\!\!\sim m_2^+ + M_2 \xrightarrow{k_{22}} \sim\!\!\sim\!\!\sim m_2^+$$

$$\sim\!\!\sim\!\!\sim m_2^+ + M_1 \xrightarrow{k_{21}} \sim\!\!\sim\!\!\sim m_1^+$$

On the basis of these equations and the foregoing assumptions, the classical, copolymer-composition equation (1) is derived that relates the ratio of mers in the copolymer to the monomer concentrations in the feed.

$$\frac{dM_1}{dM_2} = m_1/m_2 = M_1(r_1 M_1 + M_2)/M_2(r_2 M_2 + M_1), \qquad (1)$$

in which M_1 and M_2 are the monomer concentrations; m_1/m_2, the ratio of mers in the copolymer; and the monomer reactivity ratios, $r_1 = k_{11}/k_{12}$ and $r_2 = k_{22}/k_{21}$. This equation may be used in the derivative form for low-conversion copolymerizations, or in the integrated form for higher-conversion experiments.

The reactivity ratios may be evaluated by performing a series of low-conversion copolymerizations at different, monomer-feed ratios, isolating the copolymer, and determining its composition. A number of mathematical analyses have been proposed in order to provide, from the experimental data, correct values for the two unknown reactivity ratios. There is some difference of opinion as to the best method for obtaining values having quantifiable errors.[84,84a] However, several of

(83) G. Goldfinger and T. Kane, J. Polym. Sci., 3 (1948) 462–463.
(84) R. C. McFarlane, P. M. Reilly, and K. F. O'Driscoll, J. Polym. Sci., Polym. Chem. Ed., 18 (1980) 251–257.
(84a) H. Patino-Leal, P. M. Reilly, and K. F. O'Driscoll, J. Polym. Sci., Polym. Lett. Ed., 18 (1980) 219–227.

the accepted methods give values of r_1 and r_2 that are usually satisfactory for practical purposes, as, in most cases, the errors are probably no greater than the uncertainty of the experimental method itself.

The largest body of information on copolymerization relates to radical copolymerization of vinyl monomers. However, cationic copolymerization has been extensively studied, and some additional complexities are found to be present in such systems. The copolymerization behavior is determined not only by the monomers employed, but also by the catalyst; this is not surprising, because the anion derived from the catalyst remains in proximity to the growing center, and can affect the approach of monomer. Ionic copolymerization is also more sensitive to the experimental conditions than radical copolymerization, and monomer reactivities can be altered substantially by a change in the temperature or the solvent. The range of monomers that undergo ionic copolymerization is also much narrower than that reacting by the radical mechanism. These differences are, in part, a consequence of the great range of Lewis acidities and basicities found among the catalysts, monomers, and solvents. Undoubtedly, leveling effects, and competition between monomer and solvent, influence the course of many of these processes, although a detailed explanation of the various differences cannot usually be given.

Only a limited amount of information is available on the copolymerization of oxygen heterocycles, and this can only serve as a general background in the investigation of anhydro sugar copolymerizations. It is more useful to consider the general principles on which binary copolymerization theory is based. It appears probable that, in some ionic copolymerizations, the copolymer composition has been influenced by the mode of initiation, as well as by the propagation reactions. This problem seems unlikely to be found in the copolymerization of closely related monomers, such as anhydro sugars having the same ring-size. Nevertheless, copolymerization data obtained on polymers of very low molecular weight should be viewed with scepticism. The assumption that the reactivity of a growing chain is determined by its terminal, active mer is eminently reasonable in the formation of $(1 \rightarrow 6)$-linked polysaccharides, but will not necessarily be valid for copolymers linked through the secondary positions, especially C-2. In these, there is a possibility that the penultimate unit may influence the propagation by special, steric interactions; if that is so, more than four propagation reactions will be involved, and the simple binary statistics will no longer hold.

The simple, copolymer-composition equation should also be applied only to those cases in which the stereoselectivity is very high. If partial racemization occurs during copolymerization of anhydro sugar derivatives, more than one propagating chain-end may be characteristic of each of the mers. There may, for example, be both a trialkyloxonium ion and a carbocation; more than four propagation reactions are then involved. Should the carbocation and trialkyloxonium ion interconvert, the mechanistic situation becomes reminiscent of stereoblock polymerization by the anionic mechanism, which may be described by the Coleman–Fox treatment.[85] With the added complication of two monomers, however, it is doubtful that the system could be rigorously defined. Finally, it should be noted that the copolymer-composition equation is based on the premise that the monomers react to form copolymer only, but, in many stereoselective polymerizations, an oligomeric fraction is formed as well. To the extent that this process occurs, calculations of reactivities may be erroneous.

Because of these complications, data from copolymerizations should be used with caution. Nevertheless, it appears that, for the copolymerization of 1,6-anhydroaldopyranoses, the major premises are met (as well as for radical copolymerization of vinyl monomers).

3. Polymerization of 1,6-Anhydroaldo-pyranoses and -furanoses

The polymerization of unsubstituted 1,6-anhydro-β-D-glucopyranose was first observed by Pictet,[21] and was further investigated by Pringsheim and Schmalz.[86] Irvine and Oldham[23] tried unsuccessfully to prepare a linear polysaccharide by polymerization of 2,3,4-tri-O-methyl-β-D-glucopyranose. Interest in the polymerization again arose[22,87,88] in 1959, and unsuccessful attempts were made to polymerize fully substituted derivatives under melt conditions similar to those used earlier with unsubstituted 1,6-anhydro-β-D-glucopyranose.[28]

The first successful polymerizations were obtained in solution, with cationic initiators, under conditions typical of ring-opening polymerization. Brederek and Hutten polymerized the perbenzyl ether and peracetate of levoglucosan, using, as initiators, various carbonium ions formed *in situ* from organic halides and silver perchlorates. The products were apparently not stereoregular, but were definitely poly-

---

(85) B. D. Coleman and T. G. Fox, *J. Chem. Phys.*, 38 (1963) 1065–1075.
(86) H. Pringsheim and K. Schmalz, *Ber.*, 55 (1922) 3001–3007.
(87) J. DaS. Carvalho, W. Prins, and C. Schuerch, *J. Am. Chem. Soc.*, 81 (1959) 4054–4058.
(88) M. L. Wolfrom, A. Thompson, and R. B. Ward, *J. Am. Chem. Sco.*, 81 (1959) 4623–4625.

meric. This important work was not published by the investigators, but was reported in 1963 in a review article.[26]

The utilization of products from wood and cellulose was under investigation at the Institute of Wood Chemistry of the Latvian Academy of Sciences during the same period, and R. Ya. Pernikis[27] undertook a study of the polymerization of levoglucosan and some derivatives as a doctoral-research project under the supervision of V. V. Korshak. They chose to investigate the use of catalysts effective in ring-opening polymerization of oxygen heterocycles. Levoglucosan was polymerized under the action of a number of Lewis acids in 1,4-dioxane at ~90°. The products were branched and of mixed anomeric configuration, and were similar to those obtained elsewhere. Polymerization of 1,6-anhydro-2,3,4-tri-O-methyl-β-D-glucopyranose was conducted at 20–50° in toluene with boron trifluoride etherate; the products were of low viscosity, but crystalline. Polymerizations at lower temperature gave polymers having a higher viscosity. The investigators proposed that water was present as cocatalyst in the polymerization of the trimethyl ether. Initiation was presumed to occur by protonation of the anhydro ring, with [BF_3OH] as the counter-ion. The propagation process was postulated to proceed by way of a trialkyloxonium ion, both with the free anhydro sugar and its trimethyl ether.[27] No stereochemical conclusions were drawn from the mechanism, and the optical activity of the fully methylated polysaccharide was not recorded. However, the later development of the polymerization of anhydro sugars depended critically on the discoveries in Bredereck's and the Latvian laboratories.

The polymerization of 1,6-anhydro-2,3,4-tri-O-methyl-β-D-glucopyranose with boron trifluoride etherate in toluene was confirmed by Tu and Schuerch,[89] although the molecular weights of the products were found to be much lower than claimed by the Soviet authors. They also observed that the polymer had an extremely high, positive rotation, and therefore had an overwhelming preponderance of α-linkages. The postulate by the Soviet authors of a cocatalyst was also confirmed in this work, as the addition of a small proportion of water or isopropyl alcohol appeared to enhance, slightly, the stereoregularity under their conditions. Further work in Latvia demonstrated that a product of maximum viscosity was obtained at a polymerization temperature of −20° with the same monomer and catalyst, and that the effectiveness of the catalyst decreased abruptly at lower temperatures.[90]

(89) C.-C. Tu and C. Schuerch, *Polym. Lett.*, 1 (1963) 163–165.
(90) V. V. Korshak, V. A. Sergeev, Ya. A. Surna, and R. Ya. Pernikis, *Vysokomol. Soedin.*, 5 (1963) 1593–1596.

A systematic investigation was also undertaken in Syracuse, in order to optimize the conditions of polymerization. At room temperature, the boron trifluoride-catalyzed polymerization was found to give relatively low conversions, as well as products of low molecular weight, when conducted in benzene or toluene, and dichloromethane as the solvent gave even lower values. When the catalyst was changed to phosphorus pentafluoride, and the polymerization performed in dichloromethane, the stereoregularity of the polymerization was much lower at room temperature, but, at $-78°$, the polymerization was again stereoselective and the molecular weight was dramatically increased.[91] The exact temperature at which the polymerization changed from stereoselective to relatively random appears not to have been determined exactly, but, with phosphorus pentafluoride, the temperature range is probably[92] around $-40°$.

No further work was done to investigate whether or not a cocatalyst was advantageous with boron trifluoride as the catalyst, but addition of a small proportion of water to the phosphorus pentafluoride was found to result in products of lower degree of polymerization (d.p.), and, perhaps, slightly lower optical rotations. It was later found that traces of ethanol were present [from recrystallation of the monomer 1,6-anhydro-2,3,4-tri-O-benzyl-β-D-glucopyranose (GlcAnBzl$_3$) in the early experiments], and that products of higher molecular weight could be obtained if this was completely removed by recrystallization from aprotic solvents.[93]

Little work was conducted on the role of the solvent. The standard conditions for most of the subsequent work involved polymerizations in dichloromethane at $-60°$ to $-78°$. However, a few experiments using sulfur dioxide as the solvent, with phosphorus pentafluoride as the catalyst, at $-78°$, gave, from GlcAnBzl$_3$, a polymer of low d.p. having only a slightly lower specific rotation.[92] Whether this was the result of slightly lower stereoregularity in the polymer, or contamination with a low-d.p. fraction of different tacticity, was not established.

Among the ethers of levoglucosan that have by now been tested for polymerizability are the trimethyl,[91] triethyl,[91] tribenzyl,[91-97] 2,4-di-O-

(91) E. R. Ruckel and C. Schuerch, *J. Org. Chem.*, 31 (1966) 2233–2236.
(91a) E. R. Ruckel and C. Schuerch, *Biopolymers*, 5 (1967) 515–523.
(92) J. Zachoval and C. Schuerch, *J. Am. Chem. Soc.*, 91 (1969) 1165–1169.
(93) T. Uryu and C. Schuerch, *Macromolecules*, 4 (1971) 342–345.
(94) J. W. P. Lin and C. Schuerch, *J. Polym. Sci., Part A1*, 10 (1972) 2045–2060.
(95) C. Schuerch and T. Uryu, *Macromol. Synth.*, 4 (1972) 151–155.
(96) T. Uryu, H. Tachikawa, K. I. Ohaku, K. Terui, and K. Matsuzaki, *Makromol. Chem.*, 178 (1977) 1929–1940.
(97) T. Uryu, K. Ito, K. I. Kobayashi, and K. Matsuzaki, *Makromol. Chem.*, 180 (1979) 1509–1510.

TABLE II

Polymerization of 1,6-Anhydro Sugar Derivatives

Monomer	Catalyst monomer (mole%)	Monomer solvent (g/100 mL)	Temperature (degrees)	Yield (%)	Time (hours)	[α]D (degrees)	[η]	D.p.	References
Derivative of 1,6-anhydro-β-D-glucopyranose									
tri-O-acetyl-	10–25	20	0			138–178	0.04	—	26,92
tri-O-benzyl-	0.8	50	−60	77	0.66	114	1.08	1800	26,91–97
tri-O-(p-bromobenzyl)-	1	30	−60	97	1	116	1.05	—	104
tri-O-ethyl-	12–20	10	−78	75		190		~150	91
tri-O-methyl-	3.7	10	−78			199	—	—	27,91
tri-O-(p-methylbenzyl)-	1	30	−60	58		103	0.40	—	99–103
3-O-acetyl-2,4-di-O-benzyl-	9	30	−60	45	40	132	0.16		105
2,4-di-O-benzyl-3-O-(2-butenyl)-	1	30	−60	70	0.17	118	1.25		98
Perbenzylated derivative of 1,6-anhydro-									
β-D-allopyranose	10		−60	25	75	134		~4	106
β-D-altropyranose			40					~3	106
α-D-galactofuranose	20–30	100		50	125	+8 to −17			93,94,101,107,108
β-D-galactopyranose	1–25	100	−60	70	24	104	0.6	400	94,101,102,107,109
β-D-mannopyranose	1	50	−60	90	3	58	2.6		
cellobiose	10	50	−60	70	100	77		$\bar{M}_n$ 5800	110
maltose	20	100	−60	70	100	98		$\bar{M}_n$ 11,000	111

benzyl-3-O-crotyl,[98] tri-(p-methylbenzyl),[99-103] tri-p-xylyl, and tri-(p-bromobenzyl).[104] All polymerize well, to afford polymers of high molecular weight at -60 to $-78°$ with phosphorus pentafluoride (see Table II). The tris(trimethylsilyl) ether, however, failed to polymerize under a variety of conditions.[91]

No triester of levoglucosan was found[92] that polymerized at temperatures much below 0°. At $-78°$, the triacetate complexed with phosphorus pentafluoride, and, at high concentration, precipitated from solution.[91] The tris(monofluoroacetate) failed to polymerize under a variety of conditions.[91] The trinitrate polymerized at 0°, but the product was not fully characterized.[92] Polymerization of the triacetate proceeded to reasonable conversions with a number of catalysts at 0°, but the viscosity and the stereoregularity of these polymers were low.[92] In a simple, copolymerization experiment, it was demonstrated that the low polymerizability of levoglucosan triacetate was due not only to a failure to initiate but also to sluggish propagation.[92]

Several catalysts and initiator systems have been tested for the polymerization of GlcAnBzl$_3$, including the following Lewis acids: boron trifluoride and its etherate, phosphorus pentafluoride, titanium tetrachloride, and antimony pentachloride and pentafluoride. Several cationic initiators have also been used, including (triphenylmethyl) antimony hexachloride, 2,3,4,6-tetra-O-acetyl-D-glucopyranosyl hexafluorophosphate, acetyl hexafluorophosphate, pentamethylbenzyl hexafluorophosphate (most of which were generated *in situ*), and triethyl-

---

(98) H. Ito and C. Schuerch, *J. Am. Chem. Soc.*, 101 (1979) 5797–5806.
(99) J. W. P. Lin and C. Schuerch, *Macromolecules*, 6 (1973) 320–324.
(100) W. H. Lindenberger and C. Schuerch, *J. Polym. Sci., Polym. Chem. Ed.*, 11 (1973) 1225–1235.
(101) K. Kobayashi and C. Schuerch, *J. Polym. Sci., Polym. Chem. Ed.*, 15 (1977) 913–926.
(102) K. Kobayashi, R. Eby, and C. Schuerch, *Biopolymers*, 16 (1977) 415–426.
(103) H. Ito and C. Schuerch, *J. Polym. Sci., Polym. Chem. Ed.*, 16 (1978) 2217–2224.
(104) H. Ito, *J. Polym. Sci., Polym. Lett. Ed.*, 19 (1981) 43–47.
(105) K. Kobayashi, H. Sumitomo, and A. Yasui, *Macromolecules*, 12 (1979) 1019–1023.
(105a) K. Kobayashi, unpublished results.
(106) K. Matsuzaki, *U. S.–Jpn. Semin. Synth. Polysaccharides Their Biochem. Functions*, Syracuse, N. Y., 1979.
(107) H. Ito, V. Marousek, and C. Schuerch, *J. Polym. Sci., Polym. Chem. Ed.*, 17 (1979) 1299–1307.
(108) T. Uryu, H. Libert, J. Zachoval, and C. Schuerch, *Macromolecules*, 3 (1970) 345–349.
(109) J. M. Fréchet and C. Schuerch, *J. Am. Chem. Soc.*, 91 (1969) 1161–1164.
(110) B. Veruovic and C. Schuerch, *Carbohydr. Res.*, 14 (1970) 199–206.
(111) V. Masura and C. Schuerch, *Carbohydr. Res.*, 15 (1970) 65–72.

oxonium salts with various anions. Phosphorus pentafluoride gave a polymer of the highest molecular weight having the highest specific rotation, and some other fluorinated catalysts or initiators also gave stereoregular polysaccharides. Stable cations cause polymerization at higher temperatures than does phosphorus pentafluoride, but cations having perchlorate counter-ions, which were introduced by Bredereck and Hutten,[26] appear not to have been tested under conditions now known to give stereoregular products.[92] Chlorosulfonic acid has been reported to give a stereoregular polymer from tri-O-methyl-levoglucosan.[112] This seems to be the only reported, stereoselective polymerization with a nonfluorinated catalyst, or counter-ion.

Each of the fluorinated catalysts has an optimum temperature-range for stereoselective polymerization. At the lower temperatures, the rate of propagation and yield of polymer decrease dramatically. At higher temperatures, the molecular weight of the polymer produced becomes lower, presumably because chain transfer or termination processes increase in importance. At still higher temperatures, the stereoregulation is lost, and the low-d.p. polymer produced has a mixed, anomeric configuration.

In general, a polymer of the highest molecular weight and specific rotation is formed when the catalyst is used at low catalyst:monomer ratios.[92] For the polymerization of $GlcAnBzl_3$, the minimum proportion of catalyst effective in inducing polymerization is slightly less than 1 mol%. Monomers having tri-O-p-xylyl substituents generally require somewhat higher catalyst concentrations; this is due to a sluggish initiation, rather than to a significantly lower rate of propagation. Presumably, p-xylyl ether oxygen is slightly more basic than benzyl ether oxygen. The Lewis-acid catalyst undoubtedly complexes with both the ring-oxygen atom and the ether oxygen atoms. If the basicity of an ether oxygen atom is enhanced, the complexation equilibrium is shifted away from the ring-oxygen atom, and, as a result, initiation is slower.[103] The fact that polymerization does not proceed at concentrations of catalyst much less than 1 mol% also suggests complexation of ether oxygen atoms.

The concentration of the monomer also affects the course of the polymerization in a predictable way. Because propagation is a reaction between a propagating, mer cation and monomer, the rate is substantially enhanced at high monomer concentrations, and the molecular weight of the polysaccharide is higher. This effect is most noticeable in the polymerization of sluggish monomers, and concentrations of monomer as high as 50% are often advantageous.

(112) A. Klemer and C. L. Apostolides, *Carbohydr. Res.*, 22 (1972) 432–435.

In the polymerization of oxygen heterocycles, it is not uncommon to observe a simultaneous, chain-degradation process; this has also been observed[96] in the polymerization of GlcAnBzl$_3$. On treatment of the stereoregular polymer with phosphorus pentafluoride, the molecular weight decreased, and its distribution broadened. On the basis of this observation, the time of polymerization was limited to 40 min, and it proved possible to prepare this polymer at the highest viscosity obtained to date.

A mechanism based on that proposed for the polymerization of oxolane and other heterocycles serves as a satisfactory, general statement of the processes occurring during polymerization of 1,6-anhydroaldopyranoses.[91,92,101]

a. There is an unproductive complexation of Lewis acid with ester or ether functions on the mer unit. The complexation equilibrium competes with productive complexation on the ring-oxygen atom, and determines how much catalyst is available for initiation.

b. Initiation may involve reaction of a Lewis acid with a cocatalyst, such as water or other hydroxylic solvent, to form a protonic acid.

$$HOH + BF_3 \rightleftarrows H^+ BF_3OH^-$$

However, initiation proceeds readily in the absence of hydroxylic solvents, and transfer processes are also minimized in their absence. It is, therefore, advantageous to operate under strictly aprotic conditions. The initiation process with phosphorus pentafluoride can then be formulated as taking place in three steps: (*i*) an initial complexation with the ring-oxygen atom, resulting in partial electron-deficiency at C-1; (*ii*) attack by the ring-oxygen atom of a second monomer on the electron-deficient C-1, with simultaneous bond-breaking to form a terminal -CH$_2$-O-PF$_5^-$ anion and an active trialkyloxonium ion; and (*iii*) transfer of fluoride anion to phosphorus pentafluoride, to give an uncharged, terminal -CH$_2$-O-PF$_4$, and a hexafluorophosphate counterion that can migrate and remain close to the trialkyloxonium ion during propagation (see Scheme 3).

c. Propagation consists in the repeated attack on monomer ring-oxygen atoms by C-1 of the trialkyloxonium ion of the growing chain-end, and migration of phosphorus hexafluoride anion with the changing cation. The conformational changes indicated in Scheme 3 are based on evidence from copolymerization studies to be described later.

d. Termination or transfer processes clearly occur during the polymerization, but their nature has not yet been determined.

Some physical evidence in favor of this (largely speculative) mechanism comes from ^{19}F- and ^{31}P-nuclear magnetic resonance studies.[97] In the polymerization system of GlcAnBzl$_3$ and phosphorus penta-

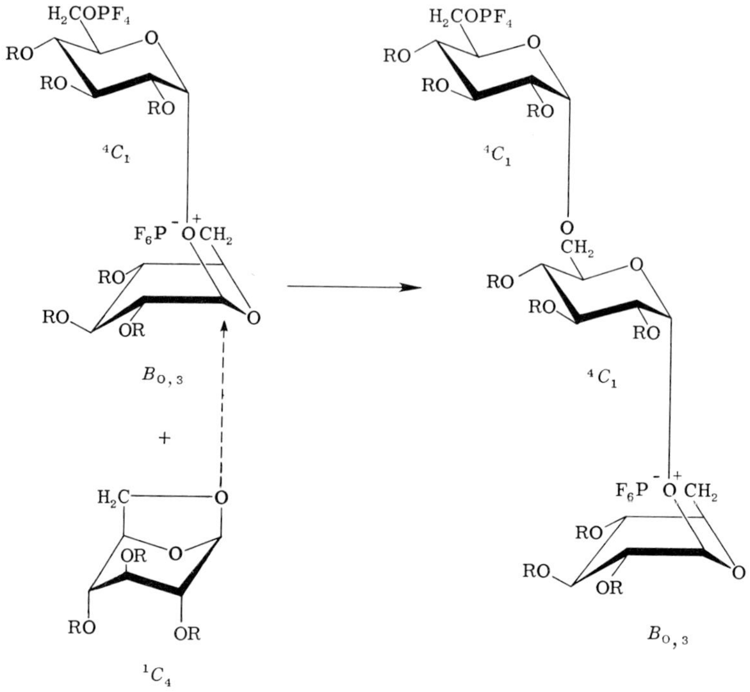

Scheme 3

fluoride in dichloromethane at $-80°$, absorptions assigned to a complex between the two compounds were observed. At $-60°$, the absorptions disappeared, and absorptions due to phosphorus hexafluoride anion replaced them. In addition, at $-80°$, absorptions identified as due to phosphoryl fluoride (POF_3) and $-CH_2-O-PF_4$ are also observed. The phosphoryl fluoride appears to be formed by the reaction of phosphorus pentafluoride with the glass of the container. At higher temperatures, $-CH_2-O-PF_4$ (the proposed substituent on C-5 of the nonreducing end-unit of the polymer) gradually disappears. These interpretations are consistent with the general mechanism proposed.

Under conditions that give products having mixed, anomeric configurations, bond breaking must precede bond formation during propagation, and the presumed intermediate is a planar carbocation[92] (see Scheme 4). Termination processes are much more severe under these circumstances, and the products have much lower molecular weights. It is conceivable either that the cation is a much more powerful electrophile, or that it much more readily eliminates a proton than the

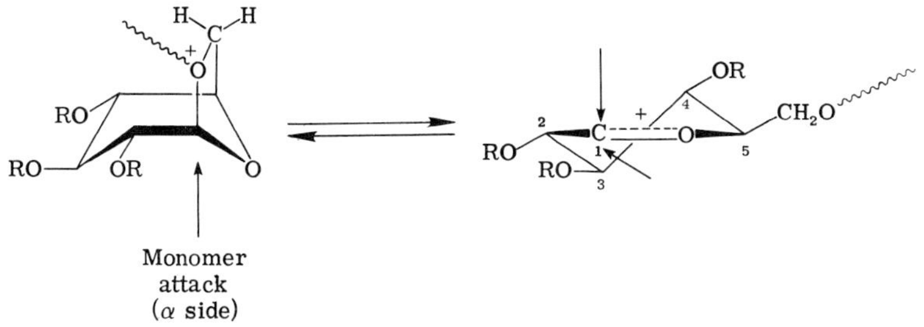

Monomer
attack
(α side)

Scheme 4

trialkyloxonium ion intermediate. No evidence is as yet available to substantiate any mechanism of termination.

Although there may be a minor, entropic contribution to the polymerization of the 1,6-anhydro-β-D-glucopyranose derivatives, the main driving-force clearly comes from the release of steric strain. These monomers are in the $^1C_4$(D) conformation in which all of the substituents are axially attached and produce corresponding, destabilizing 1,3-interactions. Were the polymerizability of the 1,6-anhydroaldopyranoses dependent solely on the ground-state energy of the monomers, it would be expected that the levoglucosan derivatives would be the most readily polymerizable 1,6-anhydroaldopyranoses, but this is not the case. Five of the eight isomeric 1,6-anhydro-2,3,4-tri-O-benzyl-β-D-aldohexopyranoses have now been tested for polymerizability, the *gluco*, *manno*,[94,101,102,107,109] and *galacto*[93,94,107,108] isomers extensively, and the *allo* and *altro* isomers[106] somewhat preliminarily. Nevertheless, the polymerizability of the five isomers is almost certainly in the order *manno* > *gluco* > *galacto* > *allo* > *altro*. The *manno*, *gluco*, and *galacto* isomers have been compared independently by a number of investigators, all of whom found, from qualitative observations, that the rate of polymerization is in the order *manno* > *gluco* > *galacto*, and the molecular weights of the polymers produced under similar, or identical, conditions lay in the same order. Copolymerization data, presented later, gave the same order for rates of propagation. Polymerization of the *allo* isomer under comparable conditions gave a stereoregular polymer, but only relatively low conversions.[106] The *altro* isomer failed to polymerize, except at high temperature and high concentration of catalyst.[106] It thus appears unlikely that the order given for the polymerizability of these monomers will be changed, even when increased experience with the last two isomers has been gained. Furthermore, the same order of reactivity has been found in the acid-catalyzed polymerization of the unsubstituted anhydro-*manno*, -*gluco*, and -*galacto* sugars.[113] Factors other than the ground-state energy of the monomer must cause the *manno* to be more reactive than the *gluco* isomer, and these will be discussed in more detail (see later in this Section).

The polymerization of esters of levoglucosan is unsatisfactory for a number of reasons.[91,92] The ester group probably complexes unproductively, and more strongly, with Lewis acids than the ether group. The electron-withdrawing character of the ester groups probably lowers the basicity of the ring-oxygen atom, and, therefore, the polymerization requires higher temperatures in order to proceed. Boron trifluoride fails to polymerize levoglucosan triacetate, and phosphorus pentafluoride does not cause polymerization at temperatures much below 0°. Under these more-drastic conditions, the reactive intermediate has less trialkyloxonium character; consequently, high termination rates result, and β attack is permitted. In addition, the ester group on C-2 may participate as a neighboring group, inhibiting α attack. With minor changes in the conditions of polymerization, the specific rotations of the polymeric esters vary over a wide range.

The results of polymerization of a number of 1,6-anhydroglycopyranose derivatives with phosphorus pentafluoride are given in Table II. With the exception of the trimethyl and triethyl ethers of levoglucosan, the values listed are generally typical of those obtained in current practice. To equal or surpass these values, it is necessary to ensure aprotic conditions, high monomer and low catalyst concentrations, a reasonably short reaction-time, and a temperature of polymerization near the optimum. In general, earlier, pioneering experiments gave stereoregular polysaccharide derivatives of lower viscosity. These results are discussed later in conjunction with a consideration of copolymerization results.

Copolymerization has been used for evaluating the reactivity of the monomeric, anhydro sugar derivatives, and also to prepare stereoregular polysaccharides of structures more complex than those of those prepared from a single monomer.[98-104,107] The procedure adopted has been first to determine the reactivity ratios of the monomers, and then to perform preparative experiments under conditions that provide polysaccharides having the desired, copolymer composition.

The method of determining monomer reactivity-ratios is well defined. A series of copolymerizations is conducted over a wide range of monomer ratios, and the polymerizations are terminated before the

---

(113) P. C. Wollwage and P. A. Seib, *J. Polym. Sci., Part A1*, 9 (1971) 2877–2892.

monomer ratio changes too greatly, usually after proceeding to 10–20% conversion. The copolymer is isolated, and carefully separated from residual monomer, and the isolated copolymers and residual monomer mixture are analyzed, to determine their compositions. These experimental data are conventionally plotted with the copolymer composition as the ordinate and the composition of the monomer feed as the abscissa. The range of monomer-feed composition in each experiment then appears as a horizontal line at the appropriate level signifying average, copolymer composition. A smooth curve drawn through the midpoints of the lines then defines the relationship between monomer feed and copolymer composition. If the initial and final monomer-feed compositions are determined for any later copolymerization, it is possible to obtain from this curve a very good estimate not only of the average copolymer composition, which can be measured, but also of the range of copolymer compositions produced from beginning to end of the reaction. These data may be sufficient for preparative purposes, and are independent of mechanistic considerations (except for the proved assumption of chain propagation).

For a detailed analysis of monomer reactivity and of the sequence-distribution of mers in the copolymer, it is necessary to make some mechanistic assumptions. The usual assumptions are those of binary, copolymerization theory; their limitations were discussed in Section III,2. There are a number of mathematical transformations of the equation used to calculate the reactivity ratios r_1 and r_2 from the experimental results. One of the earliest and most widely used transformations, due to Fineman and Ross,[114] converts equation (1) into a linear relationship between r_1 and r_2. Kelen and Tudos[115] have since developed a method in which the Fineman–Ross equation is used with redefined variables. By means of this new equation, data from a number of cationic, vinyl polymerizations have been evaluated, and the questionable nature of the data has been demonstrated in a number of them.[116] (A critique of the significance of this analysis has appeared.[117]) Both of these methods depend on the use of the derivative form of the copolymer-composition equation and are, therefore, appropriate only for low-conversion copolymerizations. The integrated

(114) M. Fineman and S. D. Ross, *J. Polym. Sci.*, 5 (1950) 259–262.
(115) J. Kelen and F. Tudos, *J. Macromol. Sci. Chem.*, 9 (1975) 1–27.
(116) J. P. Kennedy, J. Kelen, and F. Tudos, *J. Polym. Sci., Polym. Chem. Ed.*, 13 (1975) 2277–2289.
(117) R. C. McFarlane, P. M. Reilly, and K. F. O'Driscoll, *J. Polym. Sci., Polym. Lett. Ed.*, 18 (1980) 81–84.

TABLE III

Reactivity Ratios of Copolymerizations of 1,6-Anhydro Sugars,
Calculated by Various Methods

Monomer[a]		Mayo–Lewis		Kelen–Tudos		Fineman–Ross			
						(F–G)		(1/F)–(G/F)	
1	2	r_1	r_2	r_1	r_2	r_1	r_2	r_1	r_2
GlcAnXy$_3$	GlcAnBzl$_3$	1.25	1.25	1.21	1.22	1.22	1.24	1.18	1.19
GlcAnXy$_3$	GalAnBzl$_3$	1.36	0.14	1.41	0.31			1.40	0.29
GlcAnXy$_3$	ManAnBzl$_3$	0.90	11.5	0.95	9.58	0.93	9.20	0.92	9.43
GlcAnXy$_3$	Mal^{b}AnBzl$_6$	1.91	0.28	1.51	0.55	1.63	0.67	1.38	0.50
GalAnXy$_3$	ManAnBzl$_3$	0.37	38.0						
GlcAnBzl$_3$	GlcAnBbl$_3$			~2.1	~1.1	~2.3	~1.3		
GlcAnBzl$_3$	GlcAnAcBzl$_2$	3.3	0.06						
GlcAnBzl$_3$	GlcAnBzl$_2$Cro	0.70	1.75	0.83	1.63	0.83	1.63	0.83	1.63
GlcAnBzl$_3$	GalAnBzlIp	1.4	0.06						

[a] Monomers are all 1,6-anhydroaldohexose derivatives. Parent monosaccharides are indicated by the conventional abbreviations, and maltose by Mal. Other symbols: Ac, 3-O-acetyl; An, 1,6-anhydro; Bzl, benzyl; Bbl, p-bromobenzyl; Cro, 3-O-(2-butenyl); Ip, 3,4-O-isopropylidene; and Xy, p-methylbenzyl (p-xylyl). [b] 10 mole-% PF$_5$.

form should be used for copolymerization carried out to high conversions.[118]

Although the values of reactivity ratios obtained by these procedures are essentially correct, it has been emphasized that their precision is not as high as believed, due to statistical consequences of the linearization transformation.[84] It was recommended that data be gathered to establish that the Mayo–Lewis, kinetic model is the correct one, and that, thereafter, an "exact" or "approximate" design scheme be employed, in which replicate experiments are conducted at specific, mol ratios of monomers, but, to date, this has rarely been done. The reactivity ratios for a number of anhydro sugar copolymerizations have, however, been calculated by Ito and Schuerch,[103] using two Fineman–Ross expressions, the Kelen–Tudos equation, and the integrated form of the Mayo–Lewis equation. An empirical estimate of the precision of these data may be made by visual observation (see Table III). For higher precision, reference should be made to the published critiques.[84,117] No evidence has been obtained that would indicate that the basic assumptions of classical, binary-copoly-

(118) F. R. Mayo and F. M. Lewis, J. Am. Chem. Soc., 66 (1944) 1594–1601.

merization theory are inapplicable to these anhydro sugar copolymerizations.

A practical problem immediately presents itself in any investigation of copolymerization, namely, that of analyzing for the two components. Because of this analytical problem, most anhydro sugar polymerizations have been performed with different substituents on the sugars. Most of the reactions have been between an anhydro sugar substituted with benzyl groups and a second anhydro sugar having p-xylyl substituents. ^{1}H-N.m.r. spectra have then been used to determine the ratio of the two mers in the copolymer. A series of copolymerizations between $GlcAnBzl_3$ and the tri-p-xylyl ether of levoglucosan ($GlcAnXy_3$) was performed, and it was found that the two monomers were of nearly identical reactivity, and copolymerized almost azeotropically.[103] The analyses of reactivity ratios, using either of the monomers, therefore mainly reflect the influence of the structure of the anhydro sugar, and not a substituent effect. Nevertheless, the polymerizability of the two monomers is not identical. As already described, the initiation of p-xylylated monomers requires slightly higher concentrations of Lewis acid for comparable conversions, and the viscosity of the polymers produced under identical conditions tends to be lower. Apparently, the p-xylyl derivative has more tendency towards ether complexation, and towards chain transfer or termination. These differences are not serious shortcomings for studies of propagation, which is the pertinent process determining copolymer composition.

A comparison of the three binary copolymerizations of 1,6-anhydro-β-D-gluco-, -galacto-, and -manno-pyranose derivatives gives some insight into the mechanism of copolymerization, if it is assumed on this evidence that the per-p-xylyl and perbenzyl derivatives can be used interchangeably.[107]

The reactivity ratios (from Table III) are as follows.

$$\frac{k_{Gal^+Gal}}{k_{Gal^+Glc}} = 0.14 \qquad \frac{k_{Glc^+Glc}}{k_{Glc^+Gal}} = 1.36$$

$$\frac{k_{Glc^+Glc}}{k_{Glc^+Man}} = 0.90 \qquad \frac{k_{Man^+Man}}{k_{Man^+Glc}} = 11.5$$

$$\frac{k_{Gal^+Gal}}{k_{Gal^+Man}} = 0.37 \qquad \frac{k_{Man^+Man}}{k_{Man^+Gal}} = 38$$

From these values, the approximate equality can be deduced.

$$\frac{k_{Glc^+Glc}/k_{Glc^+Gal}}{k_{Gal^+Gal}/k_{Gal^+Glc}} \times \frac{k_{Man^+Man}/k_{Man^+Glc}}{k_{Glc^+Glc}/k_{Glc^+Man}} \simeq \frac{k_{Man^+Man}/k_{Man^+Gal}}{k_{Gal^+Gal}/k_{Gal^+Man}}$$

The self-propagation constants can be cancelled from this equation, to give the following relationship between cross-propagation constants.

$$\frac{k_{\text{Gal}^+\text{Glc}}}{k_{\text{Glc}^+\text{Gal}}} \times \frac{k_{\text{Glc}^+\text{Man}}}{k_{\text{Man}^+\text{Glc}}} \simeq \frac{k_{\text{Gal}^+\text{Man}}}{k_{\text{Man}^+\text{Gal}}}$$

If this approximate equality is not the result of chance, it could indicate either that the propagation rate is not primarily dependent on the monomer, but on the growing cation, or that the rate is not primarily dependent on the cation but is dependent on the monomer. The assumption that the relationship is due to chance is unlikely, and it is known that $k_{A\ A} \neq k_{A\ B}$, that is, self-propagation rates differ from cross-propagation rates. Therefore, the assumption that the propagation rate is dependent, not on the monomer, but mainly on the growing cation, is unacceptable. Consequently, it is assumed that the rate of propagation is mainly determined by the structure of the attacking monomer, and that the influence of the three propagating cations is less important.

However, the side of the monomers that approaches a cation during bond formation is the β side, and it is identical for all three monomers in the $^1C_4(\text{D})$ conformation.[101] Furthermore, the rates of homopolymerization observed for the three monomers is in the order D-mannose > D-glucose > D-galactose. These rates of homopolymerization almost certainly reflect differences in the rates of propagation, because the same order of polymer molecular weights is observed when the polymerizations are performed with equal care to minimize termination processes [$\bar{M}_n = R_{\text{prop}}/(R_{\text{term}} + R_{\text{transfer}})$]. However, the potential energy of the monomers, and the ΔH of propagation must be in the order D-glucose > D-mannose $\simeq$ D-galactose, in view of the 1,3-axial interactions present in each anhydro sugar before and after polymerization. Therefore, neither structural differences nor energetic differences of the monomers in the $^1C_4(\text{D})$ conformation explain the order of propagation rates.

Another factor must thus influence monomer reactivity, and it was proposed[101] that this is a conformational change of the monomer when it reacts with the propagating trialkyloxonium ion (see Fig. 1). The most probable change would be a movement of C-3 below the plane of the pyranose ring, producing a boat conformation, which would occur as a cation attacks the ring-oxygen atom, to convert monomer into the new, propagating cation. This movement would move the O-3 substituent away from the ring-oxygen atom, and provide space for bond formation and counter-ion. The subsequent attack of a second monomer

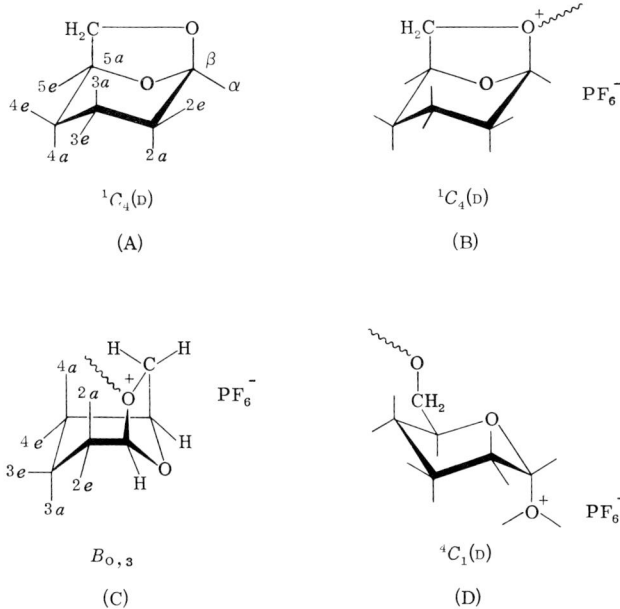

FIG. 1.—Conformational Changes During Polymerization.[101] [Substituents: benzyl, or p-xylyl on oxygen atoms indicated (other sites, hydrogen); GlcAnXy$_3$ (A + B) 2a, 3a, 4a, (C) 2e, 3e, 4e, (D) 2e, 3e, 4e; ManAnBzl$_3$ (A + B) 2e, 3a, 4a, (C) 2a, 3e, 4e, (D) 2a, 3e, 4e; GalAnBzl$_3$ (A + B) 2a, 3a, 4e, (C) 2e, 3e, 4a, (D) 2e, 3e, 4a. Reproduced, by permission, from Ref. 101.]

would then convert the new, trialkyloxonium ion into a penultimate chain-unit, with only the movement of the ring-oxygen atom to provide the $^4C_1(\text{D})$ conformation.

The conversion of a monomer into a polymer chain unit thus occurs in two steps, and the main, rate-determining feature of the process seems to involve a conversion of the monomer in the $^1C_4(\text{D})$ conformation into a trialkyloxonium ion in a boat conformation, although, of course, both steps occur on adjacent units simultaneously. During these changes, conformations develop in which the substituents become eclipsed. In the change of 1,6-anhydro-β-D-galactopyranose, the oxygen atoms on C-3 and C-4 become eclipsed, and O-4 adopts a position close to C-6 and its rigidly bonded hydrogen atoms. In contrast, 1,6-anhydro-β-D-glucopyranose can be converted into a boat with no pairs of oxygen substituents becoming eclipsed. 1,6-Anhydro-β-D-mannopyranose, in changing from the $^1C_4$ conformation to a boat conformation, also undergoes eclipsing of the oxygen on atoms C-2 and C-3, and the O-2 atom approaches the anhydro ring-oxygen atom as it is converted into an oxonium ion. That these processes do not inhibit reaction may be due to the fact that the C-1–O$^+$ bond of the oxonium ion and the C-2–O bond are dipoles in the reverse sense, and an attractive force assists the transformation.

If the conclusion is correct that the propagating cation is in a boat conformation, the α-side of all three anhydro sugars would present the same appearance to an entering monomer, and it would be understandable that the rate-determining feature would primarily involve conformational changes in the monomer. However, if the cation is in the $^1C_4(D)$ conformation, C-1 of the anhydromannose would be the least hindered. Furthermore, in converting a boat form of these trialkyloxonium ions into a $^4C_1$ unit, only D-galactose requires an eclipsed position of adjacent, large substituents, C-6 and O-4. Both of these features could affect the observed reactivities of their respective monomers in the observed direction, although the model proposed assigns them a minor role.[101]

The low reactivity of 1,6-anhydro-2,3,4-tri-O-benzyl-β-D-allopyranose is also of interest.[106] This monomer has 1,3-interaction between substituents on C-2 and C-4 in the $^1C_4(D)$ conformation. However, when conversion into a boat conformation, as proposed, occurs on reaction with a propagating cation, eclipsed bonds develop at C-2, C-3, and C-4. It is, therefore, not surprising that the D-*allo* anhydride is less reactive than those of D-mannose, D-glucose, or D-galactose. The corresponding, D-altrose derivative has only one axial substituent in addition to the anhydro ring, and these are on opposite sides of the pyranose ring, and therefore do not interact; it would be expected to be, and has proved to be, of very low polymerizability.[106]

The relative reactivities of five 1,6-anhydro-tri-O-benzyl-β-D-aldohexopyranoses and levoglucosan triacetate have already been discussed. Some other information is summarized in Tables II and III. The tris(*p*-bromobenzyl) ether of levoglucosan is only slightly more reactive in polymerization than GlcAnBzl$_3$ and, in copolymerization, is somewhat less selective.[104] Two copolymerizations of 1,6-anhydro-2,3,4-tri-O-*p*-xylyl-β-D-galactopyranose with GlcAnBzl$_3$ were performed. The copolymers produced were similar in composition to those formed from the two monomers having the benzyl and *p*-xylyl substituents exchanged.[107] Substituents on the benzyl group of benzyl ethers, therefore, produce much less influence on glycoside-forming reactions than substituents on the acid moiety of esters.

The reactivity of the trimethyl and triethyl ethers of levoglucosan has not been compared with that of the GlcAnBzl$_3$ since the earliest pioneering work, but they are probably of similar or greater reactivity.

1,6-Anhydro-2,4-di-O-benzyl-3-O-(2-butenyl)-β-D-glucopyranose polymerizes more readily than the tribenzyl ether, and copolymerization demonstrated that the monomer is more reactive in propagation processes.[98] Presumably, the smaller substituent on C-3 shields the ring-oxygen atom less than benzyl, and may also provide less inertial resistance to conformational change. In contrast, the 3-O-acetyl substituent inhibits polymerization and copolymerization for the same reasons as those suggested to explain the behavior of the triacetate. Nevertheless, polymers of the 3-O-acetyl monomer have been prepared.[105,105a] 1,6-Anhydro-2-O-benzyl-3,4-O-isopropylidene-β-D-galactopyranose has also been tested in polymerization and copolymerization. This monomer fails to homopolymerize, but it has a relative reactivity in copolymerization that is not greatly different from that of the tribenzyl ether of 1,6-anhydro-β-D-galactopyranose.[119]

The anhydrodisaccharide derivatives based on maltose[110] and cellobiose[111] [1,6-anhydro-2,3-di-O-benzyl-4-O-(2,3,4,6-tetra-O-benzyl-α- and -β-D-glucopyranosyl)-β-D-glucopyranose] have been polymerized to stereoregular products. To polymerize, these monomers require high concentrations of catalyst, presumably because of the relatively large number of unproductive, complexation sites in their structures. The stereoselectivity of the polymerization is also sensitive to the reaction conditions, and the molecular weight of the products is not high. Copolymerization of the 1,6-anhydromaltose and levoglucosan derivatives shows a lower reactivity for the former. All of these polymerizations lead to the formation of comb-shaped (1→6)-α-D-glucopyranans having pendant D-glucopyranosyl groups linked to O-4 of the backbone units. The polymer derived from cellobiose has side chains linked β to each backbone unit, that from maltose has them linked α, and the D-glucose–maltose copolymers have 4-O-α-D-glucopyranosyl groups randomly distributed along the chain, the number and sequence distribution being determined by the concentration of the feed and the relative reactivities of the two monomers.

It should be emphasized that copolymerizations that conform to the premises of binary-copolymerization theory produce copolymers of well defined structure. The kinetics of the competitive propagation-reactions determine not only the copolymer composition but also the sequence distribution. The mathematical procedures needed for calculating number-average sequence-lengths of mers, and sequence length-distributions of mers, are well known and have been

(119) T. Uryu, *U. S.–Jpn. Semin. Synth. Polysaccharides Their Biochem. Functions,* Syracuse, N. Y., 1979.

thoroughly reviewed.[81,82] They have been applied to most of the examples of copolymers already discussed. Evidence on sequence length consistent with the Mayo–Lewis treatment was found in the ^{13}C-n.m.r. spectra of GlcAnXy$_3$–ManAnBzl$_3$ copolymers.[102] The GlcAnXy$_3$ homopolymer and the ManAnBzl$_3$ homopolymer exhibit only one α-anomeric peak, at 97.8 and 98.8 p.p.m., respectively. The GlcAnXy$_3$–ManAnBzl$_3$ copolymers, however, display a third absorption, at 96.5 p.p.m., in the region of the α-anomeric carbon atom. The relative peak-heights for a series of copolymers strongly indicated that this peak represented the GlcAnXy$_3$ anomeric carbon atom linked to a ManAnBzl$_3$ unit, whereas the 97.8-p.p.m. peak represented only sequences of GlcAnXy$_3$ linked to GlcAnXy$_3$. No corresponding distinction could be made for internal and terminal anomeric peaks in the ManAnBzl$_3$ sequences. Assignment of each peak to diad fractions F_{11} 97.8 p.p.m., F_{12} 96.5 p.p.m., and ($F_{22} + F_{21}$) 98.8 p.p.m. resulted in rather good agreement with calculated diad values, based on reactivity ratios.[102] Although the precision of these data was, perhaps, not sufficient for a rigorous test of the premises, the results supported their validity.

The possibility of calculating sequence-length distributions is of interest, not only in the synthesis of heteropolysaccharides, but also in the synthesis of branched homopolysaccharides. The distribution of (1→4)-α-D-glucopyranosyl side-chains in the family of dextrans derived from anhydro-D-glucose and anhydromaltose derivatives can thus be calculated.[100] This information can be important for enzymological and immunological investigations.

The synthesis of dextrans having (1→3)-α-D-glucopyranosyl side-chains has been achieved by an alternative approach.[98] 1,6-Anhydro-2,4-di-O-benzyl-3-O-(2-butenyl)-β-D-glucopyranose (GlcAnBzl$_2$Cro) was copolymerized with GlcAnBzl$_3$, the crotyl groups were eliminated completely with potassium *tert*-butoxide, and the hydroxyl groups liberated were allowed to react with 2,3,4-tri-O-(p-methylbenzyl)-6-O-(N-phenylcarbamoyl)-1-O-tosyl-D-glucopyranose in 2,2-dimethoxyethane. The glycosidation reaction was 95% stereoselective (α), and appeared to be essentially complete. Elimination of the substituent protecting groups gave the parent family of branched dextrans. These are isomeric (in position, and distribution, of side chains) with the corresponding polysaccharides available by copolymerization of GlcAnXy$_3$ and MalAnBzl$_6$. The most probable stereochemical flaws in these two dextrans are also in two different sites. In the GlcAnBzl$_2$Cro copolymer, ~5% of the linkages introduced by glycosidation are β, whereas β linkages are most likely to be found in the main chain of the

MalAnBzl$_6$ copolymer at C-1 of the maltose units, as the stereoselectivity of that propagation step is the most sensitive to the conditions. The sequence of reactions leading to the (1→3)-branched dextran is the most ambitious, and most carefully monitored, work in this field to date.[98] However, the number of side chains introduced is somewhat in doubt, and may be less than the original, crotyl substitution.[98,120]

4. 1,6-Anhydroaldopyranans: Characterization and Applications

The products of polymerization have usually been isolated by repeated precipitation, and freeze-dried from benzene. They have generally been characterized by intrinsic viscosity, specific rotation, number-average molecular weight, and ^{1}H- and ^{13}C-n.m.r. spectra. Circular dichroism[94,98,101,121–123] and molecular-weight distributions[96,124,125] have also been studied on selected samples. The specific rotation is useful as an accurate, relative measure of the stereoregularity of homopolymers. As usually measured, without special precautions, values are reproducible to within ±1–2% on different preparations. The specific rotations of a series of GlcAnXy$_3$–GalAnBzl$_3$ copolymers have been found not to be linearly related to the mole fraction of mers in the copolymer. Instead, there was a broad minimum for the intermediate range of compositions. The nonlinearity was interpreted as meaning that each homopolymer had, in solution, a generally favored conformation that was disrupted by introducing different mer units as points of disorder into the backbone.

Circular dichroism (c. d.) spectra have been recorded for the tribenzyl ethers[122] and triacetates[94] of stereoregular (1→6)-α-D-gluco-, -manno-, and -galacto-pyranan. The c.d. bands of the acetates occurred at 210 nm, the wavelength of the acetyl $n \rightarrow \pi^*$ absorption band, and, for the D-glucan, the intensity of the band was shown to be sensitive to structural irregularities.[121] The c.d. bands of D-glucan and D-galactan were almost identical, but the change in configuration of C-2, from the equatorial acetoxyl group in the D-glucan and D-galactan to the axial acetoxyl group in the D-mannan resulted in complete reversal of the sign of the c.d. band, from negative to positive. The c.d. spectra of the

(120) I. J. Goldstein, U. S.–Jpn. Semin. Synth. Polysaccharides Their Biochem. Functions, Syracuse, N. Y., 1979.
(121) S. Mukherjee, A. Sarko, and R. H. Marchessault, Biopolymers, 11 (1972) 303–314.
(122) J. P. Merle and A. Sarko, Carbohydr. Res., 30 (1973) 390–394.
(123) A. J. Stipanovic and E. S. Stevens, Abstr. Pap. Am. Chem. Soc. Meet., 179 (1980) Carb. 45.
(124) J. P. Merle and A. Sarko, Macromolecules, 5 (1972) 132–136.
(125) T. Bluhm and A. Sarko, Macromolecules, 6 (1973) 578–581.

three benzylated polysaccharides were broad envelopes from ~210 to 270 nm, including six regularly spaced, vibrational bands. A comparison of the spectra of benzylated D-glucan and benzylated D-mannan showed a similar, complete reversal of sign at the shortest-wavelength, vibrational band only, and showed differences in intensity of the higher-wavelength bands. The lowest-vibration bands were opposite in sign to those of the corresponding acetates. Some indication of conformational effects was also found in the spectra of the benzyl ethers. C.d. spectra of GlcAnXy$_3$–ManAnBzl$_3$ copolymers show similar features. Circular dichroism has also been used as an analytical tool for determination of N-phenylcarbamate substituents on branched copolymers.[98]

The intrinsic viscosity gives an estimate of the molecular size (viscosity average) that does not always correlate with the number-average molecular weight, because of variations in the molecular-weight distribution. Molecular-weight distributions have been determined for the tribenzyl ethers of (1→6)-α-D-gluco-, -manno-, and -galactopyranans by the method of sedimentation velocity, and the variations were confirmed, because the polydispersity index $\bar{M}_w/\bar{M}_n$ ranged[125] from 1.6 to 3.2. The distributions were typical of high-conversion, cationic polymerization with spontaneous termination and transfer. Mark–Houwink exponents were also calculated.

^{1}H-N.m.r. spectra (100 MHz) are of limited utility for the benzylated homopolymers, although the absence of β-anomeric protons can be established for such spectra of perethylated D-glucan. Proton-n.m.r. spectra have primarily been used for determining mole ratios of mers in copolymers of p-xylyl and benzyl derivatives of anhydro sugars. In the spectra of GlcAnXy$_3$–ManAnBzl$_3$ copolymers, the difference in chemical shifts of methyl protons from the GlcAnXy$_3$ homopolymer also gave unequivocal evidence of copolymerization. ^{13}C-N.m.r. spectra are useful for preliminary evaluation of benzylated polymers and copolymers. In most cases, stereoselectivity can be accurately estimated from the relative peak-heights of α- and β-anomeric carbon atoms, or from the absence of the latter peaks. The relative peak-heights of aromatic C-1 absorptions give a fair estimate of mer ratios in GlcAnBzl$_3$–GlcAnBbl$_3$ copolymers. In GlcAnXy$_3$–ManAnBzl$_3$ copolymers, diad information may be derived, as already mentioned, from the anomeric region. Other features may be of interest for individual compounds, but complete assignment of peaks has not yet been accomplished for these derivatives.

The benzylated polymers have been deprotected, to give the free polysaccharides, by treatment with sodium in liquid ammonia. A few chain-breaks occur, so that the d.p. of the resultant polysaccharides is smaller than that of the original polymer. The polysaccharides that have been prepared in the free form include stereoregular (1→6)-α-D-gluco-,[91a,93-96] -manno-,[94,109] and -galacto-pyranan,[94,108] comb-shaped, low polymers derived from anhydro-maltose[110] and -cellobiose,[111] stereoregular D-glucose–D-mannose copolymers,[102] and a series of D-glucans having (1→3)-α-D-glucopyranosyl side-chains.[98] The isomeric, linear D-glucan, D-mannan, and D-galactan exhibit striking differences in solubility. The D-glucan is water-soluble, but, on drying, tends to form some ordered regions that resist dissolution. The D-mannan is dispersible in water to afford cloudy solutions, and is completely soluble, as is the D-glucan, in dimethyl sulfoxide or aqueous dimethyl sulfoxide. In contrast, the D-galactan is insoluble in both solvents, and in virtually all of the other simple, and complexing, solvents customarily used for cellulose, proteins, or other hydrogen-bonding polymers. N,N-Dimethylformamide–nitrogen dioxide and lithium hydroxide–lithium borate are the only reported exceptions, and there are difficulties in recovering pure D-galactan from these solutions.[108] If it is isolated in pure form, the D-galactan is so insoluble that derivatization is very difficult. However, if it is allowed to precipitate onto a finely divided solid from the liquid ammonia solution in which it is formed, acetylation proceeds normally.[94]

The structures of the polysaccharides have been proved by a number of physical, chemical, and enzymic methods. Elementary analysis of freeze-dried polymers invariably shows some residual water. The proportion is more variable than was originally realized, and so such physical constants as specific rotation are difficult to determine accurately. On periodate oxidation of the linear D-glucan, D-mannan,[109] and D-galactan,[108] two molecules of oxidant are consumed and one molecule of formic acid is produced per hexosyl residue. Each of the residual, oxidized polymers should have the same structure, and, as evidence of their identity, the specific rotation has been found to be the same after oxidation. For these measurements, the experimental error was largest for the highly insoluble D-galactan. Oxidation of a synthetic dextran and D-mannan that were not stereoregular gave different, and lower, values for the final value of the specific rotation. (The near identity of the circular dichroism spectra of D-glucan and D-galactan acetates also supported the concept of the stereoregularity of both.[94]) The specific rotation of the D-glucan is in the same range as that of *Leuconostoc mesenteroides* NRRL B512 dextran, and those of the D-mannan and D-galactan are higher than those of any known, natural polysaccharides based on the same sugars. The ^{1}H-n.m.r. spec-

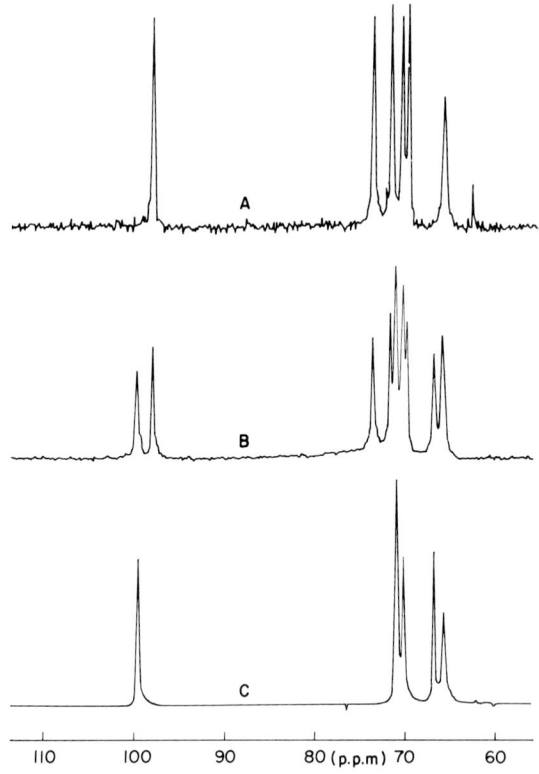

FIG. 2.—The ^{13}C-N.m.r. Spectra[102] of A, Synthetic Glucan; B, Glucomannan; and C, Mannan. (Reproduced, by permission, from Ref. 102.)

trum of the synthetic, deuterated dextran in D_2O shows a single anomeric-proton peak and an appropriate chemical-shift for equatorial hydrogen (δ 5.05), and no axial H-1 resonance is present. ^{13}C-N.m.r. spectra demonstrated the high degree of stereoregularity in linear D-glucan, D-mannan, and D-glucomannan copolymers[102] (see Fig. 2). The linear D-glucan has been used in establishing peak assignments for polymers of D-glucose.[126] The ^{13}C-n.m.r. spectra of a family of D-glucans having various numbers of randomly distributed 3-O-α-D-glucopyranosyl side-chains have been completely characterized, and compared with those of similar, naturally occurring dextrans.[98] The data obtained have been used to clarify some peak assignments in such spectra of natural dextrans. A proof of the structure of these branched

(126) P. Colson, H. J. Jennings, and I. C. P. Smith, *J. Am. Chem. Soc.*, 96 (1974) 8081–8087.

D-glucans by methylation and hydrolysis confirmed their structure, but indicated that the degree of branching in the more highly branched samples was somewhat lower than estimated by ^{13}C-n.m.r. spectroscopy.[120]

Number-average molecular weights have been determined for linear and branched D-glucans by membrane osmometry in aqueous dimethyl sulfoxide.[93,98] The values obtained showed that, on deprotection with sodium and liquid ammonia, a perbenzylated, linear D-glucan containing 980 D-glucosyl residues produces a synthetic dextran[98] of d.p. 270. On debenzylation, a similar, perbenzylated, synthetic D-glucan derivative having 10 to 12% of branches gave a branched dextran having 240 D-glucosyl residues. Occurrence of degradation in a series of deprotecting steps was shown to be more severe with the branched dextrans.[98]

Vacuum-ultraviolet, circular dichroism (v.u.c.d.) measurements have been made on films and solutions of linear D-glucan in a study of the conformation of dextran and its oligomers.[123] Film formation of the linear dextran was accompanied by crystallization, and the v.u.c.d. band was observed at 165 nm, in contrast to nonlinear dextran films displaying a band at 177 nm. The difference was ascribed to hydrogen bonding of the ring-oxygen atom in the crystalline state.

Success in preparing single crystals from linear dextran, and in obtaining electron diffraction spectra, suggests that information on the favored conformations of dextran in the solid state will soon be available.[127]

The linear dextran has also been used in investigating the mechanism of the periodate oxidation of polysaccharides. The results support the proposal that inter-residue, hemiacetal formation is a general occurrence in the later stages of the oxidation.[128]

The stereoregularity of the D-glucan has been confirmed by enzymic degradation. A random-acting dextranase, specific for α-(1→6)-linkages cleaved the polymer to D-glucose, isomaltose, and isomaltotriose, to the extent of over 92%. Calculations indicated the presence of less than 2% of flaws in the chain. The residual oligosaccharides were completely hydrolyzed to D-glucose by α-D-glucosidases. Glucoamylase, an enzyme that successively removes D-glucosyl groups from the (nonreducing) end of the chain hydrolyzed the polysaccharide to the extent of 31% in 24 hours, whereas a natural dextran having 5% of branches was hydrolyzed only[129] to the extent of 16%.

(127) H. Chanzy, C. Guizard, and A. Sarko, *Int. J. Biol. Macromol.*, 2 (1980) 149–153.
(128) M. F. Ishak and T. J. Painter, *Acta Chem. Scand.*, 25 (1971) 3875–3877.
(129) E. T. Reese and F. W. Parrish, *Biopolymers*, 4 (1966) 1043–1045.

D-Mannans of known molecular weight were degraded by an exo-D-mannanase that released single D-mannosyl groups from the (nonreducing) terminus.[130] A D-mannan of d.p. 300 was degraded to the extent of 42–43%, and one of d.p. 500, to the extent of 22–25%. By calculating the theoretical relationship between the maximum percentage of enzymic degradation and the mol% of irregularity in the polymer, according to the equation of Sarid and coworkers,[131] Tkacz and coworkers[130] estimated less than 1% of irregularity in the polymer. The same calculations have been made for the results of a degradation of the linear D-glucan to the limit of its hydrolysis (35%) by dextran-D-glucosidase, and the results again confirmed the estimate of less than 2% of flaws.[132]

Concanavalin A, a carbohydrate-binding protein, or lectin, from jack-bean meal reacts with terminal α-D-glucopyranosyl and α-D-mannopyranosyl groups. At neutral pH values, the lectin is tetrameric and tetravalent; that is, it can react with up to four α-D-glycopyranosyl end-groups. If two or more are present in a polysaccharide, the potential for a network exists on mixing the lectin and the polysaccharide. The network formation can result in precipitation (the precipitin reaction), and the extent of precipitation is related to the degree of branching. Neither synthetic, linear D-glucan[133] nor D-mannan[134,135] precipitates with concanavalin A. This was considered to constitute strong evidence for the linear nature of the synthetic dextran, but a synthetic dextran having 10 to 12% of α-D-glucopyranosyl groups linked as (single unit) branches to the backbone also failed to precipitate concanavalin A, although more-highly branched dextrans of the same type did cause precipitation.[120]

The linear polysaccharides have been used in a number of immunological reactions. The linear D-mannan was found to confer cutaneous activity in guinea pigs to a dermatophyte polysaccharide.[136] It was, however, found to be non-cross-reactive with antibodies to yeast D-mannan,[137,138] and antibodies to linear D-mannan were not cross-reactive to yeast D-mannan.[138] Anti-linear-D-mannan antibodies have been

(130) J. S. Tkacz, J. O. Lampen, and C. Schuerch, *Carbohydr. Res.*, 21 (1972) 465–472.
(131) S. Sarid, A. Berger, and E. Katchalski, *J. Biol. Chem.*, 234 (1959) 1740–1746.
(132) G. J. Walker and A. Pulkownik, *Carbohydr. Res.*, 29 (1973) 1–14.
(133) I. J. Goldstein, R. D. Poretz, L. L. So, and Y. Yang, *Arch. Biochem. Biophys.*, 127 (1968) 787–794.
(134) R. Robinson and I. J. Goldstein, *Carbohydr. Res.*, 13 (1970) 425–431.
(135) L. L. So and I. J. Goldstein, *J. Biol. Chem.*, 243 (1968) 2003–2007.
(136) S. F. Grappel, *Experientia*, 27 (1971) 329–330.
(137) W. O. Mitchell and H. F. Hasenclever, *Infect. Immun.*, 1 (1970) 61–63.
(138) W. Richter, unpublished results.

induced by injection of a linear D-mannan–albumin conjugate into rabbits. Non-cross-reactivity confirmed that the (1→6)-α-D-mannopyranan backbone of yeast D-mannans is not accessible, because of attachment of branches of other linkages [(1→2) and (1→3)]. The antilinear-D-mannan antibodies were also used for characterizing a series of D-glucomannan copolymers. When the percentage of one of the components was ≤ 15, this was reflected in the precipitation pattern with the antisera.[138]

The nature of the antigenic determinants of dextran has been elucidated with the use of the synthetic, linear D-glucan. The oligosaccharide-inhibition studies of Kabat demonstrated that the antigenic determinants of dextran correspond to three to seven (1→6)-α-D-glucopyranosyl residues.[139–141] The experiments were sometimes interpreted to indicate that the antigenic determinants were nonreducing, terminal sequences. However, Richter[142] demonstrated that synthetic, linear dextran was cross-reactive with rabbit anti-B512 dextran antiserum, and, in a battery of tests, demonstrated identity of antigenic determinants with a sample of B-512 dextran of comparable molecular weight. These findings were taken as constituting evidence that the antigenic determinants of both linear and B-512 dextran were sequences of (1→6)-linked α-D-glucopyranosyl residues arranged repetitively along the molecular chain.[142] An anti-B-512 antiserum has also been shown not to recognize synthetic, linear D-mannan, although the only structural difference in the determinant is[143] the configuration of C-2.

The binding properties of immunoglobulin binding-sites were further clarified in an investigation of dextran-reactive BALB/c mouse IgA myeloma proteins.[144] These are homogeneous immunoglobulins having single specificities. One protein had maximum complementarity for isomaltopentaose, and most of its binding energy was directed against methyl α-D-glucopyranoside, indicating the importance of the nonreducing end-group. This protein precipitated with branched dextrans high in α-D-(1→6)-linkages, but failed to precipitate synthetic, linear dextran. Another protein had maximum binding with isomaltohexaose. Most of its binding energy was directed towards isomaltotriaose, and little towards the mono- and di-sac-

(139) E. A. Kabat, *J. Am. Chem. Soc.*, 76 (1954) 3709–3713.
(140) E. A. Kabat, *J. Cell. Comp. Physiol.*, 50 (1957) 79–102.
(141) E. A. Kabat, *J. Immunol.*, 84 (1960) 82–85.
(142) W. Richter, *Int. Arch. Allergy Appl. Immunol.*, 46 (1974) 438–447.
(143) W. Richter, *Int. Arch. Allergy Appl. Immunol.*, 48 (1975) 505–512.
(144) J. Cisar, E. A. Kabat, M. M. Dorner, and J. Liao, *J. Exp. Med.*, 142 (1975) 435–459.

charides. This protein was clearly directed against nonterminal portions of the dextran chain, and precipitated linear dextran. Human antibodies appeared to be mixtures of molecules having terminal and nonterminal specificities, and different human individuals apparently have different proportions of the two classes of antibody. The conclusion that one antidextran antibody was directed solely against terminal sequences was confirmed by the method of ligand-induced, altered fluorescence, using Fab¹ fragments of the same protein.[145]

Different allergic results have also been observed in rats and man, when linear and branched dextrans are compared. The relative activities of the two kinds of dextran can be reversed when two different, allergic phenomena are studied, such as an anaphylactoid reaction following parenteral injection, and wheal and erythema production following injection into the skin. Presumably, the relative importance of terminal and nonterminal specificities is also involved in these phenomena.[146]

5. Polymerization of 1,4-Anhydroaldopyranoses

The polymerization of 1,4-anhydroaldopyranoses has not been investigated as completely as that of the 1,6-anhydroaldopyranoses, but the promising results that have been obtained suggest that, with further effort, a number of stereoregular, 1,4-anhydroaldopyranoses may become available.

The first polymerizations reported by Kops and Schuerch[147] were those of 1,4-anhydro-2,3,6-tri-O-methyl-β-D-galactopyranose and 1,4-anhydro-2,3-di-O-methyl-α-L-arabinopyranose. The latter compound was slightly contaminated with 1,4-anhydro-2,3-di-O-methyl-α-D-xylopyranose, but the course of the polymerization could nevertheless be monitored reasonably accurately. For the most part, the polymerizations were conducted at 10% concentration (g/mL) in dichloromethane, or aromatic hydrocarbons, with 1–5 mol% of phosphorus pentafluoride, or boron trifluoride etherate. At low temperature (-78 to $-97°$), the d.p. of both polymers produced was ~90; at increasing temperatures of polymerization, termination processes became more severe, and the d.p. lower. Usually, the reaction times were long (perhaps unnecessarily so), and the conversions were 50 to 90%. The specific rotations of the D-galactans prepared at -28 and $-90°$ differ by only ~10° (-85 to $-95°$), but those of the L-arabinans varied from $+6$

(145) L. G. Bennett and C. P. J. Glaudemans, *Carbohydr. Res.*, 72 (1979) 315–319.
(146) A. K. Delitheos, T. H. P. Hanahoe, and G. B. West, *Int. Arch. Allergy Appl. Immunol.*, 50 (1976) 436–445.
(147) J. Kops and C. Schuerch, *J. Polym. Sci., Part C*, 11 (1965) 119–138.

to −102°; the high, negative rotations were obtained at the higher temperatures of polymerization (−28°), and the low positive or low negative values, at low temperatures (−78 to −97°).

Structural information was obtained from degradation studies, and by comparison of specific rotations with the known values for methyl α- and β-glyco-furanosides and -pyranosides. Hydrolyses of the polymers in 4:1 oxolane–water required acid concentrations much higher than are normally needed for hydrolyzing furanosidic linkages in completely aqueous media. However, by monitoring the change in rotation during hydrolysis, it was apparent that about half of the glycosidic linkages in both polymer systems was much more acid-labile than the rest. This difference in lability to acid was also demonstrated by monitoring the rate of degradation of the polymers by methanolysis. By comparing the change in optical rotation on hydrolysis with the known rotations of methyl glycosides, it was shown that the D-galactan had mainly β-D-furanosidic linkages as the more acid-labile bonds, and largely β-D-pyranosidic linkages as the more acid-stable bonds. The structure of the L-arabinan was similar, but a variable proportion of β-L-furanosidic linkages was also present; these were the most readily hydrolyzed bonds, and gave a glycan having a high, positive specific rotation. The linkage was present in the low-temperature polymers, but less was found in those prepared at higher temperatures.

The presence of at least three different glycosidic linkages, unequivocally demonstrated by the changes in rotation during hydrolysis, requires the assumption of a carbonium-ion mechanism in this polymerization, and a detailed, conformational analysis has been proposed. Some steric control was also noted with change in the solvent.[147]

The polymerization of 1,4-anhydro-2,3,6-tri-O-benzyl-α-D-glucopyranose was reported by Micheel and coworkers.[148] The action of 15–20 mol% of phosphorus pentafluoride on this monomer gave somewhat similar results, in that at least three anomeric forms were shown to be present in polymers of molecular weight 21,000–41,000. Debenzylation of óne of the products gave a polysaccharide of mol. wt. 11,200. Partial hydrolysis to eliminate furanosidic linkages resulted in very little degradation. The main product (90%) had a molecular weight of ~9350, and the only other product isolated (5.9%) was identified as glucose. It is quite clear that the furanosidic linkages must be clustered in separate chains, or, as the authors suggested,[148] as short, terminal sequences. Polymerizations with antimony pentafluoride were less successful, and those attempted with boron trifluoride etherate failed.

(148) F. Micheel, O. E. Brodde, and K. Reinking, *Justus Liebigs Ann. Chem.*, (1974) 124–136.

Micheel and Brodde[149] also used triethyloxonium tetrafluoroborate to polymerize the same monomer, with strikingly different results. This catalyst is less active, and the optimum temperature of polymerization found was -16 to $-35°$. The polymer produced had a molecular weight of 9500, calculated by the end-group method from the results of debenzylation, methylation, hydrolysis, and gas–liquid chromatography. The specific rotation of products from a variety of polymerization conditions varied little ($[\alpha]_D^{20} + 17.6$ to $+25.8°$), and was markedly lower than that of polymers produced with Lewis acids. No evidence of furanosidic linkages was found. The partially hydrolyzed polymer produced with phosphorus pentafluoride, and the polymer produced with the oxonium ion, were converted into permethyl ethers and peracetates. A comparison of their specific rotations with those of cellulose and amylose suggested that phosphorus pentafluoride produces a polymer containing 70–75% of α-linkages, whereas the oxonium ion affords polymer primarily by inversion, and the product has $\sim 78\%$ of β-linkages; this calculation assumed a linear relationship between specific rotation and anomeric composition.

A short communication by Uryu and coworkers[150] described the polymerization of 1,5-anhydro-2,3-O-benzylidene-β-D-ribofuranose. The compound was prepared as a mixture of benzylidene isomers, and polymerized as the mixture. Polymerization proceeded well between 0 and $-40°$ with a number of Lewis acids. Phosphorus pentafluoride, boron trifluoride etherate, and stannic chloride each produced polysaccharides having specific rotations of $+40$ to $+63°$ and molecular weights of 2,000 to 20,000. Proton and ^{13}C-n.m.r. spectra indicated a mixture of structural units, presumably mainly α-furanose and α-pyranose. Polymerizations with antimony pentachloride were strikingly different. The specific rotations were negative, -59 to $-62°$, for polymers having molecular weights of 26,000 to 32,500. The ^1H- and ^{13}C-n.m.r. spectra were much sharper and simpler than those of the other glycans, and were interpreted as indicating a stereoregular polymer, presumably a (1→4)-β-D-ribopyranan. This was a striking and encouraging result, although further structural evidence is needed.

Apparently, the choice of a catalyst, the protecting groups, the temperature of polymerization, and the solvent all affect the stereoregularity of polymerization. It appears probable that the synthesis of a number of regular, (1→4)-linked aldopyranans and, perhaps, (1→5)-linked aldofuranans, should be possible in the near future.

(149) F. Micheel and O. E. Brodde, *Justus Liebigs Ann. Chem.*, (1974) 702–708.
(150) T. Uryu, K. Ito, and K. Matsuzaki, *Prepr. Short Commun. Int. Symp. Macromol.*, 26th, 1979.

6. Polymerization of 1,3-Anhydroaldopyranoses

Only preliminary data are available on the polymerization of 1,3-anhydroaldopyranoses. Both 1,3-anhydro-2,4,6-tri-O-benzyl-β-D-gluco- and -manno-pyranose have been found to polymerize with Lewis acids.[62,151] The former has been treated, under a variety of conditions, with phosphorus pentafluoride, boron trifluoride etherate, triethyloxonium hexafluorophosphate, antimony pentachloride, and silicon tetrachloride. The most favorable results to date have been obtained with phosphorus pentafluoride. Reaction is rapid and complete; yields of 60 to 70% of petroleum ether-insoluble polymer are obtained; and the rest of the monomer is converted into more-soluble oligomers. The stereoselectivity is variable, and the most regular products so far synthesized contain, at best, 80% of α-linkages. The viscosities of all of the products are low. Extended treatment of the polymers with phosphorus pentafluoride appears to alter the molecular-weight distribution. A catalyst consisting of aluminumtriethyl–water (in 1:1 mol ratio) also caused comparable polymerization with lower stereoselectivity, and zincdiethyl–water as the catalyst gave poor yields. Basic catalysts failed to cause polymerization. At present, minimal structural information is available on these polymers, and exploratory studies on alternative catalyst systems are needed.

7. Polymerization of 1,2-Anhydroaldopyranoses

3,4,6-Tri-O-acetyl-1,2-anhydro-α-D-glucopyranose (the Brigl anhydride) was subjected to thermal polymerization by Haq and Whelan[25] in 1956, and oligomeric products were obtained. Catalyzed polymerizations under melt conditions also gave low polymers.[152] After success had been achieved in the stereospecific polymerization of levoglucosan derivatives, similar methods were tried with the Brigl anhydride. Acetyl cation with a number of different counter-ions and a variety of Lewis acids were tested[153] in solution polymerizations at temperatures of −100 to +25°; again, the products were oligomeric and had specific rotations of +110 to +186°, and, in some cases, gel formation accompanied polymerization. Structural investigations demonstrated the presence of α- and β-linkages, with a preponderance of α. The gel fraction was shown to contain orthoester functions which served as cross-links. The monomeric units retained the D-

(151) H. Ito and C. Schuerch, *Macromolecules*, in press.
(152) E. J. Quinn, Ph.D. Thesis, State University College of Forestry at Syracuse University, 1962.
(153) J. Zachoval and C. Schuerch, *J. Polym. Sci., Part C*, 28 (1969) 187–195.

gluco configuration, but there were no clear trends that would suggest any method for producing high polymers, or stereoregular sequences.

Presumably, this reaction proceeds by a carbonium-ion mechanism which permits either α- or β-linkages to be formed. The preference for α-linkages may, in part, be thermodynamic, but it is more probably due to shielding of the β-side of C-1 by participation of the 6- or, more likely, the 3-acetate group. The monomer clearly acts as a multifunctional unit capable of gelation. However, no deep-seated structural changes seem to occur, as D-glucose is the only sugar formed on hydrolysis. The acetate groups are, therefore, the most probable sites of cross-linking and termination. Presumably, the carbonium ion from the Brigl anhydride is sufficiently reactive to attack acetate groups on neighboring molecules and to form acetoxonium ions. These may either eliminate protons to cause termination and oligomer formation, or else attack a 1,2-anhydro ring with the formation of an orthoacetate cross-link. Apparently, the corresponding acetoxonium ions from esters of levoglucosan are not reactive enough to open the (more stable) 1,6-anhydro ring, and, consequently, gelation does not occur.[92]

The further development of the polymerization of 1,2-anhydroaldopyranose derivatives has had to await the synthesis of monomers having ether rather than ester substituents. The first, exploratory polymerizations of 1,2-anhydro-3,4,6-tri-*O*-benzyl-β-D-mannopyranose have now been carried out.[154] Iodine, methyl borate, phosphorus pentafluoride, and triethyloxonium hexafluorophosphate all produce polymers of rather low, positive rotations (+2.4 to +8.1°). These values clearly indicate the presence of α- and β-anomeric linkages. Coordination catalysts formed by fixed proportions of water and zinc- or aluminum-alkyl caused 1,2-substituted oxiranes to polymerize stereoselectively. However, when these catalysts were tested on this monomer, the polymer yields were very low, but the higher specific rotation showed an increased preference for α-linkages. Anionic polymerization proved to be very slow, but gave the highest specific rotations. Methyllithium and lithium *tert*-butoxide gave polymers having specific rotations comparable to those produced by the complexation catalysts, and methylmagnesium bromide gave somewhat higher values. A series of experiments employing potassium *tert*-butoxide with macrocylic, crown compounds gave the best, although somewhat variable, results.

Gel-permeation chromatography of the polymers showed a rather complex, molecular-weight distribution. Usually, the cationic polymers had higher molecular weights, but in all cases, the molecular

(154) H. Yamaguchi and C. Schuerch, *Biopolymers*, 19 (1980) 297–309.

weights were probably lower than 10,000. In other epoxide polymerizations, the formation of cyclic oligomers of low molecular weight has been observed, and the possibility exists that different catalytic species in these systems produce oligomers and polymers at different sites.

A typical, cationic and a typical, anionic polymer were debenzylated with sodium and liquid ammonia, and purified and fractionated by ultrafiltration. The molecular weights of the polymers were ~1,000 after debenzylation, corresponding approximately to hexamers. The ^{13}C-n.m.r. spectrum of the anionic polymer indicated a high stereoregularity, but that of the cationic polymer was much more complex, as expected of an oligosaccharide having mixed anomeric-configurations.

The optical rotations of the debenzylated oligomers were compared with those of a series of (1→2)-α-D-mannopyranose oligomers isolated from natural sources. The specific rotations of the natural oligomers and the anionic, synthetic hexamer were plotted against the reciprocal of the degree of polymerization. The plot was reasonably linear, and the specific rotation (+61.2°) of the anionic polymer fell at its predicted position (within experimental error). The cationic polymer had a much lower specific rotation (+23°). The polymers so far produced from this monomer are thus of relatively low molecular weight, and are further degraded on debenzylation. Stereoregulation is very high in polymerizations with anionic initiators, but methods for preparing higher polymers having high stereoregularity are still needed.

The structure of the (1→2)-α-D-mannopyranan has been confirmed by methylation analysis, and its interaction with concanavalin A has been tested. Concanavalin A forms a weak complex with this oligosaccharide, and this is precipitated. Two molecules of concanavalin A must, therefore, be able to link to a sequence of two, or three, (1→2)-α-D-mannopyranosyl residues. The polymer thus gives some insight into the steric requirements, and the size of the active site, of this lectin.[120]

8. Miscellaneous Polymerizations

A few polymerizations leading to nonhydrolyzable polysaccharides have been reported. The first was a base-catalyzed polymerization of 5,6-anhydro-1,2-O-isopropylidene-3-O-methyl-α-D-glucofuranose.[11] The polymerization was initiated most successfully with solid potassium or cesium hydroxide, and the d.p. achieved was ~25. Hydrolysis gave a free, methylated, nonhydrolyzable polysaccharide, which was oxidized with hypoiodite to the poly(D-gluconic acid.)

5,6-Anhydro-1,2-O-isopropylidene-α-D-glucofuranose has also been shown to polymerize. Successful polymerizations were achieved with a number of Lewis acids,[12] to give polymers having d.p. 15 after short periods and up to 45 for long polymerization times. Loss of isopropylidene groups at 0° was observed after extended polymerizations. Spectral data were reported. The optical rotation and the distribution of molecular weight varied with the conditions of polymerization. Base-catalyzed polymerizations of the same monomer gave results[13] rather similar to those obtained with the methylated derivative.[11] Spectral data indicated greater regularity than in the cationic polymers.[12] The optical rotations were relatively constant, and gel-permeation chromatography of a one-stage and a two-stage polymerization showed the presence of some terminated and some nonterminated chains after complete reaction of the oxirane ring.

The polymerization of a variety of bicyclic acetals is under active investigation. Those corresponding to anhydrodeoxy sugars are the 6,8-dioxabicyclo[3.2.1]-octane and -octenes, and 2,7-dioxabicyclo[2.2.1]heptane. The former are related to 1,6-anhydroaldopyranoses and the latter to 1,5-anhydroaldofuranoses. These classes have been investigated most extensively by the research groups of H. Sumito, H. K. Hall, Jr., and J. Kops, and have been reviewed by Sumitomo and Okada.[155] Often, highly stereoregular polymers are obtained having configurations at the acetal center analogous to those of the synthetic polysaccharides. Some complexity is introduced into structural interpretation of the products owing to the racemic nature of the monomers. However, an optically active monomer, namely, (+)-(1R,5S)-6,8-dioxabicyclo[3.2.1]octane has been synthesized, and polymerized to a product having configurations analogous to those in the L-sugar series.[156] Furthermore, *cis*- and *trans*-hydroxylation of polymers of 6,8-dioxabicyclo[3.2.1]octene resulted in the formation of isomeric polysaccharide analogs that have been tested in a few biological systems,[155] and sulfated to afford anticoagulants.[157] These interesting researches provide an intellectual continuum from the chemistry of synthetic, aliphatic polymers to that of natural polysaccharides.

The Latvian school has continued its active research on the polymerization and application of levoglucosan and its derivatives. The main thrust of their research has been in the utilization of levoglucosan in commercial resins and polymers. A substantial part of that re-

(155) H. Sumitomo and M. Okada, *Adv. Polym. Sci.*, 28 (1978) 48–82.
(156) M. Komada, M. Okada, and H. Sumitomo, *Macromolecules*, 12 (1979) 5–9.
(157) M. Okada, H. Sumitomo, M. Hasegawa, and H. Komada, *Makromol. Chem.*, 180 (1979) 813–817.

search has involved the copolymerization of levoglucosan and its derivatives with commercial monomers. Because of the difficulties of translation, and the limitations of space, and because the present author chose to emphasize biochemical implications of this field, these contributions have not been covered. Similarly, Uryu and coworkers have investigated the copolymerization of GlcAnBzl$_3$ and ManAnBzl$_3$ with 3,3-bis(chloromethyl)oxetane, 1,3-dioxolane, and epichlorohydrin. A number of interesting observations on relative reactivities were made, and have been reported verbally.[119]

Uryu and coworkers also reported the first synthesis of a nonhydrolyzable polysaccharide by cationic, ring-opening polymerization of an oxetane ring. The monomer was 3,5-anhydro-1,2-O-isopropylidene-α-D-xylofuranose, and it formed a polymer of molecular weight of 10,000–11,000. Spectral and optical data indicated regular polymerization to a (3→5)-linked polymer of 1,2-O-isopropylidene-α-D-xylofuranose.[14]

IV. Summary and Prospects

The reactivity of the 1,6-anhydroaldopyranose derivatives in polymerization is now well understood. Because of unfavorable energetics, it is unlikely that 1,6-anhydro-ido-, -altro-, -gulo-, or -allo-pyranose will produce, directly, high polymers comparable to those of mannose, glucose, and galactose. Nevertheless, a great variety of regioselectively modified α-(1→6)-linked polysaccharides can be derived by selective reactions of the parent polymers and copolymers. Research in that direction has already been reported,[105] and more is in progress.[105a] Methods of stereoselective modification of monomers is even more fully developed[19], and polymerization of many of these compounds can lead to other unusual polysaccharides that may be of biomedical significance. Advances in this area will be most fruitful if there is collaboration between synthetic chemists and biochemists.

The variable selectivity in polymerizations of the 1,4-anhydroaldopyranoses suggests that a systematic investigation of substituent groups, catalysts, and polymerization conditions should lead to stereoregular, (1→4)-linked polysaccharides. The first reported example is already at hand.[150] As the driving force of these polymerizations derives more from ring strain than from substituent interactions, these polymerizations should proceed more generally than those employing 1,6-anhydro sugars. Similarly, the polymerization of 1,3- and 1,2-anhydro sugars depends on ring strain, and not on the configurations of vicinal carbon atoms, and should also be general. The main problems promise to be control of the configuration at C-1, and minimization of the termination and transfer processes. The optimization of these polymerizations will require creative, polymer chemistry.

If all these challenges are met, and comparable progress can be made with the furanosides, there will still be the challenge of the synthesis of regularly ordered copolymers, and *cis*-linked, rather than *trans*-linked, polysaccharides. The field can still use the good offices of "green-eyed magicians in white overalls."

THE CHEMISTRY OF MALTOSE

By Riaz Khan

*Tate & Lyle Limited, Group Research & Development,
Philip Lyle Memorial Research Laboratory, Reading, Berkshire, England*

I. Introduction	214
II. Structure and Synthesis	214
III. Glycosides	217
IV. Ethers	219
1. Trityl Ethers	219
2. Other Ethers	220
V. Esters	223
1. Acetates and Benzoates	223
2. Sulfonates	227
VI. Nucleophilic-displacement Reactions	229
1. Nucleophilic-displacement Reaction of Sulfonates	229
2. Sulfuryl Chloride Reaction	231
3. Methanesulfonyl Chloride–N,N-Dimethylformamide Reaction	233
VII. Anhydro Derivatives	235
VIII. Cyclic Acetal Derivatives	237
IX. Aminodeoxy Derivatives	239
X. Miscellaneous Compounds	242
1. Unsaturated Derivatives	242
2. 1-(2-Chloroethyl)-1-nitrosourea Derivatives	243
3. Sulfur-containing Compounds	243
4. Polymer-forming Reactions	244
5. Complex-forming Reactions	245
6. Selective-oxidation Reactions	246
7. Alkaline Degradation	247
8. Hydrolysis	248
XI. Physical Methods	249
1. ^{1}H-Nuclear Magnetic Resonance Spectroscopy	249
2. ^{13}C-Nuclear Magnetic Resonance Spectroscopy	254
3. Mass Spectrometry	258
XII. Tables of Properties of Maltose Derivatives	263

I. Introduction

During the past fifty years, a supply of cheap, petrochemical feed-stocks has been available for synthetic chemistry. These chemical feed-stocks are non-regenerable and, consequently, are likely to become scarce and expensive. To meet this challenge, a new technology based on such regenerable, natural products as carbohydrates needs to be developed. In the case of sucrose, its potential as a chemical raw-material has been demonstrated.[1] A sucrose-based surfactant[2] is already being produced on a commercial scale. Other products from sucrose that are of considerable commercial interest are noncaloric sweeteners,[3] male anti-fertility agents,[4] resins,[5] plastics,[5] and plasticizers.[5]

Maltose is another important disaccharide. It can be produced on a large scale from starch, but its potential as a raw material for chemical industry needs to be explored. In order to achieve this objective, an understanding of the fundamental chemistry of maltose is essential. The purpose of this article is, therefore, to collate information on the reactions of maltose, and to illustrate some of the physical methods that have contributed to the characterization of maltose derivatives.

II. Structure and Synthesis

Maltose (1), systematically named 4-O-α-D-glucopyranosyl-D-glucopyranose, is a reducing disaccharide. The numbering of the carbon

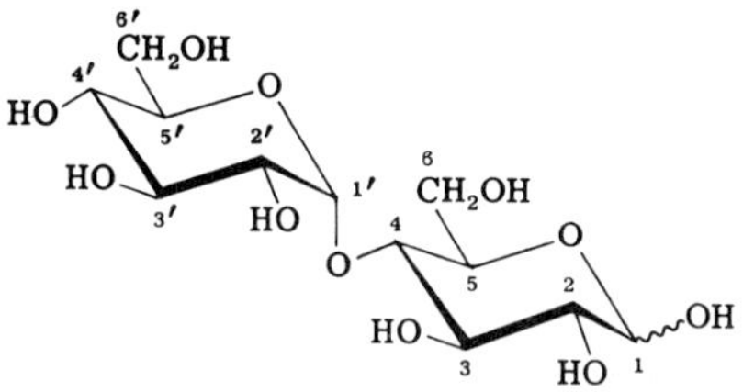

Maltose
(4-O-α-D-Glucopyranosyl-D-glucopyranose)

1

---

(1) R. Khan, *Adv. Carbohydr. Chem. Biochem.*, 33 (1976) 235–294.
(2) K. J. Parker, R. Khan, and K. S. Mufti, Br. Pat Specif. 1,399,053; U. S. Pat. 3,996,206; Ger. Offen. 2,412,374; *Chem. Abstr.*, 82 (1975) 100,608r.
(3) L. Hough, S. P. Phadnis, R. Khan, and M. R. Jenner, Br. Pat. Applic. (1976); L. Hough and S. P. Phadnis, *Nature (London)*, 263 (1976) 800; L. Hough and R. Khan, *Trends Biochem. Sci.*, (1978) 61–63.
(4) G. M. H. Waites, W. C. L. Ford, R. Khan, and H. F. Jones, Br. Pat. Applic. (1977); W. C. L. Ford and G. M. H. Waites, *J. Reprod. Fertil.*, 52 (1978) 153–157.
(5) V. Kollonitsch, *Sucrose Chemicals*, The International Sugar Research Foundation, Washington, D.C., 1970.

atoms is shown in formula 1. Its constitution was proved by methylation studies. On hydrolysis, methyl hepta-O-methylmaltoside gave[6] 1 mol of 2,3,4,6-tetra-O-methyl-D-glucose and 1 mol of 2,3,6-tri-O-methyl-D-glucose per mol. Hydrolysis of methyl hepta-O-methylmaltobionate gave equimolar proportions of 2,3,4,6-tetra-O-methyl-D-glucose and 2,3,5,6-tetra-O-methyl-D-gluconic acid.[7] These facts indicated that maltose contains a D-glucopyranosyl group linked from C-1 of the nonreducing group to O-4 of the reducing D-glucose residue. Zemplén's work on the degradation of maltose oxime supported this conclusion,[8] as did a study of the lactonization of maltobionic acid.[9] As maltose is hydrolyzed by maltase, an enzyme specific for α-D-glucosides, and as the high specific rotations of maltose and its derivatives are consistent with an α-D-linkage, maltose was designated 4-O-α-D-glucopyranosyl-D-glucopyranose. This was confirmed by synthesis, and such physical methods as X-ray crystallography, neutron diffraction, and nuclear magnetic resonance spectroscopy.

The first synthesis of octa-O-acetyl-β-maltose was reported by Lemieux[10] in 1953. When 1,2,3,6-tetra-O-acetyl-β-D-glucopyranose was heated with 3,4,6-tri-O-acetyl-1,2-anhydro-α-D-glucopyranose for 13 h at 100°, it gave a mixture from which the product was isolated, after acetylation, in 8.7% yield. Octa-O-acetyl-β-maltose has been obtained in 43.6% yield by way of condensation of 3,4,6-tri-O-acetyl-2-O-benzyl-β-D-glucopyranosyl chloride with 1,2,3,6-tetra-O-acetyl-β-D-glucopyranose.[11]

The structures of α-maltose,[12] β-maltose monohydrate,[13] and methyl

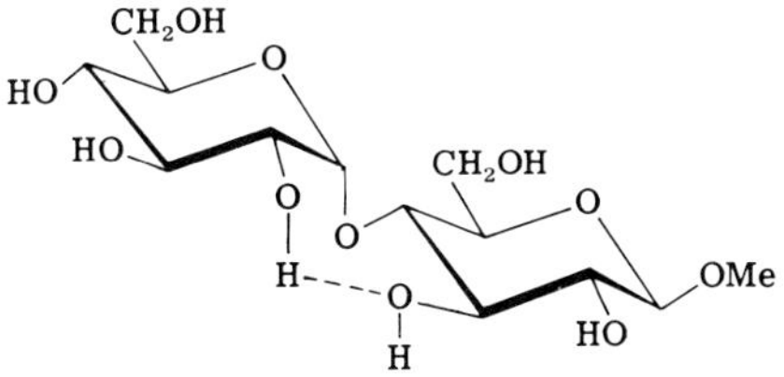

Methyl β-maltoside

2

(6) W. N. Haworth and G. C. Leitch, *J. Chem. Soc.*, 115 (1919) 809–817.
(7) W. N. Haworth and S. Peat, *J. Chem. Soc.*, (1926) 3094–3101.
(8) G. Zemplén, *Ber.*, 60 (1927) 1555–1564.
(9) P. A. Levene and H. Sobotka, *J. Biol. Chem.*, 71 (1926) 471–475.
(10) R. U. Lemieux, *Can. J. Chem.*, 31 (1953) 949–951.
(11) K. Igarashi, J. Irisawa, and T. Honma, *Carbohydr. Res.*, 39 (1975) 341–343.
(12) F. Takusagawa and R. A. Jacobson, *Acta Crystallogr., Sect. B*, 34 (1978) 213–218.
(13) G. J. Quigley, A. Sarko, and R. H. Marchessault, *J. Am. Chem. Soc.*, 92 (1970) 5834–5839.

β-maltoside[14] (**2**) were investigated by X-ray crystallography. An investigation of the molecular structure of α-maltose revealed that the D-glucopyranosyl and D-glucose moieties are slightly distorted from the $^4C_1$ to a skew conformation. Such a distortion was not observed for β-maltose[13] or methyl β-maltoside,[14] in which both moieties assume the expected $^4C_1$ conformation. The torsional angles O-4'–C-1'–O-1'–C-4 and O-1–C-4–O-1'–C-1' in α-maltose are $+172.1$ and $-177.6°$, respectively.[12] This small twist in the backbone of the molecule is such as to favor intramolecular hydrogen-bonding between O-2' and O-3. This interesting feature has been observed in all known crystal structures that contain a maltose residue, and appears to be one of the principal forces controlling conformations based on α-(1→4)-linked D-glucoses. The presence of OH-2' ------- OH-3, intramolecular hydrogen-bonding in maltose has also been supported by the results of high-resolution, nuclear magnetic resonance spectroscopy[15] (see Section XI, 1). The n.m.r. results also revealed that the OH-3 group is the donor in this interaction. This direction of hydrogen bonding in solution appears to be the opposite of that for the solid-state interaction as deduced from diffraction studies on methyl β-maltoside.[14]

A neutron-diffraction study of β-maltose monohydrate provided a more defined description of the hydrogen bonding present in the molecule.[16] With the exception of the intramolecular bond, O-2'–H-----O-3, all of the hydroxyl groups in β-maltose monohydrate are involved in both donor and acceptor hydrogen-bond interaction. Neither the ring-oxygen atoms nor the glycosidic oxygen atom are involved in hydrogen bonding. The stereochemistry of the hydroxymethyl groups in α-maltose,[12] β-maltose,[13] and methyl β-maltoside[14] has also been examined. For α-maltose, the conformational angles of the primary hydroxyl groups, C-6–O-6 and C-6'–O-6' have *gauche-trans* (–synclinal) orientations. In β-maltose and methyl β-maltoside, the α-D-glucosyl group has a *gauche-trans*, whereas the β-D-glucose residue has a *gauche-gauche* (+synclinal) orientation.

Favored conformations of maltose arrived at by free-energy calculations were in good agreement with those for the solid state and in solution.[17,18] From thermal-expansibility experiments, it has been suggested that, in aqueous solution, the maltose molecule folds, and undergoes extensive, intramolecular, hydrophobic bonding.[19]

(14) S. S. C. Chu and G. A. Jeffrey, *Acta Crystallogr.*, 23 (1967) 1038–1049.
(15) M. St.-Jacques, P. R. Sundararajan, K. J. Taylor, and R. H. Marchessault, *J. Am. Chem. Soc.*, 98 (1976) 4386–4391.
(16) M. E. Gress and G. A. Jeffrey, *Acta Crystallogr.*, Sect. B, 33 (1977) 2490–2495.
(17) C. V. Goebel, W. L. Dimpfl, and D. A. Brant, *Macromolecules*, 3 (1970) 644–654.
(18) D. A. Rees and P. J. C. Smith, *J. Chem. Soc., Perkin Trans.* 2, (1975) 836–840.
(19) J. L. Neal and D. A. I. Goring, *Can. J. Chem.*, 48 (1970) 3745–3747.

III. GLYCOSIDES

Reducing groups of sugars are generally protected by glycosidation before chemical or biological transformations are carried out at other functional groups of the molecule. The aglycon can, if desired, then be removed by acid hydrolysis, use of a cation-exchange resin (H^+ form), or hydrogenolysis, depending upon the nature of the aglycon. Several alkyl and aryl glycosides of maltose have been synthesized. Wolfrom and coworkers[20] prepared methyl β-maltoside[21] (**3**) from maltose by a modified, Koenigs–Knorr reaction, with mercuric cyanide[22] as the condensing agent instead of silver carbonate.[23] Synthesis of compound **3** has been achieved[24] by a slight modification of the process.[20] In the glycosidation reaction, mercuric cyanide was replaced by mercuric acetate, and the de-esterification step was effected by the Zemplén method. Enzymic and chemical syntheses of methyl α-maltoside (**4**) have been reported.[25–30] However, with the exception of the

3 R = OMe, R' = H
4 R = H,   R' = OMe

---

(20) M. L. Wolfrom, Y.-L. Hung, P. Chakravarty, G. U. Yuen, and D. Horton, *J. Org. Chem.*, 31 (1966) 2227–2232.
(21) J. C. Irvine and I. M. A. Black, *J. Chem. Soc.*, (1926) 862–875; T. J. Schoch, E. J. Wilson, Jr., and C. S. Hudson, *J. Am. Chem. Soc.*, 64 (1942) 2871–2872.
(22) B. Helferich and K. F. Wedemyer, *Justus Liebigs Ann. Chem.*, 563 (1949) 139–145; R. Kuhn and W. Kirschenlohr, *Chem. Ber.*, 86 (1953) 1331–1333.
(23) F. H. Newth, S. D. Nicholas, F. Smith, and L. F. Wiggins, *J. Chem. Soc.*, (1949) 2550–2553.
(24) P. L. Durette, L. Hough, and A. C. Richardson, *Carbohydr. Res.*, 31 (1973) 114–119.
(25) S. Peat, W. J. Whelan, and G. Jones, *J. Chem. Soc.*, (1957) 2490–2495.
(26) J. H. Pazur, J. M. Marsh, and T. Ando, *J. Am. Chem. Soc.*, 81 (1959) 2170–2172.
(27) S. Matsubara, *J. Biochem. (Tokyo)*, 49 (1961) 226–231.
(28) S. E. Zurabyan, V. Bilík, and Š. Bauer, *Chem. Zvesti*, 23 (1969) 923–927; *Chem. Abstr.*, 75 (1971) 6235a.
(29) Y. Inouye, K. Onodera, I. Karasawa, and Y. Nishisawa, *Nippon Nogei Kagaku Kaishi*, 26 (1952) 631–634.
(30) W. E. Dick, Jr., D. Weisleder, and J. E. Hodge, *Carbohydr. Res.*, 18 (1971) 115–123; W. E. Dick, Jr., and J. E. Hodge, *Methods Carbohydr. Chem.*, 7 (1976) 15–18.

method described by Hodge and coworkers,[30] most of them have disadvantages. The action of D-glucosyltransferases on mixtures of starch and methyl α-D-glucoside produces anomerically pure material, but the yields are low, and the method involves extensive chromatographic purification.[25,26] A claim[27] that 4 can be prepared directly from a mixture of maltose and methanolic hydrogen chloride was not substantiated when subjected to careful reexamination.[28] Methyl β-maltoside heptaacetate has been anomerized to give a mixture of the methyl α- and β-maltoside hepta-acetates.[28,29] Only one method, however, provided for selective removal of starting material.[29] An efficient synthesis of compound 4 has been achieved.[30] Treatment of ethyl 1-thio-β-maltoside with a mixture of methanol, bromine, and silver carbonate gave a mixture containing methyl α,β-maltoside and traces of maltose. The α anomer, which initially constituted 85% of the maltosides fraction, was enriched by selective oxidation of the β anomer (3) with chromic acid. Chromatographic purification gave a 65% yield of 4 of 97% anomeric purity.

Phenyl β-maltoside heptaacetate[31] was synthesized, and obtained in crystalline form, by treating octa-O-acetyl-β-maltose with phenol in the presence of p-toluenesulfonic acid.[32] The mother liquors yielded a small proportion of phenyl α-maltoside heptaacetate. Treatment of octa-O-acetyl-β-maltose with phenol and zinc chloride gave phenyl α-maltoside heptaacetate as the major product.[31,33] Crystalline phenyl α-maltoside[33] was subsequently obtained from the corresponding heptaacetate derivative by de-esterification with methanolic ammonia.[34] p-Bromophenyl β-maltoside has been prepared by an adaptation of Helferich and Griebel's method[35] for the preparation of phenyl β-lactoside.[36] Hepta-O-acetyl-α-maltosyl bromide was treated with p-bromophenol and potassium hydroxide in 50% aqueous acetone. The p-bromophenyl hepta-O-acetyl-β-maltoside was then deesterified to afford the free glycoside in 42% yield. Synthesis of benzyl β-maltoside has also been described.[37]

---

(31) E. Fischer and E. F. Armstrong, *Ber.*, 35 (1902) 3153–3155.
(32) L. Asp and B. Lindberg, *Acta Chem. Scand.*, 6 (1952) 941–946.
(33) B. Helferich and S. R. Peterson, *Ber.*, 68 (1935) 790–794.
(34) S. Matsubara, *Bull. Chem. Soc. Jpn.*, 34 (1961) 718–722.
(35) B. Helferich and B. Griebel, *Justus Liebigs Ann., Chem.*, 544 (1940) 191–205.
(36) I. C. M. Dea, *Carbohydr. Res.*, 11 (1969) 363–367.
(37) B. Helferich and W. Speicher, *Justus Liebigs Ann. Chem.*, 579 (1953) 106–112.

IV. ETHERS

1. Trityl Ethers

The value of trityl ethers of carbohydrates as synthetic intermediates is well recognized.[38] Selective tritylation of maltose and its derivatives has been investigated. The primary hydroxyl groups in maltose and many of its derivatives show a large difference in reactivity. Reaction of maltose with 1 molar equivalent of trityl chloride in pyridine, followed by acetylation and chromatographic fractionation on silica gel, gave 1,2,3,2',3',4'-hexa-O-acetyl-6,6'-di-O-tritylmaltose[39,40] (**5**), 1,2,3,6,2',3',4'-hepta-O-acetyl-6'-O-trityl-β-maltose (**6**), and 1,2,3,2',3',4',6'-hepta-O-acetyl-6-O-trityl-α-maltose (**7**) in yields of 13, 31, and 3%, respectively.[41] No attempt was made to establish the configuration at the anomeric center of **5**. The structures and the anomeric configurations of compounds **6** and **7** were established by ^{1}H-n.m.r. spectroscopy and by the structural relationship between **7** and its β-D anomer (**8**).

Compound **8** was prepared from 2,3-di-O-acetyl-1,6-anhydro-4-O-(2,3,4,6-tetra-O-acetyl-α-D-glucopyranosyl)-β-D-glucopyranose by an authentic synthesis.[41] The optical rotations of compounds **6** and **7** were not in accord with Hudson's Rules of Isorotation,[42] and this led to the synthesis, and determination of the molecular rotation, of all three anomeric pairs of 6- and 6'-mono- and 6,6'-di-O-tritylmaltose peracetates.[43] Reaction of maltose with 1 molar equivalent of trityl chloride in pyridine followed by acetylation gave, after chromatography on silica gel and precise fractional recrystallization, 1,2,3,2',3'4'-hexa-O-acetyl-6,6'-di-O-trityl-β-maltose (**9**) and its α-D anomer (**10**), **6** and its α-D anomer (**11**), **7**, and **8** in yields of 0.27, 0.41, 51.67, 2.03, 4.05, and 0.3%, respectively. Compound **11** has also been prepared by detritylation of the ditrityl ether **10** with 80% aqueous acetic acid at 100°, followed by retritylation with an equimolar proportion of the reagent, and subsequent acetylation. These results indicated that the 6'-hydroxyl group in maltose is more reactive towards tritylation than that at C-6. The isomers having the trityl group on the nonreducing group are much more dextrorotatory than

---

(38) B. Helferich, *Adv. Carbohydr. Chem.*, 3 (1948) 79–111.
(39) K. Josephson, *Justus Liebigs Ann. Chem.*, 472 (1929) 230–240.
(40) Y. Hirasaka, I. Matsunaga, K. Unemoto, and M. Sukegawa, *Yakugaku Zasshi*, 83 (1963) 966–971; *Chem. Abstr.*, 60 (1964) 4232–4233.
(41) M. L. Wolfrom and K. Koizumi, *J. Org. Chem.*, 32 (1967) 656–660.
(42) C. S. Hudson, *J. Am. Chem. Soc.*, 31 (1909) 66–86; 38 (1916) 1566–1575.
(43) K. Koizumi and T. Utamura, *Carbohydr. Res.*, 33 (1974) 127–134.

those having it on the reducing residue. Studies on the molecular rotations of compounds **6–11** suggested that changes in the orientation at the glycosidic linkage, as expressed in torsional angles about the C–O and O–C bonds, may influence the magnitude of the optical rotation of these compounds. All of the six trityl ethers gave a positive o.r.d. curve.

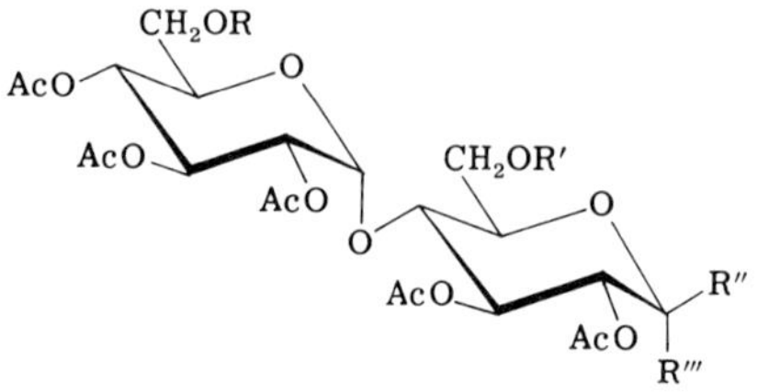

```
5   R = R' = Tr, R" = R'" = H, OAc
6   R = Tr, R' = Ac, R" = OAc, R'" = H
7   R = Ac, R' = Tr, R" = H, R'" = OAc
8   R = Ac, R' = Tr, R" = OAc, R'" = H
9   R = R' = Tr, R" = OAc, R'" = H
10  R = R' = Tr, R" = H, R'" = OAc
11  R = Tr, R' = Ac, R" = H, R'" = OAc
```

On selective tritylation, benzyl β-maltoside,[37] ethyl 1-thio-β-maltoside,[44] and methyl 3-O-(methylsulfonyl)-β-maltoside[45] give, as observed for maltose,[41,43] mainly the corresponding 6'-trityl ether. The large difference in the relative reactivities of the two primary hydroxyl groups in maltose and its derivatives has been ascribed to the non-identical stereoelectronic environment at these two positions,[46] and molecular models of the favored conformation of maltose[13] and its methyl glycoside[14] indicate that the 2'-, 3-, and 6-hydroxyl groups will be subjected to the greatest steric interaction.

2. Other Ethers

Carbohydrate ethers, particularly methyl ethers, constitute an important class of compounds. The value of methylation studies in the structural determination of sugars is well known. The 6-, 6'-, and 3-methyl ethers of phenyl α-maltoside were needed for study of their

(44) N. K. Kochetkov, E. M. Klimov, and M. U. Orchimakov, *Izv. Akad. Nauk SSSR, Ser. Khim.*, 8 (1977) 1864–1867.
(45) P. L. Durette, L. Hough, and A. C. Richardson, *J. Chem. Soc. Perkin Trans. 1*, (1974) 97–101.
(46) G. G. S. Dutton and K. N. Slessor, *Can J. Chem.*, 44 (1966) 1069–1074.

effect on the hydrolytic action of the enzyme Taka amylase A (Ref. 47). Methylation of 1,2,3,2′,3′,4′,6′-hepta-O-acetyl-β-maltose[32] (**12**, see Section V,1) with diazomethane–boron trifluoride,[48,49] a reagent that does not cause acyl migration, gave the corresponding 6-methyl ether **13** in 80% yield. Treatment of **13** with phenol in the presence of zinc chloride for 30 min at 115°, followed by de-esterification with sodium methoxide in methanol, yielded phenyl 6-O-methyl-α-maltoside (**14**). Synthesis of phenyl 6′-O-methyl-α-maltoside has been achieved from

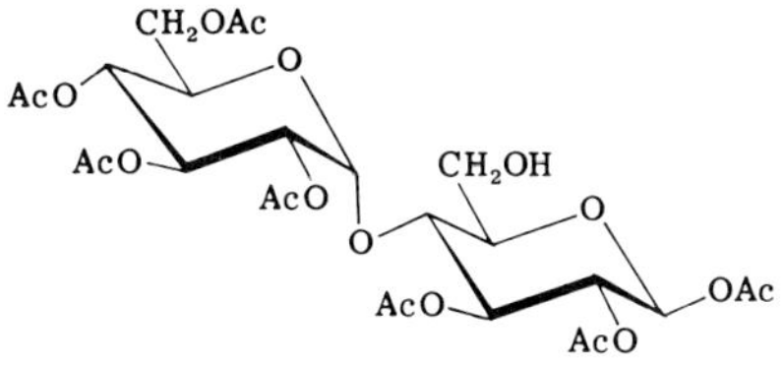

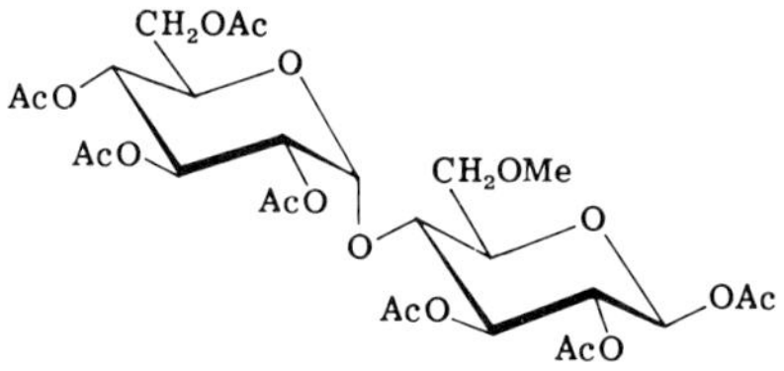

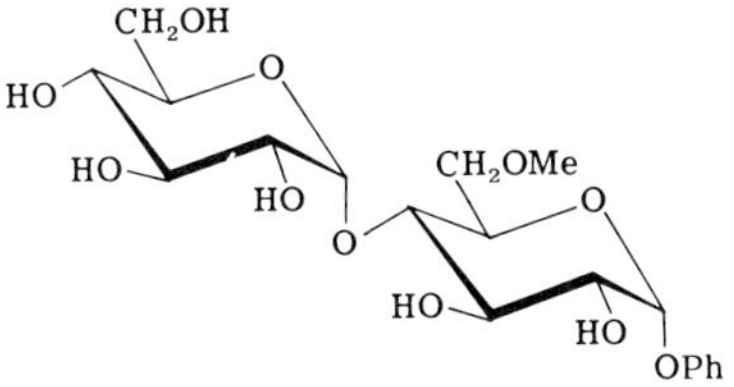

(47) H. Arita, M. Isemura, T. Ikenaka, and Y. Matsushima, *Bull. Chem. Soc. Jpn.*, 43 (1970) 818–823.
(48) E. G. Gros and S. M. Flematti, *Chem. Ind. (London)*, (1966) 1556–1557.
(49) J. O. Deferrari, E. G. Gros, and I. O. Mastronardi, *Carbohydr. Res.*, 4 (1967) 432–434; E. G. Gros and I. O. Mastronardi, *ibid.*, 10 (1969) 318–321; C. P. J. Glaudemans and H. G. Fletcher, Jr., *ibid.*, 7 (1968) 480–482; P. A. Seib, *ibid.*, 8 (1968) 101–109.

2,3,2',3',4'-penta-O-acetyl-1,6-anhydro-6'-O-trityl-β-maltose[46] by a sequence of reactions involving detritylation, methylation, glycosidation, and de-esterification.[47] Reaction of 1,2,6,2',3',4',6'-hepta-O-acetyl-β-maltose[50] (see Section V,1) with diazomethane–boron trifluoride etherate in dry acetone, followed by phenyl glycosidation and de-esterification, gave phenyl 3-O-methyl-α-maltoside.

The structures of partially benzoylated derivatives of 1,6-anhydro-4',6'-O-benzylidene-β-maltose were established by methylation studies.[51] During the course of this investigation, 2,3- and 3,3'-di- and 2,3,3'-tri-methyl ethers of 1,6-anhydro-4',6'-O-benzylidene-β-maltose were prepared by using diazomethane–boron trifluoride etherate reagent. Methyl [methyl 2,3-di-O-methyl-4-O-(2,3,4,6-tetra-O-methyl-α-D-glucopyranosyl)-β-D-glucopyranosid]uronate (**15**) has been synthe-

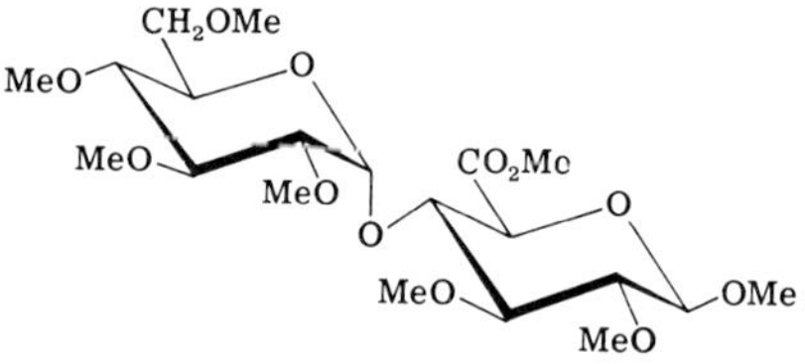

15

sized, and used as a model for the study of the base-catalyzed β-elimination of 4-O-substituted hexopyranosiduronate without degradation of the exposed reducing sugar, and of the selective, acid hydrolysis of the hex-4-enopyranosiduronate.[52] Synthesis of **15** was achieved by chromium trioxide oxidation of methyl 2,3,2',3',4',6'-hexa-O-methyl-β-maltoside (**16**), which had been prepared by the following sequence

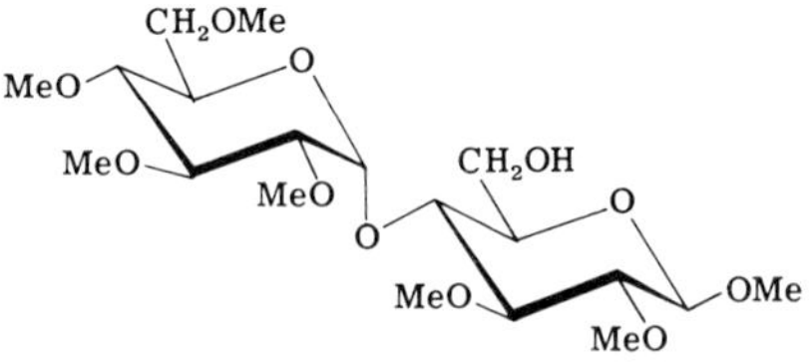

16

of reactions: methyl 4',6'-O-benzylidene-β-maltoside → methyl 2,3,6,2',3'-penta-O-acetyl-β-maltoside → methyl 2,3,6,2',3'-penta-O-

(50) W. E. Dick, B. G. Baker, and J. E. Hodge, *Carbohydr. Res.*, 6 (1968) 52–62.
(51) M. Mori, M. Haga, and S. Tejima, *Chem. Pharm. Bull.*, 24 (1976) 1173–1178.
(52) G. O. Aspinall, T. N. Krishnamurthy, and W. Mitura, *Can. J. Chem.*, 53 (1975) 2182–2188.

acetyl-4′,6′-di-O-methyl-β-maltoside → methyl 4′,6′-di-O-methyl-β-maltoside → methyl 4′,6′-di-O-methyl-6-O-trityl-β-maltoside → methyl 2,3,2′,3′,4′,6′-hexa-O-methyl-6-O-trityl-β-maltoside → **16**. Attempts to synthesize **16** from 1,6-anhydro-2,3,2′,3′,4′,6′-hexa-O-methyl-β-maltose by a reaction involving ring opening by acetolysis, conversion into the glycosyl halide, and glycosidation were unsuccessful.[52] Permethylated ethers of maltose,[6] 6′-O-tritylmaltose,[40] and 6,6′-di-O-tritylmaltose have been described.[40]

The synthesis, and the polymerization reaction, of 1,6-anhydro-2,3-di-O-benzyl-4-O-(2,3,4,6-tetra-O-benzyl-α-D-glucopyranosyl)-β-D-glucopyranoside has been reported.[53] The monomer was prepared from maltose by a sequence involving acetylation,[54] phenyl glycosidation,[32] cyclization,[55] re-acetylation, and benzylation.[56]

V. Esters

1. Acetates and Benzoates

Acetates and benzoates are widely used as characteristic derivatives of carbohydrates, and for the assay and protection of hydroxyl groups. Various maltose heptaacetates having a free hydroxyl group on C-1, C-2, C-3, C-6, or C-6′ have been synthesized. 2,3,6,2′,3′,4′,6′-Hepta-O-acetyl-β-maltose (**17**) has been prepared by hydrolysis of the corresponding glycosyl halide.[57,58] The reaction conditions may not always result in the isolation of the anomerically unsubstituted acetate, as was shown by Corbett and coworkers,[59] who obtained three

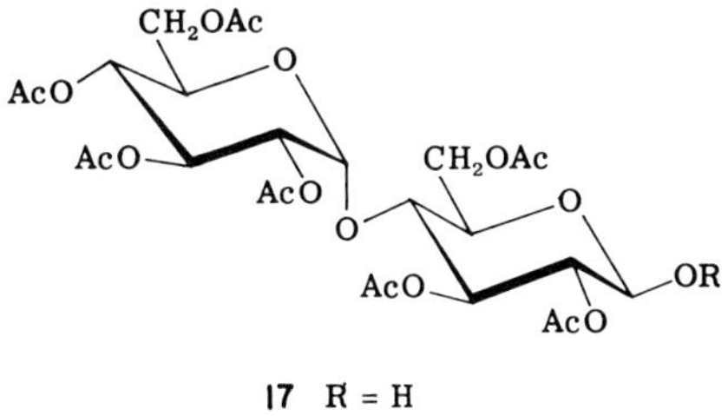

17 R = H
18 R = Ac

(53) B. Veruovic and C. Schuerch, *Carbohydr. Res.*, 14 (1970) 199–206.
(54) M. L. Wolfrom and A. Thompson, *Methods Carbohydr. Chem.*, 1 (1962) 334–335.
(55) Y. Hirasaka and I. Matsunaga, *Chem. Pharm. Bull.*, 13 (1965) 176–179.
(56) G. Zemplén, Z. Csürös, and S. (J.) Angyal, *Ber.*, 70 (1937) 1848–1856.
(57) B. Helferich and R. Steinpreis, *Chem. Ber.*, 91 (1958) 1794–1798.
(58) C. S. Hudson and R. Sayre, *J. Am. Chem. Soc.*, 38 (1916) 1867–1873.
(59) W. M. Corbett, J. Kidd, and A. M. Liddle, *J. Chem. Soc.*, (1960) 616–619.

different hepta-*O*-acetylcellobioses from a hydrolyzate of the acetylated glycosyl bromide. Hence, an alternative method of synthesis of the heptaacetates of cellobiose, maltose, melibiose, and lactose having a free hydroxyl group on C-1 was studied.[60] Treatment of octa-*O*-acetyl-β-maltose (**18**) with piperidine in oxolane for 1 h at 25° gave the heptaacetate **17** in 76% yield. Synthesis of 1,3,6,2′,3′,4′,6′-hepta-*O*-acetyl-β-maltose (**19**) from **18** has been reported[61] by way of 3,6,2′,3′,4′,6′,-hexa-*O*-acetyl-2-*O*-(trichloroacetyl)-β-maltosyl chloride[62] (**20**).

The reaction[61,62] of **18** with phosphorus pentachloride has been reinvestigated.[63] Treatment of **18** with 5 molar equivalents of phosphorus pentachloride in boiling carbon tetrachloride for 3 h under reflux gave, after chromatography on silica gel, crystalline **20** in 58% yield. The structure of **20** was corroborated by anomerization, and conversion into the β-heptaacetate **19**. Compound **20** was anomerized with tetramethylammonium chloride[64] in acetonitrile, to afford the corresponding α-chloride **21** in crystalline form. Attempts to anomerize **20** with titanium chloride[65] in chloroform were unsuccessful. The 2-chloroacetyl group in **20** was selectively removed by treatment with ammonia, to afford the crystalline β-chloride **22**, which was readily anomerized with titanium tetrachloride to yield the crystalline α-chloride **23**. Subsequent treatment of **23** with mercuric

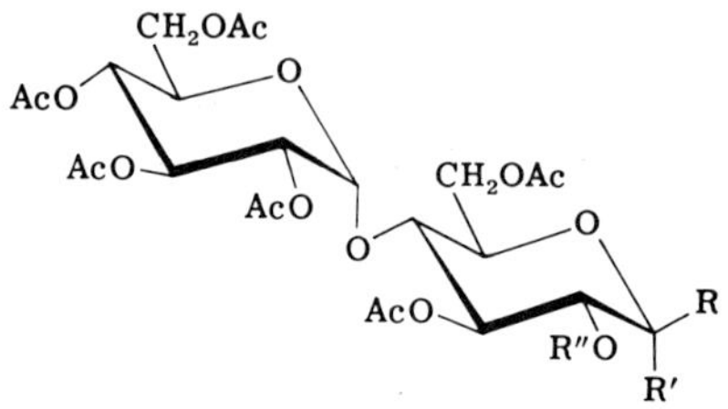

19 R = OAc, R′ = H,   R″ = H
20 R = Cl,   R′ = H,   R″ = COCl$_3$
21 R = H,   R′ = Cl, R″ = COCl$_3$
22 R = Cl,   R′ = H,   R″ = H
23 R = H,   R′ = Cl, R″ = H

(60) R. M. Rowell and M. S. Feather, *Carbohydr. Res.*, 4 (1967) 486–491.
(61) B. Koeppen, *Carbohydr. Res.*, 13 (1970) 193–198.
(62) P. Brigl and P. Mistele, *Z. Physiol. Chem.*, 126 (1923) 120–129.
(63) K. Takeo, *Carbohydr. Res.*, 48 (1976) 290–293.
(64) R. U. Lemieux and J. Hayam, *Can. J. Chem.*, 43 (1965) 2162–2173.
(65) M. L. Wolfrom, A. O. Pittet, and I. C. Gillam, *Proc. Natl. Acad. Sci. U. S. A.*, 47 (1961) 700–705.

acetate in acetic acid gave the heptaacetate **19** in 85% yield. Selective acetylation of β-maltose monohydrate with acetyl chloride in pyridine–toluene was reported[50] to give a mixture of 1,2,6,2′,3′,4′,6′-hepta-*O*-acetyl-β-maltose and **18** in the ratio of 7:3. The low reactivity of OH-3 was attributed[50] to the presence of a strong, intramolecular hydrogen-bond between O-2′ and O-3; however, this has been questioned by Hough and coworkers.[66] 1,2,3,2′,3′,4′,6′-Hepta-*O*-acetyl-β-maltose[20,32] has been prepared by the action of titanium bromide in chloroform upon 1,6-anhydro-β-maltose hexaacetate and treatment of the product with mercuric acetate in acetic acid.[67]

Selective benzoylation of β-maltose monohydrate,[68,69] methyl β-maltoside,[69] methyl 4′,6′-*O*-benzylidene-β-maltoside,[69] and 1,6-anhydro-4′,6′-*O*-benzylidene-β-maltose[51] has been studied. When benzoylation of β-maltose was performed under the conditions in which cellobiose[70] and lactose[71] are completely transformed into the corresponding octabenzoates, maltose gave[68] a mixture of 1,2,6,2′,3′,4′,6′-hepta-*O*-benzoyl-β-maltose and the octabenzoate in the ratio of 11:9. The formation of the heptabenzoate having its free hydroxyl group on C-3 has been attributed, as indicated earlier, to the presence of a strong, intramolecular hydrogen-bond between the 3- and 2′-hydroxyl groups, which has been observed in the crystal structures of maltose monohydrate[13] and methyl β-maltose monohydrate,[14] and has been postulated for maltose in dimethyl sulfoxide solution from i.r.- and n.m.r.-spectral evidence.[72] The role of intramolecular hydrogen-bonding in this case needs clarification, as it has been established for other sugars that intramolecular hydrogen-bonding enhances the rate of acetylation by acid chlorides.[73,74] A strong, OH-3---O-2′, intramolecular hydrogen-bond could, however,

---

(66) P. L. Durette, L. Hough, and A. C. Richardson, *J. Chem. Soc., Perkin Trans. 1*, (1974) 88–96.

(67) M. Mori, M. Haga, and S. Tejima, *Chem. Pharm. Bull.*, 22 (1974) 1331–1338.

(68) I. M. E. Thiel, J. O. Deferrari, and R. A. Cadenas, *Justus Liebigs Ann. Chem.*, 723 (1969) 192–197.

(69) K. Takeo and S. Okano, *Carbohydr. Res.*, 59 (1977) 379–391.

(70) J. O. Deferrari, I. M. E. Thiel, and R. A. Cadenas, *J. Org. Chem.*, 30 (1965) 3053–3055.

(71) J. O. Deferrari and I. M. E. Thiel, *Carbohydr. Res.*, 29 (1973) 141–146.

(72) B. Casu, M. Reggiani, G. G. Gallo, and A. Vigevani, *Tetrahedron*, 22 (1966) 3061–3083.

(73) K. W. Buck, J. M. Duxbury, A. B. Foster, A. R. Perry, and J. M. Webber, *Carbohydr. Res.*, 2 (1966) 122–131.

(74) A. H. Haines, *Adv. Carbohydr. Chem. Biochem.*, 33 (1976) 11–109.

cause the molecule to fold in such a way that access to HO-3 would be severely hindered.

When maltose was treated with 8 molar equivalents of benzoyl chloride in pyridine at $-40°$, it afforded octa-O-benzoyl-β-maltose, 1,2,6,2',3',4',6'-hepta-O-benzoyl-β-maltose, and 1,2,6,2',3',6'-hexa-O-benzoyl-β-maltose in yields of 3, 79, and 12%, respectively.[69] Similar treatment of methyl β-maltoside with 7 molar proportions of benzoyl chloride in pyridine gave crystalline methyl 2,6,2',3',4',6'-hexa-O-benzoyl-β-maltoside and methyl 2,6,2',3',6'-penta-O-benzoyl-β-maltoside in yields of 71 and 23%, respectively.[69] Analysis of the relative yields of the products of these reactions suggests that, after HO-3, the 4'-hydroxyl group in maltose and methyl β-maltoside is the most hindered. Benzoylation of methyl 4',6'-O-benzylidene-β-maltoside with 10 molar proportions of reagent gave 2,6,2',3'-tetra-O-benzoyl-4',6'-O-benzylidene-β-maltoside (71%) and an equimolar mixture (22%) of the 2,6,2'- and 2,6,3'-tribenzoates. These results indicated that the order of reactivity of the hydroxyl groups towards acid halides in pyridine is HO-2, HO-6 > HO-2' $\approx$ HO-3' > HO-3.

1,6-Anhydro-4',6'-O-benzylidene-β-maltose has proved to be a useful compound in selective-esterification studies,[51] as it is a rigid system having well defined, molecular geometry. Treatment of 1,6-anhydro-4',6'-O-benzylidene-β-maltose with 2 molar equivalents of benzoyl chloride in pyridine at $-10°$ gave, after chromatography on silica gel, the corresponding 2,3,2',3'-tetra-, 2,2',3'-tri-, 2,2'- and 2',3'-di-, and 2'-mono-benzoates in yields of 1.2, 2.8, 9, 7.3, and 48%, respectively. Based on these results, it was concluded that the order of reactivity of the secondary hydroxyl groups in 1,6-anhydro-4',6'-O-benzylidene-β-maltoside towards reaction with benzoyl chloride is HO-2' > HO-2 $\approx$ HO-3' > HO-3. It was expected that the equatorial hydroxyl groups, HO-2' and HO-3', would be more reactive than the axial hydroxyl groups, HO-2 and HO-3, and that this would result in the favored formation of the 2',3'-dibenzoate. However, contrary to this expectation, the 2'-monobenzoate was the major, and the 2',3'-dibenzoate, a minor, product. It is therefore obvious that the factors governing the relative reactivities of hydroxyl groups are not at present sufficiently known,[74,75] and that factors other than conformational effects must be invoked in order to explain the sequence of favored reactivity.

Selective de-esterification of octa-O-benzoyl-β-maltose has been

(75) J. Staněk, Jr., P. Chuchvalec, K. Čapek, K. Kefurt and J. Jarý, *Carbohydr. Res.*, 36 (1974) 273–282; M. E. Chacón-Fuertes, and M. Martin-Lomas, *ibid.*, 42 (1975) c4–c5.

achieved.[68,76] Treatment thereof with ammonia for 31 h at room temperature gave, after chromatography on a column of cellulose, 1,1-bis(benzamido)-1-deoxy-4-O-α-D-glucopyranosyl-D-glucitol, maltose, 6-O-benzoylmaltose, and N-benzoylmaltosylamine in yields of 20.8, 37.8, 31.6, and 0.1%, respectively. The ester group in 6-O-benzoylmaltose was found to be very resistant to ammonolysis. It was completely eliminated only after 15 days, whereas under similar reaction-conditions, the ester group of 6-O-benzoylmaltitol was eliminated after 2 days. The persistence of the 6-ester group in 6-O-benzoylmaltose has been attributed to steric hindrance towards nucleophilic attack by ammonia, together with the favorable stereoelectronic relationship of the benzoate group to the pyranose-ring oxygen atoms and the interglycosidic oxygen atom, as shown in formula **24**.

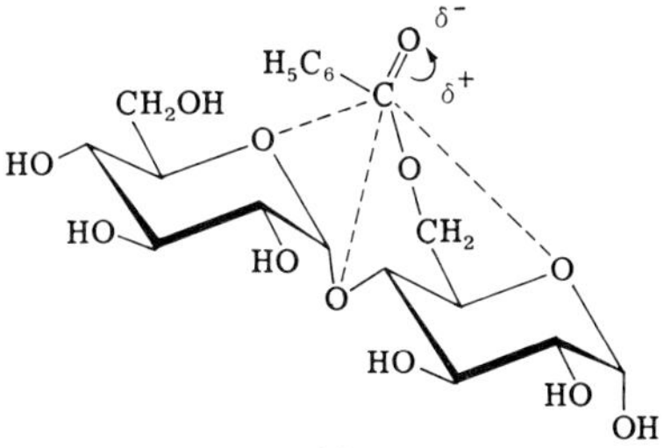

24

2. Sulfonates

The importance of sugar sulfonates as synthetic intermediates has been widely recognized.[77,78] Selective methanesulfonylation and p-toluenesulfonylation of maltose and its derivatives has been studied. Umezawa and coworkers used sequential p-toluenesulfonylation and acetylation to obtain crystalline 1,2,3,2′,3′,4′-hexa-O-acetyl-6,6′-di-O-p-tolylsulfonyl-β-maltose in 45% yield.[79] Dimolar methanesulfonylation of methyl β-maltoside followed by acetylation afforded crystalline methyl 2,3,2′,3′,4′-penta-O-acetyl-6,6′-di-O-(methylsulfonyl)-β-maltoside. The mother liquor also yielded a small proportion of a

(76) I. M. E. Thiel, J. O. Defarrari, and R. A. Cadenas, *J. Org. Chem.*, 31 (1966) 3704–3707.
(77) R. S. Tipson, *Adv. Carbohydr. Chem.*, 8 (1953) 107–215.
(78) D. H. Ball and F. W. Parrish, *Adv. Carbohydr. Chem.*, 23 (1968) 233–280; *Adv. Carbohydr. Chem. Biochem.*, 24 (1969) 139–197.
(79) S. Umezawa, T. Tsuchiya, S. Nakada, and K. Tatsuta, *Bull. Chem. Soc. Jpn.*, 40 (1967) 395–401.

mono-methanesulfonate (m.p. 153–156°; $[\alpha]_D$ +53.5°), whose structure has not yet been established.[23] On treatment with 2 molar equivalents of p-toluenesulfonyl chloride, methyl β-maltoside afforded the 6,6'-bis(p-toluenesulfonate) derivative **25** in 50% yield.[20] On the other hand, treatment with 1.1 molar equivalents of p-toluenesulfonyl chloride in pyridine gave the 6,6'-disulfonate (**25**) and the 6- (**26**) and 6'-

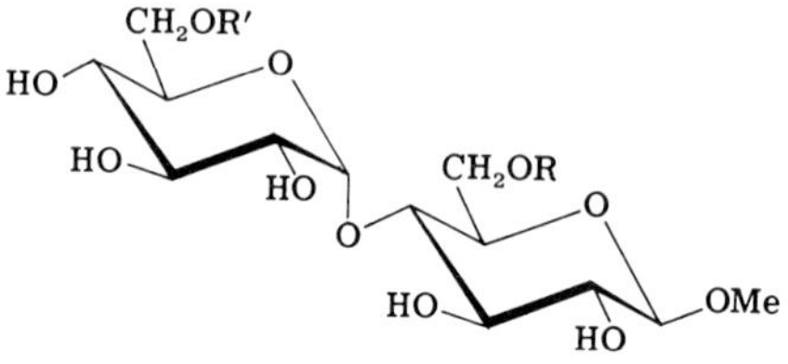

25 R = R' = Ts
26 R = Ts, R' = H
27 R = H, R' = Ts

sulfonate (**27**) in yields of 28, 1, and 18%, respectively.[80] The large difference in the reactivity of the two primary hydroxyl groups (OH-6 and OH-6') has been attributed to the non-identical, electronic environment of these two groups.[46] A similar difference in the reactivity of the two primary hydroxyl groups in maltose and its derivatives towards tritylation has already been mentioned (see Section I,1).

Selective p-toluenesulfonylation of methyl 4',6'-O-benzylidene-β-maltoside with 2 molar equivalents of the reagent in pyridine, followed by acetylation, gave a mixture of a major and a minor product, from which methyl 2,3,2',3'-tetra-O-acetyl-4',6'-O-benzylidene-6-O-p-tolylsulfonyl-β-maltoside was isolated in 66% yield.[81] Interestingly, a similar attempt at selective p-toluenesulfonylation of benzyl 4',6'-O-benzylidene-β-maltoside with a view to obtaining the 6-O-p-tolylsulfonyl derivative failed.[46] The presence of a bulky aglycon in the β-anomeric position apparently exerts a marked steric effect upon substitution at O-6.

A large number of sulfonate derivatives of maltose and its derivatives has been synthesized from 1,6-anhydro-β-maltose,[32,46,47,67] 1,6-anhydro-4',6'-O-benzylidene-β-maltose,[82] methyl 4',6'-O-benzylidene-β-malt-

---

(80) R. T. Sleeter and H. B. Sinclair, *J. Org. Chem.*, 35 (1970) 3804–3807.
(81) K. Takeo, *Carbohydr. Res.*, 69 (1979) 272–276.
(82) M. Mori, M. Haga, and S. Tejima, *Chem. Pharm. Bull.*, 23 (1975) 1480–1487.

oside,[81] and partially acylated maltose and its derivatives.[45,61,63,66,69,83,84] Sequential treatment of 2,3,2',3',4',6'-hexa-O-acetyl-1,6-anhydro-β-maltose with titanium chloride and mercuric acetate in acetic acid gave 1,2,3,2',3',4',6'-hepta-O-acetyl-β-maltose which, on p-toluenesulfonylation, afforded the corresponding 6-sulfonate.[32] Synthesis of 2,3,2',3'-tetra-O-acetyl-1,6-anhydro-4',6'-di-O-(methylsulfonyl)-β-maltose has been achieved[82] from 2,3,2',3'-tetra-O-acetyl-1,6-anhydro-4',6'-O-benzylidene-β-maltose by O-debenzylidenation followed by methanesulfonylation. From 1,2,6,2',3',4',6'-hepta-O-acetyl-β-maltose, the corresponding 3-sulfonate derivative (95%) has been prepared[66] by treatment with methanesulfonyl chloride in pyridine for 2 h at 0°.

VI. NUCLEOPHILIC-DISPLACEMENT REACTIONS

1. Nucleophilic-displacement Reactions of Sulfonates

Bimolecular, nucleophilic-displacement reactions of sulfonic esters of carbohydrates have been reviewed.[77,78]

The reaction of methyl 2,3,2',3',4'-penta-O-acetyl-6,6'-di-O-p-tolylsulfonyl-β-maltoside with an equimolar proportion of sodium iodide in acetone gave methyl 2,3,2',3',4'-penta-O-acetyl-6-deoxy-6-iodo-6'-O-p-tolylsulfonyl-β-maltoside as the major product.[80] The fact that the sulfonyloxy group on C-6 is more reactive than that on C-6' was surprising, because, in the partial p-toluenesulfonylation of methyl β-maltoside,[80] and the partial tritylation of maltose[41] and benzyl β-maltoside,[37] the hydroxyl group on C-6' is more reactive than that on C-6. When methyl 2-O-acetyl-3-O-(methylsulfonyl)-6-O-p-tolylsulfonyl-4-O-(2,3,4-tri-O-acetyl-6-O-p-tolylsulfonyl-α-D-glucopyranosyl)-β-D-glycopyranoside (28) was treated with sodium iodide under reflux for 20 h, it gave, after chromatography, methyl 2-O-acetyl-6-deoxy-6-iodo-3-O-(methylsulfonyl)-4-O-(2,3,4-tri-O-acetyl-6-deoxy-6-iodo-α-D-glucopyranosyl)-β-D-glucopyranoside (29) as the major, and methyl 2-O-acetyl-6-deoxy-6-iodo-3-O-(methylsulfonyl)-4-O-(2,3,4-tri-O-acetyl-6-O-p-tolylsulfonyl-α-D-glucopyranosyl)-β-D-glucopyranoside (30) as the minor, product.[84] These results were in agreement with an earlier observation,[80] and further indicated that the reactivity of the sulfonyloxy groups in 28 is in the order of 6 > 6' > 3.

(83) H. Arita and Y. Matsushima, J. Biochem. (Tokyo), 69 (1971) 409–413.
(84) N. K. Sharma, J. Norula, S. Mall, and R. Varadarajan, J. Inst. Chem. Calcutta, 38 (1976) 129–133.

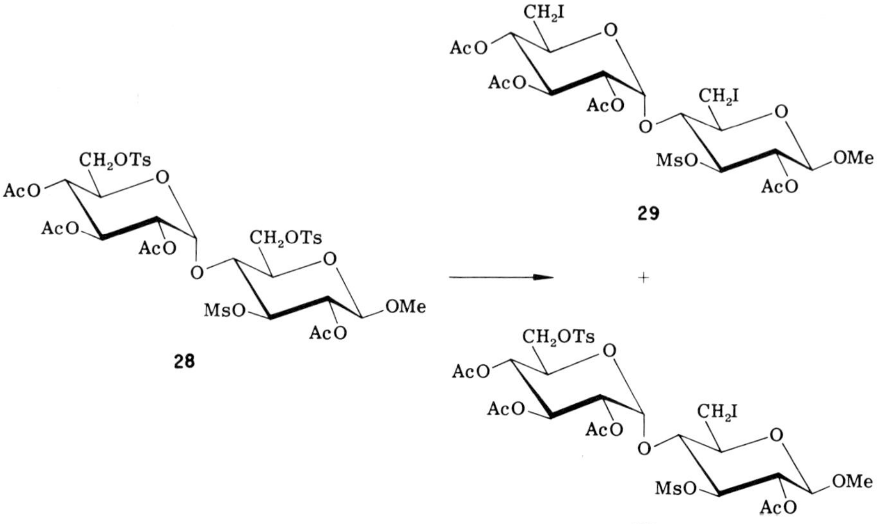

The nucleophilic-displacement reaction of the 3-sulfonyloxy group in methyl 3-O-(methylsulfonyl)-β-maltose hexaacetate,[45,66] 3-O-(methylsulfonyl)-β-maltose heptabenzoate,[69] and 3,4′,6′-tri-O-(methylsulfonyl)-β-maltose tetrabenzoate[69] has been reported. From a consideration of steric and polar factors influencing the feasibility of S_N2 reactions,[85] displacements should occur readily at C-3 in suitable derivatives of β-maltose or a β-maltoside. Attempts[66] at displacement of the sulfonate group in 3-O-(methylsulfonyl)-β-maltose heptaacetate (**31**) by azide in N,N-dimethylformamide at elevated temperatures proceeded with extensive decomposition, probably initiated by de-acetylation at O-1. Hence, formation of the methyl β-glycoside (**32**) was deemed necessary prior to nucleophilic displacement. The mesyloxy group in compound **32** underwent ready displacement with both sodium azide and sodium benzoate to afford, with inversion of configuration at C-3, the 3-azido and 3-O-benzoyl derivative, **33** and **34**, respectively. Similarly when methyl 2,6-di-O-acetyl-3-chloro-3-deoxy-4-O-(2,3,4,6-tetra-O-acetyl-α-D-glucopyranosyl)-β-D-allopyranoside reacted with sodium azide in N,N-dimethylformamide for 24 h at 120°, it gave methyl 2,6,2′,3′,4′,6′-hexa-O-acetyl-3-azido-3-deoxy-β-maltoside in 81% yield.[66] The S_N2 reaction of methyl 2,6,2′,3′-tetra-O-benzoyl-3,4′,6′-tri-O-(methylsulfonyl)-β-maltoside with sodium benzoate in hexamethylphosphoric triamide gave the expected methyl

(85) A. C. Richardson, *Carbohydr. Res.*, 10 (1969) 395–402.

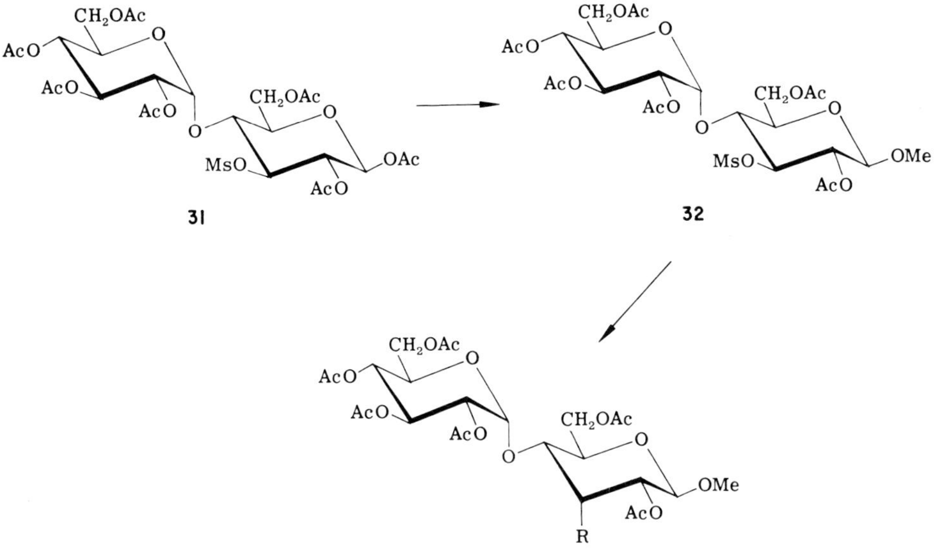

31

32

33 R = N₃
34 R = OBz

2,3,6-tri-O-benzoyl-4-O-(2,3,4,6-tetra-O-benzoyl-α-D-galactopyranosyl)-β-D-allopyranoside in 68% yield.[69]

The nucleophilic-displacement reactions of the 6- (Refs. 45 and 80), 6'- (Refs. 37, 45, and 80), and 4'- (Ref. 45) sulfonic derivatives of methyl β-maltoside, the 6-sulfonates[32,86] of β-maltose, and the 6'-sulfonate[46] of 1,6-anhydro-β-maltose have been reported.

2. Sulfuryl Chloride Reaction

The selective replacement of hydroxyl groups in carbohydrates by chlorine has been achieved by the use of sulfuryl chloride.[1,74,87,88] The reaction of sulfuryl chloride with maltose[89,90] and methyl β-maltoside[24] has been studied. Treatment of maltose with sulfuryl chloride, initially for 2 h at ~−70° and then processing at room temperature, gave, after methyl glycosidation and dechlorosulfation, a low yield

---

(86) J. Guerrera and C. E. Weil, *Carbohydr. Res.*, 27 (1973) 471–474.
(87) W. A. Szarek, *Adv. Carbohydr. Chem. Biochem.*, 28 (1973) 225–306.
(88) L. Hough, in J. L. Hickson (Ed.), "Sucrochemistry," *ACS Symp. Ser.*, 41 (1977) 9–21.
(89) H. J. Jennings and J. K. N. Jones, *Can. J. Chem.*, 40 (1962) 1408–1414.
(90) P. Colson, K. N. Slessor, H. J. Jennings, and I. C. P. Smith, *Can. J. Chem.*, 53 (1975) 1030–1037.

of methyl 6-chloro-6-deoxy-4-O-(4,6-dichloro-4,6-dideoxy-α-D-galactopyranosyl)-D-glucopyranoside[89] (**35**), but its anomeric configuration

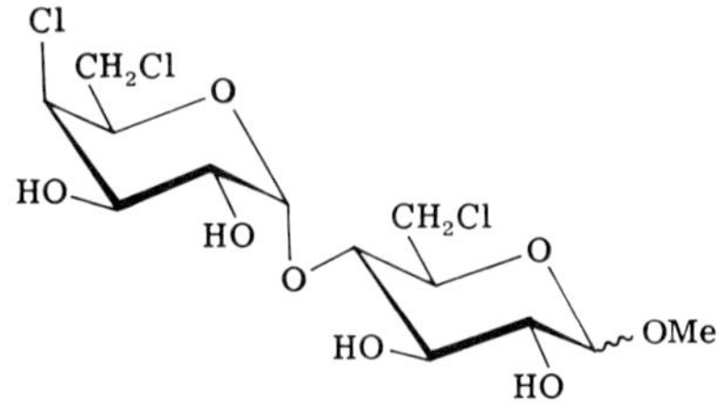

35

was not ascertained. The lack of reactivity of the 3-hydroxyl group in **35** has been attributed[24] to 1,3-steric hindrance[85,91] by the axial, anomeric substituent in its thermodynamically more stable conformation.

When maltose was treated with sulfuryl chloride in pyridine–chloroform at −70°, and the mixture processed at −10 to 0°, it gave 6,6′-dichloro-6,6′-dideoxymaltosyl chloride penta(chlorosulfate) in 80% yield.[90] Dechlorosulfation of the pentachlorosulfate with sodium iodide in acetone afforded 6,6′-dichloro-6,6′-dideoxymaltose (69.4%). The structure assigned the 6,6′-dichloride was supported by its ^{13}C-n.m.r. spectrum (see Section XI,2) and by the fact that, on hydrolysis, it gave only 6-chloro-6-deoxy-D-glucose. Methyl β-maltoside has been converted[24] into methyl 3,6-dichloro-3,6-dideoxy-4-O-(4,6-dichloro-4,6-dideoxy-α-D-galactopyranosyl)-β-D-allopyranoside, isolated in 48% yield as its acetate (**36**), on reaction with sulfuryl chloride for 2 h at

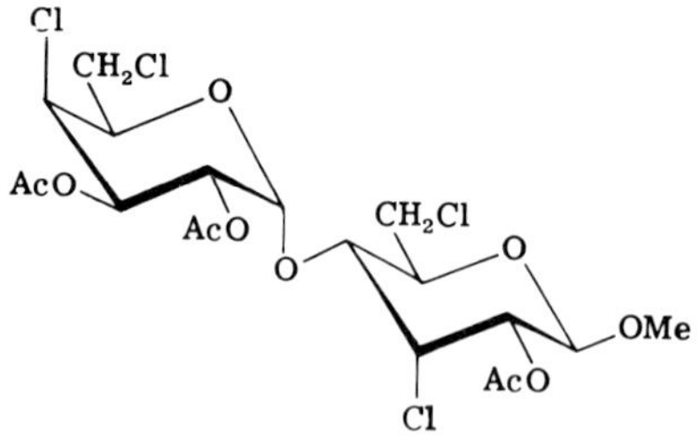

36

−70° and then for 24 h at room temperature. The substitution pattern in **36** has been rationalized in terms of stereoelectronic forces.[85,91] The conversion of the α-D-glucopyranosyl group into a 4,6-dichloro-4,6-dideoxy-D-galactopyranosyl group is in agreement with previous results

(91) A. G. Cottrell, E. Buncel, and J. K. N. Jones, *Can. J. Chem.*, 44 (1966) 1483–1491.

obtained with methyl α-D-glucopyranoside.[92] The resistance of the equatorial group on C-3' in **36** to displacement is attributed to the presence of a vicinal, axial chloro substituent on C-4, and to the β-*trans*-diaxial effect of the C-1' substituent.[85,91] In the β-D-glucopyranose residue, the C-1 substituent has the equatorial orientation, and, in addition, reaction at C-4 is prevented by the presence of the D-glucosyl group; therefore, the equatorial 3-chlorosulfate group can undergo displacement by chloride to give a 3,6-dichloro-D-allopyranose residue. The lack of reactivity at C-2 and C-2' is primarily due to the high energy, in the S$_N$2 reaction, of the transition state resulting from an unfavorable alignment of dipoles.[85]

Reaction of various, isolated hydroxyl groups of maltose[66] and methyl β-maltoside[66,69] has been examined. Treatment of 1,2,6,2',3',4',6'-hepta-*O*-acetyl-β-maltose with sulfuryl chloride afforded crystalline 1,2,6-tri-*O*-acetyl-3-chloro-3-deoxy-4-*O*-(2,3,4,6-tetra-*O*-acetyl-α-D-glucopyranosyl)-β-D-allopyranose in 84% yield.[66] Similarly, when methyl 2,6-di-*O*-benzoyl-4-*O*-(2,3-di-*O*-benzoyl-4,6-*O*-benzylidene-α-D-glucopyranosyl)-β-D-glucopyranoside was treated with the reagent, it gave the expected 3-chloro-3-deoxy-D-*allo* product in high yield.[69]

3. Methanesulfonyl Chloride—*N,N*-Dimethylformamide Reaction

Selective chlorination of carbohydrates has been achieved with the methanesulfonyl chloride–*N,N*-dimethylformamide reagent,[1,74,93–97] and an attempt to rationalize the reaction has been made.[93] Chlorination by this reagent proceeds by initial attack of the alcohol on the iminium salt [Me$_2$N$^+$=CHOMs] Cl$^-$, to afford the formiminium salt [Me$_2$N$^+$=CHOR] Cl$^-$, which then either undergoes hydrolysis to give a formic ester, or nucleophilic attack by chloride anion to give the chlorodeoxy compound (R–Cl). It was further suggested that the rate-limiting step in the chlorination is the formation of the iminium salt, and that the rate of the subsequent, nucleophilic-displacement reaction does not affect the overall rate of reaction. This hypothesis was supported by the fact that the addition of lithium chloride to the

---

(92) H. J. Jennings and J. K. N. Jones, *Can. J. Chem.*, 43 (1965) 2372–2386.
(93) M. E. Evans, L. Long, Jr., and F. W. Parrish, *J. Org. Chem.*, 33 (1968) 1074–1076.
(94) R. G. Edwards, L. Hough, A. C. Richardson, and E. Tarelli, *Tetrahedron Lett.*, (1973) 2369–2370.
(95) R. G. Edwards, L. Hough, A. C. Richardson, and E. Tarelli, *Carbohydr. Res.*, 35 (1974) 111–129.
(96) R. Khan, M. R. Jenner, and K. S. Mufti, *Carbohydr. Res.*, 39 (1975) 253–262.
(97) L. Hough and A. C. Richardson, *Pure Appl. Chem.*, 49 (1977) 1069–1084.

reaction mixture failed to accelerate the formation of chlorinated products.

Reaction of methyl β-maltoside with 30 equivalents of methanesulfonyl chloride in N,N-dimethylformamide for 8 days at 65° gave a mixture of methyl 3,6-dichloro-4-O-(6-chloro-6-deoxy-α-D-glucopyranosyl)-3,6-dideoxy-β-D-allopyranoside (37), isolated in 46% yield as its tetraacetate (38), and methyl 3,6-dichloro-3,6-dideoxy-4-O-(4,6-di-

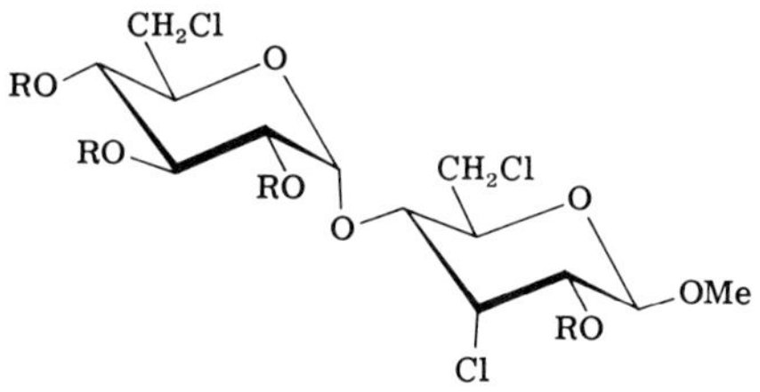

37 R = H
38 R = Ac

chloro-4,6-dideoxy-α-D-galactopyranosyl)-β-D-allopyranoside (8%, as its triacetate 36); the yield of 36 was improved by conducting the reaction at 100°. Under milder reaction-conditions, methyl 6,6'-dichloro-6,6'-dideoxy-β-maltoside (39) was obtained as the major product. The

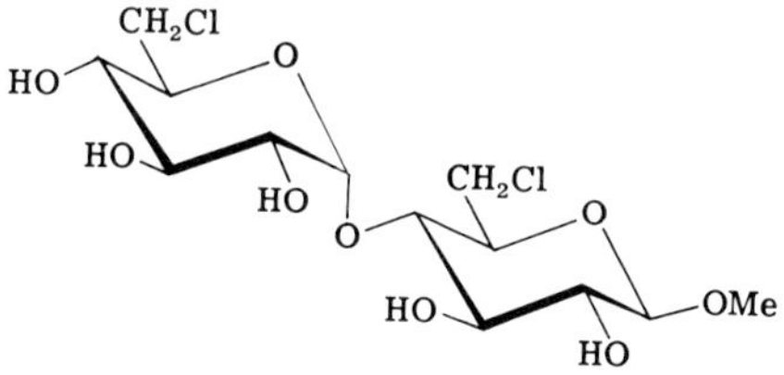

39

reaction mixture also yielded a substantial proportion of the 3,6,6'-trichloride 37. Conditions were not found that would favor the sole formation of the 6,6'-dichloride 39; in all experiments conducted, the 3,6,6'-trichloride 37 appeared before all of the starting material had been consumed. The slow appearance of the 3,6,4',6'-tetrachloride derivative revealed that the reactivity of HO-4' is considerably less than that of HO-3; this is somewhat surprising, as selective acylation studies (see Section V,1) indicated that C-3 is the most sterically hindered position in maltose, and it seems unexpected that an S_N2 transition-state should form more readily at this position than at C-4'. Similarly, the nucleophilic displacements occur more readily at O-6 than at O-6' (see Section VI, 1), although the latter is more readily esterified[50] and etherified.[41] These observations suggest that steric factors, which are important in acylation reactions, may not be significant in displacement reactions, which are more sensitive to polar effects.

VII. ANHYDRO DERIVATIVES

1,6-Anhydro-4-O-α-D-glucopyranosyl-β-D-glucopyranoside[32,67,98] (**41**) has proved to be a very useful starting-material for the production of such synthetic intermediates as 1,2,3,2',3',4',6'-hepta-O-acetyl-β-maltose,[32] 1,6-anhydro-4',6'-O-benzylidene-β-maltose,[82] and 2,3,2',3',4'-penta-O-acetyl-1,6-anhydro-β-maltose.[46]

On treatment with aqueous potassium hydroxide for 40 h on a steam bath,[32] phenyl β-maltose heptaacetate (**40**) was converted into the 1,6-anhydride **41**, isolated in 72% yield as its acetate (**42**). Compound **42**

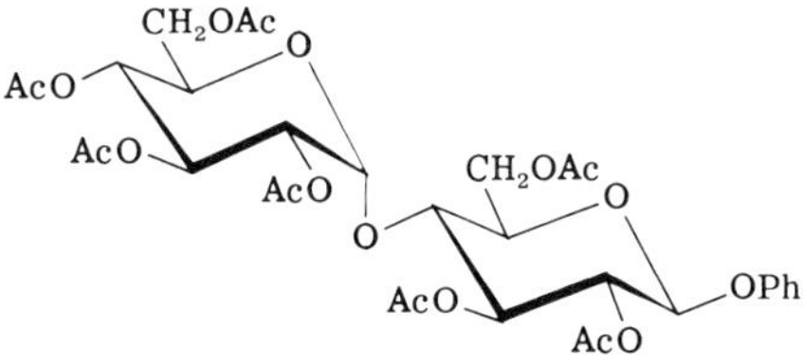

40

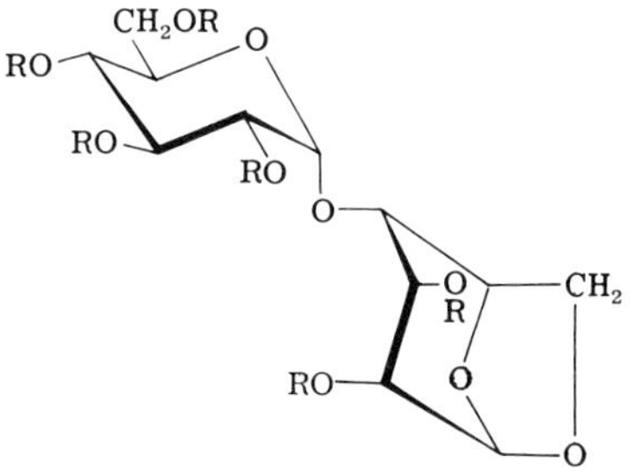

41 R = H
42 R = Ac

has also been prepared from the readily crystallizable p-chlorophenyl β-maltoside heptaacetate[99] by alkaline degradation followed by acetylation.[67] The 1,6-anhydro ring can be cleaved by acid hydrolysis, or by acetolysis, without serious disruption of the inter-residue glycosidic linkage. Generally, the 1,6-anhydro ring in maltose has been cleaved by treatment with titanium(IV) chloride or titanium(IV) bromide, to give the corresponding glycosyl halide having the 6-hydroxyl group unsubstituted. The glycosyl halide (for example, 2,3,2',3',4',6'-hexa-O-acetyl-α-maltosyl halide) can then be transformed into 1,2,3,2',3',4',6'-hepta-O-acetyl-β-maltose by treatment with mercury(II) acetate in acetic acid,[32] or into methyl 2,3,2',3',4',6'-hexa-O-acetyl-β-maltoside by reaction with mercury(II) acetate in methanol[45] (see Section III). The 220-MHz, ^{1}H-n.m.r. spectrum of **42** revealed that the

(98) P. Karrer and L. Kamienski, *Helv. Chim. Acta*, 15 (1932) 739–745.
(99) I. C. M. Dea, *Carbohydr. Res.*, 12 (1970) 297–299.

1,6-anhydride adopts the $^1C_4$ conformation[45] despite the axial glycosyl-oxy group on C-4.

Synthesis of methyl 2,3-anhydro-6-deoxy-4-O-(2,3,4-tri-O-acetyl-6-deoxy-α-D-glucopyranosyl)-β-D-allopyranoside has been achieved by the following sequence of reactions.[84] Selective acetylation of methyl 6,6'-di-O-p-tolylsulfonyl-β-maltoside with acetyl chloride in pyridine–toluene gave methyl 2,2',3',4'-tetra-O-acetyl-6,6'-di-O-p-tolylsulfonyl-β-maltoside in 85% yield. Conventional methanesulfonylation of the latter compound, followed by displacement with iodide anion, and reduction gave methyl 2-O-acetyl-6-deoxy-3-O-(methylsulfonyl)-4-O-(2,3,4-tri-O-acetyl-6-deoxy-α-D-glucopyranosyl)-β-D-glucopyranoside. On treatment with M sodium methoxide in methanol, this compound gave the 2,3-anhydride in 80% yield.

Attempts[66] to prepare the 2,3-anhydride from methyl 3-O-(methylsulfonyl)-β-maltoside, using M sodium methoxide, resulted in the corresponding 3,6-anhydride **43**. To circumvent the formation of **43**, the reaction was performed with methyl 3-O-(methylsulfonyl)-6,6'-di-O-trityl-β-maltoside. The desired product, methyl 2,3-anhydro-4-O-(2,3,4-tri-O-acetyl-6-O-trityl-α-D-glucopyranosyl)-6-O-trityl-β-D-allopyranoside (**44**), was obtained in 72% yield. Attempted detritylation of **44** by brief treatment with hydrogen bromide in glacial acetic acid, followed by conventional acetylation, gave methyl 2,6,2',3',4',6'-hexa-O-acetyl-

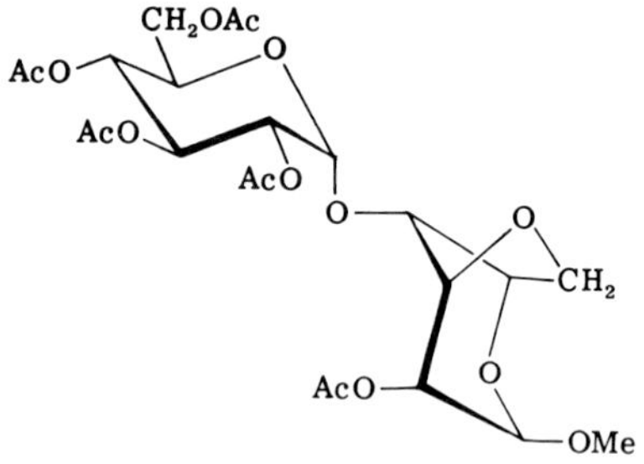

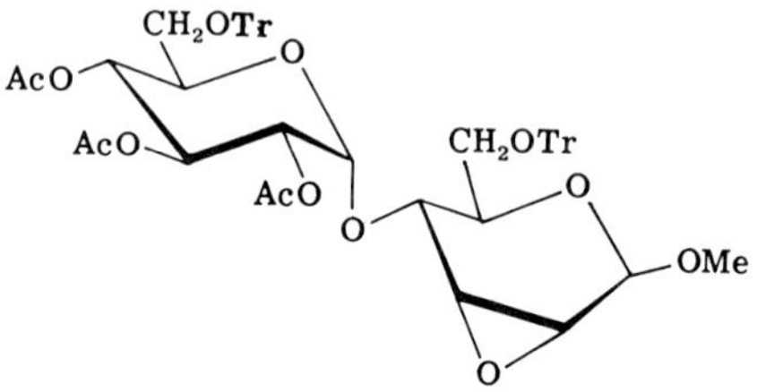

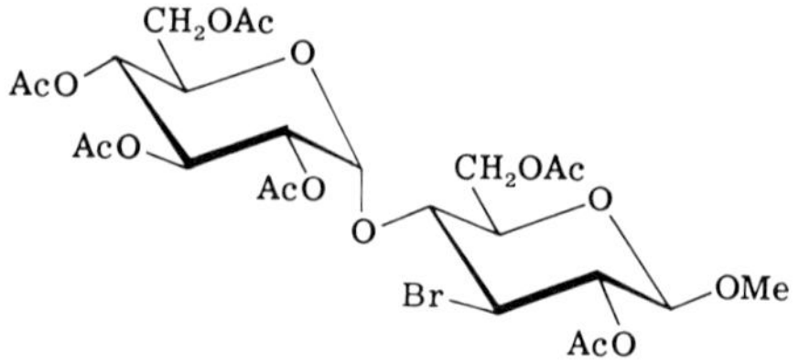

3-bromo-3-deoxy-β-maltoside (**45**) in 85% yield. Compound **45** must have arisen from diequatorial ring-opening, by attack at C-3 in **44** by the bromide anion. Had the reaction had SN2 character, the epoxide would have had to react by way of the $^5H_0$ conformation **46**, so that the 2- and 3-substituents would be held antiperiplanar in the $^1C_4$ conformation **47** first formed. However, the approach of the bromide anion to

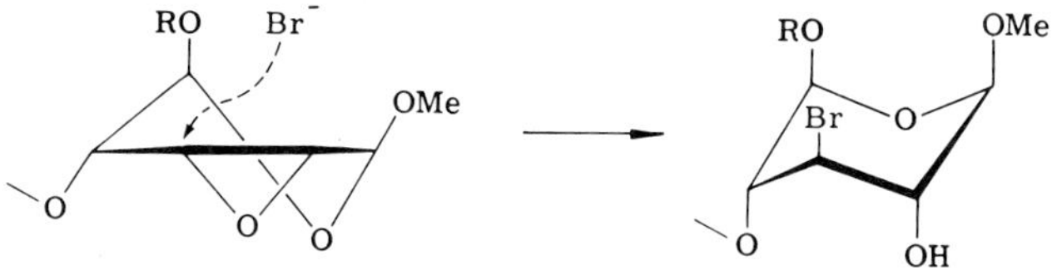

46   **47**

the rear of C-3 would be severely hindered by the axial group on C-5 (β-*trans*-axial effect[85]), and so this mechanism would be highly unlikely. Based on these considerations, it has been suggested that the ring-opening reaction proceeds by an SN1 mechanism which would be initiated by protonation of the epoxide oxygen atom, followed by heterolysis of the C-3–O bond to give the C-3 carbonium ion, the underside probably being protected by the newly formed hydroxyl group (for example, an ion pair).[66]

Synthesis of methyl 3,6:3',6'-dianhydro-β-maltoside has been achieved by alkaline alcoholysis of the 6,6'-di-O-(methylsulfonyl) or 6,6'-di-O-p-tolylsulfonyl derivative of methyl β-maltoside.[20,23] Similar treatment of methyl 2,3,2',3',4'-penta-O-acetyl-6-deoxy-6'-O-p-tolylsulfonyl-β-maltoside with ethanolic sodium hydroxide for 3.5 h at 60° afforded methyl 3',6'-anhydro-6-deoxy-β-maltoside in 90% yield.[80]

VIII. CYCLIC ACETAL DERIVATIVES

The cyclic acetals of sugars constitute an important class of compounds, and their value as synthetic intermediates is well recognized.[100] The reaction of methyl β-maltoside with benzaldehyde in the presence of zinc chloride has been reported to give the crystalline methyl 4',6'-O-benzylidene-β-maltoside.[52] Similar benzylidenation of 1,6-anhydro-β-maltose affords the corresponding 4',6'-O-benzylidene derivative in 63% yield.[82]

The reaction of sugars with a combination of 2,2-dimethoxypro-

(100) A. N. de Belder, *Adv. Carbohydr. Chem.*, 20 (1965) 219–302; *Adv. Carbohydr. Chem. Biochem.*, 34 (1977) 179–241.

pane-N,N-dimethylformamide-p-toluenesulfonic acid is known to give strained, and otherwise inaccessible, acetals.[1,101–104] However, when this combination of reagents was used for the acetonation of such (1→4)-linked, reducing disaccharides as lactose, it gave a complex mixture that was difficult to characterize.[105,106] The acetonation of lactose, maltose, and cellobiose was achieved by using a slight modification of the reagent, from which the N,N-dimethylformamide (or other, conventional, aprotic, polar solvent) was omitted.[105,106] Treatment of maltose with 2,2-dimethoxypropane in the presence of p-toluenesulfonic acid gave, after chromatography on silica gel, 2,3:5,6-di-O-isopropylidene-4-O-(4,6-O-isopropylidene-α-D-glucopyranosyl)-*aldehydo*-D-glucose dimethyl acetal (**48**) and 4-O-α-D-glucopyranosyl-2,3:5,6-di-O-isopropylidene-*aldehydo*-D-glucose dimethyl acetal (**49**) in yields of 22 and 18%, respectively.

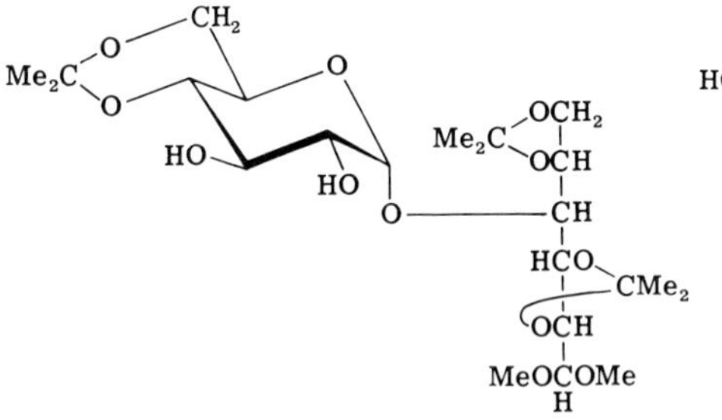

48 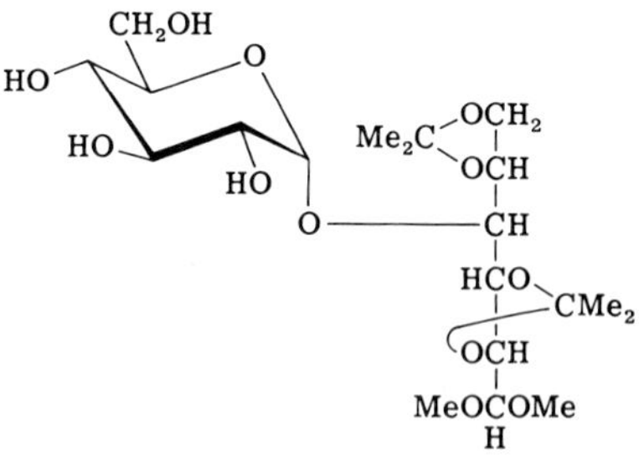 **49**

The kinetic-acetonation procedure previously reported,[107] employing alkyl isopropenyl ethers in N,N-dimethylformamide, and applied to the common pentoses and hexoses, has been extended to various hexuloses, oligosaccharides, and other sugar systems.[108] Maltose gave the 4′,6′-monoacetal, isolated as the crystalline hexaacetate, on treatment with isopropenyl methyl ether in N,N-dimethylformamide in the presence of p-toluenesulfonic acid.

(101) M. E. Evans and F. W. Parrish, *Tetrahedron Lett.*, (1966) 3805–3807.
(102) M. E. Evans, F. W. Parrish, and L. Long, Jr., *Carbohydr. Res.*, 3 (1967) 453–462.
(103) A. Hasegawa and H. G. Fletcher, Jr., *Carbohydr. Res.*, 29 (1973) 209–222.
(104) A. Hasegawa and M. Nakajima, *Carbohydr. Res.*, 29 (1973) 239–245.
(105) L. Hough, A. C. Richardson, and L. A. W. Thelwall, Br. Pat. Applic. (1978); *Carbohydr. Res.*, 75 (1979) c11–c12.
(106) L. Hough, A. C. Richardson, and G. Patel, unpublished results.
(107) J. Gelas and D. Horton, *Carbohydr. Res.*, 67 (1978) 371–387.
(108) D. Horton and J. Gelas, *Int. Symp. Carbohydr. Chem.*, 9th, London (1978); E. Fanton, J. Gelas, and D. Horton, *J. Chem. Soc. Chem. Commun.*, (1980) 21–22; J. Celas, *Adv. Carbohydr. Chem. Biochem.*, 39 (1981) 71–156.

IX. AMINODEOXY DERIVATIVES

Amino sugars are components of antibiotic substances[109] and bacterial polysaccharides,[110] and are therefore of interest. The nucleoside antibiotics amicetin, bamicetin, and plicacetin contain, as the sugar residue, a monoaminopentadeoxy disaccharide that is closely related to maltose. In view of the reported antibiotic and antitumor properties of these pyrimidine nucleosides,[111,112] the synthesis of aminodeoxy derivatives of maltose would be of interest.

Several aminodeoxy derivatives of maltose having an aminodeoxy function at C-1, C-2, C-6', C-3 and 6', or C-6 and 6' have been prepared. The synthesis of 2,3,6-tri-O-acetyl-4-O-(2,3,4,6-tetra-O-acetyl-α-D-glucopyranosyl)-β-D-glucopyranosylamine has been achieved.[113] Octa-O-acetylmaltose was converted into the corresponding glycosyl bromide by treatment with hydrogen bromide in glacial acetic acid; the bromide reacted with sodium azide in N,N-dimethylformamide to give 2,3,6-tri-O-acetyl-4-O-(2,3,4,6-tetra-O-acetyl-α-D-glucopyranosyl)-β-D-glucopyranosyl azide[114] in 60% yield, and catalytic hydrogenation of the azide gave the crude glycosylamine. This was sufficiently pure for use in a coupling reaction with 1-benzyl N-(benzyloxycarbonyl)-L-aspartate, but recrystallization or any further purification only increased decomposition of the product.[113] Formation of 1,1-bis(benzamido)-1-deoxy-4-O-α-D-glucopyranosyl-D-glucitol and N-benzoylmaltosylamine during the de-esterification of octa-O-benzoyl-β-maltose with methanolic ammonia has been mentioned in Section V,1. Based on results with benzoylated hexoses,[115,116] it has been suggested that the benzoyl group on O-3 in octa-O-benzoyl-β-maltose participates strongly in the formation of products having a nitrogen atom on C-1. The compound having a free 3-hydroxyl group, liberated by the migration, may be an intermediate step in the intramolecular migration of a second benzoyl group, giving rise to the bis(benzamido) derivative. The importance of the 3-O-benzoyl group was confirmed by the fact that, in the ammonolysis of 1,2,6,2',3',4',6'-hepta-O-benzoyl-

(109) J. D. Dutcher, *Adv. Carbohydr. Chem.*, 18 (1963) 259–308.
(110) N. Sharon, in E. A. Balazs and R. W. Jeanloz (Eds.), *The Amino Sugars*, Vol. IIA, Academic Press, New York, 1965, pp. 1–45.
(111) R. J. Suhadolnik, *Nucleoside Antibiotics*, Wiley–Interscience, New York, 1970, pp. 203–217.
(112) A. Block and C. Coutsogeorgopoulos, *Biochemistry*, 5 (1966) 33–45.
(113) D. Dunstan and L. Hough, *Carbohydr. Res.*, 23 (1972) 17–21.
(114) A. Bertho, *Justus Liebigs Ann. Chem.*, 562 (1949) 229–239.
(115) E. G. Gros, M. A. Ondetti, J. O. Sproviero, V. Deulofeu, and J. O. Deferrari, *J. Org. Chem.*, 27 (1962) 924–929.
(116) E. G. Gros and V. Deulofeu, *J. Org. Chem.*, 29 (1964) 3647–3654.

β-maltose,[68] compounds nitrogenated at C-1 could not be detected. An attempt to rationalize the formation of the bis(benzamido) derivative during ammonolysis of octa-O-acetyl-β-maltose has been made.[71] The benzoyl group on O-1 must be eliminated first, giving a free hydroxyl group at C-1. After liberation of the 1-hydroxyl group, the crowding of the benzoyl groups could favor the opening of the pyranose ring of the reducing moiety, and its 1-carbonyl group would then be rapidly attacked by ammonia, allowing facile migration of the other benzoyl groups.

After hydrolysis of carboxyl-reduced and partially desulfated heparin with hydrochloric acid, Wolfrom and coworkers[117] isolated a disaccharide that was identified[118] as 2-amino-2-deoxy-4-O-α-D-glucopyranosyl-D-glucopyranose ("maltosamine") hydrochloride, which gave a crystalline N-acetyl derivative (50). This structure was subsequently

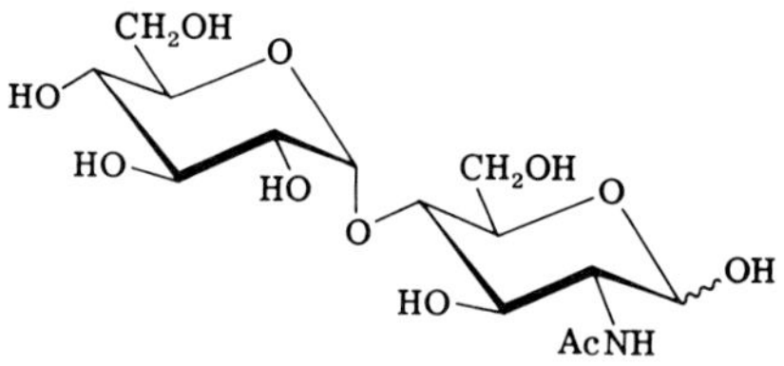

50

supported[119] by a synthesis, albeit in very low yield. However, the $[\alpha]_D$ value (+110°) reported[120] by Selinger and Schramm for a syrupy mixture containing 50 plus inorganic salts, synthesized by enzymic condensation of β-D-glucopyranosyl phosphate with 2-acetamido-2-deoxy-D-glucose, is very different from that (+39°) reported[118] for "maltosamine." In view of this discrepancy, compound 50 has been synthesized by an alternative route.[121] Treatment of benzyl 2-acetamido-3,6-di-O-benzyl-2-deoxy-α-D-glucopyranoside in anhydrous nitromethane with an excess of 2,3,4,6-tetra-O-benzoyl-1-O-(N-methylacetimidoyl)-β-D-glucopyranose in the presence of p-toluenesulfonic acid for 7 days at room temperature gave, after chromatography, 36.5%

(117) M. L. Wolfrom, J. R. Vercellotti, and G. H. S. Thomas, *J. Org. Chem.*, 26 (1961) 2160.
(118) M. L. Wolfrom, J. R. Vercellotti, and D. Horton, *J. Org. Chem.*, 27 (1962) 705–706.
(119) M. L. Wolfrom, H. S. El Khadem, and J. R. Vercellotti, *J. Org. Chem.*, 29 (1964) 3284–3286.
(120) Z. Selinger and M. Schramm, *J. Biol. Chem.*, 236 (1961) 2183–2185.
(121) M. A. M. Nassr, J.-C. Jacquinet, and P. Sinaÿ, *Carbohydr. Res.*, 77 (1979) 99–105.

of benzyl 2-acetamido-3,6-di-O-benzyl-2-deoxy-4-O-(2,3,4,6-tetra-O-benzoyl-α-D-glucopyranosyl)-α-D-glucopyranoside. Catalytic hydrogenolysis then afforded crystalline 50.

Compound 50 has been prepared in 85% yield by an alternative route. The aforementioned imidate was condensed with 2-acetamido-3-O-acetyl-1,6-anhydro-2-deoxy-β-D-glucopyranose, and the resulting disaccharide was catalytically hydrogenolyzed, the product acetylated, and the peracetate acetolyzed, to afford 2-acetamido-1,3,6-tri-O-acetyl-2-deoxy-4-O-(2,3,4,6-tetra-O-acetyl-α-D-glucopyranosyl)-α-D-glucopyranose. O-Deaectylation then gave N-acetylmaltosamine (50). The specific rotation of 50 was similar to that reported[120] by Selinger and Schramm. Thus, it appears that the disaccharide described by Wolfrom and coworkers[117] may not have been maltosamine, and that the structure of both the natural and the synthetic disaccharide require reinvestigation.

Synthesis of 6'-amino-6'-deoxymaltose has been achieved from 2,3,2',3',4'-penta-O-acetyl-1,6-anhydro-6'-O-p-tolylsulfonyl-β-maltose.[46] Treatment of the 6'-sulfonate with sodium azide in N,N-dimethylformamide gave the corresponding 6'-azidodeoxy derivative. Conventional de-esterification of the azide, followed by catalytic hydrogenation, led to the desired 6'-amino-1,6-anhydro-6'-deoxy-β-maltose. Attempts to hydrolyze the 1,6-anhydro ring selectively, under mildly acidic conditions, were not very successful, and 6'-amino-6'-deoxymaltose was obtained only as a minor product. A relatively simple procedure for the preparation of 6'-azido-6'-deoxymaltose, a precursor of the 6'-aminodeoxy derivative, has been developed through the action of *Aspergillus oryzae* amylase on 6'-azido-6'-deoxycyclohexaamylose.[122]

Nucleophilic displacement of the sulfonate groups in methyl 2,6,2',3',4'-penta-O-acetyl-3,6'-di-O-(methylsulfonyl)-β-maltoside was readily achieved with sodium azide in hexamethylphosphoric triamide, to give methyl 3-azido-4-O-(6-azido-6-deoxy-α-D-glucopyranosyl)-3-deoxy-β-D-allopyranoside pentaacetate.[45] Catalytic hydrogenation of the 3,6'-diazide, followed by acetylation, gave the 3,6'-diacetamido derivative in high, overall yield. In a similar sequence of reactions, methyl 2,3,2',3',4'-penta-O-acetyl-6,6'-di-O-p-tolylsulfonyl-β-maltoside has been transformed into the corresponding 6,6'-diacetamido derivative.[45]

(122) L. D. Melton and K. N. Slessor, Can. J. Chem., 51 (1973) 327–332.

X. Miscellaneous Compounds

1. Unsaturated Derivatives

The potential importance of unsaturated sugars as synthetic and biological intermediates is widely recognized. It is therefore surprising that only two unsaturated maltose derivatives, hexa-O-acetylmaltal[123,124] (**51**) and methyl 2,3,6-trideoxy-5-O-(2,3,4-tri-O-acetyl-

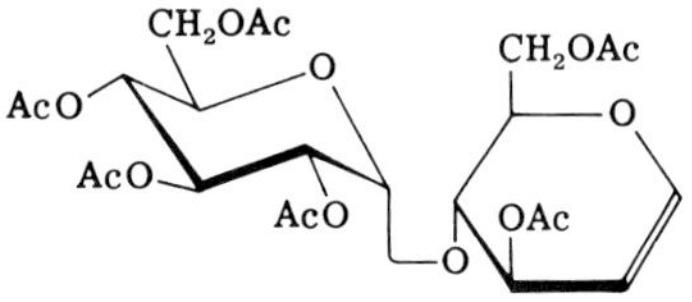

51

6-deoxy-α-D-glucopyranosyl)-β-D-*erythro*-hex-2-enopyranoside[84] (**52**),

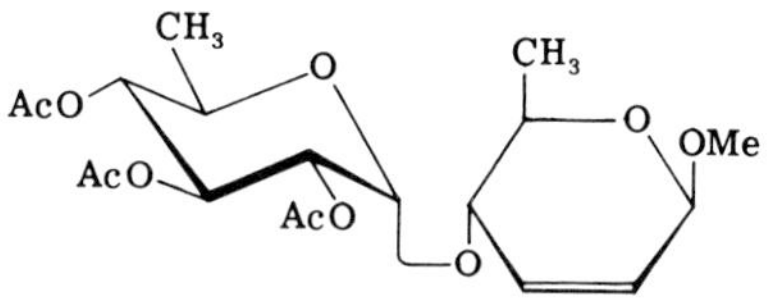

52

have so far been reported. Maltal hexaacetate (**51**) was synthesized by conventional reduction of a peracetylated maltosyl halide with zinc and acetic acid. Some of the reactions of **51** have been described.[125] Treatment of compound **51** with sodium azide in acetonitrile in the presence of boron trifluoride etherate gave, by way of allylic rearrangement, 6-O-acetyl-1,5-anhydro-3-azido-1,2,3-trideoxy-4-O-(2,3,4,6-tetra-O-acetyl-α-D-glucopyranosyl)-D-*ribo*-hex-1-enitol (**53**). The reaction of **51** with 1,3,4,6-tetra-O-acetyl-α-D-glucopyranose

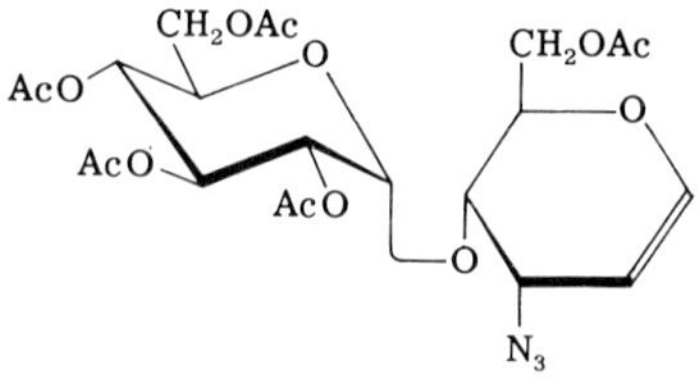

53

(123) W. N. Haworth, E. L. Hirst and R. J. W. Reynolds, *J. Chem. Soc.*, (1934) 302–303.
(124) A. M. Gakhokidze, *Zh. Obschch. Khim.*, 8 (1948) 60–67.
(125) K. Heyns and M. T. Lim, *Tetrahedron Lett.*, (1978) 891–894.

under iodoalkylation conditions gave 1,3,4,6-tetra-O-acetyl-2-O-[3,6-di-O-acetyl-2-deoxy-2-iodo-4-O-(2,3,4,6-tetra-O-acetyl-α-glucopyranosyl)]-α-D-glucopyranose. Synthesis of **52** has been achieved[84] from methyl 2,3-anhydro-6-deoxy-4-O-(2,3,4-tri-O-acetyl-6-deoxy-α-D-glucopyranosyl)-β-D-*ribo*-hexopyranoside by treatment with sodium iodide in acetone and acetic acid, followed by dehydrohalogenation with phosphoryl chloride.

2. 1-(2-Chloroethyl)-1-nitrosourea Derivatives

Synthesis of N-(2-chloroethyl)-N-nitrosocarbamoyl derivatives of six glycosylamines, including that from maltose, has been described.[126] Their antitumor activities against leukemia L1210 in mice have been determined. The reaction of hepta-O-acetyl-β-maltosylamine[114] (**54**) with 2-chloroethyl isocyanate in 1,4-dioxane gave 1-(2-chloroethyl)-3-(hepta-O-acetyl-β-maltosyl)urea (**55**) in 56% yield. De-esterification of **55** with ammonia in methanol yielded crystalline 1-(2-chloroethyl)-3-β-maltosylurea (**56**), which, on treatment with nitrogen trioxide in acetone, afforded 1-(2-chloroethyl)-3-β-maltosyl-1-nitrosourea (**57**) in 75% yield. Acetylation of **57**, or nitrosation of **55**, gave 1-(2-chloroethyl)-3-(hepta-O-acetyl-β-maltosyl)-1-nitrosourea (**58**). All of the 1-(2-chloroethyl)-1-nitrosoureas tested were active against lymphoid leukemia L1210 transported in mice.

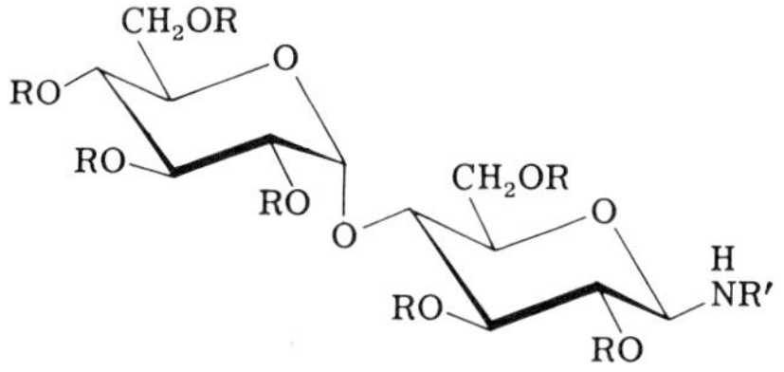

54 R = Ac, R' = H
55 R = Ac, R' = CONH(CH$_2$)$_2$Cl
56 R = H, R' = CONH(CH$_2$)$_2$Cl
57 R = H, R' = CON(NO)CH$_2$CH$_2$Cl
58 R = Ac, R' = CON(NO)CH$_2$CH$_2$Cl

3. Sulfur-containing Compounds

Ethyl 1-thio-β-maltoside heptaacetate (**59**) has proved to be a useful intermediate in the synthesis of the α-glycoside of maltose (see Section III).[30] Treatment of β-maltose octaacetate with ethanethiol and fused zinc chloride gave compound **59** in 83% yield. Synthesis of mal-

(126) T. Suami, T. Machinami, and T. Hasamatsu, *J. Med. Chem.*, 22 (1979) 247–250.

tose dialkyl dithioacetal octaacetates, has been described by Wolfrom and his coworkers.[127] 1,6-Anhydro-2,3,2',3',4',6'-hexa-O-acetyl-6-thio-β-maltose (**60**) was obtained from 2,3,2',3',4',6'-hexa-O-acetyl-6-O-p-

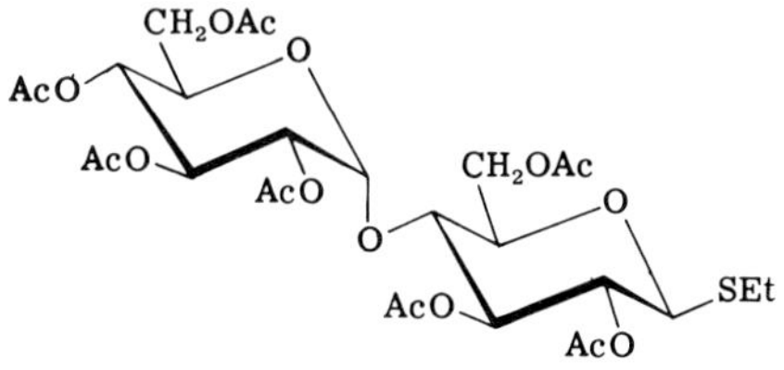

tolylsulfonyl-β-maltosyl ethylxanthate by treatment with sodium methoxide in methanol.[67] The application of bimolecular, nucleophilic substitution (S$_N$2) reactions to maltose sulfonates has been found of value in synthesizing thiomaltoses.[46,67] The reaction of 1,2,3,2',3',4',6'-hepta-O-acetyl-6-O-p-tolylsulfonyl-β-maltose with potassium thiolacetate in N,N-dimethylformamide for 1 h on a steam bath gave the corresponding 6-thioacetate derivative in 74% yield.[67] Similarly, 6'-thio derivatives of maltose and 1,6-anhydro-β-maltose have been prepared from their corresponding sulfonates.[46]

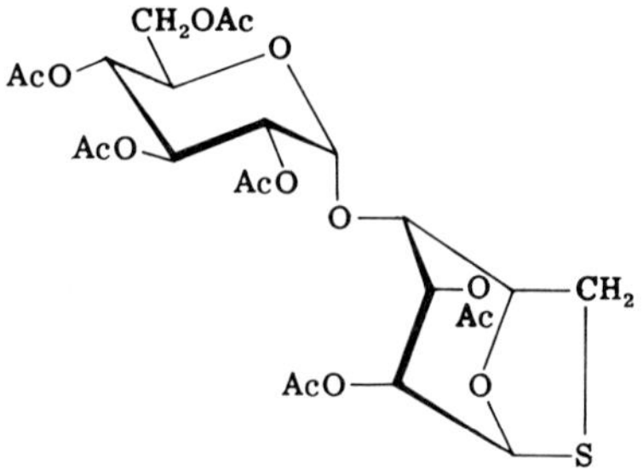

4. Polymer-forming Reactions

The condensation of benzyl 2,3,2',3'-tetra-O-acetyl-4',6'-O-benzylidene-β-maltoside with tetra-O-acetyl-α-D-glucosyl bromide in chloroform in the presence of silver oxide and anhydrous calcium sulfate gave the corresponding 6-O-β-D-glucosylmaltose.[128] Synthesis of 6-O-β-D-galactosylmaltose has been achieved by the condensation of tetra-O-acetyl-α-D-galactosyl bromide with 2,3,2',3',4'-penta-O-acetyl-1,6-anhydro-β-maltose or 2,3,2',3',4'-penta-O-acetyl-1,6-anhydro-6'-O-trityl-β-maltose.[129] The Lewis acid-catalyzed polymerization of per-O-benzylated 1,6-anhydro-β-maltose led to products of high molecular weight.[53] Based on molecular weight and the values of optical rota-

---

(127) M. L. Wolfrom, M. R. Newlin, and E. E. Stahly, *J. Am. Chem. Soc.*, 53 (1931) 4379–4383.
(128) A. Klemer, *Chem. Ber.*, 92 (1959) 218–226.
(129) C. T. Gi and S. Tejima, *Chem. Pharm. Bull.*, 25 (1977) 464–470.

tion, it was suggested that the polysaccharides have a high degree of stereoregularity in the main chain, with mainly α-D-linkages in both the main chain and branches. Such comb-shaped molecules have also been obtained by the condensation of peracetylated 1,2-O-(1-ethoxyethylidene) acetals of maltose, maltotetraose, and maltohexaose with the 2,3-di-O-(phenylcarbamoyl) derivatives of amylose and cellulose.[130].

5. Complex-forming Reactions

Complexes of maltose with urea, thiourea, imidazole, methanol, 2-oxazolidinone, N,N-dimethylformamide, and hexamethylphosphoric triamide have been described.[131] These complexes were noncrystalline and hygroscopic. Such complex-forming reactions could be valuable in the preservation of food flavors during the dehydration process. Sugars have been shown to complex with ethylenediamine.[132,133] The nature of the complex has been suggested to be that of a proton-transfer type, in which the carbohydrate moiety is the proton donor and the amine is the proton acceptor.

Complexes of alkali metals and alkaline-earth metals with carbohydrates have been reviewed in this Series,[134] and the interaction of alkaline-earth metals with maltose has been described.[135] Standard procedures for the preparation of adducts of D-glucose and maltose with the hydroxides of barium, calcium, and strontium have been established. The medium most suitable for the preparation of the adduct was found to be 80% methanol. It is of interest that the composition of the adducts, from D-glucose, maltose, sucrose, and α,α-trehalose was the same, namely, 1:1, in all cases. The value of such complex-forming reactions in the recovery of metals from industrial wastes has been recognized. Metal hydroxide–sugar complexes may also play an important biological role in the transport of metal hydroxides across cell membranes.

(130) B. Pfannemüller, G. C. Richter, and E. Husemann, *Carbohydr. Res.*, 56 (1977) 147–151.
(131) J. E. Hodge, J. A. Rendleman, Jr., and E. L. Nelson, *Cereal Sci. Today*, 17 (1972) 180–188.
(132) S. P. Moulik, A. K. Mitra, and K. K. Sen Gupta, *Carbohydr. Res.*, 19 (1971) 416–418.
(133) S. P. Moulik and A. K. Mitra, *Carbohydr. Res.*, 23 (1972) 65–74.
(134) J. A. Rendleman, Jr., *Adv. Carbohydr. Chem.*, 21 (1965) 209–271.
(135) N. Roy and A. K. Mitra, *Carbohydr. Res.*, 24 (1972) 175–179, 180–183.

6. Selective-oxidation Reactions

Oxidative, degradative reactions of maltose have been studied,[136-138] and reviewed in this Series,[139-142] and this subject will not, therefore, be discussed. The selective oxidation of primary[52,143-146] and secondary[66,147] hydroxyl groups in maltose has been investigated. The synthesis[143] of "maltouronic acid," 4-O-(α-D-glucopyranosyluronic

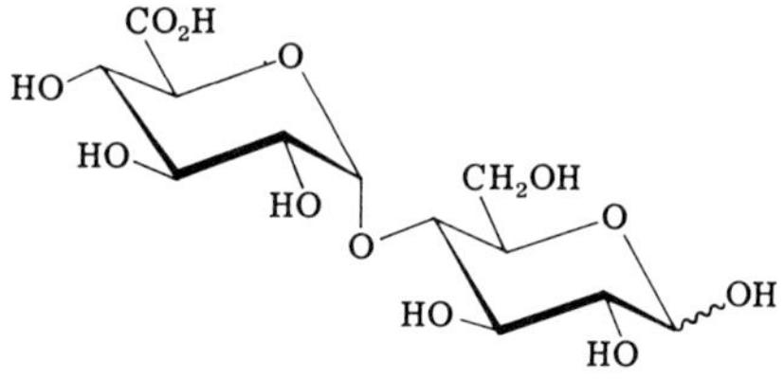

61

acid)-D-glucose (**61**), was first reported by Hirasaka,[144] who subjected benzyl β-maltoside to catalytic oxidation. The lack of reactivity at C-6 has been attributed[145] to steric hindrance caused by the benzyl group on O-1. Unambiguous synthesis of **61** was achieved by Roy and Glaudemans[145] by the following reaction-sequence: 1,6-anhydro-β-maltose hexaacetate → 1,6-anhydro-β-maltose → 1,6-anhydro-6'-O-trityl-β-maltose pentabenzoate → 1,6-anhydro-2,3,2',3',4'-penta-O-benzoyl-β-maltose pentabenzoate → 1,6-anhydro-2,3,2',3',4-penta-O-benzoyl-β-maltose → 1,6-anhydro-2,3,2',3',4'-penta-O-benzoyl-β-maturonic acid → methyl 1,6-anhydro-2,3,2',3',4'-penta-O-benzoylmalturonate → **61**. Aspinall and coworkers[52] described the synthesis of methyl [methyl 2,3-di-O-methyl-4-O-(2,3,4,6-tetra-O-methyl-α-D-glucopyranosyl)-β-D-glucopyranosid]uronate (**15**).

Microbial oxidation of maltose by *Agrobacterium tumefaciens* to give 4-O-(α-D-*ribo*-hexosyl-3-ulose)-D-glucose (**62**) has been reported.[147] Chemical oxidation of methyl, 2,6,2',3',4',6'-hexa-O-acetyl-

---

(136) R. L. Whistler and K. J. Yagi, *J. Org. Chem.*, 26 (1961) 1050–1052.
(137) H. S. Isbell and R. Schaffer, *J. Am. Chem. Soc.*, 78 (1956) 1887–1889.
(138) H. S. Isbell and R. G. Naves, *Carbohydr. Res.*, 36 (1974) c1–c4.
(139) J. W. Green, *Adv. Carbohydr. Chem.*, 3 (1948) 129–184.
(140) J. M. Bobbitt, *Adv. Carbohydr. Chem.*, 11 (1956) 1–41.
(141) A. S. Perlin, *Adv. Carbohydr. Chem.*, 14 (1959) 9–61.
(142) R. W. Bailey and J. B. Pridham, *Adv. Carbohydr. Chem.*, 17 (1962) 121–167.
(143) G. G. S. Dutton and K. N. Slessor, *Can. J. Chem.*, 42 (1964) 1110–1112.
(144) Y. Hirasaka, *Yakugaku Zasshi*, 83 (1963) 960–965; *Chem. Abstr.*, 60 (1964) 4232c.
(145) N. Roy and C. P. J. Glaudemans, *J. Org. Chem.*, 33 (1968) 1559–1562.
(146) D. Abbott and H. Weigel, *J. Chem. Soc.*, (1965) 5157–5162.
(147) S. Fukui and R. M. Hochester, *Can. J. Biochem. Physiol*, 41 (1963) 2363–2371.

β-maltoside with ruthenium tetraoxide gave the corresponding 3-keto derivative **63** in 84% yield.[66] It is of interest that **63** underwent racemization to give the C-2 epimer **64**.

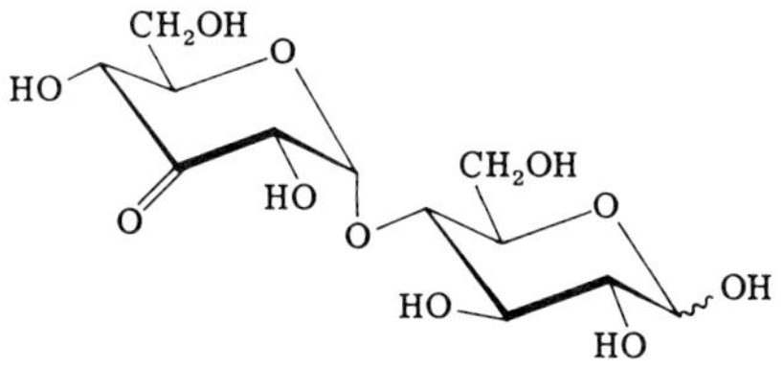

62

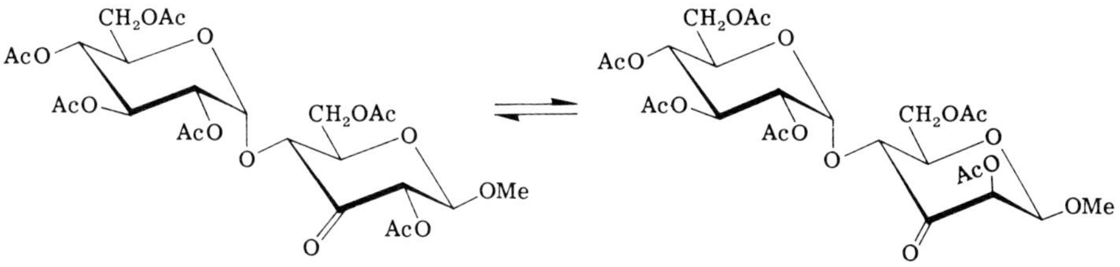

63   **64**

7. Alkaline Degradation

Alkaline degradation-reactions of sugar glycosides and polysaccharides have been reviewed in this Series.[148,149] The effect of alkali on maltose has been studied.[52,150–153] The anaerobic degradation of 4-O-substituted hexoses gives mainly 3-deoxy-2-C-(hydroxymethyl)-D-erythro-and -D-threo-pentonic acid. Methyl [methyl 2,3-di-O-methyl-4-O-(2,3,4,6-tetra-O-methyl-α-D-glucopyranosyl)-β-D-glucopyranosid]uronate (**15**) has been used as a model for the study of the base-catalyzed elimination of 4-O-substituted hexopyranosiduronates without degradation of the exposed, reducing sugars, and of the selective hydrolysis of hex-4-enopyranosiduronates.[52]

---

(148) C. E. Ballou, Adv. Carbohydr. Chem., 9 (1954) 59–95.
(149) R. L. Whistler and J. N. BeMiller, Adv. Carbohydr. Chem., 13 (1958) 289–329.
(150) G. Machell and G. N. Richards, J. Chem. Soc., (1960) 1924–1931, 1932–1938.
(151) L. Hough, J. K. N. Jones, and E. L. Richards, Chem. Ind. (London), (1954) 545–546.
(152) R. F. Burns and P. J. Somers, Cabohydr. Res., 31 (1973) 289–300, 301–309.
(153) I. Ziderman, J. Bel-Ayche, A. Basch, and M. Lewin, Carbohydr. Res., 43 (1975) 255–263.

Treatment of **15** with 1,5-diazabicyclo[5.4.0]undec-5-ene (DBU) in benzene resulted in β-elimination, but under these conditions, the liberated 2,3,4,6-tetra-O-methyl-D-glucose was further degraded, probably with formation of 3-deoxy-2,4,6-tri-O-methyl-D-*erythro*-hex-2-enopyranose. In order to prevent this degradation, the reaction of **15** with DBU was carried out in the presence of acetic anhydride. The reaction mixture gave, after chromatography on silica gel, methyl (methyl 4-deoxy-2,3-di-O-methyl-α-L-*threo*-hex-4-enopyranosid)uronate (**65**) and a mixture of the 1-acetates of 2,3,4,6-tetra-O-methyl-α- and -β-D-glucose (**66**).

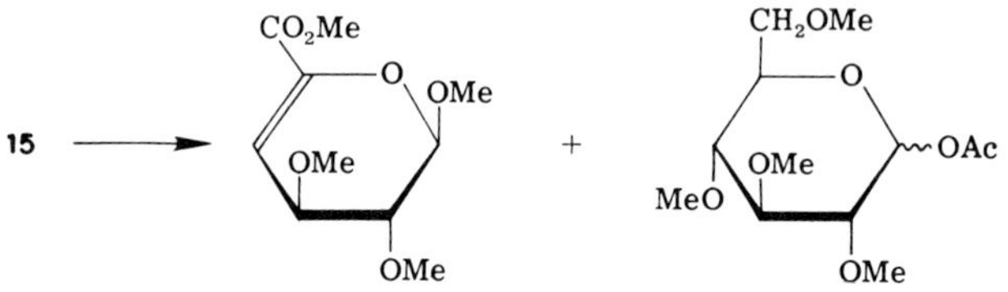

8. Hydrolysis

Maltose may be hydrolyzed by treatment with dilute, mineral acids. As with other sugars, a complex mixture of products may arise from dehydration reactions that yield such products as higher saccharides, other disaccharides, or furan derivatives.[154] The α-D-glucosidic linkage in maltose is hydrolyzed about nine times as fast as that of methyl α-D-glucopyranoside, and over twice as fast as the related, β-(1→4)-linked disaccharide.[155] Like acid hydrolysis, acetolysis can be used to fragment carbohydrate chains to afford oligosaccharides.[156] The initial acetolysis rates of several disaccharides have been compared.[157] For D-glucose disaccharides, β-linkages were cleaved faster than α-linkages, indicating anchimeric assistance from the *trans*-2-acetoxyl group. The rates of the acetolysis reaction for the various α-linked D-glucose disaccharides decreases in the order (1→6) >> (1→4) > (1→3) > (1→2). For the β-linked disaccharides, the order is (1→6) >> (1→3) > (1→2) > (1→4).

(154) M. S. Feather and J. F. Harris, *Adv. Carbohydr. Chem. Biochem.*, 28 (1973) 161–224.
(155) B. Capon, *Chem. Rev.*, 69 (1969) 407–498; J. N. BeMiller, *Adv. Carbohydr. Chem.*, 22 (1967) 25–108.
(156) R. D. Guthrie and J. F. McCarthy, *Adv. Carbohydr. Chem.*, 22 (1967) 11–23.
(157) L. Rosenfeld and C. E. Ballou, *Carbohydr. Res.*, 32 (1974) 287–298.

XI. Physical Methods

1. ¹H-Nuclear Magnetic Resonance Spectroscopy

The value of ¹H-n.m.r. spectroscopy in determining the configuration and conformation of carbohydrates is well recognized. The ¹H-n.m.r. spectra of maltose[158–160] and some of its derivatives[24,45,66,95,161–168] have been described. De Bruyn and co-workers[160] studied the ¹H-n.m.r. spectrum of maltose in deuterium oxide at 300 MHz, and the chemical shifts were compared with those of α- and β-D-glucopyranose. The resonances due to individual protons, and the position of the glycosidic linkage, were established by homo-INDOR experiments and by using increment rules. The resonances due to H-1,2,3,4,5,6a,6b in β-maltose were identified at δ 4.66, 3.28, 3.77, 3.65, 3.60, 3.92, and 3.77, respectively. The coupling constants ($J_{1,2}$ 7.8, $J_{2,3}$ 10.0, $J_{3,4}$ 8.8, $J_{4,5}$ 10.0, $J_{5,6a}$ 2.0, $J_{5,6b}$ 5.0, and $J_{6a,6b}$ − 12.0 Hz) were in agreement with the β-D-*gluco* configuration and the $^4C_1$(D) conformation for the reducing residue in β-maltose. The signals due to H-1′,2′,3′,4′,5′,6′a,6′b were observed at δ 5.41, 3.58, 3.69, 3.74, 3.86, and 3.77, respectively. The derived coupling-constants ($J_{1′,2′}$ 3.8, $J_{2′,3′}$ 10.0, $J_{3′,4′}$ 8.6, $J_{4′,5′}$ 10.0, $J_{5′,6′a}$ 2.0, $J_{5′,6′b}$ 5.0, and $J_{6′a,6′b}$ − 12 Hz) revealed the *e, a, a, a, a* arrangement of H-1′, H-2′, H-3′, H-4′, and H-5′, respectively, in agreement with the α-D-*gluco* configuration and the $^4C_1$(D) conformation for the nonreducing group. The concentration of α-maltose at equilibrium was too low to permit interpretation of the spectrum beyond H-1 (δ 5.23, $J_{1,2}$ 3.8 Hz),

---

(158) T. Usui, M. Yokoyama, N. Yamaoka, K. Matsuda, K. Tuzimura, H. Sugiyama, and S. Seto, *Carbohydr. Res.*, 33 (1974) 105–116.
(159) J. M. van der Veen, *J. Org. Chem.*, 28 (1963) 564–566.
(160) A. De Bruyn, M. Anteunis, and G. Verhegge, *Bull. Soc. Chim. Belg.*, 84 (1975) 721–734.
(161) C. G. Hellerqvist, O. Larm, and B. Lindberg, *Acta Chem. Scand.*, 25 (1971) 743–744.
(162) J. P. Kamerling, M. J. A. De Bie, and J. F. G. Vliegenthart, *Tetrahedron*, 28 (1972) 3037–3047.
(163) J. Haverkamp,. M. J. A. De Bie, and J. F. G. Vliegenthart, *Carbohydr. Res.*, 37 (1974) 111–125.
(164) D. G. Streefkerk and A. M. Stephen, *Carbohydr. Res.*, 57 (1977) 25–37.
(165) B. Casu, M. Reggiani, G. G. Gallo, and A. Vigevani, *Carbohydr. Res.*, 12 (1970) 157–170.
(166) K. Koizumi and T. Utamura, *Carbohydr. Res.*, 63 (1978) 283–287.
(167) J. O. Deferrari, I. M. E. Thiel, and R. A. Cadenas, *Carbohydr. Res.*, 26 (1973) 244–246.
(168) J. O. Deferrari, I. M. E. Thiel, and R. A., Cadenas, *An. Asoc. Quim. Argent.*, 61 (1973) 107–112.

H-1' (δ 5.41, $J_{1',2'}$ 10.0 Hz), H-2 (δ 3.57, $J_{2,3}$ 10.0 Hz), and H-2' (δ 3.58, $J_{2',3'}$ 10.0 Hz).

^{1}H-N.m.r. studies[161,162] of per-O-(trimethylsilyl) (Me$_3$Si) derivatives of oligosaccharides have proved valuable in determining the configuration of interglycosidic linkages. Per-O-(trimethylsilyl)ation is a simple procedure that, in the case of reducing sugars, results in a mixture of Me$_3$Si anomeric forms, having almost the same composition as in the starting medium, for example, water. ^{1}H-N.m.r. spectroscopy of such derivatives may, therefore, also be used for the determination of the anomeric equilibrium, provided that the quantity of each anomer present is >5%. The 100-MHz, ^{1}H-n.m.r. spectrum of octa-O-(trimethylsilyl)maltose revealed three anomeric signals, with coupling constants of $\sim$3 Hz (axial–equatorial arrangements), and one anomeric signal with a coupling constant of $\sim$7 Hz (axial–axial arrangement).[162] Based on these results, it was concluded that the configuration of the glycosidic linkage in the per-O-(trimethylsilyl) derivative was α.

This method of determining the configuration of interglycosidic linkages has been applied to other oligosaccharides. However, the approach is limited to oligosaccharides that contain monosaccharide units having axial protons on C-2 (for example, D-glucose and D-galactose). Although, in the ^{1}H-n.m.r. spectra of Me$_3$Si derivatives of sugars, the signals due to anomeric protons are well separated from the rest of the proton resonances, the nonanomeric-proton resonances are not well resolved. Hence, for the determination of the conformation of monosaccharide residues in oligosaccharides, the Me$_3$Si derivatives are not suitable. For this purpose, peracetylated derivatives are preferred. The nonanomeric protons of peracetylated derivatives of oligosaccharides are spread over a large δ-range ($\sim$2 p.p.m.), making their identification possible. However, for determination of the configuration of the glycosidic linkages, the peracetylated derivatives are unsuitable, because the anomeric proton on C-1 of the glycosyl units appears in the region of nonanomeric protons, and this makes identification difficult.[166,168–171]

The ^{1}H-n.m.r. spectrum of octa-O-acetyl-β-maltose (67) at 100 MHz has been described.[165] The resonances due to H-1,2,3,4,5,6a,6b were observed at δ 5.75, 4.97, 5.31, 4.04, 4.11, 4.23, and 4.49, respectively. The first-order coupling-constants ($J_{1,2}$ 7.8, $J_{2,3}$ 9.0, $J_{3,4}$ 7.8, $J_{4,5}$ 8.0, $J_{5,6a}$

(169) W. W. Binkley, D. Horton, and N. S. Bhacca, *Carbohydr. Res.*, 10 (1969) 245–258.
(170) H. Friebolin, G. Keilich, and E. Siefert, *Org. Magn. Reson.*, 2 (1970) 457–465.
(171) G. Keilich, E. Siefert, and H. Friebolin, *Org. Magn. Reson.*, 3 (1971) 31–36.

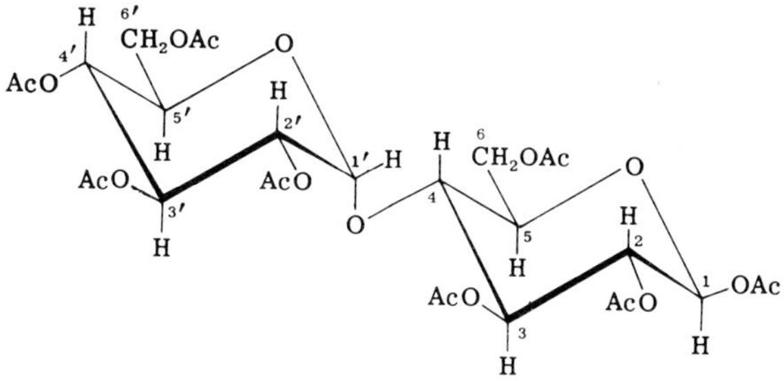

67

$3.9, J_{5,6b}\ 3.9, J_{6a,6b}\ -12.2$ Hz) confirmed the β-D-*gluco* configuration and the $^4C_1(\text{D})$ conformation for the reducing moiety in **67**. The signals due to H-1',2',3',4',5',6'a,6'b appeared at δ 5.41, 4.86, 5.38, 5.05, 4.18, 4.23, and 4.49, respectively. The first-order coupling-constants ($J_{1',2'}$ 3.9, $J_{2',3'}$ 9.0, $J_{4',5'}$ 9.7, $J_{5',6'a}$ 3.9, $J_{5',6'b}$ 3.9, $J_{6'a,6'b}$ − 12.2 Hz) were in agreement with the α-D-*gluco* configuration and the $^4C_1(\text{D})$ conformation for the nonreducing moiety in **67**. The resonances due to H-1' and H-4 in **67**, compared with the chemical-shift values for H-1 and H-4 in 1,2,3,4,6-penta-*O*-acetyl-α-D-glucopyranose, are shifted upfield dramatically. This was explained on the basis that the acetyl groups at O-1 and O-4 have been replaced by the glycosidic linkage. Besides these, upfield shift for the signals due to H-2, H-2', H-3, and H-3' in **67** is also observed; in particular, the "internal" protons, H-2' and H-3, are the most affected.

High-resolution, ^{1}H-n.m.r. spectroscopy of maltose derivatives has been extensively studied by Hough and his coworkers.[24,45,66,95] The structure of methyl 2-*O*-acetyl-3,6-dichloro-3,6-dideoxy-4-*O*-(2,3-di-*O*-acetyl-4,6-dichloro-4,6-dideoxy-α-D-galactopyranosyl)-β-D-allopyranoside (**36**) was confirmed by its 220-MHz, ^{1}H-n.m.r. spectrum.[24] The *allo* configuration in the $^4C_1(\text{D})$ conformation of the reducing residue of the molecule was established on the basis of the first-order coupling-constants ($J_{1,2}$ 7.7, $J_{2,3}$ 3.3, $J_{3,4}$ 2.8, $J_{4,5}$ 9.1, $J_{5,6}$ 2.5, and $J_{5,6'}$ 3.4 Hz). The *galacto* configuration in the $^4C_1(\text{D})$ conformation of the nonreducing group of the disaccharide was also clearly indicated by the resonances of H-1'–4'. The H-1' signal appeared as a doublet ($J_{1,2}$ 3.6 Hz), indicating that H-1' and H-2' are gauche-disposed. The H-2' and H-3' signals appeared at lowest field, as pairs of strongly coupled quartets ($J_{2',3'}$ 10.8, $J_{3',4'}$ 3.2 Hz), which indicated that H-2' and H-3' are attached to the same carbon atoms as the acetoxyl groups. The H-4' signal appeared as a narrow quartet having $J_{4',5'}$ 1.3 Hz, indicative of an axial 4-substituent as, in this case, both hydrogen atoms bear an antiperiplanar relationship to vicinal, electronegative substituents (O-5' and Cl-4', respectively).

The structure of methyl 2-O-acetyl-3,6-dichloro-3,6-dideoxy-4-O-(2,3,4-tri-O-acetyl-6-chloro-6-deoxy-α-D-glucopyranosyl)-β-D-allopyranoside (38) was supported[95] by its ^{1}H-n.m.r. spectrum at 220 MHz. Comparison of the spectrum with that of methyl 6,6'-dichloro-6,6'-dideoxy-β-maltoside indicated that the resonance position of the protons on the nonreducing ring had been unaltered, except for the H-1' resonance, which had undergone a diamagnetic shift of 0.47 p.p.m. On the other hand, the resonances due to protons on the reducing ring had been markedly perturbed. Significantly, the paramagnetic shifts experienced by H-1 and H-5 (0.48 and 1.15 p.p.m., respectively) were characteristic of the deshielding influence of a *syn*-axial chlorine atom at C-3. This was confirmed by the appearance of the H-3 resonance as a narrow triplet (J ~3 Hz) at δ 4.68, and by the H-2 and H-4 resonances, which appeared as doublets (J ~3 and ~9 Hz).

^{1}H-N.m.r. spectroscopy has proved of immense value in identifying the positions of isolated hydroxyl groups in maltose derivatives.[50,61,66] In the ^{1}H-n.m.r. spectrum (220 MHz) of methyl 2,6,2',3',4',6'-hexa-O-acetyl-β-maltoside, the H-3 resonance appeared as a wide triplet ($J_{2,3} = J_{3,4} = 9.4$ Hz) at ~1.6 p.p.m. to higher field of the corresponding resonance for methyl β-maltoside heptaacetate, in accord with attachment of H-3 to the carbon atom bearing the less-deshielding hydroxyl group.[66] The resonances due to the other protons on the reducing ring were also moved upfield with respect to the corresponding resonances for the peracetylated derivative, but to a smaller extent. The upfield shift of resonances due to protons on adjacent carbon atoms bearing methylsulfonyloxy and azidodeoxy groups, in comparison with the corresponding acetoxyl substituents, has also been observed.[45,66]

Comparison of the ^{1}H-n.m.r. spectrum of 3-O-(methylsulfonyl)-β-maltose heptaacetate (31) with that of β-maltose octaacetate showed only one divergence, ~0.34 p.p.m. upfield of the H-3 resonance, indicative of the location of the methylsulfonyloxy group on C-3. A similar comparison of ^{1}H-n.m.r.-spectral parameters as between methyl 4'-O-(methylsulfonyl)-β-maltoside hexaacetate and methyl β-maltoside heptaacetate revealed that the methylsulfonyloxy group in the former compound is located on C-4', as the H-4' resonance appeared ~0.3 p.p.m. to higher field than the H-4' resonance for the heptaacetate.[66] In the ^{1}H-n.m.r. spectrum of methyl 3-azido-3-deoxy-β-maltoside heptaacetate, the H-3 resonance appeared at 1.70 p.p.m. to higher field of the corresponding resonance of methyl β-maltoside heptaacetate, indicative of the presence of the less-deshielding, azido group on C-3. Comparison of the ¹H-n.m.r. spectrum of the azide with that of methyl β-maltoside heptaacetate indicated that, besides the H-3 resonance, the H-4 resonance is the only one affected by the change in the nature of the substituent on C-3 (an upfield shift of 0.35 p.p.m.), which suggests that it lies within the shielding cone of the anisotropic, azido group.[66]

High-resolution, ¹H-n.m.r. spectra of trityl ethers of maltose and its derivatives have been described.[45,66,166] In the ¹H-n.m.r. spectrum of methyl 2,2′,3′,4′-tetra-O-acetyl-3-O-(methylsulfonyl)-6,6′-di-O-trityl-β-maltoside (**68**), the chemical shifts of methine protons (H-2, H-2′, H-

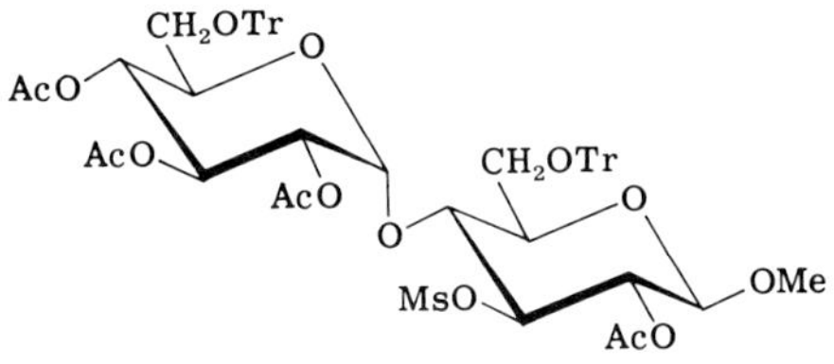

68

3′, and H-4′) were indicative of adjacent acetoxyl groups. One of the acetyl methyl groups resonated at abnormally high field (δ 1.67), most probably due to the 4′-acetoxyl group's lying in the shielding cone of one of the aromatic rings of the 6′-O-trityl group.[45] This assignment was subsequently confirmed by Koizumi and Utamara[166] by using an analog of methyl 6′-O-trityl-β-maltoside hexaacetate (**69**) having a trideuterioacetoxyl group at C-4′ (**70**). The ¹H-n.m.r. spectrum (60 MHz) of **70** was identical in all respects with that of **69**, except that one three-proton singlet, at δ 1.71, disappeared for the former, and the signal could thus be unambiguously assigned to the 4′-acetoxyl signal.

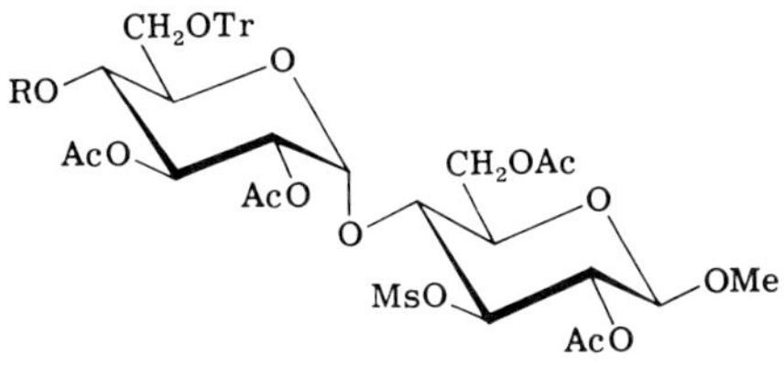

69 R = Ac
70 R = COCD₃

2. ^{13}C-Nuclear Magnetic Resonance Spectroscopy

The value of ^{13}C-n.m.r. spectroscopy for the determination of the structure of carbohydrates is now well recognized.[172-175] ^{13}C-N.m.r. spectra of maltose and its derivatives have been reported. In the ^{13}C-n.m.r. spectrum of maltose, the ^{13}C signals due to C-1', C-1α, and C-1β were readily identified at lowest field. The ^{13}C nuclei bearing the primary hydroxyl groups (C-6, C-6') appeared at the higher-field end of the spectrum. The assignment of the C-4 signals of the molecule were obvious, because of the downfield shift that occurs for a ^{13}C nucleus on glycosidation of its attached hydroxyl group. The rest of the spectrum was tentatively assigned by using the chemical-shift correlation-technique,[176-179] which involves comparisons of the chemical-shift values with those of the monomer components. More-definitive methods of ^{13}C assignments involve chemical substitution,[90,178,179] selective, heteronuclear, spin decoupling,[180-182] or selective spin-labelling, either by ^{13}C (see Refs. 183-186) or ^{2}H (see Refs. 187-194).

(172) G. Kotowycz and R. U. Lemieux, *Chem. Rev.*, 73 (1973) 669-698.
(173) N. K. Wilson and J. B. Stothers, *Top. Stereochem.*, 8 (1974) 1-158.
(174) A. S. Perlin, *Int. Rev. Sci., Org. Chem. Ser. Two, Carbohydrates*, 7 (1976) 1-34.
(175) B. Coxon, in C. K. Lee (Ed.), *Developments in Food Carbohydrate-2*, Applied Science, London, 1980, pp. 351-390.
(176) D. E. Dorman and J. D. Roberts, *J. Am. Chem. Soc.*, 92 (1970) 1355-1361; 93 (1971) 4463-4472.
(177) T. Uusi, N. Yamaoka, K. Matsuda, K. Tuzimura, H. Sugiyama, and S. Seto, *J. Chem. Soc., Perkin Trans. 1*, (1973) 2425-2432.
(178) P. Colson, H. J. Jennings, and I. C. P. Smith, *J. Am. Chem. Soc.*, 96 (1974) 8081-8087.
(179) See Ref. 163.
(180) D. Y. Gagnaire, F. R. Taravel, and M. R. Vignon, *Carbohydr. Res.*, 51 (1976) 157-168.
(181) J. Feeny, P. Shaw, and P. J. S. Pauwels, *J. Chem. Soc. Chem. Commun.*, (1970) 554-555.
(182) N. S. Bhacca, F. W. Wehrli, and N. M. Fischer, *J. Org. Chem.*, 38 (1973) 3618-3622.
(183) G. Excoffier, D. Y. Gagnaire, and F. R. Taravel, *Carbohydr. Res.*, 56 (1977) 229-238.
(184) D. Y. Gagnaire, R. Nardin, F. R. Taravel, and M. R. Vignon, *Nouv. J. Chim.*, 1 (1977) 423-430.
(185) T. E. Walker, R. E. London, T. W. Whaley, R. Barker, and N. A. Matwiyoff, *J. Am. Chem. Soc.*, 98 (1976) 5807-5813.
(186) J. A. Schwarcz and A. S. Perlin, *Can. J. Chem.*, 50 (1972) 3667-3676.
(187) A. S. Perlin and B. Casu, *Tetrahedron Lett.*, (1969) 2921-2924.
(188) H. J. Koch and A. S. Perlin, *Carbohydr. Res.*, 15 (1970) 403-410.
(189) P. A. J. Gorin, *Can. J. Chem.*, 52 (1974) 458-461.
(190) P. A. J. Gorin, *Carbohydr. Res.*, 39 (1975) 3-10.

The ^{13}C-n.m.r. spectrum of maltose has been unambiguously assigned (see Table I) by using selective proton-irradiation, ^{13}C selective spin-decoupling, and the isotopic chemical-shift of hydroxyl-bearing carbon atoms that is induced by deuterium.[194] The $\alpha:\beta$ ratio corresponding to the mutarotational equilibrium was equal to the value obtained for D-glucose. Hence, the signals due to C-1α, C-1β, C-1', C-6β, C-6α, and C-6' were readily assigned; the last two signals overlapped. By analogy with cellobiose, the resonances at 78.9, 78.6, and 101.1 p.p.m. were assigned to C-4α, C-4β, and C-1', respectively. In comparison with cellobiose, the resonances due to C-1' and C-4α,β in maltose appeared at higher field. The resonances for C-2α, C-3α, and C-5α were distinguished from those for C-2β, C-3β, and C-5β on the basis of the $\alpha:\beta$ anomeric ratio. The signals at 71.4, 74.0, and 76.0 p.p.m. were assigned to C-5α, C-5', and C-5β, respectively, by reason of a β-isotopic effect. Selective spin-decoupling of H-2', H-2α, and H-2β permitted the assignment of C-2', C-2α, and C-2β. Selective labelling of maltose with ^{13}C at C-1' confirmed the assignment for C-2' (73.1 p.p.m.), because of coupling with C-1'. The existence of a geminal coupling, ^2J (C-1'–C-5') ~ 1.5 Hz, which is characteristic of a 1,2-*cis* linkage[184,185] confirmed that the resonance at 74.0 p.p.m. was due to C-5'.

Deuteration has been used extensively in order to facilitate the analysis of ^{13}C-n.m.r. spectra[187–193] Replacement of a ^{13}C-bound proton by a deuterium atom usually causes a substantial diminution in the intensity of the α- carbon signal in the spectrum[188] and a shift of the signal of a β-carbon atom,[195] thereby identifying these signals; or, it may demonstrate the presence, or absence, of coupling between a proton and other various atoms of the molecule. Until 1977, use of C-deuteration as an assignment technique was limited, because of the complexity then associated with the synthesis of isotopically labelled oligosaccharides. Koch and Stuart[191] showed that carbon-attached hydrogen atoms of hydroxymethyl groups of various carbohydrate derivatives undergo ready ^{1}H–^{2}H exchange in deuterium oxide in the presence of

---

(191) H. J. Koch and R. S. Stuart, *Carbohydr. Res.*, 59 (1977) c1–c6.
(192) F. Balza, N. Cyr, G. K. Hamer, A. S. Perlin, H. J. Koch, and R. S. Stuart, *Carbohydr. Res.*, 59 (1977) c7–c11.
(193) G. K. Hamer, F. Balza, N. Cyr, and A. S. Perlin, *Can. J. Chem.*, 56 (1978) 3109–3116.
(194) A. Heyraud, M. Rinaudo, M. Vignon, and M. Vincendon, *Biopolymers*, 18 (1979) 167–185.
(195) G. E. Maciel, P. D. Ellis, and D. C. Hofer, *J. Phys. Chem.*, 71 (1967) 2160–2164.

TABLE I

^{13}C Chemical-Shift[a] Assignments for Some Maltose Derivatives

Sugar moiety	^{13}C Atom	Compounds			
		α-Maltose[194]	β-Maltose[194]	Methyl β-maltoside[177]	Benzyl β-maltoside[202]
D-Glucose residue	C-1	93.2	97.2	104.4	102.2
	C-2	72.7	75.4	74.6	73.9
	C-3	74.5	77.5	77.8	77.3
	C-4	78.9	78.6	78.7	78.2
	C-5	71.4	76.0	76.1	75.7
	C-6	62.0	62.2	62.3	61.7
D-Glucosyl group	C-1	101.1	101.1	101.1	100.8
	C-2	73.2	73.1	74.3	74.1
	C-3	74.3	74.3	74.6	74.1
	C-4	70.8	70.8	70.9	70.5
	C-5	74.0	74.0	73.4	72.8
	C-6	62.0	62.0	62.3	61.9

[a] Chemical shifts in p.p.m. downfield from external tetramethylsilane.

Raney nickel catalyst. Methyl α-D-glucopyranoside was fully deuterated at carbon atoms 2, 3, 4, and 6, to afford **71**, by treatment with

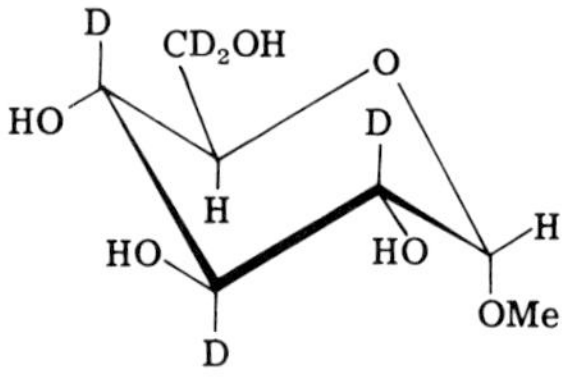

71

deuterium oxide for 18 h under reflux in the presence of Raney nickel catalyst prewashed with deuterium oxide.[191] The ^{1}H-coupled ^{13}C spectrum of **71** revealed a signal for C-5 as a doublet of doublets ($^1J_{C-5,H-5}$ 142 Hz, $^3J_{C-5,H-1}$ 6.5 Hz), whereas, prior to deuteration, the signal is a complex multiplet. Hence, the simplified splitting-pattern provides a value of 6.5 Hz for vicinal coupling between an sp^3-hybridized, ^{13}C atom and an *anti*-oriented ^{1}H atom, through oxygen (^{13}C–O–C–^{1}H pathway), information that had not previously been accessible.[186] The ^{1}H–^{2}H exchange-reaction[193] with methyl β-maltoside was complete after 24 h, to give **72**. The ^{13}C-n.m.r. spectrum revealed six signals at-

TABLE I (continued)

	Compounds				
Octa-O-acetyl-β-maltose[180]	6'-Chloro-6'-deoxy-α-maltose[90]	6'-Chloro-6'-deoxy-β-maltose[90]	6,6'-Dichloro-6,6'-dideoxy-α-maltose[90]	6,6'-Dichloro-6,6'-dideoxy-β-maltose[90]
91.35	93.0	96.9	93.2	97.0
71.10	72.4	75.1	72.4	75.1
75.10	74.4	77.3	74.2	77.1
72.90	78.3	78.1	78.9	78.6
73.05	71.0	75.0	69.7	74.0
62.70	62.1	61.9	46.1	45.6
95.85	100.8	100.8	101.0	101.0
70.20	73.4	73.4	72.9	72.9
69.40	73.6	73.6	73.6	73.6
68.20	71.2	71.2	71.2	71.2
68.70	72.8	72.8	72.8	72.8
61.65	45.4	45.4	45.6	45.6

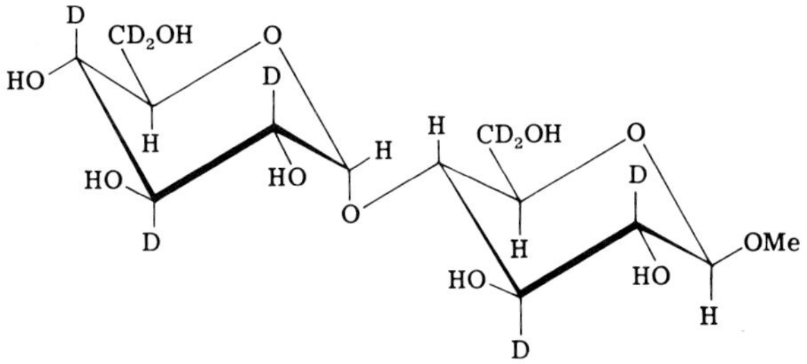

72

tributable to the four carbon atoms engaged with the oxygen atoms on the pyranoid ring and in the two glycosidic bonds, namely, C-1, C-5, C-1', and C-5', and C-4 and CH_3, plus a residual signal due to slow exchange of H-2'. On the basis of these results, the earlier assignments for[176] C-3' and C-5', and for[176,177] C-2 and C-5 of the maltoside should be reversed.

The possibility of inversion of configuration during the $^1H-^2H$ exchange-reaction has been considered.[191,192] However, the inversion process is much slower than deuteration, and it is likely to become significant only during prolonged exchange-treatments. Purification, with an ion-exchange resin, of the products of the deuteration reaction is strongly advisable, in order to ensure removal of paramagnetic metal ions that may cause broadening of the ^{13}C signals. Carbohydrates having groups sensitive to hydrogenation cannot be used directly, and, therefore, reducing sugars are converted into glycosides before the ^{1}H–^{2}H exchange treatment.

The value of chlorinated and alditol derivatives of D-glucose and related disaccharides in the assignment process has been demonstrated.[90] Chlorination of C-6' of the nonreducing group of maltose had, as expected, little effect on the chemical shifts of the carbon atoms of the reducing residue (see Table I). The increased shielding of C-6', by 16.3 p.p.m., on chlorination is comparable to that observed for D-glucose (α, −16.1 p.p.m.; β, −16.7 p.p.m.), but a corresponding effect on C-5' was not observed. The glycosidic link at C-1' must be responsible for this decreased sensitivity. The small displacement of C-2', C-3', and C-4', by −0.4, −0.5, and +0.7 p.p.m., respectively, was attributed to the chlorination effect. In the ^{13}C-n.m.r. spectrum of 6,6'-dichloro-6,6'-dideoxymaltose, the chemical-shift values for the nonreducing group were similar to those found for that of 6'-chloro-6'-deoxymaltose. The changes in chemical shift of C-6 (α, −15.9 p.p.m.; β, −16.2 p.p.m.) were comparable to those found for that of 6-chloro-6-deoxy-D-glucose. Similarly, an increased shielding was found for C-5 (α, −1.5 p.p.m.; β, −1.7 p.p.m.), in contrast to the "anomalous" behavior for the nonreducing group. The distinction between the C-6 resonances of the α and β anomers of the reducing residue is enhanced, relative to maltose, as was found for D-glucose.

3. Mass Spectrometry

Mass spectrometry has become an important technique in the determination of the structure of carbohydrates.[196] Fragmentation patterns of disaccharide derivatives can indicate the type of linkage between monosaccharide residues.[196–198] High-resolution, electron-impact, mass spectrometry of maltose derivatives has been studied.[24,52,86,95] For methyl 2,3,2',3',4'-penta-O-acetyl-6,6'-dichloro-6,6'-dideoxy-β-maltoside (**73**), a probable pattern of fragmentation has been suggested[95] (see Scheme 1). Compound **73** underwent cleavage of the two bonds leading from the interglycosidic oxygen atom (⤬O⤬) at m/z 307

---

(196) N. K. Kochetkov and O. S. Chizhov, *Adv. Carbohydr. Chem.*, 21 (1966) 39–93.
(197) O. S. Chizhov, N. V. Molodtsov, and N. K. Kochetkov, *Carbohydr. Res.*, 4 (1967) 273–276.
(198) N. K. Kochetkov, O. S. Chizhov, and N. V. Molodtsov, *Tetrahedron*, 24 (1968) 5587–5593.

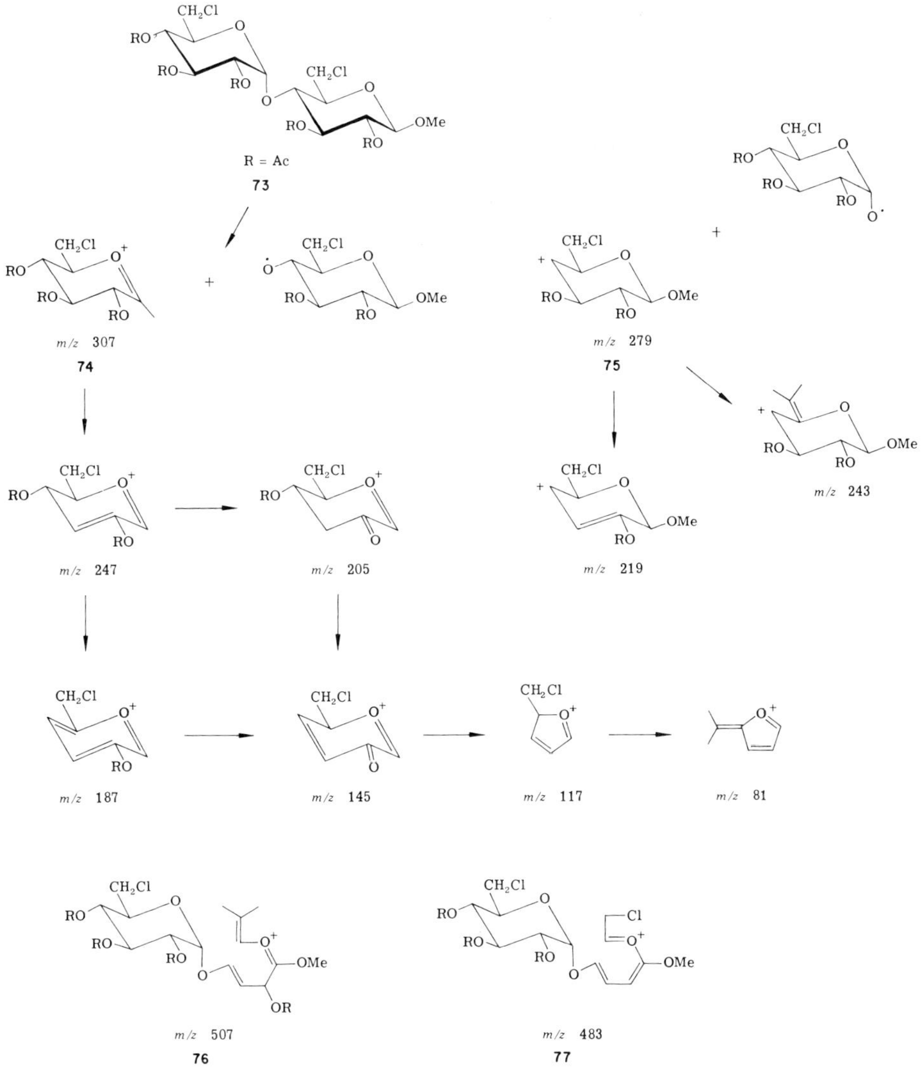

Scheme 1

(**74**) and 279 (**75**), in the ratio of 4:1. The oxycarbonium ion **74** fragmented in the usual way, revealing the sequential loss of (*i*) acetic acid, (*ii*) acetic acid or ketene, (*iii*) ketene or acetic acid, (*iv*) carbon monoxide, and hydrogen chloride. The ion **75** lost either acetic acid (*m/z* 219) or hydrogen chloride (*m/z* 243). No molecular-ion peak was observed. However, fragments of low intensity were noted at *m/z*

507 and 483, corresponding to **76** (M⁺−OAc−HCl) and **77** (M⁺−OAc−OAc), of which an analog of the former had been observed[24] in the mass spectrum of methyl 2-O-acetyl-3,6-dichloro-3,6-dideoxy-4-O-(2,3-di-O-acetyl-4,6-dichloro-4,6-dideoxy-α-D-galactopyranosyl)-β-D-allopyranoside (**36**).

In the mass spectrum of **36**, the major pathway was initiated, as expected, by cleavage of the interglycosidic linkage, to give the glycosyl carbonium ions **78** (m/z 283) of the nonreducing group of the mole-

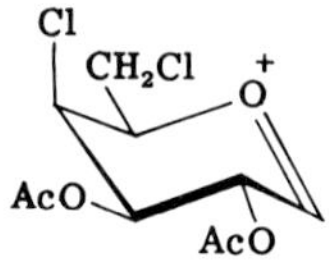

m/z 283

78

cule, which then underwent sequential loss of acetic acid, ketene, hydrogen chloride, carbon monoxide, and hydrogen chloride. The first stage, the loss of acetic acid, suggested the presence of an acetoxyl group on C-3′, and the following loss of ketene was indicative of a 2′-acetoxyl group. These results were consistent with the structure depicted for **36**. The oxycarbonium ion (**79**) at m/z 255 originated from

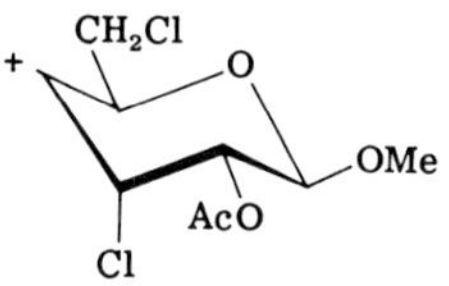

m/z 255

79

the reducing residue of the disaccharide, and was formed either by cleavage of the C-4−O bond, or by cleavage of the C-1′−C-2′ bond, followed by rearrangement of the 3′-acetoxyl group to C-1′ and cleavage of the C-1′−O-5 bond, to give fragment **80** (m/z 343). The loss of

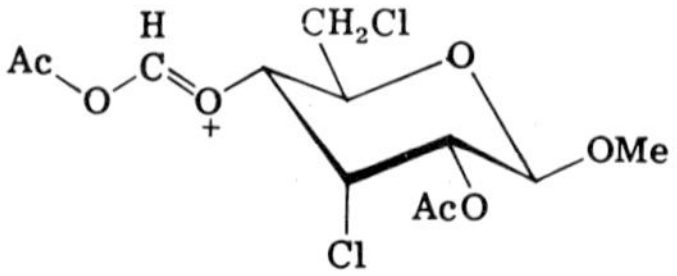

m/z 343

80

CH_3CO_2CHO would afford fragment **79**. This structural analysis demonstrates the potential of the technique for structural studies of oligosaccharides.

The mass spectrum of methyl 2,3,6,2',3'-penta-O-acetyl-4',6'-di-O-methyl-β-maltoside has been studied.[52] Although only a weak molecular ion at m/z 594 was observed, the fragmentation ions at m/z 563 (M–$CH_3O\cdot$), 562 (M–CH_3OH), 535 (M–$CH_3COO\cdot$), 534 (M–CH_3COOH), 503 (M–$CH_3O\cdot$–CH_3COOH), 502 (M–CH_3OH–CH_3COOH), and 475 (M–$CH_3COO\cdot$–CH_3COOH) provided indirect evidence of the molecular weight. The fragmentation ion at m/z 549 (M–$CH_3OCH_2\cdot$) suggested the presence of a primary methoxyl substituent. The major fragmentation-pattern was followed, as expected, by cleavage of the two bonds attached to the interglycosidic oxygen atom, giving rise to ions **81** (m/z 275) and **82** (m/z 303). The

presence of ion **81** suggested that the nonreducing group in the disaccharide contained two methoxyl groups, and the successive formation of ions at m/z 215 and 173, which was supported by the appearance of the relevant, metastable peaks, was consistent with the proposed location of the O-methyl and O-acetyl substituents.

Chemical ionization (c.i) mass spectra, employing ammonia or isobutane as the reagent gas, have been reported[199] for peracetylated maltose and other oligosaccharides. In all cases, the disaccharide acetate–ammonium complex (a.c.i.) was the base peak in the spectrum (**83**, m/z 696), above the reagent-gas region (m/z 60). The most striking feature of the spectra of the oligosaccharide studied was the preponderance of even-electron ions with even mass. As virtually all of the ions obtained under these conditions are even-electron ions, the even-mass ions must contain a nitrogen atom. Ways in which nitrogen-containing fragments could have arisen, such as thermolysis followed by attachment of NH_4^+, have been suggested. The glycosyl ions (m/z 331) resulting from cleavage of the interglycosidic linkage of the disaccharide peracetate were also observed as the major fragments. Metastable transi-

(199) R. C. Dougherty, J. D. Roberts, W. W. Binkley, O. S. Chizhov, V. I. Kadentsev, and A. A. Solv'yov, *J. Org. Chem.*, 39 (1974) 451–455.

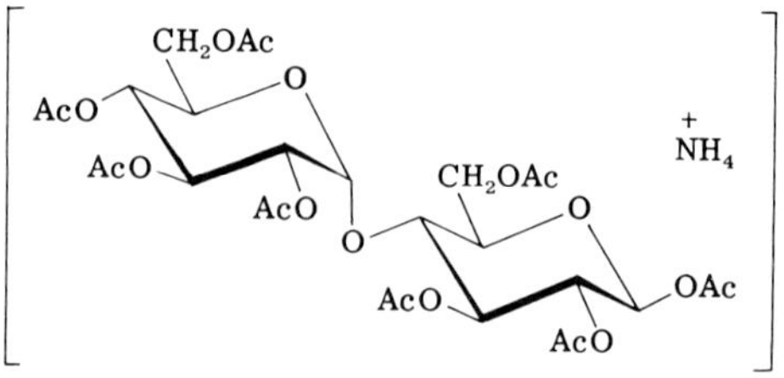

[Structure **83**, m/z 696, ammonium adduct of per-acetylated disaccharide]

$\downarrow$ −AcONH$_4$

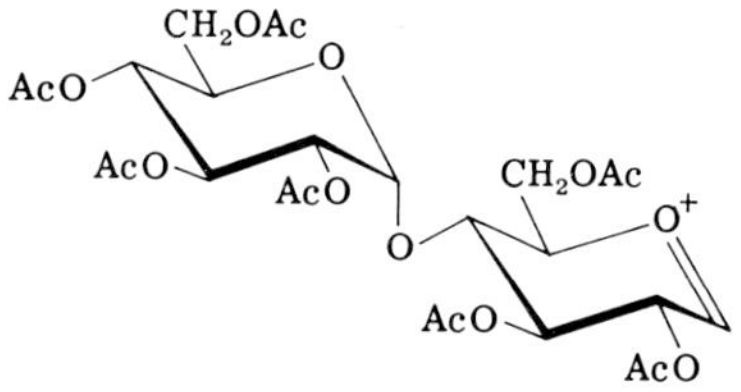

[Structure **84**, m/z 619, glycosyl oxocarbenium ion]

tions indicated that either the molecular a.c.i. (**83**) or the ion **84** (m/z 619) could be the precursor of this ion. Further fragmentation of the glycosyl ion proceeded in a manner precisely analogous to the electron-impact mass spectra of the disaccharide acetates. Sequential elimination of acetic acid and ketene, often with corresponding metastable ions, accounted for the odd-mass ions between m/z 331 and 109. The relatively simple spectra obtained with chemical ionization, together with the formation of molecular and high-mass ions, indicates the potential of this technique for structural determination of oligosaccharides.

The potential of the negative chemical-ionization (n.c.i.) technique for obtaining valuable structural information on very small samples of underivatized oligosaccharides has been demonstrated.[200] The n.c.i. spectra of several mono-, di-, tri-, and tetra-saccharides were recorded, using methane as the reagent gas, and direct-probe insertion of the samples into the ion source.[201] Fairly intense, molecular ions, M$^-$, were observed in each case. A small number of fragment ions were also observed, but could not be interpreted. Use of dichlorodifluoro-

(200) A. K. Ganguly, *J. Chem. Soc. Chem. Commun.*, (1979) 148–149.
(201) A. L. C. Smit, M. A. J. Russetto, and F. H. Field, *Anal. Chem.*, 48 (1976) 2042–2044.

methane as the reagent gas in n.c.i. mass spectrometry has proved valuable for studying oligosaccharides, especially those having a number of free hydroxyl groups.[200] Apparently, the larger the number of sites for attachment of Cl^-, the higher the probability of the formation of an anion. Larger fragments are thus more prominent than the smaller fragments in the n.c.i. mass spectra.

XII. Tables of Properties of Maltose Derivatives

Tables II–XV constitute a list of most of the known, characterized derivatives of maltose. The names of the solvents used for measuring the specific rotations are abbreviated as follows: A, acetone; B, benzene; C, chloroform; E, ethanol; M, methanol; P, pyridine; and W, water.

TABLE II

Glycosides

Compound	M.p. (degrees)	$[\alpha]_D$ (degrees)	(solvent)	References
α-Maltoside				
isoamyl	152–155	+143	W	27
tert-butyl	188–190	+153	W	27
cyclohexyl	112–115	+211	W	27
isopropyl	177–179	+166	W	27
methyl	201–202	+183	W	27
	—	+147	W	30
	—	+176	W	28
hepta-O-acetyl-	66–69	+134	C	28
1-naphthyl	156–158	+192	W	27
hepta-O-acetyl-	192–194	+185	C	27
phenyl	202–204	+211	W	34
hepta-O-acetyl-	184–184.5	+170.2	C	34
β-Maltoside				
benzyl	148–149			143
hepta-O-acetyl-	124–125			143
p-bromophenyl		+34.5	W	36
hepta-O-acetyl-	172.5–173.5	+44.5	C	36
	104–106	+80.7	W	68
methyl (monohydrate)	108–109	+78	W	20
	110–111	+81	W	23
hepta-O-acetyl-	122–124	+53	C	20,23
phenyl	96	+34	W	31
hepta-O-acetyl-	154–155	+42	C	32
	157–158			31

Table III

Trityl Ethers

Compound	M.p. (degrees)	[α]$_D$ (degrees)	(solvent)	References
α-Maltose				
6-O-trityl-				
hepta-O-acetyl-	189–190	+100	C	41
hexa-O-acetyl-	191.8–192.0	+88	C	43
6'-O-trideuterioacetyl-	190.4–191.0	+87	C	43
6'-O-trityl-				
hepta-O-acetyl-	161–161.6	+131	C	43
hexa-O-acetyl-				
6-O-trideuterioacetyl-	160.6–161.0	+131.3	C	43
6,6'-di-O-trityl-				
hexa-O-acetyl-	224–225.4	+98	C	43
β-Maltose				
6-O-trityl-				
hepta-O-acetyl-	116–118	+36	C	41
6'-O-trityl-				
hepta-O-acetyl-	164–164.5	+96	C	41
	162–163	+86.3		40
penta-O-acetyl-	141–142	+40.8		46
1,6-anhydro-	101–102	+55	C	46,47
6,6'-di-O-trityl-				
hexa-O-acetyl-	126.4–126.8	+59	C	43
Maltose				
6,6'-di-O-trityl-	149–152	+68.4		40
hexa-O-acetyl-	215–217	+88		40
	222–223	+88	C	41
Benzyl β-maltoside				
6-O-trityl-				
hexa-O-acetyl-	169–170	+62.1	C	37
Methyl β-maltoside				
6'-O-trityl-				
hexa-O-acetyl-	165–168	+88	C	45
penta-O-acetyl-3-O-(methylsulfonyl)-	100–103	+74	C	45
6,6'-di-O-trityl-				
tetra-O-acetyl-3-O-(methylsulfonyl)-	118–122	+51	C	66
Ethyl 1-thio β-maltoside				
6'-O-trityl-				
hexa-O-acetyl-	101–103	+71	C	44

Table IV

Methyl Ethers

Compound	M.p. (degrees)	$[\alpha]_D$ (degrees)	(solvent)	References
β-Maltose				
2-O-methyl-, hepta-O-acetyl-	136–138	+82	C	204
3-O-methyl-				
hepta-O-acetyl-	115–116	+77.2	C	50
1,6-anhydro-	syrup	+74.6	M	51
tri-O-benzoyl-4′,6′-O-benzylidene-	140–142	+74.5	C	51
6-O-methyl-, hepta-O-acetyl-	135–136	+60.7	C	47
6′-O-methyl-, penta-O-acetyl-1,6-anhydro-	164	+20	C	47
3,4′-di-O-methyl-, hexa-O-benzoyl-	97–99	+73.1	C	69
3,3′-di-O-methyl-, 1,6-anhydro-di-O-benzoyl-4′,6′-O-benzylidene-	126–128	+71.2	C	51
2,3,3′-tri-O-methyl-, 1,6-anhydro-2-O-benzoyl-4′,6′-O-benzylidene-	105–106	+65.5	C	51
Methyl β-maltoside				
4′,6′-di-O-methyl-	154–155	+70.5	M	52
penta-O-acetyl-	131–132			52
6-O-trityl-	105	+49	C	52
3,2′-di-O-methyl-				
tri-O-benzoyl-4′,6′-O-benzylidene-	189–190	+61.4	C	69
3,3′-di-O-methyl-				
tri-O-benzoyl-4′,6′-O-benzylidene-	209–210	+83.6	C	69
2,3,2′,3′,4′,6′-hexa-O-methyl-	syrup	+67	C	52
6-O-trityl-	syrup	+46.5	C	52
hepta-O-methyl-	syrup	+81.4	W	6
Phenyl α-maltoside				
3-O-methyl-	syrup	+199	W	47
6-O-methyl-	syrup	+168	W	47
6′-O-methyl-	syrup	+183	W	47

(202) E. E. Lee and J. O. Wood, *Carbohydr. Res.*, 75 (1979) 317–321.
(203) C. S. Hudson and J. M. Johnson, *J. Am. Chem. Soc.*, 37 (1915) 1276–1280.
(204) H. Arita, T. Ikenaka, and Y. Matsushima, *J. Biochem. (Tokyo)*, 69 (1971) 401–407.
(205) R. A. Cadenas and J. O. Deferrari, *J. Org. Chem.*, 28 (1963) 2613–2616.
(206) D. H. Brauns, *J. Am. Chem. Soc.*, 5 (1929) 1820–1831.
(207) D. S. Genghof, C. F. Brewer, and E. J. Hehre, *Carbohydr. Res.*, 61 (1978) 291–299.
(208) J. O. Deferrari, personal communication.
(209) V. N. Nigam, C. A. Brailovsky, and C. Chopra, *Cancer Res.*, 38 (1978) 3315–3321.

TABLE V

Acetates and Benzoates

Compound	M.p. (degrees)	[α]D (degrees)	(solvent)	References
Maltose				
6-O-benzoyl-	140–145	+130 → +105.5	C	68
hepta-O-acetyl-	192–193	+120.2	C	65
α-Maltose				
octa-O-acetyl-	127	+122.1	C	124
	125	+122.8		203
β-Maltose				
3,6,2′,3′,4′,6′-hexa-O-acetyl-	155–157	+116	C	61
1,2,3,2′,3′,4′-hexa-O-acetyl-	—	+111		40
2,3,6,2′,3′,4′,6′-hepta-O-acetyl-	183–184	+70	C	57
	181	+67.8	C	58
	188	+84 → +144	P	60
1,3,6,2′,3′,4′,6′-hepta-O-acetyl-	166.5–168.5	+86.7	C	61
	168–169	+85.9	C	63
1,2,6,2′,3′,4′,6′-hepta-O-acetyl-	—	+77.3	C	47
	—	+87.5		50
1,2,3,2′,3′,4′,6′-hepta-O-acetyl-	141–143	+66	C	67
	140–141	+65		32,55
	140–142			47,83
1,2,3,6,2′,3′,4′-hepta-O-acetyl-	183–184	+58.3		40
	177–178	+65	C	41
octa-O-acetyl-	152–157	+62.9	C	11
	155–158	+57.1	C	55
	159–160	+62.5	C	54
	159	+62.8	C	72
3,4′-di-O-acetyl-1,2,6,2′,3′,6′-hexa-O-benzoyl-	175–179	+76.7	C	69
1,2,6,2′,3′,6′-hexa-O-benzoyl-	—	+58.4	C	69
1,2,6,2′,3′,4′,6′-hepta-O-benzoyl-	141–142	+46.1	C	68,69
octa-O-benzoyl-	191–193	+66	C	68
	193–194	+67.7	C	69
Benzyl β-maltoside				
2,3,4,2′,3′,6′-hexa-O-acetyl-	162–164	+24.9	C	37
Methyl β-maltoside				
2,3,6,2′,3′-penta-O-acetyl-	152–153	+41	C	52
3,6,2′,3′,4′,6′-hexa-O-acetyl-	157–158	+85.3	C	63
2,6,2′,3′,4′,6′-hexa-O-acetyl-	75–77	+61.0	C	66
2,3,6,2′,3′,4′-hexa-O-acetyl-	187–190	+49	C	45
hepta-O-acetyl-	124–126	+52.6	C	68
	132–133	+61	C	63
	123–124	+53.5	C	23
2,6,2′,3′-tetra-O-benzoyl-	183–184	+81.4	C	69
2,6,2′,3′,6′-penta-O-benzoyl-		+72.2	C	69
2,6,2′,3′,4′,6′-hexa-O-benzoyl-	114–115	+56.8	C	69
hepta-O-benzoyl-	93–95	+80.1	C	68

TABLE V (continued)

Compound	M.p. (degrees)	[α]D (degrees)	(solvent)	References
4-O-α-D-Glucopyranosyl-β-D-allopyranoside				
octa-O-acetyl-	154–155	+45.2	C	69
Methyl 4-O-α-D-glucopyranosyl-β-D-allopyranoside				
hepta-O-acetyl-	72–74	+49	C	66
hepta-O-benzoyl-	syrup	+107	C	69
hexa-O-acetyl-3-O-benzoyl-	75–78	+30	C	66
Methyl 4-O-α-D-galactopyranosyl-β-D-allopyranoside	215–216	+112.3	W	69
4-O-α-D-Galactopyranosyl-D-glucopyranose	227–229 (dec.)	+159.6	W	82
4-O-α-D-Galactopyranosyl-α-D-glucopyranose				
octa-O-acetyl-	syrup	+117	C	82

TABLE VI

Sulfonates

Compound	M.p. (degrees)	[α]D (degrees)	(solvent)	References
β-Maltose				
2-O-(methylsulfonyl)-				
hepta-O-acetyl-	165–166	+66.9	C	61
3-O-(methylsulfonyl)-				
hepta-O-acetyl-	152–153	+65	C	66
hepta-O-benzoyl-	156–157	+37.1	C	69
6-O-(methylsulfonyl)-				
hepta-O-acetyl-	175–176	+58.4	C	67
	173–174	+63.6	C	83
hexa-O-acetyl-, 1-(dimethylamino)dithiocarbamate	189.5–190.5	+77.8	C	67
3,4'-di-O-(methylsulfonyl)-				
hexa-O-benzoyl-	125–126	+46.8	C	69
4',6'-di-O-(methylsulfonyl)-				
tetra-O-acetyl-1,6-anhydro-	199–200	+40.8	C	82
6-O-p-tolylsulfonyl-				
hepta-O-acetyl-	145–147.5	+58.9	C	67
	138–141	—		47
	140–141	+50		32
hexa-O-acetyl-, 1-ethylxanthate	178–179	+70.3	C	67

Continued

TABLE VI (continued)

Compound	M.p. (degrees)	$[\alpha]_D$ (degrees)	(solvent)	References
6'-O-p-tolylsulfonyl-				
hepta-O-acetyl-	syrup	+100	C	47
penta-O-acetyl-1,6-anhydro-	170–171	+51.5	C	46
	170–171	+55.7	C	47
4',6'-di-O-p-tolylsulfonyl-				
tetra-O-acetyl-1,6-anhydro-	180–181	+36.2	C	82
6,6'-di-O-p-tolylsulfonyl-				
hepta-O-acetyl-	190–192	+82	C	79
Benzyl β-maltoside				
6-O-p-tolylsulfonyl-				
hexa-O-acetyl-	135–137	+35.4	C	37
Methyl β-maltoside				
2-O-(methylsulfonyl)-				
hexa-O-acetyl-	132–133	+61	C	63
3-O-(methylsulfonyl)-				
hexa-O-acetyl-	73–76	+56	C	66
4'-O-(methylsulfonyl)-				
hexa-O-acetyl-	syrup	+44	C	45
6'-O-(methylsulfonyl)-				
hexa-O-acetyl-	154.5–156	+58	C	45
3,4'-di-O-(methylsulfonyl)-				
penta-O-benzoyl-	109–110	+82.6	C	69
3,6'-di-O-(methylsulfonyl)-				
penta-O-acetyl-	syrup	+53	C	45
6,6'-di-O-(methylsulfonyl)-				
penta-O-acetyl-	175–176	+56.5	C	23
3,4',6'-tri-O-(methylsulfonyl)-				
tetra-O-benzoyl-	110–112	+85	C	69
4',6'-di-O-(methylsulfonyl)-				
tetra-O-acetyl-6-O-p-				
tolylsulfonyl-	193–195	+48.7	C	81
3-O-(methylsulfonyl)-				
tetra-O-acetyl-6,6'-di-O-p-				
tolylsulfonyl-	184–185	+55		84
	184–185.5	+55	C	45
6-O-p-tolylsulfonyl-	154–155	—		80
hexa-O-acetyl-	143–144	+55.3	C	80
tetra-O-acetyl-4',6'-O-				
benzylidene-	197–198	+20.3	C	81
6'-O-p-tolylsulfonyl-				
penta-O-acetyl-6-deoxy-	143–144	+43.2	C	80
6,6'-di-O-p-tolylsulfonyl-	120–121	+46.3	C	80
	125–126			84
	124–126	+45	C	20
2,2',3',4'-tetra-O-acetyl-		+85	C	84
penta-O-acetyl-	186–189	+60.4	C	80
Phenyl α-maltoside				
6-O-(methylsulfonyl)-				
hexa-O-acetyl-	190–192	+158	C	83

TABLE VII

Halides

Compound	M.p. (degrees)	[α]$_D$ (degrees)	(solvent)	References
Maltose				
6'-bromo-6'-deoxy-	syrup	+105	W	122
6'-chloro-6'-deoxy-	syrup	+116	W	122
6'-deoxy-6'-iodo-	110–112	+89	W	122
hepta-O-acetyl-	syrup	+102		47
6,6'-dichloro-6,6'-dideoxy-	syrup	+99.6	W	90
β-Maltose				
6'-bromo-6'-deoxy- tetra-O-acetyl-1,6-anhydro-4'-O-benzoyl-	syrup	−5	C	83
6-chloro-6-deoxy- hepta-O-acetyl-	138–139	+50	P	47
6'-chloro-6'-deoxy- penta-O-acetyl-1,6-anhydro-	129	+53.8	C	47
6-deoxy-6-iodo- hepta-O-acetyl-	129–131			47
6'-deoxy-6'-iodo- hepta-O-acetyl-	88–90	+50	C	32
	84–86	+51	C	67
1,6-anhydro- penta-O-acetyl-	194–195	+39	C	46
tetra-O-acetyl- 4'-O-p-tolylsulfonyl-	syrup	+24.3	C	82
4'-O-(methylsulfonyl)-	182–183	+36	C	82
Methyl β-maltoside				
6-bromo-6-deoxy-	166–167 (dec.)	+69.1	W	81
hexa-O-acetyl-	128–129.5	45.1	C	81
tetra-O-acetyl-4',6'-O-benzylidene-	215–217	+13.6	C	81
3-bromo-3-deoxy- hexa-O-acetyl-	85–88 → 154.5–156	+48	C	66
6-chloro-6-deoxy-	182–183	+72.6	W	81
hepta-O-acetyl-	126–127	+43.8	C	81
tetra-O-acetyl-4',6'-O-benzylidene-	218.5–220	+11.6	C	81
6,6'-dichloro-6,6'-dideoxy- penta-O-acetyl-	180–182	+58.0	C	95
6'-deoxy-6'-iodo- hexa-O-acetyl-	170–171	+52.4	C	80
6-deoxy-6-iodo- hexa-O-acetyl-	129–130	+47.7	C	80,81
tetra-O-acetyl-4',6'-O-benzylidene-	217–218.5	+20.5	C	81
penta-O-acetyl-6'-O-p-tolylsulfonyl-	203–204	+50.1	C	80

Continued

TABLE VII (continued)

Compound	M.p. (degrees)	[α]$_D$ (degrees)	(solvent)	References
6,6'-dideoxy-6,6'-diiodo-				
penta-O-acetyl-	196–197	+48.2	C	81
tetra-O-acetyl-3-O-				
(methylsulfonyl)-	syrup	+43.4	A	84
3,6,6'-trideoxy-3-iodo-				
2',3',4'-tri-O-acetyl-	syrup	+50.2	A	84
Phenyl α-maltoside				
6'-chloro-6'-deoxy-				
hexa-O-acetyl-	133–134	+186.9	C	47
6-deoxy-6-fluoro-	syrup	+227.3	W	83
6-deoxy-6-iodo-	174 (dec.)	+128	W	47
6'-deoxy-6'-iodo-	syrup	+130	W	47
Benzyl β-maltoside				
6-deoxy-6-iodo-				
hexa-O-acetyl-	114–116	+31	C	37
4-O-α-D-Glucopyranosyl-β-D-				
allopyranose				
3-chloro-3-deoxy-	121–130	+155	W	66
2,6,2',3',4',6'-hexa-O-acetyl-	188–193	+115	C	66
hepta-O-acetyl-	154–157	+96	C	66
Methyl 4-O-α-D-glucopyranosyl-β-D-allopyranoside				
3-chloro-3-deoxy-	175–176	+121	M	69
	178–179	+120	M	66
hexa-O-acetyl-	182–184	+89	C	66
tetra-O-benzoyl-4',6'-O-				
benzylidene-	214–215	+65.9	C	69
3,6,6'-trichloro-3,6,6'-trideoxy-				
tetra-O-acetyl-	202–204	+88.5	C	95
Methyl 4-O-α-D-galactopyranosyl-β-D-allopyranoside				
3,6,4',6'-tetrachloro-3,6,4',6'-tetradeoxy-				
tri-O-acetyl-	161–162	+118.9	C	24
	158–160	+114	C	95

TABLE VIII

Maltosyl Halides

Compound	M.p. (degrees)	[α]$_D$ (degrees)	(solvent)	References
α-Maltosyl bromide				
2,3,6,2',3',4'-hexa-O-acetyl-	—	+116.4	C	44
hepta-O-acetyl-	113–114	+180.4	C	124
	112–113	+180.1	C	206
	84	—		31
hexa-O-acetyl-				
6-O-(methylsulfonyl)-	—	+159	C	67
penta-O-acetyl-				
6,6'-di-O-p-tolylsulfonyl-	141–142	+164	C	79
α-Maltosyl chloride				
3,6,2',3',4',6'-hexa-O-acetyl-	72–73	+162.2	C	63
hepta-O-acetyl-	123–124			61
	118–120	+159.5	C	206
hexa-O-acetyl-				
2-O-(methylsulfonyl)-	160–162	+148.3	C	61
2-O-(trichloroacetyl)-	124–125	+142.1	C	63
β-Maltosyl chloride				
3,6,2',3',4',6'-hexa-O-acetyl-	143–144	+80.5	B	63
2-O-(trichloroacetyl)-	133–134	+57.3	B	63
α-Maltosyl fluoride				
hepta-O-acetyl-	174–175	+111.1	C	206
	172–174	+110.2	C	207
β-Maltosyl fluoride				
hepta-O-acetyl-	130–131	+80.1	C	207
α-Maltosyl iodide				
hepta-O-acetyl-	62–63			206

Table IX

Anhydro Derivatives

Compound	M.p. (degrees)	$[\alpha]_D$ (degrees)	(solvent)	References
β-Maltose				
1,6-anhydro-	156–158	+108	W	46
	154–156	+92	W	55
	132–137	+76.6	W	82
2,3,2′,3′-tetra-O-acetyl-	177–179	+39.1	C	82
2,3,2′,3′,4′-penta-O-acetyl-	82–83	+43.4	C	46,47
hexa-O-acetyl-	181.5–183.5	+48.9	C	55
	182–184	+49	C	67
	182–183	+48	C	32
2′-O-benzoyl-	176–177	+86	W	51
2,2′-di-O-benzoyl-	syrup	+120.5	M	51
2,3′-di-O-benzoyl-	syrup	+100	M	51
hexa-O-benzyl-	77–78	+7.6	C	53
Methyl β-maltoside				
3,6-anhydro-				
hexa-O-acetyl-	159–165	+29	C	66
3,6:3′,6′-dianhydro-	95–101	−66	W	20,23
tri-O-acetyl-	218–219	+25.4	C	23
tri-O-p-tolylsulfonyl-	—	—		20
Phenyl α-maltoside				
3,6-anhydro-	syrup	+120	W	83
3′,6′-anhydro-	194–195	+136	W	47
4-O-α-D-Galactopyranosyl-β-D-glucopyranose				
1,6-anhydro-	197–200	+104.7	W	82
hexa-O-acetyl-	132–133	+54.5	C	82
tetra-O-acetyl-4′,6′-di-O-benzoy	—	+58.8	C	82
4-O-α-D-Glucopyranosyl-D-glucitol				
hepta-O-acetyl-1,5-anhydro-	125–127	+83.3	C	67
Methyl 4-O-α-D-glucopyranosyl-β-D-allopyranoside				
tri-O-acetyl-2,3-anhydro-6,6′-di-O-trityl-	117–120	+120	C	66

TABLE X

Cyclic Acetals

Compound	M.p. (degrees)	[α]D (degrees)	(solvent)	References
Maltose				
4′,6′-O-isopropylidene-hexa-O-acetyl-	90–101	+32	C	108
Methyl β-maltoside				
4′,6′-O-benzylidene-	140–141	+47.6	E	52
penta-O-acetyl-	200–201	+25.4	C	52
2,6,2′-tri-O-benzoyl-	178–181	—		69
2,6,3′-tri-O-benzoyl-	213–216	—		69
2,6,2′,3′-tetra-O-benzoyl-	196–197	+53.3	C	69
β-Maltose				
4′,6′-O-benzylidene-				
1,6-anhydro-	224–226	+56.4	W	82
tetra-O-acetyl-	215–216	+20.0	C	82
2,3,3′-tri-O-acetyl-	110–112	+48.1	C	51
2,3-di-O-acetyl-2′,3′-di-O-benzoyl-	109–111	+46.1	C	51
3,3′-di-O-acetyl-2,2′-di-O-benzoyl-	113–115	+87.0	C	51
3-O-acetyl-2,2′,3′-tri-O-benzoyl-	115–117	+74.7	C	51
2′-O-benzoyl-	—	+64.2	C	51
2,2′-di-O-benzoyl-	185–187	+56.2	C	51
2′,3′-di-O-benzoyl-	228–230	+65.1	C	51
2,3-di-O-methyl-	189–191	+59.6	C	51
2,2′,3′-tri-O-benzoyl-	syrup	+69.1	C	51
tetra-O-benzoyl-	197–199	+75.3	C	82
	197–200	—		51
2,3:5,6-Di-O-isopropylidene-aldehydo-D-glucose dimethyl acetal				
4-O-α-D-glucopyranosyl-	syrup	+191.7	C	105
tetra-O-acetyl-	109–110	+85	C	105
4-O-(4,6-O-isopropylidene-α-D-glucopyranosyl)-	132–133	—		105
di-O-acetyl-	syrup	+66	C	105
di-O-(methylsulfonyl)-	70–71	+55	C	105
2′-O-p-tolylsulfonyl-	68–69	+66.5	C	105
di-O-p-tolylsulfonyl-	69–70	+48.0	C	105

TABLE XI

Deoxy Derivatives

Compound	M.p. (degrees)	[α]$_D$ (degrees)	(solvent)	References
Maltose				
2-deoxy-	182	+30.4		124
hepta-O-acetyl-	167	—		124
6'-deoxy-	syrup	+112.7	W	46
hepta-O-acetyl-	183–185	+64	C	46
	85–87	+85	C	47
β-Maltose				
6-deoxy-				
hepta-O-acetyl-	163–164.5	+17	C	47
	164–166	+64.1	C	67
1,6-anhydro-6'-deoxy-				
penta-O-acetyl-	141–142	+44.5	C	46
	142–143	+51.5	C	82
tetra-O-acetyl-				
4'-O-benzoyl-	syrup	0	C	82
4'-O-(methylsulfonyl)-	184–186	+34	C	82
4'-O-p-tolylsulfonyl-	syrup	+16.5	C	82
Benzyl β-maltoside				
6-deoxy-	syrup	+40	W	37
hexa-O-acetyl-	162–163	+29.1	C	37
Methyl β-maltoside				
6-deoxy-	194–195	+74.6	W	81
hexa-O-acetyl-	121–122	+46.6	C	80,81
penta-O-acetyl-6'-O-benzoyl-	196–197	—		80
3',6'-anhydro-	—	—		80
6'-deoxy-				
hexa-O-acetyl-	176–177	—		80
6,6'-dideoxy-				
penta-O-acetyl-	186–187	+50	C	80
Phenyl α-maltoside				
6-deoxy-	syrup	+176	W	47
6'-deoxy-	syrup	+170	W	47
4-O-α-D-Galactopyranosyl-β-D-glucopyranose				
1,6-anhydro-6'-deoxy-	192–194	+101.3	W	82
penta-O-acetyl-	140–141	+52.2	C	82
tetra-O-acetyl-4'-O-benzoyl-	syrup	+76.1	C	82
4-O-α-D-Galactopyranosyl-α-D-glucopyranose				
6'-deoxy-	syrup	+128.3	W	82
hepta-O-acetyl-	186–189	+137.5	C	82
Methyl 2,3,6-trideoxy-4-O-(6-deoxy-α-D-glucopyranosyl)-β-D-erythro-hexopyranoside	118–119	+34.6		84
4-O-α-D-Glucopyranosyl-D-glucitol				
6-deoxy-				
hexa-O-acetyl-1,5-anhydro-	130–131	+79.6	C	67

TABLE XII

Nitrogen-containing Compounds

Compound	M.p. (degrees)	$[\alpha]_D$ (degrees)	(solvent)	References
Maltose				
6'-amino-6'-deoxy-	—	+88	W	46
6,6'-diamino-6,6'-dideoxy-	138–140 (dec.)	+98	W	79
1,6,6'-triazido-1,6,6'-trideoxy-	65–67	+98	W	79
penta-O-acetyl-	78–79 (dec.)	+81	C	79
1,6,6'-triacetamido-1,6,6'-trideoxy-	124–125 (dec.)	+79	W	79
penta-O-acetyl-	146–147	+81	C	79
β-Maltose				
1-azido-1-deoxy-				
hepta-O-acetyl-	90–91	+52.4	C	113
	91	+53	C	114
1-amino-1-deoxy-				
hepta-O-acetyl-	191.5	+73.7	C	114
1-N[1-benzyl N-(benzyloxycarbonyl)-L-aspart-4-oyl]-				
hepta-O-acetyl-	95–98	+68	C	113
1-N-(L-aspart-4-oyl)-	239–240	+7	W	113
6'-azido-6'-deoxy-				
hepta-O-acetyl-	151–152	+47.9	C	46
Methyl β-maltoside				
3-azido-3-deoxy-				
hexa-O-acetyl-	86–89	+30	C	66
4'-azido-4'-deoxy-				
hexa-O-acetyl-	106–108	+21	C	45
6-azido-6-deoxy-				
hexa-O-acetyl-	100–102	+73.7	C	81
	107–110	+73	C	45
tetra-O-acetyl-4',6'-O-benzylidene-	139–140	+62.2	C	81
6'-azido-6'-deoxy-				
hexa-O-acetyl-	145–146	+68	C	45
6,6'-diazido-6,6'-dideoxy-				
penta-O-acetyl-	149–150.5	+75	C	45
6,6'-diacetamido-6,6'-dideoxy-				
penta-O-acetyl-	syrup	+46	C	45
Methyl 4-O-α-D-glucopyranosyl-β-D-allopyranoside				
3-azido-3-deoxy-				
hexa-O-acetyl-	102–104	+51	C	66
3-acetamido-3-deoxy-				
hexa-O-acetyl-	syrup	+49	C	66
3,6'-diazido-3,6'-dideoxy-				
penta-O-acetyl-	158–160	+69	C	45

Continued

TABLE XII (continued)

Compound	M.p. (degrees)	[α]D (degrees)	(solvent)	References
3,6'-diacetamido-3,6'-dideoxy-				
penta-O-acetyl-	syrup	+46	C	45
1,1-Bis(acetamido)-1-deoxy-4-O-α-				
D-glucopyranosyl-D-glucitol	84	+80.9	W	205
2,3,5,6,2',3',4',6'-octa-O-acetyl-	95	+46.6	C	205
1,1-Bis(benzamido)-1-deoxy-4-O-				
α-D-glucopyranosyl-D-glucitol	225	+42.6	P	76
2,3,5,6,2',3',4',6'-octa-O-acetyl-	130	+31.9	C	76

TABLE XIII

Sulfur-containing Compounds

Compound	M.p. (degrees)	[α]D (degrees)	(solvent)	References
Maltose				
6'-thio-	—	+137.6	W	46
α-Maltose				
6-S-acetyl-6-thio-				
hepta-O-acetyl-	177–178	+138.4	C	67
6'-thio-				
hepta-O-acetyl-	73–75	+113	C	46
β-Maltose				
1-thio-, sodium salt	amorphous	+90.7	W	67
1-S-acetyl-1-thio-				
hepta-O-acetyl-	152–153	+69	C	67
6-thio-				
1,6-anhydro-	—	+66.1	M	67
hexa-O-acetyl-	135–137	+82	C	67
6-S-acetyl-6-thio-				
hepta-O-acetyl-	130–131	+72	C	67
6'-S-acetyl-6'-thio-				
1,6-anhydro-	219–220	+34.6	C	47
Maltose				
diethyl dithioacetal				
octa-O-acetyl-	122–122.5	+87.5	C	127
β-Maltosyl ethylxanthate				
2,3,2',3',4',6'-hexa-O-acetyl-	155–156	+86.8	C	67
6-O-(methylsulfonyl)-	143–144	+73.1	C	67
hepta-O-acetyl-	83–85	+75.4	C	67
β-Maltoside				
ethyl 1-thio-	149–152	+55	M	30
2,3,6,2',3',4'-hexa-O-acetyl-	182–183	+42.8	C	44
hepta-O-acetyl-	135.5–136.5	+50.1	C	30

TABLE XIV

Carboxylic Derivatives

Compound	M.p. (degrees)	$[\alpha]_D$ (degrees)	(solvent)	References
4-O-(α-D-Glucopyranosyluronic acid)-D-glucose	powder	+108.1		40
	—	+116	W	143
methyl ester	196–197	+73.7		40
sodium salt, H₂O	106–109 (dec.)	+94		40
hepta-O-acetyl-	202–203	+72.2		40
methyl ester	196–197	+73.7		40
	197–198	+77	C	143
4-O-(α-D-Glucopyranosyluronic acid)-D-glucuronic acid	—	+73		40
hexa-O-acetyl-	—	+104.2		40
methyl ester	—	+110		40
disodium salt	166–168 (dec.)	+140		40
Benzyl 4-O-(α-D-glucopyranosyluronic acid)-β-D-glucopyranoside	190–193	+45.4		40
methyl ester	162–163	+42.4		40
hexa-O-acetyl- methyl ester	164–165	+33.8	C	143
barium salt		+30	W	143
4-O-α-D-Glucopyranosyl-D-glucuronic acid		+73.5	W	55
hepta-O-acetyl-	186–186.5	+64.2	C	55
methyl ester	85 → 120 (dec.)	+65.9	C	55
sodium salt	126–129 (dec.)	+112.6	W	55
barium salt	149–150 (dec.)	+90.9	W	55
Methyl [methyl di-O-methyl-4-O-(tetra-O-methyl-α-D-glucopyranosyl)-β-D-glucopyranosid]uronate		+71	C	52
Maltobionic acid		+104.1 → +106		137
lithium salt, 3 H₂O	106.5–108	+97.3	W	137
		+96.8	W	136

TABLE XV

Miscellaneous Derivatives

Compound	M.p. (degrees)	$[\alpha]_D$ (degrees)	(solvent)	References
4-O-α-D-Galactopyranosyl-D-glucopyranose	227–229	+159.6	W	82
4-O-α-D-Glucopyranosyl-D-allopyranose		+118.9	W	69
3-O-α-D-Glucopyranosyl-D-arabinose	172	+16.5	W	124
monohydrate	121	+56.9 → +47		136
hexa-O-acetyl-	153	−42	C	124
phenylosazone	195–200 (dec.)			136
3-O-α-D-Glucopyranosyl-D-ribofuranose	70–72	+59.5		84
β-Maltose				
3-O-(phenylcarbamoyl)-hepta-O-acetyl-	174.5–175.5	+58.7	C	50
1-O-(phenylcarbamoyl)-hepta-O-acetyl-	146–148	+60.1	C	50
β-Maltosyl				
N,N-dimethyldithiocarbamate		+46.7	W	67
hepta-O-acetyl-	132–133	+79.3	C	67
hexa-O-acetyl-6-O-p-tolylsulfonyl-	154–155	+73.4	C	67
Methyl 4-O-α-D-glucopyranosyl-β-D-*ribo*-hexopyranosid-3-ulose				
hexa-O-acetyl-	112–113.5	+109	C	66
Methyl 4-O-α-D-glucopyranosyl-β-D-*arabino*-hexopyranosid-3-ulose				
hexa-O-acetyl-	75–78	+28	C	66
Maltal	176	+1.16	W	124
hexa-O-acetyl-	132–136	−22.5	C	124
	131–132	+68	C	123
Methyl-4-O-α-D-glucopyranosyl-β-D-*erythro*-hex-2-enopyranoside				
tri-O-acetyl-2,3,6,6'-trideoxy-	102–103	+11.8		84
1-(2-Chloroethyl)-3-β-maltosylurea	108–110 (dec.)	+68	W	126
(Hepta-O-acetyl-β-maltosyl)urea	72	+74	C	126
β-Maltosyl-1-nitrosourea	96 (dec.)	+60	W	126
(Hepta-O-acetyl-β-maltosyl)-1-nitrosourea	95–96 (dec.)	+63.8	C	126
Maltononitrile				
octa-O-benzoyl-	90–92	+97.7	C	208
Maltose				
tetra-O-palmitoyl-	—	—		209

CHEMISTRY AND BIOCHEMISTRY OF D- AND L-FUCOSE

By Harold M. Flowers

*Department of Biophysics,
The Weizmann Institute of Science, Rehovot, Israel*

I. Introduction 280
   1. General Remarks 280
   2. Occurrence 280
   3. Isolation 282
   4. Biological Significance 282
II. Chemistry 283
   1. Steric and Electronic Effects of the 5-Methyl Compared to the
      5-(Hydroxymethyl) Group 283
   2. Synthesis of D- and L-Fucose 285
   3. Analytical Detection and Determination 289
   4. Substituents and Protecting Groups 290
   5. Relative Reactivities of Hydroxyl Groups 290
   6. Glycoside Formation by Acid Equilibration 293
   7. Synthesis of Complex Glycosides 294
III. Fucose in Glycans 301
   1. General Occurrence 301
   2. Glycosides Containing D-Fucose 302
   3. Polysaccharides 302
   4. Glycoproteins 304
   5. Glycolipids 307
IV. Immunological Aspects of Complex Fucans 311
   1. Bacterial Antigens 311
   2. Plant Antigens 312
   3. Blood-Group Antigens 313
V. Metabolism 316
   1. Biosynthesis 316
   2. Degradation 324
VI. Clinical 329
   1. Effect of Free Sugars 329
   2. Urinary Oligosaccharides 330
   3. Malignancy 331
   4. Fucosidosis 332
   5. Other Illnesses 333
   6. Fucose and the Immune-Defense System 334

VII. Concluding Remarks 335
VIII. Tables of Some Properties of the Fucoses and Their Derivatives 337

I. INTRODUCTION

1. General Remarks

The fucoses (6-deoxygalactoses) are found in a wide variety of natural products from many different sources, and probably occur in all higher organisms. Interest in their chemistry and biochemistry has expanded considerably with elucidation of the structures of many complex glycans containing the sugar that have been found in micro-organisms, plants, and animals. However, there are no reviews available on the chemistry of the fucoses, although it was briefly dealt with in chapters on cardiac glycosides, by Elderfield[1] in Volume 1 of this Series (1945) and by Reichstein and Weiss[2] in 1962. An article by Hanessian[3] in 1966 dealt mainly with the synthesis, purification, and spectral analysis of deoxy sugars, among them the fucoses. Reviews have since appeared[4,5] that were devoted to the biochemistry and biological importance of fucolipids.

The object of this chapter is twofold: to discuss aspects of the chemistry of the fucoses, especially developments in this field, and to consider their metabolism and biochemistry, and point out possible biological functions and medical applications of the fucoses and their compounds.

2. Occurrence

Both optical isomers of fucose are found widely. In 1890, the L enantiomer was first obtained[6] crystalline from the hydrolyzate of a polysaccharide. It occurs in many bacterial and plant glycosides and polysaccharides,[7–11] at times sulfated,[11,12] in oligosaccharides of human

(1) R. C. Elderfield, *Adv. Carbohydr. Chem.*, 1 (1945) 147–173.
(2) T. Reichstein and E. Weiss, *Adv. Carbohydr. Chem.*, 17 (1962) 65–120.
(3) S. Hanessian, *Adv. Carbohydr. Chem.*, 21 (1966) 143–207.
(4) S.-I. Hakomori, *Prog. Biochem. Pharmacol.*, 10 (1975) 167–196.
(5) J. M. McKibbin, *J. Lipid Res.*, 19 (1978) 131–147.
(6) A. Günther and B. Tollens, *Ber.*, 23 (1890) 2585–2586.
(7) B. Lindberg and S. Svensson, *MTP Int. Rev. Sci., Ser. One, Carbohydr.*, 7 (1973) 319–344.
(8) K. Jann and B. Jann, in I. W. Sutherland (Ed.), *Surface Carbohydrates of the Prokaryotic Cell*, Academic Press, New York, 1977, pp. 247–287.

milk,[13,14] and in many glycolipids[4,5,9] and glycoproteins,[15,16] including several families of blood-group antigens.[15]

D-Fucose also occurs widely in plants. It was isolated from hydrolyzates of the complex glycoside convolvulin, and characterized as a crystalline phenylhydrazone[17]; chartreusin[18] contains D-fucose and D-digitalose (3-O-methyl-D-fucose), and D-fucose has also been isolated from a number of other cardiac glycosides, such as ledienoside[19] and cheirotoxin[20].

Methyl ethers of the fucoses are found in naturally occurring glycosides and polysaccharides. Thus, D-digitalose occurs in chartreusin,[18] curacose (4-O-methyl-D-fucose) was obtained from the antibiotic curamycin produced by *Streptomyces cura-coi*,[21] and labilose (2,3-di-O-methyl-D-fucose) was isolated[22] after (aqueous) acid hydrolysis of the methanolyzate of labilomycin, an antimycobacterial, antibiotic substance. Demethylation of labilose with hydrobromic acid gave free D-fucose, which was characterized. Structural assignments were made by nuclear magnetic resonance (n.m.r.) analysis of the methyl glycosides.

2-O-Methyl-L-fucose occurs as a constituent of plum-leaf polysaccharides,[23] linseed mucilage,[24] soybean polysaccharides,[25] and gum tragacanth.[26]

---

(9) S. G. Wilkinson, in Ref. 8, pp. 97–175.
(10) G. O. Aspinall and A. M. Stephen, *MTP Int. Rev. Sci., Ser. One, Carbohydr.*, 7 (1973) 285–317.
(11) E. Percival, *Methods Carbohydr. Chem.*, 1 (1962) 195–198.
(12) B. Larsen, A. Haug, and T. J. Painter, *Acta Chem. Scand.*, 20 (1966) 219–230.
(13) H. H. Baer, *Fortschr. Chem. Forsch.*, 3 (1958) 822–910.
(14) A. Kobata, in M. I. Horowitz and W. Pigman (Eds.), *The Glycoconjugates*, Vol. 1, Academic Press, New York, 1977, pp. 423–440.
(15) K. O. Lloyd, *MTP Int. Rev. Sci., Ser. Two, Carbohydr.*, 7 (1976) 251–281.
(16) R. Kornfeld and S. Kornfeld, *Annu. Rev. Biochem.*, 45 (1976) 217–237.
(17) E. Votoček and F. Valentin, *Collect. Czech. Chem. Commun.*, 2 (1930) 36–46.
(18) L. H. Sternbach, S. Kaiser, and M. W. Goldberg, *J. Am. Chem. Soc.*, 80 (1958) 1639–1647.
(19) R. Zelnik and O. Schindler, *Helv. Chim. Acta*, 40 (1957) 2110–2129.
(20) J. A. Moore, C. Tamm, and T. Reichstein, *Helv. Chim. Acta*, 37 (1954) 755–770.
(21) O. L. Galmarini and V. Deulofeu, *Tetrahedron*, 15 (1961) 75–86.
(22) E. Akita, K. Maeda, and H. Umezawa, *J. Antibiot., Ser. A*, 16 (1963) 147–151; 17 (1964) 37–38, 200–217.
(23) J. D. Anderson, P. Andrews, and L. Hough, *Chem. Ind. (London)*, (1957) 1453.
(24) K. Hunt and J. K. N. Jones, *Can. J. Chem.*, 40 (1962) 1266–1279.
(25) G. O. Aspinall, K. Hunt, and J. M. Morrison, *J. Chem. Soc., C*, (1967) 1080–1086.
(26) G. O. Aspinall and J. Baillie, *J. Chem. Soc.*, (1963) 1702–1714.

3. Isolation

Free D- and L-fucose have been isolated from acid hydrolyzates of natural products,[11] and purified by means of such crystalline derivatives as the 2-methyl-2-phenylhydrazone,[11] 2-benzyl-2-phenyl- and 2,2-diphenyl-hydrazone,[17] (p-bromophenyl)hydrazone,[19] and phenylhydrazone.[11,20,27] The values for the constants of the crystalline, free sugar that have been published from different laboratories are in the region of m.p. 145°, and $[\alpha]_D$ $-153 \rightarrow -76°$ (L) or $+153 \rightarrow +76°$ (D).

4. Biological Significance

The widespread occurrence of the fucoses in different organisms and various glycans suggests diverse biological functions. D-Fucose is generally limited to plant products and microbial, antibiotic substances, whereas plant polysaccharides and animal glycans usually contain the L enantiomer. L-Fucose is either the immunodominant sugar of many complex, carbohydrate antigens, or its presence may increase the strength of the antigenic response. Addition of the free sugar may influence living cells, and the L-fucose content of animal glycans may change under pathological conditions, including transformation of cells in culture, and cancer *in vivo*. Schemes have been put forward to indicate the biosynthesis of L-fucose, and the oxidative degradation of L- and D-fucose. However, no information is as yet available on the biosynthesis of the D enantiomer, and on whether it is an intermediate or a side-product in pathways of biosynthesis of other deoxy sugars, for example, the dideoxy sugars often found together with it in cardiac glycosides. The biological methylation of D-fucose in plant systems has also not been monitored.

exo-Fucosidases have been characterized that catalyze the hydrolytic removal of D- or L-fucose from the nonreducing termini of glycans, and L-fucosyltransferases are known for the transfer of L-fucose to such substrates. However, no *endo*-fucosidases have as yet been found to attack internal, fucosyl residues.

Only, a few sugars are found in Nature in the form of both optical isomers (for example, galactose and arabinose). It is intriguing to speculate on the special conditions that determine the presence of the rarer form (for instance, the limiting of D-fucose to certain plant and microbial products), and on the specific pathways involved in their metabolism.

---

(27) W. A. P. Black, W. J. Cornhill, E. T. Dewar, and F. N. Woodward, *J. Sci. Food Agric.*, 4 (1953) 85–91.

II. CHEMISTRY

1. Steric and Electronic Effects of the 5-Methyl Compared to the 5-(Hydroxymethyl) Group

Fucose is a 6-deoxyhexose having the *galacto* configuration, that is, it has *cis*-hydroxyl groups on C-3 and C-4. The most stable conformation is the $^4C_1$ (D) or $^1C_4$ (L), in which bulky C-2, 3, and 5 substituents are equatorially (*eq.*) oriented. Differences in reactivity between galactose and fucose are derived from steric and electronic effects of the 5-substituent: the hydroxymethyl and the methyl group, respectively. Thus, fucopyranosides are much more labile than galactopyranosides to acid hydrolysis: methyl 6-deoxy-α-D-galactopyranoside is hydrolyzed at approximately six times the rate of methyl α-D-galactopyranoside.[28] There has been considerable discussion of the mechanism of hydrolysis of glycosiduronic acids,[29,30] and of the possibility that the inductive effect of the 5-carboxyl group is responsible for the marked decline in the rate of hydrolysis at low pH of glucosiduronic acids compared with glucosides. By analogy, it might be supposed that the inductive effect of CH_2OH-5 would be appreciable, and opposite in direction to that of CH_3-5, thus explaining the greater lability of fucopyranosides compared to galactopyranosides in aqueous acid. However, it is not, in fact, clear that inductive effects are of such importance in these reactions, as rate changes may be relatively large, even though the 5- and 1-substituents are separated by three singly-bonded atoms. It was suggested that polar effects are actually of minor importance in this matter,[31] and that it is the conformational factors that are more significant.[30]

The accepted mechanism of hydrolysis of glycopyranosides involves intermediate, oxocarbonium-ion formation at C-1, with attainment of a half-chair conformation,[32] and it has been postulated[33] that the rate of hydrolysis of glycopyranosides will be affected by the relative ease of attainment of this conformation, in which C-2, C-1, the ring O, and C-5 lie in one plane. Bulky 5-substituents may be ex-

---

(28) W. G. Overend, C. W. Rees, and J. S. Sequiera, *J. Chem. Soc.*, (1962) 3429–3440.
(29) B. Capon, *Chem. Ber.*, 69 (1969) 407–498.
(30) D. Keglević, *Adv. Carbohydr. Chem. Biochem.*, 36 (1979) 57–134.
(31) T. E. Timell, W. Enterman, F. Spencer, and E. J. Soltes, *Can. J. Chem.*, 43 (1965) 2296–2305.
(32) W. G. Overend, in W. Pigman and D. Horton (Eds.), *The Carbohydrates: Chemistry and Biochemistry*, 2nd edn., Vol. IA, Academic Press, New York, 1972, pp. 279–353.
(33) J. T. Edward, *Chem. Ind. (London)*, (1955) 1102–1104.

pected to hinder this transformation by increasing the steric barrier, so that more steric hindrance to the necessary conformational change will be afforded by CH_2OH than by CH_3. However, the polar nature of the substituents cannot be ignored, as mutual, electronic interactions between OH-4 and the 5-substituent may be different when the latter is CH_2OH (−I effect) on the one hand or CH_3 (+I effect) on the other. In this connection, it is interesting[32] that methyl α-D-fucopyranoside is hydrolyzed more rapidly than methyl α-L-arabinopyranoside in 2 M hydrochloric acid at 60°. Were steric factors alone responsible for the relative rates of hydrolysis of these two glycosides, it might be supposed that the attainment of the half-chair conformation would be easier for the arabinopyranoside (5-H,H) than for the fucopyranoside (5-H, CH_3).

In studies on the anomerization of acetylated aldopyranoses in acetic anhydride–acetic acid–sulfuric acid at 25°, it was found[34] that the peracetates of 6-deoxy-D-glucopyranose anomerize about ten times as fast as those of D-glucopyranose.

In the formation of glycosides by the Fischer process (equilibration in the presence of an acid catalyst), fucopyranosides are apparently formed more rapidly than galactopyranosides. In the presence of mineral acids, fucose attains equilibrium in boiling methanol within 6–8 h, whereas a reaction time of 20 h is customary for D-galactose. In reactions catalyzed by cation-exchange resins, maximum pyranoside formation is attained by 8 h for L-fucose,[35a] whereas D-galactose requires[35b] 12–24 h.

There are no striking differences in interatomic distances apparent from crystal analyses of β-D-galactose and α-L-fucose, and the C-5 to C-6 distances are essentially the same.[36]

Apart from possible effects produced on the reactivity at C-1 by the 5-substituent, the vicinal OH-4 may be more obviously affected. The comparatively lower reactivity of OH-4 in galactopyranosides towards a variety of reagents is well known,[37] and has been ascribed to its axial orientation. However, although glycosylation reactions at OH-4 of galactopyranosides are often difficult,[38] variations in reagents and reac-

(34) W. A. Bonner, *J. Am. Chem. Soc.*, 81 (1959) 1448–1452.
(35) (a) D. F. Mowery, Jr., *Carbohydr. Res.*, 43 (1975) 233–238; (b) R. H. Pater, P. A. Coelho, and D. F. Mowery, Jr., *J. Org. Chem.*, 38 (1973) 3272–3277.
(36) F. Longchambon, J. O. Ohannessian, D. Avenel, and A. Neuman, *Acta Crystallogr. Sect. B*, 31 (1975) 2623–2627; see G. A. Jeffrey and M. Sundaralingam, *Adv. Carbohydr. Chem. Biochem.*, 34 (1977) 350, 351, 372.
(37) A. H. Haines, *Adv. Carbohydr. Chem. Biochem.*, 33 (1976) 11–109.
(38) H. M. Flowers and D. Shapiro, *J. Org. Chem.*, 30 (1965) 2041–2043; H. M. Flowers, *Carbohydr. Res.*, 4 (1967) 312–317.

tion conditions may afford excellent yields of disaccharide products.[39] The question arises as to the possible effects of the vicinal CH_2OH group on the reactivity of OH-4. In this respect, selective tribenzoylation[40] of methyl α-D-gluco-, -galacto-, and -manno-pyranosides demonstrated the lowest substitution at OH-4, even when it was *eq.* (D-*gluco* configuration), or *eq.* in the presence of (a more reactive) *axial* (*ax.*) OH-2 (D-*manno* configuration).

Similarly, peracetylation of D-mannose with acetic anhydride–sodium acetate gave an appreciable yield of 1,2,3,6-tetra-*O*-acetyl-β-D-mannopyranose (29%), as well as the β-pentaacetate[41] (60%).

The effect of replacing CH_2OH-5 by CH_3 on the reactivity of OH-4 in these compounds may be speculated about, and this topic will be discussed later.

2. Synthesis of D- and L-Fucose

a. D-Fucose.—D-Fucose has been synthesized by several different methods, all based on a common approach: conversion of CH_2OH-5 in D-galactopyranose into CH_3-5.

(i) **From the 6-Iodo Derivative (3).** The first-described synthesis[42] of D-fucose (**7**) commenced from the readily available 1,2:3,4-di-*O*-isopropylidene-α-D-galactopyranose (**1**), which was converted into the 6-*p*-toluenesulfonate (**2**). Treatment of **2** with sodium iodide, and reduction of the resulting iodide (**3**) with Raney nickel, followed by acid hydrolysis, afforded the free sugar (**7**) (see Scheme 1).

Ultraviolet irradiation[43] of a methanolic solution of compound **3** in the presence of alkali led to its rapid, almost quantitative, conversion into 1,2:3,4-di-*O*-isopropylidene-α-D-fucopyranose (**5**); an excellent yield of **5** was also obtained[44] by irradiation of a solution of 6-*O*-acetyl-1,2:3,4-di-*O*-isopropylidene-α-D-galactopyranose (**6**) in aqueous hexamethylphosphoric triamide.

(ii) **Directly from the 6-*p*-Toluenesulfonate (2).** With the introduction of metallic organoborohydride reducing agents, it was found pos-

---

(39) D. D. Cox, E. K. Metzner, and E. J. Reist, *Carbohydr. Res.*, 62 (1978) 245–252.
(40) A. C. Richardson and J. M. Williams, *Tetrahedron*, 23 (1967) 1369–1378.
(41) E. Lee, A. Bruzzi, E. O'Brien, and P. S. O'Colla, *Carbohydr. Res.*, 71 (1979) 331–334.
(42) K. Freudenberg and K. Raschig, *Ber.*, 60 (1927) 1633–1636.
(43) W. W. Binkley and R. W. Binkley, *Carbohydr. Res.*, 11 (1969) 1–8.
(44) J.-P. Pete, C. Portella, C. Monneret, J.-C. Florent, and Q. Khuong-Huu, *Synthesis*, (1977) 774–776.

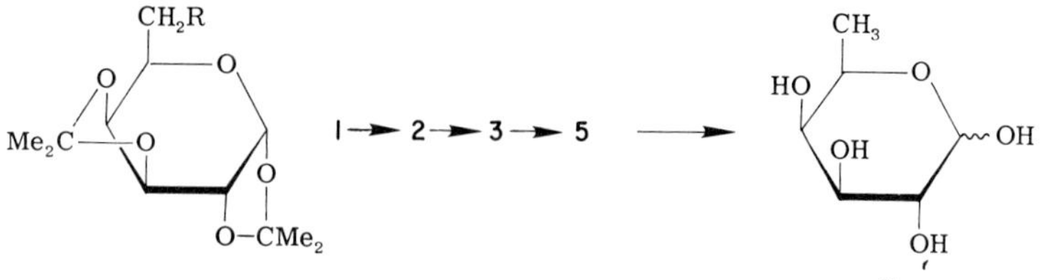

1	R = OH
2	R = OTs
3	R = I
4	R = Cl
5	R = H
6	R = OAc

Scheme 1

sible[45] to reduce **2** directly to the deoxy compound **5** by means of lithium aluminum hydride.

(iii) From the 4,6-Benzylidene Acetal. Treatment of methyl 4,6-*O*-benzylidene-α-D-galactopyranoside (**8**) with *N*-bromosuccinimide and an excess of barium carbonate in boiling carbon tetrachloride gave[46] an almost quantitative yield of methyl 4-*O*-benzoyl-6-bromo-6-deoxy-α-D-galactopyranoside (**10**), which was converted into methyl α-D-fucopyranoside (**11**) in good yield by deacylation and reduction (see Scheme 2).

Methyl β-D-fucopyranoside (**12**) was prepared similarly[47] from methyl 4,6-*O*-benzylidene-β-D-galactopyranoside (**9**).

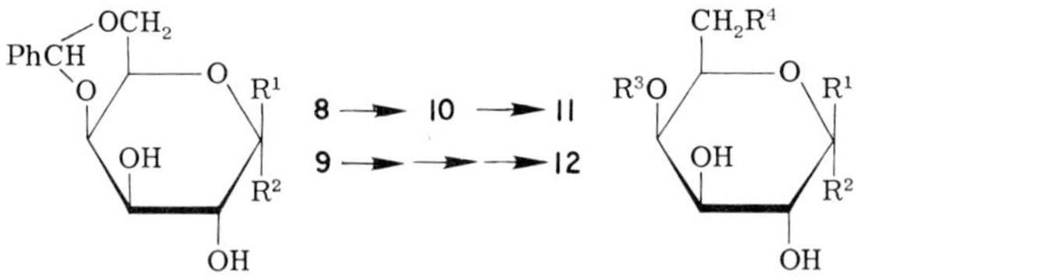

8	R^1 = H, R^2 = OMe
9	R^1 = OMe, R^2 = H
10	R^1 = H, R^2 = OMe, R^3 = Bz, R^4 = Br
11	R^1 = R^3 = R^4 = H, R^2 = OMe
12	R^1 = OMe, R^2 = R^3 = R^4 = H

Scheme 2

(45) H. Schmid and P. Karrer, *Helv. Chim. Acta*, 32 (1949) 1371–1378.
(46) S. Hanessian, *Carbohydr. Res.*, 2 (1966) 86–88.
(47) K. Eklind, P. J. Garegg, and B. Gotthammer, *Acta Chem. Scand.*, 29 (1975) 633–634.

(iv) **By Use of Triphenylphosphite Methiodide.** In the presence of relatively hindered, secondary hydroxyl groups, the primary hydroxyl group reacts preferentially with triphenylphosphite methiodide to give a 6-iodo derivative. Thus, methyl 3,4-*O*-isopropylidene-β-D-galactopyranoside (**13**) gave[48] the deoxyiodo compound **14** in 60% yield, and this was converted into **7** by reduction with Raney nickel, followed by acid hydrolysis. Similar treatment of **1**, however, gave **3** in only 20.5% yield.

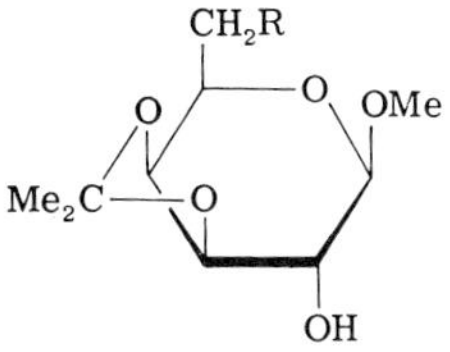

13 R = OH
14 R = I

(v) **By Way of the 6-Chloro Derivative.** The primary hydroxyl group in **1** can also be replaced by chlorine in good yield by using triphenylphosphine in boiling carbon tetrachloride,[49] or (chloromethylene)dimethyliminium chloride.[49,50] The resulting 6-chloro-6-deoxy compound is reduced to **5** by activated Raney nickel, or lithium aluminum hydride.

(vi) **By Way of the 6-Bromo Derivative.** Reaction of penta-*O*-acetyl-β-D-galactopyranose with an excess of hydrogen bromide causes replacement of both OH-1 and OH-6 by bromine.[51] Treatment of the product with methanol in the presence of silver carbonate, followed by catalytic reduction of CH_2Br to CH_3, catalytic *O*-deacetylation, and acid hydrolysis, afforded **7**.

(vii) **By Way of a Hexosid-4-ulose.** A novel and convenient synthesis[52] of D-fucose was based on analogy with the accepted pathway for the biosynthesis of L-fucose from guanosine 5'-(D-mannosyl diphosphate) (see later). Oxidation of methyl α(or β)-D-galactopyranoside with platinum and oxygen apparently led to selective oxidation of

---

(48) N. K. Kochetkov and A. I. Usov, *Tetrahedron*, 19 (1963) 973–983.
(49) S. Hanessian and N. R. Plessas, *J. Org. Chem.*, 34 (1961) 2163–2170.
(50) S. Hanessian, M. M. Ponpipom, and P. Lavallée, *Carbohydr. Res.*, 24 (1972) 45–56.
(51) H. H. Schlubach and E. Wagenitz, *Ber.*, 65 (1932) 304–308.
(52) O. Gabriel, *Carbohydr. Res.*, 6 (1968) 111–117.

OH-4 (*ax*). The presumed, intermediate keto derivatives were not isolated, but were directly reduced by bubbling hydrogen into the reaction vessel, and the resulting mixture of products was hydrolyzed. Free **7** was isolated in 15 and 35% overall yields from the α- and β-D-galactopyranosides, respectively. It was postulated that the labile, intermediate glycosid-4-uloses eliminated water betwen C-5 and C-6 to give hex-5-enosid-4-ulose derivatives which were reduced preferentially by hydrogen in the presence of platinum to the D-*galacto* configuration. When the hydrogen–platinum catalyst was replaced by sodium borohydride for the reduction stage, fucose could not be isolated.

b. L-Fucose.—The methods described for the synthesis of D-fucose from D-galactose are not practical for the L enantiomer, owing to the rarity of L-galactose. Two different approaches for the synthesis of L-fucose have been elaborated.

(i) From L-Arabinose. A very involved synthesis of L-fucose has been described[53] that resulted in an overall yield of 1% from L-arabinose, by way of nine intermediates, including L-mannose and 5-deoxy-L-lyxose.

(ii) From D-Galactose. A much simpler method was devised[54] that utilized inversion from the D- to the L-*galacto* configuration by formal exchange of C-1 for C-6. This was achieved by reduction of the 1-aldehyde group to a methyl group, followed by oxidation of the primary 5-(hydroxymethyl) group to aldehyde.

D-Galactose was converted by ethanethiol and hydrochloric acid into crystalline D-galactose diethyl dithioacetal, which was acetonated with acetone–zinc chloride. The product (**15**) was reduced to the L-fucitol derivative (**16**) with Raney nickel. The overall yield of **16** was 29%, and it was characterized as the crystalline 6-*p*-toluenesulfonate **17**. Oxidation of **16** by the Pfitzner–Moffatt reagent[55] proceeded readily and, after *O*-deacetonation, and purification of the product by chromatography on a column of silica gel, L-fucose (**18**; 13% overall yield from D-galactose) and L-fucitol (**19**; 1% yield) were isolated (see Scheme 3).

(53) A. Tanimura, *Eisei Shikenjo Hokoku*, 77 (1959) 123–125; *Chem. Abstr.*, 55 (1961) 12,306.
(54) M. Dejter-Juszynski and H. M. Flowers, *Carbohydr. Res.*, 28 (1973) 144–146; for the underlying principle involved, see R. S. Tipson, *J. Biol. Chem.*, 125 (1938) 341–344.
(55) K. E. Pfitzner and J. G. Moffatt, *J. Am. Chem. Soc.*, 87 (1965) 5670–5678.

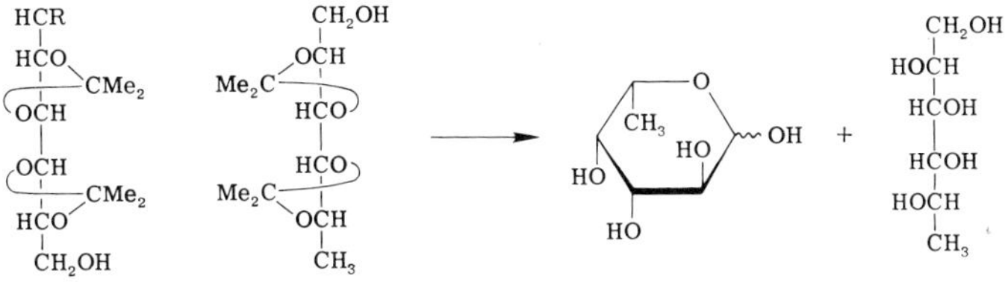

CHEMISTRY AND BIOCHEMISTRY OF D- AND L-FUCOSE

[Scheme 3 structures:]

15 R = (SEt)$_2$
16 R = H$_2$
17 R = H, OTs

Scheme 3

L-Fucitol has also been converted[56] into **18** by oxidation of **16** with potassium permanganate to the substituted fuconate, which was reduced (after deacetonation and conversion into the lactone) with sodium amalgam to syrupy **18**, characterized as its diphenylhydrazone.

3. Analytical Detection and Determination

a. General Methods for Deoxy Sugars.

—(i) Fucose is determined colorimetrically by a general method[57] for 6-deoxyhexoses that involves the chromophores produced by reaction of reagents with the 5-methyl-2-furaldehyde chromogen formed by treatment of the free sugar or its glycosides with sulfuric acid. The two commonest reagents used are (**1**) cysteine, with reading at 380 and 412–414 nm to correct for contaminating hexoses; and (**2**) mercaptoacetic acid, with reading at 400 and 430 nm.

The yellow color that develops serves also as a qualitative test for the presence of 6-deoxyhexoses, including fucose, in samples.

(ii) On oxidation by periodate, free 6-deoxyhexoses give acetaldehyde, and this can be absorbed in sodium hydrogensulfite solution, and determined by means of the color developed with *p*-phenylphenol.[58]

b. Specific Enzymic Methods.

—(i) A very sensitive, specific method for detection and estimation of free L-fucose is based on its oxidation[59] by L-fucose dehydrogenase (EC 1.1.1.122) in the presence of NAD$^+$. The formation of NADH is monitored spectrophotometrically at 340 nm.

(56) S. Akiya and S. Suzuki, *Yakugaku Zasshi*, 74 (1954) 1296–1298.
(57) Z. Dische, *Methods Carbohydr. Chem.*, 1 (1962) 501–503.
(58) M. N. Gibbons, *Analyst (London)*, 80 (1955) 268–276.
(59) C. Tsay and G. Dawson, *Anal. Biochem.*, 78 (1977) 423–427.

(ii) A less specific enzyme, D-galactose dehydrogenase, (EC 1.1.1.48) will catalyze[60] the NAD⁺-dependent oxidation of D-galactose, L-arabinose, or D-fucose to the corresponding aldonic acid, with formation of NADH.

c. Other Methods.—The introduction of chromatographic methods,[60a] for example, gas–liquid chromatography (g.l.c.), g.l.c.–mass spectrometry (m.s.), high-performance liquid chromatography (h.p.l.c.), and thin-layer chromatography (t.l.c.) enables identification of derivatives of fucose, and there are methods for determination of group and anomeric configurations. The n.m.r. spectra of numerous derivatives have been analyzed, but, as these determinations have already been treated extensively (see, for example, Refs. 10, 61–63), they will not be discussed here.

4. Substituents and Protecting Groups

The hydroxyl groups in fucose undergo the usual reactions, leading to a considerable variety of acetals, esters, ethers, and glycosides. Tables of properties of derivatives of fucose are appended that indicate some physical contents, and provide literature references for details of relevant reactions. Some rather unexpected findings on the relative reactivities of hydroxyl groups have now been reported, and these will be discussed. In addition, considerable progress has been made during the past few years in the synthesis of oligosaccharides of L-fucose, and these reactions will be discussed.

5. Relative Reactivities of Hydroxyl Groups

a. Conformational Effects: The Lower Reactivity of Axial OH-4.—Selective dibenzoylation[64] of methyl α-L-fucopyranoside (**20**) with benzoyl chloride–pyridine at a low temperature gave the amorphous 2,3-dibenzoate in high yield (80–85%); methyl 3-O-benzoyl-α-L-fucopyranoside was also isolated, in 6% yield. [Similar treatment of methyl 6-deoxy-α-L-mannopyranoside (α-L-rhamnopyranoside) gave the corresponding 2,3-dibenzoate in 50% yield, although, in this compound,

---

(60) A. S. L. Hu and S. Grant, *Anal. Biochem.*, 25 (1968) 221–227.
(60a) H. Rauvala, J. Finne, T. Krusius, J. Kärkkäinen, and J. Järnefelt, *Adv. Carbohydr. Chem. Biochem.*, 38 (1981) 389–416.
(61) J. Lönngren and S. Svensson, *Adv. Carbohydr. Chem. Biochem.*, 29 (1974) 41–106.
(62) J. J. Marshall, *Adv. Carbohydr. Chem. Biochem.*, 30 (1974) 257–370.
(63) B. Lindberg, J. Lönngren, and S. Svensson, *Adv. Carbohydr. Chem. Biochem.*, 31 (1975) 185–240.
(64) A. C. Richardson and J. M. Williams, *Tetrahedron*, 23 (1967) 1641–1646.

OH-4 is equatorial.] The same authors obtained similar effects on selective tribenzoylation of methyl α-D-galactopyranoside[40]; that is, they observed lower reactivity of OH-4 (ax).

Partial methylation[65] of **20** with methyl iodide–silver oxide gave the 2,3-, 2,4-, and 3,4-dimethyl ethers of **20** in the ratios of 11:4.5:1, indicating that OH-2 is the most reactive, followed by OH-3. (However, direct observation of product ratios resulting from disubstitutions do not give true estimations of the relative reactivities of hydroxyl groups.[37]) Analysis of the monomethyl ether fraction produced in this reaction showed a high proportion of methyl 2-O-methyl-α-L-fucopyranoside (43.5%) and the ratio of 2- to 3-methyl ether was almost 5:1. No 4-monomethyl ether of **20** was detected, again indicating the much greater reactivity of OH-2 in **20**. In methyl β-L-fucofuranoside also, OH-2 is by far the most reactive hydroxyl group.

b. Apparently Enhanced Reactivity of OH-4.—Partial benzylation,[66] with α-chlorotoluene (benzyl chloride)–potassium hydroxide in 1,4-dioxane–toluene, of methyl 2-O-benzyl-α-L-fucopyranoside (**22**), obtained from **20** by way of the syrupy 3,4-isopropylidene acetal **21**, afforded the 2,3- (**23**) and 2,4- (**24**) dibenzyl ethers in equimolar. ratio, indicating similar reactivities for OH-3 and OH-4 in **22** (see Scheme 4). However, dibenzylation of **20** gave **24** and methyl 3,4-di-O-benzyl-α-L-fucopyranoside (**25**) in the ratio of 3:2, with no evidence for the presence of any of the 2,3-di-ether (**23**). Apparently, OH-4 in compound **22** is no less reactive than OH-3, but it may be more

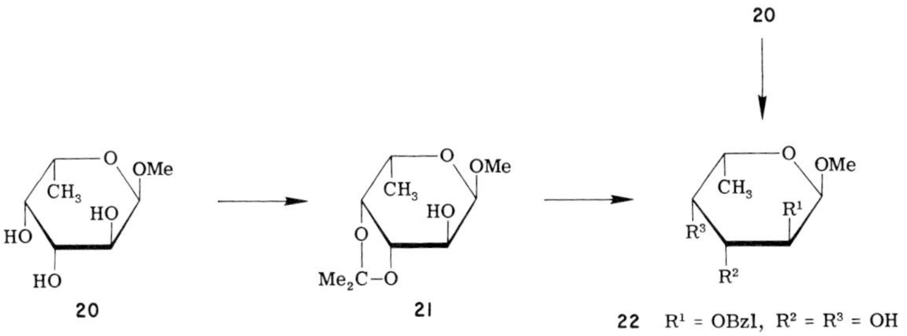

20

22	R^1 = OBzl, R^2 = R^3 = OH
23	R^1 = R^2 = OBzl, R^3 = OH
24	R^1 = R^3 = OBzl, R^2 = OH
25	R^1 = OH, R^2 = R^3 = OBzl
	Bzl = $PhCH_2$

Scheme 4

(65) J. G. Gardiner and E. E. Percival, *J. Chem. Soc.*, (1958) 1414–1418.
(66) M. Dejter-Juszynski and H. M. Flowers, *Carbohydr. Res.*, 28 (1973) 61–74.

reactive in compound **20**, in which its reactivity is also high compared with that of OH-2.

Confirmation of these results was provided by a different laboratory[67] in studies with allyl α-L-fucopyranoside. Partial benzylation of this compound, under conditions similar to those described previously,[66] also gave the 2,4- and 3,4-dibenzyl ethers in the ratio of 3:2. In both publications, the structures assigned were primarily confirmed by methylation of unreacted hydroxyl groups with methyl iodide–sodium hydride in dimethyl sulfoxide,[68] followed by catalytic debenzylation, and isolation and characterization of the known methyl ethers.

A similar reactivity of OH-4, greater than might have been expected from conformational factors, was exhibited[69] in the selective benzylation of methyl 2,6-di-O-benzyl-α(and β)-D-galactopyranoside with benzyl bromide–sodium hydride in N,N-dimethylformamide. Much more 2,4,6- than 2,3,6-tribenzyl ether was formed, with rather higher selectivity being exhibited by the α- than the β-glycoside. However, the primary hydroxyl group (OH-6) of this compound was more reactive than OH-4.

c. Tritylation.—Treatment of methyl α-D-fucopyranoside with 2 molar proportions of chlorotriphenylmethane in pyridine at room temperature gave[70] ~70% of monotrityl ethers, which were separated into the 2- and 3-trityl ether (3:2 ratio). There was no evidence for substitution at OH-4.

Resolution of these conflicting results of selective-substitution reactions requires assumption of different reaction-mechanisms. Acylation and tritylation reactions involve attack by electrophilic reagents in a weak base on nucleophilic hydroxyl groups. In these examples, the relative reactivities of free hydroxyl groups are, presumably, a function of their relative nucleophilicities, and conformational effects are important, especially when bulky substituents are to be introduced. However, in reactions involving strong bases, carbohydrate anions will be formed as the reactive species. The relative acidities of the different hydroxyl groups and the reactivities of the carbohydrate anions, including conformational factors, will then determine the pattern of early substitution. After monosubstitution has occurred, other factors will also intervene,[37] as vicinal hydroxyl groups will be affected by the substituent groups.

(67) P. J. Garegg and T. Norberg, *Carbohydr. Res.*, 52 (1976) 235–240.
(68) S.-I. Hakomori, *J. Biochem. (Tokyo)*, 55 (1964) 205–208.
(69) H. M. Flowers, *Carbohydr. Res.*, 31 (1975) 245–251.
(70) T. Otake and T. Sonobe, *Bull. Chem. Soc. Jpn.*, 49 (1976) 1050–1054.

6. Glycoside Formation by Acid Equilibration

a. Mineral Acids.—Treatment of fucose with boiling, methanolic hydrogen chloride leads to an equilibrium mixture from which methyl α-D- or α-L-fucopyranoside can be isolated, depending upon the enantiomer used initially.[71,72] Treatment of L-fucose with methanolic hydrogen chloride (0.8%) for 66 h at room temperature led[65] to attainment of equilibrium, as shown by no further change in optical rotation. Syrupy methyl β-L-fucofuranoside (44%), and crystalline methyl α-L-fucofuranoside (21%), α-L-fucopyranoside (10%), and β-L-fucopyranoside (20%) were isolated after separation of the mixture of products on a column of cellulose. At room temperature, the equilibrium mixture obtained is rich in furanosides which, at higher temperatures, are converted into pyranosides with favored, acid-catalyzed anomerization to α-pyranosides.

b. Ion-exchange Resins.—A convenient method for the preparation of methyl α-L-fucopyranoside utilizes strongly acidic, ion-exchange resins in boiling methanol.[73] After removal of the resin by filtration, and evaporation of the solvent, the desired product may be obtained pure, in ~40% yield, as a crystalline product from ethanol. Recycling of the mother liquors permits isolation of additional batches of product, and total yields of the order of 90% have been obtained by repeating, three or four times,[74] the equilibration with resin and the isolation.

The kinetics of the reaction have been monitored[35a] polarimetrically, and by separation of the products, after various intervals, by g.l.c. of their per(trimethylsilyl) ethers. L-Fucose dissolved completely in boiling methanol within 30 min, even in the absence of the resin, and the initial solution contained ~5% of L-fucofuranoses, 38% of α-L-fucopyranose, and 57% of β-L-fucopyranose. Immediately upon adding the resin, glycoside formation occurred. After 1 min, all four isomers appeared, in approximately equimolar ratios and the furanoside content attained a maximum after 15 min (18% of α- and 34% of β-furanoside, with 23% of α- and 21% of β-pyranoside). The final equilibrium mixture (after 12 h at the b.p., with the resin) contained ~6% of α- and 13% of β-fucofuranoside, and 54% of α- and 27% of β-fucopyranoside.

(71) R. C. Hockett, F. P. Phelps, and C. S. Hudson, *J. Am. Chem. Soc.*, 61 (1939) 1658–1660.
(72) H. B. MacPhillamy and R. C. Elderfield, *J. Org. Chem.*, 4 (1939) 150–161.
(73) U. Zehavi and N. Sharon, *J. Org. Chem.*, 37 (1972) 2141–2145.
(74) H. M. Flowers, unpublished data.

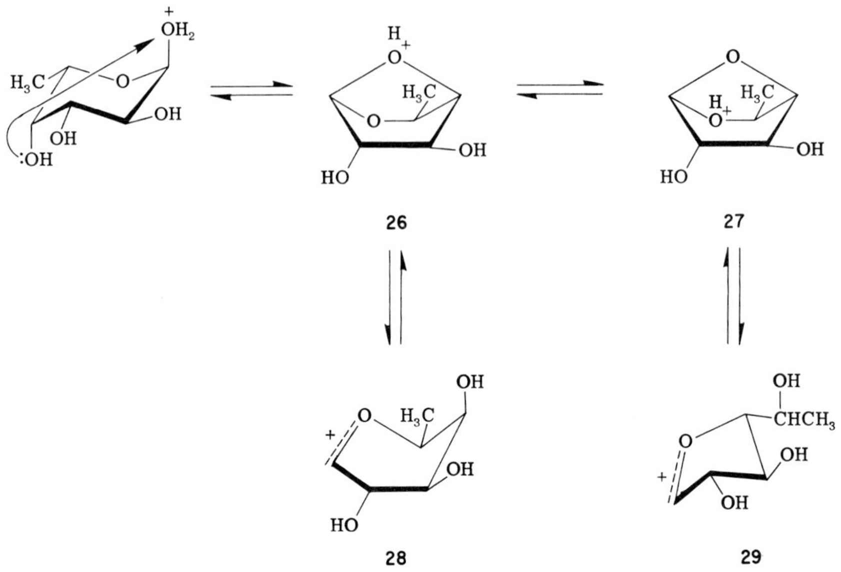

Scheme 5

The ratios of the products isolated supported a mechanism involving initial formation, from the free sugar, of a bicyclic cation (**26**), protonated on the ring-oxygen atom attached to C-4, which was in equilibrium with the bicyclic cation **27**, protonated on the ring-oxygen atom attached to C-5 (see Scheme 5). These species were rapidly converted into pyranose (**28**), or furanose (**29**), carbonium ions, which were the intermediates reacting directly with methanol to afford pyranosides or furanosides, respectively, by S_N1 processes. This behavior contrasts with that of D-galactose, for which the β-D-furanoside and α-D-pyranoside are initially formed.[35b] It was postulated that, in this case, the proposed bicyclic intermediates could be stabilized by double hydrogen-bonding between the 6-hydroxyl group and the two ring-oxygen atoms; this would then be the species attacked by methanol without prior formation of carbonium ions, leading to steric control, and formation of the products initially observed. Of course, as the reaction proceeds, equilibration occurs, so that the final product-ratio reflects neither the initial composition nor the mechanism of formation of initial products.

7. Synthesis of Complex Glycosides

a. General.—Synthesis of di- and oligo-saccharides, and glycosides of acid-labile alcohols, as well as glycosides of a particular stereochemistry (β-L or β-D) cannot be satisfactorily achieved by equilibration reactions between O-nucleophiles and free fucose, but necessitate, instead, the preparation from fucose of intermediate, reactive electrophiles that are suitably activated at C-1 and protected elsewhere in the molecule.[75] Protected fucopyranosyl halides have generally been used, although orthoesters have also been employed with some success. Although naturally occurring oligosaccharide derivatives of D-fucose have been characterized, interest in synthetics has been concentrated on the L enantiomer, from which a considerable number of di- and oligo-saccharides of biological importance have been prepared.

Because many of the reactive intermediates needed for these condensation reactions are derived from peracetates of fucose, it is of interest first to examine methods for their preparation.

b. **Tetra-O-acetylfucopyranoses.**—Crystalline α- and β-tetraacetates of both D- and L-fucopyranose have been prepared.[76–79] Difficulties were encountered in obtaining the crystalline β-L-tetraacetate originally described[79] as being obtained by heating L-fucose with acetic anhydride–sodium acetate, and an indirect method was employed instead.[80] Alternatively, treatment of the free hexose with acetic anhydride–pyridine afforded a mixture of the α- and β-tetraacetates, from which the α anomer could be readily separated and crystallized.[77]

c. **Tri-O-acetyl-α-L-fucopyranosyl Chloride (30).**—A variety of methods has been utilized[77] to prepare this halide, including treatment of crystalline tetra-O-acetyl-β-L-fucopyranose with acetyl chloride–acetic anhydride, or a mixture of the α- and β-tetraacetates with titanium tetrachloride, or hydrogen chloride in acetic anhydride–acetic acid. It was found unnecessary to isolate and purify the tetraacetate before conversion into the chloride, and acetylation of L-fucose with acetic anhydride, in the presence of either sodium acetate or zinc chloride, followed by saturation of the solution with hydrogen chloride, led to formation of the desired product, which could be obtained crystalline (if necessary, after purification by chromatography on a column of silica gel).

d. **Tri-O-acetyl-α-D(and L)-fucopyranosyl Bromide.**—Direct conversion[76] of D- and L-fucose, or reaction of the crude β-L-tetraacetate

(75) H. M. Flowers, *Methods Enzymol.*, 50 (1978) 93–121.
(76) G. A. Levy and A. McAllan, *Biochem. J.*, 80 (1961) 433–435.
(77) D. H. Leaback, E. C. Heath, and S. Roseman, *Biochemistry*, 8 (1969) 1351–1359.
(78) M. L. Wolfrom and J. A. Orsino, *J. Am. Chem. Soc.*, 56 (1934) 985–987.
(79) O. Westphal and H. Feier, *Chem. Ber.*, 89 (1956) 582–588.
(80) H. S. Prihar and E. J. Behrman, *Biochemistry*, 12 (1973) 997–1002.

with hydrogen bromide–acetic acid,[81] provided the corresponding tri-O-acetyl-α-D(or L)-fucopyranosyl bromide, which was not crystallized. However, the L enantiomer (**31**) has now been crystallized,[82] and converted into the syrupy 1,2-(ethoxyethylidene) acetal (by treatment with s-collidine in dry ethanol) for use as a reagent in the stereospecific synthesis of β-L-fucopyranosyl phosphate.

e. Benzylated Fucopyranosyl Halides.—Methyl α-L-fucopyranoside was converted[83] into crystalline 2,3,4-tri-O-benzyl-1-O-(p-nitrobenzoyl)-β-L-fucopyranose by benzylation, followed by partial hydrolysis with acid, and p-nitrobenzoylation in pyridine of the 2,3,4-tri-O-benzyl-L-fucose in the hydrolyzate. Treatment of the product with hydrogen bromide in dichloromethane led to precipitation of p-nitrobenzoic acid and formation of 2,3,4-tri-O-benzyl-α-L-fucopyranosyl bromide (**32**), which could not be crystallized.

Other syrupy, benzylated bromides bearing O-acetyl or O-p-nitrobenzoyl groups were also prepared and characterized,[66,84] as well as a partially benzylated chloride.[85] Acid hydrolysis of **22** gave crystalline 2-O-benzyl-α-L-fucose, which was acylated with p-nitrobenzoyl chloride–pyridine. Treatment[84] of the resulting tris(p-nitrobenzoate) with hydrogen bromide–dichloromethane led to precipitation of p-nitrobenzoic acid and formation of syrupy 2-O-benzyl-3,4-di-O-(p-nitrobenzoyl)-α-L-fucopyranosyl bromide (**33**).

A very useful intermediate in the synthesis of oligosaccharides was derived[86] from crystalline tri-O-benzyl-α-L-fucopyranosyl chloride

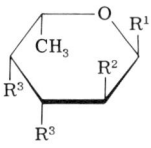

30 R^1 = Cl, R^2 = R^3 = OAc
31 R^1 = Br, R^2 = R^3 = OAc
32 R^1 = Br, R^2 = R^3 = OBzl
33 R^1 = Br, R^2 = OBzl,
    R^3 = p-$O_2NC_6H_4CO_2$
34 R^1 = Cl, R^2 = R^3 = OBzl

(81) H. M. Flowers, A. Levy, and N. Sharon, *Carbohydr. Res.*, 4 (1967) 189–195.
(82) J.-H. Tsai and E. J. Behrman, *Carbohydr. Res.*, 64 (1978) 297–301.
(83) M. Dejter-Juszynski and H. M. Flowers, *Carbohydr. Res.*, 18 (1971) 219–226.
(84) M. Dejter-Juszynski and H. M. Flowers, *Carbohydr. Res.*, 23 (1972) 41–45.
(85) E. M. Acton, K. J. Ryan, and A. E. Luetzow, *J. Med. Chem.*, 20 (1977) 1362–1371.
(86) J. R. Pougny, J.-C. Jacquinet, M. Nassr, D. Duchet, M.-L. Milat, and P. Sinaÿ, *J. Am. Chem. Soc.*, 99 (1977) 6762–6763.

(**34**), prepared by the reaction of tri-*O*-benzyl-α-L-fucopyranose with thionyl chloride. On treatment of **34** with a mixture of *N*-methylacetamide, silver oxide, *N,N*-diisopropylethylamine, and powdered molecular sieves, it was converted into 2,3,4-tri-*O*-benzyl-1-*O*-(*N*-methylacetimidyl)-β-L-fucopyranose (**35**), which was isolated crystalline in excellent yield.

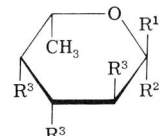

35   R = OBzl

f. 2,3,4-Tri-*O*-(chlorosulfonyl)-α- and β-L-fucopyranosyl Chloride.
—Reaction of L-fucose with sulfuryl chloride in pyridine at low temperature gave mainly the β-chloride (**37**), which was isolated as a crystalline solid[87,88] in ~50% yield. (There is a conflict between the authors of Refs. 87 and 88 as to the m.p. of the product.) At the same time,[88] a small amount of the α-chloride (**36**) was also obtained as a crystalline solid.

36 R^1 = Cl, R^2 = H,
         R^3 = OSO$_2$Cl
37 R^1 = H, R^2 = Cl,
         R^3 = OSO$_2$Cl

g. Fucopyranosides by Condensation Between Activated Fucopyranosyl Derivatives and Nucleophiles.—(i) The preparation of simple glycosides by reaction of peracetylated halides of D- and L-fucose with alcohols or phenols has been described.[76] D-Fucopyranosyl di- and oligo-saccharides have not been prepared from such halides. However, a great deal of effort has been devoted to the synthesis of complex L-fucopyranosides.

(87) M.-E. Rafestin, D. Delay, and M. Monsigny, *Can. J. Chem.*, 52 (1974) 210–212.
(88) J.-R. Pougny, P. Sinaÿ, and G. Hajduković, *Carbohydr. Res.*, 34 (1974) 351–360.

The first description of synthetic disaccharides of L-fucose appeared in 1967, when 2-O-α-L-fucopyranosyl-L-fucose[81] and -D-galactose[89] were obtained stereospecifically by reaction between syrupy **31** and suitably protected sugar nucleophiles. The configuration of the products was rather unexpected at the time, as the postulated mechanism of the Koenigs–Knorr reaction implied participation between the acyl group on O-2 and the intermediate carbonium ion at C-1, sterically controlling attack of the nucleophile and leading to formation of the *trans*-1,2- (that is, the β-L-)fucopyranoside (see the discussion in Ref. 75). Later attempts[90,91] to utilize the same bromide for the preparation of other disaccharides led to formation of the expected β-L-linked anomers, and it would appear that the structure of the nucleophile also has some effect on the stereochemistry of the product of this condensation reaction.[92]

As L-fucose is in the pyranosyl form and the α configuration in almost all, naturally occurring, complex carbohydrates, there is a challenge to synthesize oligosaccharides of this type. A variety of approaches has been developed in different laboratories in order to achieve this aim, and to attain reactions having high degrees of stereospecificity.

(ii) α-L-Fucopyranosides. The fundamental approach was to eliminate possible participation of the 2-substituent in the halide. If an S_N2 type of reaction occurs with the nucleophile, use of a β-L halide would provide an α-L disaccharide, whereas S_N1 reactions would be expected to yield mixtures of anomers, with the possibility of affecting their ratios by suitable control of the reaction conditions and reagents.[75]

Halide **32** was used first, in nitromethane–benzene solution in the presence of an equimolar proportion of mercuric cyanide. It was found[83] that **32** gave a disaccharide at O-6 of a protected derivative of 2-acetamido-2-deoxy-D-glucose in good yield, with a certain degree of stereoselectivity (α:β ratio 7:3). Complete stereospecificity was attained, leading to the desired α-linked disaccharides, when the partially acylated bromide **33** was condensed with a number of different nucleophiles.[66,83,84,93–95] A mechanism was postulated[83] to explain the

(89) A. Levy, H. M. Flowers, and N. Sharon, *Carbohydr. Res.*, 4 (1967) 305–311.
(90) E. S. Rachaman and R. W. Jeanloz, *Carbohydr. Res.*, 10 (1968) 429–434.
(91) M. Dejter-Juszynski and H. M. Flowers, *Carbohydr. Res.*, 37 (1974) 75–79.
(92) H. M. Flowers, *Carbohydr. Res.*, 74 (1979) 177–185.
(93) M. Dejter-Juszynski and H. M. Flowers, *Carbohydr. Res.*, 30 (1973) 287–292.
(94) M. Dejter-Juszynski and H. M. Flowers, *Carbohydr. Res.*, 44 (1975) 308–312.
(95) K. L. Matta, E. A. Z. Johnson, and J. J. Barlow, *Carbohydr. Res.*, 32 (1974) 396–399, 418–422.

steric control obtained in these reactions involving participation between the 4-O-acyl group and C-1.

Halide 33 was also employed[96] for synthesizing α-linked trisaccharides of L-fucose by reaction with protected nucleophiles that gave β-linked anomers with compound 31.

(iii) The fully benzylated bromide 32 could also be utilized to give good steric control under suitable conditions. It was argued[97] that perbenzylated hexopyranosyl bromides should be ideal halides for preparing α-linked disaccharides. Addition of an excess of the halide ion should lead to formation, *in situ*, of the more-reactive β-glycopyranosyl halide, which might be considered to react by way of an ion pair to give an α-glycoside preferentially (see also Ref. 75). (This approach might be compared to the silver perchlorate-catalyzed, glycosidation reactions, which give stereoselective α-disaccharide formation.[98]) This type of reaction ("halide ion-catalyzed glycosidation") was employed in the successful synthesis of a number of disaccharides, including α-L-Fuc-(1→3)-D-Glc,[97,99] α-L-Fuc-(1→2)-D-Gal,[100] and α-L-Fuc-(1→4)-2-acetamido-2-deoxy-D-glucose.[101] The general procedure employed was to stir a mixture of 32, tetraethylammonium bromide, N,N-diisopropylethylamine, molecular sieves, and the protected nucleophile in dichloromethane or dichloroethane for several days at room temperature. The ratios of the reactants were varied in different preparations, from approximately equivalent amounts of 32 and the nucleophile to a two, or three, times excess of 32; the yields were often very good, and the stereospecificity was high.

Use of these complex, saccharide nucleophiles permitted the synthesis of tri- and tetra-saccharides, and several blood-group ABH antigenic determinants were prepared in this way; for example, α-L-Fuc-(1→2)-β-Gal-(1→3)-R (Ref. 102), α-L-Fuc-(1→3)-[β-Gal-(1→4)]-R (Ref.

(96) H. M. Flowers, unpublished data.
(97) R. U. Lemieux, K. B. Hendricks, R. V. Stick, and K. James, *J. Am. Chem. Soc.*, 97 (1975) 4056–4062.
(98) K. Igarashi, J. Irasawa, and T. Honma, *Carbohydr. Res.*, 39 (1975) 341–343.
(99) Abbreviations: Fuc = fucopyranosyl, Gal = D-galactopyranosyl, Glc = D-glucopyranosyl, Man = mannopyranosyl, Xyl = D-xylopyranosyl, GlcA = D-glucopyranosyluronic acid, GalA = D-galactopyranosyluronic acid, GlcNAc = 2-acetamido-2-deoxy-D-glucopyranosyl, and GalNAc = 2-acetamido-2-deoxy-D-galactopyranosyl.
(100) R. U. Lemieux and H. Driguez, *J. Am. Chem. Soc.*, 97 (1975) 4069–4075.
(101) J.-C. Jacquinet and P. Sinaÿ, *Carbohydr. Res.*, 42 (1975) 251–258.
(102) J.-C. Jacquinet and P. Sinaÿ, *Tetrahedron*, 32 (1976) 1693–1697.

103), α-L-Fuc-(1→4)-[β-Gal-(1→3)]-R (Refs. 104 and 105), α-L-Fuc-(1→2)-[α-Gal-(1→3)]-D-Gal,[100] α-L-Fuc-(1→2)-β-Gal-(1→4)-[α-L-Fuc-(1→3)]-R (Ref. 106), α-GalNAc-(1→3)-[α-L-Fuc-(1→2)]-β-Gal-(1→3)-R (Ref. 107), and α-Gal-(1→3)-[α-L-Fuc-(1→2)]-β-Gal-(1→3)-R (Ref. 107), where R = 2-acetamido-2-deoxy-D-glucose.

In all of the syntheses of tri- and tetra-saccharides thus far described, emanating from several different laboratories, fucosylation was performed on a suitably protected nucleophile. A bromide of a fucose-containing disaccharide was not prepared, although use of such a compound would be feasible in principle, as was shown in one case of a synthetic glycoside (see later).

(iv) The introduction of the imidate procedure has permitted replacement of the syrupy α-bromide **32** by the crystalline β-imidate (**35**), which was found to react satisfactorily with nucleophiles in the presence of p-toluenesulfonic acid, to give good yields of oligosaccharides, including[108] the B-antigenic trisaccharide α-L-Fuc-(1→2)-[α-Gal-(1→3)]-D-Gal.

(v) **Other Reagents.** Several glycosides have been prepared from the chlorosulfonyl chloride **37**, including p-nitrobenzyl α-L-fucopyranoside[87] and methyl α- and β-L-fucopyranoside.[88] In the methanolysis of **37**, in chloroform solution, the stereochemistry of the condensation was somewhat affected by the reaction conditions. On employing silver carbonate and calcium sulfate in excess, the reaction was complete within 16 h, and the ratio of the methyl α- to β-L-fucopyranosides obtained was ~9:1. De(chlorosulfonyl)ation was attained by treatment with a weakly basic, anion-exchange resin. If, however, the reaction was performed in the presence of silver carbonate plus a catalytic amount of iodine, glycosidation was complete within 15 min and was accompanied by hydrolytic removal of the chlorosulfonyl groups, providing the methyl L-fucopyranosides directly (α:β ratio ~6:1). The α-L chloride **36** was less reactive than the β-L, and, after 4 days, the methyl L-fucopyranosides were obtained (α:β ratio ~1:6). Starting from either chloride (**36** or **37**), the yields were excellent.

(103) J.-C. Jacquinet and P. Sinaÿ, J. Chem. Soc. Perkin Trans. 1, (1979) 314–318.
(104) R. U. Lemieux and H. Driguez, J. Am. Chem. Soc., 97 (1975) 4063–4068.
(105) J.-C. Jacquinet and P. Sinaÿ, J. Chem. Soc. Perkin Trans. 1, (1979) 319–322.
(106) J.-C. Jacquinet and P. Sinaÿ, J. Org. Chem., 42 (1977) 720–724.
(107) H. Paulsen and C. Koláv̌, Chem. Ber., 112 (1979) 3190–3202.
(108) J.-C. Jacquinet and P. Sinaÿ, Tetrahedron, 35 (1979) 365–371.

III. FUCOSE IN GLYCANS

1. General Occurrence

D-Fucose is usually found, in relatively simple glycosides comprising only a few sugar units, in plants and some micro-organisms. The L enantiomer is, however, much more diverse in the nature of its chemical bonding. It may occur in such homopolymers as fucoidans containing sulfated α-L-fucopyranosyl residues or groups involved in (1→2)-, (1→3)-, or (1→4)-linkages. It may occur in bacterial or polysaccharide structures, where it is frequently located at nonreducing ends of chains, but is at times found inside oligosaccharide chains in the form of Man-(1→3)- (Refs. 109 and 110), GlcA-(1→4)- (Refs. 111 and 112), Xyl-(1→3)- (Ref. 113), GlcA-(1→3)- (Refs. 113 and 114), and GalA-(1→2)-L-Fuc (Ref. 115) entities. In M-antigen from strains of *Salmonella typhimurium, Escherichia coli,* and *Aerobacter cloacae,* a hexasaccharide repeating-unit is found, part of which is a -(1→3)-β-L-Fuc-(1→4)-α-L-Fuc-(1→3)-β-Glc portion.[116] In animal glycoproteins, L-fucose has so far only been identified at nonreducing termini; this situation had also applied to glycolipids, but internally situated L-fucosyl residues have now been identified in certain glycolipids obtained from aquatic invertebrates,[117] in which the 2-acetamido-2-deoxy-D-glucosyl-L-fucose unit probably occurs.

In glycoproteins, L-fucose and sialic acid are often mutually exclusive at some sites of substitution, and both sugars occur at nonreducing termini.[118] In fact, they have been assumed to act as "chain-stoppers" in the biosynthetic process, controlling the extent of chain elongation. Nevertheless, from the jelly coat of the eggs of the sea ur-

---

(109) T. Makamoto and H. Yamaguchi, *Carbohydr. Res.*, 59 (1977) 614–621.
(110) S. V. Tomshich, R. P. Gershkova, Yu. N. El'kin, and Yu. S. Ovodov, *Eur. J. Biochem.*, 65 (1976) 193–199.
(111) G. O. Aspinall and K. S. Jiang, *Carbohydr. Res.*, 38 (1975) 247–255.
(112) C. Erbing, L. Kenne, B. Lindberg, J. Lönngren, and I. W. Sutherland, *Carbohydr. Res.*, 50 (1976) 115–120.
(113) E. (E.) Percival, *Carbohydr. Res.*, 17 (1971) 121–126.
(114) K. Jann, B. Jann, and K. F. Schneider, *Eur. J. Biochem.*, 5 (1968) 456–465.
(115) K. Jann, B. Jann, F. Ørskov, I. Ørskov, and O. Westphal, *Biochem. Z.*, 342 (1965) 1–22.
(116) P. J. Garegg, B. Lindberg, T. Onn, and I. W. Sutherland, *Acta Chem. Scand.*, 25 (1971) 2103–2108.
(117) T. Hori, O. Hasaka, and M. Sugita, *Proc. Int. Symp. Glycoconjugates, 5th,* Thieme, Stuttgart, 1979, pp. 46–47.
(118) Z. Dische, *Ann. N. Y. Acad. Sci.*, 106 (1963) 259.

chin *Pseudocentrotus depressus* was isolated[119] a disaccharide that was formulated as Fuc-(1→4)-N-glycolylneuraminic acid [Fuc-(1→4)-NeuGc]. It occurs as part of a glycoprotein, but its position in the saccharide chain is not yet known.

2. Glycosides Containing D-Fucose

Most D-fucopyranosides occur in the cardiac glycosides in plants as glycoside or oligosaccharide derivatives of steroids.[2] A considerable number of these glycosides has been isolated, and their structures have been determined. Steroid glycosides of D-fucose, 3-O-methyl-D-fucose (D-digitalose), and 2,3-di-O-methyl-D-fucose, have been identified, including ledienoside,[19] panstroside,[120] digipronin,[121] strospeside,[122] strebloside,[123] and kamaloside[123]; disaccharide derivatives include that from *Digitalinum verum*[122] and gracibioside[124] [β-Glc-(1→4)-(3-O-Me)-β-D-Fuc], cheiroside A (Ref. 20) [β-Glc-(1→4)-β-D-Fuc], and the antibiotic chartreusin[18] (a disaccharide of D-fucose and D-digitalose). A number of trisaccharides have also been identified, including odorotrioside G (Ref. 125) [β-Glc-(1→6)-β-Glc-(1→4)-(3-O-Me)-β-D-Fuc], gitostin[126] [β-Glc-(1→4)-β-Glc-(1→4)-(3-O-Me)-β-D-Fuc], and neogitostin[126] [β-Glc-(1→6)-β-Glc-(1→4)-(3-O-Me)-β-D-Fuc]. The 4-methyl and 2,3-dimethyl ethers[21,22] of D-fucose, which have been isolated from hydrolyzates of antibiotic substances (for example, from *Steptomyces* spp.), are also presumably present as glycosides or oligosaccharides, but their complete structures have not yet been reported.

3. Polysaccharides

The presence of fucose or of methyl ethers of fucose has been reported in many polysaccharides, but the configuration of the sugar has not always been established, as identification has often been based solely on chromatographic data. Great advances have been made with the introduction of new techniques for the elucidation of structures of

---

(119) K. Hotta, M. Kurokawa, and S. Isaka, *J. Biol. Chem.*, 248 (1973) 629–631.
(120) J. P. Rosselet, A. Hunger, and T. Reichstein, *Helv. Chim. Acta*, 34 (1951) 2143–2147.
(121) R. Tschesche, G. Lipp, and G. Grimmer, *Ann.*, 606 (1957) 160–166.
(122) W. Rittel, A. Hunger, and T. Reichstein, *Helv. Chim. Acta*, 35 (1952) 434–441.
(123) M. P. Khare, O. Schindler, and T. Reichstein, *Helv. Chim. Acta*, 45 (1962) 1534–1546.
(124) J. P. Rosselet and T. Reichstein, *Helv. Chim. Acta*, 36 (1953) 787–801.
(125) A. Rheiner, A. Hunger, and T. Reichstein, *Helv. Chim. Acta*, 35 (1952) 687–716.
(126) A. Okano, *Chem. Pharm. Bull.*, 6 (1958) 178–182.

oligo- and poly-saccharides.[10,63,127] The plant polysaccharides are often highly branched structures having L-fucopyranosyl units either at chain ends or within the chains, and, apparently, D-fucosyl units are not found in them. It also appears that the α-anomeric configuration of L-fucose is the only one found in plants. In some cases, methylated fucopyranosyl units occur; for example, 2-O-methyl-L-fucose has been identified in polysaccharides from Victoria-plum leaf[128] and sugar-beet pectin.[129] Roots of grasses secrete hydrophilic polysaccharides that may serve as protective lubricants, and these exuded glycans may be rich in fucose.[130,131]

Seaweed polysaccharides form especially complex structures[132,133] with residues of D-xylose, L-fucose, and hexuronic acids, and with sulfate groups, and, apparently, covalently bound protein. The results of methylation studies[133] indicated that doubly and, possibly, triply substituted fucopyranosyl residues were present.

β-GlcA-(1→4)-L-Fuc was isolated from partial, acid hydrolyzates of soybean polysaccharide,[134] the pectin of leaves of Tussilago,[135] and tragacanthic acid,[136] and α-L-Fuc-(1→2)-D-Xyl was obtained from soybean polysaccharide.[25]

Capsular antigens from various, Gram-negative bacteria contain L-fucose,[137,138] and complex, branched glycans from some fungi also contain[138] L-fucopyranosyl residues. The fucomannan from the fungus *Fomes annosus* was purified[139] by precipitation with H-agglutinin from eel serum, which reacts specifically with terminal, α-L-fucopyranosyl groups. From *Polyporus* spp. were isolated[140–142] fucogalac-

(127) G. O. Aspinall, in Ref. 15, pp. 201–222.
(128) P. Andrews and L. Hough, *J. Chem. Soc.*, (1958) 4476–4483.
(129) P. Andrews, L. Hough, D. B. Powell, and B. M. Woods, *J. Chem. Soc.*, (1959) 774–779.
(130) P. J. Harris and D. H. Northcote, *Biochem. J.*, 120 (1970) 479–491.
(131) K. Wright and D. H. Northcote, *Biochem. J.*, 139 (1974) 525–534.
(132) B. Larsen, *Acta Chem. Scand.*, 21 (1967) 1395–1396.
(133) A. J. Mian and E. (E.) Percival, *Carbohydr. Res.*, 26 (1973) 147–161.
(134) G. O. Aspinall, W. J. Cottrell, J. V. Egan, J. M. Morrison, and J. N. C. Whyte, *J. Chem. Soc., C*, (1967) 1071–1080.
(135) E. Haaland, *Acta Chem. Scand.*, 26 (1972) 2322–2328.
(136) G. O. Aspinall, D. B. Davies, and R. N. Fraser, *J. Chem. Soc., C*, (1967) 1086–1088.
(137) S. G. Wilkinson, in Ref. 8, pp. 97–175.
(138) B. Lindberg and S. Svensson, in Ref. 7, pp. 319–344.
(139) P. A. J. Gorin, J. F. T. Spencer, and A. J. Finlayson, *Carbohydr. Res.*, 16 (1971) 161–166.
(140) H. Björndal and B. Lindberg, *Carbohydr. Res.*, 10 (1969) 79–85.
(141) H. Björndal and B. Wågstrom, *Acta Chem. Scand.*, 23 (1969) 1560–1566, 3313–3320.
(142) K. Axelsson and H. Björndal, *Acta Chem. Scand.*, 23 (1969) 1815–1817.

tans having α-D-(1→6)-linked poly(galactopyranosyl) chains substituted by α-L-Fuc-(1→2) or α-Man-(1→3)-α-L-Fuc-(1→2) side chains.

4. Glycoproteins

Both O-and N-glycosyl chains of glycoproteins may contain L-fucose bound to chain ends of the carbohydrate.[16,143,144] The secreted, blood-group substances[15] provide a good example of the former type, and the latter type is exemplified by many other secreted[16] and membrane[145] glycoproteins.

In all glycoproteins that have thus far been studied, the L enantiomer is the fucose identified, and it has always been located at nonreducing positions in the α-anomeric configuration substituting (1→2) on Gal, or (1→3), (1→4), or (1→6) on GlcNAc. The external location of the fucopyranosyl groups makes them labile to the exo-L-fucosidases that are present in most higher organisms, and the afucoglycoproteins are usually good acceptors for fucosyltransferases (see later). Hence, there is a possibility that changes in the fucose content of glycoproteins may occur in pathological states. An extensive literature on this matter has appeared,[146] and attempts have been made to monitor these changes as a means of diagnosing pathological states, including cancer. In this respect, the incorporation of L-fucose into animal glycoproteins without its apparently undergoing transformation into other sugars has permitted controlled labelling of these materials (but, see Ref. 147).

It has been argued that the glycoprotein composition of the cell surface is a determinant in cell behavior,[148,149] and it is clear that many properties of cells may be affected by the action of glycosidases on cell-surface glycoproteins. In this respect, changes in their electrophoretic mobility, in the regulation of the cell cycle, and in cell–cell interactions, and the effects on lymphocyte stimulation of cells, their immunogenicity, their interaction with hormones, and their onco-

---

(143) R. Montgomery, in Ref. 7, pp. 213–249.
(144) R. G. Spiro, *Adv. Protein Chem.*, 27 (1974) 349–467.
(145) R. C. Hughes, *Membrane Glycoproteins*, Butterworth, London, 1976.
(146) L. Warren, C. A. Buck, and G. P. Tuszynski, *Biochim. Biophys. Acta*, 516 (1978) 97–127.
(147) J. Y. Chan, N. A. Nurokoro, and H. Schachter, *J. Biol. Chem.*, 250 (1975) 6185–6190.
(148) C. W. Lloyd, *Biol. Rev. Cambridge Philos. Soc.*, 50 (1975) 325–350.
(149) M. C. Glick and H. M. Flowers, in M. I. Horowitz and W. Pigman (Eds.), *The Glycoconjugates*, Vol. 2, Academic Press, New York, 1978, pp. 337–384.

genicity have all been reported,[150] and they point to the possible role of the carbohydrate chains of surface glycoproteins as information markers.

Examination[151] of the glycoproteins of transformed cells shows differences from their normal counterparts, and the role of fibronectin[152] has been especially intriguing and controversial. However, apart from the considerable decrease in fibronectin reported for many cells after transformation, gel electrophoresis has revealed changes in the molecular weight of a number of other glycopeptide chains. It is not yet clear whether these changes are due to the action of glycosidases, or to the altered activity of glycosyltransferases on transformation.

The difficulties inherent in the isolation of intact glycoproteins from cell membranes in order to examine their structures led to an approach[153] based on the selective removal of glycopeptides from cell surfaces by proteolytic enzymes. It was found that there were differences between the glycopeptides isolated from normal cells and those from virally transformed lines derived from these cells. After pronase digestion of cells grown in media containing either 2-amino-2-deoxy-D-[^{14}C]glucose or L-[^{3}H]fucose, followed by treatment with sialidase, different elution patterns were observed on columns of Sephadex G-50 gel. Certain glycopeptides that increased markedly on transformation were shown to be correlated to tumorigenesis, but others disappeared in virus-transformed and chemically transformed[154,155] cells; they appear to be a good marker for tumorigenesis. It was also observed that the typical glycopeptides apparently characteristic of surfaces of transformed cells also occur in other organelles, including mitochondria, nuclei, and endoplasmic reticulum,[156] leading to the conclusion that glycoproteins of all membranes (plasma and intracellular) are affected by transformation.

Treatment[157] of several transformed lines of cultured cells with cyclic AMP caused morphological changes, so that the cells behaved more like their normal counterparts. At the same time, the pronase-derived fucopeptides were more similar to those of normal cells.

(150) H. M. Flowers and N. Sharon, *Methods Enzymol.*, 48 (1979) 30–95.
(151) K. M. Yamada and J. Pouyssegur, *Biochimie*, 60 (1978) 1221–1233.
(152) A. Vaheri and D. F. Mosher, *Biochim. Biophys. Acta*, 516 (1978) 1–25.
(153) C. A. Buck, M. C. Glick, and L. Warren, *Biochemistry*, 9 (1970) 4567–4576.
(154) M. C. Glick, Z. Rabinowitz, and L. Sachs, *J. Virol.*, 13 (1974) 967–974.
(155) L. A. Smets, W. P. van Beek, and H. van Rooij, *Int. J. Cancer*, 18 (1976) 462–468.
(156) T. Muramatsu, M. Ogata, and N. Koide, *Biochim. Biophys. Acta*, 444 (1976) 53–68.
(157) R. M. Roberts, A. Walker, and J. J. Cetorelli, *Nature (New Biol.)*, 244 (1973) 86–89.

The predominant, surface glycopeptide from a line of virally transformed cells in culture, labelled with L-[^{14}C]fucose, has been isolated[158] after trypsinization of the cells, and extensively digested by pronase; a partial structure was proposed for its saccharide portion. It was suggested that α-L-Fuc is attached to GlcNAc of the di-N-acetylchitobiosyl portion attached to asparagine in N-glycosyl chains, but which GlcNAc unit was substituted was not established, nor was the position of substitution clear, as methylation analysis was not performed.

The Sephadex G-50-gel patterns of the pronase glycopeptides analyzed in these and similar experiments[159] indicated an elevation in molecular weight upon transformation of the cells. However, the single L-fucose unit in this carbohydrate structure cannot account for the apparent increase in molecular weight; in fact, similarly bound fucose in not markedly different quantity seems to occur in the typical glycopeptide peak of non-transformed cells, and it would appear that the differences in molecular weight between the two sets of glycopeptides is due to the degree of branching and the lengths of chains *external* to the mannose core of the glycopeptide. The only sugar that is strikingly elevated in the "transformed" glycopeptides is sialic acid,[160] and, after sialidase treatment, the differences in the gel elution-patterns between the glycopeptides of the two cell-types disappear.[161]

Experiments with L cells,[162] using labelled D-glucose, 2-amino-2-deoxy-D-glucose, L-valine, and L-leucine, gave indications of incorporation of radioactive label into membrane glycoproteins, with little, if any, turnover in the plasma membrane. However, in pulse-chase experiments using HeLa cells[163] with [^{3}H]fucose labelling, decay in membrane glycoprotein was demonstrated, with a half-life similar to that expected in a chase where the medium is not changed. A kinetic analysis of glycoprotein metabolism in HeLa cells using [^{3}H]fucose labelling confirmed[164] that very little free fucose ($\sim 2\%$) occurs in the cells, and negligible amounts were degraded oxidatively to carbon dioxide. Nevertheless, there was exchange between the internalized, free fucose and membrane fucosyl glycoproteins. It was concluded

(158) U. V. Santer and M. C. Glick, *Biochemistry*, 18 (1979) 2533–2540.
(159) S.-I. Ogata, T. Muramatsu, and A. Kobata, *Nature (London)*, 259 (1976) 580–582.
(160) L. Warren, J. P. Fuhrer, C. A. Buck, and E. F. Warborg, Jr., *Miami Winter Symp.*, 8 (1974) 1–16.
(161) L. Warren, J. P. Fuhrer, and C. A. Buck, *Fed. Proc. Fed. Am. Soc. Exp. Biol.*, 32 (1973) 80–85.
(162) L. Warren and M. C. Glick, *J. Cell Biol.*, 37 (1969) 729–746.
(163) R. L. Kaufman and V. Ginsburg, *Exp. Cell Res.*, 50 (1968) 127–132.
(164) P. D. Yurchenco and P. H. Atkinson, *Biochemistry*, 16 (1977) 944–953.

that there were two glycoprotein pools in the cells: the majority of the incorporated [³H]fucose entered the plasma-membrane glycoprotein-pool and ~50% of this [³H]fucose was exchanged in 23 h, 75–80% of it appearing as free fucose, and 20–25% as glycopeptide or glycoprotein fucose.

Exchange of labeled fucose between plasma and intracellular membranes of the cell may complicate analysis of membrane glycoproteins, and this possibility stresses the importance of clearly separating plasma from internal membranes before studying the effects of external conditions upon glycoproteins of the cell.

5. Glycolipids

The two main classes of fucose-containing glycolipids are lipopolysaccharides of Gram-negative bacteria and animal glycosphingolipids. Although these materials are structurally different, they do have some similarities: both are heteroglycans and amphiphilic molecules, usually dissolving in water; in both classes, it is the L enantiomer of fucose that has been identified, almost always in the α-L configuration; and recognition properties, including reaction with specific antibodies and lectins, may be due to their fucosyl units (see later.)

a. Bacterial Lipopolysaccharides.—The lipopolysaccharide (LPS) is a characteristic component of the outer membrane of Gram-negative, bacterial cell-envelopes, and is responsible for the toxicity and much of the antigenic specificity of these organisms. It is a receptor for bacteriophages,[165] and is therefore important for phage-typing of bacteria. Some capsular antigens (polysaccharides), which occur in many organisms,[137,138] also contain L-fucose and they may be bound covalently to the LPS, making differentiation difficult.

The LPS proper almost always contains[9] an internal β-(1→6)-linked disaccharide of 2-amino-2-deoxy-D-glucose, partially substituted by long-chain fatty acids, to which is attached a heteroglycan. Fucose occurring in such saccharide chains is always situated in the outer section (O-antigen), often, but not always, at nonreducing termini. Examples of internal L-Fuc residues include the O-antigenic side-chains of *Yersinia* (*Pasteurella*) spp., which contain[166,167] repeating units of

---

(165) A. A. Lindberg, *Annu. Rev. Microbiol.*, 27 (1973) 205–241.
(166) C. J. Hellerqvist, B. Lindberg, K. Samuelsson, and R. R. Brubaker, *Acta Chem. Scand.*, 26 (1972) 1389–1393, 1394–1398.
(167) K. Samuelsson, B. Lindberg, and R. R. Brubaker, *J. Bacteriol.*, 117 (1974) 1010–1016.

→2)-α-Man-(1→3)-α-L-Fuc-(1→, substituted in the Man portions, and *Salmonella* spp., which contain[168] GalNAc-(1→4)-L-Fuc-D-Gal units.

Apart from L-fucose, 3-amino-3,6-dideoxy-D-galactose[169] and the D (Ref. 170) and L (Ref. 171) enantiomers of 2-amino-2,6-dideoxygalactose (fucosamine) have been identified in bacterial LPS. 2-Acetamido-4-amino-2,4,6-trideoxy-D-galactopyranose has been isolated[171a] from the capsular polysaccharide of *Streptococcus pneumoniae*.

b. Animal Glycolipids.—(i) Occurrence. Fucosphingolipids[5] occur in erythrocytes and secretory tissues of several mammalian species, and their antigenic properties in particular have been studied (see later). Other aspects of considerable interest are their cellular location and their role in tumor-cell biology. All of the fucolipids thus far identified are ceramide oligosaccharides containing a number of different sugar residues, apart from the sole example of a fucosylceramide[172] (**38**).

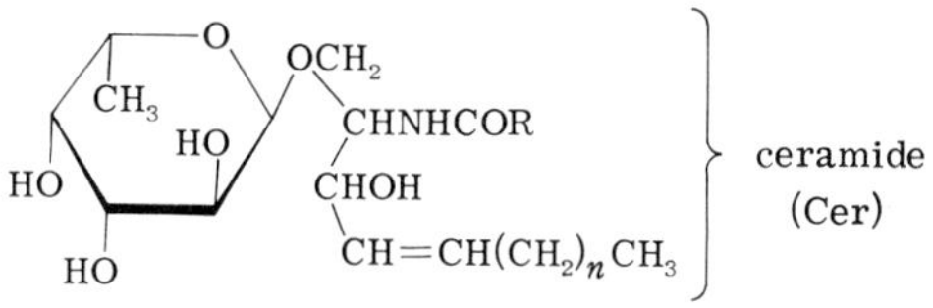

$n = 12, 14,$ etc.

38

α-L-Fuc-(1→2)-β-Gal-(1→3)-β-GalNAc-(1→4)-β-Gal-(1→4)-β-Glc-Cer
                                              |
                                              R

39   R = N-acetylneuraminyl (NeuAc)
40   R = N-glycolylneuraminyl (NeuGc)

α-NeuAc-(2→3)-β-Gal-(1→4)-β-GlcNAc-(1→3)-β-Gal-(1→4)-β-Glc-Cer
                    3
                    ↑
                    1
                 α-L-Fuc

41

(168) D. Lüderitz, D. A. R. Simmons, and O. Westphal, *Biochem. J.*, 97 (1965) 820–826.
(169) S. G. Wilkinson, *J. Gen. Microbiol.*, 70 (1972) 365–369.
(170) M. J. Crumpton and D. A. L. Davies, *Biochem. J.*, 70 (1958) 729–736.
(171) D. T. Drewry, K. C. Symes, G. W. Gray, and S. G. Wilkinson, *Biochem. J.*, 149 (1975) 93–106.
(171a) B. Lindberg, B. Lindqvist, J. Lönngren, and D. A. Powell, *Carbohydr. Res.*, 78 (1980) 111–117.
(172) K. Watanabe, T. Matsubara, and S.-I. Hakomori, *J. Biol. Chem.*, 251 (1976) 2385–2387.

They can be quite complex, with as many as 30 or more sugar units present,[173] and, except for rare examples[117] isolated from some aquatic invertebrates, which may also contain O-methylated fucose, only have (nonreducing) terminal fucosyl groups. They rarely contain sialic acid, although some fucosylated gangliosides have been characterized, including **39** (Ref. 174), **40** (Ref. 174), and **41** (Ref. 175).

(ii) **Structure Determination.** Techniques for extraction, purification, and structural determination of these lipids have been well elaborated.[5] Their solubility in water facilitates some steps of the extraction and purification processes. However, complicating factors are their amphiphilic properties, leading to micelle formation in aqueous media, and their high polarity, causing them to be bound very strongly to silica gel, so that suitable rates of migration, using silica in t.l.c. or column chromatography, of most of the members of this family can only be obtained after peracetylation. Sugar compositions, sequences, anomeric configurations, and positions of substitution have all been determined by the usual methods of carbohydrate chemistry and biochemistry, including n.m.r.-[176] and mass-spectral[177] analysis of intact glycolipids and their degradation products, and the use of specific glycosidases (to be discussed later).

(iii) **Fucolipids: Tumorigenesis and Ontogeny.** In addition to their antigenic properties, the evolutionary significance of which is speculative, other biological functions of fucolipids are also not clear, although some suggestions have been made.[4,5] The fucolipids are generally found in rapidly proliferating tissues, such as gastrointestinal epithelium and bone marrow, and are usually absent from more-slowly dividing cells, such as brain, myocardium, or kidney, but occur in significant proportions in erythrocytes. In oncogenesis,[4,178] where simpler carbohydrate chains of sphingolipids are often evident, the fucosyl units frequently become deleted; on the other hand, chains may actually be extended in ontogeny.[178] They may be synthesized more rapidly at higher cell-densities,[179] and the accumulation of glycosphin-

---

(173) M. Dejter-Juszynski, N. Harpaz, H. M. Flowers, and N. Sharon, *Eur. J. Biochem.*, 83 (1978) 363–373.
(174) H. Wiegandt, *Z. Physiol. Chem.*, 354 (1973) 1049–1056.
(175) H. Rauvala, *J. Biol. Chem.*, 251 (1976) 7517–7520.
(176) K.-E. Folk, K.-A. Karlsson, and B. E. Samuelsson, *Arch. Biochem. Biophys.*, 192 (1979) 164–202.
(177) H. Egge, *Chem. Phys. Lipids,* 21 (1978) 349–360.
(178) K. Watanabe and S.-I. Hakomori, *J. Exp. Med.*, 144 (1976) 644–653.
(179) S. Steiner, M. Gacto, and M. R. Steiner, in L. A. Witting (Ed.), *Glycolipid Methodology*, Am. Oil Chem. Soc. Press, Champaign, Ill., 1976, pp. 141–157.

golipids less complex than normal in experimental, tumor models is now well documented.[180,181]

In the biosynthesis of blood-group active A and B glycolipids from precursor H glycolipid, 2-acetamido-2-deoxygalactosyl- and galactosyl-transferases, respectively, catalyze the addition of sugar units to the nonreducing termini of the carbohydrate chains. There was a significant decline[182] in this transferase activity in adenocarcinoma tissue from A, B, and AB individuals, compared to normal mucosa from the same subjects, while the 2-acetamido-2-deoxygalactosidase activity did not seem to be affected. In some human adenocarcinomas, there was deletion of A and B glycolipids,[183] and a less fucosylated glycolipid accumulated, leading to change in the blood-group specificity. Cultured cells transformed by oncornaviruses incorporated radioactively labelled fucose differently than normal cells; more fucose was incorporated into lipids migrating more rapidly in t.l.c., while the less-mobile lipids incorporated much less fucose.[184] Similar differences in labelling patterns were observed with human cancer-cells compared with normal, human-embryonic lung-cells.[185] Cells infected with temperature-sensitive viruses had normal, fucolipid-labelling patterns, and normal morphology at nonpermissive temperature, but were abnormal at a permissive temperature.[186]

The structural changes involved in the formation of these abnormal fucolipids have not yet been determined, but pulse-chase experiments with labelled and unlabelled fucose indicated[179] that the chromatographically least-mobile fucolipid was the stable end-product of the cell, and was formed from more-mobile precursors. Like the transformed cells, normal-cell cultures at low density did not synthesize this fucolipid as rapidly as cells at higher density, but rates of synthesis in transformed cells were not affected by cell density; that is, there was a relationship between fucolipid metabolism, contact-inhibition, and transformation. The simplest fucoglycolipid (fucosylceramide) was actually found to accumulate in some human-colon tumors,[177] and

(180) P. H. Fishman and R. O. Brady, *Science*, 194 (1976) 906–915.
(181) S.-I. Hakomori and W. W. Young, Jr., *Scand. J. Immunol.*, 7, Suppl., 6 (1978) 97–117.
(182) K. Stellner, S.-I. Hakomori, and G. A. Warner, *Biochem. Biophys. Res. Commun.*, 55 (1973) 439–445.
(183) S.-I. Hakomori and R. W. Jeanloz, in D. Aminoff (Ed.), *Blood and Tissue Antigens*, Academic Press, New York, 1970, p. 149.
(184) S. Steiner, P. J. Brennan, and J. L. Melnick, *Nature (New Biol.)*, 245 (1973) 19–21.
(185) S. Steiner and J. L. Melnick, *Nature (London)*, 251 (1974) 717–719.
(186) S. M. Steiner, J. L. Melnick, S. Kit, and K. D. Somers, *Nature (London)*, 248 (1974) 682–684.

the quantity present was apparently proportional to the degree of malignancy of the tumor. The isolation of this glycoside, and also a hexaosylceramide having Le^b activity (from human-colonic adenocarcinoma[187]) containing 2 moles of fucose per mole, would seem to indicate that fucosyltransferase activity may not be depressed in certain tumors, and that the building of more-complex chains in these examples may be prevented by lack of other transferase activities.

Cell division is stimulated by lectins (for a review, see Ref. 188) and by other saccharide-directed reagents, for example, sodium periodate and D-galactose oxidase after desialylation,[189] but there is no evidence that it is the fucopyranosyl residues that are the specific receptors in these instances. The L-fucose-specific lectins from, for example, *Lotus tetragonolobus* and *Ulex europeus* are not mitogenic.

It would seem, then, that the carbohydrate portions of glycolipids or glycoproteins, or both, serve a regulatory role in cell division of tissues capable of rapid proliferation. Fucolipids also seem to be involved in these control processes, although no specific biological effect has been suggested for the fucopyranosyl units within these glycans.

IV. Immunological Aspects of Complex Fucans

1. Bacterial Antigens

The external location of bacterial glycans, in the cell envelope or capsule, makes them important in the recognition and immune response of higher organisms to invading bacteria, and their immunology has been studied for many years,[190] following the appreciation of their importance in 1915. Rational immunization techniques based on these investigations became of vital medical importance, but then, with the advent of antibiotics, interest in this approach to treatment faded. However, with the discovery of resistant strains of bacteria, renewed appreciation of its value has arisen. Attention focused on bacterial saccharide antigens, with the correlation of the serological classification of *Salmonella* spp., based on the Kauffman–White

(187) B. Siddiqui, J. S. Whitehead, and Y. S. Kim, *J. Biol. Chem.*, 253 (1978) 2168–2175.
(188) H. Lis and N. Sharon, in M. Sela (Ed.), *The Antigens*, Vol. 4, Academic Press, New York, 1977, pp. 429–529.
(189) A. Novogrodsky and E. Katchalski, *Proc. Natl. Acad. Sci. USA*, 70 (1973) 1824–1827.
(190) M. Heidelberger, *Fortschr. Chem. Org. Naturst.*, 18 (1960) 503–536.

scheme, with the structure of the bacterial-surface glycans.[191,192] Antigenic specificity and the antigen–antibody reaction were carefully studied, leading to a clearer understanding of immunodominant sugars and antigenic determinants.[191] Artificial antigens were prepared[193,194] by coupling oligosaccharide determinants to protein, and were injected into animals in order to prepare antibodies that reacted with the active glycans and also with whole bacteria.

The structural features of a large number of bacterial antigens have now been analyzed, and their immunological determinants defined. It has become clear that a major immunological effect is expressed by nonreducing sugar groups at chain ends, although internally situated residues can also play a role, and a number of unusual dideoxy sugars have been identified as important antigenic determinants in bacteria. However, these studies have not implicated L-fucosyl residues as being of major importance.

On the other hand, examination of a large number of Gram-negative bacteria has demonstrated antigens shared between them and mammalian erythrocytes,[195] lending strong support to an immunogenetic hypothesis of isoantibody origin. Out of 282 aerobic, Gram-negative strains examined, ~50% were A-, B-, or H-active, and, of these, 40% were H-active, as shown by interaction with eel lectin, which is specific for α-fucopyranosyl linkages.

Similar results were obtained by other workers on additional species,[196] although they showed a much lower proportion of H-active bacteria.

2. Plant Antigens

Some plant glycosides and polysaccharides cross-react[195,197] with reagents for blood-group antigens, and interesting cross-reactions were also shown by methylated-fucose haptens and L-fucose-specific lectins; for example, it was found that 2-O-methyl-L-fucose was immunodominant in the Japanese yew (*Taxus cuspidata*), and that the 3-O-

---

(191) O. Lüderitz, A. M. Staub, and O. Westphal, *Bacteriol. Rev.*, 30 (1966) 192–255.
(192) O. Lüderitz, O. Westphal, A. M. Staub, and H. Nikaido, in G. Weinbaum, S. Kadis, and S. J. Aijl (Eds.), *Microbial Toxins*, Vol. 4, Academic Press, New York, 1971, p. 145.
(193) W. F. Goebel, F. H. Babers, and O. T. Avery, *J. Exp. Med.*, 60 (1934) 85–94.
(194) W. F. Goebel, O. T. Avery, and F. H. Babers, *J. Exp. Med.*, 60 (1934) 599–617.
(195) G. F. Springer, *Prog. Allergy*, 15 (1971) 9–77.
(196) G. W. Drach, W. P. Reed, and R. C. Williams, Jr., *J. Lab. Clin. Med.*, 79 (1972) 38–46.
(197) G. F. Springer, P. R. Desai, and B. Kolecki, *Biochemistry*, 3 (1964) 1076–1085.

methyl- and 2,3-di-O-methyl-D-Fuc in cardiac glycosides also reacted with the eel lectin. Cross-reactivity was even shown by the 3-O-methyl-Gal occurring in certain plant polysaccharides, and structural features for this reactivity were discussed.

3. Blood-Group Antigens

Gene-dependent antigens of ABH and related types occur on human erythrocytes (and those of other species), and also on many other cells.[198] Their chemistry and biochemistry have been reviewed frequently during the past two decades (see, for example, Refs. 199 and 200), and speculations have been made as to their biological significance.[199] It was pointed out that the ABH antigens do not appear to have a specific function essential for cell vitality, as there are some individuals ("Bombay" type) who lack these antigens and yet seem to be completely normal. At the same time, the immunological relationship between ABH activity and various Gram-negative pathogens[195] indicates possible advantages of developing antibodies to these substances within the vertebrate body. There does seem to be some correlation[201] between ABH types and the tendency towards certain pathological states with, for example, A individuals more prone to gastric carcinoma and pernicious anemia, and a slightly greater tendency of group O (H)-individuals to become afflicted with gastric ulcer.

Secreted glycoproteins with ABH and related blood-group activities have been intensively studied, and biochemical and genetic factors in their interrelationships have been determined.[15,199,200] There have also been considerable advances in our knowledge of erythrocyte antigens of glycolipid and glycoprotein nature.[202-205] Earlier work concentrated on the secreted glycoprotein antigens, and it was shown that hydrolysis of ABH antigens with dilute acid, which apparently only involved removal of fucosyl units, led to a marked decrease in their ABH activities.[206] The use of lectins revealed that eel lectin reacted specifically

(198) A. E. Szulman, Annu. Rev. Med., 17 (1966) 307–322.
(199) S.-I. Hakomori and A. Kobata, in Ref. 188, pp. 79–140.
(200) W. M. Watkins, in D. M. N. Surgenor (Ed.), The Red Blood Cell, 2nd edn., Vol. 1, Academic Press, New York, 1974, pp. 293–360.
(201) L. H. Muschel, Bacteriol. Rev., 30 (1966) 427–441.
(202) J. Koscielák, A. Piasek, H. Gorniak, and A. Gardas, Eur. J. Biochem., 37 (1973) 214–225.
(203) T. Krusius, J. Finne, and H. Rauvala, Eur. J. Biochem., 92 (1978) 289–300.
(204) S. Takasaki, K. Yamashita, and A. Kobata, J. Biol. Chem., 253 (1978) 6086–6091.
(205) J. Järnefelt, J. Rush, Y.-T. Li, and R. A. Laine, J. Biol. Chem., 253 (1978) 8006–8009.
(206) E. A. Kabat, Bacteriol. Rev., 13 (1949) 189–202.

with H and Le antigens, and the lectin from *Ulex europeus*, which is bound most strongly by H substance (but not by Le^a substance), showed highest specificity for α-L-Fuc-(1→2)-Gal linkages. Immune antisera also indicated that α-L-Fuc was the immunodominant part of H, Le^a, and Le^b determinants. The selective removal of L-fucose by α-L-fucosidases (see later) destroyed H and Le^a specificities. It was demonstrated that A and B activities are respectively due to (nonreducing) terminal α-GalNAc and α-Gal groups, but these were accentuated by the concomitant presence of neighboring α-L-Fuc-(1→2)-Gal units in the saccharide chains. Careful characterization of one of the partially purified α-L-fucosidases employed showed[207] that it was strictly specific for α-L-Fuc-(1→2)-Gal, and controlled, alkaline degradation of secreted ABH-glycoproteins permitted the isolation of fucose-containing oligosaccharides whose structures were established by conventional means[15] (for example, the structure proposed for the largest carbohydrate chain from the H-active glycoprotein from human ovarian-cyst, namely, **42**).

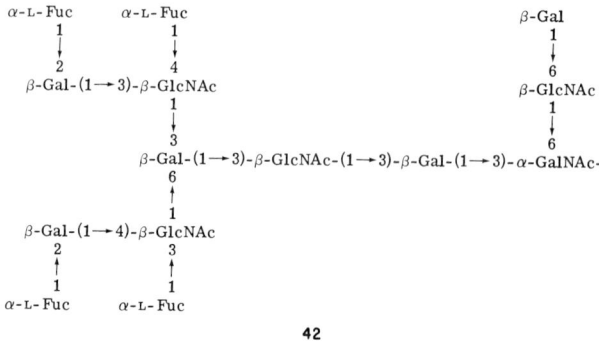

42

Glycosphingolipid antigens, extracted from erythrocyte membranes, have the same immunodominant groups as the cross-reacting, secreted glycoprotein antigens, with α-L-Fuc again being immuno-

```
        Fuc-(1→2)-Gal-(1→4)-GlcNAc
                    1
                    ↓
                    3
            Gal-(1→4)-[GlcNAc-(1→3)-Gal]₇-GlcCer
                    6
                    ↑
                    1
        Fuc-(1→2)-Gal-(1→4)-GlcNAc
```

43

(207) M. E. A. Pereira, E. C. Kisailus, F. Gruezo, and E. A. Kabat, *Arch. Biochem. Biophys.*, 185 (1978) 108–115.

dominant for H and Le glycolipids. Amongst characterized products was a glycolipid (**43**) from human H(O)-erythrocytes.[208] However, blood-group activity was also detected in glycopeptide fractions prepared from human erythrocytes, and structure **44** was suggested[205] for a poly(glycosyl)peptide having H-activity.

α-L-Fuc-(1→2)-β-Gal-(1→4)-β-GlcNAc(1→3)-β-Gal-(1→4)-β-GlcNAc-(1→$\overline{x}$?)-(Man)$_{2-3}$-GlcNAc

44

Similar fucolipids were also isolated from hog gastric-mucosa, for example, fucolipids **45(a–d)** having[209] (A + H) activity, but human sal-

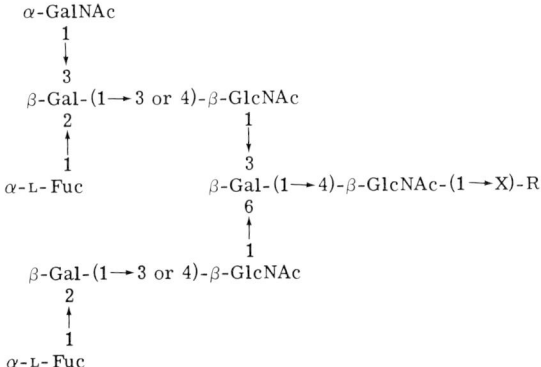

45

a X = 3, R = β-Gal-(1→4)-β-Gal-(1→4)-Glc-Cer
b X = 4, R = β-GlcNAc-(1→3)-β-Gal-(1→4)-Glc-Cer
c X = 3, R = β-Gal-(1→4)-β-Gal-(1→4)-Glc-Cer
   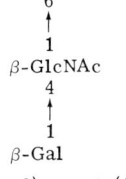
d X = 4, R = β-GlcNAc-(1→3)-β-Gal-(1→4)-Glc-Cer
                                        6
                                        ↑
                                        1
                                    β-GlcNAc
                                        4
                                        ↑
                                        1
                                    β-Gal

(208) A. Gardas, *Eur. J. Biochem.*, 68 (1976) 177–183.
(209) B. L. Slomiany and A. Slomiany, *Eur. J. Biochem.*, 90 (1978) 39–49.

ivary ABH antigens were found to be entirely glycoprotein in nature.[210] It is clear that the H-determinant is α-L-Fuc-(1→2)-Gal, irrespective of whether the saccharide chains are attached to protein or to sphingolipid.

The Lewis-type antigens are closely related to the ABH family, and investigations similar to those just described showed that the immunodominant sugar in the Le blood-groups was L-fucose attached either α-(1→4) to GlcNAc, giving Le^a activity, or α-(1→2) to Gal in addition to α-(1→4) to GlcNAc, producing Le^b activity. In this respect, it should be noted that α-Fuc-(1→3)-GlcNAc linkages, with apparently no special antigenic properties, occur in blood-group-active glycoproteins. This linkage can readily be distinguished from the Le^a determinant of α-Fuc-(1→4)-GlcNAc by its reaction with the lectin from *Lotus tetragonolobus*.[211]

Other blood-group determinants, such as human P and I, and the J-determinant from cattle, are carbohydrate in nature, but are deficient in, or devoid of, L-fucose, which apparently does not play an immunological role in these antigens.[199]

V. METABOLISM

1. Biosynthesis

a. L-Fucose.—There had been a suggestion that L-fucose might be derived biosynthetically from D-galactose (a common dietary sugar and biosynthetic product) by reduction at C-1 to CH_3 and oxidation of CH_2OH to CHO (compare, the chemical synthesis[54]). However, it was demonstrated[212-214] that certain bacteria are able to convert specifically labeled D-[^{14}C]glucose into L-[^{14}C]fucose without inversion, or cleavage of the carbon chain. In humans, after administration of D-[6-^{14}C]glucose, it was found[215] that the L-fucose present in milk was asymmetrically labelled, with most of the label in C-5 and C-6, again indicating conversion from D-glucose into L-fucose without rupture of the chain.

---

(210) B. L. Slomiany and A. Slomiany, *Eur. J. Biochem.*, 85 (1978) 249–254.
(211) L. Rovis, B. Anderson, E. A. Kabat, F. Greuzo, and J. Liao, *Biochemistry*, 12 (1973) 5340–5354.
(212) J. F. Wilkinson, *Nature (London)*, 180 (1957) 995.
(213) S. Segal and Y. J. Topper, *Biochim. Biophys. Acta*, 25 (1957) 419–420.
(214) E. C. Heath and S. Roseman, *J. Biol. Chem.*, 230 (1958) 511–519.
(215) S. Segal and Y. J. Topper, *Biochim. Biophys. Acta*, 42 (1960) 147–151.

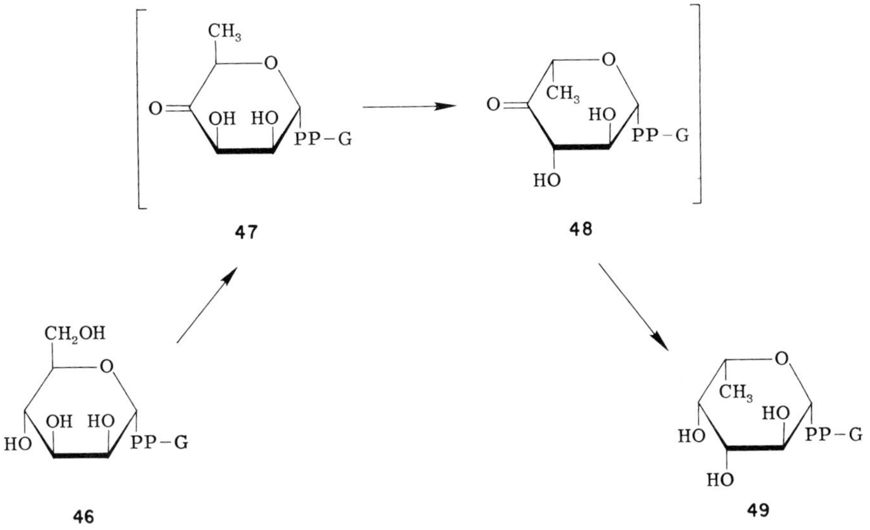

CHEMISTRY AND BIOCHEMISTRY OF D- AND L-FUCOSE

Scheme 6

PP–G = guanosine 5'-diphosphate group

The isolation of a guanosine nucleotide containing fucose, and presumed to be guanosine 5'-(L-fucosyl diphosphate) (GDP-Fuc), from *Aerobacter aerogenes*[216] and ewe's milk[217] suggested its possible implication in the biosynthesis of L-fucose. Subsequently, crude extracts of *Aerogenes* were shown[218] to catalyze the conversion of guanosine 5'-(D-mannosyl diphosphate) (GDP-Man, **46**) into GDP-Fuc. Detailed investigation of this process showed[219] that $NADP^+$ was an essential cofactor, and that an oxidation–reduction process was involved; OH-4 was oxidized to a ketone group, and the CH_2OH was converted into CH_3 (**47**).

Epimerization of **47** at C-3 and C-5 to afford **48**, followed by reduction of the 4-ketone to hydroxyl by NADPH led to GDP-Fuc (**49**) (see Scheme 6).

Similar pathways were later elucidated for the biosynthesis of thy-

(216) V. Ginsburg and H. N. Kirkman, *J. Am. Chem. Soc.*, 80 (1958) 3481.
(217) R. Denamur, G. Fauconneau, and G. Guntz, *C. R. Acad. Sci.*, Ser. A, 246 (1958) 2820–2823.
(218) V. Ginsburg, *J. Am. Chem. Soc.*, 80 (1958) 4426.
(219) V. Ginsburg, *J. Biol. Chem.*, 235 (1960) 2196–2201; 236 (1961) 2389–2393.

midine(dTDP)-L-rhamnose from dTDP-D-glucose,[220-222] and GDP-L-galactose from GDP-Man.[223]

Various mammalian-tissue preparations were also found[224] to catalyze the formation of GDP-Fuc from GDP-Man; both NAD^+ and an NADPH-generating system were necessary for maximum activity.

b. Glycosyltransferases and Biosynthesis of Glycans.—Early interest in glycosyltransferases arose out of the scheme postulated[225,226] for the genetic basis of blood-group ABH and Le specificity, depending on the control of transfer, catalyzed by specific glycosyltransferases, of glycosyl groups to a growing chain. The necessary glycosyltransferases were soon located in a variety of tissues, their specificities determined, and their ability to convert carbohydrate structures into those having A, B, H, or Le activities demonstrated. Later, great interest was shown in the characterization of glycosyltransferases in connection with studies on the tissue localization, and mechanism of biosynthesis, of other animal glycoproteins.

Similarly, the biosynthesis of glycosphingolipids is apparently a process of stepwise addition of single glycosyl groups, catalyzed by specific glycosyltransferases, to a growing chain starting from ceramide,[227] and this pathway has been established both for neutral glycolipids of the blood-group active type[228,229] and for gangliosides.[180,229]

c. Enzymes Catalyzing Transfer of L-Fucose to Glycoproteins.—One of the problems encountered in examining the specificities of glycosyltransferases, including fucosyltransferases, has been their particulate nature and the difficulty in freeing them of other activities and of substrates that might serve as endogenous acceptors. A crude, particulate preparation from hog gastric-mucosa catalyzed[230] transfer of L-fucose from GDP-[^{14}C]Fuc to blood-group H-substance, with for-

---

(220) L. Glaser and S. Kornfeld, *J. Biol. Chem.*, 236 (1961) 1795–1799.
(221) R. Okazaki, T. Okazaki, and J. L. Strominger, *Fed. Proc. Fed. Am. Soc. Exp. Biol.*, 20 (1961) 85.
(222) J. H. Pazur and E. W. Shuey, *J. Biol. Chem.*, 236 (1961) 1780–1785.
(223) J.-C. Su and W. Z. Hassid, *Biochemistry*, 1 (1962) 468–480.
(224) D. W. Foster and V. Ginsburg, *Biochim. Biophys. Acta*, 54 (1961) 376–378.
(225) R. Ceppellini, *Ciba Found. Symp. Biochem. Human Genet.*, (1959) 242–263.
(226) W. M. Watkins and W. T. J. Morgan, *Vox Sang.*, 4 (1959) 97–119.
(227) S. Roseman, *Chem. Phys. Lipids*, 5 (1970) 270–297.
(228) S. Basu, M. Basu, J. R. Moskal, J.-L. Chien, and D. A. Gardner, in Ref. 179, pp. 123–139.
(229) G. Dawson, in Ref. 149, pp. 255–284.
(230) A. P. Grollman and D. P. Marcus, *Biochem. Biophys. Res. Commun.*, 25 (1966) 542–548.

mation of α-L-Fuc-(1→2)-Gal linkages. It was shown that "secretor" status does, indeed, have an enzymic basis, as proposed by the biosynthetic scheme[225,226] for blood-group-active glycoproteins, as defatted milk from secretors catalyzed transfer of L-fucosyl groups to a variety of simple acceptors,[231] producing α-L-Fuc-(1→2)-Gal linkages of the M-specificity type, whereas the milk of nonsecretors was inactive. Interestingly, both kinds of donor produced in the milk additional fucosyltransferase activities in much larger quantities, and these catalyzed the formation of α-L-Fuc-(1→4)-GlcNAc linkages (of the Le type) and α-L-Fuc-(1→3)-Glc linkages of unknown significance. Crude, particulate fractions of human submaxillary-glands and stomach-tissues also catalyzed the transfer[232] of L-fucosyl groups to a variety of acceptors, and they contained three separate, enzymic activities, leading to formation of the same three types of linkage just described; although the enzyme forming α-L-Fuc-(1→2)-Gal is absent from the submaxillary glands of nonsecretors, it *does* occur in stomach tissues of nonsecretors and in serum of all normal, ABH donors, irrespective of their ABH secretor or nonsecretor status.[233]

"Bombay" (Oh) individuals show no A, B, or H activity, and cannot accept the glycosyl groups necessary for development of these activities. It was shown[234] that treatment of erythrocytes from these individuals with GDP-Fuc and a fucosyltransferase that had been prepared from the gastric juice of normal, O individuals, indeed failed to transform them into O-type cells. However, after prior incubation with sialidase, the "Bombay" erythrocytes became acceptors of L-fucosyl groups, leading to expression of H-antigenic activity. The resulting, H-active cells could now accept GalNAc from GDP-GalNAc, catalyzed by a suitable transferase, leading to A-type cells; that is, sialic acid units apparently mask the acceptor site for L-fucose.

The presence of three separate, fucosyltransferase activities in human milk was confirmed, and their pH optima and substrate requirements were defined by use of simple oligosaccharides of known structure.[235] It was demonstrated that increased affinity for each of

---

(231) L. Shen, E. P. Grollman, and V. Ginsburg, *Proc. Natl. Acad. Sci. USA*, 59 (1968) 224–230.
(232) M. A. Chester and W. M. Watkins, *Biochem. Biophys. Res. Commun.*, 34 (1969) 835–842.
(233) H. Schenkel-Brunner, M. A. Chester, and W. M. Watkins, *Eur. J. Biochem.*, 30 (1972) 269–277.
(234) H. Schenkel-Brunner, R. Prohaska, and H. Tuppy, *Eur. J. Biochem.*, 56 (1975) 591–594.
(235) J.-P. Prieels, T. Beyers, and R. L. Hill, *Biochem. Soc. Trans.*, 5 (1977) 838–839.

these three enzymes was correlated to increase in the chain-length of the substrate.

Other glycoprotein acceptors for L-fucose are of interest in the study of fucosyltransferases. It was found that there are present, in Triton X-100 extracts of HeLa cells,[236] two transferases having different pH optima and metal requirements. One of them catalyzed transfer of L-fucosyl groups to Gal residues of asialoafuco-porcine submaxillary mucin, although not to asialofetuin or α_1-acid-glycoproteins, and the other fucosylated GalNAc residues of asialoagalactofetuin. Both enzymes were membrane-bound, and were partially purified by passing solutions of them containing Triton X-100 through columns of Sephadex G-200, followed by centrifugation for 4 h at 300,000 g. Both enzymes were inactive towards several mono- and di-saccharides, and acceptors of high molecular weight seemed to be necessary.

The biosynthesis of glycoproteins in various cell-types was monitored histologically[237] by injection of L-[³H]fucose into rats, followed by radioautography of different cells after suitable periods of growth, and sacrifice of the animals. The pattern of transfer of [³H]fucose from the initially labelled, Golgi apparatus into lysosomes, secretory materials, and plasma membranes demonstrated that fucosyl groups were added to growing glycoprotein chains in the Golgi bodies; that is, the carbohydrate chains of the glycoproteins were completed there.

The inverse relationship existing between the contents of L-fucose and sialic acid in many secreted glycoproteins has already been mentioned.[118] Part of the mutually exclusive presence of these two sugars might be explained as due to competition for the same site on the acceptor, for example, substitution of Gal residues. However, the typical positions of linkage of sialic acid in glycans [(2→3) or (2→6) to Gal, or to GalNAc] are not those occupied by L-Fuc [(1→2) to Gal, and (1→3, 4, or 6) to GlcNAc]. It has now been demonstrated[238] that occupation of a site by one of these sugar residues may inhibit the entry of the other sugar into a neighboring, unoccupied site. Thus, the asparagine-linked oligosaccharides of asialotransferrin can either be sialylated to form α-NeuAc-(2→6)-β-Gal-(1→4)-GlcNAc, or fucosylated to afford β-Gal-(1→4)-[α-L-Fuc-(1→3)]-GlcNAc. Sialylation of asialotransferrin prevents subsequent addition of fucose by the fucosyltransferase, and fucosylation of the asialoglycoprotein blocks addition of NeuAc. It

(236) H. B. Bosmann, A. Hagopian, and E. H. Eylar, *Arch. Biochem. Biophys.*, 128 (1968) 470–481.
(237) G. Bennett, C. P. Leblond, and A. Haddad, *J. Cell Biol.*, 60 (1974) 258–284.
(238) J. C. Paulson, J.-P. Prieels, L. R. Glasgow, and R. L. Hill, *J. Biol. Chem.*, 253 (1978) 5617–5624.

was concluded that a sequence such as α-NeuAc-(2→6)-β-Gal-(1→4)-[α-L-Fuc-(1→3)]-GlcNAc is not possible in these asparagine-linked oligosaccharides.

There is often an apparent correlation between the L-fucose content of serum glycoprotein and pathological manifestations (see later), and part of the differences in the levels found in normal and pathological tissues is due to changed fucosyltransferase activities. Elevated fucosyltransferase activities were found in a variety of malignant tissues,[239–242] and it has been suggested[243] that decrease in human-serum, fucosyltransferase activity may be monitored as a convenient indicator of successful, tumor therapy. It should be noted, however, that many malignant tissues grow more rapidly than the surrounding, normal tissues, so that elevated glycosyltransferase activities would not be unexpected. In cases of neoplasia in cancer patients, notably those with tumor metastatic to liver, elevated levels of sialyl-, galactosyl-, and fucosyl-transferases were all found.[241] However, specificity in the enhanced activity of fucosyltransferases was evidenced; for example, the α-L-Fuc-(1→3)-transferase activity to GlcNAc was especially elevated[240] in human serum [rather than α-L-Fuc-(1→2) to Gal] in cases of neoplasia; in infectious hepatitis, the α-L-Fuc-(1→3)-transferase activity was low.

On the other hand, in patients having untreated, acute myelogenous leukemia, the plasma fucosyltransferase activity to GlcNAc (A) was normal, whereas the activities of sialyltransferase and Gal-transferase were slightly increased, and that of fucosyltransferase to Gal (B) was elevated to 3 times that of normal controls.[243] In patients having drug-induced, disease remission, the plasma activity of A was markedly increased, but that of B was correspondingly decreased. In six established strains of spontaneously metastasizing, rat-mammary tumors and four nonmetastasizing strains, using asialofetuin (C) and asialoagalactofetuin (D), respectively, as acceptors, it was found that the level of fucosyltransferase activity to C (involving, presumably, addition to Gal) was comparable in both groups of tumors, but the activity towards D (GlcNAc acceptor) was raised 6–7-fold in the metastasizing

---

(239) P. Khilanani, T.-H. Chou, P. L. Loman, and D. Kessel, *Cancer Res.*, 37 (1977) 2557–2559.
(240) C. H. Bauer, E. Köttgen, and W. G. Reutter, *Biochem. Biophys. Res. Commun.*, 76 (1977) 488–494.
(241) D. Kessel, E. Sykes, and M. Henderson, *J. Natl. Cancer Inst.*, 59 (1977) 29–32.
(242) C. H. Bauer, W. G. Reutter, K. P. Erhart, E. Köttgen, and W. Gercke, *Science*, 201 (1978) 1232–1233.
(243) D. Kessel, T.-H. Chou, and P. Khilanani, *Biochem. Soc. Trans.*, 6 (1978) 187–190.

tumors. The authors speculated that increased activity of the GlcNAc-fucosyltransferase in the group of metastasizing tumors might have reflected faster synthesis, and shedding of fucosylated glycoprotein antigens, whereas the synthesis of similar molecules proceeded in the nonmetastasizing cells at a much lower rate, as these compounds were not readily lost from the cell surface. Interestingly, between the different groups of cells, there was little difference in fucosidase activities, which were very low at pH 6.3–7.4, and had an optimal pH of 4–5. An important precaution in these experiments[244] was the matching of the two groups of tumors according to their growth rates, to ensure that differences in the transferase activity were not a reflection of differential rates of growth. In fact, it was found that the specific activities of the transferases did not vary significantly when the tumors were collected at different periods of growth, or when homogenates were prepared from tumors of various sizes.

The plasma fucosyltransferase to Gal produces α-L-Fuc-(1→2)-Gal linkages, and is the H-dependent, blood-group-specific enzyme. Its selective inhibition by thiol-blocking agents has permitted its discrimination[245] from other fucosyltransferases.

d. Fucosyltransferases for Glycolipid Substrates.—L-Fucose has been transferred[228,246] from GDP-Fuc to suitable glycosphingolipid acceptors by using crude, enzyme preparations to produce H-active fucolipids, and the reaction was more specific for glycosylceramide than for oligosaccharide or glycoprotein acceptors. Precursor glycolipids were also converted into Lewis specific lipids by the action of crude fucosyltransferases from human, gastric mucosa.[247]

Differences in fucolipid content as between normal and transformed cells, and changes on ontogenesis, have already been discussed, and these effects are presumably due to variations in fucosyltransferase activities, with possible specificity differences, namely, α-(1→2) to Gal or (1→3 or 4) to GlcNAc. However, the enzymes involved in these cells have not yet been separated and purified, and information on the separate enzyme-activities involved is almost nonexistent.

Little research has as yet been published on the biochemistry of incorporation of L-fucose into bacterial glycans. Most of the studies have been made on the genetics of the biosynthesis of colanic acid in Gram-

(244) S. K. Chatterjee and U. Kim, *J. Natl. Cancer Inst.*, 61 (1978) 151–162.
(245) D. Kessel and T.-H. Chou, *Biochem. J.*, 181 (1979) 767–769.
(246) S. Basu, M. Basu, and J.-L. Chien, *J. Biol. Chem.*, 250 (1975) 2956–2962.
(247) R. Prohaska, H. Schenkel-Brunner, and H. Tuppy, *Eur. J. Biochem.*, 84 (1978) 161–166.

negative bacteria. Colanic acid is the capsular polysaccharide of *Escherichia coli, Salmonella,* and *Enterobacter cloacae* spp.; it contains[248] a repeating hexasaccharide (**50**) consisting of D-glucose, D-ga-

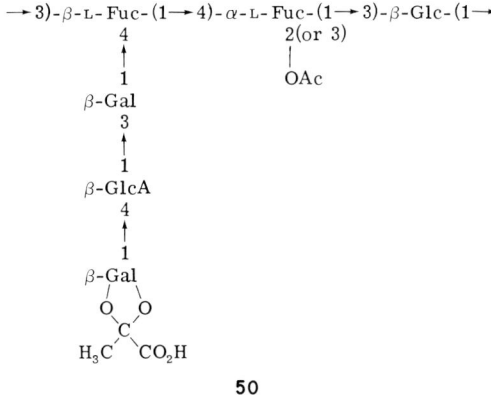

50

lactose, L-fucose, and D-glucuronic acid in the ratios of 1:2:2:1, and it has been demonstrated[249] that membrane fractions of a mucoid (*lon*) mutant of *E. coli* that produces the polysaccharide catalyze the incorporation of fucose from GDP-Fuc into a lipid, as well as into polymer. The glycolipid formed had the structure (Fuc_2, Glc_2)-Glc-P-P-lipid, and the lipid was found to be identical to undecaprenol, demonstrating the similarity of this system to the one involved in the incorporation of abequose into lipopolysaccharide,[250] and also to pathways of biosynthesis of bacterial glycans in general, including peptidoglycan,[251] lipopolysaccharide O-antigens,[252] and teichoic acids.[253] During the production[130,131] of slime polysaccharide by membranes of the root-cap cells of maize (*Zea mays*), fucolipids, including a polar, polyprenol lipid, were synthesized.[254] A microsomal fucosyltransferase was isolated that catalyzed the transfer of L-fucose to polyprenol phosphate, and a second transferase was identified in microsomal, but

(248) I. W. Sutherland, *Biochem. J.*, 115 (1969) 935–945; *Eur. J. Biochem.*, 23 (1971) 582–587.
(249) J. G. Johnson and D. B. Wilson, *J. Bacteriol.*, 129 (1977) 225–236.
(250) M. J. Osborn and I. M. Weiner, *J. Biol. Chem.*, 243 (1968) 2631–2639.
(251) J. S. Anderson, M. Matsuhashi, M. A. Haskin, and J. L. Strominger, *Proc. Natl. Acad. Sci. USA*, 53 (1965) 881–889.
(252) A. Wright, M. Dankert, P. Fennessey, and P. W. Robbins, *Proc. Natl. Acad. Sci. USA*, 57 (1967) 1798–1803.
(253) R. G. Anderson, L. J. Douglas, H. Hussey, and J. Baddiley, *Biochem. J.*, 136 (1973) 871–876.
(254) J. R. Green and D. H. Northcote, *Biochem. J.*, 178 (1979) 661–671.

especially in Golgi, fractions of the cells catalyzing the incorporation of L-fucose into polysaccharides and glycoproteins.[255]

2. Degradation

a. Glycosidases in General.—The isolation of glycosidases from a large variety of sources, and the investigation of their catalysis of the hydrolysis of many different substrates is an active field of study, and interest has especially developed in certain aspects of medical application, for example, in the treatment of enzyme-deficiency disorders. A useful collection of articles on techniques of isolation and purification of glycosidases,[256] and a review of their properties and their application to the study of complex carbohydrates and cell surfaces,[150] have been published.

b. α-L-Fucosidases.—Early interest in fucosidases developed from the finding that extracts of certain bacteria degrade blood-group-active glycoproteins, with liberation of monosaccharides and removal of ABH activity. Application of crude extracts that contained a mixture of glycosidases in the presence of specific inhibitors facilitated elucidation of the immunological determinants of ABH substances.[200] It was found that many extracts contained α-L-fucosidase activity that hydrolyzed simple glycosides,[76] in addition to more-complex glycans, permitting their assay by use of chromogenic or fluorogenic substrates. However, it should be realized that some fucosidases, like other glycosidases, may act only on simple, but not on complex, substrates, whereas others may attack the more-complex substrates preferentially.[150,256,257] All of the α-L-fucosidases characterized are specific for the anomeric and D or L configuration of the sugar; they also seem to be quite specific for the hexopyranose conformation. However, they differ in specificity towards the structure of the next saccharide residue to which the α-L-Fuc group is attached. Thus, α-L-fucosidases have been isolated (and purified) from the marine mollusc *Chamelea gallina* L.,[258] and from rat liver[259] and human[260] lysosomes, that hydrolyze α-L-Fuc-(1→2)-Gal- and α-L-Fuc-(1→3 or 4)-GlcNAc-linkages. Whereas the mollusc enzyme acts on glycopeptides and desialylated glycoproteins containing the necessary structures, the rat-liver en-

---

(255) J. R. Green and D. H. Northcote, *J. Cell Sci.*, 40 (1979) 235–244.
(256) V. Ginsburg (Ed.), *Methods Enzymol.*, 28 (1972) 699–1000.
(257) A. Kobata, *Anal. Biochem.*, 100 (1979) 1–14.
(258) A. Reglero and J. A. Cabezas, *Eur. J. Biochem.*, 66 (1976) 379–387.
(259) D. J. Opheim and O. Touster, *Methods Enzymol.*, 50 (1978) 505–510.
(260) G. Dawson and G. C. Tsay, *Arch. Biochem. Biophys.*, 184 (1977) 12–23.

zyme was reported not to act on intact glycoproteins, and the human enzyme hydrolyzed H-active glycolipids in the presence of detergents. All of these enzymes hydrolyzed p-nitrophenyl α-L-fucopyranoside.

In contrast, many sources provide α-L-fucosidases of higher specificity, especially towards α-L-Fuc-(1→2)-Gal and α-L-Fuc-(1→3 or 4)-GlcNAc substrates. A highly purified enzyme from *Aspergillus niger*[261] acted on α-L-Fuc-(1→2)-Gal only, and not even on α-L-Fuc-(1→2)-Glc. Similarly, specific activity towards α-L-Fuc-(1→2)-Gal was purified from *Clostridium perfringens*,[262] *Trichomonas foetus*,[263] rat cerebral-cortex,[264] and almond emulsin.[265] These sources[262-265] also provided pure samples of fucosidases acting on α-L-Fuc-(1→3 or 4)-GlcNAc linkages

Certain studies have sought to correlate the immunological identities of α-L-fucosidases from various mammalian sources, with a view to their utilization in replacement therapy in cases of lysosomal disorders involving this enzyme.[266]

Inhibitors of some of these enzymes have been reported. Usually, L-fucose inhibits at higher concentrations as do certain aldonolactones.[258] L-Fucono-1,4-lactone was found to be a specific inhibitor for porcine-kidney α-L-fucosidase,[267] although it had been reported[257] that the crude enzymes from rat epididymis and ox liver were not inhibited by this lactone.

c. β-D-Fucosidases.—β-D-Fucosidase activity towards simple glycosides has been reported in many, diverse extracts, including those from mammalian tissues,[268,269] the snail *Helix pomatia*,[270] and almond emulsin.[271] The available evidence would appear to indicate that β-D-fucosidase activity cannot be defined separately from β-D-galactosidase activity. Thus, a human-liver enzyme was reported to be capable

(261) O. P. Bahl, *J. Biol. Chem.*, 245 (1970) 299–304.
(262) D. Aminoff and K. Furukawa, *J. Biol. Chem.*, 245 (1970) 1659–1669.
(263) M. A. Chester, A. D. Yates, and W. M. Watkins, *Eur. J. Biochem.*, 69 (1976) 583–592.
(264) H. B. Bosmann and B. A. Hemsworth, *Biochim. Biophys. Acta*, 242 (1971) 152–171.
(265) M. Ogata-Arakawa, T. Muramatsu, and A. Kobata, *Arch. Biochem. Biophys.*, 181 (1977) 353–358.
(266) T. Alam and A. Balasubramanian, *Biochim. Biophys. Acta*, 524 (1978) 373–384.
(267) G. Ya. Vidershaim and E. I. Rozenfel'd, *Biokhimiya*, 34 (1969) 398–403.
(268) G. A. Levvy, *Nature (London)*, 187 (1960) 1027.
(269) M. L. Llanillo, N. Perez, and J. A. Cabezas, *Int. J. Biochem.*, 8 (1977) 557–564.
(270) A. Marnay, *Experientia*, 20 (1964) 441.
(271) D. E. Walker and B. Axelrod, *Arch. Biochem. Biophys.*, 187 (1978) 102–107.

of hydrolyzing 4-methylumbelliferyl β-D-gluco-, -galacto-, -fuco-, and -xylo-pyranosides.[272] Although it was reported[270] that the β-D-fucosidase and β-D-galactosidase activities of *Helix pomatia* could be separated by differential heat-inactivation, the latter activity being the more labile, other investigators[269,271] could not differentiate several glycosidase activities in extracts from other sources. Thus, inhibition studies of the rabbit-liver[269] and almond-emulsin[271] enzymes showed that they had concomitant β-D-fuco-, -D-gluco-, and -D-galacto-sidase activities, and good evidence was educed for a single catalytic site for these three activities in the almond enzyme. In both cases, however, the catalytic constants indicated that fucosides were the best substrates.

d. α-D-Fucosidases.—Separation of α-D-fucosidase and α-D-galactosidase activities was also not attained in mammalian[273] and microbial enzymes,[274] although this problem has been less widely investigated than that of the β-D-specific enzyme. Although α-D-galactosidase and α-D-fucosidase activities of the enzyme from Streptomyces[274] were both inhibited by D-galactose, D-fucose inhibited α-D-fucosidase but not α-D-galactosidase activity.

Both activities were inhibited to the same extent in the enzyme from human and pig kidney by D-fucose, D-galactose, and D-galactono-1,4-lactone.[273]

e. β-L-Fucosidases.—β-L-Fucopyranosides are rare in Nature, although they occur in some bacteria.[248] A number of glycosidase activities were isolated from rat cerebral-cortex,[264] including an enzyme [that acted on *p*-nitrophenyl β(but not α)-L-fucopyranoside] which was purified and characterized. There is no information available on the wider substrate-specificity of this enzyme, and whether, like β-D-fucosidases, it can act on other structures, such as β-L-galactopyranosides.

f. L-Fucose Degradation.—**(i) Introduction.** Complex, fucose-containing polymers are synthesized by living organisms. Free L-fucose is incorporated into such polymers without prior conversion into other sugars.[275–277] Various oligosaccharides containing L-fucose have been

(272) M. A. Chester, B. Hultberg, and P.-A. Öckerman, *Biochim. Biophys. Acta*, 429 (1976) 517–526.
(273) G. Ya. Vidershaim and E. M. Beyer, *Arch. Biochem. Biophys.*, 182 (1977) 335–342.
(274) K. Oishi and K. Aida, *Agric. Biol. Chem.*, 39 (1975) 2129–2135.
(275) J. W. Coffey, O. N. Miller, and O. Z. Sellinger, *J. Biol. Chem.*, 239 (1964) 4011–4017.
(276) J. G. Bekesi and R. J. Winzler, *J. Biol. Chem.*, 242 (1967) 3873–3879.
(277) J. M. Sturgess, E. Minaker, M. M. Mitranic, and M. A. Moscarello, *Biochim. Biophys. Acta*, 320 (1973) 123–132.

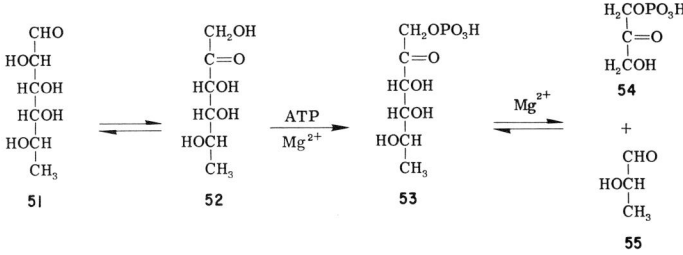

Scheme 7

isolated from mammalian urine, and characterized.[278,279] In pathological cases, more-complex oligosaccharides may be excreted,[280-282] owing to depressed activity of α-L-fucosidases acting on the complex polymers. However, there is also evidence that free L-fucose can be oxidized in the animal body.[212]

(ii) **Bacterial Metabolism.** An induced pathway for degradation of L-fucose in *Escherichia coli* has been described,[283] involving isomerization of the aldehyde form 51 to 6-deoxy-L-*lyxo*-2-hexulose ("L-fuculose," 52), followed by phosphorylation to 53, and splitting of 53 by "L-fuculose" 1-phosphate aldolase to afford 1,3-dihydroxy-2-propanone 1-phosphate (54) and L-lactaldehyde (55) (see Scheme 7). Other bacteria have been found to utilize L-fucose[284,285] by a similar pathway. However, a bacterial L-fucose (D-arabinose) dehydrogenase has been described that causes the oxidation of L-fucose to L-fuconolactone.[286]

(iii) **Mammalian Metabolism.** A soluble, NAD^+-dependent dehydrogenase purified from pork liver, and characterized,[287] oxidizes L-fucose (51) to 3-deoxy-L-*threo*-2-hexulosonate (58), D-arabinose to

(278) A. Lundblad, *Methods Enzymol.*, 50 (1978) 226–235.
(279) A. Lundblad, J. Lundstein, N. E. Nordén, S. Sjöblad, S. Svensson, P.-A. Öckerman, and M. Gehlhoff, *Eur. J. Biochem.*, 83 (1978) 513–521.
(280) G. Strecker, C. Trentesaux-Chauvet, A. Poitau, and J. Montreuil, *Biochimie*, 58 (1976) 805–814.
(281) G. Strecker, T. Riazi-Farzard, B. Fourmet, S. Bouquelet, and J. Montreuil, *Biochimie*, 58 (1976) 815–825.
(282) M. Nishigaki, K. Yamashita, I. Matsuda, S. Arashima, and A. Kobata,*J. Biochem. (Tokyo)*, 84 (1978) 823–834.
(283) M. A. Ghalambor and E. C. Heath,*J. Biol. Chem.*, 237 (1962) 2427–2433.
(284) R. G. Eagon,*J. Bacteriol.*, 82 (1961) 548–550.
(285) D. C. Old and R. P. Mortlock,*J. Gen. Microbiol.*, 101 (1977) 341–344.
(286) K. Yamanaka, K. Tzumori, and K. Matsumoto,*Kagawa Daigaku Nogakubu Gakujutsu Hokoku*, 27 (1976) 221–231; *Chem. Abstr.* 86 (1977) 85,229e.
(287) N. A. Nurokoro and H. Schachter,*J. Biol. Chem.*, 250 (1975) 6185–6190.

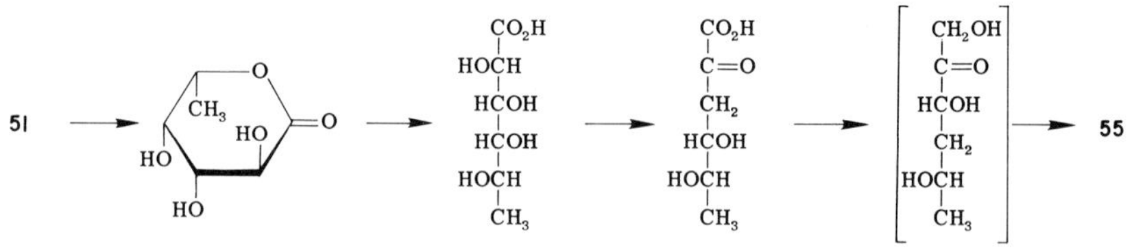

```
       CO₂H           CO₂H          ⎡ CH₂OH ⎤
       |              |             |  |    |
       HOCH           C=O           |  C=O  |
       |              |             |  |    |
51 →  [structure] → HCOH    →     CH₂    → HCOH   →  55
       HCOH           HCOH          |  CH₂  |
       |              |             |  |    |
       HOCH           HOCH          |  HOCH |
       |              |             |  |    |
       CH₃            CH₃           ⎣  CH₃  ⎦

       56             57            58           59
```

Scheme 8

3-deoxy-D-*glycero*-2-pentulosonate, and L-galactose[147] to 3-deoxy-L-*threo*-2-hexulosonate. A dehydrogenase oxidizes L-fucose to the lactone **56**, which is hydrolyzed to the free acid **57**, and the product is converted by a hydrolyase into the ketoaldonate **58**. The ketoaldonate can be further oxidized to lactate[147] by a reaction requiring NAD⁺, and evidence was provided that the enzyme "2-keto-3-deoxy-L-fuconate": NAD⁺-oxidoreductase was involved, giving an unstable, intermediate hexulose (**59**) which could not isolated (see Scheme 8).

The authors pointed out that such a metabolic pathway may complicate the results of biosynthetic studies in certain species, as it would provide a route for the oxidation of L-fucose and its eventual provision of carbon atoms to other sugars. This metabolic route does not occur in rats, but it is known[215] that humans can oxidize L-[1-¹⁴C]fucose to carbon dioxide. Only pig liver and kidney were found to contain appreciable levels of the major enzymes needed in this pathway, whereas heart, stomach, intestine, submaxillary gland, lung, and brain were deficient in or devoid of them.

g. **D-Fucose Metabolism.**—D-Fucopyranosyl derivatives and D-fucopyranosidases occur widely in Nature. The pathway of the metabolism of D-fucose has, however, only been studied in bacteria. A pseudomonad has been described[288–291] that can utilize D-fucose as the sole source of carbon and of energy. The free sugar is oxidatively degraded to pyruvate and D-lactaldehyde by steps analogous to those already described for the degradation of L-fucose in mammals. In the first stage (oxidation to D-fucono-1,5-lactone[288]), the NAD⁺-dependent dehydrogenase involved, after purification, showed broad specificity, and a number of different aldohexoses were good substrates. In fact, D-glucose was superior to D-fucose (higher V_{max} and lower K_m). Al-

(288) A. S. Dahms and R. L. Anderson, *J. Biol. Chem.*, 247 (1972) 2222–2227.
(289) A. S. Dahms and R. L. Anderson, *J. Biol. Chem.*, 247 (1972) 2228–2232.
(290) A. S. Dahms and R. L. Anderson, *J. Biol. Chem.*, 247 (1972) 2233–2237.
(291) A. S. Dahms and R. L. Anderson, *J. Biol. Chem.*, 247 (1972) 2238–2241.

though D-mannose, D-altrose, D-galactose, and 2-deoxy-D-*arabino*-hexose were satisfactory substrates, only D-glucose and D-fucose (among the sugars tested) were effective inducers of the enzyme. The D-fuconolactone formed enzymically was spontaneously hydrolyzed to D-fuconic acid.

A different aldose dehydrogenase, also having a low (broad) specificity, was purified[289] from the same organism; it oxidized D-fucose to the 1,4-lactone, and both NAD^+ and $NADP^+$ could serve as coenzymes, the L-*arabino* configuration at C-2 to C-4 seeming to be necessary in the substrate. In fact, L-arabinose was a better substrate than D-fucose. The enzyme was induced by growth of the organism on D-fucose, D-galactose, or L-arabinose. D-Fucono-1,4-lactone could be isolated from incubation mixtures, and was hydrolyzed by a lactonase also present, and purified from cell extracts to afford D-fuconic acid.

An aldonic acid dehydratase was partially purified,[290] and found to convert D-fuconate into 3-deoxy-D-*threo*-2-hexulosonate, and L-arabinonate into the corresponding 3-deoxy-L-*glycero*-2-pentulosonate. It did not act on L-fuconate or on D-galactonate.

Finally, the 3-deoxy-D-*threo*-2-hexulosonate was split[291] by an aldolase into pyruvate and 3-deoxy-D-glyceraldehyde, the L-*arabino* derivative being a rather better substrate than that derived from D-fucose.

VI. CLINICAL

1. Effect of Free Sugars

Various studies have shown the toxicity of various free monosaccharides towards mammalian cells. L-Fucose was found to inhibit the growth of some cells in culture, but only at high doses,[292] and to inhibit the growth of implanted, mammary tumors in rats.[293] In an examination of the effect of 93 carbohydrates on cell proliferation,[294] D-fucose was one of the 42 that were toxic or growth-inhibitory, but the effect of L-fucose was not reported. The inhibitory effect of L-fucose in implanted, mammary tumors[293] was considered to be accompanied by its incorporation into serum glycoproteins, and this would offer a possible explanation for some of its biological properties.

(292) R. P. Cox and B. M. Gesner, *Cancer Res.*, 28 (1968) 1162–1172.
(293) D. E. Wolfe, J. M. Roseman, E. E. Miller, M. H. Seltzer, and F. E. Rosato, *J. Surg. Oncol.*, 3 (1971) 73–77.
(294) R. L. Burns, P. G. Rosenberger, and R. J. Klobe, *J. Cell. Physiol.*, 88 (1976) 307–316.

2. Urinary Oligosaccharides

The occurrence of L-fucosides in normal urine, and changes in their structure in pathological cases, have already been mentioned.[278–282] Normal, human urine contains a large number of different glycoproteins, glycopeptides, and oligosaccharides.[295] Small amounts of fucoglycoproteins having blood-group activities have been identified in urine, but only partially characterized[296,297]; a large number of fucoglycopeptides lacking ABH activity have also been isolated, and partially characterized.[295] They constitute a very heterogeneous group, with molecular weights ranging from 2,000 to 10,000.

The interesting glycoside β-D-Glc-$(1\rightarrow3)$-α-L-Fuc-$(1\rightarrow3)$-L-threonine was isolated[298] from the urine of both secretors and nonsecretors belonging to different ABH and Lewis blood-groups. It was later synthesized[67] by a two-step procedure; first, an unstable, syrupy, benzylated bromide of a disaccharide of D-glucose and L-fucose was prepared, and this bromide was then condensed with N-(benzyloxycarbonyl)-L-threonine benzyl ester by the halide ion-catalyzed process.[97] Protecting groups were removed by hydrogenolysis.

A large number of L-fucose-containing oligosaccharides having ABH or Lewis specificities have been isolated from human urine,[295,298] and characterized. During the later stages of pregnancy and lactation, the urine also contains some of the characteristic, milk oligosaccharides, and it is considered[295] that, as the structure of the latter is dependent on the secretor and blood-group status of the individual,[299] so are the urinary oligosaccharides most probably determined.

Different oligosaccharides are excreted in pathological urine, and a number of them have been characterized. After ingestion of myo-inositol, especially by secretors, large amounts of α-L-Fuc-myo-inositol are excreted; α-L-Fuc-$(1\rightarrow2)$-D-Glc appears in normal urine (and is galactosylated in galactosuric patients), and a large variety of other fucose-containing di-, tri-, tetra-, penta-, and higher saccharides appear in normal urine. The urine of diabetics, lactosuric subjects, and patients having fucosidosis is especially enriched in these oligosaccharides. In a patient with fucosidosis, a decasaccharide was identified as a minor component[282]; in a different laboratory,[300] a

---

(295) A. Lundblad, in Ref. 14, pp. 441–458; see also, E. H. F. McGale, *Adv. Carbohydr. Chem. Biochem.*, 24 (1969) 435–452.
(296) A. Lundblad and I. Berrgård, *Biochim. Biophys. Acta*, 148 (1967) 146–150.
(297) R. Bourrillon and Y. Goussault, *Carbohydr. Res.*, 8 (1968) 175–184.
(298) P. Hallgren, A. Lundblad, and S. Svensson, *J. Biol. Chem.*, 250 (1975) 5312–5314.
(299) A. Kobata, in Ref. 14, pp. 423–440.
(300) G. C. Tsay, G. Dawson, and S.-S. J. Sung, *J. Biol. Chem.*, 251 (1976) 5852–5859.

decasaccharide possibly having the same structure was characterized as the major component of water-soluble storage-products isolated from the tissues and urine of fucosidosis patients. The glycoside α-L-Fuc-(1→6)-β-GlcNAc-asparagine, which also accumulates, forms part of the structural grouping occurring in many secreted glycoproteins.

Variations in the quantities and structures of urinary oligosaccharides may offer a diagnostic test for pathological conditions. They reflect a deficiency of α-L-fucosidase activity, leading to excretion of more of the fucosylated saccharides than in normal subjects.

3. Malignancy

In a number of cancer patients, serum fucose levels showed marked variations from normal. After some early, contradictory correlations of serum glycoprotein–fucose levels and the malignancy of tumors,[301] a vast amount of literature has accumulated that specifically shows a higher fucose content in the serum glycoproteins of the cancer patients, and the diagnostic value of its estimation.[302-307] Elevated fucosphingolipid levels were also reported in cases of malignancy and transformation, including the accumulation of the novel fucosylceramide in some human-colon tumors, mentioned previously.[172] A suggestion was made[308] that the cytotoxic action of 6-thioguanine on neoplasia might be due to its inhibition of incorporation of fucose into cellular glycoproteins, caused by its depression of formation of GDP-Fuc.

Fucosyltransferase levels were, however, found to be elevated in patients having various neoplasia[239-242] (see Section V,1,c), and assay of the level affords a means of diagnosis of neoplasia in these patients, and of ascertaining the success of surgery, chemotherapy, or radiation.[242] However, it should be recalled (see Section III,5,b,iii) that

(301) R. A. L. Macbeth and G. McBride, *Cancer Res.*, 25 (1965) 1779–1780.
(302) T. K. Dutta, U. Sengupta, and B. D. Gupta, *Indian J. Cancer*, 13 (1976) 262–266.
(303) N. Shaug and T. Hofstad, *Scand. J. Dent. Res.*, 85 (1977) 142–148.
(304) T. Tatsumura, H. Sato, A. Mori, Y. Kemori, K. Yamamoto, G. Fukatanai, and S. Kuno, *Cancer Res.*, 37 (1977) 4101–4103.
(305) R. Koehler, K. Mainzer, and J. Fuehr, *Aerztl. Lab.*, 24 (1978) 40–45.
(306) M. K. Wallack, A. S. Brown, E. F. Rosato, S. Rubin, J. L. Johnson, and F. Rosato, *J. Surg. Oncol.*, 10 (1978) 39–44.
(307) T. P. Waalkes, C. W. Gehrke, D. C. Tormey, K. B. Woo, K. C. Kuo, J. Snyder, and H. Hansen, *Cancer*, 41 (1978) 1871–1882.
(308) J. S. Lazo, K. M. Hurang, and P. C. Sartorelli, *Cancer Res.*, 37 (1977) 4250–4255.

certain fucolipids are found in normal, but rapidly dividing, tissues, rather than in cells that divide more slowly.

4. Fucosidosis

Genetic disorders, leading to the accumulation of glycolipids and glycosaminoglycans, and caused by deficiency of lysosomal enzymes have been studied intensively (for reviews, see Refs. 309–313). A specific disorder caused by α-L-fucosidase deficiency was first reported[314] in 1966, and shown to lead to accumulation of glycolipids and glycoproteins (proteoglycans). Fucolipid ("H-antigen," α-L-Fuc-(1→2)-Gal-GlcNAc-Gal-Glc-Cer) was characterized in biopsied liver from a patient with fucosidosis,[315,316] and an asparagine-linked fucoglycopeptide was found to be stored in cultured fibroblasts[314] and urine[279] from such patients. The structure of a decasaccharide (**60**) accumulating in

$$
\begin{array}{l}
\alpha\text{-L-Fuc-}(1\rightarrow 2)\text{-}\beta\text{-Gal-}(1\rightarrow 4)\text{-}\beta\text{-GlcNAc-}(1\rightarrow 2)\text{-}\alpha\text{-Man} \\
\underset{\downarrow}{1} \\
3 \\
\beta\text{-Man-}(1\rightarrow 4)\text{-GlcNA} \\
6 \\
\uparrow \\
1 \\
\alpha\text{-L-Fuc-}(1\rightarrow 2)\text{-}\beta\text{-Gal-}(1\rightarrow 4)\text{-}\beta\text{-GlcNAc-}(1\rightarrow 2)\text{-}\alpha\text{-Man}
\end{array}
$$

60

tissues and urine of patients was established,[300] and it was accompanied by some 2-acetamido-2-deoxy-6-O-α-L-fucopyranosyl-D-glucose. Similarities were drawn[317] between the accumulation of fucosaccharide and fucolipid in fucosidosis, and of glycolipid in gangliosidosis. Other fuco-oligosaccharides and fucose-rich glycopeptides

(309) H. G. Hers and F. van Hoof (Eds.), *Lysosomes and Storage Diseases*, Academic Press, New York, 1973.
(310) R. J. Desnick, S. R. Thorpe, and M. B. Fiddler, *Physiol. Rev.*, 56 (1976) 57–99.
(311) A. Dorfman and R. Matalon, *Proc. Natl. Acad. Sci. USA*, 73 (1976) 630–637.
(312) M. Cantz and J. Gehler, *Humangenetik*, 32 (1976) 233–255.
(313) R. O. Brady, *Annu. Rev. Biochem.*, 47 (1978) 687–713.
(314) P. Durand, C. Borrone, and J. Della Cella, *Lancet*, (1966) 1313; *J. Pediatr.*, 75 (1969) 665–674.
(315) G. Dawson and J. W. Spranger, *N. Engl. J. Exp. Med.*, 285 (1971) 122.
(316) G. C. Tsay and G. Dawson, *Biochem. Biophys. Res. Commun.*, 63 (1975) 807–814.
(317) G. C. Tsay and G. Dawson, *J. Neurochem.*, 27 (1976) 733–740.

were also characterized from the urine of a fucosidosis patient.[281] The increased excretion of keratan sulfate (which contains nonreducing, terminal α-L-Fuc groups) was reported in two fucosidosis patients[318]; they also exhibited profound, α-L-fucosidase deficiency. Family studies of α-L-fucosidase levels were made for fucosidosis patients,[319] and changed kinetic constants, as well as marked deficiency of the enzyme, were noted for the pathological cases. Cultured, skin fibroblasts of fucosidosis patients also showed marked depression in α-L-fucosidase activity towards p-nitrophenyl and 4-methylumbelliferyl α-L-fucopyranoside.[320] Increased activity of α-L-fucosidase was observed after incubation of cell lysates with sialidase, an observation that may indicate hindrance to attack of the enzyme on glycoconjugates by vicinal sialic acid groups in the substrate molecule; the hindrance to fucosyltransferases exhibited by vicinal sialyl groups[234] may be compared.

5. Other Illnesses

The physiological basis of cystic fibrosis (CF) has still not been clarified. Attempts have been made to correlate derangements in the metabolism of fucose with the appearance of the disorder. Conflicting reports have been made concerning the appearance of increased incorporation of fucose into cultured CF fibroblasts.[321–323] In a study[324] indicating significantly higher incorporation of fucose into glycoprotein acceptors from saliva of patients, no difference was found in either their fucosyltransferase or their fucosidase activities, in comparison with those of normal controls. On the other hand, other investigators reported[325] that the activity of α-L-fucosidase was decreased in the serum of CF patients, compared to age-matched con-

---

(318) H. Greiling, H. W. Stuhlsatz, M. Cantz, and J. Gehler, *J. Clin. Chem. Clin. Biochem.*, 16 (1978) 329–334.
(319) J. Troest, M. C. M. Van der Heijden, and G. E. J. Staal, *Clin. Chim. Acta*, 73 (1976) 329–346.
(320) N. G. Beratis, B. M. Turner, G. Labadie, and K. Hirschhorn, *Pediatr. Res.*, 11 (1977) 862–866.
(321) S. Wood, *Clin. Genet.*, 10 (1976) 183–186.
(322) P. B. Davis and H. C. Smith, *IRCS Med. Sci. Libr. Compend.*, 5 (1977) 353; *Chem. Abstr.*, 87 (1977) 116,095k.
(323) T. F. Scanlin and M. C. Glick, *Pediatr. Res.*, 11 (1977) 463.
(324) A. K. Guha, K. M. Kutty, R. K. Chandra, and R. C. Way, *Clin. Biochem.*, 10 (1977) 153–155.
(325) T. F. Scanlin, S. S. Matacic, and M. C. Glick, *Clin. Chim. Acta*, 91 (1979) 197–202.

trols, and that it was elevated in skin fibroblasts; other glycosidases were not so affected. The authors proposed a faulty regulation of the intra- and extra-cellular activities of α-L-fucosidase in the disease, and suggested that CF is an "enzyme distribution disease." It is, therefore, possible that CF is also a metabolic disorder of glycosidases, rather than of glycosyltransferases.

Elevated levels of fucose in the glycoproteins of patients with ulcers,[326] diabetes,[327] and other illnesses have been reported, but the significance of these findings is not yet clear.

6. Fucose and the Immune-Defense System

Carbohydrate receptors have been implicated in certain structures of the H-2 complex, and there was apparently specific inhibition of the killing, by T-lymphocytes, of mouse target-cells by p-nitrophenyl α-L-fucopyranoside, but not by p-nitrophenyl glycosides of other sugars.[328] The binding of anti-Ia antibodies to B-lymphocytes was inhibited by a number of monosaccharides and milk oligosaccharides, and the Ia.7 region was apparently characterized by L-fucose, and by the α-L-Fuc-(1→2)-Gal type of linkage.[329] It was proposed that at least some of the genes in the I region of the H-2 complex code for glycosyltransferases.[330] However, other investigators did not find results consistent with this hypothesis, and postulated[331] that it is the protein portion of the Ia glycoprotein antigens that determines their antigenic activity.

Macrophages are activated or inhibited by various factors (lymphokines). In guinea-pigs, it was found that α-L-Fuc inhibits the macrophage migration–inhibition reaction, and α-L-fucosidase causes macrophages to become transiently refractory to the action of macrophage, migration–inhibition factor (MIF), thus suggesting that α-L-Fuc may

---

(326) P. D. Rabinovitch and N. I. Domracheva, *Vopr. Med. Khim.*, 23 (1977) 478–482; *Chem. Abstr.*, 87 (1977) 131,730g.

(327) C. Sitadevi, M. V. V. Krishnamahan, K. Suryanarayan, B. Swanny, and G. Chittipantabu, *Clinician*, 41 (1977) 143–151.

(328) P. E. Trefts and J. A. Alhadeff, in W. H. Fishman and B. Sell (Eds.), *Oncodevelopmental Gene Expression*, Academic Press, New York, 1976, pp. 517–522.

(329) C. R. Parish, D. C. Jackson, and I. F. C. McKenzie, in H. O. McDevitt (Ed.), *Ir Genes and Ia Antigens, Proc. Ir Gene Workshop, 3rd.*, Academic Press, New York, 1978, pp. 243–253.

(330) C. R. Parish, D. C. Jackson, and I. F. C. McKenzie, *Immunogen*, 3 (1976) 455–463.

(331) J. H. Freed and S. G. Nathenson, in Ref. 329, pp. 263–273.

be part of a macrophage, cellular receptor-site for MIF. However, the action of L-fucose seems to be more general than inhibition of MIF activity, as it (as well as L-rhamnose) also inhibits macrophage-chemotactic and neutrophil-chemotactic factors.[332] There was some evidence[333] that glycolipids might be receptors for both MIF and macrophage-activation factor. None of the common gangliosides were implicated, but treatment of guinea-pig, peritoneal macrophages with liposomes containing protein-free preparations of bovine-brain gangliosides, or water-soluble glycolipids extracted from guinea-pig macrophages, enhanced their responsiveness to MIF. This enhanced response was abolished by prior incubation of the macrophages with fucose-binding lectins (*Lotus tetragonolobus* and *Ulex europeus* lectins), suggesting that the MIF-binding component donated by the liposomes might be a fucose-containing glycolipid. However, using disaccharides containing α-L-Fuc-(1→2, 3, and 6)-linkages, no preferential interaction of any linkage type with MIF could be revealed.[334]

VII. CONCLUDING REMARKS

The widespread occurrence of fucose in glycans and glycoconjugates has aroused great interest in its chemistry and biochemistry. The chemical reactions performed on simpler derivatives have usually not elicited unexpected results, and numerous acetals, esters, ethers, glycosides, and other products have been characterized by the usual methods. N.m.r.- and mass-spectral data are also available for many of these compounds. Comparison of the reactivity, towards certain reagents, of some derivatives of fucose with similar derivatives of galactose and arabinose indicates that steric factors alone cannot explain the relative effects of CH_3-5 and CH_2OH-5 on reactions at OH-4 and C-1, but that electronic effects must also play an important role.

There have been striking advances in the synthesis of α-linked oligosaccharides, including L-fucopyranosides, of biological interest, and artif'cial antigens have been prepared[335] that offer medical applications. Although the construction of complex heteroglycans by stepwise processes is still a formidable challenge, the simpler possibility

(332) A. Amsden, V. Euran, T. Yoshida, and S. Cohen, *J. Immunol.*, 120 (1978) 542–549.
(333) G. Poste, R. Kirsh, and I. J. Fidler, *Cell. Immunol.*, 44 (1979) 71–88.
(334) G. Poste, H. Allen, and K. L. Matta, *Cell. Immunol.*, 44 (1979) 89–98.
(335) R. U. Lemieux, *Chem. Soc. Rev.*, 7 (1978) 423–452.

of synthesizing fragments having biological activity, and artificial analogs, is now available, and will, presumably, be explored considerably in the near future.

Pathways of biosynthesis, and biological degradation, of L-fucose have been elucidated, and the enzymes involved in the transfer of L-fucose to suitable acceptors, and in its hydrolytic removal from complex saccharides, have been characterized. Considerable specificity is exhibited by α-L-fucosyltransferases and α-L-fucosidases with respect to the anomeric configuration of the linkage formed, or hydrolyzed, and also, frequently, as regards the nature of the sugar substituted and the position of substitution by the fucopyranosyl group. Such specificity apparently does not apply to α- and β-D-fucosidases, which usually exhibit additional glycosidase activities.

Accumulation of fucosylated derivatives in pathological cases, and excessive excretion of similar compounds in the urine, generally reflects deficiency of α-L-fucosidase or depression of its activity, and study of these compounds can serve as a diagnostic tool.

In the biochemistry of D-fucose and its derivatives, there are larger lacunae in our knowledge. The biosynthesis of D-fucose has not been examined, so that it is not yet known how this sugar is produced in the organisms that contain it (mainly plants and some micro-organisms). In plants, it occurs especially in the form of steroid glycosides, whereas, in micro-organisms, it has been particularly located in antibiotic substances. It has not been identified in animals.

The metabolic degradation of D-fucose has been investigated in bacteria, and oxidative steps elucidated that closely resemble those undergone by the L enantiomer in the animal body.

The biosynthesis of polymers of L-fucose of high molecular weight occurs in the plant kingdom, and is considered to proceed by transfer of L-fucopyranosyl groups from GDP-Fuc. A similar process could in, principle, be postulated for formation of D-fucans. The conversion of GDP-D-Man into GDP-Fuc serves as a preliminary stage in the biosynthesis of L-fucans, and its hydrolysis affords free L-fucose. GDP-L-Man has not been identified, and the absence of a pathway for formation of D-fucose by way of a putative GDP-D-Fuc, and the lack of synthesis of complex D-fucans, may be related. In addition, the author is unaware of any information on D-fucosyltransferases. The mechanism of methylation of D-fucose is also unknown. It is obvious that research on the biosynthesis of the fucose-containing, cardiac glycosides of plants and fucose-containing *Streptomyces* antibiotics could provide valuable information on these questions.

VIII. Tables of Some Properties of the Fucoses and Their Derivatives

The following abbreviations are used in Tables I and II: A = acetone, B = benzene, C = chloroform, D = dichloromethane, E = ethanol, M = methanol, P = pyridine, W = water, eq = equilibrium rotation value, f = furanoside, p = pyranoside, and R = 2-acetamido-2-deoxy-D-glucose; see also Ref. 99.

TABLE I

Properties of D-Fucose and its Derivatives

Compound	M.p. (°C)	$[\alpha]_D$ (degrees)	Rotation solvent	References
Free, Acyclic, and Glycosides				
Fucitol	153.5	−4.6	10% 1,4-dioxane	336
α-Fucofuranose, tetra-O-benzoyl-	88–93	+52.3	C	17
Fucono-1,4-lactone	104–106	−71	W	76
α-Fucopyranose, tetra-O-acetyl-	92–93	+143	A	76
		+129	C	76
Fucose	140–145	+76 (eq)	W	337
(*p*-bromophenyl)hydrazone	175–177	+7.8	M	19
diphenylformazan	163	—	—	338
2,2-diphenylhydrazone	187–188	—	—	18
phenylhydrazone	173–175	0	W	20
tetra-O-acetyl-	142–144	+32	E–W	338
4-Aminophenyl 1-thio-β-*p*-	75–78	−76.6	M	339
Benzyl α-*p*-	161.5–162.5	+196.6	M	340
Methyl α-*p*-	158–159	+207.8	M	341
Methyl β-*p*-	121.5–122.5	−16.4	W	342
2-O-acetyl-	135	−15.2	C	343
Methyl β-*p*, tri-O-acetyl-	99–100	−6.6	C	343
tri-O-benzoyl-	163–164	+190	C	49
2,3-di-O-acetyl-4-O-benzoyl-α-*p*-	76–77	+187	C	344
4-O-benzoyl-2,3-di-O-mesyl-α-*p*-	135–136	+121	C	49
4-O-benzoyl-2,3-di-O-mesyl-β-*p*-	—	+10.4	C	49
3-O-tosyl-β-*p*-	135	+33.9	C	343
2,4-Dinitrophenyl β-*p*-	135–136	−87	M	345
tri-O-acetyl-	169–172	+74	C	345
p-Nitrophenyl α-*p*-	193–194	+329	A	76
tri-O-acetyl-	169–170	+231	C	76
p-Nitrophenyl β-*p*-	186–188	−96	A	76
		−85	E	76
1-thio-β-*p*-	177–178	−113	M	339
tri-O-acetyl-	139–140	4.3	C	339
Acetals, Ethers				
Fuconic acid, amide, 2,3-di-O-methyl-	145–146	+15.7	M	346
phenylhydrazide, 2,3-di-O-methyl-	137–142	+18.0	M	346

Continued

TABLE I (continued)

Compound	M.p. (°C)	$[\alpha]_D$ (degrees)	Rotation solvent	References
Fucono-1,4-lactone, 3-O-methyl-	137–138	−92.5 → −74.9 (eq)	W	347
Fucopyranose, 1,2:3,4-di-O-isopropylidene-	37	−52.4	melt	42
3,4-O-isopropylidene-	110–111	+71 (eq)	W	348
2-O-methyl-	150–153	+79 (eq)	W	348
Fucose, 2-O-benzyl-	164	+66.3 (eq)	W	349
2-benzyl-2-phenylhydrazone	141	−22.5 → −17.4 (eq)	M	349
diethyl dithioacetal	96	−4.6	E	349
tri-O-acetyl-	62–63	+4.5	E	349
O-isopropylidene-	—	+15.3	E	349
dimethyl acetal	93–94	+11.1	W	349
4,5-O-isopropylidene-	—	+21.1	E	349
3-O-methyl-	—	+7.4	M	349
Fucose, dimethyl acetal, 4,5-O-isopropylidene-	95	+25.9	W	349
2-O-methyl-	155–161	+73 → +87 (eq)	W	72
	155–157	+88 (eq)		350
3-O-methyl-	99–101	+104.5 (eq)	W	351
	122–126	+105.2 (eq)	W	352
phenylosazone	178–179	+0.5	—	353
4-O-methyl-	131–132	+82	W	21
	132–133	+103		354
p-tolylsulfonylhydrazone	134	−16 → −3.3 (eq)	P	21
2,3-di-O-methyl-	79–80	+109.5	W	355
2,4-di-O-methyl-	131.5–132	+146 → +100 (eq)	W	356
2,3,4-tri-O-methyl-	—	+106	W	357
2,3,5-tri-O-methyl-	—	−36.8	E	350
Fucosylamine, N-phenyl-, 2,4-di-O-methyl-	154	−86 → +61 (eq)	E	356
2,3,4-tri-O-methyl-	133–135	+76 (eq)	E	357
Methyl 2-O-benzyl-3,4-O-isopropylidene-α-p-	—	+97.2	C	349
β-p-	—	+35.9	C	342
		+10	M	342
3,4-di-O-(methylsulfonyl)-	153	+60.7	M	342
3,4-di-O-p-tolylsulfonyl-	165–166	+56.6	C	342
3-O-benzyl-2,4-di-O-methyl-α-p-	—	+137	M	356
2,3-di-O-benzyl-β-p-	—	+31.5	C	354
4-O-methyl-β-p-	—	+8.5	C	354
4-O-p-tolylsulfonyl-α-p-	157.5–158.5	+18.0	C	358
3,4-O-benzylidene-β-p-	97–99	+2	C	47
3,4-O-isopropylidene-α-p-	38–39	+166.5	W	349

TABLE I (continued)

Compound	M.p. (°C)	[α]$_D$ (degrees)	Rotation solvent	References
Methyl 3,4-O-isopropylidene-α-p-,				
2-O-p-tolylsulfonyl-	186–188	+158	C	341
-β-p-	64–66	+24.5	C	343
		+13.2	M	343
2-O-methyl-α-p-	—	+174	W	197
3-O-methyl-α-p-	98.5–100.5	+198.5	M	197
-β-p-	108–110	+9.9	M	197
4-O-methyl-β-p-	144–145	−14.6	M	354
2,3-di-O-methyl-α-p-	—	−190	A	350
2,3-di-O-methyl-β-p-	163–167	+15.2	A	346
2,4-di-O-methyl-α-p-	86	+165	M	356
tri-O-methyl-α-p-	97–99	+207	A	350
β-p-	93–98	+112	W	357
2-O-trityl-α-p-	68–70	+55.5	C	70
3,4-di-O-acetyl-	172–174	+38.0	C	70
3-O-trityl-α-p-	71–74	+81.4	C	70

(336) E. Votoček and J. Bulíř, Z. Zuckerind., 30 (1906) 333–339; Chem. Zentralbl., 1 (1906) 1818–1819.
(337) A. Fogh, I. Lundt, and C. Pedersen, Acta Chem. Scand., Ser. B, 30 (1976) 624–626.
(338) V. Zsoldos, A. Messmer, I. Pinter, and A. Neszmélyi, Carbohydr. Res., 62 (1978) 105–116.
(339) R. H. Shah and O. P. Bahl, Carbohydr. Res., 32 (1974) 15–23.
(340) K. Heyns, A. L. Baron, and H. Paulsen, Chem. Ber., 97 (1964) 921–925.
(341) B. Iselin and T. Reichstein, Helv. Chim. Acta, 29 (1946) 508–512.
(342) H. Paulsen and H. Redlich, Chem. Ber., 107 (1974) 2992–3012.
(343) S. A. S. Al Janobi, J. G. Buchanan, and A. R. Edgar, Carbohydr. Res., 35 (1974) 151–164.
(344) L. M. Jeffersen, I. Lundt, and C. Pedersen, Acta Chem. Scand., 27 (1973) 3579–3585.
(345) F. Ballardie, B. Capon, J. D. Sutherland, D. Cocker, and M. Sinott, J. Chem. Soc., Perkin Trans. 1, (1973) 2418–2419.
(346) M. P. Khare, O. Schindler, and T. Reichstein, Helv. Chim. Acta, 45 (1962) 1547–1551.
(347) S. Morgenlie, Carbohydr. Res., 41 (1975) 77–83.
(348) O. T. Schmidt, W. Mayer, and A. Distelmaier, Ann., 555 (1943) 26–41.
(349) O. T. Schmidt and E. Wernicke, Ann., 558 (1947) 70–80.
(350) G. F. Springer and P. Williamson, Biochem. J., 85 (1962) 282–293.
(351) O. T. Schmidt and E. Wernicke, Ann., 556 (1944) 179–186.
(352) O. Schindler and T. Reichstein, Helv. Chim. Acta, 36 (1953) 370–376.
(353) H. Hegedüs and T. Reichstein, Helv. Chim. Acta, 38 (1955) 1133–1146.
(354) E. G. Gros, Carbohydr. Res., 2 (1966) 56–62.
(355) T. Fujikawa, Carbohydr. Res., 38 (1974) 325–327.
(356) P. P. Singh and G. A. Adams, Carbohydr. Res., 13 (1970) 229–234.
(357) S. P. James and F. Smith, J. Chem. Soc., (1945) 746–748.
(358) C. L. Stevens, P. Blumbergs, F. A. Danther, D. H. Otterbach, and K. G. Taylor, J. Org. Chem., 31 (1966) 2822–2929.

TABLE II

Properties of L-Fucose and its Derivatives

Compound	M.p. (°C)	[α]$_D$ (degrees)	Rotation solvent	References
Free, Acyclic, and Glycosides				
Fucal	70–72	+10.4	C	359
3,4-di-O-acetyl-	47–49	+9.9	C	359
Fucitol	153–154	+4.7	10% borax	360
1-O-benzoyl-	177–178	+4.3	P	361
2,3,4,5-tetra-O-acetyl-	116–117	+18.6	C	361
penta-O-acetyl-	127	+20.5	C	362
penta-O-benzoyl-	149–150	−5.96	C	361
2,3,4,5-tetra-O-acetyl-	92–94	−15	C	362
1-O-trityl-	152	−18	C	362
Fuconic acid, phenylhydrazide	203–204	—	—	363
Fucono-1,4-lactone	104–106	+74	W	76
α-Fucopyranose, 2,3,4-tri-O-acetyl-	117	−118	E	80
tetra-O-acetyl-	92	−120	C	78
β-Fucopyranose, 2,3,4-tri-O-acetyl-	102–103	−5.2 → −77 (eq)	E	80
tetra-O-acetyl-	172	−39	C	79
Fucose	145	−121 → −76 (eq)	W	364
2-(barium sulfate)	—	−49.6	W	365
3-(barium sulfate)	—	−41.0	W	365
4-(barium sulfate)	—	−51.2	W	365
benzylphenylhydrazone	172–173	+9.1	P	363
diethyl dithioacetal	167–168.5	—	—	366
diphenylhydrazone	198	—	—	363
methylphenylhydrazone	177	+3.6	P	363
phenylosazone	178	−70	P–E	367
β-Fucosylamine	146–147 (dec.)	−50	W	368
N-(6-aminohexanoyl)-	170–171	+0.5	W	368
N-(6-benzamidohexanoyl)-	169–170	+1.5	W	368
Allyl α-D-	158–160	−216	W	369
p-Aminophenyl α-D-	175	−204	M	79
1-thio-α-D-	138–139	−357	M	339
β-D-	65–78	+66.0	M	339
Benzyl β-D-	—	+46	W	81
2-O-acetyl-	112–113	+45.0	C	370
tri-O-acetyl-	141–145	+41	C	81
2-O-acetyl-3,4-di-O-(methylsulfonyl)-	132–133	+19.1	C	370

TABLE II (continued)

Compound	M.p. (°C)	$[\alpha]_D$ (degrees)	Rotation solvent	References
Methyl α-f-	127–128	−108	W	65
β-f-	—	+112	W	65
tri-O-acetyl-	71	+75.5	M	65
α-p-	154	−197	W	71
	158–159	−191	W	65
2-O-acetyl-	82–83	−196	C	73
tri-O-acetyl-	67	−150	C	71
2-O-acetyl-3,4-di-O-(methyl-sulfonyl)-	144	−168	C	73
β-p-	121–123	+14.2	W	71
	126–127	+10.5		65
tri-O-acetyl-	97–99	+7.1	C	71
2-(barium sulfate)	103	+71.5	W	365
Methyl 2,3-di-O-benzoyl-α-p-	—	−187	C	64
tri-O-benzoyl-	147–148	−260	C	64
2,3-di-O-benzoyl-4-O-(methyl-sulfonyl)-α-p-	162–164	−159	C	64
tri-O-(chlorosulfonyl)-α-p-	86	−142.5	C	88
β-p-	99–100	+8	C	77
	101.5	−20	C	88
2-O-p-tolylsulfonyl-α-p-	158	−85	C	371
p-Nitrobenzyl α-p-	149–151	−142	E	87
p-Nitrophenyl α-p-	194–196	−317	A	76
	196–197	—	—	79
tri-O-acetyl-	178	−222	C	79
β-p-	186–188	+95	E	77
2-O-acetyl-	165–167	+2.0	A	77
1-thio-α-p-	224–226	−382	M	339
tri-O-acetyl-	133–134	−273.5	C	339
β-p-	177–178	+116.0	M	339
tri-O-acetyl-	141–141.5	+4.7	C	339
Acetals, Ethers				
Fucitol, 2,3:4,5-di-O-isopropylidene-	59–60	+11.7	E	361
1-O-acetyl-	46–47	+26.1	C	361
1-O-benzoyl-	56–58	+18.7	C	361
1-O-p-tolylsulfonyl-	78–79	+19.7	C	361
Fuconic acid, amide, 3-O-methyl-	176–180	+16.4	W	372
Fucono-1,4-lactone, 3-O-methyl-	136–140	+20 → −74 (eq)	W	372
Fucopyranose, 1,2:3,4-di-O-isopropylidene-	37	+62.1	melt	346
Fucose, 2-O-benzyl-	168–170	−64.5	W	84
2,3-di-O-benzyl-	—	+12	C	66

Continued

TABLE II (continued)

Compound	M.p. (°C)	$[\alpha]_D$ (degrees)	Rotation solvent	References
2,4-di-O-benzyl-	133–135	−75.5	C	66
3,4-di-O-benzyl-	—	−72	C	66
2,3,4-tri-O-benzyl-	102–103	−26.5	C	83
2-O-methyl-	149–150	−75	W	23
	153.5–155	−91	W	350
3-O-methyl-	110	−97	W	65
phenylosazone	172–176	—	W	372
4-O-methyl-	—	−76.0	W	66
2,3-di-O-methyl-	75–76	−101	W	197
2,4-di-O-methyl-	—	−97	W	65
3,4-di-O-methyl-	82	−118	W	371
2,3,4-tri-O-methyl-	63–64	—	—	26
	36–37	−184 → −128 (eq)	W	348
anilide	133–134	—	—	24
Allyl 3,4-di-O-benzyl-α-p-	85–86	−126	C	67
tri-O-benzyl	86–87	−92	C	67
Benzyl 2-O-acetyl-3,4-O-isopropylidene-β-p-	96–96.5	+1.3	C	370
3,4-O-isopropylidene-α-p-	—	−126	C	92
β-p-	—	+3	C	81
Methyl 2-O-acetyl-3,4-O-isopropylidene-α-p-	97–98	−187	C	44
	100–101	−230	B	73
3,4-di-O-acetyl-2-O-benzyl-α-p-	—	−77	C	66
2-O-methyl-α-p-	—	−133	C	66
2,3-di-O-acetyl-4-O-methyl-α-p-	93–95	−138	C	66
3,4-di-O-acetyl-2-O-trityl-α-p-	208–210	−37.5	C	362
2,4-di-O-acetyl-3-O-methyl-α-p-	115–117	−177	C	66
2-O-benzyl-α-p-	79–81	−118	C	66
2,3-di-O-benzyl-	78–80	−57.8	C	66
2,4-di-O-benzyl-	—	−61.7	C	66
3,4-di-O-benzyl-	92–94	−57	C	66
tri-O-benzyl-α-p-	—	−20.0	C	83
3,4-O-isopropylidene-α-p-	—	−160	W	371
2-O-p-tolylsulfonyl-	182	−146	C	372
β-p-	58–62	−21.0	C	372
2-O-p-tolylsulfonyl-	96–98	−26.4	C	373
2-O-methyl-β-p-	88–92	−10.9	M	348
2-O-methyl-α-p-	—	−196	W	66
	—	−179	W	197
β-p-	98–99	+17.15	M	197
3-O-methyl-α-p-	76–78	−200	C	66
4-O-methyl-α-p-	134–136	−192	W	66

TABLE II (continued)

Compound	M.p. (°C)	$[\alpha]_D$ (degrees)	Rotation solvent	References
2,3-di-O-methyl-α-p-	49–51	−190	W	373
	—	−189.5	A	197
β-p-	78	−1.23	A	197
3,4-di-O-methyl-α-p-	100	−213	W	371
2-O-p-tolylsulfonyl-α-p-	103	−84	C	371
tri-O-methyl-α-p-	97–98	−209	W	348
β-p-	101.5–102.5	−21	W	348
Glycopyranosyl Halides, and Miscellaneous				
β-Fucopyranose tri-O-benzyl-1-O-(N-methyl-acetimidyl)-	89–90	−67	B	86
Fucopyranosyl bromide				
tri-O-acetyl-α-	64–66	−265	C	82
tri-O-benzyl-α-	—	−120	C	83
2-O-benzyl-3,4-di-O-(p-nitrobenzoyl)-α-	—	−272	C	84
Fucopyranosyl chloride				
tri-O-acetyl-α-	65	−215	C	77
tri-O-benzyl-α-	72–73	−169	D	86
tri-O-(chlorosulfonyl)-α-	71–72	−171	C	88
β-	112.5	−29	C	88
	128	−30	C	87
Disaccharides				
α-L-Fuc-(1→2)-L-Fuc	193–195	−168.5	W	92
β-	152–154	−53 → −50 (eq)	W	92
α-L-Fuc-(1→3)-L-Fuc	198–200	−191	W	374
α-L-Fuc-(1→4)-L-Fuc	—	−170	W	374
α-L-Fuc-(1→2)-D-Gal	—	−57	W	89
benzylphenylhydrazone	163.5	—	—	375
α-L-Fuc-(1→6)-D-Gal	—	−67	W	97
α-L-Fuc-(1→2)-D-Glc	—	−65.5	W	376
α-L-Fuc-(1→3)-D-Glc	—	−86	W	97
α-L-Fuc-(1→3)-R	218–220	−60 → −74 (eq)	W	93
β	148–150	+8 → +1 (eq)	W	93
α-L-Fuc-(1→4)-R	194–196	−77 → −99 (eq)	M–W	101
α-L-Fuc-(1→6)-R	—	+31 → +29 (eq)	E–W	90
α-L-Fuc-(1→2)-D-Xyl	185–190	−61	W	26
β-Glc-(1→2)-L-Fuc	—	−71	W	376
octaacetate	228–230	−59	C	377

Continued

TABLE II (continued)

Compound	M.p. (°C)	$[\alpha]_D$ (degrees)	Rotation solvent	References
β-GlcA-(1→4)-L-Fuc	—	−61	W	25
β-Xyl-(1→3)-L-Fuc	—	−66.5	W	132
Benzyl 2-O-α-L-Fuc-β-L-Fucp	242–244	−81.8	95% E–W	81
2-O-β-L-Fuc-α-L-Fucp	181–183	−140	M	92
Benzyl 2-O-α-L-Fuc-β-D-Galp	205–207	−97.8	W	89
Methyl 2-O-α-L-Fuc-α-L-Fucp	190–192	−227	M	92
2-O-α-L-Fuc-β-L-Fucp	210–212	−91.4	M	92
2-O-β-L-Fuc-α-L-Fucp	236–238	−88.9	M	92
2-O-β-L-Fuc-β-L-Fucp	188–190	−1.2	M	92
2-O-α-L-Fuc-α-D-Galp	202–204	+0.9	W	92
2-O-β-L-Fuc-α-D-Galp	226–228	+115	W	92
Oligosaccharides				
α-L-Fuc-(1→2)-β-Gal-(1→3)-R		−19.4	M	107
α-L-Fuc-(1→2)-β-Gal-(1→4)-R		−46.5	W	102
α-L-Fuc-(1→3)-[β-Gal-(1→4)]-R	140–142	−33	W	105
α-L-Fuc-(1→4)-[β-Gal-(1→3)]-R	—	−45.1	W	104
α-L-Fuc-(1→2)-β-Gal-(1→4)-[α-L-Fuc-(1→3)]-R	214–216	−113 → −124.5 (eq)	10% M–W	106
α-L-Fuc-(1→2)-[α-Gal-(1→3)]-β-Gal-(1→3)-R		+33.3	M	107
α-L-Fuc-(1→2)-[α-GalNAc-(1→3)]-β-Gal-(1→3)-R		+53.8	M	107
α-L-Fuc-(1→2)-α-Gal-(1→3)-D-Gal		+35.2	W	100
α-L-Fuc-(1→2)-β-Gal-(1→4)-D-Glc	230–231	−57	W	378
phenylosazone	217	−29	W	378
phenyl-p-tolylsulfonylosazone	205	−73	W	378
Lacto-N-fucopentaose I [α-L-Fuc-(1→2)-β-Gal-(1→3)-GlcNAc-(1→3)-β-Gal-(1→4)-D-Glc]	214–216	−16.3	W	379
Lacto-N-fucopentaose II {α-L-Fuc-(1→4)-[β-Gal-(1→3)]-β-GlcNAc-(1→3)-β-Gal-(1→4)-D-Glc}	213–215	−30.4	W	380
Lacto-N-difucohexaose II {α-L-Fuc-(1→4)-[β-Gal-(1→3)-β-GlcNAc-(1→3)-β-Gal-(1→4)-[α-L-Fuc-(1→3)]-D-Glc}	218–220	−68.8	W	381

(359) B. Iselin and T. Reichstein, *Helv. Chim. Acta*, 27 (1944) 1200–1203.
(360) E. Votoček and R. Potmesil, *Ber.*, 46 (1913) 3653–3655.
(361) A. T. Ness, R. M. Hann, and C. S. Hudson, *J. Am. Chem. Soc.*, 64 (1942) 982–985.
(362) R. C. Hockett and D. F. Mowery, Jr., *J. Am. Chem. Soc.*, 63 (1943) 403–409.
(363) A. Müther and B. Tollens, *Ber.*, 37 (1904) 306–311.

(364) B. Tollens and F. Rorive, *Ber.*, 42 (1909) 2009-2012.
(365) P. F. Forrester, P. F. Lloyd, and C. H. Stuart, *Carbohydr. Res.*, 49 (1976) 175-184.
(366) E. Votoček and V. Veselý, *Ber.*, 47 (1914) 1515-1519.
(367) K. Freudenberg and K. Raschig, *Ber.*, 62 (1929) 373-383.
(368) S. Blumberg, J. Hildesheim, J. Yariv, and K. J. Wilson, *Biochim. Biophys. Acta*, 264 (1972) 171-176.
(369) V. Hořejši and J. Kocourek, *Biochim. Biophys. Acta*, 297 (1973) 346-351.
(370) A. Liav (Levy) and N. Sharon, *Carbohydr. Res.*, 30 (1973) 109-126.
(371) E. E. Percival and E. G. V. Percival, *J. Chem. Soc.*, (1950) 690-691.
(372) H. Paulsen and W. Koebernick, *Chem. Ber.*, 110 (1977) 2127-2145.
(373) J. Conchie and E. G. V. Percival, *J. Chem. Soc.*, (1950) 827-832.
(374) R. H. Côté, *J. Chem. Soc.*, (1959) 2248-2254.
(375) R. Kuhn, H. H. Baer, and A. Gauhe, *Ann.*, 611 (1957) 242-249.
(376) P. A. J. Gorin and J. F. T. Spencer, *Can. J. Chem.*, 39 (1961) 2274-2281.
(377) V. P. Rege, T. J. Painter, W. M. Watkins, and W. T. J. Morgan, *Nature (London)*, 204 (1964) 740-742.
(378) R. Kuhn, H. H. Baer, and A. Gauhe, *Chem. Ber.*, 88 (1955) 1135-1146.
(379) R. Kuhn, H. H. Baer, and A. Gauhe, *Chem. Ber.*, 89 (1956) 2514-2523.
(380) R. Kuhn, H. H. Baer, and A. Gauhe, *Chem. Ber.*, 91 (1958) 364-374.
(381) R. Kuhn and A. Gauhe, *Chem. Ber.*, 93 (1960) 647-651.

THE UTILIZATION OF DISACCHARIDES AND SOME OTHER SUGARS BY YEASTS[1]

By James A. Barnett[2]

*School of Biological Sciences,
University of East Anglia, Norwich, England*

I. Introduction 347
II. Glycoside Structure and Hydrolysis 349
III. The Utilization of Glycosides Hydrolyzed Outside the Plasmalemma 353
   1. β-D-Fructofuranosides 355
   2. α-D-Galactopyranosides 374
IV. The Utilization of Glycosides Hydrolyzed Inside the Plasmalemma 378
   1. The Entry of Glycosides into Yeasts 379
   2. α-D-Glucopyranosides 381
   3. β-D-Glucopyranosides 391
   4. β-D-Galactopyranosides 394
V. The Requirement of Oxygen for Utilizing Glycosides and D-Galactose 397
VI. Addendum 401
   1. Glycosides Hydrolyzed Outside the Plasmalemma 401
   2. Glycosides Hydrolyzed Inside the Plasmalemma 402

I. Introduction

"Biochemists as well as geneticists should use only S288C or 'isogenic' Saccharomyces strains in their investigations. Regrettably, this has not been the rule in the past."
(Fink, 1970)[3]

Although Fink's injunction might seem exceedingly funny to many microbiologists, geneticists may accept it as sound advice. Genetical analysis is a powerful method for elucidating mechanisms responsible for variability, including the variability in nutritional versatility. How-

---

(1) This article derives from a talk given at the Symposium on Yeasts, in memory of the late Professor Maurice Ingram, at the University of Newcastle-upon-Tyne on 17 July 1979.
(2) The writer thanks Dr. A. P. Sims for helpful criticisms and suggestions, and Dr. R. W. Payne for computing the information in Table III.
(3) G. R. Fink, *Methods Enzymol.*, 17A (1970) 59–78.

ever, this approach to yeast nutrition has been restricted, because only two species, *Saccharomyces cerevisiae* and *Schizosaccharomyces pombe*, have been subjected to extensive genetical analysis, and both utilize relatively few exogenous organic compounds.

At present, only *Saccharomyces cerevisiae* is of major industrial importance. Yet some of the other four- or five-hundred yeast species could certainly assume a considerable role in industry, as a source of protein, for producing ethanol, and for carrying out other chemical transformations, such as those involved in synthesizing chiral precursors of important natural products. The following kinds of research will further such ends. (*i*) Some of the more versatile yeasts should be characterized genetically. (*ii*) Artificial means of manipulating yeasts genetically should be developed further, in order to make a wider range of yeasts available; it may no longer be essential to choose yeasts having an amenable, natural system of genetic recombination, and the useful physiological characteristics found in various strains that have been little studied could be transferred, by gene cloning, to a commonly used, but less versatile strain. (*iii*) In addition, more must be found out about physiological differences between yeasts. The present article is concerned with one aspect of these differences, namely, the ability to use glycosides.

In Volume 32, Barnett[4] surveyed certain aspects of this subject, including the utilization by yeasts of the monosaccharide components of the common glycosides. The similarities and differences between yeasts in their capacity to use disaccharides and other glycosides will be discussed, and attempts made to consider the following matters: (*i*) the location of hydrolysis; (*ii*) the substrate specificity of the glycosidases; (*iii*) the substrate specificity of the carriers of the glycosides; (*iv*) the affinity of the glycosidases and of the carriers for their substrates; (*v*) the regulation, by induction or repression, of these glycosidases and carriers; (*vi*) the numbers, in any given yeast strain, of glycosidases and carriers of glycosides having overlapping specificities; and (*vii*) the effects of changes in the environment, such as those of pH and oxygen concentration. The coverage of publications is fairly thorough, but only selected topics will be discussed at all fully.

Most of the work described herein has been conducted with very few kinds of yeast, chiefly *Saccharomyces cerevisiae*. The names of yeasts used here are those given by Barnett and his colleagues.[5] When

(4) J. A. Barnett, *Adv. Carbohydr. Chem. Biochem.*, 32 (1976) 125–234.
(5) J. A. Barnett, R. W. Payne, and D. Yarrow, *A Guide to Identifying and Classifying Yeasts*, Cambridge University Press, Cambridge, 1979.

the name differs from that used by the author whose work is cited, his name for the yeast is given in parentheses. Organisms referred to as "baker's" or "brewer's" yeast are here called *Saccharomyces cerevisiae*, brewing bottom-yeasts now being included in this species. However, in many publications, evidence is not given for the identity, or even the microbiological homogeneity, of the yeast. The names used for enzymes are those given by the Nomenclature Committee of the International Union of Biochemistry.[6] The *Tentative Rules for Carbohydrate Nomenclature*[7] have been obeyed.

II. Glycoside Structure and Hydrolysis

The first step in the utilization, by a yeast, of the most common glycosides is either (*i*) their passage intact across the plasmalemma into the cytosol, or (*ii*) their hydrolysis outside the plasmalemma and then entry of some or all of the hydrolytic products into the cell. Inside or outside the plasmalemma, hydrolysis is usually the first catabolic change undergone by a glycoside. Hydrolysis is catalyzed by a specific glycosidase (*O*-glycosyl hydrolase; see Table I), which usually can also catalyze glycosyl transfers.

In any glycoside, at least one monosaccharide group, the "glycon," is combined through an oxygen atom of the hemiacetal hydroxyl group to another residue, the "aglycon" (see formula 1). The aglycon of an oligosaccharide consists of a residue composed of one or more other monosaccharide residues. The linkage between the glycon and the aglycon may be of the α or β configuration.

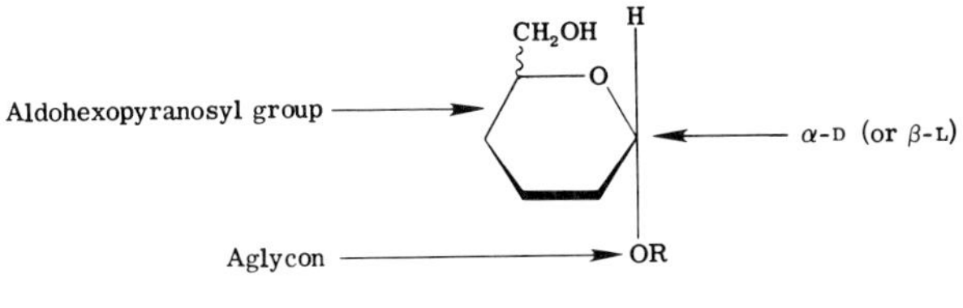

Aldohexopyranosyl group — α-D (or β-L)

Aglycon — OR

1

By using $H_2^{18}O$, hydrolysis has been shown to involve splitting of the bond between the glycosyl group and the oxygen atom of the

---

(6) *Enzyme Nomenclature 1978. Recommendations of the Nomenclature Committee of the International Union of Biochemistry on the Nomenclature and Classification of Enzymes*, Academic Press, New York, 1979.
(7) Tentative Rules for Carbohydrate Nomenclature, *Eur. J. Biochem.*, 21 (1971) 455–477.

TABLE I. Certain Glycosidases of Yeasts[6]

Enzyme Commission number	Name used herein	Other names	Reaction catalyzed
3.2.1.1	Alpha amylase	diastase	endohydrolysis, in a random manner, of $(1\rightarrow4)$-α-D-glucosidic linkages in polysaccharides containing three or more $(1\rightarrow4)$-α-D-linked D-glucosyl residues
3.2.1.3	$(1\rightarrow4)$-α-D-Glucan glucohydrolase	exo-$(1\rightarrow4)$-α-D-glucosidase, glucoamylase	chiefly, hydrolysis of successive, terminal $(1\rightarrow4)$-α-D-glucosyl groups from the nonreducing ends of polysaccharide chains
3.2.1.7	Inulinase	$(2\rightarrow1)$-β-D-fructan fructanohydrolase, inulase	hydrolysis of $(2\rightarrow1)$-β-D-fructosyl links in inulin
3.2.1.10	Oligo-$(1\rightarrow6)$-D-glucosidase	isomaltase, dextrin 6-α-glucanohydrolase	hydrolysis of $(1\rightarrow6)$-α-D-glucosyl links in isomaltose, panose, and gluco-oligosaccharides produced from starch by alpha amylase
3.2.1.20	α-D-Glucosidase	α-D-glucoside glucohydrolase, maltase	hydrolysis of terminal, nonreducing, $(1\rightarrow4)$-linked α-D-glucosyl groups; oligosaccharides are hydrolyzed much faster than polysaccharides
3.2.1.21	β-D-Glucosidase	β-D-glucoside glucohydrolase, cellobiase	hydrolysis of terminal, nonreducing, β-D-glucosyl groups of β-D-glucopyranosides; shows broad substrate specificity
3.2.1.22	α-D-Galactosidase	α-D-galactoside galactohydrolase, melibiase	hydrolysis of terminal, nonreducing, α-D-galactosyl groups of α-D-galactopyranosides
3.2.1.23	β-D-Galactosidase	β-D-galactoside galactohydrolase, lactase	hydrolysis of terminal, nonreducing, β-D-galactosyl groups of β-D-galactopyranosides
3.2.1.26	β-D-Fructofuranosidase	β-D-fructofuranoside fructohydrolase, invertase	hydrolysis of terminal, nonreducing, β-D-fructofuranosyl groups of β-D-fructofuranosides
3.2.1.28	α,α-Trehalase		α,α-trehalose + $H_2O \rightarrow$ 2 D-glucose

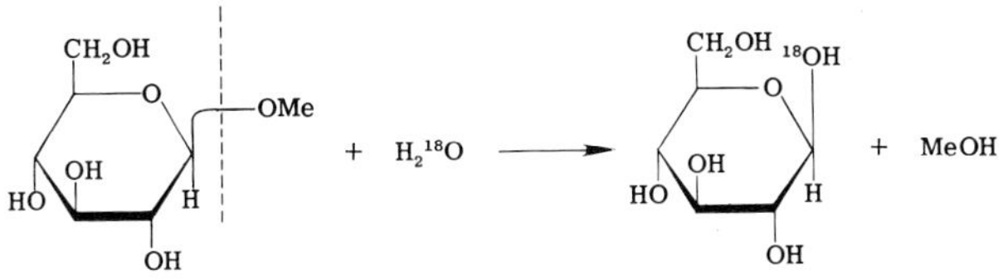

Scheme 1

Methyl β-D-glucopyranoside + $H_2^{18}O$ ⟶ β-D-[1-^{18}O]Glucopyranose + MeOH (1)

bridge, for both α- and β-D-glucopyranosides,[8] β-D-fructofuranosides,[9] and o-nitrophenyl β-D-galactopyranoside.[10] In many, although not all, glycosidase reactions, the configuration of the glycoside is retained, without a Walden[11] inversion, by the sugar formed from the glycon[12,13]; the direction of mutarotation provides the evidence for the initial configuration of that sugar. Examples include β-D-fructofuranosidase, which liberates β(not α)-D-fructofuranose,[9,14] β-D-galactosidase, which liberates β-D-galactose,[15] and α-D-glucosidase and β-D-glucosidase, which produce α- and β-D-glucopyranose, respectively.[16] Hehre and coworkers[17] studied this mechanism further. Thus, the hydrolysis of methyl β-D-glucopyranoside to give β-D-glucose may be represented as shown in Scheme 1. Gottschalk[18] found that baker's yeast (*Saccharomyces cerevisiae* ?) could not ferment the pyranose form of β-D-fructose; consumption of this sugar was rate-limited by its conversion into β-D-fructofuranose. In the utilization of a D-glucopyranoside, the phosphorylation of D-glucose liberated by hydrolysis is probably catalyzed by hexokinase (EC 2.7.1.1), and this reaction seems to occur at about the same rate for both α- and β-D-glucose[18–21] (for a review,

(8) C. A. Bunton, T. A. Lewis, D. R. Llewellyn, H. Tristram, and C. A. Vernon, *Nature (London)*, 174 (1954) 560.
(9) D. E. Koshland and S. S. Stein, *J. Biol. Chem.*, 208 (1954) 139–148.
(10) K. Wallenfels and O. P. Malhotra, *Adv. Carbohydr. Chem.*, 16 (1961) 239–298.
(11) P. Walden, *Ber.*, 26 (1893) 210–215.
(12) D. E. Koshland, *Biol. Rev. Cambridge Philos. Soc.*, 28 (1953) 416–436.
(13) D. E. Koshland, in W. D. McElroy and B. Glass (Eds.), *The Mechanism of Enzyme Action*, Johns Hopkins Press, Baltimore, 1954, pp. 608–641.
(14) B. Andersen and H. Degn, *Acta Chem. Scand.*, 16 (1962) 215–220.
(15) K. Wallenfels and G. Kurz, *Biochem. Z.*, 335 (1962) 559–572.
(16) E. F. Armstrong, *J. Chem. Soc.*, 83 (1903) 1305–1313.
(17) E. J. Hehre, D. S. Genghof, H. Sternlicht, and C. F. Brewer, *Biochemistry*, 16 (1977) 1780–1787.
(18) A. Gottschalk, *Aust. J. Exp. Biol. Med. Sci.*, 21 (1943) 133–137.
(19) J. M. Bailey, P. H. Fishman, and P. G. Pentchev, *J. Biol. Chem.*, 243 (1968) 4827–4831.
(20) M. Salas, E. Viñuela, and A. Sols, *J. Biol. Chem.*, 240 (1965) 561–568.
(21) B. Wurster and B. Hess, *Eur. J. Biochem.*, 36 (1973) 68–71.

see Ref. 22). E. H. Fischer and Stein[23] and Nisizawa and Hashimoto[24] have reviewed the mechanisms of glycosidase action.

The movement of glycosides and most of their hydrolytic products across the plasmalemma into the cell generally occurs by means of carriers associated with the plasmalemma. Without such carriers, the plasmalemma is effectively impermeable to sugars[25]; but some molecules, such as methanol,[25a] may enter cells by simple diffusion, as described for the stoneworts (algae) *Chara australis* and *Nitella translucens*.[26]

The common glycosides considered here (for reviews, see Refs. 27 and 28) are those of D-glucose, D-galactose, and D-fructose. Table II gives their structures. Each of these glycosides is formed by the reaction of the anomeric (carbonyl) carbon atom of an aldohexose (C-1) or 2-hexulose ("ketohexose") (C-2) with an alcohol. The aglycon may be formed (*i*) by a monohydric alcohol, such as methanol, to give, for example, methyl α- or β-D-glucopyranoside, as in the converse of reaction (*1*), (*ii*) by a phenolic compound, such as phenol itself, as in phenyl α- or β-D-glucopyranoside, or (*iii*) by another sugar (in the oligosaccharides), such as a second molecule of D-glucose, as in maltose, cellobiose, or α,α-trehalose.

D-Aldohexopyranose

2

D-2-Hexulofuranose

3

(22) S. J. Benkovic and K. J. Schray, *Adv. Enzymol. Relat. Areas Mol. Biol.*, 44 (1976) 139–164.
(23) E. H. Fischer and E. A. Stein, in P. D. Boyer, H. Lardy, and K. Myrbäck (Eds.), *The Enzymes*, 2nd edn., Vol. 4, Academic Press, New York, 1960, pp. 301–312.
(24) K. Nisizawa and Y. Hashimoto, in W. Pigman and D. Horton (Eds.), *The Carbohydrates: Chemistry and Biochemistry*, 2nd edn., Vol. IIA, Academic Press, New York, 1970, pp. 241–300.
(25) E. J. Conway and M. Downey, *Biochem. J.*, 47 (1950) 347–355.
(25a) J. R. Quayle, personal communication, 1978.
(26) J. Dainty and B. Z. Ginzburg, *Biochim. Biophys. Acta*, 79 (1964) 122–128.
(27) J. H. Pazur, in W. Pigman and D. Horton (Eds.), *The Carbohydrates: Chemistry and Biochemistry*, 2nd edn., Vol. IIA, Academic Press, New York, 1970, pp. 69–137.
(28) W. G. Overend, in W. Pigman and D. Horton (Eds.), *The Carbohydrates: Chemistry and Biochemistry*, 2nd edn., Vol. IA, Academic Press, New York, 1972, pp. 279–353.

α,α-Trehalose is an example of a double glycoside, in which the anomeric carbon atom of a monosaccharide has reacted with the hydroxyl group of the anomeric carbon atom of a second monosaccharide. As both reducing groups are substituted, these double glycosides are nonreducing sugars. α,α-Trehalose is composed of two D-glucose monomers, but double glycosides may be formed by two unlike monomers, such as D-glucose and D-fructose, as in sucrose. Sucrose is simultaneously both a β-D-fructofuranoside and an α-D-glucopyranoside, and so it can be hydrolyzed either by a β-D-fructofuranosidase or an α-D-glucosidase.

Certain synthetic glycosides have been used extensively for investigating the mechanisms by which naturally occurring sugars are utilized. (i) Chromogenic substrates are exemplified by p-nitrophenyl α-D-glucopyranoside. On hydrolysis, this glycoside liberates p-nitrophenol (λ_{max} 400 nm) which can be measured directly (see, for example, Refs. 29–34). (ii) Some non-metabolized 1-thioglycosides (S-glycosyl derivatives), often radioactively labelled, have been used as analogs of metabolizable sugars for measuring the kinetics of the carriers that take those sugars into the cells. Thus, for example, Kepes[35] and subsequent workers used 1-thio-β-D-galactopyranosides as analogs of lactose.

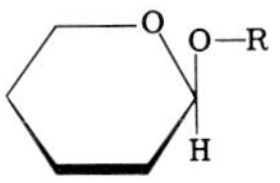

Glycoside   1-Thioglycoside
4   5

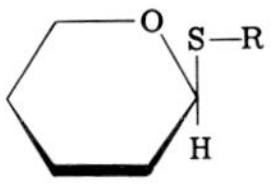

where R is the aglycon

III. The Utilization of Glycosides Hydrolyzed Outside the Plasmalemma

Of the glycosidases that are important for hydrolyzing exogenously supplied glycosides, β-D-fructofuranosidase (invertase) and α-D-galactosidase (melibiase) activities are well known to occur outside the

---

(29) K. Aizawa, *J. Biochem. (Tokyo)*, 30 (1939) 89–100.
(30) K. Aizawa, *Enzymologia*, 6 (1939) 321–324.
(31) J. Lederberg, *J. Bacteriol.*, 60 (1950) 381–392.
(32) J. D. Duerksen and H. [O.] Halvorson, *J. Biol. Chem.*, 233 (1958) 1113–1120.
(33) H. [O.] Halvorson and L. Ellias, *Biochim. Biophys. Acta*, 30 (1958) 28–40.
(34) H. [O.] Halvorson, *Methods Enzymol.*, 8 (1966) 559–562.
(35) A. Kepes, *Biochim. Biophys. Acta*, 40 (1960) 70–84.

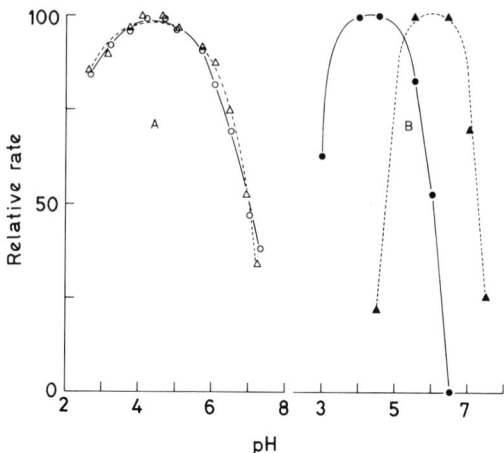

FIG. 1.—Relative Rates of Activity Against Glycosides at Various pH Values. [A. Hydrolysis of sucrose by β-D-fructofuranosidase of *Saccharomyces cerevisiae*: ○, intact cells and △, toluene-treated cells (results of Wilkes and Palmer[40]). B. Maltose: ●, fermentation by intact baker's yeast; ▲, hydrolysis by α-D-glucosidase (results of Hestrin[41]; see also Ref. 42).]

plasmalemmata of yeasts. There is also evidence that some yeasts may have external β-D-glucosidase activity[36] and external α,α-trehalase[37,38]; probably, other such glycosidases of certain yeasts will eventually be found to act externally. These external hydrolases are usually retained by the cell wall, but in some cases, they are released into the suspending medium. This must occur with polysaccharidases, which, in some instances, may also act on oligosaccharides.

The following kinds of observation have provided evidence of external glycosidase activity. (*i*) The idea of similarity of activity *in vivo* and *in vitro* has long been recognized; for example, by von Euler and Kullberg[39] in 1911. The effect, on the glycosidase activity of intact cells, of varying the pH is the same as that for the enzyme in solution (see Fig. 1).[40,43–45] (*ii*) Inhibitors of transport across the plasmalemma,

(36) J. G. Kaplan and W. Tacreiter, *J. Gen. Physiol.*, 50 (1966) 9–24.
(37) S. Janda and M. von Hedenström, *Arch. Microbiol.*, 101 (1974) 273–280.
(38) W. N. Arnold, *Curr. Microbiol.*, 2 (1979) 109–112.
(39) H. [von] Euler and S. Kullberg, *Z. Physiol. Chem.*, 71 (1911) 14–30.
(40) B. J. Wilkes and E. T. Palmer, *J. Gen. Physiol.*, 16 (1932) 233–242.
(41) S. Hestrin, *Wallerstein Lab. Commun.*, 11 (1948) 193–207.
(42) R. Weidenhagen, *Z. Zuckerind.*, 80 (1930) 374–383.
(43) H. von Euler and S. Heintze, *Z. Physiol. Chem.*, 108 (1919) 165–185.
(44) R. Willstätter and F. Racke, *Ann.*, 425 (1921) 1–135.
(45) K. Myrbäck and E. Willstaedt, *Ark. Kemi*, 8 (1956) 367–374.

such as uranyl ions (UO_2^{2+}) have little effect on glycoside hydrolysis by intact cells, although greatly lessening the rate of utilization of sugar.[46-49] (iii) Hexose liberated by external glycosidase activity can be trapped by adding hexokinase exogenously; the hexose phosphates formed do not enter the cells across the plasmalemma.[50,51] (iv) As the cell wall disintegrates when spheroplasts are formed, soluble glycosidase is simultaneously liberated into the suspending fluid[52,53]; the resulting spheroplasts are correspondingly low in glycosidase activity.[53-55] (v) Isolated cell-walls, obtained by mechanically disintegrating the yeast, contain much β-D-fructofuranosidase.[53] (vi) Cells that utilize a glycoside form spheroplasts that cannot do so, although they can use D-glucose.[53]

1. β-D-Fructofuranosides

In 1860, Berthelot[56] prepared β-D-fructofuranosidase from brewer's yeast by alcoholic precipitation. Since then, the β-D-fructofuranosidase activity of yeasts, particularly of *Saccharomyces cerevisiae*, has been extensively studied, chiefly with sucrose or raffinose as the substrate, because chromogenic β-D-fructofuranosides have not been available. β-D-Fructofuranosidases have been reviewed by Neuberg and Mandl,[57] Myrbäck,[58] and Lampen[59]; see also, Ref. 60.

(46) E. S. G. Barron, J. A. Muntz, and B. Gasvoda, *J. Gen. Physiol.*, 32 (1948) 163-178.
(47) D. J. Demis, A. Rothstein, and R. Meier, *Arch. Biochem. Biophys.*, 48 (1954) 55-62.
(48) A. Rothstein, *Symp. Soc. Exp. Biol.*, 8 (1954) 165-201.
(49) A. Rothstein, in G. C. Ainsworth and A. S. Sussman (Eds.), *The Fungi: An Advanced Treatise*, Vol. I, Academic Press, New York, 1965, pp. 429-455.
(50) A. Sols and G. de la Fuente, in A. Kleinzeller and A. Kotyk (Eds.), *Membrane Transport and Metabolism*, Publishing House of the Czechoslovak Academy of Sciences, Praha, 1961, pp. 361-377.
(51) G. de la Fuente and A. Sols, *Biochim. Biophys. Acta*, 56 (1962) 49-62.
(52) M. Burger, E. E. Bacon, and J. S. D. Bacon, *Nature (London)*, 182 (1958) 1508.
(53) D. D. Sutton and J. O. Lampen, *Biochim. Biophys. Acta*, 56 (1962) 303-312.
(54) J. Friis and P. Ottolenghi, *C. R. Trav. Lab. Carlsberg*, 31 (1959) 259-271.
(55) J. Friis and P. Ottolenghi, *C. R. Trav. Lab. Carlsberg*, 31 (1959) 272-281.
(56) M. P. E. Berthelot, *C. R. Acad. Sci.*, 50 (1860) 980-984.
(57) C. Neuberg and I. Mandl, in J. B. Sumner (Ed.), *The Enzymes*, Vol I, Part 1, Academic Press, New York, 1950, pp. 527-550.
(58) K. Myrbäck, in P. D. Boyer, H. Lardy, and K. Myrbäck (Eds.), *The Enzymes*, 2nd edn., Vol. 4, Academic Press, New York, 1960, pp. 379-396.
(59) J. O. Lampen, in P. D. Boyer (Ed.), *The Enzymes*, 3rd edn., Vol. 5, Academic Press, New York, 1971, pp. 291-305.
(60) A. Goldstein and J. O. Lampen, *Methods Enzymol.*, 42 (1975) 504-511.

a. **β-D-Fructofuranosidase Substrate Specificity.**—β-D-Fructofuranosidase catalyzes the hydrolysis of compounds, such as **6**, having a terminal, unsubstituted β-D-fructofuranosyl group.[58] Substitution in the β-D-fructofuranosyl group, as in melezitose (see Table II) prevents hy-

TABLE II

The Structures of Some D-Glycosides

Glycoside	Structure
α-D-Glucopyranosides	
Maltose 4-O-α-D-Glucopyranosyl-D-glucopyranose	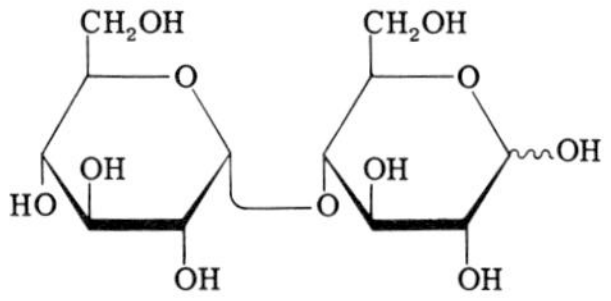
Sucrose β-D-Fructofuranosyl α-D-glucopyranoside	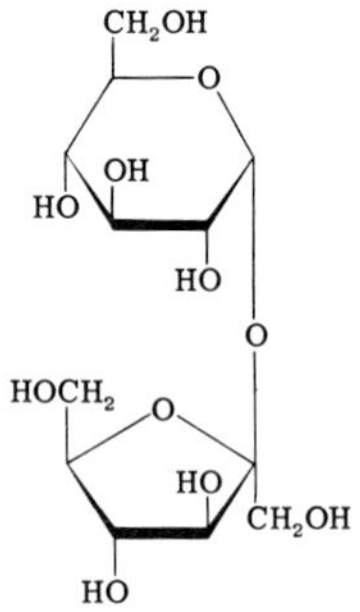
Methyl α-D-glucopyranoside	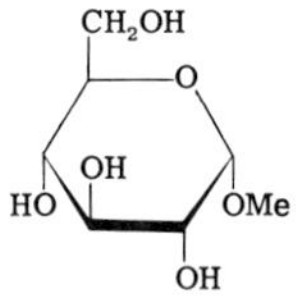
Isomaltose 6-O-α-D-Glucopyranosyl-D-glucopyranose	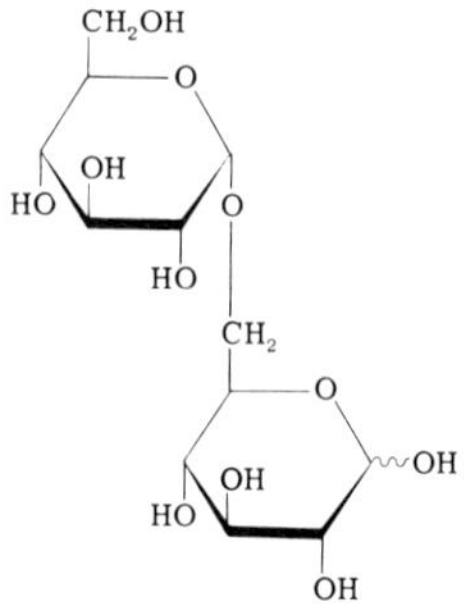

TABLE II (continued)

Glycoside	Structure
Turanose 3-O-α-D-Glucopyranosyl-D-fructopyranose	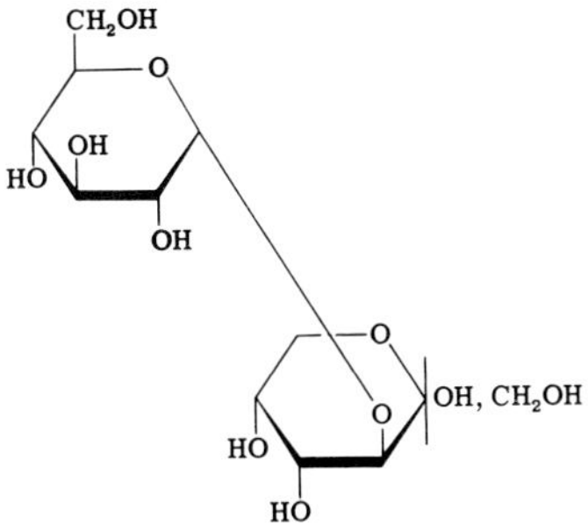
Ethyl 1-thio-α-D-glucopyranoside (α-TEG)	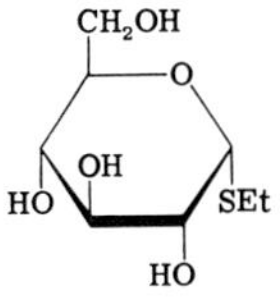
p-Nitrophenyl α-D-glucopyranoside	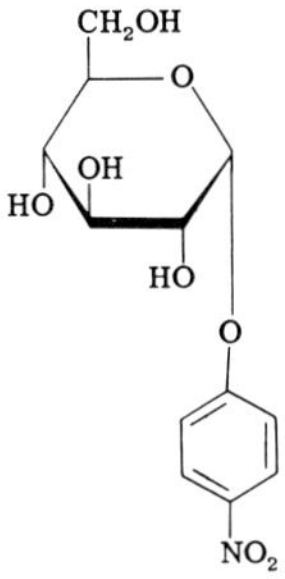
Maltotriose O-α-D-Glucopyranosyl-(1→4)-O-α-D-glucopyranosyl-(1→4)-D-glucopyranose	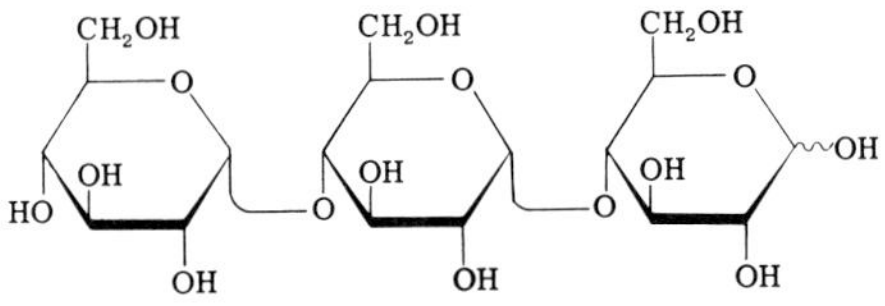
Melezitose O-α-D-Glucopyranosyl-(1→3)-O-β-D-fructofuranosyl α-D-glucopyranoside	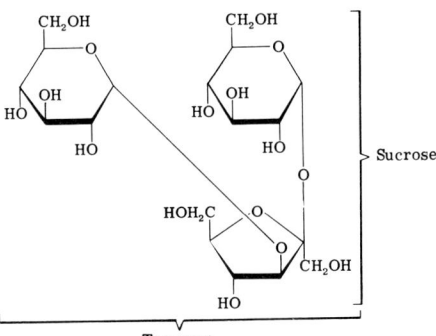

(continued)

TABLE II (continued)

Glycoside	Structure
α,α-Trehalose α-D-Glucopyranosyl α-D-glucopyranoside	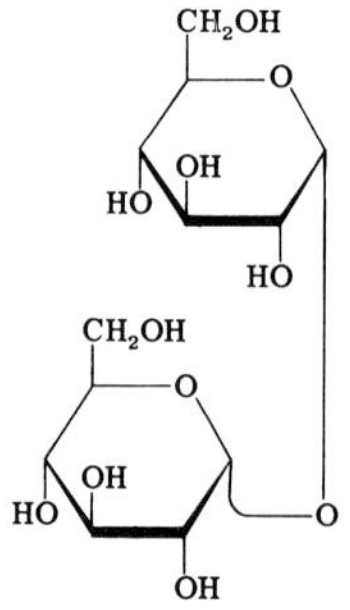
β-D-Glucopyranosides	
Cellobiose 4-O-β-D-Glucopyranosyl-D-glucopyranose	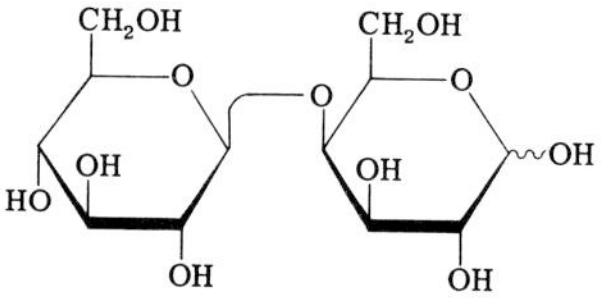
Salicin 2-(Hydroxymethyl)phenyl β-D-glucopyranoside	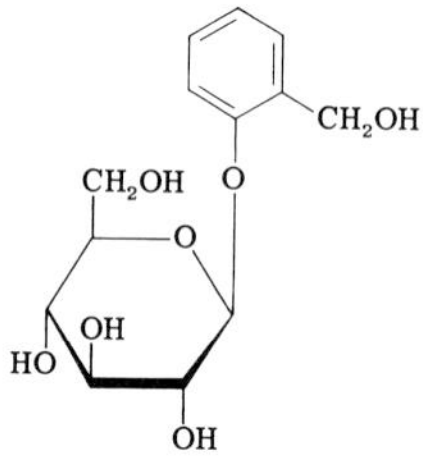
Arbutin 4-Hydroxyphenyl β-D-glucopyranoside	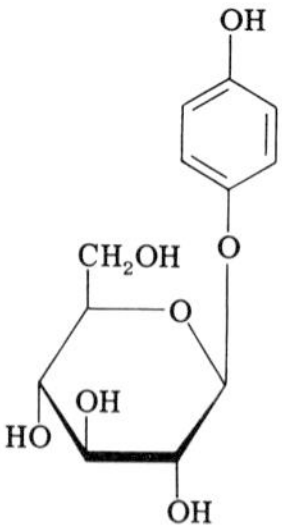
Esculin 7-Hydroxycoumarin-6-yl β-D-glucopyranoside	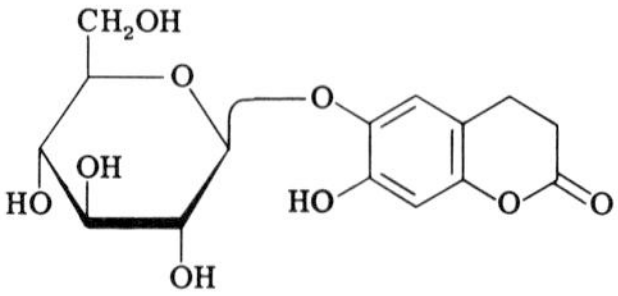

TABLE II (*continued*)

Glycoside	Structure
Amygdalin *O*-β-D-Glucopyranosyl-(1→6)-β-D-glucopyranosyl-OCH(CN)Ph	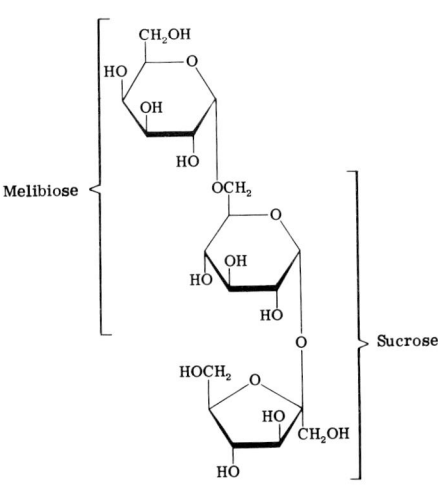
D-Galactopyranosides	
Melibiose 6-*O*-α-D-Galactopyranosyl-D-glucopyranose	
Raffinose *O*-α-D-Galactopyranosyl-(1→6)-α-D-glucopyranosyl β-D-fructofuranoside	
Lactose 4-*O*-β-D-Galactopyranosyl-D-glucopyranose	

359

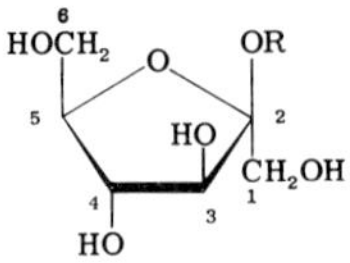

A β-D-fructofuranoside

6

drolysis.[61–63] On the other hand, the characteristics of the afructon have less effect on hydrolysis, although this is slower with a larger afructon, possibly because of steric hindrance. Accordingly, Hudson and coworkers[62] found the following relative rates of hydrolysis by purified β-D-fructofuranosidase from baker's yeast (*Saccharomyces cerevisiae* ?).

	Molecular weight of afructon	Relative rate of hydrolysis
Sucrose	179	100
Raffinose	341	23
Stachyose	503	7

However, GrootWassink and Fleming[64] reported a higher rate of hydrolysis of stachyose than raffinose for a commercial preparation of β-D-fructofuranosidase of *Saccharomyces cerevisiae*.

Weidenhagen[65] found no hydrolysis of either methyl or phenyl β-D-fructofuranoside by β-D-fructofuranosidase (? of *Saccharomyces cerevisiae*); but Purves and Hudson,[66] like Schlubach and Rauchalles,[67] reported that this enzyme, presumed by Gottschalk[68] to be from "yeast," hydrolyzed methyl β-D-fructofuranoside, although only 7% as fast as for sucrose. Purified β-D-fructofuranosidase of *Saccharomyces cerevisiae* hydrolyzed ethyl β-D-fructofuranoside and N-*p*-nitrophenyl-β-D-fructofuranosylamine according to Baseer and Shall,[69] who suggested

(61) R. Kuhn and G. E. von Grundherr, *Ber.*, 59 (1926) 1655–1664.
(62) M. Adams, N. K. Richtmyer, and C. S. Hudson, *J. Am. Chem. Soc.*, 65 (1943) 1369–1380.
(63) H. Negoro, *Hakko Kogaku Zasshi*, 51 (1973) 879–886.
(64) J. W. D. GrootWassink and S. E. Fleming, *Enzyme Microb. Technol.*, 2 (1980) 45–53.
(65) R. Weidenhagen, *Z. Zuckerind.*, 82 (1932) 912–922.
(66) C. B. Purves and C. S. Hudson, *J. Am. Chem. Soc.*, 56 (1934) 702–707.
(67) H. H. Schlubach and G. Rauchalles, *Ber.*, 58 (1925) 1842–1850.
(68) A. Gottschalk, in W. Ruhland (Ed.), *Handbuch der Pflanzenphysiologie*, Vol. VI, Springer-Verlag, Berlin, 1958, pp. 87–124.
(69) A. Baseer and S. Shall, *Int. J. Biochem.*, 2 (1971) 503–506.

TABLE III

Abilities of 439 Species of Yeast to Utilize Certain Glycosides, D-Glucose, and D-Galactose, Aerobically or Anaerobically[a]

Carbohydrate	Aerobic growth[b]			Anaerobic fermentation[b]		
	+	−	?	+	−	?
D-Glucose	439	0	0	207	148	84
D-Galactose	221	148	70	49	310	80
Sucrose	258	152	29	80	295	64
Maltose	231	178	30	24	343	72
Methyl α-D-glucopyranoside	139	226	74	5	303	131
Melezitose	167	227	45	5	335	99
α,α-Trehalose	274	94	71	27	254	158
Cellobiose	243	135	61	9	290	140
Salicin	222	136	81	0	0	439
Arbutin	240	132	67	0	0	439
Melibiose	52	365	22	9	402	28
Lactose	51	342	46	3	429	7
Raffinose	144	256	39	47	343	49
Inulin	20	384	35	0	0	439
Starch	78	312	49	0	0	439

[a] Information computed by R. W. Payne from the compilation of Barnett and co-workers.[5] These results differ from those given by Barnett,[4] chiefly because (i) newly described species are included, and (ii) many previously described species have been combined. The figures in the Table are the numbers of species. [b] Key: the symbol +, sugar utilized; −, sugar not utilized. Under the symbol ? are those species for which a definite + or − cannot be given, for one or more of the following reasons: (i) some strains of the species are + and others −; (ii) utilization was given an equivocal qualification, such as "weak," "slow," or "delayed"; or (iii) there are no results for that species in the compilation.

that the latter compound might be a useful chromogenic substrate, giving, on hydrolysis, a significant decrease in absorbance at 350 nm and an increase at 380 nm.

b. Inulinase.—Some yeast β-D-fructofuranosidases appear not to cleave inulin, and only 20 yeast species have been unequivocally reported to utilize inulin (see Table III); these are all among the 144 species capable of using both sucrose and raffinose[5] (see also, Ref. 70 and Table IV). Inulin is composed of long molecules, each consisting

(70) J. A. Barnett, *J. Gen. Microbiol.*, 99 (1977) 183–190.

TABLE IV

Species of Yeast That Utilize Inulin[a]

Candida brassicae	Hansenula beijerinckii
Candida kefyr	Hansenula jadinii
Candida macedoniensis	Hansenula lynferdii
Candida membranaefaciens	Hansenula petersonii
Candida pseudotropicalis	Kluyveromyces bulgaricus
Candida salmanticensis	Kluyveromyces marxianus
Candida utilis	Lipomyces kononenkoae
Debaryomyces castellii	Lipomyces starkeyi
Debaryomyces phaffii	Lipomyces tetrasporus
Debaryomyces polymorpha	Pichia guilliermondii

[a] Abstracted from the compilation of Barnett and coworkers.[5]

of ~30 (2→1)-linked β-D-fructofuranosyl residues, ending in a sucrose residue[71] (formula 7), and so inulin has[72,73] a molecular weight of ~5000.

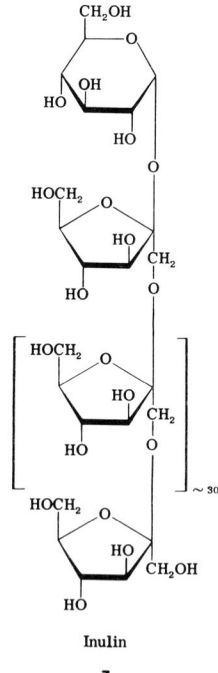

Inulin

7

(71) E. L. Hirst, *Proc. Chem. Soc., London*, (1957) 193–204.
(72) W. N. Haworth, E. L. Hirst, and E. G. V. Percival, *J. Chem. Soc.*, (1932) 2384–2388.
(73) S. R. Carter and B. R. Record, *J. Chem. Soc.*, (1939) 664–675.

Snyder and Phaff[74] examined the action on inulin of the β-D-fructofuranosidase from a strain of *Kluyveromyces marxianus* (*Saccharomyces fragilis*). This yeast fermented (that is, anaerobically formed CO_2 from) inulin and D-glucose at about the same rate. The β-D-fructofuranosidase of this yeast contrasted with that of *Saccharomyces cerevisiae*. The approximate ratio of rate of hydrolysis of raffinose to that of inulin was 3.5:1 (also found for *Kluyveromyces lactis* by Yurkevich and his colleagues[75]); on the other hand, commercial β-D-fructofuranosidase gave a corresponding ratio of 5,800:1, approximating that (5,000 to 8,000) found for yeast (? *Saccharomyces cerevisiae*) β-D-fructofuranosidase by Weidenhagen,[65,76] and ~3,000:1 for an inulin-utilizing strain of *Saccharomyces cerevisiae*.[50,51] The β-D-fructofuranosidase of *Kluyveromyces marxianus* also hydrolyzed bacterial levan; consequently, this enzyme attacked both β-(2→6)- and β-(2→1)-linkages. The pH optima for hydrolysis were different for sucrose (4.3) and inulin (5.1), as Takahashi and Soutome[77] found for the same species, and Hudson and coworkers[62] for β-D-fructofuranosidase of baker's yeast. Nonetheless, Snyder and Phaff[74] considered that their partly purified enzyme was homogeneous. Yurkevich and his colleagues[75,78,79] were also unable to separate the sucrase and inulinase activities of the β-D-fructofuranosidases of *Kluyveromyces marxianus* (*fragilis*), *Kluyveromyces lactis*, and *Saccharomyces cerevisiae* (including *Saccharomyces paradoxus*). However, Schlubach and Grehn[80] found that streptomycin inhibited the activity of a highly purified, yeast β-D-fructofuranosidase against inulin, but not against sucrose.

Snyder and Phaff[74] classified their enzyme from *Kluyveromyces marxianus* as inulinase (EC 3.2.1.7) rather than β-D-fructofuranosidase (EC 3.2.1.26). Their inulinase hydrolyzed inulin from the D-fructosyl end of the polymer, by an endwise action.[81,82] Release of the en-

(74) H. E. Snyder and H. J. Phaff, *Antonie van Leeuwenhoek J. Microbiol. Serol.*, 26 (1960) 433–452.
(75) V. V. Yurkevich, N. S. Kovaleva, and Kh. Kh. Baker, *Fiziol. Rast.*, 19 (1972) 937–945; *Chem. Abstr.*, 78 (1973) 25,723m.
(76) R. Weidenhagen, *Naturwissenschaften*, 20 (1932) 254.
(77) M. Takahashi and S. Soutome, *Utsunomiya Daigaku Nogakubu Gakujutsu Hokoku Tokushu*, 9 (1975) 95–98; *Chem. Abstr.*, 83 (1975) 39,304r.
(78) V. V. Yurkevich and N. S. Kovaleva, *Biol. Nauki* (*Moscow*), 15 (1972) 98–103; *Chem. Abstr.*, 77 (1972) 111,240n.
(79) V. V. Yurkevich and N. S. Kovaleva, *Dokl. Akad. Nauk SSSR*, 207 (1972) 1233–1235; *Chem. Abstr.*, 81 (1974) 59,786m.
(80) H. H. Schlubach and M. Grehn, *Ann.*, 647 (1961) 51–53.
(81) H. E. Snyder and H. J. Phaff, *J. Biol. Chem.*, 237 (1962) 2438–2441.
(82) G. Avigad and Š. Bauer, *Methods Enzymol.*, 8 (1966) 621–628.

zyme from the yeast cells into the suspending medium seemed to be inducible (see Ref. 237). Working with another strain of the same species (CBS 397), the type strain of *Saccharomyces fragilis*), Davies[83] found that its β-D-fructofuranosidase was repressed by D-glucose: with 60 or 100 mM D-glucose in the growth medium, the sucrose-hydrolyzing activity formed in the cells was only 0.4% of that with 10 mM D-glucose, and was similarly repressed by D-galactose. The inulinase of another strain of *Kluyveromyces marxianus* (*fragilis*) has been studied[64] with a view to commercial exploitation; some of the characteristics of this enzyme are shown in Table V.

An inulin-utilizing yeast, *Candida kefyr*, also releases into the medium an enzyme having both inulinase and other β-D-fructofuranosidase activity,[63,84,85] similar to that of *Kluyveromyces marxianus* (see Table V). Release of the β-D-fructofuranosidase into the medium seems a necessary condition for inulin utilization, because the cell walls, of *Saccharomyces cerevisiae* for example, probably exclude[86] compounds having molecular weights >700, inulin in particular.[25] Can this observation be reconciled with the passage, through the cell walls of certain yeasts, of small protein molecules[87,88] having molecular weights <14,000, of bovine serum albumin,[89] probably having a molecular weight of 65,000, or of β-D-fructofuranosidase, molecular weight 270,000 (Ref. 90)? Furthermore, from "killer" yeasts, toxin having a molecular weight of ~100,000 can penetrate the walls.[91] Scherrer and coworkers[86] suggested that sufficiently large openings are formed in the cell wall in response to the presence of high concentrations of the protein, in media of low ionic strength (as Ottolenghi[89] described). Only certain strains of *Saccharomyces cerevisiae* can utilize inulin[92]; this utilization probably depends on (*i*) liberation through the cell wall of β-D-fructofuranosidase, and (*ii*) its specificity for inulin. Some yeasts release much of their β-D-fructofuranosidase

(83) A. Davies, *J. Gen. Microbiol.*, 14 (1956) 109–121.
(84) H. Negoro and E. Kito, *Hakko Kogaku Zasshi*, 51 (1973) 96–102.
(85) H. Negoro and E. Kito, *Hakko Kogaku Zasshi*, 51 (1973) 103–110.
(86) R. Scherrer, L. Louden, and P. Gerhardt, *J. Bacteriol.*, 118 (1974) 534–540.
(87) F. Schlenk and C. R. Zydek-Cwick, *Arch. Biochem. Biophys.*, 138 (1970) 220–225.
(88) F. Schlenk, *Biochim. Appl.*, 17 (1970) 89–103.
(89) P. Ottolenghi, *C. R. Trav. Lab. Carlsberg*, 36 (1967) 95–111.
(90) N. P. Neumann and J. O. Lampen, *Biochemistry*, 6 (1967) 468–475.
(91) R. B. Wickner, *Bacteriol. Rev.*, 40 (1976) 757–773.
(92) J. P. van der Walt, in J. Lodder (Ed.), *The Yeasts, A Taxonomic Study*, 2nd edn., North-Holland Publishing Co., Amsterdam, 1970, pp. 555–718.

TABLE V
Characteristics of β-D-Fructofuranosidases Prepared from Various Yeasts

	Relative rate of hydrolysis[a]				K_m (mM)[b]		
	Sucrose	Raffinose	Stachyose	Inulin	Sucrose	Raffinose	Stachyose
Saccharomyces cerevisiae	100	16	—	0.006	47	69	286
Candida utilis	100	8	—	0.009	—	—	—
Candida kefyr	100	12	—	2	—	—	—
Kluyveromyces marxianus (fragilis)[b]	100	28	24	—	9.4	6.1	6.8

[a] Results for Saccharomyces cerevisiae, Candida utilis, and Candida kefyr are from Ref. 63. [b] Results are from Ref. 64.

into the suspending medium,[93–98] and this characteristic may prove to be highly associated with the ability to use inulin.

c. β-D-Fructofuranosidase Kinetics.—For the β-D-fructofuranosidases of *Saccharomyces cerevisiae*, estimates of the Michaelis constant (K_m) of 16 to 40 mM sucrose[98–108] and 81 to 292 mM raffinose have been made.[102,103,106] For other species, the following values for K_m of β-D-fructofuranosidases have been published: (*i*) *Candida utilis*, 9 mM sucrose[98]; (*ii*) *Debaryomyces hansenii*, 17 mM sucrose[109]; (*iii*) *Kluyveromyces lactis*, some strains of which utilize inulin,[110] 25 mM sucrose, ~3 mM inulin[75]; (*iv*) *Kluyveromyces marxianus* (*Saccharomyces fragilis*), 11 mM sucrose, 8 mM raffinose[77]; and *Zygosaccharomyces* (*Saccharomyces*) *rouxii*, 83 mM raffinose.[111]

d. β-D-Fructofuranosidase Location.—Many studies on *Saccharomyces cerevisiae* and *Kluyveromyces marxianus* have shown that β-D-fructofuranosidase is synthesized within the cell and exported across the plasmalemma. Some authors have suggested that the enzyme is held in the periplasmic space,[112–115] of width ~10 nm and volume[116] ~1 fL, between the plasmalemma and the cell wall. Other workers

(93) A. Guilliermond, *The Yeasts*, Wiley, New York, 1920.
(94) L. J. Wickerham, *Arch. Biochem. Biophys.*, 76 (1958) 439–448.
(95) R. G. Dworschack and L. J. Wickerham, *Arch. Biochem. Biophys.*, 76 (1958) 449–456.
(96) R. G. Dworschack and L. J. Wickerham, *Appl. Microbiol.*, 9 (1961) 291–294.
(97) L. J. Wickerham and R. G. Dworschack, *Science*, 131 (1960) 985–986.
(98) E. T. Reese, R. Birzgalis, and M. Mandels, *Can. J. Biochem. Physiol.*, 40 (1962) 273–283.
(99) R. Kuhn, *Z. Physiol. Chem.*, 125 (1923) 28–92.
(100) M. V. Tracey, *Biochim. Biophys. Acta*, 77 (1963) 147–149.
(101) J. Hoshino, T. Kaya, and T. Sato, *Plant Cell Physiol.*, 5 (1964) 495–506.
(102) S. Gascón, N. P. Neumann, and J. O. Lampen, *J. Biol. Chem.*, 243 (1968) 1573–1577.
(103) B. Berggren, *Ark. Kemi*, 32 (1970) 167–180.
(104) A. Baseer and S. Shall, *Biochim. Biophys. Acta*, 250 (1971) 192–202.
(105) P. Ottolenghi, *Eur. J. Biochem.*, 18 (1971) 544–552.
(106) S. Yamamoto, S. Nagasaki, and H. Kubo, *Kochi Daigaku Gakujutsu Kenkyu Hokoku*, 19 (1970) 163–170; *Chem. Abstr.*, 76 (1972) 31,646p.
(107) D. K. Kidby, *J. Gen. Microbiol.*, 84 (1974) 343–349.
(108) K. Toda and M. Shoda, *Biotechnol. Bioeng.*, 17 (1975) 481–497.
(109) B. Johnson, E. Merdinger, and C. Lange, *Trans. Ill. State Acad. Sci.*, 67 (1974) 131–138.
(110) J. P. van der Walt, in J. Lodder (Ed.), *The Yeasts, A Taxonomic Study*, 2nd edn., North-Holland Publishing Co., Amsterdam, 1970, pp. 316–378.
(111) W. N. Arnold, *J. Bacteriol.*, 120 (1974) 886–894.
(112) J. B. Best, *J. Cell. Comp. Physiol.*, 46 (1955) 29–52.
(113) M. Burger, E. E. Bacon, and J. S. D. Bacon, *Biochem. J.*, 78 (1961) 504–511.

have held that external β-D-fructofuranosidase is located within the cell wall itself, held simply by physical restraint,[117,118] or, alternatively, attached to the wall either by D-mannose phosphoric diester bridges or by hydrogen bonds.[59,119]

Most of the β-D-fructofuranosidase of "fresh baker's yeast"[115] or of brewer's yeast[120] may be released into solution by mechanical disruption of the cells. Intact cells of *Kluyveromyces marxianus* (*Saccharomyces fragilis*), treated with 2-mercaptoethanol, may also release most of their β-D-fructofuranosidase.[117,118,121,122] So the unbound enzyme, retained in enzyme-impermeable structures within the wall, is probably released by the reduction, and hence breakdown, of disulfide bridges in mannan–protein complexes of the outer region of the cell wall (see Fig. 2). Sommer and Lewis[123] and Smith and Ballou[124] found that a more powerfully reducing sulfhydryl compound, 1,4-dithiothreitol, acts similarly on *Saccharomyces cerevisiae*. Nagasaki and colleagues[125] reported the use of 2-mercaptoethanol to extract β-D-fructofuranosidase from four species of *Saccharomyces* and from *Trigonopsis variabilis*, a yeast that does not utilize sucrose, raffinose, or inulin.[126]

In common with other exoenzymes, external β-D-fructofuranosidase is associated with much carbohydrate.[90,119,127–130] Thus, the enzyme of

---

(114) N. M. Mityushova and A. M. Ugolev, *Dokl. Akad. Nauk SSSR*, 195 (1970) 503–506; *Chem. Abstr.*, 74 (1971) 73,018z.
(115) W. N. Arnold, *J. Bacteriol.*, 112 (1972) 1346–1352.
(116) W. N. Arnold, *Physiol. Chem. Phys.*, 5 (1973) 117–123.
(117) D. K. Kidby and R. Davies, *J. Gen. Microbiol.*, 61 (1970) 327–333.
(118) D. K. Kidby and R. Davies, *Biochim. Biophys. Acta*, 201 (1970) 261–266.
(119) J. O. Lampen, *Antonie van Leeuwenhoek J. Microbiol. Serol.*, 34 (1968) 1–18.
(120) N. J. Williams and A. Wiseman, *Biochem. Soc. Trans.*, 1 (1974) 1299–1301.
(121) R. Davies and P. A. Elvin, *Biochem. J.*, 93 (1964) 8P–9P.
(122) R. Davies, *Abh. Dtsch. Akad. Wiss. Berlin, Kl. Med.*, 6 (1967) 195–198; *Chem. Abstr.*, 70 (1969) 65,506j.
(123) A. Sommer and M. J. Lewis, *J. Gen. Microbiol.*, 68 (1971) 327–335.
(124) W. L. Smith and C. E. Ballou, *Biochem. Biophys. Res. Commun.*, 59 (1974) 314–321.
(125) S. Nagasaki, H. Matsuoka, and S. Yamamoto, *Kochi Daigaku Gakujutsu Kenkyu Hokoku*, 19 (1970) 171–175; *Chem. Abstr.*, 76 (1972) 44,658u.
(126) W. C. Slooff, in J. Lodder (Ed.), *The Yeasts, A Taxonomic Study*, 2nd edn., North-Holland Publishing Co., Amsterdam, 1970, pp. 1353–1357.
(127) E. H. Fischer, L. Kohtès, and J. Fellig, *Helv. Chim. Acta*, 34 (1951) 1132–1138.
(128) K. Myrbäck and W. Schilling, *Enzymologia*, 29 (1965) 306–314.
(129) N. P. Neumann and J. O. Lampen, *Biochemistry*, 8 (1969) 3552–3556.
(130) J. O. Lampen, S. C. Kuo, and F. R. Cano, in J. R. Villanueva, I. García-Acha, S. Gascón, and F. Uruburu (Eds.), *Yeast, Mould and Plant Protoplasts*, Academic Press, London, 1973, pp. 143–156.

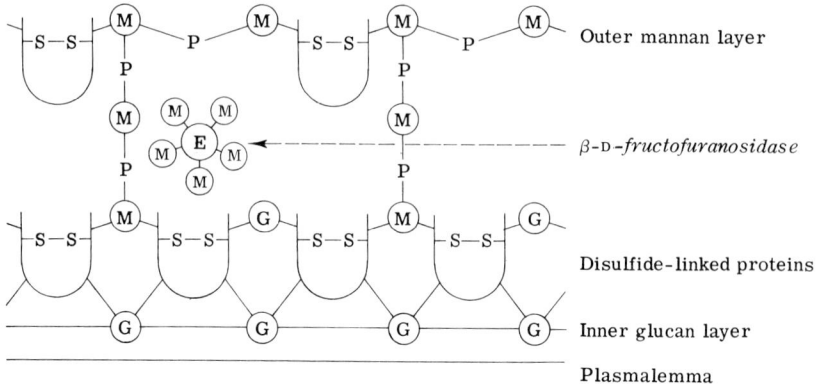

FIG. 2.—Diagram of Hypothetical Structure of Yeast Cell-wall. [Phosphoric diester links are represented by —P—; Ⓔ is β-D-fructofuranosidase; Ⓖ is D-glucan; Ⓜ is D-mannan; and S is sulfur. From Ref. 117; printed here by permission of Cambridge University Press.]

Saccharomyces cerevisiae is a mannoprotein of molecular weight 270,000, containing 50 to 70% of D-mannose,[90,102,131] and the β-D-fructofuranosidase of *Candida utilis* seems to be similar.[132] The mannan appears to be attached to the protein as 18 to 20 asparagine-linked units of polysaccharide, that is, 9 to 10 units per subunit of protein.[133,134] An endo-2-acetamido-2-deoxy-β-D-glucosidase has been used[134] to remove, by cleaving the unit of di-*N*-acetylchitobiose, the mannan that is linked to asparagine. Thus, a subunit of the protein, having a molecular weight of ~60,000, is released with mannan units that are terminated at the reducing end by a single 2-acetamido-2-deoxy-D-glucose residue.[133] Lehle and his colleagues[135] held that, in the biosynthesis of β-D-fructofuranosidase in *Saccharomyces cerevisiae*, the mannan core is added to the protein as a unit and, after this, there is a stepwise elaboration of the outer chain, by the addition of single mannose units (see also Refs. 136–138).

Waheed and Shall[139] worked with a heterogeneous, commercial

(131) B. Andersen and O. S. Jørgensen, *Acta Chem. Scand.*, 23 (1969) 2270–2276.
(132) M. Iizuka, H. Chiura and T. Yamamoto, *Agric. Biol. Chem.*, 42 (1978) 1207–1211.
(133) A. L. Tarentino, T. H. Plummer, and F. Maley, *J. Biol. Chem.*, 249 (1974) 818–824.
(134) R. B. Trimble and F. Maley, *J. Biol. Chem.*, 252 (1977) 4409–4412.
(135) L. Lehle, R. E. Cohen, and C. E. Ballou, *J. Biol. Chem.*, 254 (1979) 12,209–12,218.
(136) L. Lehle and W. Tanner, *Biochim. Biophys. Acta*, 539 (1978) 218–229.
(137) L. Lehle and W. Tanner, *Eur. J. Biochem.*, 83 (1978) 563–570.
(138) T. Nakajima and C. E. Ballou, *Proc. Natl. Acad. Sci. USA*, 72 (1975) 3912–3916.
(139) A. Waheed and S. Shall, *Enzymologia*, 41 (1971) 291–303.

preparation of β-D-fructofuranosidase from *Saccharomyces cerevisiae*. Having separated six fractions chromatographically, these authors found that the specific activity against sucrose, in units per mg of protein, was linearly associated with the carbohydrate content (%, w/w) of the enzyme. On the other hand, later authors found that removal of the D-mannan affects neither the activity nor the stability of the enzyme.[133,140] Having purified the D-mannan–protein β-D-fructofuranosidases of three strains of *Saccharomyces cerevisiae* whose walls differ in the structure of the D-mannan, Smith and Ballou[140] found that the structure of each β-D-fructofuranosidase D-mannan is immunochemically similar to that of the cell wall of the corresponding strain only. Mutations affecting the structure of the one also produced similar changes in the other. Smith and Ballou[124] suggested that the enzyme contains disulfide bonds, and that the role of these and the attached carbohydrate is to retain the β-D-fructofuranosidase in the cell wall, so producing a physiologically advantageous concentration of enzyme.[54] In addition, the bound carbohydrate may play a part in promoting folding of the β-D-fructofuranosidase into its most stable conformation.[141] After removing ~90% of the carbohydrate enzymically, the catalytic properties were almost unchanged, but the carbohydrate-free form was less resistant to freezing and thawing, to incubation at 50°, to acidic conditions, and to trypsin digestion.

Using gel filtration on columns of Sephadex G-200, Gascón and Ottolenghi[142] discovered in *Saccharomyces cerevisiae* a form of β-D-fructofuranosidase of low molecular weight, and predicted correctly that this form would be found to be free from carbohydrate. This enzyme occurred within the protoplast; its molecular weight (135,000) and specific activity were similar to those of the protein moiety of the external enzyme. The two β-D-fructofuranosidases gave the same K_m for sucrose, the same K_m for raffinose, and the same pH optimum (3.5 to 5.5) for enzymic activity; but their pH–stability curves differed, the internal enzyme being reversibly inactivated under acidic conditions, that is, below pH 5.

At one time, the small, carbohydrate-free, internal β-D-fructofuranosidase was considered to be a precursor of the large, external enzyme.[143,144] However, two kinds of experimental evidence led Lampen

(140) W. L. Smith and C. E. Ballou, *Biochemistry*, 13 (1974) 355–361.
(141) F. K. Chu, R. B. Trimble, and F. Maley, *J. Biol. Chem.*, 253 (1978) 8691–8693.
(142) S. Gascón and P. Ottolenghi, *C. R. Trav. Lab. Carlsberg*, 36 (1967) 85–93.
(143) P. Beteta and S. Gascón, *FEBS Lett.*, 13 (1971) 297–300.
(144) F. Moreno, A. G. Ochoa, S. Gascón and J. R. Villanueva, *Eur. J. Biochem.*, 50 (1975) 571–579.

and his colleagues[102,145,146] to question the precursory role of the internal β-D-fructofuranosidase of low molecular weight. First, they reported that the two forms of enzyme (large and small) differ in their amino acid composition. Secondly, these authors carried out "pulse-chase" experiments: when cells that were synthesizing β-D-fructofuranosidase were grown in the presence of ^{3}H-labelled amino acids, the large enzyme not only incorporated more total radioactivity, but also developed much greater specific activity than the small enzyme. Then, when protoplasts, labelled in this way, were incubated with unlabelled amino acids for 60 minutes, the tritiated large β-D-fructofuranosidase became diluted with new unlabelled enzyme, whereas the specific activity of the small enzyme changed relatively little. Hence, Lampen and his colleagues[146] concluded that the small, internal, carbohydrate-free β-D-fructofuranosidase is not a precursor of the external enzyme. In addition, a mutant revertant, from a β-D-fructofuranosidase-less strain, having fully regained the ability to form the large enzyme, produced only traces of the small enzyme.[147] Furthermore, whereas the large form is synthesized throughout the cell-cycle, the small enzyme is produced only at budding. In the presence of tunicamycin, the formation of the large, but not the small, enzyme ceased, and the latter did not accumulate in the cells.[148]

With Sephadex G-200, Gascón and coworkers[144] found many intermediates between light and heavy β-D-fructofuranosidase, apparently reflecting the sequential addition of D-mannose to a light form of the enzyme, culminating in the heavy form. More than one pool of the enzyme may occur within the plasmalemma: Holbein and his colleagues[149] reported an internal pool, associated with the plasmalemma, containing much β-D-fructofuranosidase of intermediate molecular weights, in addition to a soluble, internal pool of the small enzyme. Gascón and his colleagues found that the addition of 3-[2-(3,5-dimethyl-2-oxocyclohexyl)-2-hydroxyethyl]pentanedioimide (cycloheximide) stopped synthesis *de novo* of β-D-fructofuranosidase,[150] but glycosylation continued, the heavy enzyme appearing to accumulate at the expense of the carbohydrate-free form.[144] Holbein

(145) S. Gascón and J. O. Lampen, *J. Biol. Chem.*, 243 (1968) 1567–1572.
(146) J. O. Lampen, S. Kuo, F. R. Cano, and J. S. Tkacz, *Ferment. Technol. Today, Proc. Int. Ferment. Symp., 4th*, (1972) 819–824; *Chem. Abstr.*, 84 (1976) 27,875k.
(147) B. B. Abrams, R. Hackel, T. Mizunaga, and J. O. Lampen, *J. Bacteriol.*, 135 (1978) 809–817.
(148) G. Gallili and J. O. Lampen, *Biochim. Biophys. Acta*, 475 (1977) 113–122.
(149) B. E. Holbein, C. W. Forsberg, and D. K. Kidby, *Can. J. Microbiol.*, 22 (1976) 989–995.
(150) P. Liras and S. Gascón, *Eur. J. Biochem.*, 23 (1971) 160–165.

and Kidby[151,152] also uncoupled the synthesis of the protein of β-D-fructofuranosidase from its secretion, this time by damaging the plasmalemma of spheroplasts by the action of trypsin (EC 3.4.4.4). Secretion was stopped before synthesis, giving an increase of over fivefold in the concentration of small enzyme in the soluble, internal pool, as well as some small and intermediate enzyme associated with membrane fractions.

Babczinski and Tanner[153] have reported an additional isoenzyme of β-D-fructofuranosidase which they considered to be a precursor of the main enzyme. This new enzyme is associated with crude, membrane fractions of the cells; it has a molecular weight of 190,000, and appears transiently, as < 5% of the total β-D-fructofuranosidase, when the yeast is actively synthesizing the external enzyme. From studies on the effects of D-glucose repression and of cycloheximide, Sentandreu and coworkers (see Ref. 154) also considered that such an enzyme, associated more with internal membranes than with plasmalemma, might be, at least in part, a precursor of the external β-D-fructofuranosidase in the process of secretion.

Theories explaining the secretion of β-D-fructofuranosidase across the plasmalemma have included the involvement of vacuolar fragments, eliminated with the enzyme by a kind of reversed pinocytosis,[143] or of other vesicles.[146,155] Such theories depend particularly on techniques of fractionating cells, and on correct identification of isolated organelles. Matile and his colleagues[156] found D-mannan synthetase activity to be associated with the plasmalemma; consequently, Meyer and Matile[157] suggested that glycosylation of β-D-fructofuranosidase occurs in the plasmalemma and is coupled with transport of the enzyme from the cytosol to the cell-wall space. The findings of Holbein and Kidby[152] lend support to this suggestion.

e. **Regulation of β-D-Fructofuranosidase.**—There are reports, some of them contradictory, of hexose repression of both synthesis and secretion of β-D-fructofuranosidase in *Saccharomyces cerevisiae*. Gascón and Ottolenghi[142] found that the concentration of the external

(151) B. E. Holbein and D. K. Kidby, *Can. J. Microbiol.*, 23 (1977) 202–208.
(152) B. E. Holbein and D. K. Kidby, *Can. J. Microbiol.*, 25 (1979) 528–534.
(153) P. Babczinski and W. Tannner, *Biochim. Biophys. Acta*, 538 (1978) 426–434.
(154) L. Rodríguez, T. Ruiz, J. R. Villanueva, and R. Sentandreu, *Curr. Microbiol.*, 1 (1978) 41–44; *Chem. Abstr.*, 89 (1978) 159,858v.
(155) R. A. Holley and D. K. Kidby, *Can. J. Microbiol.*, 19 (1973) 113–117.
(156) M. Cortat, P. Matile, and F. Kopp, *Biochem. Biophys. Res. Commun.*, 53 (1973) 482–489.
(157) J. Meyer and P. Matile, *Arch. Microbiol.*, 103 (1975) 51–55.

enzyme varies by a factor of ~1000 with the concentration of exogenous D-glucose and, consistent with this finding, Meyer and Matile[158] reported abrupt derepression of secretion at <6 mM D-glucose. However, according to Sentandreu and coworkers,[159] and consistent with observations by Dodyk and Rothstein[160] and Gascón and Ottolenghi,[161] β-D-fructofuranosidase is synthesized only in media that are <60 mM in D-glucose; the inhibition and kinetics of this D-glucose repression resemble the effects of cycloheximide, in which synthesis, but not secretion or catalysis, of β-D-fructofuranosidase is prevented. Synthesis of β-D-fructofuranosidase was not repressed by sucrose, except when its hydrolysis produced a high concentration of exogenous hexose. Working with a mutant of *Saccharomyces cerevisiae* that hyperproduced β-D-fructofuranosidase, Hackel and N. A. Khan[162] confirmed that D-glucose represses synthesis. Some of their results were as follows.

Growth substrate (and initial concentration) (mM)		Relative specific activity of β-D-fructofuranosidase in crude extract	
D-Glucose	Sucrose	Original strain	Mutant
28	—	16	94
111	—	2	100
278	—	2	52
555	—	2	11
—	58	24	82

Genetic analysis has established the existence of at least five genes, designated *SUC1* to *SUC5*, the presence of any one of which is sufficient to allow the synthesis of a β-D-fructofuranosidase and the utilization of raffinose or sucrose[163,164] (for a review, see Ref. 165). The enzyme formed in response to the presence of any one of the *SUC* genes appears functionally similar to each of the others.[164] These genes are probably concerned with determining the primary amino acid se-

(158) J. Meyer and P. Matile, *Biochem. Physiol. Pflanz.*, 166 (1974) 377–385.
(159) M. V. Elorza, J. R. Villanueva, and R. Sentandreu, *Biochim. Biophys. Acta*, 475 (1977) 638–651.
(160) F. Dodyk and A. Rothstein, *Arch. Biochem. Biophys.*, 104 (1964) 478–486.
(161) S. Gascón and P. Ottolenghi, *C. R. Trav. Lab. Carlsberg*, 39 (1972) 15–24.
(162) R. A. Hackel and N. A. Khan, *Mol. Gen. Genet.*, 164 (1978) 295–302.
(163) Ø. Winge and C. Roberts, *C. R. Trav. Lab. Carlsberg*, 25 (1952) 141–171.
(164) P. Ottolenghi, *C. R. Trav. Lab. Carlsberg*, 38 (1971) 213–221.
(165) R. K. Mortimer and D. C. Hawthorne, in A. H. Rose and J. S. Harrison (Eds.), *The Yeasts*, Vol. 1, Academic Press, London, 1969, pp. 385–460.

quence of both internal and external β-D-fructofuranosidases,[166] possibly in addition to regulatory functions.[162,167]

Yurkevich and his colleagues presented evidence of some novel controls on the synthesis and secretion of β-D-fructofuranosidase by strains of *Saccharomyces cerevisiae*. Addition of D-mannitol to give a concentration of 0.5 M in the medium increased the concentration of β-D-fructofuranosidase in *Saccharomyces (globosus) cerevisiae*. Similar results were obtained with D-glucitol and D-xylose. As this increase was inhibited by 36 μM cycloheximide, Yurkevich and Khamani[168] concluded that a change to a hypertonic medium caused an increase in *de novo* biosynthesis of β-D-fructofuranosidase.

Heterothallic, but not homothallic, strains of *Saccharomyces cerevisiae* (including *Saccharomyces carlsbergensis*, *Saccharomyces paradoxus*, and *Saccharomyces chevalieri*) released some β-D-fructofuranosidase into the suspending medium[169]; this release was subject to feedback control, as exogenously supplied, active (but not inactivated) β-D-fructofuranosidase diminished the release.[170] Secretion of β-D-fructofuranosidase by spheroplasts was unaffected by exogenous enzyme,[171] but the regulatory mechanism was restored when the cell walls had partly regenerated. In 1955, Ingram[172] drew attention to the adsorption of β-D-fructofuranosidase from solution by yeast cells; subsequently, Galcheva-Gargova and Yurkevich[173] showed that [^{125}I]-labelled β-D-fructofuranosidase becomes bound to intact yeast-cells and to cell-wall material, but not to protoplasts. Yurkevich and Galcheva-Gargova[174] suggested that this regulation is mediated by transcriptional control of synthesis of β-D-fructofuranosidase: after addition of enzyme, incorporation of [^{14}C]uracil into RNA remained unchanged, but that of L-[^{14}C]valine into protein decreased.

(166) M. K. Grossmann and F. K. Zimmermann, *Mol. Gen. Genet.*, 175 (1979) 223–229.
(167) R. A. Hackel, *Mol. Gen. Genet.*, 140 (1975) 361–370.
(168) V. V. Yurkevich and D. Khamani, *Biol. Nauki (Moscow)*, 15 (1972) 101–104; *Chem. Abstr.*, 78 (1973) 55,181a.
(169) V. V. Yurkevich and T. I. Naumova, *Biol. Nauki (Moscow)*, 16 (1973) 95–98; *Chem. Abstr.*, 78 (1973) 121,056s.
(170) V. V. Yurkevich, N. V. Shamshurina, and D. Khamani, *Biol. Nauki (Moscow)*, 16 (1973) 97–100; *Chem. Abstr.*, 79 (1973) 39,929r.
(171) V. V. Yurkevich and Z. I. Galcheva-Gargova, *Zh. Obshch. Biol.*, 36 (1975) 414–420; *Chem. Abstr.*, 83 (1975) 144,343a.
(172) M. Ingram, *An Introduction to the Biology of Yeasts*, Pitman, London, 1955, p. 57.
(173) Z. I. Galcheva-Gargova and V. V. Yurkevich, *Dokl. Akad. Nauk SSSR Ser. Biokhim.*, 225 (1975) 446–449; *Chem. Abstr.*, 84 (1976) 56,357y.
(174) V. V. Yurkevich and Z. I. Galcheva-Gargova, *Dokl. Akad. Nauk SSSR Ser. Biokhim.*, 225 (1975) 971–973; *Chem. Abstr.*, 84 (1976) 86,600q.

f. Delayed Utilization of Sucrose.—Certain yeasts utilize sucrose, but only after a delay of three or four weeks[175]; this delayed utilization seems to result from physiological changes in the cells, and not from selection of mutants.[176] These yeasts have been found notably among *Zygosaccharomyces* (*Saccharomyces*) *rouxii*, *Zygosaccharomyces* (*Saccharomyces*) *bisporus*, and *Schizosaccharomyces octosporus*, many strains of which tolerate high concentrations of sugar[175-181] of at least up to 4 M hexose. Arnold and his colleagues[111,182,183] investigated the delay in a strain of *Zygosaccharomyces rouxii*. Electron micrographs[182] showed many organelles in the periplasmic zones of cells of this yeast that were in the stationary phase of growth. Accordingly, these authors suggested that these periplasmic bodies are short-lived and might be involved in translocating β-D-fructofuranosidase into the space between the plasmalemma and the cell-wall. With aging, the bodies disintegrate, and release the enzyme. However, these results do not explain the observation that, although strains of *Zygosaccharomyces rouxii* may utilize sucrose after a delay, van der Walt[92] found that none of the 55 strains he examined could use raffinose.

2. α-D-Galactopyranosides

Whereas ~60% of yeast species can use sucrose, and 33% can use raffinose, only ~12% of the 439 species considered in Table III unequivocally utilize melibiose [O-α-D-galactopyranosyl-(1→6)-α-D-glucopyranose]. The best known of these melibiose-using yeasts are the bottom (lager) yeasts, formerly called *Saccharomyces carlsbergensis*,[184,185] subsequently *Saccharomyces uvarum*,[92] and now, *Saccharomyces cerevisiae*.[5] (Consequently, at present, *Saccharomyces*

(175) M. P. Scarr, *J. Gen. Microbiol.*, 5 (1951) 704–713.
(176) D. Pappagianis and H. J. Phaff, *Antonie van Leeuwenhoek J. Microbiol. Serol.*, 22 (1956) 353–370.
(177) M. P. Scarr, *Proc. Soc. Appl. Bacteriol.*, 16 (1953) 119–127.
(178) M. Ingram, *Symp. Soc. Gen. Microbiol.*, 7 (1957) 90–133.
(179) M. Ingram, in A. H. Cook (Ed.), *The Chemistry and Biology of Yeasts*, Academic Press, New York, 1958, pp. 603–633.
(180) M. Ingram, *Rev. Ferment. Ind. Aliment.*, 14 (1959) 23–33.
(181) H. Ōnishi, *Adv. Food Res.*, 12 (1963) 53–94.
(182) W. N. Arnold, R. G. Garrison, and K. S. Boyd, *Appl. Microbiol.*, 28 (1974) 1047–1054.
(183) W. N. Arnold and R. G. Garrison, *J. Bacteriol.*, 137 (1979) 1386–1394.
(184) N. M. Stelling-Dekker, *Verh. K. Ned. Akad. Wet. Amsterdam Afd. Natuurk.*, Reeks 2, 28 (1931) 1–547.
(185) J. Lodder and N. J. W. Kreger-van Rij, *The Yeasts, A Taxonomic Study*, North-Holland Publishing Company, Amsterdam, 1952.

cerevisiae includes yeasts that are melibiose-utilizing and -non-utilizing.) Accordingly, unlike top (ale) yeasts, bottom yeasts have long been known to be a good source of α-D-galactosidase.[186,187] This enzyme has certain features in common with β-D-fructofuranosidase; it is also a glycoprotein, and acts outside the plasmalemma.[50,51,55,188–190] Cartledge and Lloyd[191] fractionated subcellular particles from late-exponential-phase *Saccharomyces (carlsbergensis) cerevisiae*, and found that, of all the hydrolases assayed, the α-D-galactosidase had the most heterogeneous distribution.

In 1972, Dey and Pridham[192] reviewed the α-D-galactosidases. Those from yeasts seem to hydrolyze melibiose to D-galactose and D-glucose, or raffinose to D-galactose and sucrose[193,194] (see Scheme 2). Gascón and his colleagues[194] found the molecular weight of α-D-galactosidase from *Saccharomyces (carlsbergensis) cerevisiae* to be 300,000, and that the enzyme contained 57% of carbohydrate, composed primarily of D-mannose. Their estimates of K_m were 6 mM p-nitrophenyl α-D-galactopyranoside, 6 mM melibiose, and 135 mM raffinose, the pH optimum lying between 4 and 5. Sugars having the same configuration at C-2, C-3, and C-4 as D-galactose, such as D-fucose and L-arabinose, competitively inhibit the enzyme. Wakabayashi and Nishizawa[195] found that α-D-galactosidase from bottom yeast hydrolyzed phenyl, *m*-methylphenyl, *p*-methylphenyl, and *p*-nitrophenyl α-D-galactopyranoside at similar rates, and that these differed from those for methyl α-D-galactopyranoside and melibiose.

Over 20 years ago, Winge and his colleagues at the Carlsberg Laboratorium made various hybrids of *Saccharomyces (carlsbergensis) cerevisiae* which, according to their genotype, could utilize all, part,

---

(186) E. Fischer and P. Lindner, *Ber.*, 28 (1895) 3034–3039.
(187) A. Bau, *Chem.-Ztg., Chem. Appar.*, 19 (1895) 1873–1874.
(188) G. de la Fuente and A. Sols, *Abstr. Int. Congr. Biochem., 4th, Vienna*, (1958) 131.
(189) S. Gascón, P. S. Lazo, F. Moreno, and A. G. Ochoa, in J. R. Villanueva, I. García-Acha, S. Gascón, and F. Uruburu (Eds.), *Yeast, Mould and Plant Protoplasts*, Academic Press, London, 1973 pp. 157–166.
(190) P. S. Lazo, A. G. Ochoa, and S. Gascón, *Eur. J. Biochem.*, 77 (1977) 375–382.
(191) T. G. Cartledge and D. Lloyd, *Biochem. J.*, 126 (1972) 755–757.
(192) P. M. Dey and J. B. Pridham, *Adv. Enzymol. Relat. Areas Mol. Biol.*, 36 (1972) 91–130.
(193) V. I. Isajev, *Chem. Listy*, 21 (1927) 101–107, 141–147, 191–202; *Chem. Abstr.*, 22 (1928) 97.
(194) P. S. Lazo, A. G. Ochoa, and S. Gascón, *Arch. Biochem. Biophys.*, 191 (1978) 316–324.
(195) K. Wakabayashi and K. Nishizawa, *Seikagaku*, 27 (1955) 662–665; *Chem. Abstr.*, 55 (1961) 1743a.

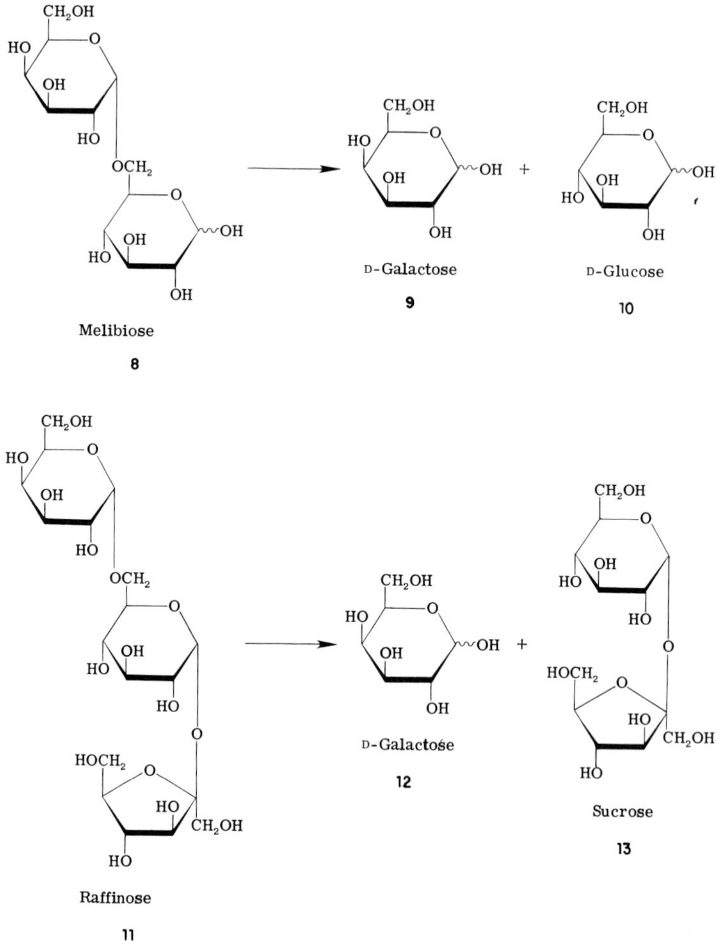

Scheme 2

or none of the raffinose molecule[196-198] (see Table VI). Thus, (i) yeasts lacking genes for both α-D-galactosidase (*MEL*) and β-D-fructofuranosidase (*SUC*) did not hydrolyze raffinose; (ii) yeasts having *SUC* genes, but no *MEL* genes, utilized the D-fructofuranosyl group, so that melibiose remained; (iii) yeasts having a *MEL* gene, but no *SUC* gene, hydrolyzed raffinose to sucrose plus D-galactose, and these were uti-

(196) Ø. Winge and C. Roberts, *Nature (London)*, 177 (1956) 383–384.
(197) Ø. Winge and C. Roberts, *C. R. Trav. Lab. Carlsberg*, 25 (1957) 419–459.
(198) M. Losada, *C. R. Trav. Lab. Carlsberg*, 18 (1957) 460–482.

TABLE VI

Hydrolysis and Utilization of Raffinose by Cultures of *Saccharomyces uvarum* of Various Genotypes[198]

	Genotype							
	MAL	MEL	SUC	MEL GAL	MEL MAL	MEL SUC	MEL MAL GAL	MEL SUC GAL
Enzymes present[a]								
α-D-Glucosidase	+	−	−	−	+	−	+	−
α-D-Galactosidase	−	+	−	+	+	+	+	+
β-D-Fructofuranosidase	−	−	+	−	−	+	−	+
?Galactokinase[b]	−	−	−	+	−	−	+	+
Sugars produced by hydrolysis of raffinose[c]								
D-Fructose	0	0	PU	0	PU	PU	PU	PU
D-Glucose	0	0	0	0	PU	PU	PU	PU
D-Galactose	0	P	0	PU	P	P	PU	PU
Sucrose	0	P	0	P	0	0	0	0
Melibiose	0	0	P	0	0	0	0	0

[a] Key: +, enzyme present; −, enzyme not present. [b] The gene *GAL*, for D-galactose utilization, may refer to one or more genes controlling enzymes of the Leloir pathway (see Ref. 4). [c] Key: 0, sugar not produced; P, sugar produced, but not utilized by yeast; PU, sugar produced, and utilized, by yeast.

lized only if genes *MAL* and *GAL* were present, respectively. The *MAL* genes are concerned with the synthesis of α-D-glucosidase (which can cleave sucrose, both an α-D-glucopyranoside and a β-D-fructofuranoside). The *GAL* genes subserve the utilization of D-galactose.

Table VI instances eight possible gene combinations affecting the utilization of raffinose. Correspondingly, yeasts of different species utilize different parts of the raffinose molecule[199] or none; Barnett[4] exemplified each kind of utilization for species of *Saccharomyces* described by van der Walt.[92]

The genetical regulation of the utilization of melibiose may be quite complex: a single locus, *GAL 3* is concerned pleiotropically in the initiation of induction of utilizing melibiose, D-galactose, and maltose.[200] Utilization of melibiose and production of α-D-galactosidase are also controlled by three regulatory genes, *i, c* and *GAL 4*, concerned with at least three enzymes of the Leloir pathway by which D-galactose is converted into D-glucose 6-phosphate (for a survey, see Ref. 4). Kew and Douglas[200] also found D-galactose to be a non-metabolized (gratuitous) inducer of α-D-galactosidase in strains lacking the Leloir pathway.

IV. THE UTILIZATION OF GLYCOSIDES HYDROLYZED INSIDE THE PLASMALEMMA

Having found that extracts from dried brewer's yeast readily split maltose to D-glucose, Emil Fischer held that yeasts always hydrolyze disaccharides before fermenting them,[186,201,202] but this view was disputed by Richard Willstätter in the nineteen-twenties (for a review, see Ref. 203). Willstätter and Oppenheimer[204] found that some lactose-using yeasts ferment lactose more rapidly than D-glucose, D-galactose, or a mixture of the two. From such observations, Willstätter concluded that the first catabolic event is not always hydrolysis. Although, early in the twentieth century, Prior and Schulze[205] and Slator[206] considered the possibility that different sugars may enter yeasts at different rates, these two views, initial hydrolysis on the one hand, and "direct" fer-

(199) N. J. W. Kreger-van Rij, *Symp. Soc. Gen. Microbiol.*, 12 (1962) 196–211.
(200) O. M. Kew and H. C. Douglas, *J. Bacteriol.*, 125 (1976) 33–41.
(201) E. Fischer, *Ber.*, 27 (1894) 3479–3483.
(202) E. Fischer, *Z. Physiol. Chem.*, 26 (1898) 60–87.
(203) J. Leibowitz and S. Hestrin, *Adv. Enzymol.*, 5 (1945) 87–127.
(204) R. Willstätter and G. Oppenheimer, *Z. Physiol. Chem.*, 118 (1922) 168–188.
(205) E. Prior and H. Schulze, *Z. Angew. Chem.*, 14 (1901) 208–215.
(206) A. Slator, *J. Chem. Soc.*, 93 (1908) 217–242.

mentation on the other, were a subject of dispute until Gottschalk[207,208] suggested that entry into the cell can be the rate-limiting step in the catabolism of a disaccharide, and this is the current view.

1. The Entry of Glycosides into Yeasts

Thus, for those glycosides that are hydrolyzed within the plasmalemma, the first step in utilization is their movement into the cell across the plasmalemma, by means of carriers associated with the plasmalemma. These carriers (or permeases) take the glycosides into the cytoplasm from solution between the cell wall and the plasmalemma. The importance of the carriers is emphasized by cases of crypticity, when the yeasts fail to utilize a sugar, despite their possessing all of the necessary intracellular enzymes. For example, some yeasts cannot use maltose, although they can use D-glucose, and contain high concentrations of α-D-glucosidase. The carriers provide a means of controlling entry, and hence also, in part, of controlling utilization of the glycosides.

The carriers are like enzymes in that they (a) are proteins that form complexes with their substrates, (b) have various degrees of substrate specificity, including stereospecificity (cells may be equipped with more than one carrier having overlapping specificities), and (c) show saturation kinetics. Classical, Michaelis–Menten relationships between substrate concentration, maximum initial velocity (V_{max}), and K_m, are used to analyze uptake, but carriers may also show allosteric cooperativity. Like many initial catabolic enzymes, transport carriers often have wide substrate specificities. This apparent broad specificity may be accentuated, as the number of substrates for a carrier is more properly compared with that for an enzyme, in addition to structurally related competitive inhibitors; this is because the transporting system may depend solely on the binding of the carrier to its substrate, instead of on its chemical transformation. Many of the carriers are inducible or repressible, or both, and may be under direct and separate genetic control. This genetic control is probably also separable from that concerning other processes of utilizing the glycoside taken in by that carrier. Hence, for any such sugar utilization, it may be possible to obtain a series of single-locus mutants, each locus affecting a carrier or catabolic enzyme.

The following considerations have been important for studying the

---

(207) A. Gottschalk, *Wallerstein Lab. Commun.*, 12 (1949) 55–67.
(208) A. Gottschalk, in K. Myrbäck (Ed.), *The Enzymes*, Vol. I. Part 1, Academic Press, New York, 1950, pp. 551–582.

uptake of sugars and glycosides. (*i*) Sugars that are not metabolized by a yeast may nonetheless enter its cells. For example, L-sorbose is taken up, but not catabolized, by *Saccharomyces cerevisiae*.[209] Consequently, many studies have been made with non-metabolized, substituted sugars, such as thio sugars (see, for example, Refs. 210 and 211), in order to avoid the problem of separating the effects of transport from those of metabolism. (*ii*) Transport is often studied by using inhibitors of either metabolism or transport (for a review, see Ref. 212). Certain compounds that specifically inhibit the transport of sugars do so by competing externally for combination with the carriers, or by so modifying them that internal release is impaired. By analogy with enzymes, transport of a given sugar may be inhibited by other sugars that are transported on the same carrier, by sugar analogs that are bound but not released by the carrier, or by structurally dissimilar compounds. (*iii*)As it is difficult to measure the affinity of a carrier for many sugars directly, such measurement is often achieved by testing the capacity of a sugar to inhibit the uptake of non-metabolized sugars. The extent of inhibition provides a measure of the affinity of the inhibiting sugar for the carrier: the inhibitor constant, K_i, is the dissociation constant of the inhibitor–carrier complex.

a. Kinds of Transport.—Carrier-mediated movement of glycosides (or monosaccharides) across the plasmalemma of yeasts involves the combination of the glycoside or sugar with the carrier protein on one side of the plasmalemma, and then release of the glycoside or sugar into the cytoplasm on the other side. Such movement is described either as *facilitated diffusion*, when the movement requires no metabolic energy, or *active transport*, if it involves the expenditure of metabolic energy. Sugars entering yeast cells by active transport may be accumulated within the cells to a concentration over a hundred times the external level. For yeasts, the subject has been reviewed by various authors.[49,213–217]

(209) V. P. Cirillo, *Trans. N. Y. Acad. Sci.*, 23 (1961) 725–734.
(210) H. Okada and H. O. Halvorson, *Biochim. Biophys. Acta*, 82 (1964) 538–546.
(211) H. Okada and H. O. Halvorson, *Biochim. Biophys. Acta*, 82 (1964) 547–555.
(212) V. P. Cirillo, in R. M. Hochster, M. Kates, and J. H. Quastel (Eds.), *Metabolic Inhibitors: A Comprehensive Treatise*, Vol. III, Academic Press, New York, 1972, pp. 47–68.
(213) V. P. Cirillo, *Annu. Rev. Microbiol.*, 15 (1961) 197–218.
(214) C. Divies and J. N. Morfaux, *Ann. Technol. Agric.*, 17 (1968) 355–377.
(215) H. Suomalainen and E. Oura, in A. H. Rose and J. S. Harrison (Eds.), *The Yeasts*, Vol. 2, Academic Press, London, 1971, pp. 3–74.
(216) D. H. Jennings, *Trans. Br. Mycol. Soc.*, 62 (1974) 1–24.
(217) A. Kotyk and K. Janáček, *Cell Membrane Transport*, 2nd edn., Plenum Press, New York, 1975.

One kind of active transport, namely *group translocation*, occurs in bacteria (for reviews, see Refs. 218–223), and some workers consider that this also takes place in yeasts (see, for example, Refs. 224 and 225); by this means, uptake of a sugar is directly coupled to its phosphorylation, and the sugar is released into the cytoplasm as a phosphate.

Energy for active transport may also come from a direct interaction between the carrier and ATP or another source of "high energy." In some cases, uptake of a sugar by a yeast seems to depend on the simultaneous, stoichiometric uptake of protons by the carrier of that sugar; the free energy produced by the transport of protons along their transmembrane, concentration gradient (an electrochemical, potential gradient) drives the transport of the sugars. The protons are then pumped out of the cell by another carrier, which may be associated with ATPase activity as a source of energy. Most of the research on the energetics of active transport has been conducted with bacteria, in which proton-coupled transport is of widespread importance (for a review, see Ref. 226). Eddy[227] reviewed the subject for micro-organisms, including yeasts, and Deák[228] surveyed proton–sugar symport in 22 yeast species.

2. α-D-Glucopyranosides

a. The Uptake of α-D-Glucopyranosides by Yeasts.—This process has been studied almost exclusively with a number of different strains of *Saccharomyces cerevisiae*. As with enzymes, there appear to be various, distinct, multiple forms of carrier for α-D-glucopyranosides, but earlier descriptions of the uptake systems have not been reinvestigated. There may be a maltose-specific carrier: Harris and Thompson[229] recorded that maltose is taken up by yeast grown on maltose,

---

(218) S. Roseman, in L. E. Hokin (Ed.), *Metabolic Pathways*, 3rd edn., Vol. 6, Academic Press, New York, 1972, pp. 41–89.
(219) S. Roseman, in J. F. Woessner and F. Huijing (Eds.), *The Molecular Basis of Biological Transport*, Academic Press, New York, 1972, pp. 181–215.
(220) S. Roseman, in G. Semenza and E. Carafoli (Eds.), *Biochemistry of Membrane Transport*, FEBS Symp. No. 42, Springer-Verlag, Berlin, 1977, pp. 582–597.
(221) H. L. Kornberg, *Proc. R. Soc. London, Ser. B*, 183 (1973) 105–123.
(222) H. L. Kornberg, *Symp. Soc. Exp. Biol.*, 27 (1973) 175–193.
(223) A. Kepes, *Biochimie*, 55 (1973) 693–702.
(224) J. van Steveninck, *Biochim. Biophys. Acta*, 203 (1970) 376–384.
(225) H. T. A. Jaspers and J. van Steveninck, *Biochim. Biophys. Acta*, 406 (1975) 370–385.
(226) F. M. Harold, *Curr. Top. Bioenerg.*, 6 (1977) 83–149.
(227) A. A. Eddy, *Curr. Top. Membr. Transp.*, 10 (1978) 279–359.
(228) T. Deák, *Arch. Microbiol.*, 116 (1978) 205–211.
(229) G. Harris and C. C. Thompson, *Biochim. Biophys. Acta*, 52 (1961) 176–183.

TABLE VII

Estimates of Apparent Michaelis Constants for the Entry of α-D-Glucopyranosides into *Saccharomyces cerevisiae*

Growth substrate	Maltose (mM)	Methyl α-D-glucopyranoside (mM)	1-thio-α-D-glucopyranoside (mM)	Ethyl α-D-glucopyranoside (mM)	References	Comments
Maltose	80	80	—	—	240	9–15 mg (dry wt) of yeast per mL, hence unlikely to be fully aerobic; no correction for loss of $^{14}CO_2$
Maltose	2.4 (K_i)	120	—	—	241	~3 mg (dry wt) of yeast
Methyl α-D-glucopyranoside	—	—		1.8	211	active transport
D-Glucose	—	—		50	210	facilitated diffusion
Maltose	4	—	—	—	242	estimated from rate of anaerobic fermentation, assuming entry to be rate-limiting
D-Glucose	50	—	—	—	242	
Complex medium	—	8	—	—	224	active transport; no catabolism of substrate (see also, Ref. 239)
Complex medium for inducing protoplasts	—	—	—	15	238	*Saccharomyces carlsbergensis* protoplasts; active transport "induced" with 100 mM maltose + 17 mM D-glucose

but not by that grown on methyl α-D-glucopyranoside. From competition experiments, Hautera and Lövgren[230] concluded that maltose and ethyl 1-thio-α-D-glucopyranoside (α-TEG) enter on the same carrier. However, the findings of Okada and Halvorson[211] indicated that α-TEG enters by the carrier for methyl α-D-glucopyranoside; yeasts generally unable to take up that D-glucoside can nevertheless ferment maltose.[231-233] Harris and Thompson[234] suggested that maltotriose penetrates *Saccharomyces cerevisiae* by yet another carrier having a high substrate-specificity. Uptake is often described[210,211,229,235,236] as "inducible" (Ref. 237), but, as comparisons have usually been made solely between yeasts grown on D-glucose and those grown on an α-D-glucopyranoside, the carrier may well be derepressible (compare Ref. 238). Entry into repressed (or non-induced) cells is by facilitated diffusion,[210] but, after exposure to maltose, methyl α-D-glucopyranoside, or α-TEG, by active transport.[211,238,239] The carriers responsible for active transport seem to have a higher affinity for their substrates than those for facilitated diffusion (see Table VII). Okada and Halvorson[210,211] studied the uptake of α-TEG, which is not metabolized. In D-glucose-grown yeast, α-TEG entered, but was not concentrated by, the cells; entry was unaffected by 10 mM sodium azide, but inhibited by D-glucose and methyl α-D-glucopyranoside. Yeast grown on the last could concentrate α-TEG by a factor of about three from a 5 mM solution,

(230) P. Hautera and T. Lövgren, *J. Inst. Brew.*, London, 81 (1975) 309–313.
(231) D. C. Hawthorne, *Heredity*, 12 (1958) 273–284.
(232) G. Terui, H. Okada, and Y. Oshima, *Technol. Rep. Osaka Univ.*, 9 (1959) 237–259.
(233) T. Takahashi and Y. Ikeda, *Z. Vererbungsl.*, 90 (1959) 66–73; *Chem. Abstr.*, 55 (1961) 26,114g.
(234) G. Harris and C. C. Thompson, *J. Inst. Brew.*, London, 66 (1960) 293–297.
(235) H. O. Halvorson, H. Okada, and J. Gorman, in J. F. Hoffman (Ed.), *The Cellular Functions of Membrane Transport*, Prentice-Hall, Englewood Cliffs, New Jersey, 1964, pp. 171–191.
(236) J. H. Slavenburg, R. van Wijk, and G. Zoutewelle, *Proc. K. Ned. Akad. Wet.*, Ser. C, 75 (1972) 55–66.
(237) Some enzymes and carriers are synthesized only in response to the presence of the sugar, or of a structurally similar compound; these enzymes and carriers are said to be "inducible." Contrariwise, enzyme synthesis may be repressed by an increase in the concentration of ATP, or of some other metabolite. Induction and repression of enzymes and carriers provide two important kinds of control in metabolic regulation.
(238) R. A. de Kroon and V. V. Koningsberger, *Biochim. Biophys. Acta*, 204 (1970) 590–609.
(239) R. Brocklehurst, D. Gardner, and A. A. Eddy, *Biochem. J.*, 162 (1977) 591–599.
(240) G. Avigad, *Biochim. Biophys. Acta*, 40 (1960) 124–134.
(241) J. J. Robertson and H. O. Halvorson, *J. Bacteriol.*, 73 (1957) 186–198.
(242) C. P. M. Görts, *Biochim. Biophys. Acta*, 184 (1969) 299–305.

and by a factor of up to a hundred from solutions containing <0.5 mM α-TEG.[210,211,238,239] This accumulation was inhibited by 0.8 mM sodium azide. Halvorson and his colleagues[211,235] suggested that active transport depends on an inducible coupling between the energy metabolism of the cell and the carrier responsible for facilitated diffusion. This suggestion is not inconsistent with a subsequent view, namely, that a carrier responsible for the active transport of a sugar by proton symport may, when deprived of the proton gradient, subserve facilitated diffusion. This change could also involve decreased velocity of transport (compare Ref. 243), and, perhaps, a change in the affinity for the sugar (compare Ref. 244). Active transport of maltose is inhibited by 2,4-dinitrophenol, sodium fluoride, sodium azide, and carbonyl cyanide m-chlorophenylhydrazone.[245]

Van Steveninck proposed[224] a different mechanism for methyl α-D-glucopyranoside which was taken up by the maltose carrier. This uptake, he suggested, involves phosphorylation of the D-glucoside (group translocation), and then dephosphorylation to afford nonphosphorylated, intracellular D-glucoside. When the yeast was fed methyl α-D-[^{14}C]glucopyranoside, the D-glucoside phosphate was labelled before the free, internal D-glucoside. However, Eddy and his colleagues,[239] working with the same strain of yeast, questioned the validity of these findings. They suggested that the chromatographic system used was inadequate to separate D-glucose from methyl α-D-glucopyranoside. Kotyk and Michaljaničová also found[246] that this D-glucoside was taken up without phosphorylation.

About one equivalent of protons was taken up with each equivalent of methyl α-D-glucopyranoside[239] or maltose[247]; α-TEG was also absorbed with protons. In addition, Serrano[245] and Palacios and Serrano[248] found maltose transport to be coupled to the electrochemical, proton gradient, but that entry was independent of the intracellular concentration of ATP.

b. Yeast α-D-Glucopyranoside Hydrolases.—About half the yeast species can utilize maltose, one third methyl α-D-glucopyranoside, and one third melezitose (see Table III). Of 253 strains of yeasts of

(243) A. Seaston, G. Carr, and A. A. Eddy, *Biochem. J.*, 154 (1976) 669–676.
(244) R. D. Simoni and P. W. Postma, *Annu. Rev. Biochem.*, 44 (1975) 523–554.
(245) R. Serrano, *Eur. J. Biochem.*, 80 (1977) 97–102.
(246) A. Kotyk and D. Michaljaničová, *Biochim. Biophys. Acta*, 332 (1974) 104–113.
(247) A. A. Eddy, R. Philo, P. Earnshaw, and R. Brocklehurst, in G. Semenza and E. Carafoli (Eds.), *Biochemistry of Membrane Transport*, FEBS-Symposium No. 42, Springer-Verlag, Berlin, 1977, pp. 250–260.
(248) J. Palacios and R. Serrano, *FEBS Lett.*, 91 (1978) 198–201.

miscellaneous genera that used methyl α-D-glucopyranoside, and of 206 that used melezitose, all could utilize maltose.[4] In addition, maltotriose, which constitutes 15 to 20% of the fermentable carbohydrate of brewer's wort,[249] is utilized by a number of strains of yeast now called *Saccharomyces cerevisiae* (including *Saccharomyces pastorianus*, *Saccharomyces logos*, *Saccharomyces bayanus*, *Saccharomyces willianus*, *Saccharomyces carlsbergensis*, *Saccharomyces heterogenicus*, and *Saccharomyces diastaticus*), *Zygosaccharomyces* (*Saccharomyces*) *rouxii*, *Torulaspora delbrueckii* (*Saccharomyces fermentati*), and *Zygosaccharomyces* (*Saccharomyces*) *florentinus*.[250,251] The order in which a brewer's yeast uses the wort sugars depends on how that yeast has been cultivated.[252] Some yeasts can also use turanose or palatinose.[51,253,254]

Many yeasts seem to hydrolyze sucrose by means of an α-glucosidase. Barnett[4,70] reported that 128 (26%) out of 497 strains of miscellaneous genera could use sucrose, but not raffinose, and at least 124 of these yeasts could use maltose as well.

(i) Substrate Specificity of Yeast Hydrolases Acting on α-D-Glucopyranosides.—Gottschalk[68,208] and Larner[255] have reviewed the α-D-glucopyranoside hydrolases. Those considered in these paragraphs are classified by the Enzyme Commission[6] as follows. (*i*) α-D-Glucosidase (EC 3.2.1.20) hydrolyzes terminal, (1→4)-linked α-D-glucopyranosyl groups, releasing α-D-glucose; oligosaccharides are hydrolyzed rapidly, and polysaccharides slowly, or not at all. (*ii*) Oligo-(1→6)-D-glucosidase (EC 3.2.1.10) hydrolyzes (1→6)-α-D-glucopyranosyl linkages in isomaltose and in the gluco-oligosaccharides ("dextrins") produced from starch and glycogen by alpha amylase. (*iii*) Alpha amylase (EC 3.2.1.1) randomly hydrolyzes (1→4)-α-D-glucopyranoside linkages in starch, glycogen, other D-glucans, and gluco-oligosaccharides. (*iv*) Glucoamylase (EC 3.2.1.3) hydrolyzes terminal, (1→4)-linked groups successively from nonreducing ends of chains, particularly of polysaccharides. However, it is not clear whether the yeast α-D-glucopyranosidases can all be satisfactorily classified in this way.

(249) I. C. MacWilliam, *J. Inst. Brew., London*, 74 (1968) 38–54.
(250) R. B. Gilliland, *C. R. Trav. Lab. Carlsberg*, 26 (1956) 139–148.
(251) D. J. Millin and D. G. Springham, *J. Inst. Brew., London*, 72 (1966) 388–393.
(252) J. A. Budd, *Eur. J. Appl. Microbiol.*, 3 (1977) 267–272.
(253) D. S. Hwang and C. C. Lindegren, *Nature (London)*, 203 (1964) 791–792.
(254) D. S. Hwang and C. C. Lindegren, *Nature (London)*, 205 (1965) 880–883.
(255) J. Larner, in P. D. Boyer, H. Lardy, and K. Myrbäck (Eds.), *The Enzymes*, 2nd edn., Vol. 4, Academic Press, New York, 1960, pp. 369–378.

Neuberg[256] and Neuberg and Marx[257] found that α-D-glucosidase does not hydrolyze raffinose, and suggested that this is because C-6 of the D-glucosyl residue of the α-D-glucoside is blocked by the linkage with the D-galactosyl group. Phillips[258] confirmed Neuberg's finding with α-D-glucosidase of *Saccharomyces cerevisiae* (see also, Refs. 68 and 259). However, sucrose (an unsubstituted α-D-glucopyranoside, as well as a β-D-fructofuranoside) was shown by Weidenhagen[42,260,261] to be hydrolyzed by α-D-glucosidase at pH 6.9, but not at pH 4.7, which is optimal for β-D-fructofuranosidase. Indeed, *Candida tropicalis*, which utilizes sucrose and not raffinose,[262] gave an unfractionated extract that hydrolyzed sucrose, maltose, and methyl α-D-glucopyranoside, but not raffinose.[263] The same yeast produced an alpha amylase that was active against maltose and isomaltose, but not against sucrose.[264,265] Furthermore, Eaton and his colleagues[266] described the utilization of sucrose by means of an α-D-glucosidase in a strain of *Saccharomyces cerevisiae* lacking β-D-fructofuranosidase, and Avigad reported[240] the accumulation of sucrose by a non-sucrose-utilizing strain of this yeast. *Candida solani* also utilizes sucrose, but not raffinose, apparently hydrolyzing the sucrose by means of an internal enzyme.[267] It would be interesting to study the specificity of the sucrose carriers in these yeasts.

In other yeasts, there is evidence of enzymes that hydrolyze maltose but not sucrose; this applies to *Schizosaccharomyces octosporus*,[268,269] *Schizosaccharomyces pombe*,[270] *Candida tropicalis*,[264,265] *Candida*

---

(256) C. Neuberg, *Biochem. Z.*, 3 (1907) 519–534.
(257) C. Neuberg and F. Marx, *Biochem. Z.*, 3 (1907) 535–538.
(258) A. W. Phillips, *Arch. Biochem. Biophys.*, 80 (1959) 346–352.
(259) D. French, *Adv. Carbohydr. Chem.*, 9 (1954) 149–184.
(260) R. Weidenhagen, *Naturwissenschaften*, 16 (1928) 654–655.
(261) R. Weidenhagen, *Ergeb. Enzymforsch.*, 2 (1933) 90–103.
(262) N. van Uden and H. Buckley, in J. Lodder (Ed.), *The Yeasts, A Taxonomic Study*, 2nd edn., North-Holland Publishing Co., Amsterdam, 1970, pp. 893–1087.
(263) T. Sawai, *Bot. Mag. (Tokyo)*, 69 (1956) 177–185.
(264) T. Sawai, *J. Biochem. (Tokyo)*, 45 (1958) 49–56.
(265) T. Sawai, *J. Biochem. (Tokyo)*, 48 (1960) 382–391.
(266) N. A. Khan, F. K. Zimmermann, and N. R. Eaton, *Mol. Gen. Genet.*, 123 (1973) 43–50.
(267) E. K. Novák, *Acta Microbiol. Acad. Sci. Hung.* 10 (1963) 7–10.
(268) E. Hofmann, *Biochem. Z.*, 272 (1934) 417–425.
(269) E. K. Novák, F. Kevei, B. Oláh, and J. Zsolt, *Acta Microbiol. Acad. Sci. Hung.*, 19 (1972) 39–42.
(270) S. Chiba and T. Shimomura, *Agric. Biol. Chem.*, 29 (1965) 540–547.

TABLE VIII

Substrate Specificity of Enzymes of *Saccharomyces cerevisiae*
Having α-D-Glucosidase Activity[a]

	Enzyme					
	α-D-Glucosidase (EC 3.2.1.20)			Oligo-(1→6)-D-glucosidase (EC 3.2.1.10)		
Substrate		Relative rate of hydrolysis[b,c]	K_i (mM)[c]		Relative rate of hydrolysis[b,d]	K_i (mM)[e]
Sucrose	+	80	23	−		170
Maltose	+	57	81	−		100
Methyl α-D-glucopyranoside	−	2	400	+	11	60
Maltotriose	+			−		
p-Nitrophenyl α-D-glucopyranoside	+	60	0.25 (K_m)	+		1.2 (K_m)
Turanose	+	87	9.5	−		
Phenyl α-D-glucopyranoside	+	100	0.3	+	100	
Isomaltose	−	0	11	+	9	

[a] These findings come from (i) [c] Halvorson and Ellias,[33] and Halvorson[34] for *Saccharomyces cerevisiae (italicus)*, (ii) [d]Matsusaka and coworkers[273] for brewer's yeast, and (iii) [e]Gorman and Halvorson.[274] See also, Needleman and coworkers.[275] [b] The relative rates were calculated from values for V. Further observations on the specificity of these enzymes were given by Spiegelman and coworkers,[276] Cook and Phillips,[277] Lövgren and Siro,[278] and Legler and Lotz[279]; Khan and coworkers reported[280,281] hydrolysis of sucrose by oligo-(1→6)-D-glucosidase. Further estimates of K_m and K_i were given by Halvorson and coworkers,[282] Lövgren,[283] Wandrey and coworkers,[284] and Siro and Lövgren.[285]

(stellatoidea) albicans,[271] and *Saccharomyces (logos) cerevisiae*.[272] The specificities of α-D-glucosidase and oligo-(1→6)-D-glucosidase from *Saccharomyces cerevisiae* are given in Table VIII. Spencer-Martins and van Uden[286] reported that *Lipomyces kononenkoae* produces

(271) E. K. Novák, F. Kevei, B. Oláh, and J. Zsolt, *Acta Biol. Acad. Sci. Hung.*, 16 (1965) 137–140.
(272) S. Chiba, T. Saeki, and T. Shimomura, *Agric. Biol. Chem.*, 37 (1973) 1823–1829.
(273) K. Matsusaka, S. Chiba, and T. Shimomura, *Agric. Biol. Chem.*, 41 (1977) 1917–1923.
(274) J. Gorman and H. [O.] Halvorson, *Methods Enzymol.*, 8 (1966) 562–565.
(275) R. B. Needleman, H. J. Federoff, T. R. Eccleshall, B. Buchferer, and J. Marmur, *Biochemistry*, 17 (1978) 4657–4661.

(i) an alpha amylase having an optimum pH of 5.5 and K_m of 0.2 M maltotriose, and (ii) a glucoamylase, optimum pH 4.5, with values for K_m of 1 mM for both maltose and p-nitrophenyl α-D-glucopyranoside.

(ii) **Molecular Weights of α-D-Glucopyranoside Hydrolases.**—Estimates of the molecular weights of α-D-glucosidases from *Saccharomyces cerevisiae*[33,34,275,279,280,285] have ranged from 44,000 to 85,000. Chiba and coworkers reported[272,287] that the α-D-glucosidase activity of one strain of *Saccharomyces (logos) cerevisiae* is associated with a glycoprotein of molecular weight 270,000, containing 50% of carbohydrate. Perhaps this was an external enzyme, as Novák and his colleagues suggested[269] for the α-D-glucosidase of *Schizosaccharomyces octosporus*. Khan and Eaton[280] gave the molecular weight of oligo-(1→6)-D-glucosidase of *Saccharomyces cerevisiae* as 65,000. *Lipomyces kononenkoae* produces an alpha amylase of molecular weight 38,000 and a glucoamylase of molecular weight 81,500 (Ref. 286).

(iii) **Regulation of α-D-Glucosidase and Oligo-(1→6)-D-Glucosidase.**—Both α-D-glucosidases and oligo-(1→6)-D-glucosidases of yeasts are controlled by induction and by catabolite repression. Even within *Saccharomyces cerevisiae*, there seem to be differences of inducer specificity between strains. Spiegelman and his colleagues reported[276] that *Saccharomyces cerevisiae* grown on, or adapted to, maltose but not to D-glucose, formed both α-D-glucosidase and oligo-(1→6)-D-glucosidase. The latter enzyme hydrolyzed methyl α-D-glucopyranoside, but intact yeast cells did not do so. Spiegelman and Halvorson[288] found that methyl α-D-glucopyranoside, although it is not a substrate of α-D-glucosidase, induced that enzyme. On the other

---

(276) S. Spiegelman, M. Sussman, and B. Taylor, *Fed. Proc. Fed. Am. Soc. Exp. Biol.*, 9 (1950) 120.
(277) A. H. Cook and A. W. Phillips, *Arch. Biochem. Biophys.*, 69 (1957) 1–9.
(278) T. Lövgren and M. R. Siro, *Acta Acad. Abo., Ser. B*, 32 (1972) No. 4, 1–10.
(279) G. Legler and W. Lotz, *Z. Physiol. Chem.*, 354 (1973) 243–254.
(280) N. A. Khan and N. R. Eaton, *Biochim. Biophys. Acta*, 146 (1967) 173–180.
(281) N. A. Khan and R. H. Haynes, *Mol. Gen. Genet.*, 118 (1972) 279–285.
(282) H. O. Halvorson, S. Winderman, and J. Gorman, *Biochim. Biophys. Acta*, 67 (1963) 42–53.
(283) T. Lövgren, *Acta Acad. Abo., Ser. B*, 32 (1972) No. 5, 1–10.
(284) C. Wandrey, W. Hönig, and M. Kula, *Eur. J. Appl. Microbiol.*, 3 (1977) 257–265.
(285) M. R. Siro and T. Lövgren, *Acta Chem. Scand., Ser. B*, 32 (1978) 447–451.
(286) I. Spencer-Martins and N. van Uden, *Eur. J. Appl. Microbiol. Biotechnol.*, 6 (1979) 241–250.
(287) T. Saeki, S. Chiba, and T. Shimomura, *Agric. Biol. Chem.*, 39 (1975) 551–552.
(288) S. Spiegelman and H. O. Halvorson, *J. Bacteriol.*, 68 (1954) 265–273.

hand, Halvorson and coworkers[235] reported that maltose induced α-D-glucosidase, and methyl α-D-glucopyranoside induced oligo-(1→6)-D-glucosidase, and not *vice versa*. Furthermore, Ouwehand and van Wijk found[289] that maltose induced both α-D-glucosidase and oligo-(1→6)-D-glucosidase, whereas methyl α-D-glucopyranoside or α-TEG induced the latter but not the former enzyme. In another strain of *Saccharomyces (oviformis) cerevisiae*, maltose induced α-D-glucosidase, but methyl α-D-glucopyranoside induced both enzymes; α-TEG was a poor inducer with this yeast.[290] Using protoplasts of *Saccharomyces (carlsbergensis) cerevisiae*, van Wijk and coworkers[291] found that the induction of α-D-glucosidase by maltose depended on the concentration of exogenous D-glucose. Although the lag between the time of adding maltose and that of the start of synthesis of the enzyme was independent of the concentration of D-glucose, the maximum rate of synthesis occurred in the presence of ~8 mM D-glucose.

Further publications also described findings on the α-D-glucosidase activity of *Saccharomyces cerevisiae*[292-306] and *Schizosaccharomyces pombe*.[307]

Mortimer and Hawthorne[165] and Barnett[4] summarized work on the complex, genetical control of α-D-glucosidase and oligo-(1→6)-D-glu-

(289) J. Ouwehand and R. van Wijk, *Mol. Gen. Genet.*, 117 (1972) 30–38.
(290) B. Axelrod, H. Ashe, T. Fukui, and D. E. Lewis, *Rec. Chem. Prog.*, 28 (1967) 121–133.
(291) R. van Wijk, J. Ouwehand, T. van den Bos, and V. V. Koningsberger, *Biochim. Biophys. Acta*, 186 (1969) 178–191.
(292) A. I. Oparin, N. S. Gel'man, and I. G. Zhukova, *Biokhimiya*, 20 (1955) 571–575; *Chem. Abstr.*, 50 (1956) 7217c.
(293) S. Chiba, S. Sugawara, T. Shimomura, and Y. Nakamura, *Agric. Biol. Chem.*, 26 (1962) 787–793.
(294) Y. Z. Frohwein and J. Leibowitz, *Enzymologia*, 24 (1962) 211–229.
(295) D. J. Millin, *J. Inst. Brew., London*, 69 (1963) 389–393.
(296) M. Burger, E. Oura, and H. Suomalainen, *Suom. Kemistil. B*, 38 (1965) 285–289.
(297) R. van Wijk, *Proc. K. Ned. Akad. Wet., Ser. C*, 71 (1968) 60–79.
(298) R. Hartlief and V. V. Koningsberger, *Biochim. Biophys. Acta*, 166 (1968) 512–531.
(299) R. G. Bell, *Can. J. Biochem.*, 47 (1969) 677–684.
(300) R. van Wijk, K. W. van de Poll, and G. A. G. Speziali, *Proc. K. Ned. Akad. Wet., Ser. B*, 73 (1970) 357–371.
(301) J. Ouwehand, *Proc. K. Ned. Akad. Wet., Ser. C*, 75 (1972) 434–440.
(302) H. Suomalainen, J. Dettwiler, and E. Sinda, *Process Biochem.*, 7 (1972) 16–19.
(303) J. H. Slavenburg and R. van Wijk, *Proc. K. Ned. Akad. Wet., Ser. C*, 76 (1973) 382–391.
(304) H. L. Lai and B. Axelrod, *Biochim. Biophys. Acta*, 391 (1975) 121–128.
(305) N. R. Eaton and F. K. Zimmermann, *Mol. Gen. Genet.*, 148 (1976) 199–204.
(306) N. A. Khan and A. Greener, *Mol. Gen. Genet.*, 150 (1977) 107–108.
(307) G. Schlanderer and H. Dellweg, *Eur. J. Biochem.*, 49 (1974) 305–316.

cosidase in *Saccharomyces cerevisiae*. The polymeric series of *MAL* genes control the ability to form inducible and repressible α-D-glucosidase; a yeast having *MAL4* has constitutive α-D-glucosidase that is resistant to catabolite repression.[308] Constitutive α-D-glucosidase has also been reported[309,310] for mutants at the *MAL6* locus. In addition, Khan[311] isolated a mutant of *Saccharomyces cerevisiae* that was partly constitutive for oligo-(1→6)-D-glucosidase, and Schamhart and coworkers[312] reported a revertant, from a yeast not utilizing maltose, that resisted catabolite repression for α-D-glucosidase, oligo-(1→6)-D-glucosidase, β-D-fructofuranosidase, and succinate dehydrogenase. Derkanosov and coworkers[313] selected ultraviolet-produced mutants having an α-D-glucosidase activity 6 to 10 times that of the parent yeast. These mutants were useful for fermenting molasses.

c. α,α-Trehalose.—Over 60% of the known species of yeast, including many strains of *Saccharomyces cerevisiae*, can utilize exogenous α,α-trehalose as the sole source of carbon (see Table III). Even yeasts that do not use α,α-trehalose may contain α,α-trehalase,[314] as α,α-trehalose is a storage carbohydrate in many yeasts,[4,315,316] and may constitute up to 15% of the dry weight of baker's yeast.[317–319] α,α-Trehalase acts within the plasmalemma of *Saccharomyces cerevisiae*,[320] but, perhaps, externally with *Rhodotorula glutinis*[37] and *Torulopsis glabrata*.[38] The enzyme of *Saccharomyces cerevisiae* is specific for α,α-trehalose,[314,321,322] hydrolyzing each molecule to two of D-glucose, al-

(308) F. K. Zimmermann, N. A. Khan, and N. R. Eaton, *Mol. Gen. Genet.*, 123 (1973) 29–41.
(309) A. M. A. ten Berge, G. Zoutewelle, and R. B. Needleman, *Mol. Gen. Genet.*, 131 (1974) 113–121.
(310) A. M. A. ten Berge, G. Zoutewelle, K. W. van de Poll, and H. P. J. Bloemers, *Mol. Gen. Genet.*, 125 (1973) 139–146.
(311) N. A. Khan, *Mol. Gen. Genet.*, 133 (1974) 363–365.
(312) D. H. J. Schamhart, A. M. A. ten Berge, and K. W. van de Poll, *J. Bacteriol.*, 121 (1975) 747–752.
(313) N. I. Derkanosov, A. I. Ivanov, M. M. Sabel'nikova, and E. S. Samoilenko, *Fermentn. Spirt. Prom.*, (1973) Part 8, 18–20; *Chem. Abstr.*, 80 (1974) 46,483q.
(314) G. Avigad, O. Ziv, and E. Neufeld, *Biochem. J.*, 97 (1965) 715–722.
(315) W. E. Trevelyan, in A. H. Cook (Ed.), *The Chemistry and Biology of Yeasts*, Academic Press, New York, 1958, pp. 369–436.
(316) D. J. Manners, in A. H. Rose and J. S. Harrison (Eds.), *The Yeasts*, Vol. 2, Academic Press, London, 1971, pp. 419–439.
(317) G. Tanret, *C. R. Acad. Sci.*, 192 (1931) 1056–1058.
(318) K. Myrbäck, *Sven. Kem. Tidskr.*, 48 (1936) 55–61; *Chem. Abstr.*, 30 (1936) 4897^5.
(319) K. Myrbäck and B. Örtenblad, *Biochem. Z.*, 288 (1936) 329–337.
(320) N. O. Souza and A. D. Panek, *Arch. Biochem. Biophys.*, 125 (1968) 22–28.
(321) K. Myrbäck, *Ergeb. Enzymforsch.*, 10 (1949) 168–190.
(322) P. J. Kelly and B. J. Catley, *Anal. Biochem.*, 72 (1976) 353–358.

TABLE IX

Estimates of Values of K_m and Optimum pH for Yeast α,α-Trehalase (EC 3.2.1.28)

Kind of yeast	K_m (mM α,α-trehalose)	Optimum pH	References
Baker's	—	5–6	329
Baker's	0.41	5.7	322
Baker's	0.5	5.5	323
Top	0.91	5.35	331a
Bottom	—	6.2–7.4	329
Saccharomyces (carlsbergensis) cerevisiae	4.5	6.7–6.9	331
Saccharomyces cerevisiae	10.2	6.9	314
Candida tropicalis	—	4–5.5	330
Debaryomyces hansenii	0.2–2	5	109

though Panek and Souza[323] also recorded activity against raffinose. Because nearly all yeasts that utilize melezitose also use α,α-trehalose, Barnett[70] suggested that most yeast α,α-trehalases might be found to hydrolyze melezitose too. Elbein[324] has surveyed the metabolism of α,α-trehalose.

α,α-Trehalase has been isolated from various yeasts[325–331] and has been purified from baker's yeast.[322,323] Estimates of the K_m and optimum pH for this enzyme are given in Table IX. Krüger and Hess[331] separated two forms of α,α-trehalase, having molecular weights of 310,000 and 78,000, from Saccharomyces (carlsbergensis) cerevisiae.

3. β-D-Glucopyranosides

In 1976, Barnett[4] reviewed the utilization of β-D-glucopyranosides by yeasts; β-D-glucosidases (EC 3.2.1.21) have been reviewed by Vei-

---

(323) A. D. Panek and N. O. Souza, *J. Biol. Chem.*, 239 (1964) 1671–1673.
(324) A. D. Elbein, *Adv. Carbohydr. Chem. Biochem.*, 30 (1974) 227–256.
(325) E. Fischer, *Ber.*, 28 (1895) 1429–1438.
(326) A. Kalanthar, *Z. Physiol. Chem.*, 26 (1898) 88–101.
(327) A. Bau, *Wochschr. Brau.*, 16 (1899) 305–306.
(328) P. Lindner, *Wochschr. Brau.*, 28 (1911) 61–64.
(329) K. Myrbäck and B. Örtenblad, *Biochem. Z.*, 291 (1937) 61–69.
(330) T. M. Lukes and H. J. Phaff, *Antonie van Leeuwenhoek, J. Microbiol. Serol.*, 18 (1952) 323–335.
(331) J. Krüger and B. Hess, *Arch. Mikrobiol.*, 61 (1968) 154–158.
(331a) J. E. Courtois, F. Petek, and M. Kolahi-Zanouzi, *Bull. Soc. Chim. Biol.*, 44 (1962) 735–743.

bel[332] and Larner.[255] About 55% of the species currently accepted can use cellobiose, arbutin, and salicin (see Table III), and other non-utilizing species, such as *Saccharomyces cerevisiae* (as early as 1845, Piria[333] recorded that beer yeast does not split salicin), may contain β-D-glucosidase. Certain yeasts can hydrolyze phlorizin (phloretin-2-yl β-D-glucopyranoside).[334]

Detailed studies have been made on the β-D-glucosidases of the following: (*i*) *Rhodotorula minuta* (misnamed *Saccharomyces cerevisiae*),[32,335,336] (*ii*) *Kluyveromyces* (*Saccharomyces*) *lactis*,[337–344] (*iii*) *Kluyveromyces* (*fragilis*) *marxianus*,[345,346] (*iv*) *Kluyveromyces dobzhanskii*,[346] (*v*) a hybrid of *Kluyveromyces marxianus* and *Kluyveromyces dobzhanskii*,[346–351] (*vi*) a yeast of unknown identity (see Ref. 4) called "*Saccharomyces cerevisiae*,"[36,352–356] (*vii*) *Pichia* (*Candida*) *guilliermondii*,[357] and (*viii*) *Debaryomyces polymorphus* (*Pichia polymorpha*).[358] Table X lists some characteristics of these β-D-glucosi-

(332) S. Veibel, in J. B. Sumner and K. Myrbäck (Eds.), *The Enzymes*, Vol. I, Part 1, Academic Press, New York, 1950, pp. 583–620.
(333) R. Piria, *Ann.*, 56 (1845) 35–77.
(334) N. P. Jayasankar, R. J. Bandoni, and G. H. N. Towers, *Phytochemistry*, 8 (1969) 379–383.
(335) J. D. Duerksen and H. [O.] Halvorson, *Biochim. Biophys. Acta*, 36 (1959) 47–55.
(336) S. W. Tanenbaum, G. C. Burke, and S. M. Beiser, *Biochim. Biophys. Acta*, 54 (1961) 439–447.
(337) A. Herman and H. [O.] Halvorson, *J. Bacteriol.*, 85 (1963) 895–900.
(338) A. Herman and H. [O.] Halvorson, *J. Bacteriol.*, 85 (1963) 901–910.
(339) G. L. Marchin and J. D. Duerksen, *J. Bacteriol.*, 96 (1968) 1181–1186.
(340) G. L. Marchin and J. D. Duerksen, *J. Bacteriol.*, 96 (1968) 1187–1190.
(341) G. L. Marchin and J. D. Duerksen, *J. Bacteriol.*, 97 (1969) 237–243.
(342) M. A. Tingle and H. O. Halvorson, *Biochim. Biophys. Acta*, 250 (1971) 165–171.
(343) M. [A.] Tingle and H. O. Halvorson, *J. Bacteriol.*, 110 (1972) 196–201.
(344) M. [A.] Tingle and H. O. Halvorson, *Genet. Res.*, 19 (1972) 27–32.
(345) C. Neuberg and E. Hofmann, *Biochem. Z.*, 256 (1932) 450–461.
(346) L. W. Fleming and J. D. Duerksen, *J. Bacteriol.*, 93 (1967) 135–141.
(347) A. S. L. Hu, R. Epstein, H. O. Halvorson, and R. M. Bock, *Arch. Biochem. Biophys.*, 91 (1960) 210–219.
(348) A. M. MacQuillan and H. O. Halvorson, *J. Bacteriol.*, 84 (1962) 23–30.
(349) A. M. MacQuillan and H. O. Halvorson, *J. Bacteriol.*, 84 (1962) 31–36.
(350) J. D. Duerksen and L. W. Fleming, *Biochem. Biophys. Res. Commun.*, 12 (1963) 169–174.
(351) L. W. Fleming and J. D. Duerksen, *J. Bacteriol.*, 93 (1967) 142–150.
(352) J. G. Kaplan, *J. Cell Biol.*, 19 (1963) 38A.
(353) J. G. Kaplan, *Fed. Proc. Fed. Am. Soc. Exp. Biol.*, 23 (1964) 210.
(354) J. G. Kaplan, *J. Gen. Physiol.*, 48 (1965) 873–886.
(355) A. N. Inamdar and J. G. Kaplan, *Fed. Proc. Fed. Am. Soc. Exp. Biol.*, 25 (1966) 508.
(356) A. N. Inamdar and J. G. Kaplan, *Can J. Biochem.*, 44 (1966) 1099–1108.
(357) W. W. Roth and V. R. Srinivasan, *Prep. Biochem.*, 8 (1978) 57–71.
(358) T. G. Villa, V. Notario and J. R. Villanueva, *FEMS Microbiol. Lett.*, 6 (1979) 91–94.

TABLE X

Specificities of β-D-Glucosidases of Four Yeasts[a]

Substrate or inducer	Induction ratio[b]		Relative rate of hydrolysis[c]					Inhibition constant K_i (mM)			Michaelis constant, K_m (mM)	
	Rh	Kl	Rh	Kl	U	D(i)	D(ii)	Rh	Kl	U	D(i)	D(ii)
Cellobiose	1.9	1.4	6	—	6	72	25	150	—	—	22	40
Salicin	2.7	5.5	44	63	73	100	100	2.6	0.32	1.0	0.94	0.52
Arbutin	1.1	3.3	100	100	100	—	—	1.7	0.26	0.22	—	—
Aesculin	1.0	6.4	34	51	20	—	—	0.014	0.059	—	—	—
Amygdalin	1.0	6.3	0	0	47	—	—	14	1.0	7.7	—	—
Methyl β-D-glucopyranoside	72	—	18	—	27	—	—	56	—	—	—	—
Phenyl β-D-glucopyranoside	2.4	—	88	—	113	—	—	1.6	—	0.27	—	—
p-Nitrophenyl β-D-glucopyranoside	—	—	67	—	267	136	80	0.08[K_m]	0.08[K_m]	0.095[K_m]	0.72	0.21
Phenyl 1-thio-β-D-glucopyranoside	1.1	—	0	—	—	—	—	4.6	3.3	—	—	—
Methyl 1-thio-β-D-glucopyranoside	60	—	0	—	—	—	—	0.88	—	—	—	—
Ethyl 1-thio-β-D-glucopyranoside	58	—	0	—	—	—	—	0.22	—	—	—	—
D-Glucose	0	—	—	—	—	—	—	8.5	8.8	6.7	—	—
Control (no addition)	1.0	1.0	—	—	—	—	—	—	—	—	—	—

[a] Rh, *Rhodotorula minuta*, Refs. 32 and 335; Kl, *Kluyveromyces* interspecific hybrid, Ref. 347; U, unidentified yeast (Refs. 355 and 356); D(i), D(ii), two β-D-glucosidases of a strain of *Debaryomyces polymorphus* (Ref. 358). [b] Induced enzymic activity: uninduced enzymic activity. [c] The relative rates are calculated from values of V.

dases. Those of *Kluyveromyces marxianus* and *Kluyveromyces dobzhanskii* were similar to that of their hybrid. That of *Pichia guilliermondii* gave a K_m of 125 μM p-nitrophenyl β-D-glucopyranoside. All had pH optima lying between 6 and 6.8. The enzymes were carbohydrate-free, and had molecular weights of ~300,000, except for that of the inducible β-glucosidase of *Pichia guilliermondii*, which was 48,000. Herman and Halvorson[337] presented evidence for the presence of two β-D-glucosidases of different substrate specificities in *Kluyveromyces lactis*: the first hydrolyzed p-nitrophenyl β-D-glucopyranoside, arbutin, salicin, and aesculin; the second hydrolyzed cellobiose. Methyl β-D-glucopyranoside induced the first, but not the second, enzyme.[338] For β-D-glucosidases of three strains of *Kluyveromyces lactis*, the K_m value for p-nitrophenyl β-D-glucopyranoside was ~0.2 mM, and the inhibition constants were 1.1 to 3.8 mM salicin, 3.3 to 22 mM amygdalin, and 116 mM methyl β-D-glucopyranoside. Kaplan and Tacreiter[36] also considered that their unidentified yeast, when cellobiose-grown, contained two β-D-glucosidases: an internal enzyme (see Table X) and a "surface β-glucosidase" that differed from the internal enzyme by hydrolyzing methyl β-D-glucopyranoside and cellobiose.

4. β-D-Galactopyranosides

Although <12% of the species of yeast use lactose (see Table III), this utilization is of commercial interest, as β-D-galactosidase (EC 3.2.1.23) from yeasts may be used to lower the lactose content of dairy products, such as whey from cheese-making (for example, see Ref. 359) or milk itself.[360]

The work of Monod and his colleagues on the genetic regulation of induction and repression of enzyme synthesis has produced a vast glut of publications on the β-D-galactosidase of *Escherichia coli* (for reviews, see Refs. 10 and 361). On the other hand, β-D-galactosidase was probably first studied in yeasts[201,362-364]; in 1904, Heinze and Cohn[365] reviewed aspects of the early work.

However, relatively little has been published on the β-D-

(359) R. R. Mahoney and J. R. Whitaker, *J. Food Sci.*, 43 (1978) 584–591.
(360) C. Tosi, South African Pat. 72 08, 897 (1973); *Chem. Abstr.*, 82 (1975) 29,928n.
(361) K. Wallenfels and R. Weil, in P. D. Boyer (Ed.), *The Enzymes*, 3rd edn., Vol. II, Academic Press, New York, 1972 pp. 617–663.
(362) M. W. Beyerinck, *Zentralbl. Bakteriol., Parasitenkd.*, 6 (1889) 44–48.
(363) M. W. Beyerinck, *Arch. Neerl. Sci. Exactes Nat.*, 24 (1891) 369–442.
(364) E. Fischer, *Ber.*, 27 (1894) 2985–2993.
(365) B. Heinze and E. Cohn, *Z. Hyg. Infektionskr.*, 46 (1904) 286–366.

galactosidase of yeasts. The work has chiefly been concerned with two species, namely, *Kluyveromyces marxianus* (*Saccharomyces fragilis*)[51,204,345,366–377] and *Kluyveromyces* (*Saccharomyces*) *lactis*.[344,378–382] Myrbäck and Vasseur[383] also studied the β-D-galactosidase activities of *Candida pseudotropicalis* (*Torula cremoris*) and *Candida kefyr* (*Torula lactosa*). In addition, some yeasts, including strains of *Saccharomyces cerevisiae*, that do not use lactose, nonetheless appear to contain some β-D-galactosidase.[384–386]

Yeast β-D-galactosidase hydrolyzes lactose and *o*-nitrophenyl β-D-galactopyranoside (see Table XI), and also fluroan-3,6-diyl di(β-D-ga-

TABLE XI

Characteristics of β-D-Galactosidases (EC 3.2.1.23) of Two *Kluyveromyces* Species

Species	Michaelis constant (mM)		Optimum pH	Molecular weight
	ONPβGal[a]	Lactose		
K. marxianus[b]	2.5 to 4[c]	14 to 24[d]	6.2 to 7[e]	~200,000[f]
	5.6 μM[g]	—	6.7 to 7.1[g]	>500,000[g]
K. lactis	1.2 to 1.6[h]	12 to 17[i]	6.9 to 7.3[h]	135,000[i]

[a] ONPβGal = *o*-nitrophenyl β-D-galactopyranoside. [b] Includes yeasts variously named *Kluyveromyces fragilis*, *Saccharomyces fragilis*, and *Candida pseudotropicalis* (for nomenclature, see Refs. 5 and 387). [c] Refs. 375, 377, 387, and 388. [d] Refs. 375, 377, and 387. [e] Refs. 204, 367, 369, 371, 375, 377, 388, and 389. [f] Refs. 359 and 375. [g] Ref. 374. [h] Refs. 378 and 381. [i] Ref. 381.

(366) E. Hofmann, *Biochem. Z.*, 256 (1932) 462–474.
(367) R. Caputto, L. F. Leloir, and R. E. Trucco, *Enzymologia*, 12 (1948) 350–355.
(368) E. Hofmann and H. Scheck, *Biochem. Z.*, 319 (1949) 522–528.
(369) B. van Dam, J. G. Revallier-Warffemius, and L. C. van Dam-Schermerhorn, *Neth. Milk Dairy J.*, 4 (1950) 96–114.
(370) H. R. Roberts and E. F. McFarren, *J. Dairy Sci.*, 36 (1953) 620–632.
(371) R. Davies, *J. Gen. Microbiol.*, 37 (1964) 81–98.
(372) G. Szabó and R. Davies, *J. Gen. Microbiol.*, 37 (1964) 99–112.
(373) G. Szabó and J. Rózsa, *Acta Microbiol. Acad. Sci. Hung.*, 12 (1965) 91–102.
(374) A. K. Kulikova, A. S. Tikhomirova, and R. V. Feniksova, *Biokhimiya*, 37 (1972) 405–409.
(375) T. Uwajima, H. Yagi, and O. Terada, *Agric. Biol. Chem.*, 36 (1972) 570–577.
(376) R. R. Mahoney, T. A. Nickerson, and J. R. Whitaker, *J. Dairy Sci.*, 58 (1975) 1620–1629.
(377) R. R. Mahoney and J. R. Whitaker, *J. Food Biochem.*, 1 (1977) 327–350; *Chem. Abstr.*, 89 (1978) 72,362e.
(378) L. Biermann and M. D. Glantz, *Biochim. Biophys. Acta*, 167 (1968) 373–377.
(379) J. Yashphe and H. O. Halvorson, *Science*, 191 (1976) 1283–1284.
(380) I. P. Ferrero, C. Rossi, M. P. Landini, and P. P. Puglisi, *Biochem. Biophys. Res. Commun.*, 80 (1978) 340–348.

lactopyranoside), which Yashphe and Halvorson[379] used for assaying the enzymic activity in individual cells of *Kluyveromyces lactis*. Table XI gives various estimates of the dissociation constants found for β-D-galactopyranosidases of *Kluyveromyces marxianus* and *Kluyveromyces lactis*; that of *Sporobolomyces singularis* has given a K_m of 50 mM *o*-nitrophenyl β-D-galactopyranoside.[390]

In 1899, Dienert reported[391] that the extracts of lactose-fermenting yeasts hydrolyzed lactose, but only after the yeasts had been incubated with that sugar and, from subsequent work, yeast β-D-galactosidase appears to be either inducible or derepressible.[371,380,381] Although D-galactose might well induce β-D-galactosidase,[371] Wendorff and coworkers[392] found low yields of β-D-galactosidase in *Kluyveromyces marxianus* (*Saccharomyces fragilis*) grown on D-galactose; working with 41 strains of that yeast, Mahoney and coworkers[376] obtained the maximum yield of β-D-galactosidase early in the stationary phase of growth, and Shahani and coworkers have reported[393] a strain that is constitutive for β-D-galactosidase.

Work that may prove of immense industrial significance is the isolation from *Kluyveromyces lactis*, by Dickson and Markin,[394] of the gene for β-D-galactosidase by use of recombinant-DNA techniques. The DNA from the yeast was partly digested with endonuclease, and the product was joined to a plasmid DNA by means of DNA ligase. The ligated DNA was then successfully used to transform a lactose-nega-

---

(381) R. C. Dickson, L. R. Dickson, and J. S. Markin, *J. Bacteriol.*, 137 (1979) 51–56.
(382) V. Jirku, J. Turkova, A. Kuchynkova, and V. Krumphanzl, *Eur. J. Appl. Microbiol. Biotechnol.*, 6 (1979) 217–222.
(383) K. Myrbäck and E. Vasseur, *Z. Physiol. Chem.*, 277 (1943) 171–180.
(384) E. Hofmann, *Biochem. Z.*, 265 (1933) 209–212.
(385) P. P. Puglisi, I. P. Ferrero, and A. Algeri, *G. Microbiol.*, 18 (1970) 57–67; *Chem. Abstr.*, 77 (1972) 98,657g.
(386) J. Rossi and G. Trovarelli, *Ann. Microbiol. Enzimol.*, 26 (1976) 197–207; *Chem. Abstr.*, 89 (1978) 18,909a.
(387) J. Lodder (Ed.), *The Yeasts, A Taxonomic Study*, 2nd edn., North-Holland Publishing Co., Amsterdam, 1970.
(388) W. L. Wendorff and C. H. Amundson, *J. Milk Food Technol.*, 34 (1971) 300–306; *Chem. Abstr.*, 75 (1971) 84,655b.
(389) S. A. de Bales and F. J. Castillo, *Appl. Environ. Microbiol.*, 37 (1979) 1201–1205.
(390) J. A. Blakely and S. L. MacKenzie, *Can. J. Biochem.*, 47 (1969) 1021–1025.
(391) F. Dienert, *C. R. Acad. Sci.*, 129 (1899) 63–64.
(392) W. L. Wendorff, C. H. Amundson, and N. F. Olson, *J. Milk Food Technol.*, 33 (1970) 451–455; *Chem. Abstr.*, 74 (1971) 30,735p.
(393) K. M. Shahani, M. G. Carrancedo, and A. Kilara, *J. Milk Food Technol.*, 38 (1975) 208–211; *Chem. Abstr.*, 83 (1975) 126,845m.

tive mutant of *Escherichia coli*. According to Dickson and Markin,[394] there are at least seven, unlinked genes involved in lactose utilization. One of these codes for β-D-galactosidase, one for a carrier, and the other five appear to subserve the regulation of concentration of enzyme.

V. The Requirement of Oxygen for Utilizing Glycosides and D-Galactose

In 1940, Kluyver and Custers[395] confirmed earlier reports that certain yeasts can utilize certain disaccharides aerobically, but not anaerobically, although these yeasts can use one or more of the component hexoses anaerobically[396]; this is the Kluyver effect.[397] Table III shows this phenomenon to be widespread: only a minority of fermenting yeasts that can utilize a given glycoside aerobically can also use it anaerobically. For example, only 24 of 231 (~1 in 10) maltose-using species ferment maltose, and the ratio for lactose is ~1 in 17; by contrast, about half the species can ferment D-glucose.

Considering only those yeasts that are capable of fermenting D-glucose anaerobically, the effect must result either from (i) the requirement by most of these yeasts for oxygen for active transport of the glycoside across the plasmalemma to an internal glycosidase,[398] or (ii) the activity of the glycosidase.[395] Sims and Barnett[397] showed that oxygen is necessary for transport and not for enzymic activity. If facilitated diffusion were the usual means by which yeasts take up glycosides, there would be no apparent reason for any oxygen requirement for transport. The notion of this effect was also applied[397] to D-galactose (in addition to the glycosides), because the Leloir pathway by which D-galactose is transformed into D-glucose 6-phosphate (an interme-

---

(394) R. C. Dickson and J. S. Markin, *Cell*, 15 (1978) 123–130.
(395) A. J. Kluyver and M. T. J. Custers, *Antonie van Leeuwenhoek J. Microbiol. Serol.*, 6 (1940) 121–162.
(396) Many yeasts can utilize certain sugars anaerobically. The most obvious sign of this anaerobic utilization, here called "fermentation," is the formation of carbon dioxide. Unlike some bacteria, which can ferment a wide range of organic compounds, yeasts can ferment only D-glucose, D-fructose, D-mannose, D-galactose, and their glycosides. Apparently, all species of yeast that ferment any sugar at all can ferment D-glucose, D-fructose, and D-mannose. Furthermore, except for cytochrome-deficient mutants, a yeast that uses a sugar anaerobically can also use it aerobically (see Ref. 4).
(397) A. P. Sims and J. A. Barnett, *J. Gen. Microbiol.*, 106 (1978) 277–288.
(398) J. A. Barnett, in G. C. Ainsworth and A. S. Sussman (Eds.), *The Fungi: An Advanced Treatise*, Vol. III, Academic Press, New York, 1968, pp. 557–595.

diate in the glycolytic pathway) involves no net oxidation (see Ref. 4). Hence, there seems to be no reason for the catabolism of D-galactose to differ from that of D-glucose in its oxygen requirements. Of 229 species that ferment D-glucose,[70,399] all of the strains of 97 species showed the Kluyver effect for at least one glycoside or for D-galactose.

Using four such yeasts [*Candida utilis, Kluyveromyces (fragilis) marxianus, Zygosaccharomyces fermentati (Saccharomyces montanus),* and *Kluyveromyces thermotolerans (Torulopsis dattila)*], Sims and Barnett[397] studied the utilization of maltose, cellobiose, and D-galactose. These authors examined the different responses of these yeasts to anaerobic conditions (see Table XII) with respect to enzymic activity, uptake of sugar, and production of carbon dioxide. Measurement of the production of carbon dioxide established the rapidity, and the reversibility, of the effects of changes of the concentration of oxygen. For example, within a few seconds of adding air to *Candida utilis* which had been incubated anaerobically with maltose, there was a ninefold increase in the output of carbon dioxide, and this was immediately linear. When *Candida utilis*, incubated with cellobiose, was made anaerobic, there was a decline in the production of carbon dioxide if cellobiose was still present, but not when D-glucose was added. Hence, the decline could not have resulted from deactivation of any glycolytic enzyme. Furthermore, over this period, there was no corresponding loss of β-D-glucosidase activity. The rate of anaerobic output of carbon dioxide by *Kluyveromyces thermotolerans*, in the presence of D-galactose, declined during anaerobic incubation; and nystatin, which makes the plasmalemma generally permeable,[400] increased that rate. On using a non-metabolized, labeled analog of D-galactose, namely, D-[1-^{3}H]fucose, Sims and Barnett[397] found that anaerobiosis immediately decreases the rate of transport to about one quarter of the aerobic rate. Subsequently, Barnett and Sims[401] studied the uptake of [^{14}C]methyl 1-thio-β-D-galactopyranoside, a non-metabolized analog of lactose, by *Debaryomyces polymorphus*, a yeast that shows the Kluyver effect with lactose, and by a strain of *Kluyveromyces marxianus* that ferments lactose. With both yeasts, active transport was abolished immediately by changing from aerobic to anaerobic conditions. However, the nature of the difference between the two yeasts under anaerobic conditions has not yet been elucidated.

There are a few, well authenticated reports of the Kluyver effect

(399) J. A. Barnett and R. J. Pankhurst, *A New Key to the Yeasts*, North-Holland Publishing Co., Amsterdam, 1974.
(400) J. Béchet and J. M. Wiame, *Biochem. Biophys. Res. Commun.*, 21 (1965) 226–234.
(401) J. A. Barnett and A. P. Sims, in preparation.

TABLE XII

The Utilization of Certain Glycosides and D-Galactose by Some Yeast Species That Show the Kluyver Effect for at Least One Substrate[a]

Species	Substrate									
	Gal	Suc	Mal	Cel	Tre	Lac	Mel	Mlz	MeG	Raf
Candida chilensis	K	K	K	?	K	K	–	?	?	–
Candida entomophila	?	+	K	?	?	K	?	?	?	?
Candida utilis	–	+	K	K	?	–	–	K	?	+
Citeromyces matritensis	–	+	K	–	?	–	–	–	?	+
Debaryomyces polymorphus[b]	?	+	?	?	?	K	K	?	–	+
Hansenula holstii	?	K	K	?	?	–	–	?	?	–
Kluyveromyces marxianus[c]	+	+	–	K	–	+	–	–	–	+
Kluyveromyces thermotolerans[d]	K	+	K	–	K	–	–	K	?	?
Metschnikowia pulcherrima	?	K	K	K	K	–	–	K	K	–
Pachysolen tannophilus	K	–	–	?	–	–	–	–	–	–
Pichia wickerhamii	–	K	K	?	K	–	–	?	?	–
Torulopsis gropengiesseri	K	+	–	K	–	–	–	–	–	K
Wingea robertsii	K	+	+	K	+	–	–	K	+	+
Zygosaccharomyces fermentati[e]	+	+	+	K	+	–	–	?	?	+

[a] Abstracted from Ref. 397, with corrections from Ref. 5. Key: K, Kluyver effect, that is, fermentation negative and aerobic growth positive; +, fermentation and aerobic growth both positive; –, fermentation and aerobic growth both negative; ?, doubt as to how the results should be categorized. Gal, D-galactose; Suc, sucrose; Mal, maltose; Cel, cellobiose; Tre, α,α-trehalose; Lac, lactose; Mel, melibiose; Mlz, melezitose; MeG, methyl α-D-glucopyranoside; Raf, raffinose. [b] Refers to strains formerly called *Debaryomyces cantarellii*. [c] Refers to strains formerly called *Kluyveromyces fragilis*. [d] Refers to strains formerly called *Torulopsis dattila*. [e] Synonymous with *Saccharomyces montanus*.

with glycosides that are typically hydrolyzed outside the plasmalemma. Such anomalies have been examined and confirmed[402] in the writer's laboratory for *Torulopsis gropengiesseri* with raffinose, and *Candida entomophila* with melibiose. In both cases, however, the yeast fermented the glycoside for a short period of time, and this fermentation involved an external glycosidase. The anomaly occurs because, for some unknown reason, the fermentation is not sustained beyond a few hours.

The findings concerning the Kluyver effect throw some light on the utilization of glycosides by yeasts in general. For example, according

(402) J. A. Barnett, T. J. Kitchell, M. V. San Romão, and A. P. Sims, in preparation.

TABLE XIII

The Anaerobic Utilization of Sucrose by 137 Species That Ferment D-Glucose and Use Sucrose Aerobically: Association with Utilization of Raffinose[a]

Fermentation of sucrose	Aerobic growth on raffinose			
	+	−	?	Σ
+	62	7	11	80
−	4	26	1	31
?				26
			Σ	137

[a] Figures are number of species. From a compilation of 439 yeast species.[5]

to Barnett and his colleagues,[5] 137 species use sucrose aerobically and also ferment D-glucose. Of these 137, 80 species also ferment sucrose, and 31 do not; that is, 31 species show the Kluyver effect for sucrose (see Table XIII). Hence, those 31 species under aerobic conditions might be expected to transport sucrose intact into the cells and, unlike the 80 that ferment sucrose, be devoid of external β-D-fructofuranosidase. Sucrose would be hydrolyzed internally, probably by an α-D-glucosidase. The finding (see Table XIII) that 26 of these 31 species cannot use raffinose is consistent with this interpretation; by contrast, only 7 of the 80 sucrose-fermenting species were reported to be unable to use raffinose. This suggests that >20% of the species of yeast that utilize sucrose do so by means of an internal α-D-glucosidase, not by an external β-D-fructofuranosidase. Some support is given to this view by the association between the abilities to utilize sucrose and maltose.[70]

In addition, from information about the Kluyver effect, it is tempting to speculate further about the nature of glycoside transport. In a table of utilization of glycosides and D-galactose by 97 species, all known strains of which show the Kluyver effect[397] (abstracted for 14 species in Table XII), the most remarkable feature is the absence of any pattern; instead, there is a striking individuality within the yeasts in their responses to each of the substrates. A possible interpretation of these results is as follows. (i) The transporting carriers differ greatly from species to species, or even within species, both in their substrate specificity and in whether or not metabolic energy is necessary for entry of the substrate. (ii) Most yeasts that hydrolyze glycosides internally concentrate them by active transport. (*iii*) In yeasts, active transport operates aerobically only. (*iv*) Those yeasts that can ferment glycosides can transport them across the plasmalemma by facilitated diffusion and, perhaps, compensate for the low, internal concentration of glycoside by achieving a high concentration of the appropriate glycosidase. This suggestion contrasts with that of Serrano[245] that active transport compensates for the low affinity of the yeast glycosidases for their substrates, and it should not be difficult to investigate experimentally. Serrano[245] made the following points: (*i*) unlike glycosides, hexoses enter *Saccharomyces cerevisiae* by facilitated diffusion; and (*ii*) the first, intracellular reaction of a hexose is catalyzed by a kinase of low K_m value (~0.1 mM). He suggested that the high K_m values of the glycosidases (~10 mM) necessitate active transport, in order to provide a sufficiently high concentration of substrate for fast catabolism. If Serrano's views prove to be correct, the anaerobic transport of maltose into *Saccharomyces cerevisiae* may be expected to be active. Alternatively, it might be found that anaerobic conditions derepress the synthesis of glucosidase, and so achieve a high concentration of the enzyme. It will be interesting to see whether the chief difference between fermenting and nonfermenting yeasts is the presence or absence of facilitated diffusion for D-glucose.

VI. ADDENDUM

1. Glycosides Hydrolyzed Outside the Plasmalemma

Consistent with the glycosylation of β-D-fructofuranosidase occurring at the plasmalemma[157] is the finding that a highly purified preparation of plasmalemma from *Saccharomyces cerevisiae* glycosylated endogenous proteins.[403] This glycosylation involved glycolipid intermediates, with uridine 5'-(2-acetamido-2-deoxy-D-glucosyl diphosphate) as the sugar donor.

Two yeasts that utilize inulin have been examined, namely, *Candida salmanticensis*[404] and *Debaryomyces polymorphus (cantarellii)*.[405] The former released inulinase into the culture medium and, when purified, this enzyme gave the following values of K_m: 17 mM inulin, 43 mM sucrose, and 2 mM raffinose, with an optimum at pH ~4 for all three

---

(403) G. W. Welton-Verstegen, P. Boer, and E. P. Steyn-Parvé, *J. Bacteriol.*, 141 (1980) 342–349.
(404) J. P. Guiraud, C. Viard-Gaudin, and P. Galzy, *Agric. Biol. Chem.*, 44 (1980) 1245–1252.
(405) I. Beluche, J. P. Guiraud, and P. Galzy, *Folia Microbiol. (Prague)*, 25 (1980) 32–39.

substrates. Crude extracts of *Debaryomyces polymorphus* contained more than one β-D-fructofuranosidase, or inulinase, with optima at pH 4 and 6. Activity of one of these enzymes appeared to be induced when the yeast grew on inulin, but not on sucrose or raffinose. Another enzyme was constitutive, and not repressible by D-glucose.

2-Deoxy-D-*arabino*-hexose ("2-deoxy-D-glucose") stimulated *Saccharomyces cerevisiae* to synthesize β-D-fructofuranosidase,[406] but inhibited the formation of α-D-galactosidase. Adding D-xylose, which this yeast does not utilize,[5] stimulated the synthesis of β-D-fructofuranosidase, even in the presence of D-glucose that, alone, would have repressed the synthesis. In a search for yeasts having high α-D-galactosidase activity,[407] one strain of *Schwanniomyces occidentalis* (*alluvius*) was found to have particularly high activity of this enzyme.

2. Glycosides Hydrolyzed Inside the Plasmalemma

Further evidence has been obtained[408] on the role of the *MAL* genes of *Saccharomyces cerevisiae* as structural genes that have additive effects on the specific activity of α-D-glucosidase. The *hex2-3* mutation is responsible for a high specific activity for phosphorylating D-glucose and D-fructose;[409] in the presence of maltose, yeast carrying *hex2-3* accumulates a high, intracellular concentration of D-glucose (of at least 150 mM, compared with 2.5 mM in wild-type cells), possibly because of uncontrolled uptake of maltose.[410]

In *Lipomyces starkeyi*, which utilizes starch as the sole source of carbon for growth,[5] alpha amylase is associated with cell walls, and is formed abundantly during exponential growth on starch or maltose,[411] but to only a small extent on sucrose or D-glucose. Mutants of *Lipomyces kononenkoae*, that were derepressed for producing alpha amylase, were selected[412] from a culture medium containing starch and 2-deoxy-D-*arabino*-hexose. This method of selecting derepressed mutants was developed by Zimmermann and Scheel.[413] Selection depends on two particular characteristics of 2-deoxy-D-*arabino*-hexose,

---

(406) F. Moreno, P. Herrero, F. Parra, and S. Gascón, *Cell. Mol. Biol.*, 25 (1979) 1–6.
(407) I. V. Ulezlo, O. M. Zaprometova, and A. M. Bezborodov, *Prikl. Biokhim. Mikrobiol.*, 16 (1980) 347–350.
(408) D. B. Mowshowitz, *J. Bacteriol.*, 137 (1979) 1200–1207.
(409) K.-D. Entian and F. K. Zimmerman, *Mol. Gen. Genet.*, 177 (1980) 345–350.
(410) K.-D. Entian, *Mol. Gen. Genet.* 179 (1980) 169–175.
(411) G. Moulin and P. Galzy, *Agric. Biol. Chem.* 43 (1979) 1165–1171.
(412) N. van Uden, C. Cabeça-Silva, A. Madeira-Lopes, and I. Spencer-Martins, *Biotechnol. Bioeng.*, 22 (1980) 651–654.
(413) F. K. Zimmermann and I. Scheel, *Mol. Gen. Genet.*, 154 (1977) 75–82.

which (*i*) unlike D-glucose, is not used for growth,[4] and (*ii*) like D-glucose, represses the synthesis of a number of catabolic enzymes.[414]

The production has been studied[415–419] of (1→4)-α-D-glucan glucohydrolase (glucoamylase) by *Hansenula beckii* (*Endomycopsis bispora*).

Differences have been investigated[420] in the ability of certain yeasts to grow on various β-D-glucopyranosides as the sole source of carbon. These differences are exemplified as follows (+ = growth, − = no growth).

Yeast	β-D-Glucopyranoside		
	Cellobiose	Salicin	Arbutin
Debaryomyces polymorphus strain CBS 4349	+	−	−
Rhodotorula minuta strain CBS 2179	+	+	−
Candida tsukubaensis strain CBS 6389	+	−	+

The findings were as follows. (*i*) Washed, starved suspensions of each yeast could respire, at similar rates, each of the three β-D-glucopyranosides mentioned, as well as *p*-nitrophenyl β-D-glucopyranoside and methyl β-D-glucopyranoside. (*ii*) A combination of causes was responsible for failures to grow on the phenolic β-D-glucopyranosides, including differences in the degree of inhibition of growth of each yeast by the aglycons 2-(hydroxymethyl)phenol (saligenin) and 4-hydroxyphenol (hydroquinone).

Physiological studies have been made[421] on a strain of *Kluyveromyces lactis* that grows faster on lactose than on D-glucose (compare, Ref. 204). β-D-Galactosidase is formed during exponential growth, requiring the constant presence of an inducer, namely, lactose, D-galactose, or β-D-galactopyranosyl-(1→4)-D-gluconic acid. Certain inducers of the *lac* operon of *Escherichia coli*, such as methyl, isopropyl,

(414) I. Witt, R. Kronau, and H. Holzer, *Biochim. Biophys. Acta*, 118 (1966) 522–537.
(415) H. Ruttloff, A. Täufel, and F. Zickler, *Z. Allg. Mikrobiol.*, 19 (1979) 195–201.
(416) K. Winkler, F. Zickler, and H. Ruttloff, *Z. Allg. Mikrobiol.*, 19 (1979) 577–584.
(417) H. Ruttloff, K. Winkler, and F. Zickler, *Z. Allg. Mikrobiol.*, 19 (1979) 277–281.
(418) K. Winkler, H. Ruttloff, and F. Zickler, *Z. Allg. Mikrobiol.*, 19 (1979) 517–522.
(419) A. Täufel, K. Winkler, F. Zickler, and H. Ruttloff, *Z. Allg. Mikrobiol.*, 19 (1979) 349–356.
(420) T. Dick, A. P. Sims, and J. A. Barnett, in preparation.
(421) R. C. Dickson and J. S. Markin, *J. Bacteriol.*, 142 (1980) 777–785.

and phenyl 1-thio-β-D-galactopyranoside and β-D-galactopyranosyl-(1→6)-D-glucose, entered the yeast cells, but did not induce their synthesis of β-D-galactosidase. This enzyme was partly repressed by D-glucose which, however, by contrast with *Escherichia coli,* did not prevent the yeast from taking up, or utilizing, lactose.

AFFINITY CHROMATOGRAPHY OF MACROMOLECULAR SUBSTANCES ON ADSORBENTS BEARING CARBOHYDRATE LIGANDS

By John H. Pazur

Paul M. Althouse Laboratory, The Pennsylvania State University, University Park, Pennsylvania

I. Introduction 405
II. General Considerations 408
III. Activation of Supports 412
    1. Cellulose 412
    2. Agarose 413
    3. Poly(acrylamide) 416
    4. Porous Glass 417
    5. Other Supports 419
IV. Derivatives of Carbohydrates 419
    1. Types of Derivatives 419
    2. Chemical Synthesis 421
    3. Enzymic Synthesis 429
V. Coupling Reactions 430
    1. For Supports Activated with Cyanogen Bromide 430
    2. For Supports Activated with Bisoxirane 431
    3. For Supports Activated with Divinyl Sulfone 432
    4. Reductive Amination 433
    5. Other Reactions 434
VI. Applications 437
    1. Antibodies 437
    2. Enzymes 441
    3. Lectins 444
    4. Myeloma Proteins 446

I. Introduction

Affinity chromatography has become the method of choice for purifying macromolecular substances and for determining the nature of the stereospecific interactions of these substances.[1,2] The method has

---

(1) J. Porath and T. Kristiansen, in H. Neurath and R. L. Hill (Eds.), *The Proteins*, 3rd edn., Vol. I, Academic Press, New York, 1975, pp. 95–178.
(2) I. M. Chaiken, *Anal. Biochem.*, 97 (1979) 1–10.

been used successfully for the purification of antibodies, coenzymes, enzymes, enzyme-inhibitors, hormone-receptors, lectins, nucleic acids, and associated substances.[3,4] It has also been used for determining the dissociation constants of complexes, and the interactions of various proteins with ligands of low and high molecular weight.[5,6] Briefly, affinity chromatography is a type of column chromatography in which a selective adsorption of a biological substance occurs on a ligand that is attached to an insoluble support, and which is specific for the substance. In the initial reaction, the biological substance forms an insoluble complex with the ligand, and is retained on the affinity adsorbent. In subsequent reactions, the complex is dissociated by a solvent system of altered composition, and the desired substance is recovered in the eluate from the adsorbent.

Many types of compounds have been attached to insoluble supports, providing affinity adsorbents having different types of ligands. These adsorbents have been used for the purification of a wide spectrum of biological macromolecules.[1,3,4] Some carbohydrates or derivatives thereof that have been used for preparing affinity adsorbents are D-galactose,[7-9] 2-acetamido-2-deoxy-D-glucose,[10,11] 2-acetamido-2-deoxy-D-galactose,[12-14] 2-amino-2-deoxy-D-galactose,[15,16] D-glucose,[12,17] L-fucose,[18-20] D-mannose,[12,21] 1-thio-D-galactose,[22] D-glucuronic

(3) P. Cuatrecasas and C. B. Anfinsen, *Annu. Rev. Biochem.*, 40 (1971) 259–278.
(4) O. Hoffmann-Ostenhof, M. Breitenbach, F. Koller, D. Kraft, and O. Scheiner, *Affinity Chromatography*, Pergamon Press, New York, 1978.
(5) P. Brodelius and K. Mosbach, *Anal. Biochem.*, 72 (1976) 629–636.
(6) D. Eilat and I. M. Chaiken, *Biochemistry*, 18 (1979) 790–795.
(7) L. Wofsy and B. Burr, *J. Immunol.*, 103 (1969) 380–382.
(8) C. E. Hayes and I. J. Goldstein, *J. Biol. Chem.*, 249 (1974) 1904–1914.
(9) J. H. Pazur, K. L. Dreher, and L. S. Forsberg, *J. Biol. Chem.*, 253 (1978) 1832–1837.
(10) J. H. Shaper, R. Barker, and R. L. Hill, *Anal. Biochem.*, 53 (1973) 564–570.
(11) R. Lotan, A. E. S. Gussin, H. Lis, and N. Sharon, *Biochem. Biophys. Res. Commun.*, 52 (1973) 656–662.
(12) V. Hořejší and J. Kocourek, *Methods Enzymol.*, 34B (1974) 361–367.
(13) P. Vretblad, *Biochim. Biophys. Acta*, 434 (1976) 169–176.
(14) R. Uy and F. Wold, *Anal. Biochem.*, 81 (1977) 98–107.
(15) A. K. Allen and A. Neuberger, *FEBS Lett.*, 50 (1975) 362–364.
(16) B. A. Baldo, W. H. Sawyer, R. B. Stick, and G. Uhlenbruck, *Biochem. J.*, 175 (1978) 467–477.
(17) J. H. Pazur, Y. Tominaga, K. L. Dreher, L. S. Forsberg, and B. M. Romanic, *J. Carbohydr. Nucleos. Nucleot.*, 5 (1978) 1–14.
(18) I. Matsumoto and T. Osawa, *Methods Enzymol.*, 34B (1974) 329–331.
(19) V. Hořejší and J. Kocourek, *Biochim. Biophys. Acta*, 297 (1973) 346–351.
(20) H. J. Allen and E. A. Z. Johnson, *Carbohydr. Res.*, 58 (1977) 253–265.
(21) N. Fornstedt and J. Porath, *FEBS Lett.*, 57 (1975) 187–191.
(22) E. Steers, Jr., P. Cuatrecasas, and H. B. Pollard, *J. Biol. Chem.*, 246 (1971) 196–200.

acid,[7,23] lactose,[7,9,24] maltose,[25] isomaltose,[17] melibiose,[8,25] chitobiose,[25,26] lacto-N-fucohexaose,[27] streptococcal cell-wall glycans,[23,28] enzymically degraded starch,[14] mannan,[29] and blood-group substances.[30]

In the preparation of affinity adsorbents, it is necessary to attach the ligand residues to insoluble supports by chemical bonds or physical interactions. Adsorbents having carbohydrate ligands have been prepared for the most part by chemical reaction of carbohydrate derivatives with the support material. Methods for the preparation of some types of carbohydrate derivatives are described in Section IV. Methods for the preparation of other derivatives of carbohydrates that may be suitable for synthesizing affinity adsorbents have been described in an article on neoglycoproteins.[31]

Many materials have been tested in order to determine their suitability for use in the preparation of affinity adsorbents.[32] However, relatively few materials have been found to possess desirable properties. In the latter group are agarose, cellulose, poly(acrylamide) beads, and porous-glass beads. Generally, the coupling of a support material with a ligand occurs by reaction of an activated form of the support with a derivative of the ligand. The activation of a support involves the introduction of chemically reactive groups into the material. This activation is achieved by chemical modification of the functional groups of the support substance.

Polysaccharides that have been modified chemically, or altered physically, have been used as adsorbents for affinity chromatography. The modification of the structure of polysaccharides has been achieved by introducing cross-linkages between the chains of the polymer and bifunctional reagents. The alteration of the properties of polysaccharides by physical means can be effected by embedding the polysaccharide in a network of the support material. The molecular in-

(23) J. H. Pazur, K. L. Dreher, and R. L. Kubrick, *Fed. Proc.*, 39 (1980) 1633.
(24) Y. D. Kim and F. Karush, *Immunochemistry*, 10 (1973) 365–371.
(25) R. J. Baues and G. R. Gray, *J. Biol. Chem.*, 252 (1977) 57–60.
(26) C. M. T. Kieda, F. M. Delmotte, and M. L. P. Monsigny, *Proc. Natl. Acad. Sci. USA*, 74 (1977) 168–172.
(27) A. M. Jeffrey, D. A. Zopf, and V. Ginsburg, *Biochem. Biophys. Res. Commun.*, 62 (1975) 608–613.
(28) K. Eichmann and J. Greenblatt, *J. Exp. Med.*, 133 (1971) 424–441.
(29) B. V. McCleary, *Phytochemistry*, 17 (1978) 651–653.
(30) T. Kristiansen, L. Sundberg, and J. Porath, *Biochim. Biophys. Acta.*, 184 (1969) 93–98.
(31) C. P. Stowell and Y. C. Lee, *Adv. Carbohydr. Chem. Biochem.*, 37 (1980) 225–281.
(32) J. Porath, *Methods Enzymol.*, 34B (1974) 13–30.

teractions between the polysaccharide and the support substance hold the polysaccharide in the network. Dextran, glycogen, guaran, and pectin have been modified by these methods, and have been used for the purification of lectins and enzymes that bind carbohydrate moieties.[33–37]

Affinity adsorbents having carbohydrate ligands have been used for the isolation and purification of many types of macromolecular substances. To illustrate, several types of antibodies, enzymes, lectins, and myeloma proteins have been obtained in highly purified form. All of the substances that have been purified by this procedure exhibit specificity for a particular carbohydrate moiety. It is the purpose of this article to assemble information on the methods for preparing affinity adsorbents having carbohydrate ligands, and to illustrate the use of these adsorbents for the purification of representative, macromolecular substances.

II. General Considerations

Affinity chromatography is possible because of the stereospecific interactions that can occur between the functional groups of one compound and the complementary groups of another. For affinity chromatography, adsorbents are needed, and such adsorbents can be prepared by attaching one of the interacting compounds to an insoluble support. The support, having the immobilized ligand, is then used to adsorb the second compound as an insoluble complex, a process that separates the desired substance from impurities. Subsequently, this substance is released from the adsorbent by dissociation of the complex with a suitable solvent, and the substance is recovered from the column in pure form. The theoretical aspects of biospecific interactions and affinity chromatography have been discussed.[32] In the discussion, consideration was given to the role of the distribution coefficient and the effect of the molecular-sieving process on affinity chromatography.[38,39] On the basis of these theoretical considerations, it is now possible to predict the type of results that may be obtained in various types of affinity experiments. Guidelines for making such predictions have been given.[40]

(33) B. B. L. Agrawal and I. J. Goldstein, *Biochim. Biophys. Acta*, 147 (1967) 262–271.
(34) N. M. Young and M. A. Leon, *Carbohydr. Res.*, 66 (1978) 299–302.
(35) K. Sutoh, L. Rosenfeld, and Y. C. Lee, *Anal. Biochem.*, 79 (1977) 329–337.
(36) L. Rexová-Benková and V. Tibenský, *Biochim. Biophys. Acta*, 268 (1972) 187–193.
(37) J. J. Marshall and W. Woloszczuk, *Carbohydr. Res.*, 61 (1978) 407–417.
(38) T. C. Laurent and J. Killander, *J. Chromatogr.*, 14 (1964) 317–330.
(39) J. Porath, *Nature (London)*, 218 (1968) 834–838.
(40) D. J. Graves and Y.-T. Wu, *Methods Enzymol.*, 34B (1974) 140–162.

The interaction between biological macromolecules can be of various degrees of complexity. These interactions may be of high specificity, as in antigen–antibody complexes, or of low specificity, as in protein–lipid associations. The complexes that are formed may be very stable, or highly unstable. These properties of a complex must be considered in selecting an affinity adsorbent to be used for the purification of a specific substance.

Affinity adsorbents may be prepared by attaching, by covalent linkages, a derivative of the ligand to an activated, insoluble support. Adsorbents may also be prepared by the entrapment of the ligand substance in the network of the support material. Adsorbents prepared by the latter method are less stable, and tend to lose ligand residues by leakage. Such affinity adsorbents eventually lose their effectiveness, and then need to be replaced. Another disadvantage of affinity adsorbents prepared by entrapment methods is the lowered capacity of such materials. The capacity of an adsorbent is determined by the number of reactive ligand-residues attached to the adsorbent. In adsorbents prepared by physical methods, a significant fraction of the ligand residues may be inactive, due to shielding of the residues by the matrix material. These residues do not participate in the affinity chromatography process. Affinity adsorbents having covalent linkages contain few shielded ligand-residues, and generally have high capacities. Only minor losses of ligand residues occur by leakage from such adsorbents.

The total number of ligand residues in an adsorbent can be measured by analysis of the adsorbent by a method specific for a particular ligand. Adsorbents having carbohydrate ligands attached to non-carbohydrate supports[41] can be analyzed by a general method for carbohydrates,[42] those having hexosamine ligands can be analyzed by a method specific for hexosamines,[43] and those having deoxyhexose ligands can be analyzed by reagents specific for deoxyhexoses.[44] Indirect methods can be used to determine the amount of ligand attached to a support. In such methods, the amount of ligand that remains in the coupling solution at completion of a reaction is determined by chromatographic and colorimetric methods.[17] The difference between this value and the initial value is a measure of the amount of ligand that has been coupled to the support. Radio-isotopes have also been used to determine the amount of ligand coupled to a support. In this

(41) G. R. Gray, *Arch. Biochem. Biophys.*, 163 (1974) 426–428.
(42) M. Dubois, K. A. Gilles, J. K. Hamilton, P. A. Rebers, and F. Smith, *Anal. Chem.*, 28 (1956) 350–356.
(43) W. T. J. Morgan and L. A. Elson, *Biochem. J.*, 28 (1934) 988–995.
(44) Z. Dische, *Methods Carbohydr. Chem.*, 1 (1962) 488–494.

method, a radioactive ligand is employed in the coupling reaction, and the radioactivities of the ligand solution and the affinity adsorbent are determined. From the radioactivity values, the number of ligand residues attached to the support can be calculated.[14,45]

For the successful application of affinity chromatography, a proper selection of support material is extremely important. The ideal support-material should have the properties listed next. It should be insoluble in the solvent system being employed. It should have good permeability. It should be stable to chemical reagents. It should be reactive with the activating reagents. It should be resistant to microbial and enzymic decomposition. It should be hydrophilic in nature. It should be readily coupled to the ligand. Agarose, cellulose, poly(acrylamide) beads, and porous-glass beads are materials that possess characteristics desirable in insoluble supports. Agarose is, perhaps, the most widely used material for preparing affinity adsorbents. Many of the reactions for activating supports have been developed with agarose, and such reactions are, for the most part, applicable for activating other support materials having structural features similar to those of agarose.

The fundamental reactions of affinity chromatography are the formation of an insoluble complex between a ligand and the desired substance, and the subsquent dissociation of the complex by a suitable solvent. The formation of the complex is regulated by the nature of the ligand group, the structure of the desired substance, and the types of interactions that can occur. The dissociation of the complex is achieved by altering the conformation of the substance being isolated, or by using the displacement-elution technique. A substance having an altered conformation will not bind to the ligand, and will be released from the complex on the adsorbent. Changes in conformation of the adsorbed substance may be effected by altering the pH, increasing the ionic strength, or adding a denaturing agent to the solvent. If biological activity is to be retained in the substance isolated, it is necessary to establish that an irreversible, conformational change has not been induced in the substance. A displacement elution of an adsorbed substance is effected with a solution of the ligand, or of a structural analog. The soluble and the immobilized form of the ligand then compete for the desired substance. In this competition, the substance binds with the soluble ligand, and is eluted from the affinity adsorbent. A diagram showing the formation of an insoluble complex and the routes for the dissociation of the complex[46] is reproduced in Fig. 1.

(45) P. Cuatrecasas and I. Parikh, *Biochemistry*, 11 (1972) 2291–2299.
(46) P. O'Carra, S. Barry, and T. Griffin, *Methods Enzymol.*, 34B (1974) 108–126.

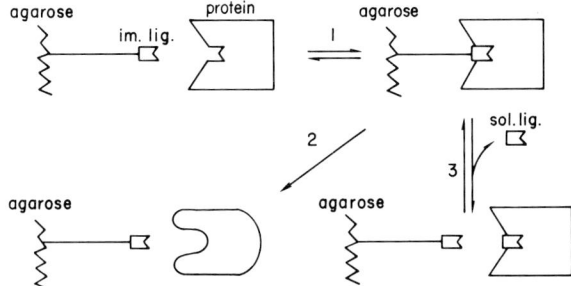

FIG. 1.—Diagrammatic Representation of the Formation of an Insoluble Complex (1) Between a Protein and an Immobilized Ligand (im. lig.) Attached to Agarose and (2) the Dissociation of the Complex, Due to a Conformational Change in the Protein, or (3) Due to Displacement Elution with the Soluble Ligand (sol. lig.) (see Ref. 46).

In the development of the techniques of affinity chromatography, it was found that the formation of a complex between a ligand and a desired substance did not always occur on an affinity adsorbent.[47,48] As a result, a purification of the substance was not possible. The affinity adsorbents used in such studies contained ligand residues attached directly to the support material. It was proposed that, in such adsorbents, steric hindrance interfered with the adsorption of the desired substance.[48] The steric hindrance was due to the proximity of the functional groups of the support and of the ligand residues, and an interaction occurred between these groups, rather than with the desired substance. Steric-hindrance effects could be minimized by using affinity adsorbents that contained a spacer group between the ligand and the support.[48] Compounds that function as spacer residues are long-chain hydrocarbons containing reactive groups on the terminal carbon atoms. In the preparation of affinity adsorbents having spacer groups, the spacer compound can first be attached to the activated support, and then coupled to the ligand substance, or the spacer compound can first be attached to the ligand, and then coupled to the support. Although steric hindrance may influence the type of separation that can be obtained on affinity adsorbents having spacer groups, other factors may also be involved. Thus, hydrophobic interactions of the spacer group with the substance being isolated may be responsible for separations that are obtained with these types of adsorbents.[46]

The structural features of compounds that have been used to intro-

(47) P. Cuatrecasas, M. Wilchek, and C. B. Anfinsen, *Proc. Natl. Acad Sci. USA*, 61 (1968) 636–643.
(48) P. Cuatrecasas, *Adv. Enzymol.*, 36 (1972) 29–89.

duce spacer groups into adsorbents are a linear hydrocarbon chain of two to ten carbon atoms, with an amino group at one terminus and an amino or carboxyl group at the other. The amino group of such compounds will react with the imido carbonate groups of the insoluble support activated with cyanogen bromide, to yield a product containing hydrocarbon chains having a reactive group on the terminal carbon atom. If a diamine is the spacer compound, a support derivative having a free amino group at the terminal carbon atom of the spacer will be produced. If an amino- and carboxyl-containing compound is the spacer, a derivative having a carboxyl group at the terminal carbon atom of the spacer will be formed. The derivatives obtained by coupling of diamines to supports can react further with other reagents, such as succinic anhydride, N-hydroxysuccinimide, and N-(bromoacetoxy)succinimide, to yield adsorbents having spacer arms of larger dimensions.[49] Coupling of carbohydrate derivatives to supports having spacer groups is effected by the carbodiimide method described in Section V.

III. Activation of Supports

1. Cellulose

Cellulose is a polysaccharide of high molecular weight that is composed of D-glucosyl residues joined by β-D-(1→4) linkages. It is insoluble in water, and highly resistant to chemical and enzymic degradation. It is available in large amounts and is inexpensive. Cellulose and cellulose derivatives can be coupled with ligands by a number of methods, to yield a variety of affinity adsorbents. An adsorbent of cellulose and serum albumin was prepared by treating cellulose with p-nitrobenzyl chloride (α-chloro-p-nitrotoluene), reducing the product to O-(p-aminobenzyl)cellulose, converting this derivative into the diazonium salt, and coupling the salt to bovine serum albumin. The coupled product was used to isolate antibodies specific for bovine serum albumin by the methodology of affinity chromatography.[50] The diazonium derivative of the O-(p-aminobenzyl)cellulose may be coupled to other proteins and glycoproteins. Coupling to glycoproteins yields affinity adsorbents containing carbohydrate units. Such adsorbents may be used to isolate substances that combine with the carbohydrate units of the glycoprotein, and should be especially useful for

---

(49) P. Cuatrecasas, *J. Biol. Chem.*, 245 (1970) 3059–3065.
(50) D. H. Campbell, E. Luescher, and L. S. Lerman, *Proc. Natl. Acad. Sci. USA*, 37 (1951) 575–578.

studies on the glycoproteins of cell-surface receptors and cell membranes.

Many types of derivatives of cellulose have been used to prepare affinity adsorbents.[51] O-(2-Aminoethyl)cellulose and O-(carboxymethyl)cellulose can be coupled to carbohydrates bearing carboxyl or primary amino groups by the carbodiimide method.[52,53] Cellulose derivatives bearing amino groups can be coupled to reducing carbohydrates by reductive-amination reactions[25] in which the Schiff base produced in the initial reaction is reduced with sodium cyanoborohydride.[54]

Cellulose possesses some disadvantages for use as an insoluble support. The microstructure of cellulose and cellulose derivatives interferes with the permeation of substances through affinity adsorbents of these materials. Consequently, long periods are needed for separating and isolating substances on such columns. The linear chains of cellulose mask some ligand residues, and the adsorbent will have a low capacity. The amount of the substance that can be purified on such columns may be significantly diminished. Affinity adsorbents prepared from cellulose may exhibit nonspecific adsorption, and the purification of the desired substance could be difficult to achieve.

2. Agarose

Agarose is used most extensively for the preparation of affinity adsorbents. Agarose is the linear component of agar,[55] and is composed of repeating units of β-D-galactosyl-$(1\rightarrow4)$-3,6-anhydro-L-galactose joined by α-L-$(1\rightarrow3)$ linkages.[56] Agarose is isolated from agar by fractionation methods based on differences in solubilities of the linear and branched components.[55] Whereas the linear component possesses properties desirable in chromatographic material, the branched component does not. Agarose has been prepared in spherical form by special techniques,[57,58] and these spheres possess properties desirable in support materials. The spheres are stable, yield good gels, and are highly permeable to macromolecular substances. The properties of

---

(51) R. J. Boegman and M. J. Crumpton, *Biochem. J.*, 120 (1970) 373–379.
(52) H. G. Khorana, *Chem. Ind. (London)*, (1955) 1087–1088.
(53) J. C. Sheehan and G. P. Hess, *J. Am. Chem. Soc.*, 77 (1955) 1067–1068.
(54) R. F. Borch, M. D. Bernstein, and H. D. Durst, *J. Am. Chem. Soc.*, 93 (1971) 2897–2904.
(55) C. Araki and K. Arai, *Bull. Chem. Soc. Jpn.*, 30 (1957) 287–293.
(56) C. Araki, *Bull. Chem. Soc. Jpn.*, 29 (1956) 543–544.
(57) S. Hjertén, *Biochim. Biophys. Acta*, 79 (1964) 393–398.
(58) S. Bengtsson and L. Philipson, *Biochim. Biophys. Acta*, 79 (1964) 399–406.

the agarose spheres can be further modified by cross-linking the agarose chains with such bifunctional reagents as epichlorohydrin,[59] bisoxirane (1,2:3,4-diepoxybutane)[59] or divinyl sulfone.[60] Agarose of different particle-sizes and various numbers of cross linkages can be purchased commercially. Such preparations are very satisfactory for preparing affinity adsorbents.

For use in the preparation of affinity adsorbents, agarose must be converted into an activated form by modification of the functional groups of the polymer by chemical methods. Reaction with cyanogen bromide has been used most widely for activating agarose.[61] In the initial step of this reaction, cyanide groups are attached to a hydroxyl group of agarose, to yield a product having cyanic ester groups.[62,63] A further reaction occurs with adjacent hydroxyl groups of the agarose, to yield a product having cyclic imidocarbonate groups.[64] The structures of the products having the different types of groups are shown in Equation 1.

$$\begin{array}{c}\text{agarose}\\|\\-\text{OH}\\|\\-\text{OH}\end{array} + \text{BrCN} \longrightarrow \begin{array}{c}\text{agarose}\\|\\-\text{O}-\text{C}\equiv\text{N}\\|\\-\text{OH}\end{array} \longrightarrow \begin{array}{c}\text{agarose}\\|\\-\text{O}\\\diagdown\\\text{C}=\text{NH}\\\diagup\\-\text{O}\end{array} \quad (1)$$

Products containing other structural groups may be formed during the reaction of cyanogen bromide with agarose; for example, a small proportion of the carbamate group can be detected by infrared measurements.[62]

The reaction of agarose with cyanogen bromide is conducted under alkaline conditions.[32] Initially, sodium hydroxide was used[61] for maintaining an alkaline pH, but a sodium carbonate buffer has since been used.[65] Agarose that has been activated with cyanogen bromide may be used immediately in coupling reactions, or it may be stored in the dry state at low temperature. In order to retain the activated groups, the agarose should be dried in the presence of such stabilizers as lactose and dextran. Dried preparations of cyanogen bromide-activated agarose are stable for long periods, and are available from several commercial suppliers.

(59) J. Porath, J.-C. Janson, and T. Låås, *J. Chromatogr.*, 60 (1971) 167–177.
(60) J. Porath, T. Låås, and J.-C. Janson, *J. Chromatogr.*, 103 (1975) 49–62.
(61) R. Axén, J. Porath, and S. Ernback, *Nature*, 214 (1967) 1302–1304.
(62) R. Axén and P. Vretbland, *Acta Chem. Scand.*, 25 (1971) 2711–2716.
(63) J. Kohn and M. Wilchek, *Biochem. Biophys. Res. Commun.*, 84 (1978) 7–14.
(64) R. Axén and S. Ernback, *Eur. J. Biochem.*, 18 (1971) 351–360.
(65) S. C. March, I. Parikh, and P. Cuatrecasas, *Anal. Biochem.*, 60 (1974) 149–152.

Agarose may be converted into a reactive form by the bisoxirane method.[66] Bisoxirane contains two epoxide rings, and, under appropriate reaction-conditions, one epoxide ring reacts with a hydroxyl group of agarose but the other does not. The product is an agarose derivative containing a hydrophilic side-chain having an epoxide ring. This epoxide ring can react with a ligand substance that contains a hydroxyl or an amino group, to yield an affinity adsorbent. Some bisoxirane molecules may react with two hydroxyl groups located on different agarose chains, and cross-linkages will then be produced in the agarose. Cross-linkages in these materials are desirable, as such materials possess altered physical properties and yield superior, affinity, adsorbents. Directions for preparing the bisoxirane derivatives of agarose are recorded in the literature.[66] The chemical reaction is shown in Equation 2.

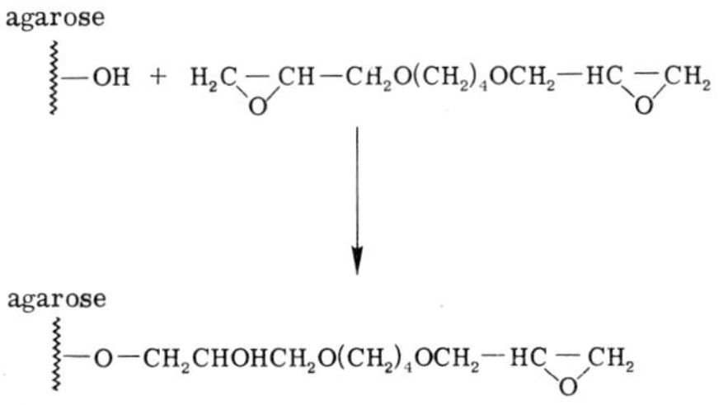

$$\text{agarose}-\text{OH} + H_2C\underset{O}{-}CH-CH_2O(CH_2)_4OCH_2-HC\underset{O}{-}CH_2$$

$$\downarrow$$

$$\text{agarose}-O-CH_2CHOHCH_2O(CH_2)_4OCH_2-HC\underset{O}{-}CH_2 \quad (2)$$

Agarose may be activated by reaction with divinyl sulfone, yielding a product having vinylsulfonylethyl ether groups.[21] This product is an activated form of agarose, and the vinyl group of the product can be coupled with appropriate ligands. The preparation of derivatives of agarose and divinyl sulfone has been described,[20,21] and the reaction is illustrated in Equation 3.

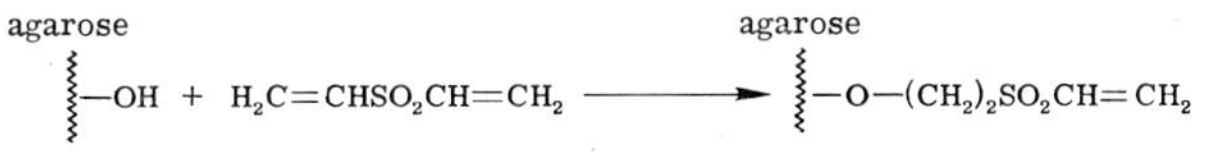

$$\text{agarose}-\text{OH} + H_2C=CHSO_2CH=CH_2 \longrightarrow \text{agarose}-O-(CH_2)_2SO_2CH=CH_2 \quad (3)$$

Agarose has been converted into an activated form by reaction with a new reagent, 1,1′-carbonylbis(imidazole).[67] The reaction of this reagent with agarose is shown in Equation 4.

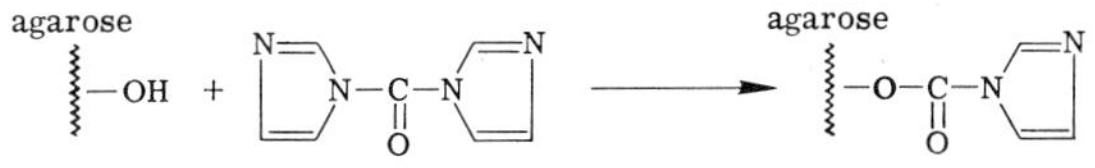

$$\text{agarose}-\text{OH} + \underset{}{\text{Im}}-\underset{\underset{O}{\parallel}}{C}-\underset{}{\text{Im}} \longrightarrow \text{agarose}-O-\underset{\underset{O}{\parallel}}{C}-\underset{}{\text{Im}} \quad (4)$$

(66) L. Sundberg and J. Porath, *J. Chromatogr.*, 90 (1974) 87–98.
(67) G. S. Bethell, J. S. Ayers, W. S. Hancock, and M. T. W. Hearn, *J. Biol. Chem.*, 254 (1979) 2572–2574.

It has been stated[67] that affinity adsorbents prepared from agarose activated with the bis(imidazole) reagent possess a number of advantages over products activated with other reagents. Thus, agarose that has been activated by bis(imidazole) and coupled to a ligand yields a product that does not contain ionic groups. In contrast, agarose activated with cyanogen bromide and coupled to a ligand yields an adsorbent having ionizable groups. The ionizable groups are responsible for nonspecific adsorption of substances on the affinity adsorbents. Nonspecific adsorption is undesirable in affinity chromatography, as the substance being isolated may be contaminated with such materials. Ligands that are attached to bis(imidazole)-activated agarose are attached more tightly than ligands that are attached to cyanogen bromide-activated agarose. Accordingly, a loss of ligand residues by leakage from affinity adsorbents prepared by the former method is not likely, but a loss of ligand residues from adsorbents prepared by the latter method can be appreciable. Loss of ligand residues from an affinity adsorbent is undesirable, as such losses lessen the capacity of the adsorbent.

The methods just discussed for activating agarose may equally well be used with other support materials that contain glycol or hydroxyl groups. Thus, cellulose, glycogen, guaran, and some types of porous glass have been activated by the cyanogen bromide and the 1,1'-carbonylbis(imidazole) methods.

3. Poly(acrylamide)

Poly(acrylamide) is a polymer of acrylamide consisting of a linear chain of ethylene units having amide side-groups on alternate carbon atoms. Poly(acrylamide) having a linear structure is soluble in water, and therefore unsuitable for chromatographic use. Insoluble poly(acrylamide) may be prepared from acrylamide and the cross-linking reagent, N,N'-methylenebis(acrylamide) by copolymerization of these substances. In the copolymerization, cross-linkages are formed between the linear polymers. Poly(acrylamide) can be prepared with a different number of cross-linkages by controlling the proportion of the bifunctional reagent and the reaction time. Different types of poly(acrylamide) beads can be prepared from poly(acrylamides) having different numbers of cross-linkages. Many types of poly(acrylamide) beads may be purchased commercially. Beads having the lowest number of cross-linkages are the most porous, and are used to prepare affinity adsorbents for macromolecular substances. Beads having the highest number of cross-linkages are the least porous, and are used to prepare affinity adsorbents for substances of low molecular weight.

Poly(acrylamide) beads possess a number of advantages over other types of support materials that have been used to prepare affinity adsorbents.[68] These advantages are the low, nonspecific adsorption, the high stability to heat, the high resistance to chemical degradation, the low susceptibility to enzymic degradation, and the low extent of leakage of ligand residues.[68] Poly(acrylamide) beads react with chemical reagents to yield other types of derivatives suitable for coupling to ligands. The 2-aminoethyl derivative, the hydrazide, the succinoyl hydrazide, the glutaroyl hydrazide, the 6-(bromoacetamido)hexanoyl, and the 4-(2-aminoethyl)glutaramoyl hydrazide are examples of such derivatives.[68] The (2-aminoethyl) derivative of poly(acrylamide) has been used in coupling reactions of many types. This derivative is synthesized by the reaction of poly(acrylamide) with ethylenediamine at elevated temperatures. The details of the preparation and isolation of (2-aminoethyl)poly(acrylamide) have been published.[69] The reaction is shown in Equation 5.

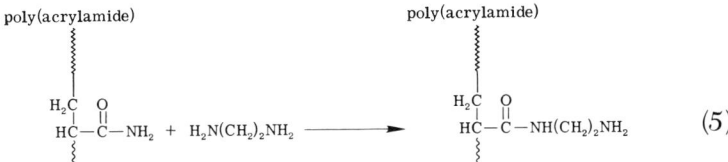

Some derivatives of poly(acrylamide), such as the (2-aminoethyl) derivative, may be dried by lyophilization, but other derivatives should not be dried. Some derivatives may be stored in the wet state in buffer solution at low temperature, but others may not be so stored. In general, poly(acrylamide) derivatives should be used immediately after preparation. It is especially important that poly(acrylamide) derivatives should not be subjected to alternate freezing and thawing, as such treatment will cause fragmentation of the poly(acrylamide) beads.

4. Porous Glass

Porous glass is now increasingly used for the preparation of affinity adsorbents.[70] By a special technique, the glass may be fabricated into beads of different sizes and porosities. The technique involves heating the glass to high temperatures, to separate the boric acid from the silica phases, and leaching with dilute acid to remove the boric acid. The residual material consists primarily of silica. On being cooled,

---

(68) J. K. Inman, *Methods Enzymol.*, 34B (1974) 30–58.
(69) J. K. Inman and H. M. Dintzis, *Biochemistry*, 8 (1969) 4074–4082.
(70) H. H. Weetall and A. M. Filbert, *Methods Enzymol.*, 34B (1974) 59–72.

this silica forms glass beads. Beads of different pore-sizes can be obtained by controlling the temperature of heating, the time of leaching, and additional treatment of the silica residue with dilute alkali.[70]

Porous glass to be used as support material must be treated with an organosilane that contains an organic functional group at one end and a silylalkoxyl group at the other. The latter group of the organosilane reacts with the silica of the glass beads. The other functional group of the organosilane is available for reacting with the appropriate ligand. The organosilane that is generally used for treating porous-glass beads is (3-aminopropyl)triethoxysilane. The reaction of this reagent with a glass bead is shown diagrammatically in Equation 6.

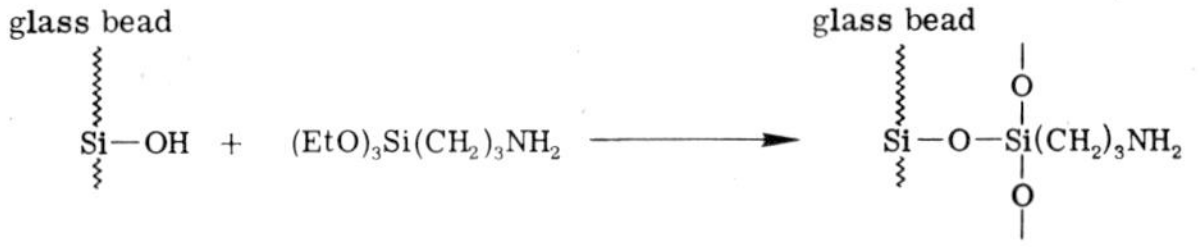

$$\text{glass bead}-\text{Si}-\text{OH} + (\text{EtO})_3\text{Si}(\text{CH}_2)_3\text{NH}_2 \longrightarrow \text{glass bead}-\text{Si}-\text{O}-\underset{\underset{\text{O}}{|}}{\overset{\overset{\text{O}}{|}}{\text{Si}}}(\text{CH}_2)_3\text{NH}_2 \qquad (6)$$

Other derivatives of porous-glass beads have been prepared by reaction of the silanized beads with appropriate reagents. Derivatives having aldehyde, alkylamine, arylamine, carboxyl, and hydrazide groups have been synthesized.[70] Procedures have also been developed for coupling proteins to derivatives of porous-glass beads.[71] Finally, new derivatives of porous glass have been introduced; these include glycophase and bis(imidazole)-activated glycophase.[67] Glycophase contains glycerol residues attached to the organosilane of porous-glass beads. The essential structure of this support is shown in 1.

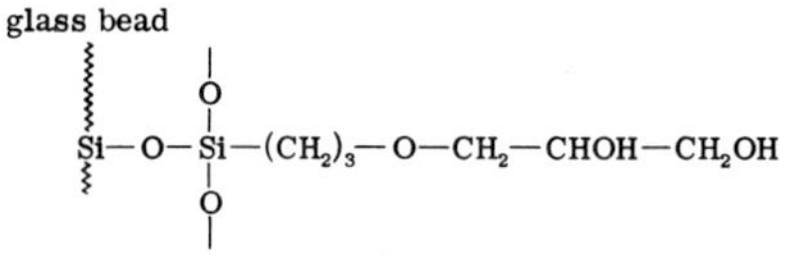

$$\text{glass bead}-\text{Si}-\text{O}-\underset{\underset{\text{O}}{|}}{\overset{\overset{\text{O}}{|}}{\text{Si}}}-(\text{CH}_2)_3-\text{O}-\text{CH}_2-\text{CHOH}-\text{CH}_2\text{OH}$$

1

Glycophase contains glycol units, and can be activated with cyanogen bromide under alkaline reaction-conditions. The activated glycophase can then be coupled to appropriate carbohydrate derivatives. Glycophase will also react with 1,1'-carbonylbis(imidazole) to yield a product having a reactive 1-imidazolylcarbamate group. The 1-imidazolylcarbamate group can be replaced by the amino group of carbohydrate derivatives, to yield affinity adsorbents having carbohydrate ligands.

(71) H. H. Weetall and L. S. Hersh, *Biochim. Biophys. Acta*, 185 (1969) 464–465.

5. Other Supports

Sparingly soluble polysaccharides have been used as affinity adsorbents, and examples of such compounds include dextran,[33,34] starch,[18] guaran,[35] pectin,[36] and glycogen.[37] The utility of a polysaccharide as an affinity adsorbent can be considerably enhanced by introducing cross-linkages between the chains of the polysaccharide. Epichlorohydrin, 1,6-diaminohexane, divinyl sulfone, and bisoxirane have been used as cross-linking reagents. Dextrans having different numbers of cross-linkages have been especially useful for chomatography. Such material was used in early experiments on the purification of concanavalin A from jack-bean meal by the methodology of affinity chromatography.[72] Starch to which monosaccharides have been attached by the epichlorohydrin reaction has been used for isolating lectins by chromatography.[18] Guaran has been used to prepare an affinity adsorbent suitable for isolating lectins that bind D-galactose. The guaran was entrapped in a poly(acrylamide) gel during the polymerization process.[35] Pectin has been cross-linked with epichlorohydrin, and the product used as an affinity adsorbent for the purification of a pectinase.[36] Glycogen has been activated with cyanogen bromide, the activated product cross-linked with 1,6-diaminohexane, and the product used to isolate an amylase.[37]

IV. Derivatives of Carbohydrates

1. Types of Derivatives

The p-aminophenyl glycosides have been most widely used for preparing affinity adsorbents having carbohydrate ligands. Because these glycosides contain a free amino group, they can be readily coupled to the activated groups of support materials. The p-aminophenyl glycosides of D-glucose, D-galactose, and oligosaccharides of these units were synthesized many years ago for use in the preparation of carbohydrate–protein conjugates.[73,74] The conjugates were used, in turn, to prepare vaccines for immunizing rabbits, from which antisera that contained anti-glycosyl antibodies were then obtained.[74,75] The p-aminophenyl glycosides of many other monosaccharides have been synthesized, and used to prepare affinity adsorbents. Some of these

---

(72) B. B. L. Agrawal and I. J. Goldstein, *Biochem. J.*, 96 (1965) 23c–25c.
(73) W. F. Goebel and O. T. Avery, *J. Exp. Med.*, 50 (1929) 521–531.
(74) F. H. Babers and W. F. Goebel, *J. Biol. Chem.*, 105 (1934) 473–480.
(75) S. M. Beiser, G. C. Burke, and S. W. Tanenbaum, *J. Mol. Biol.*, 2 (1960) 125–132.

glycosides may be purchased from commercial suppliers, and others may be synthesized by adaptation of standard methods.[76]

The glycosides that are commercially available are p-aminophenyl 2-acetamido-2-deoxy-α-D-galactopyranoside, p-aminophenyl 2-acetamido-2-deoxy-β-D-glucopyranoside, p-aminophenyl α-L-fucopyranoside, p-aminophenyl β-L-fucopyranoside, p-aminophenyl α-D-galactopyranoside, p-aminophenyl β-D-galactopyranoside, p-aminophenyl α-D-glucopyranoside, p-aminophenyl β-D-glucopyranoside, p-aminophenyl α-D-mannopyranoside, p-aminophenyl 1-thio-β-L-fucopyranoside, p-aminophenyl 1-thio-β-D-galactopyranoside, p-aminophenyl 1-thio-β-D-glucopyranoside, and p-aminophenyl β-lactoside. In addition, the p-nitrophenyl glycosides of other carbohydrates are commercially available, and these can be readily converted into the p-aminophenyl derivative by catalytic reduction with hydrogen and platinum oxide,[77] palladized charcoal and sodium borohydride,[78] or alkaline sodium hydrosulfite ($Na_2S_2O_4$).[79]

Glycosides having different aglycons that react with activated groups of the support material have also been synthesized, and used in the preparation of affinity adsorbents. Such compounds are the aminoalkyl glycosides,[80] the allyl glycosides,[19,81] the 1-[2-(p-aminophenyl)ethyl]amino-1-deoxyalditols,[27] the N-acylglycosylamines,[11] p-aminophenyl 1-thioglycosides,[22] and the glycosides of 8-(ethoxycarbonyl)octanol.[82] Coupling of the amino-containing compounds to activated groups of the support occurs by the reactions already discussed. Coupling of derivatives having carboxyl groups to the amino groups of supports is effected by the carbodiimide method,[52,53] or by the mixed-anhydride method.[83,84] Derivatives of carbohydrates having a carboxyl group containing C-1 have been prepared by mild oxidation of the carbohydrate with hypoiodite.[85,86] Derivatives of carbohydrates having a

---

(76) W. G. Overend, in W. Pigman and D. Horton (Eds.), *The Carbohydrates*, 2nd edn., Vol. IA, Academic Press, New York, 1972, pp. 279–353.
(77) V. Voorhees and R. Adams, *J. Am. Chem. Soc.*, 44 (1922) 1397–1405.
(78) T. Neilson, H. C. S. Wood, and A. G. Wylie, *J. Chem. Soc.*, (1962) 371–372.
(79) R. Bloch and M. M. Burger, *FEBS Lett.*, 44 (1974) 286–289.
(80) R. Barker, K. W. Olsen, J. H. Shaper, and R. L. Hill, *J. Biol. Chem.*, 247 (1972) 7135–7147.
(81) R. T. Lee and Y. C. Lee, *Carbohydr. Res.*, 37 (1974) 193–201.
(82) R. U. Lemieux and H. Driguez, *J. Am. Chem. Soc.*, 97 (1975) 4069–4075.
(83) S. W. Tanenbaum, M. Mage, and S. M. Beiser, *Proc. Natl. Acad. Sci. USA*, 45 (1959) 922–929.
(84) Y. Arakatsu, G. Ashwell, and E. A. Kabat, *J. Immunol.*, 97 (1966) 858–866.
(85) S. Moore and K. P. Link, *J. Biol. Chem.*, 133 (1940) 293–311.
(86) R. Schaffer and H. S. Isbell, *Methods Carbohydr. Chem.*, 2 (1963) 11–12.

carboxyl group containing C-6 occur naturally, or can be synthesized by oxidation methods.[87] Procedures for coupling carbohydrate derivatives to support materials are discussed in Section V.

2. Chemical Synthesis

a. **Aminophenyl Glycosides.**—The fundamental reaction for the preparation of *p*-aminophenyl glycosides is the condensation of the per-*O*-acetylglycosyl halide, or the fully acetylated carbohydrate, with *p*-nitrophenol under basic reaction-conditions,[88] or in the presence of zinc chloride as the catalyst.[89] The glycosides are isolated by extraction procedures, deacetylated with base, if necessary, and reduced to the *p*-aminophenyl glycoside by a catalytic method. Many *p*-aminophenyl glycosides have been prepared by these reaction routes, including the α- and β-glycosides of D-glucose, D-galactose, and L-fucose, and the β-glycosides of 2-acetamido-2-deoxy-D-galactose, 2-amino-2-deoxy-D-glucose, D-glucuronic acid, lactose, maltose, and cellobiose. The reaction sequence for the preparation of *p*-aminophenyl β-D-galactoside is shown in Scheme 1.

The *p*-aminophenyl glycosides of 1-thio sugars have been used to prepare affinity adsorbents for purifying enzymes.[4,22] Glycosides of 1-thio sugars can be prepared by condensing the per-*O*-acetylglycosyl halide with the lithium salt of *p*-nitrobenzenethiol in alkaline solution.[90] The product can be isolated from the reaction mixture by extraction, and purified by standard methods. Thioglycosides can also be synthesized by treating the peracetylated monosaccharides and *p*-nitrobenzenethiol in the presence of zinc chloride as a catalyst.[91] The nitro derivatives were deacetylated, and the products reduced to the amino derivatives. The *p*-aminophenyl glycosides of 1-thio-L-fucose, 1-thio-D-galactose, and 1-thio-D-glucose are available commercially. Thio derivatives of other carbohydrates can be prepared by the reaction routes already indicated, or by alternative, reaction routes described in the literature.[92]

It will be noted in Scheme 1 that the per-*O*-acetylated 1-halide is needed in the preparation of *p*-aminophenyl β-D-galactoside. The halide is obtained from the peracetylated carbohydrate, which can be

(87) C. L. Mehltretter, *Methods Carbohydr. Chem.*, 2 (1963) 29–31.
(88) M. Seidman and K. P. Link, *J. Am. Chem. Soc.*, 72 (1950) 4324.
(89) O. Westphal and H. Feier, *Chem. Ber.*, 89 (1956) 582–588.
(90) B. Capon, P. M. Collins, A. A. Levy, and W. G. Overend, *J. Chem. Soc.*, (1964) 3242–3254.
(91) J. Schneider, H. H. Liu, and Y. C. Lee, *Carbohydr. Res.*, 39 (1975) 156–159.
(92) D. Horton and D. H. Hutson, *Adv. Carbohydr. Chem.*, 18 (1963) 123–199.

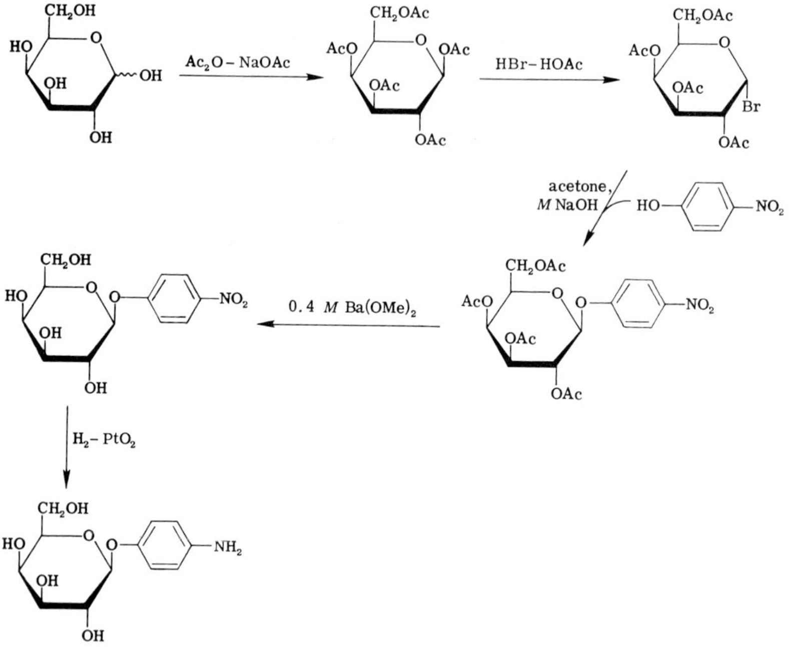

Scheme 1

prepared by the standard methods of carbohydrate chemistry.[93] However, for the individual compounds,[94] the per-*O*-acetylglycosyl halides must be formed by specialized methods. The 1-bromides can be prepared from the peracetylated carbohydrates by treating the compound with hydrogen bromide in acetic acid.[95] Under these conditions, the acetoxyl group on C-1 is replaced by a bromine atom, and the 1-bromide is obtained. The 1-bromides can also be prepared in a one-step reaction by acetylating the carbohydrate with acetic anhydride, and sequentially treating the derivative with phosphorus, bromine, and water.[96]

The per-*O*-acetylated 1-chlorides of hexosamines can be prepared by treating the hexosamine with an excess of acetyl chloride at elevated temperature.[97] Under these conditions, acetylation of the hexosamine occurs at all positions except C-1, at which a chlorine atom

(93) M. L. Wolfrom and A. Thompson, *Methods Carbohydr. Chem.*, 2 (1963) 211–215.
(94) L. J. Haynes and F. H. Newth, *Adv. Carbohydr. Chem.*, 10 (1955) 207–256.
(95) L. J. Haynes and A. R. Todd,*J. Chem. Soc.*, (1950) 303–308.
(96) M. Bárczai-Martos and F. Kőrösy, *Nature*, 165 (1950) 369.
(97) D. Horton, *Methods Carbohydr. Chem.*, 6 (1972) 282–285.

becomes attached. Tetra-*O*-acetyl-D-glucosyl chloride can be synthesized by treating penta-*O*-acetyl-D-glucose with titanium tetrachloride.[98] The D-glucose derivative can also be prepared by heating the pentaacetate with hydrogen chloride, phosphorus pentachloride, or aluminum trichloride.[98]

The condensation of phenol with a per-*O*-acetylglycosyl halide, or with the peracetates of carbohydrates, generally yields a mixture of the α- and β-glycosides. The ratio of the anomers in a reaction mixture depends on a variety of factors, including the structure of the carbohydrate reactant, the reaction conditions, the stability of the anomeric glycosides, the trans orientation of the substituents at carbon atoms 1 and 2, and the electronic environment around the anomeric carbon atom.[99,100] Some condensations of aglycons and carbohydrate derivatives occur with almost complete inversion of the anomeric configuration, and others, with almost complete retention of the anomeric configuration. The relative importance of the factors that influence the extent of inversion, or retention, of anomeric configuration is discussed in considerable detail in a chapter in a treatise.[76]

If both anomers of the glycoside are obtained in a reaction, it is necessary to separate them. One separation method that has been used is the preferential extraction[101] of the anomers with a volatile solvent. After extraction of the individual glycoside, the solvent is removed by evaporation, and the glycoside may be obtained in crystalline form. A second method of separation utilizes fractional recrystallization, and it may be possible by proper selection of the solvent to obtain both anomers in crystalline form. A third method utilizes chromatography for separating the anomers, and the pure anomers may be obtained from appropriate fractions from the column. The anomeric configuration of the anomers which have been obtained in pure form will need to be determined. Such determinations can be made by measurement of physical constants, from the n.m.r. spectra, and from the susceptibility of the anomer to enzymes of known specificity.

b. Aminoalkyl Glycosides.—The aminoalkyl glycosides have been synthesized by condensing per-*O*-acetylglycosyl halides with 6-(trifluoroacetamido)-1-hexanol in the presence of mercuric cyanide as the

(98) R. U. Lemieux, *Methods Carbohydr. Chem.*, 2 (1963) 223–224.
(99) R. S. Tipson, *J. Biol. Chem.*, 130 (1939) 55–59.
(100) R. U. Lemieux and T. L. Nagabhushan, *Methods Carbohydr. Chem.*, 6 (1972) 487–496.
(101) R. Barker, C.-K. Chiang, I. P. Trayer, and R. L. Hill, *Methods Enzymol.*, 34B (1974) 317–327.

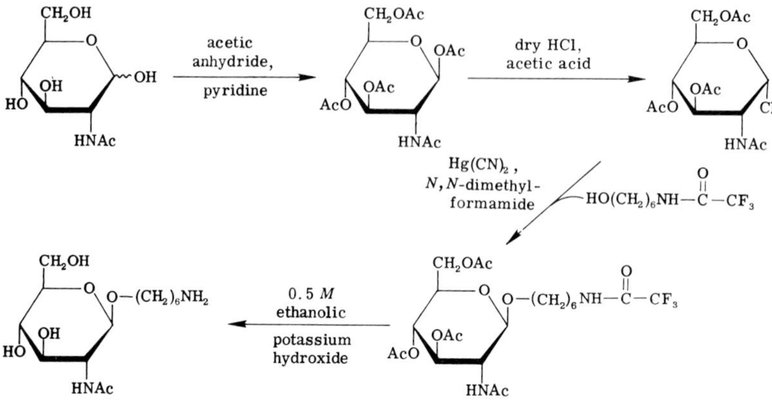

Scheme 2

catalyst.[80] The reaction sequence for preparing 6-amino-1-hexyl 2-acetamido-2-deoxy-α-D-glucopyranoside is shown in Scheme 2.

The reactions shown in Scheme 2 can be utilized with minor modifications to synthesize the 6-amino-1-hexyl glycoside of other monosaccharides. The derivatives, and the reagents, needed for the synthesis of the aminoalkyl glycosides are indicated in Scheme 2. The directions and precautions that should be taken at the various steps of the synthesis have also been reported.[80] Some modifications in the method were later described for the preparation of other glycosides of this type.[101] The 6-amino-1-hexyl glycosides of 2-acetamido-2-deoxy-D-galactopyranose, D-glucose, D-galactose, and D-xylose have been prepared by this method.

c. Glycosyl (6-Aminohexyl Diphosphates).—Affinity adsorbents having hexosyl phosphate groups attached would be useful for isolating antibodies having specificity for the hexosyl phosphate part. Antibodies having such specificity are known,[102,103] and they present a unique type of specificity. Affinity adsorbents bearing hexosyl phosphate attachments can be prepared by attaching glycosyl (6-aminohexyl diphosphates) to agarose activated with cyanogen bromide. The preparation of α-D-galactosyl (6-aminohexyl diphosphate) has been achieved,[80] and the reaction sequence is shown in Scheme 3.

The essential reaction in this Scheme is the addition of the imidazole derivative of 6-(trifluoroacetamido)hexyl phosphate to a per-O-

---

(102) J. H. Pazur, A. Cepure, J. A. Kane, and W. W. Karakawa, *Biochem. Biophys. Res. Commun.*, 43 (1971) 1421–1428.

(103) P. N. Lipke, W. C. Raschke, and C. E. Ballou, *Carbohyd. Res.*, 37 (1974) 23–35.

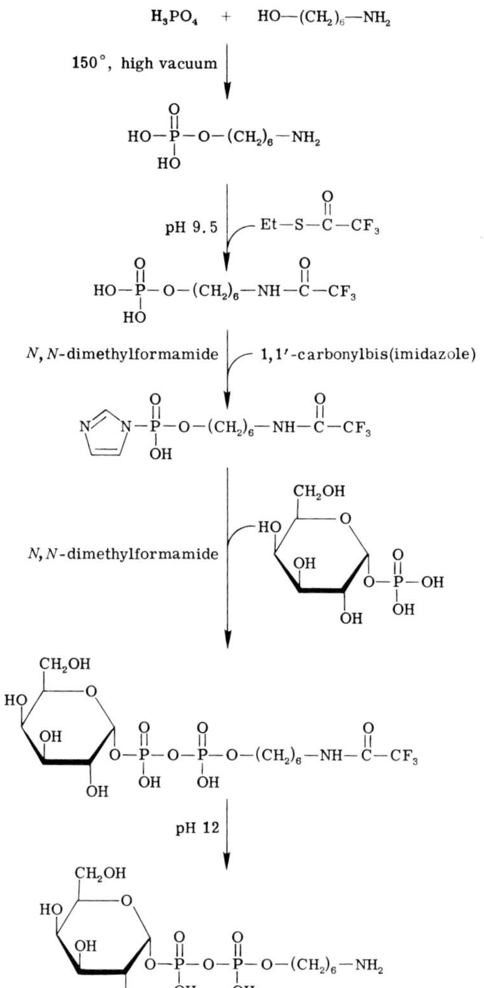

Scheme 3

acetyl-hexosyl phosphate. The reaction product contains a diphosphate linkage. The product was isolated by solvent extraction, and the protecting groups were removed by de-esterification. α-D-Galactosyl (6-aminohexyl diphosphate) was obtained as the product in this series of reactions. Some of the reactions shown in Scheme 3 are sensitive to water and amines, and extra precautions should therefore be taken to remove these impurities from the reagents used. The reaction sequence may be used to synthesize 6-aminohexyl phosphate derivatives of other monosaccharides, provided that the aldosyl phosphate is available.

d. N-Acylglycosylamines.—Glycosylamines can be synthesized by treating reducing carbohydrates with anhydrous, liquid ammonia.[104] These derivatives are generally not stable, and not suitable for preparing affinity adsorbents. However, N-acylglycosylamines can be used to prepare affinity adsorbents, and they can be synthesized from the per-O-acetylglycosyl chloride[97] by conversion of the derivative into an azide,[105] and thence into an acetylated glycosylamine.[106] This product is treated with 6-(benzyloxycarbonylamino)hexanoic acid in the presence of isobutyl chloroformate and triethylamine, to yield an acetylated N-acylglycosylamine.[106] The acetyl groups are removed by de-esterification, and the benzyloxycarbonyl group is removed reductively. The product can be coupled to activated agarose. The reaction sequence for the synthesis of 2-acetamido-N-(6-aminohexanoyl)-2-deoxy-β-D-glucosylamine is shown in Scheme 4.

e. 1-[2-(p-Aminophenyl)ethyl]amino-1-deoxyalditols.—Reducing oligosaccharides react with a diamine in the presence of sodium borohydride to yield stable glycosylamines containing a free amino group. Such glycosylamines can be coupled to activated agarose, to yield affinity adsorbents. The reaction is performed by dissolving the oligosaccharide in a diamine containing sodium borohydride.[27] In the first step of the reaction, water is eliminated between the hemiacetal hydroxyl group of the oligosaccharide and one amino group of the diamine, to yield an unstable product having a carbon–nitrogen bond. In the next step, the compound is reduced by the sodium borohydride, to yield a stable alditol derivative. The reaction sequence for cellotriose and 2-(p-aminophenyl)ethylamine[107] is shown in Scheme 5.

Derivatives of this type have been prepared from 2-(p-aminophenyl)ethylamine and the oligosaccharides from milk,[27] and the products were used to prepare affinity adsorbents for isolating antiglycosyl antibodies directed against the oligosaccharides from milk. Affinity adsorbents prepared by this route contain one glycosyl residue fewer than the original oligosaccharide, because the reducing residue of the original oligosaccharide has been reduced. As a consequence, the utility of affinity adsorbents prepared by this method may be limited.

(104) H. S. Isbell and H. L. Frush, *J. Org. Chem.*, 23 (1958) 1309–1319.
(105) C. H. Bolton and R. W. Jeanloz, *J. Org. Chem.*, 28 (1962) 3228–3230.
(106) H. Lis, R. Lotan, and N. Sharon, *Methods Enzymol.*, 34B (1974) 341–346.
(107) J. H. Pazur and L. S. Forsberg, unpublished results (1980).

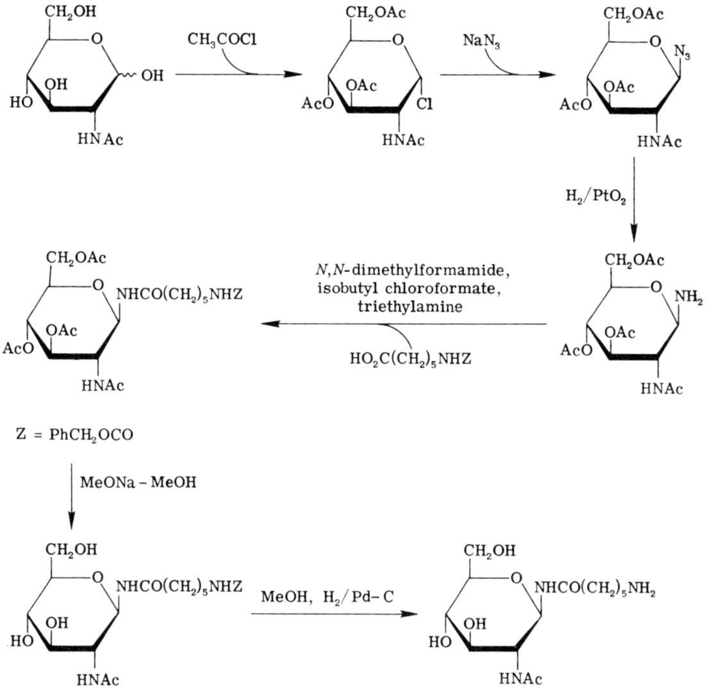

Scheme 4

f. Allyl Glycosides.—The allyl glycosides of carbohydrates have been used in two ways for synthesizing affinity adsorbents. In the first method, the glycosides can be copolymerized with acrylamide and N,N'-methylenebis(acrylamide).[12] In the polymerization process, the allyl group of the glycoside becomes attached to the ethylene units of poly(acrylamide). The polymeric material contains glycosyl residues that function as the ligands of the affinity adsorbent. Carbohydrates that have been attached to poly(acrylamide) by this method are α-L-fucose, α-D-galactose, α-D-glucose, β-D-glucose, D-mannose, 2-acetamido-2-deoxy-α-D-glucose, and β-lactose.[12]

In the second method, the allyl glycosides may be converted into derivatives containing free amino groups, and these derivatives can be attached to activated supports.[81] In the conversion, the allyl glycoside reacts with 2-aminoethanethiol, which adds to the double bond of the allyl group to yield a 3-(2-aminoethylthio)propyl glycoside. Glycosides of this type have been prepared from derivatives of D-glucose, D-galactose, and 2-acetamido-2-deoxy-D-glucose.[81] Coupling of these

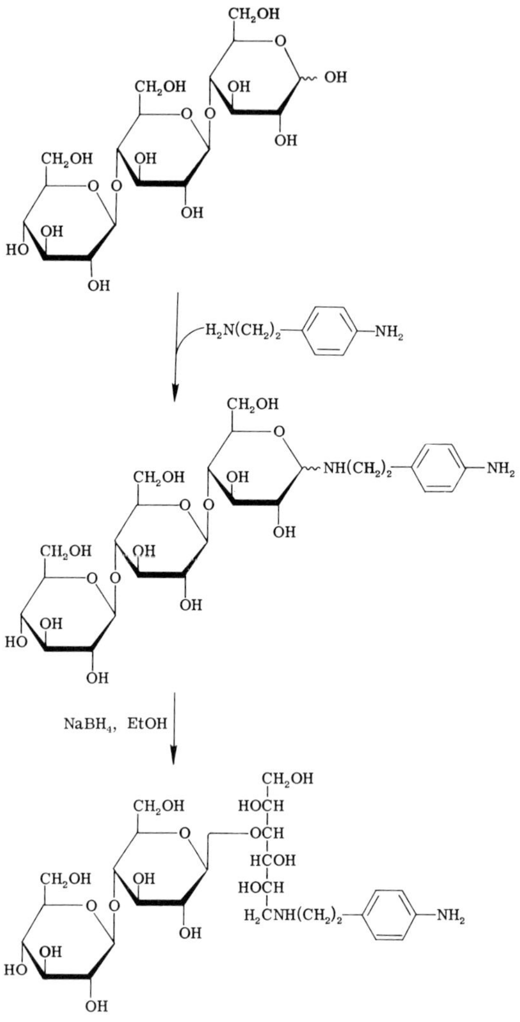

Scheme 5

derivatives to activated supports occurs at the amino group of the aglycon and the activated groups of the support.

The allyl glycosides have been synthesized from allyl alcohol and the free carbohydrate, or the per-O-acetylglycosyl halide in the presence of acid, or mercury cyanide, as the catalyst. The condensation of the alcohol occurs with both anomers of the carbohydrate, and a mixture of the α- and β-glycosides is obtained. The anomers can be separated by preferential, solvent extraction,[12] or by chromatography on ion-exchange columns.[81] The condensation of allyl alcohol with the per-O-acetylglycosyl halides in the presence of mercuric cyanide as the catalyst occurs stereospecifically, to yield a single glycoside. The acetylated glycoside can be isolated by extraction with solvent, and deacetylated to yield the allyl glycoside. The glycoside is isolated by crystallization, and the pure compound can be used for the preparation of affinity adsorbents.

3. Enzymic Synthesis

The p-aminophenyl glycosides of D-gluco-oligosaccharides having (1→4)- and (1→6)-α-D linkages have been synthesized by enzymic procedures. Macerans amylase from *Bacillus macerans* was used to synthesize the p-aminophenyl glycosides of maltose and malto-oligosaccharides.[17] Cyclomaltohexaose and p-aminophenyl β-D-glucoside were, respectively, the substrate and the cosubstrate for this enzyme. In this enzymic reaction, an α-D-(1→4) linkage of the cyclomaltohexaose is opened by the enzyme, and a maltohexaosyl–enzyme complex is formed.[108] The maltohexaosyl moiety is then transferred, with retention of configuration, to O-4 of p-aminophenyl β-D-glucoside to yield, initially, a p-aminophenyl maltoheptaoside. However, the enzyme also effects a redistribution of the D-glucosyl units at the end remote from the p-aminophenyl group of the maltoheptaoside, to produce a series of homologous glycosides.[109] These glycosides can be separated on paper chromatograms, and identified by the use of specific spray-reagents,[110] and by their R_F values.[111] The individual glycosides can be isolated by preparative paper-chromatography.[112]

The p-aminophenyl glycosides of isomaltose and other (1→6)-α-D-linked D-gluco-oligosaccharides were prepared from maltose and p-aminophenyl β-D-glucoside by use of a D-glucosyltransferase from *Aspergillus niger*.[113] This enzyme removes a terminal unit from maltose, to afford a D-glucosyl–enzyme complex. The D-glucosyl group is transferred, with retention of configuration, to O-6 of p-aminophenyl β-D-glucopyranoside, to yield p-aminophenyl isomaltoside. Additional transfers to the isomaltose derivative can occur, to yield a series of p-aminophenyl glycosides. Transfer of a D-glucosyl group can also occur

(108) D. French, M. L. Levine, E. Norberg, P. Nordin, J. H. Pazur, and G. M. Wild, *J. Am. Chem. Soc.*, 76 (1954) 2387–2390.
(109) J. H. Pazur, J. M. Marsh, and T. Ando, *J. Am. Chem. Soc.*, 81 (1959) 2170–2172.
(110) I. S. Bhatia, M. S. Bhatia, S. Singh, and K. L. Bajaj, *Ann. Chim.*, (1976) 7–8.
(111) D. French and G. M. Wild, *J. Am. Chem. Soc.*, 75 (1953) 2612–2616.
(112) J. H. Pazur, *J. Biol. Chem.*, 205 (1953) 75–80.
(113) J. H. Pazur and T. Ando, *Arch. Biochem. Biophys.*, 93 (1961) 43–49.

to other D-glucosyl compounds, and a mixture of products is obtained during action of the enzyme on maltose and p-aminophenyl β-D-glucopyranoside.[17] The p-aminophenyl D-glycosides in the reaction mixture were isolated by preparative paper-chromatography, and the isomaltose derivative was used to synthesize an affinity adsorbent bearing isomaltosyl units.

Other types of transferring enzymes may be useful for preparing p-aminophenyl glycosides. The D-galactosyltransferase from *Saccharomyces fragilis* effects transfer of D-galactosyl groups from lactose[114] and o-nitrophenyl β-D-galactopyranoside[115] to the cosubstrates D-glucose, D-fructose, D-galactose, and 2-amino-2-deoxy-D-glucose. It is probable that p-aminophenyl glycosides of these cosubstrates may also function as acceptor molecules for this enzyme, and, if so, the p-aminophenyl glycosides of the oligosaccharides O-D-galactosyl-D-glucose, O-D-galactosyl-D-fructose, O-D-galactosyl-D-galactose, or O-D-galactosyl-(2-amino-2-deoxy-D-glucose) would be produced. A D-galactosyltransferase from milk has been shown to transfer D-galactosyl groups from uridine 5′-(D-galactosyl diphosphate) to acceptor compounds.[80] A transfer of D-galactosyl groups from the D-galactosyl ester of the nucleotide to 2-acetamido-2-deoxy-D-glucose units that were already attached to agarose was observed.[80] This enzyme probably possesses the ability to transfer D-galactosyl groups to other acceptor molecules attached to agarose. Such transfers would yield affinity adsorbents containing ligands of various oligosaccharides.

V. Coupling Reactions

1. For Supports Activated with Cyanogen Bromide

Many types of affinity adsorbent have been prepared by coupling ligands containing primary amino groups to agarose activated with cyanogen bromide.[32] The structures for two types of active groups in agarose activated by cyanogen bromide have been shown in Equation 1. Coupling reactions occur between these groups and the amino groups of the ligand molecules. The structure of the linkages by which the ligand is attached to the support material has been deduced from infrared and other types of measurements.[1,62] The structures of the products from p-aminophenyl β-D-galactopyranoside and cyanogen bromide-activated agarose are shown in **2** and **3**.

(114) J. H. Pazur, J. M. Marsh, and C. L. Tipton, *J. Biol. Chem.*, 233 (1958) 277–279.
(115) J. H. Pazur, M. Shadaksharaswamy, and A. Cepure, *Arch. Biochem. Biophys.*, 94 (1961) 142–147.

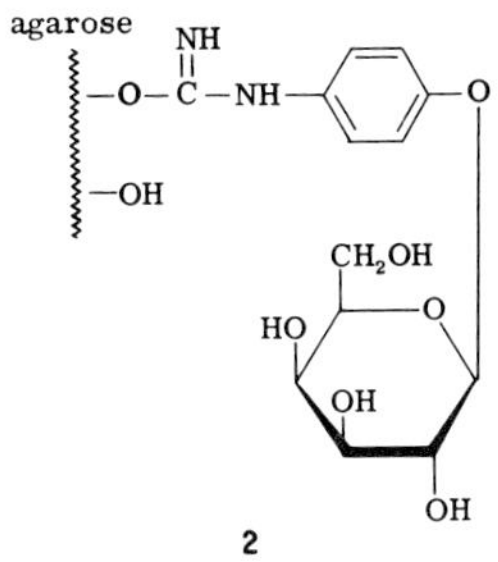

The cyanic ester group of activated agarose also reacts with p-aminophenyl β-D-galactopyranoside, to yield a product[63] of the same structure as that shown in 3. Other carbohydrate derivatives having amino groups, such as 6-amino-1-hexyl glycosides, 1-[2-(p-aminophenyl)ethyl]amino-1-deoxyalditols, and N-acylglycosylamines, react with activated agarose, to yield products having linkages like those shown in 2 and 3.

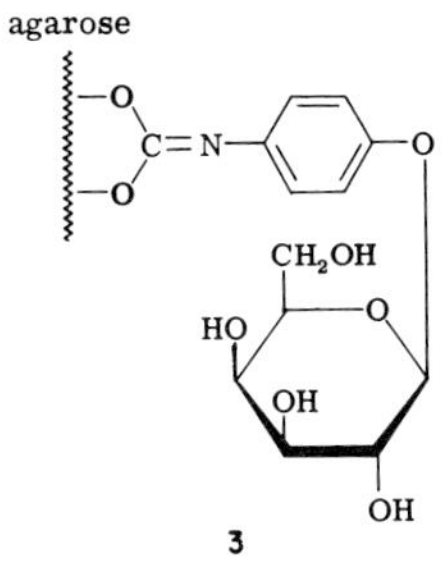

The conditions for attaching ligands to activated agarose vary, depending on the nature of the ligand and the support. A method recommended for coupling ligands to activated agarose utilizes sodium carbonate buffer to maintain an alkaline pH during the reaction.[116] At completion of the coupling reaction, the excess of the ligand derivative is removed by thoroughly washing the agarose with the buffer. The unreacted, activated groups in the agarose are blocked, to eliminate interferences in the affinity chromatography process. Such groups may be blocked by treating the agarose with 2-aminoethanol, and then thoroughly washing the adsorbent with the buffer in order to remove the excess of the reagent.

2. For Supports Activated with Bisoxirane

Support materials have been converted into activated forms by reaction of the material with bisoxirane. Bisoxirane is a bifunctional reagent having two epoxide rings, and one of the epoxide rings reacts with a hydroxyl group of the support, but the other epoxide ring is available for coupling to a ligand.[66] The coupling to a carbohydrate may occur to any of its hydroxyl groups, but the primary alcohol group is the favored point of coupling.[14] The carbohydrates that have been coupled to bisoxirane-activated supports are 2-acetamido-2-deoxy-D-glucose,[13] 2-acetamido-2-deoxy-D-galactose,[13] methyl α-D-mannopyranoside,[14] lactose,[14] and enzymically modified starch.[14] The struc-

(116) I. Parikh, S. March, and P. Cuatrecasas, *Methods Enzymol.*, 34B (1974) 77–102.

ture of the coupled product of 2-acetamido-2-deoxy-D-glucose and bisoxirane-activated agarose is shown in **4**.

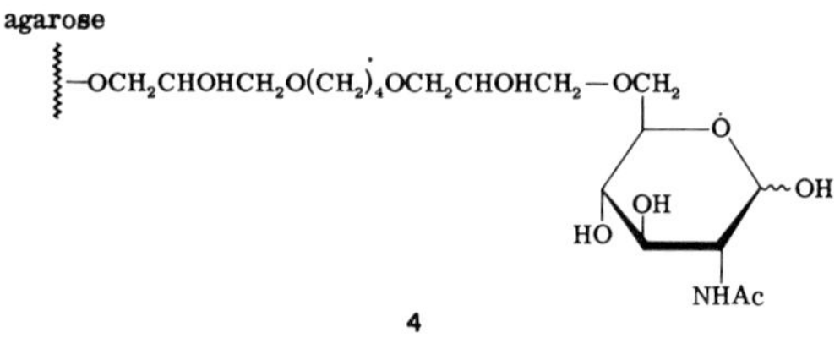

$$\text{agarose}—OCH_2CHOHCH_2O(CH_2)_4OCH_2CHOHCH_2—OCH_2-\underset{\underset{NHAc}{HO}}{\underset{|}{\bigcirc}}-OH$$

4

The coupling of the carbohydrates to supports activated by bisoxirane is performed in slightly alkaline media.[66] The alkalinity should be maintained close to neutrality, in order to minimize isomerization of the carbohydrates and the loss of carbohydrate units. In **4**, the primary alcohol group of the carbohydrate residue is linked to the oxygen atom of the original epoxide ring attached to the activated support. The evidence for this type of linkage has come primarily from data educed by periodate oxidation.[14] If the primary alcoholic group is involved in the formation of the insoluble complex, such a complex could not be formed with this type of affinity adsorbent. Accordingly, affinity adsorbents prepared by this method may not be suitable for some types of affinity chromatography.

3. For Supports Activated with Divinyl Sulfone

Supports activated with divinyl sulfone, like those activated with bisoxirane, can be coupled to underivatized carbohydrates.[21] The supports activated with divinyl sulfone couple to the hemiacetal hydroxyl group of the carbohydrate by the vinyl group. Accordingly affinity adsorbents prepared by this method will contain glycosyl units. The structure of the reaction product of L-fucose with agarose activated with divinyl sulfone is shown in **5**.

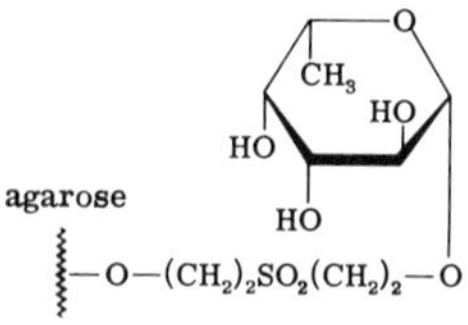

$$\text{agarose}—O—(CH_2)_2SO_2(CH_2)_2—O-\text{fucosyl}$$

5

The reaction of the vinyl groups of activated supports with ligand residues is not quantitative. The excess of the ligand must be removed by washing the adsorbent with buffer. Unreacted groups of the support need to be blocked, to prevent interference with the chromatographic process. Such unreacted groups can be blocked by treating the adsorbent with 2-mercaptoethanol for several hours, and removing the excess of the 2-mercaptoethanol by washing with a suitable buffer. The resulting product is suitable for use as an affinity adsorbent.

4. Reductive Amination

Reducing oligosaccharides have been attached to support materials, by means of primary amino groups, by the reductive-amination method.[25] The amination is performed under slightly alkaline conditions, and the reduction is effected with sodium cyanoborohydride.[54] Support materials that have been used in this method are 2-aminoethylpoly(acrylamide) or O-(2-aminoethyl)cellulose. The reducing oligosaccharide and support material are allowed to react for a long time, and then yield a product, a Schiff base, by elimination of water between the amino group of the support material and the aldehyde group of the carbohydrate. This Schiff base is reduced by the sodium cyanoborohydride to a 1-deoxyalditol-1-yl–N derivative. Under the reaction conditions, sodium cyanoborohydride does not reduce the aldehyde group of the carbohydrate. Reductive amination has been used for coupling carbohydrates to proteins, to yield neoglycoproteins[31]; these have been used for studies on (a) the transport of glycoproteins through membranes,[117,118] (b) the role of carbohydrates in the stability of glycoproteins,[119] and (c) the nature of the immune response to synthetic, carbohydrate–protein antigens.[120,121]

Because reduction of the oligosaccharide occurs in the reductive-amination reaction, affinity adsorbents prepared by this route contain one glycosyl residue fewer than the original oligosaccharide. Adsorbents having such ligands may have low utility. The structure of the product obtained from lactose and 2-aminoethylpoly(acrylamide) by the reductive-amination route is shown in **6**.

(117) C. J. A. Van Den Hamer, A. G. Morell, I. H. Scheinberg, J. Hickman, and G. Ashwell, *J. Biol. Chem.*, 245 (1970) 4397–4402.
(118) M. J. Krantz, N. A. Holtzman, C. P. Stowell, and Y. C. Lee, *Biochemistry*, 15 (1976) 3963–3968.
(119) J. H. Pazur and N. N. Aronson, Jr., *Adv. Carbohydr. Chem. Biochem.*, 27 (1972) 301–341.
(120) O. T. Avery and W. F. Goebel, *J. Exp. Med.*, 54 (1931) 437–447.
(121) R. U. Lemieux, D. A. Baker, and D. R. Bundle, *Can. J. Biochem.*, 55 (1977) 507–512.

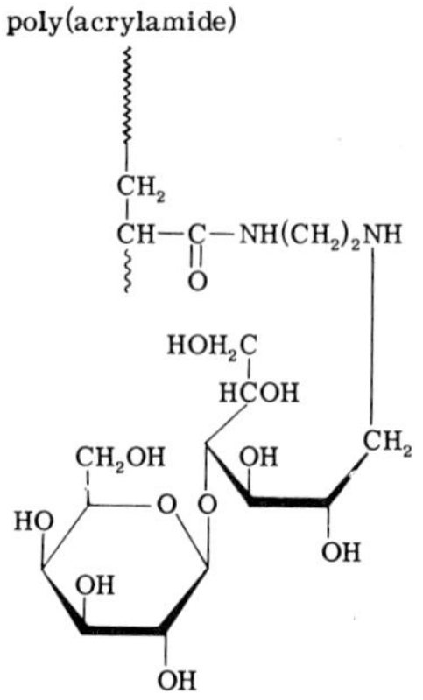

$$\begin{array}{c}
\text{poly(acrylamide)}\\
|\\
CH_2\\
|\\
CH-C-NH(CH_2)_2NH-\\
\parallel\\
O
\end{array}$$

6

In the reductive amination method, it is important to use reagents and reactants of high purity. The carbohydrate sample should not contain carbohydrate contaminants, as, if they were present, an adsorbent containing more than one type of ligand would be obtained. The sodium cyanoborohydride should be of high purity, because reduction of the Schiff base may not occur with impure preparations. Special procedures have been developed for purifying sodium cyanoborohydride, and these should be employed.[25]

5. Other Reactions

Special types of affinity adsorbents are required for purifying some types of substances that cannot be purified on the conventional, affinity adsorbents.[47,48] These special adsorbents contain spacer groups that link the ligand residues to the support material. Bifunctional reagents have been used as spacer groups, and two such compounds are 1,6-diaminohexane and 6-aminohexanoic acid. The amino group of the compound reacts with the imido carbonate group of cyanogen bromide-activated supports. A hydrocarbon chain of six carbon atoms, having an amino group on, or a carboxyl group as, the terminal carbon atom, becomes attached to the support. The reactive group of the spacer arm can be coupled to ligands by a variety of methods. Other bifunctional reagents may be used in order to attach spacer groups to activated supports.

A procedure widely used for coupling ligands to supports containing spacer groups is the carbodiimide method.[52,53] Many different carbohydrate derivatives have been coupled to supports in this way. A variety of diimides has been employed in these couplings, and these include dicyclohexylcarbodiimide (DCC), 3-(3-dimethylaminopropyl)-1-ethylcarbodiimide (DAEC) and its hydrochloride (EDC), and 1-cyclohexyl-3-(2-morpholinoethyl)carbodiimide metho-*p*-toluenesulfonate (CMC).

Carbohydrate derivatives that have been coupled to supports having carboxyl groups on the spacer arm are the *p*-aminophenyl glycosides, the glycosylamines, and other derivatives containing amino groups. Carbohydrate derivatives that contain aldonic acids or uronic acids have been coupled to supports having free amino groups on the spacer arm. The reaction series for coupling melibiose to 2-aminoethylpoly(acrylamide) by the carbodiimide method is shown in Scheme 6.

The aminophenyl glycosides can be coupled to activated supports by other routes. Thus, the glycosides may be converted into diazo-

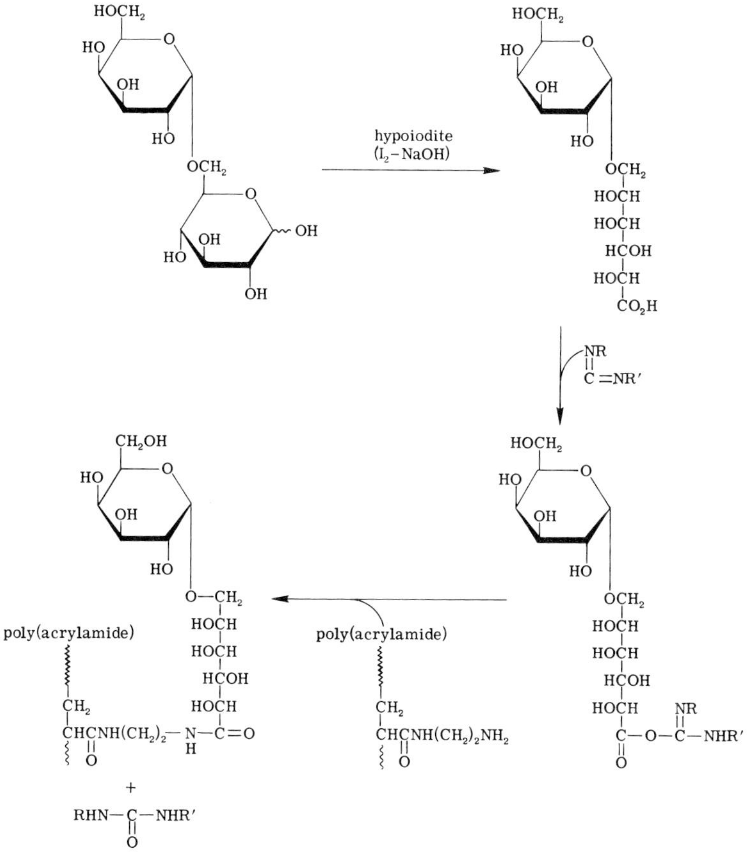

Scheme 6

nium salts by reaction of the glycoside with sodium nitrite and hydrochloric acid. The diazonium salt is then coupled to appropriate supports by the diazo reaction.[68,89] Finally, glycosides can be coupled to support materials that have been activated by 1,1′-carbonyl(bisimidazole). In this coupling reaction, the imidazole group of the activated support is displaced by the amino group of the glycoside, to yield an adsorbent bearing carbohydrate units. The product of the reaction of p-aminophenyl β-D-galactopyranoside with agarose activated by means of the bisimidazole reagent is shown in 7.

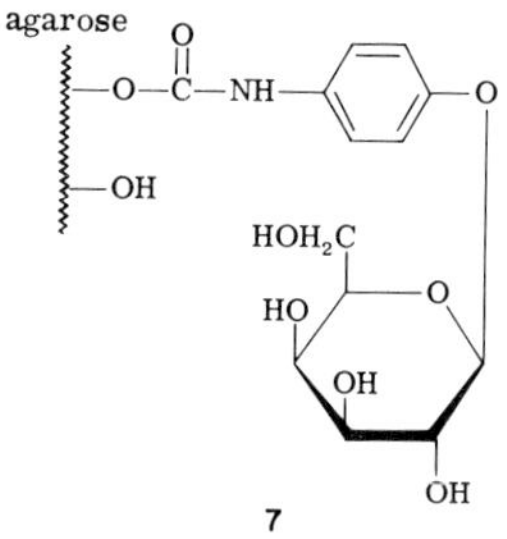

It has been stated that the affinity adsorbents prepared from the bis(imidazole)-activated supports are superior to those prepared from supports activated by other methods.[67] Some of the advantages of affinity adsorbents prepared from bis(imidazole)-activated supports are given in Section III.

A coupling reaction that has been used to prepare affinity adsorbents bearing carbohydrate ligands is the copolymerization of an allyl glycoside with acrylamide and N,N′-methylenebis(acrylamide). The copolymer contains many glycosyl residues attached by the allyl group to the ethylene units of the poly(acrylamide), and possesses many of the properties desirable in affinity adsorbents. The future availability of allyl glycosides will determine the utility of this method for preparing other types of affinity adsorbents.

Affinity adsorbents prepared from polysaccharides have been used in the isolation of enzymes and lectins. Polysaccharides used for this purpose have been modified by chemical or physical means. Chemical modification involves the introduction of cross-linkages between the polymeric chains, to yield products having properties suitable for use in chromatography. Dextran,[33,34] guaran,[35] pectin,[36] and glycogen[37] have been treated with cross-linking reagents, and the products have been used as affinity adsorbents. Polysaccharides have also been treated with cross-linking reagents in the presence of monosaccharides.[18] In the ensuing reaction, monosaccharide units become attached to the polysaccharide chains, and the resulting product can be used as an affinity adsorbent. Polysaccharides have been coupled to supports by attaching the polysaccharide to a protein by the cyanuric chloride reaction,[122] and then coupling the conjugate to cyanogen bromide-activated agarose.[123]

VI. APPLICATIONS

1. Antibodies

Antibodies that are induced by carbohydrate antigens and that combine with carbohydrate residues of these antigens have been designated anti-glycosyl antibodies.[124] The carbohydrate antigens may be glycans,[125,126] glycolipids,[127] glycoproteins,[128,129] or synthetic, carbohydrate–protein conjugates.[130–132] Glycans and glycolipids in bacterial cell-walls are the type-specific substances, and are the basis of a scheme for the serological identification of pathogenic bacteria.[133] The molecular structures of several bacterial, carbohydrate antigens have been elucidated by methylation analysis, gas–liquid chromatography, and mass spectrometry.[134,135] It was observed that many of these antigens consist of a main chain to which are attached numerous oligosaccharide side-chains. These oligosaccharide chains function as the immunodeterminant groups of the antigens. This type of information was needed before affinity adsorbents for isolating anti-glycosyl antibodies could be devised and prepared.

The isolation of anti-glycosyl antibodies by affinity chromatography

(122) R. J. Fiedler, C. T. Bishop, S. F. Goppell, and F. Blank, *J. Immunol.*, 105 (1970) 265–267.
(123) J. B. Winfield, J. H. Pincus, and R. C. Mage, *J. Immunol.*, 108 (1972) 1278–1287.
(124) J. H. Pazur, K. B. Miller, K. L. Dreher, and L. S. Forsberg, *Biochem. Biophys. Res. Commun.*, 70 (1976) 545–550.
(125) M. Heidelberger and O. T. Avery, *J. Exp. Med.*, 40 (1924) 301–316.
(126) R. M. Krause, *Bacteriol. Rev.*, 27 (1963) 369–380.
(127) O. Lüderitz, A. M. Staub, and O. Westphal, *Bacteriol. Rev.*, 30 (1966) 192–255.
(128) J. Krupey, P. Gold, and S. O. Freedman, *J. Exp. Med.*, 128 (1968) 387–398.
(129) W. D. Terry, P. A. Henkart, J. E. Coligan, and C. W. Todd, *Transplant. Rev.*, 20 (1974) 100–129.
(130) O. T. Avery and W. F. Goebel, *J. Exp. Med.*, 50 (1929) 533–550.
(131) F. Karush, *J. Am. Chem. Soc.*, 79 (1957) 3380–3384.
(132) R. U. Lemieux, D. R. Bundle, and D. A. Baker, *J. Am. Chem. Soc.*, 97 (1975) 4076–4083.
(133) R. C. Lancefield, *J. Exp. Med.*, 57 (1933) 571–595.
(134) B. Lindberg and J. Lönngren, *Methods Enzymol.*, 50 (1978) 3–40.
(135) J. H. Pazur and L. S. Forsberg, *Carbohydr. Res.*, 60 (1978) 167–178.

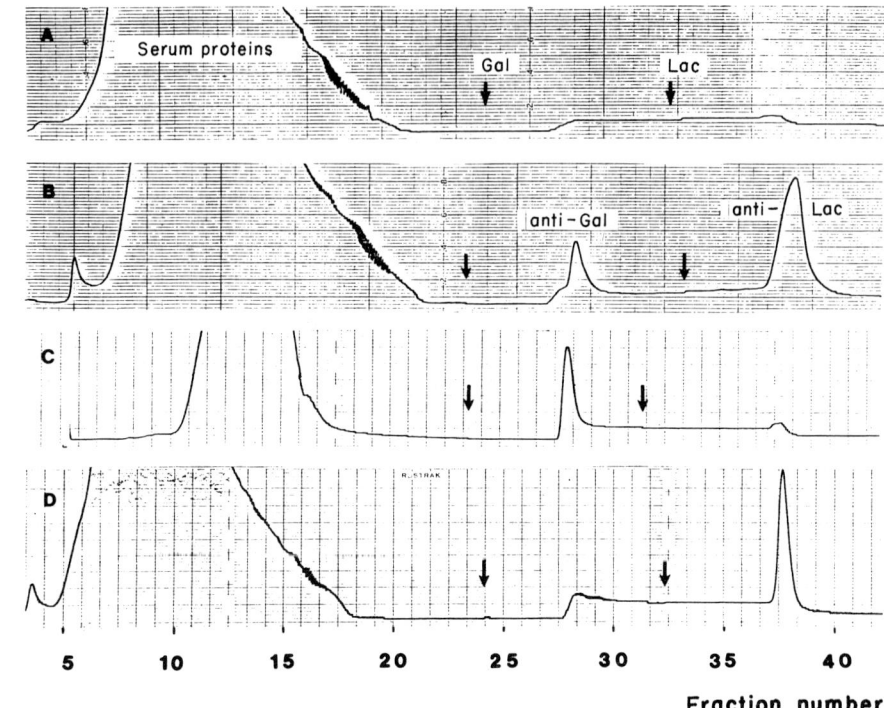

FIG. 2.—Elution Patterns for Samples of Rabbit Antisera from Lactosyl-agarose. [Initial elution with 0.02 M phosphate buffer, pH 7.2, in saline; subsequent elution with 0.5 M D-galactose (first arrow), and with 0.5 M lactose (second arrow). Preimmune serum (A), anti-S. *faecalis* serum (B), anti-Gal-BSA serum (C), and anti-Lac-BSA serum (D) (see Ref. 143).]

was first achieved with antibodies directed against synthetic, carbohydrate antigens.[7] The method has been used for isolating some novel anti-glycosyl antibodies directed against natural carbohydrate antigens.[9] The natural carbohydrate antigen employed in these studies was a glycan of D-glucose and D-galactose having side chains of lactosyl units that function as the immunodeterminant groups.[135,136] This novel antigen is present in the cell wall of some strains of group D *Streptococci*. The affinity adsorbent used for isolating the anti-glycosyl antibodies was lactosylagarose. The data from the affinity chromatography experiments are shown in Fig. 2.

It will be noted in Fig. 2B that u.v.-absorbing components were

(136) J. H. Pazur, J. S. Anderson, and W. W. Karakawa, *J. Biol. Chem.*, 246 (1971) 1793–1798.

eluted with galactose and with lactose solutions. The material that eluted with galactose has been called anti-galactose antibodies, and that which eluted with lactose, anti-lactose antibodies. Some properties of these antibodies are considered later. Fig. 2A shows the elution pattern for the preimmune serum, and 2C and D, the elution pattern for anti-galactose serum albumin (S.A.) serum and for anti-lactose S.A. serum.

Anti-glycosyl antibodies directed against other types of natural carbohydrate antigens have been investigated. Thus, anti-dextran antibodies were observed in the serum of humans immunized with the dextran. Such antibodies have been used for investigating the nature of the antigen–antibody reaction, the size of the combining site of an antibody, and the nature of the immune response.[137–139] Affinity chromatography has not been utilized extensively for the purification of anti-dextran antibodies. Other types of anti-glycosyl antibodies have been purified by this technique, and these are anti-D-glucose,[17] anti-D-glucuronic acid,[23] anti-2-acetamido-2-deoxy-D-glucose,[28,107] anti-2-amino-2-deoxy-D-glucose,[107] anti-isomaltose,[17] anti-cellobiose,[107] anti-chitobiose,[26] anti-lactosyl-N-fucohexaose,[140] and anti-sialyl oligosaccharides.[141] The experimental data for the isolation of anti-lacto-N-fucohexaose antibodies are recorded in Fig. 3.

Purified, anti-glycosyl antibodies that have been examined to date by gel electrophoresis and by electrofocusing techniques have been found to consist of multiple, protein components.[9,142] That the proteins of each set were antibodies directed against the antigen was established by agar diffusion tests. Thus, antibodies directed against a single, structural unit of an antigen can occur in multimolecular forms. The sets of antibodies induced by a single antigen, and combining with the same structural unit of this antigen, have been termed isoantibodies.[142] Not only natural carbohydrate antigens but also synthetic carbohydrate–protein conjugates induce the synthesis of isoantibodies.[143,144] The biosynthesis of anti-glycosyl isoantibodies appears to

(137) E. A. Kabat and D. Berg, *J. Immunol.*, 70 (1953) 514–532.
(138) E. A. Kabat, *J. Am. Chem. Soc.*, 76 (1954) 3709–3713.
(139) E. A. Kabat, G. M. Turino, A. B. Tarrow, and P. H. Maurer, *J. Clin Invest.*, 36 (1957) 1160–1170.
(140) D. A. Zopf, C.-M. Tsai, and V. Ginsburg, *Arch. Biochem. Biophys.*, 185 (1978) 61–71.
(141) D. F. Smith and V. Ginsburg, *J. Biol. Chem.*, 255 (1980) 55–89.
(142) J. H. Pazur and K. L. Dreher, *Biochem. Biophys. Res. Commun.*, 74 (1977) 818–824.
(143) J. H. Pazur, K. L. Dreher, and D. R. Bundle, *ACS Symp. Ser.*, 88 (1979) 102–115.
(144) D. R. Bundle, *Can. J. Biochem.*, 57 (1979) 367–371.

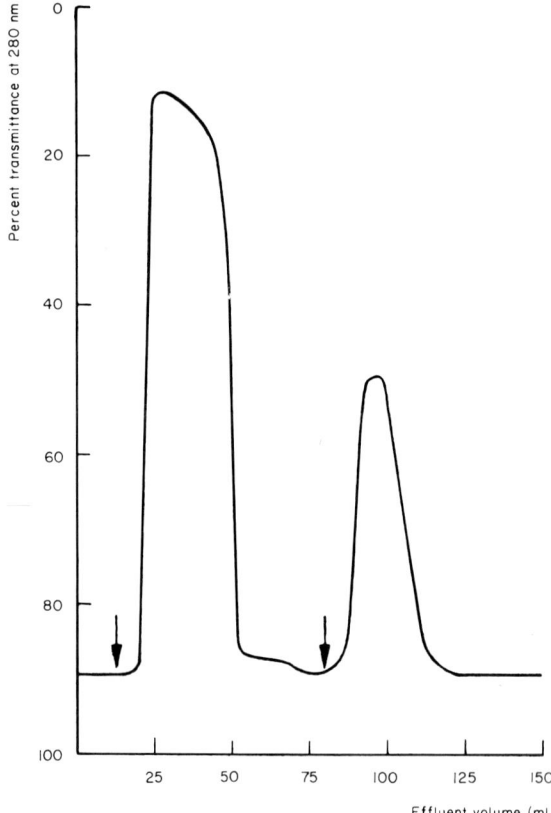

FIG. 3.—Elution Pattern for Goat Anti-lacto-N-difucohexaose Serum from Lacto-N-difucohexaosyl-agarose. [Initial elution with phosphate buffer in saline (first arrow), followed by elution with lacto-N-difucohexaose solution (second arrow) (see Ref. 27).]

be a common feature of the immune response to carbohydrate antigens. The biological significance of the synthesis of isoantibodies still needs to be elucidated.

Anti-glycosyl antibodies isolated from the sera of some animals immunized with bacterial carbohydrate antigens are of a restricted heterogeneity, that is, they occur in relatively few, multimolecular forms.[145,146] Such antibodies have been used in studies on the structure of antibodies,[147] and on the genetic regulation of synthesis of anti-

(145) R. M. Krause, *Ann. N. Y. Acad. Sci.*, 190 (1971) 276–284.
(146) F. W. Chen, A. D. Strosberg, and E. Haber, *J. Immunol.*, 110 (1973) 98–106.
(147) L. Hood, K. Eichmann, H. Lackland, R. M. Krause, and J. J. Ohms, *Nature*, 228 (1970) 1040–1044.

bodies.[148,149] Affinity chromatography-techniques have been used for isolating antibodies of restricted heterogeneity.[150]

Anti-glycosyl antibodies that are directed at the carbohydrate units of glycoproteins have been detected, and such antibodies are of potential value for studying cell-membrane processes. Affinity chromatography has been used to isolate antibodies directed against fetuin, a glycoprotein containing many types of monosaccharide residues.[151] These antibodies were isolated on an adsorbent of fetuin attached to cyanogen bromide-activated agarose. The antibodies were tested by inhibition methods with carbohydrates known to be constituents of fetuin. The tests revealed that these antibodies were inhibited by sialic acid, D-galactose, 2-acetamido-2-deoxy-D-galactose, and D-mannose. These carbohydrate units of the glycoprotein function as immunodeterminant groups of the antigen, and stimulate the immune system to produce anti-glycosyl antibodies.

2. Enzymes

Affinity chromatography on adsorbents having carbohydrate ligands has been used to purify enzymes of the carbohydrase group. In such experiments, carbohydrates that are inhibitors, or substrate analogs, of the enzyme have been used as the ligands on the affinity adsorbents. Because of the similarity in structure of the inhibitor, or analog, and the substrate, the enzyme combines with these compounds, to afford an insoluble complex. The enzyme molecule in this complex is inactive, owing to the nature of the chemical bond between the ligand and the adsorbent, and is retained on the adsorbent. The enzyme is then eluted from the adsorbent by changing the pH of the solvent, or by using the displacement-elution technique. Examples of enzymes that have been purified by chromatography on appropriate, affinity adsorbents are β-D-galactosidase,[22] galactosyltransferase,[80] α-D-galactosidase,[152] 2-acetamido-2-deoxy-β-D-glucosidase,[153] β-D-mannanase,[29]

---

(148) K. Eichmann and T. J. Kindt, *J. Exp. Med.*, 134 (1971) 532–552.
(149) J. A. Gally and G. M. Edelmen, *Annu. Rev. Genet.*, 6 (1972) 1–46.
(150) M. Freedman, H. Yeger, J. Milandre, and M. Slaughter, *Immunochemistry*, 12 (1975) 137–147.
(151) B.-A. Sela, J. L. Wang, and G. M. Edelman, *Proc. Natl. Acad. Sci. USA*, 72 (1975) 1127–1131.
(152) N. Harpaz, H. M. Flowers, and N. Sharon, *Biochim. Biophys. Acta*, 341 (1974) 213–221.
(153) M. E. Rafestin, A. Obrenovitch, A. Oblin, and M. Monsigny, *FEBS Lett.*, 40 (1974) 62–66.

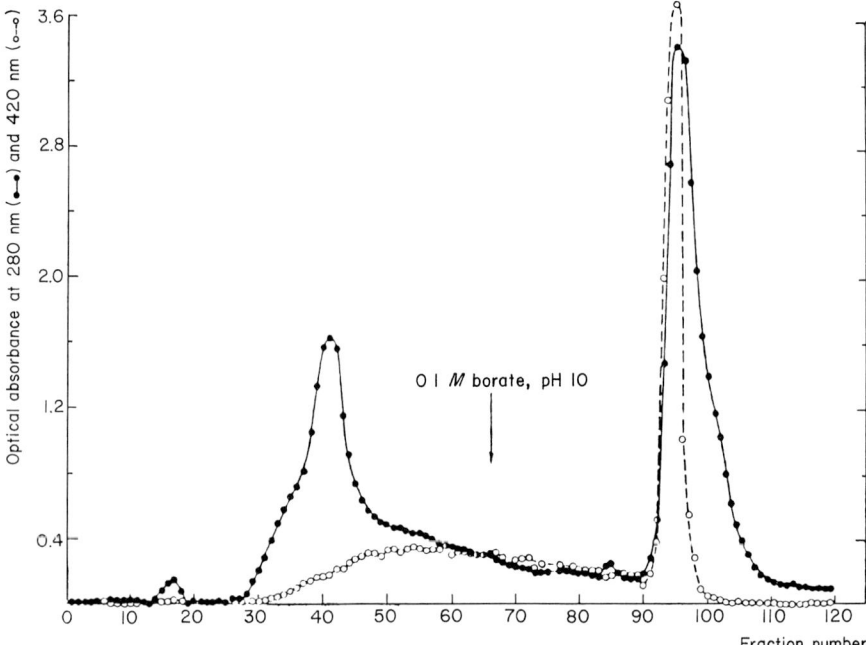

Fig. 4.—Elution Pattern for β-D-Galactosidase from 1-Thio-β-D-galactosyl-agarose. [Initial elution with 0.05 M Tris·HCl buffer, pH 7.5, followed by elution with 0.1 M sodium borate buffer, pH 10 (arrow). Protein was measured at 280 nm (●—●), and enzymic activity at 420 nm (○---○) (see Ref. 22).]

alpha amylase,[14,36] and sialyltransferase.[154] Data for purification[22] of β-D-galactosidase on 1-thio-D-galactosyl-agarose are shown in Fig. 4.

It will be noted in Fig. 4 that the elution of the enzyme was effected with a buffer of high pH. The enzyme obtained from this column was active, and, on analysis by gel electrophoresis, yielded a single protein band.[22] In a later study on β-D-galactosidase from the same source, it was found that the enzyme could be purified on affinity adsorbents that did not contain a D-galactosyl unit.[46] It was proposed that the purification of β-D-galactosidase on such adsorbents is due to nonspecific interaction of the enzyme with the hydrophobic portion of the spacer group of the affinity adsorbent.[46] Additional studies are in order for clarification of the nature of the affinity adsorption occurring during the purification of β-D-galactosidase by this technique.

Affinity adsorbents can be prepared from some compounds that are

(154) J. C. Paulson, W. E. Beranek, and R. L. Hill, *J. Biol. Chem.*, 252 (1977) 2356–2362.

substrates of an enzyme, provided that the enzyme possesses a low specificity for the compound.[14,29] Under these conditions, the adsorption and the elution of the enzyme are effected before the ligand residues are removed from the adsorbent. Affinity adsorbents suitable for purifying enzymes can be prepared by attaching the substrate to support material in a way that lessens the susceptibility of the substrate to the enzyme.[36,37] Enzymes that have been purified by the latter technique include alpha amylase from the pancreas,[14] polygalactosiduronase from *Aspergillus niger*,[36] β-D-mannanase from lucerne,[29] and alpha amylase from *Helix pomatia*.[37]

The purification of the pancreatic alpha amylase was effected on an affinity adsorbent prepared from enzymically degraded starch plus agarose activated with bisoxirane.[14] The fractions from the affinity column were analyzed for protein components by u.v. absorbance, and for alpha amylase activity by incubating the fractions with starch and measuring the increase in reducing sugars. The results are shown in Fig. 5.

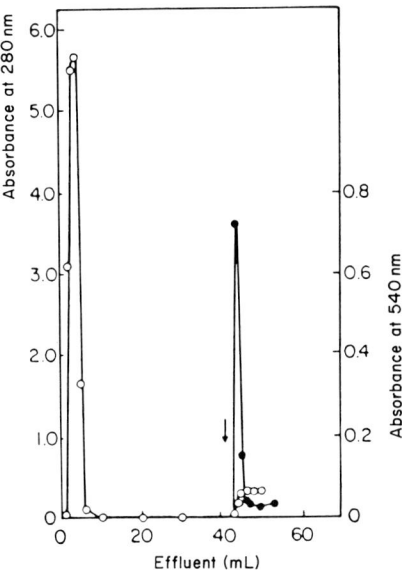

FIG. 5.—Elution Pattern for Bovine Pancreatic Amylase from an Adsorbent of Agarose plus Enzymically Modified Starch. [Initial elution with 0.05 M sodium hydrogencarbonate buffer, pH 8.5, containing M sodium chloride, followed by elution with 1% modified starch in 0.05 M acetate buffer, pH 5.5, and containing M sodium chloride (arrow). Absorbance at 280 nm (○—○), and alpha amylase activity (●—●) (see Ref. 14.]

3. Lectins

Lectins are special types of proteins or glycoproteins that display affinities for carbohydrate residues that may be single residues, or segments of residues, of simple and complex carbohydrates.[155] A well known lectin, isolated many years ago[156] from jack-bean meal, is concanavalin A. This lectin was later purified to homogeneity by affinity chromatography,[157] and the carbohydrate specificity of the lectin was determined. It was found that the lectin interacts with carbohydrate polymers that possess segments of the following residues: α-D-glucopyranosyl, 2-acetamido-2-deoxy-α-D-glucopyranosyl, α-D-mannopyranosyl, or α-D-fructofuranosyl. These segments must occupy terminal positions in the carbohydrate polymer in order that binding may occur.[158]

A large number of lectins having specificity for different carbohydrate residues has been isolated since the advent of affinity chromatography.[159] Lectins are known that possess specificity for single residues, as, for example, D-glucose, L-fucose, D-galactose, 2-acetamido-2-deoxy-D-galactose, and 2-acetamido-2-deoxy-D-glucose, having specificity for segments of carbohydrate residues as already indicated, or for an array of carbohydrate residues. A list of lectins, biological sources, and carbohydrate specificities has been compiled.[159]

In the purification of lectins, special adsorbents have been employed. The adsorbents have been prepared by attaching specific, carbohydrate ligands to activated supports.[41,106,157] An affinity adsorbent prepared from chitobiose and (2-aminoethyl)-Bio-Gel was used to purify the 2-acetamido-2-deoxy-D-glucose-binding lectin from wheat germ. The data for the purification of this lectin are shown in Fig. 6.

Another example of the preparation of a special adsorbent is the attachment of L-fucose to agarose activated with divinyl sulfone. This affinity adsorbent was prepared, and used in the purification of two different types of lectin.[20] The results on the purification of one lectin on this adsorbent are shown in Fig. 7.

Lectins have been used for purifying glycoproteins present in cell membranes. The lectin of wheat germ was immobilized on agarose ac-

---

(155) H. Lis and N. Sharon, *Annu. Rev. Biochem.*, 42 (1973) 541–574.
(156) J. B. Sumner and S. F. Howell, *J. Bacteriol.*, 32 (1936) 227–237.
(157) B. B. L. Agrawal and I. J. Goldstein, *Methods Enzymol.*, 28B (1972) 313–318.
(158) I. J. Goldstein and C. E. Hayes, *Adv. Carbohydr. Chem. Biochem.*, 35 (1978) 127–340.
(159) T. Kristiansen, *Methods Enzymol.*, 34B (1974) 331–341.

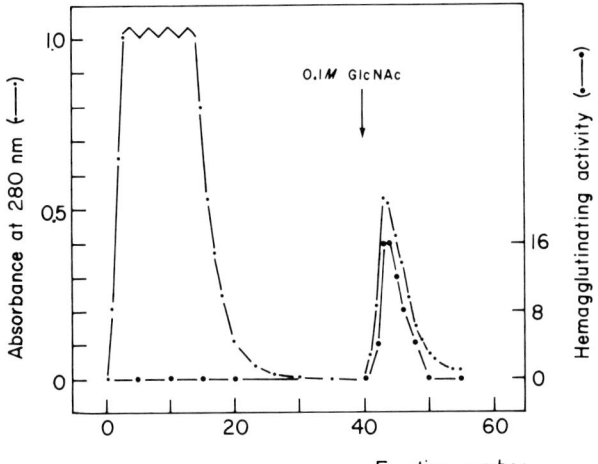

Fig. 6.—Elution Pattern for Wheat-germ Lectin from an Adsorbent Prepared from N,N'-Diacetylchitobiose and (2-Aminoethyl)-Bio-Gel by Reductive Amination. [Initial elution with phosphate buffer in saline, followed by elution with 0.1 M 2-acetamido-2-deoxy-D-glucose in phosphate buffer and saline (arrow) (see Ref. 25).]

tivated with cyanogen bromide. This adsorbent was used to isolate, in a highly purified state, a glycoprotein from erythrocyte membranes.[160] Procedures of this type are applicable to the purification of other glycoproteins if a lectin that binds the glycoprotein is available.

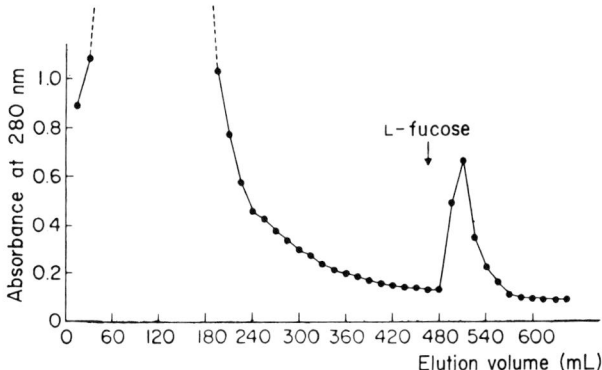

Fig. 7.—Elution Pattern for *Lotus* Extracts from L-Fucosyl-agarose. [Initial elution with 0.01 M sodium phosphate in saline, followed by elution with 0.1 M L-fucose in 0.01 M phosphate buffer in saline (arrow) (see Ref. 20).]

(160) W. L. Adair and S. Kornfeld, *J. Biol. Chem.*, 249 (1974) 4696–4704.

4. Myeloma Proteins

Myeloma proteins occur in animal tumors of various types. The synthesis of these proteins can be induced in some experimental animals by injection of mineral oil, or the implantation of plastic discs.[161] Myeloma proteins were shown to possess antibody activity, with specificity for different types of substances.[161] Some of these proteins were found to combine with such carbohydrates as bacterial, cell-wall polysaccharides, lipopolysaccharides, dextrans, levans, and galactans.[162] Myeloma proteins that combine with galactans or dextrans possess anti-galactan activity,[163,164] or anti-dextran activity,[165–167] and have been studied most extensively.

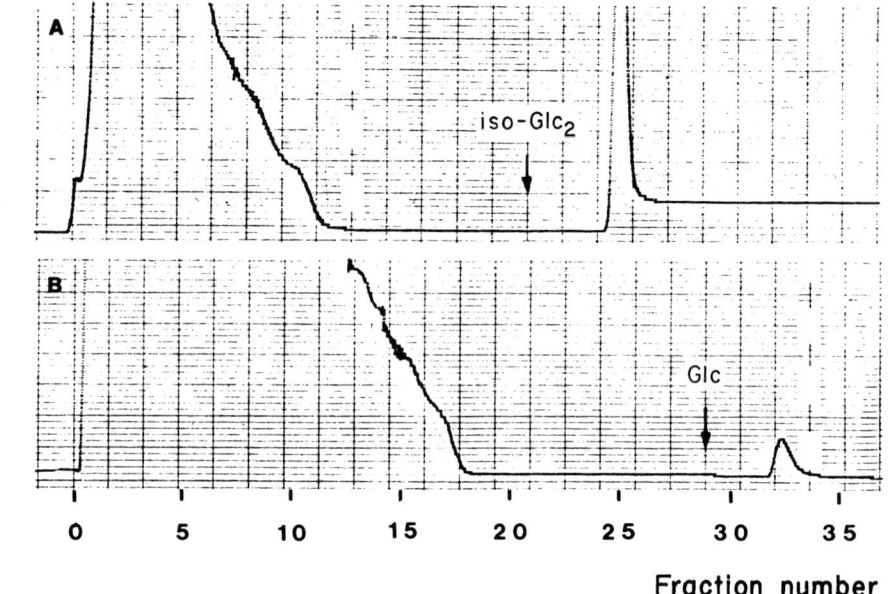

FIG. 8.—Elution Patterns for Mouse-myeloma Proteins and Rabbit Anti-Glc-BSA Serum from Isomaltosyl-agarose. [Elution with 1% isomaltose (IsoGlc$_2$) solution in (A) and with 10% D-glucose (Glc) solution in (B) in 0.02 M phosphate buffer, pH 7.2, in saline (see Ref. 17).]

(161) M. Potter, *Fed. Proc.*, 29 (1970) 85–91.
(162) M. Potter and C. P. J. Glaudemans, *Methods Enzymol.*, 28B (1972) 388–395.
(163) B. N. Manjula, C. P. J. Glaudemans, E. Mushinski, and M. Potter, *Carbohydr. Res.*, 40 (1975) 137–142.
(164) K. Eichmann, G. Uhlenbruck, and B. A. Baldo, *Immunochemistry*, 13 (1976) 1–6.
(165) N. M. Young, I. B. Jocius, and M. A. Leon, *Biochemistry*, 10 (1971) 3457–3460.
(166) J. Cisar, E. A. Kabat, J. Liao, and M. Potter, *J. Exp. Med.*, 139 (1974) 159–179.
(167) J. Cisar, E. A. Kabat, M. M. Dorner, and J. Liao, *J. Exp. Med.*, 142 (1975) 435–459.

Several types of affinity adsorbents having carbohydrate ligands have been utilized to purify anti-carbohydrate, myeloma proteins. One adsorbent was prepared from p-aminophenyl 1-thio-β-D-galactopyranoside and agarose containing a spacer arm.[163] A second adsorbent was prepared from guaran by cross-linking the chains of guaran by means of epichlorohydrin.[164] Both of these adsorbents were used to purify anti-galactan, myeloma proteins. A third adsorbent was prepared from p-aminophenyl β-isomaltoside and agarose activated with cyanogen bromide; this adsorbent was used to isolate W3129 anti-isomaltose myeloma protein.[17] As this myeloma protein combines with isomaltose units, the protein can be adsorbed on an isomaltosyl-agarose column, and eluted with an isomaltose solution. The results showing the elution pattern of these myeloma proteins are reproduced in Fig. 8A; Fig. 8B shows the results of elution of anti-glucose antibodies from the same adsorbent.

Agar-diffusion tests with the anti-isomaltose, myeloma protein and several dextran preparations showed that the protein does indeed combine with a number of dextrans. Results of gel electrophoresis of the sample showed that the purified myeloma protein consisted of several multimolecular forms.[107] This observation is at variance with a report that the W3129 myeloma protein isolated from the same type of serum was homogeneous.[168] Additional studies are needed in order to establish whether the anti-isomaltose, myeloma protein is synthesized as a homogeneous protein or in multimolecular form.

(168) L. G. Bennett and C. P. J. Glaudemans, *Carbohydr. Res.*, 72 (1979) 315–319.

AUTHOR INDEX

Numbers in parentheses are footnote reference numbers and indicate that an author's work is referred to although his name is not cited in the text.

A

Abbott, D., 246
Abdulla, R. F., 150
Abdun-Nur, A. R., 135
Abe, T., 34
Abrams, B. B., 370
Acher, A. J., 158
Acton, E. M., 296
Adair, W. L., 445
Adams, G. A., 338
Adams, M., 360, 363(62)
Adams, R., 420
Adcock, B. G., 118
Agrawal, B. B. L., 408, 419(33), 436(33), 444
Ahluwalia, R., 75
Aida, K., 326
Aizawa, K., 353
Akita, E., 281, 302(22)
Akiya, S., 289
Aksnes, G., 27
Alam, T., 325
Albrecht, H. P., 39
Albriktsen, P., 27
Albutt, A. D., 102
Alexander, B. H., 169
Alfrey, T., Jr., 176, 196(81)
Algeri, A., 395(385), 396
Alhadeff, J. A., 334
Al Janobi, S. A. S., 337(343), 339
Al-Jeboury, F. S., 26, 27(77)
Allen, A. K., 406
Allen, H., 335
Allen, H. J., 406, 415(20), 444(20), 445(20)
Allen, S., 44
Allerton, R., 45
Allison, J. B., 116
Al-Masoudi, N., 77
Al-Masudi, N. A., 95
Al-Radhi, A. K., 166
Al-Timari, U. S., 95
Amen, K.-L., 54
Aminoff, D., 325

Amsden, A., 335
Amundson, C. H., 395(388), 396
Andersen, B., 351, 368
Anderson, B., 316
Anderson, C. B., 107, 152(151)
Anderson, J. D., 281, 342(23)
Anderson, J. S., 323, 438
Anderson, R. C., 80
Anderson, R. G., 323
Anderson, R. L., 328, 329(289,290,291)
Ando, T., 217, 218(26), 429
Andrews, P., 281, 303, 342(23)
Anfinsen, C. B., 406, 411
Angyal, S. J., 16, 18, 21(26a), 22, 36, 46, 75, 83, 84(61,62), 158, 161, 162(35), 223
Ansell, E.G., 56
Anteunis, M., 249
Apostolides, C. L., 184
Arakatsu, Y., 420
Araki, C., 413
Araki, Y., 76, 80, 88, 92(81), 93(81), 94(43,44)
Arashima, S., 327, 330(282)
Arentzen, R., 39
Arita, H., 171, 221, 222(47), 228(47), 229, 264(47), 265(47), 266(47), 267(47,83), 268(37), 269(47,83), 270(47,83), 272(47,83), 274(47), 276(47)
Armstrong, E. F., 218, 263(31), 271(31), 351
Arnold, W. N., 354, 366(115,116), 367, 374(111), 390(38)
Aronson, N. N., Jr., 433
Artamonova, T. S., 118
Ashby, E. C., 122
Ashe, H., 389
Ashwell, G., 420, 433
Asp, L., 218, 221(32), 223(32), 225(32), 228(32), 229(32), 231(32), 235(32), 263(32), 266(32), 267(32), 269(32), 272(32)
Aspinall, G. O., 222, 223(52), 237(52),

449

246(52), 247(52), 258(52), 265(52), 266(52), 273(52), 277(52), 280(10), 281, 290(10), 301, 303(10,25), 342(26), 343(26), 344(25)
Atavin, A. S., 154
Atkinson, P. H., 306
Atlani, P., 81
Avenel, D., 284
Avery, O. T., 312, 419, 433, 437
Avigad, G., 363, 382(240), 383, 386(240), 390, 391(314)
Axelrod, B., 325, 326(271), 389
Axelsson, K., 303
Axen, R., 414, 430(62)
Ayers, J. S., 415, 416(67), 418(67), 436(67)
Ayupova, Z. L., 118

B

Babczinski, P., 371
Babers, F. H., 312, 419
Babouz, A., 99
Bacon, E. E., 355, 366
Bacon, J. S. D., 355, 366
Baddiley, J., 62, 116, 323
Badding, V. G., 122, 135(198), 136(196)
Badger, R. A., 138
Baer, H. H., 281, 343(375), 344(378,379, 380), 345
Baganz, H., 96
Baggett, N., 19, 22, 23(56), 26(56), 27(77)
Bahl, O. P., 325, 337(339), 339, 340(339), 341(339)
Bailey, J. M., 351
Bailey, R. W., 246
Bailey, W. J., 137
Baillie, J., 281, 342(26), 343(26)
Bajaj, K. L., 429
Baker, B. G., 222, 225(50), 234(50), 252(50), 265(50), 266(50), 278(50)
Baker, B. R., 168
Baker, D. A., 433, 437
Baker, Kh. Kh., 363, 366(75)
Balanina, I. V., 88, 95(82)
Balasubramanian, A., 325
Baldo, B. A., 406, 446, 447(164)
Balkau, D., 144, 145(278,279), 146(276)
Ball, D. H., 227, 229(78)
Ballard, J. M., 26, 32

Ballardie, F., 337(345), 339
Ballou, C. E., 26, 59, 60(266), 247, 367, 368, 369, 424
Balza, F., 254(192,193), 255, 256(193), 257(192)
Bandoni, R. J., 392
Barczai-Martos, M., 422
Barker, G. R., 59, 60(270), 121
Barker, R., 254, 255(185), 406, 420, 423, 424(80), 430(80), 441(80)
Barker, S. A., 14, 20, 21(7), 64, 72
Barlow, J. J., 35, 298
Barnett, J. A., 348, 361, 362, 374(5), 377(4), 378, 385(4), 389, 390(4), 391, 392(4), 395(5), 397(4), 398(4,70), 399(397), 400(70,397), 402(5), 403(4)
Barnett, W. E., 48
Baron, A. L., 337(340), 339
Barron, E. S. G., 355
Barry, S., 410, 411(46), 442(46)
Barton, D. H. R., 85, 86
Bartuska, V. J., 44
Basch, A., 247
Baseer, A., 360, 366
Basu, M., 318, 322(228)
Basu, S., 318, 322(228)
Batey, J. F., 167
Bau, A., 375, 391
Bauer, C. H., 321, 331(240,242)
Bauer, Š., 217, 218(28), 263(28), 363
Baues, R. J., 407, 413(25), 433(25), 434(25), 445(25)
Baum, G., 34
Bazouin, A., 136
Beaucage, S. L., 64, 65(304), 66(304,309), 67(309)
Béchet, J., 398
Bedell, S. F., 80
Behrman, E. J., 295, 296, 340(80), 343(82)
Beiser, S. M., 392, 419, 421
Bekesi, J. G., 326
Bel-Ayche, J., 247
Bell, D. J., 56, 57
Bell, R. G., 389
Beluche, I., 401
BeMiller, J. N., 44, 247, 248
Bender, M. L., 34
Bengtsson, S., 413
Benkovic, S. J., 352

AUTHOR INDEX

Bennett, G., 320
Bennett, L. G., 204, 447
Beránek, J., 37
Beranek, W. E., 442
Beratis, N. G., 333
Berg, D., 439
Berger, A., 202
Bergelson, L. D., 34
Berggren, B., 366
Berlin, K. D., 138
Bernasconi, C., 95
Bernstein, M. D., 413, 433(54)
Berrang, B., 146
Berrgård, I., 330
Berthelot, M. P. E., 355
Bertho, A., 239, 243(114), 275(114)
Bertolini, M., 35
Best, J. B., 366
Beteta, P., 369, 371(143)
Bethell, G. S., 415, 416(67), 418(67), 436(67)
Beveridge, R. J., 16, 18, 21(26a), 93, 161, 162(35)
Beyer, E., 96
Beyer, E. M., 326
Beyerinck, M. W., 394
Beyers, T., 319
Bezborodov, A. M., 402
Bhacca, N. S., 250, 254
Bhatia, I. S., 429
Bhatia, M. S., 429
Bhattacharjee, S. S., 79, 124, 125(214), 126(214), 127(214,215), 128, 130(214), 132(214)
Bhattacharya, A., 159
Bianchi, G., 139
Bickelhaupt, F., 154
Biermann, L., 395
Bilik, V., 217, 218(28), 263(28)
Binkley, R. W., 285
Binkley, W. W., 250, 261, 285
Birzgalis, R., 366
Bishop, C. T., 437
Björndal, H., 303
Black, I. M. A., 217
Black, W. A. P., 282
Blakely, J. A., 396
Blank, F., 437
Blessmann, M., 22
Bloch, R., 420

Block, A., 239
Bloemers, H. P. J., 390
Blomberg, C., 154
Bluhm, T., 197, 198(125)
Blumberg, S., 340(368), 345
Blumbergs, P., 338(358), 339
Bobbitt, J. M., 246
Bochkov, A. F., 172
Bock, K., 114, 115, 117, 120
Bock, R. M., 392, 393(347)
Boegman, R. J., 413
Boer, P., 401
Boger, J., 33
Bognár, R., 119
Bohrer, J. J., 176, 196(81)
Bolker, H. I., 136
Bolton, C. H., 426
Bonner, T. G., 22, 25, 26, 27(80), 44
Bonner, W. A., 284
Bôrbâs, A., 29
Borch, R. F., 413, 433(54)
Bornstein, J., 80
Borrone, C., 332
Bosmann, H. B., 320, 325, 326(264)
Boss, R., 50
Bouchu, M. N., 95
Bouquelet, S., 327, 330(281), 333(281)
Bourne, E. J., 14, 20, 21(7), 22, 25, 26, 27(80), 29, 36(96), 44, 72
Bourrillon, R., 330
Boyd, K. S., 374
Boznyi, Ya. V., 172
Brady, R. F., Jr., 14, 16, 17, 24(19), 71
Brady, R. O., 310, 318(180), 332
Brailovsky, C. A., 265, 278(209)
Brandstetter, H. H., 67
Brant, D. A., 216
Braun, E., 164
Brauns, D. H., 265, 271(206)
Bredereck, H., 47, 159, 180(26), 182(26), 184
Breitenbach, M., 406, 421(4)
Brennan, P. J., 310
Brewer, C. F., 265, 271(207), 351
Brewer, C. L., 15
Brewster, J. H., 19
Brigl, P., 35, 171, 224
Brimacombe, J. S., 23, 64, 165, 166, 168
Brinkmeyer, R. S., 150
Brockhaus, M., 63

Brocklehurst, R., 383, 384(239)
Brodde, O. E., 205, 206
Brodelius, P., 406
Brown, A. S., 331
Brown, D. M., 36, 59
Brown, H. C., 19, 137
Brown, J. H., Jr., 136
Brown, J. R., 58
Brown, R. K., 122, 123, 124(200,201,203), 127(201,206), 133, 136(208)
Brubaker, R. R., 307
Bruce, G. T., 22
Bruzzi, A., 285
Buchanan, J. G., 27, 79, 116, 337(343), 339
Buchferer, B., 387, 388(275)
Buck, C. A., 304, 305, 306
Buck, K. W., 21, 22, 23(46,56), 26(56), 225
Buckley, H., 386
Budd, J. A., 385
Buendia, J., 133
Bukhari, M. A., 22
Buliř, J., 337(336), 339
Bullock, C., 166, 167
Buncel, E., 232, 233(91)
Bundle, D. R., 433, 437, 438(143), 439
Bunton, C. A., 351
Burdon, J., 25
Burger, M., 355, 366, 389
Burger, M. M., 420
Burgers, P. M. J., 42
Burke, G. C., 392, 419
Burns, R. F., 247
Burns, R. L., 329
Burr, B., 406, 407(7), 438(7)
Burton, H., 45
Butler, K., 29
Byrne, K. J., 129, 136

C

Cabeça-Silva, C., 402
Cabezas, J. A., 324, 325(258), 326(269)
Cadenas, R. A., 31, 32(111,112), 225, 227(68), 240(68), 249, 250(167,168), 263(68), 265, 266(68), 276(76,205)
Calas, R., 136
Calinaud, P., 134, 136(240), 154(240)

Campbell, D. H., 412
Cano, F. R., 367, 370, 371(146)
Cantz, M., 332, 333
Čapek, K., 29, 30, 226
Capon, B., 15, 19, 248, 283, 337(345), 339, 421
Caputto, R., 395
Carr, G., 384
Carrancedo, M. G., 396
Carter, H. E., 60
Carter, S. R., 362
Cartledge, T. G., 375
Caruthers, M. H., 41
Carvalho, J. DaS., 179
Castillo, F. J., 395(389), 396
Casu, B., 225, 249, 250(165), 254, 255(187), 266(72)
Catley, B. J., 390
Ceppellini, R., 318, 319(225)
Cepure, A., 424, 430
Černý, M., 115, 158, 161(19), 162(19), 163, 164(19), 211(19)
Cetorelli, J. J., 305
Chacón-Fuertes, M. E., 27, 226
Chaiken, I. M., 405, 406
Chakravarty, P., 217, 225(20), 228(20), 237(20), 263(20), 268(20), 272(20)
Chan, J. Y., 304, 328(147)
Chandra, R. K., 333
Chaney, A., 58
Chanzy, H., 201
Chapleur, Y., 106
Chari, V. M., 35
Charpovskii, Yu. A., 136
Chatterjee, S. K., 322
Chen, C. H., 98
Chen, F. W., 440
Chênevert, R., 82
Chernetsky, V. N., 172
Chernousova, N. N., 124
Chester, M. A., 319, 325, 326
Chiang, C.-K., 423
Chiba, S., 386, 387, 388, 389
Chien, J.-L., 318, 322(228)
Chirkunova, S. K., 116
Chittipantabu, G., 334
Chiura, H., 368
Chizhov, O. S., 258, 261
Chonan, M., 103
Chopra, C., 265, 278(209)
Chou, T.-H., 321, 322, 331(239)

Christol, H., 99
Chu, F. K., 369
Chu, S. S. C., 216, 220(14), 225(14)
Chuchvalec, P., 226
Cirillo, V. P., 380
Cisar, J., 203, 446
Clasper, P., 133
Cleare, P. J. V., 26, 27(80)
Clegg, W., 77
Cleveland, J. H., 118
Clode, D. M., 14, 15(7c), 19(7c), 26(7c), 72, 142, 143(275), 144(275)
Cocker, D., 337(345), 339
Coelho, P. A., 284, 294(35b)
Coffey, J. W., 326
Cohen, R. E., 368
Cohen, S., 335
Cohn, E., 394
Coleman, B. D., 179
Coleman, G. H., 161, 163(30)
Coleman, R. A., 137
Coligan, J. E., 437
Collins, P. M., 16, 89, 90(84,85), 91(85), 92(84,85), 93(84), 421
Colonge, J., 133
Colson, P., 200, 231, 232(90), 254(90), 257(90), 258(90), 269(90)
Compton, J., 168
Conant, R., 51, 52(216)
Conchie, J., 342(373), 343(373), 345
Conway, E. J., 352, 364(25)
Cook, A. F., 41, 43
Cook, A. H., 387, 388
Cook, W. H., 61
Copeland, C., 15, 74
Corbett, W. M., 223
Corcoran, R. J., 33
Corey, E. J., 50, 64
Cornhill, W. J., 282
Cort, L. A., 96, 118(105)
Cortat, M., 371
Cosgrave, D. J., 59
Coste, J., 99
Costisella, B., 98
Côté, R. H., 343(374), 345
Cottier, L., 95
Cottrell, A. G., 232, 233(91)
Cottrell, W. J., 303
Courtois, J. E., 59, 391
Coutsogeorgopoulos, C., 239
Cox, D. D., 285

Cox, R. P., 329
Coxon, B., 254
Craciunescu, E., 87
Croxall, W. J., 76
Crumpton, M. J., 413
Csűrös, Z., 45, 223
Cuatrecasas, P., 406, 410, 411, 412, 414, 420(22), 421(22), 431, 434(47,48), 441(22), 442(22)
Cunningham, J., 49, 50(204), 51
Curtis, E. J. C., 22
Custardoy, V., 23
Custers, M. T. J., 397
Cyr, N., 254(192,193), 255, 256(193), 257(192)
Czégény, I., 133

D

Dahlhoff, W. V., 53, 54, 55(225,228)
Dahms, A. S., 328, 329(289,290,291)
Daignault, R. A., 122, 135(198)
Dainty, J., 352
Danilov, S. N., 88, 95(82)
Dankert, M., 323
Danther, F. A., 338(358), 339
Das, B. C., 63
Dasgupta, F., 80
David, S., 15
Davies, A., 364, 367, 368(117)
Davies, D. A. L., 308
Davies, D. B., 303
Davies, R., 395, 396(371)
Davis, H. A., 122, 123(204), 136(208)
Davis, P. B., 333
Dawson, G., 289, 318, 324, 330, 332(300)
Dea, I. C. M., 218, 235, 263(36)
Deák, T., 381
deBales, S. A., 395(389), 396
deBelder, A. N., 14, 15(4,5), 71, 73(2,3), 237
DeBie, M. J. A., 249, 250(162), 254(163)
deBoer, T. J., 97, 99
Debost, J. L., 77
DeBruyn, A., 249
Defaye, J., 63, 158
Deferrari, J. O., 31, 32(111,112), 221, 225, 227(68), 239, 240(68,71), 249,

250(167,168), 263(68), 265, 266(68), 276(76,205), 278(208)
Degn, H., 351
Dejter-Juszynski, M., 288, 291, 292(66), 296(66), 298(66,83,84), 309, 316(54), 341(66,84), 342(66,83), 343(83,84,93)
Dekker, C. A., 59, 60(269)
deKroon, R. A., 382(238), 383, 384(238)
de la Fuente, G., 355, 363(50,51), 375(50,51), 385(51), 395(51)
Delay, D., 297, 300(87), 341(87), 343(87)
de Leeuw, H. P. M., 38
Delitheos, A. K., 204
Della Cella, J., 332
Dellweg, H., 389
Delmotte, F. M., 407, 439(26)
Demerseman, P., 81
DeMicheli, C., 139
Demis, D. J., 355
Denamur, R., 317
Derkanosov, N. I., 390
de Rooij, J. F., M., 38, 69
Desai, P. R., 312, 339(197), 342(197), 343(197)
Descotes, G., 95
Desjobert, A., 59
Deslongchamps, P., 81, 82
Desnick, R. J., 332
Dettwiler, J., 389
Deulofeu, V., 31, 239, 281, 302(21), 338(21)
Dewar, E. T., 282
Dewar, J., 56
Dewey, R. S., 34
Dey, P. M., 375
DiCesare, P., 78, 117
Dick, T., 403
Dick, W. E., Jr., 18, 217, 218(30), 222, 225(50), 234(50), 243(30), 252(50), 263(30), 265(50), 266(50), 276(30), 278(50)
Dickson, L. R., 395(381), 396
Dickson, R. C., 395(381), 396, 397, 403
Diehl, H. W., 22, 121
Dienert, F., 396
Dimler, R. J., 158, 169
Dimpfl, W. L., 216
Diner, U. E., 122, 123(201,205,206,207), 127(201,206)
Dintzis, H. M., 417, 436(69)

Dische, Z., 289, 301, 320(118), 409
Distelmaier, A., 24, 82, 338(348), 339, 342(348), 343(348)
Divies, C., 380
Diwadkar, A. B., 74
Dodd, G. H., 67
Dodyk, F., 372
Domaschke, L., 96
Domracheva, N. I., 334
Dorfman, A., 332
Dorman, D. E., 254, 257(176)
Dorner, M. M., 203, 446
Dougherty, R. C., 261
Douglas, H. C., 378
Douglas, L. J., 323
Doukas, H. M., 122
Downey, M., 352, 364(25)
Drach, G. W., 312
Dreher, K. I., 406, 407(9,17), 409(17), 429(17), 430(17), 437, 438(9,143), 439(9,17,23), 446(17), 447(17)
Drewry, D. T., 308
Driguez, H., 299, 300(100), 344(100, 104), 420
Drolshagen, G., 30
Drummond, G. I., 61, 62(279)
Drummond, P. E., 80
Dubois, M., 409
Duchet, D., 296, 343(86)
Duerksen, J. D., 353, 392(32), 393(32, 335)
Duncan, W. P., 118
Dunstan, D., 239, 275(113)
Durand, P., 332
Durette, P. L., 217, 220, 225, 229(66), 230(45,66), 231(24,45), 232(24), 233(66), 235(45), 236(45,66), 237(66), 241(45), 246(66), 247(66), 249(24,45, 66), 251(24,45,66), 252(45,66), 253(45,66), 258(24), 260(24), 264(45, 66), 266(45,66), 267(66), 268(45,66), 269(66), 270(24,66), 272(66), 275(45, 66), 276(45), 278(66)
Durham, L. J., 114
Durst, H. D., 413, 433(54)
Dusseau, C. H. V., 99
Dutcher, J. D., 239
Dutta, T. K., 331
Dutton, G. G. S., 220, 222(46), 228(46), 229(46), 231(46), 235(46), 241(46),

244(46), 246, 263(143), 264(46), 268(46), 269(46), 272(46), 274(46), 275(46), 277(143)
Duxbury, J. M., 225
Dworschack, R. G., 366

E

Eagon, R. G., 327
Earnshaw, P., 384
Eastham, A. M., 174
Eaton, N. R., 386, 287(280), 388, 389, 390
Eby, R., 47, 163, 170, 182(102), 183, 187(102), 188(102), 196(102), 199(102), 200(102)
Eccleshall, T. R., 387, 388(275)
Eckstein, F., 43
Eddy, A. A., 381, 383, 384(239)
Edelmen, G. M., 441
Edgar, A. R., 79, 337(343), 339
Edmonds, C. G., 136
Edward, J. T., 283
Edwards, R. G., 233, 249(95), 251(95), 252(95), 258(95), 269(95), 270(95)
Egan, J. V., 303
Egge, H., 309, 310(177)
Egyed, J., 81
Eichmann, K., 407, 439(28), 440, 441, 446, 447(164)
Eilat, D., 406
Eitelman, S. J., 147, 149(286)
Eklind, K., 102, 286, 338(47)
Elad, D., 80, 87, 94(42)
Elbein, A. D., 391
Elderfield, R. C., 280, 293, 338(72)
Eliel, E. L., 122, 123, 129, 135(198), 136(196)
El Khadem, H. S., 240
El'Kin, Yu. N., 301
Ellias, L., 353, 387, 388(33)
Ellis, J. E., 138
Ellis, P. D., 255
Elorza, M. V., 372
Elrick, D. E., 58
El Sawi, M. M., 115
Elson, L. A., 409
Elvers, J., 106, 107(146)
Elvin, P. A., 367

Emoto, S., 15, 18, 154
Engberts, J. B. F. N., 97
Ennor, K. S., 56, 57(243), 58(243,244)
Enterman, W., 283
Entian, K.-D., 402
Entwistle, D. W., 64, 65(304), 66(304)
Epstein, R., 392, 393(347)
Erbing, C., 301
Erhart, K. P., 321, 331(242)
Ernback, S., 414
Esclamadou, C., 136
Euran, V., 335
Evans, M. E., 19, 24(38), 27, 77, 78, 79, 83, 84(61,62), 233, 238
Excoffier, G., 254
Eylar, E. H., 320

F

Failla, D. L., 96, 98(108)
Falconer, E. L., 57
Fanton, E., 74(16), 75, 76(16), 238, 273(108)
Farkas, I., 119
Farkas-Szabó, I., 119
Fauconneau, G., 317
Faure, A., 95
Feather, M. S., 30, 224, 248, 266(60)
Federoff, H. J., 387, 388(275)
Feeny, J., 254
Feier, H., 295, 340(79), 341(79), 421
Fellig, J., 367
Feniksova, R. V., 395
Fennessey, P., 323
Ferrero, I. P., 395, 396
Ferrier, R. J., 14, 15(7b), 16, 53, 54
Fiddler, M. B., 332
Fidler, I. J., 335
Fiedler, R. J., 437
Field, F. H., 262
Filbert, A. M., 417, 418(70)
Fineman, M., 189
Fink, A. L., 33
Fink, G. R., 347
Finlayson, A. J., 127, 303
Finnan, J. L., 48
Finne, J., 290, 313
Fischer, E., 71, 218, 263(31), 271(31), 375, 378(186), 391, 394

Fischer, E. H., 352, 367, 394(201)
Fischer, N. M., 254
Fischer, J. C., 154
Fishman, P. H., 310, 318(180), 351
Flematti, S. M., 221
Fleming, B., 136
Fleming, L. W., 392
Fleming, S. E., 360, 364(64), 365(64)
Fletcher, H. G., Jr., 22, 34, 45, 73, 121, 161, 165, 221, 238
Florent, J.-C., 104, 105(144), 106(144), 107(147), 285, 342(44)
Flowers, H. M., 284, 288, 291, 292(66), 293, 295, 296(66), 298(66,75,81,83, 84), 299(75), 304, 305, 309, 316(54), 318(149), 324(150), 340(81), 341(66, 84), 342(66,81,83,92), 343(83,84,89, 92,93), 344(81,89,92), 441
Fogh, A., 337(337), 339
Folk, K -E., 309
Fontaine, J., 43
Fontaine, T. D., 122
Fontana, A., 35
Ford, W. C. L., 214
Fornstedt, N., 406, 415(21), 432(21)
Forrest, T. P., 98
Forrester, P. F., 340(365), 345
Forsberg, C. W., 370
Forsberg, L. S., 406, 407(9,17), 409(17), 426, 429(17), 430(17), 437, 438(9, 135), 439(9,17,107), 446(17), 447(17, 107)
Fort, G., 56
Foster, A. B., 19, 20(41), 21, 22, 23(46, 56), 26(56), 27(77), 225
Foster, D. W., 318
Fourmet, B., 327, 330(281), 333(281)
Fox, J. J., 29, 36, 59, 153
Fox, T. G., 179
Frainnet, E., 136
Franke, F., 67
Fraser, R. N., 303
Fraser-Reid, B., 45, 80, 94(45)
Fréchet, J. M. J., 79, 182(109), 183, 199(109)
Freed, J. H., 334
Freedman, M., 441
Freedman, S. O., 437
Fréhel, D., 81, 82
Freiberg, J., 98

French, D., 386, 429
Freudenberg, K., 164, 285, 338(42), 340(367), 345
Fridland, S. V., 116
Friebolin, H., 250
Friedrich, E. C., 107, 152(151)
Friis, J., 355, 369(54), 375(55)
Frisch, K. C., 174
Frohwein, Y. Z., 28, 389
Frush, H. L., 426
Frye, G. H., 16
Fuchs, E. F., 63
Fügedi, P., 130, 131(227), 132, 133
Fuehr, J., 331
Fuhrer, J.-P., 306
Fujikawa, T., 100, 338(355), 339
Fukatanai, G., 331
Fukui, S., 246
Fukui, T , 389
Furukawa, J., 174
Furukawa, K., 325

G

Gabriel, O., 287
Gacto, M., 309, 310(179)
Gaertner, K., 48
Gagnaire, D. Y., 35, 254, 255(184), 257 (180)
Gakhokidze, A. M., 242, 266(124), 271(124), 274(124), 278(124)
Galcheva-Gargova, Z. I., 373
Gallili, G., 370
Gallo, G. G., 225, 249, 250(165), 266(72)
Gally, J. A., 441
Galmarini, O. L., 281, 302(21), 338(21)
Galzy, P., 401, 402
Gandolfi, R., 139
Ganguly, A. K., 262, 263(200)
Garbarino, J. A., 86
Garcia, Z., 150
Gardas, A., 313 , 315
Gardiner, J. G., 291, 293(65), 341(65), 342(65)
Gardner, D. A., 318, 322(228), 383, 384 (239)
Garegg, P. J., 102, 286, 292, 301, 330(67), 338(47), 342(67)
Garrett, C. S., 17
Garrison, R. G., 374

Gascón, S., 366, 368(102), 369, 370(102, 144), 371(143), 372, 375, 402
Gasvoda, B., 355
Gaudry, M., 97
Gauhe, A., 343(375), 344(378,379,380, 381), 345
Gaylord, N. G., 121
Gehler, J., 332, 333
Gehlhoff, M., 327
Gehrke, C. W., 331
Gelas, J., 14, 73, 74(16,17), 75(10,13), 76(16), 77, 81, 97, 118, 134, 136(240), 154(240), 238, 273(108)
Gelman, N. S., 389
Gemal, A. L., 81
Genghof, D. S., 265, 271(207), 351
Gent, P. A., 50, 51(212), 52(216,217)
Gercke, W., 321, 331(242)
Gerhardt, P., 364
Gershkova, R. P., 301
Gesner, B. M., 329
Ghalambor, M. A., 327
Gharia, M. M., 80
Gi, C. T., 244
Gibbons, M. N., 289
Gigg, J., 49, 50(205)
Gigg, R., 49, 50(201,204,205), 51(207, 212), 52(207,216,217), 130, 137
Gillam, I. C., 224, 266(65)
Gilles, K. A., 409
Gilliland, R. B., 385
Ginsburg, V., 306, 317, 318, 319, 324, 407, 420(27), 426(27), 439, 440(27)
Ginzburg, B. Z., 352
Gladieux, N., 104, 105(144), 106(144)
Glantz, M. D., 395
Glaser, L., 318
Glasgow, L. R., 320
Glaudemans, C. P. J., 35, 161, 204, 221, 246, 446, 447(163)
Glavis, F. J., 76
Glick, M. C., 304, 305, 306, 318(149), 333
Goebel, C. V., 216
Goebel, W. F., 312, 419, 433, 437
Goering, H. L., 34
Görts, C. P. M., 382(242), 383
Goethals, E., 174
Gold, P., 437
Goldberg, I. H., 20, 47(45)
Goldberg, M. W., 281, 302(18), 337(18)
Goldfinger, G., 177
Golding, B. T., 67
Goldstein, A., 355
Goldstein, I. J., 159, 179(22), 197, 201(120), 202(120), 209(120), 406, 407(8), 408, 419(33), 436(33), 444
Golova, O. P., 159, 164, 180(27), 182(27)
Gonzalez, A., 23
Goodman, L., 44, 168
Goosen, A., 99
Goppell, S. F., 437
Gorin, D. A. I., 216
Gorin, P. A. J., 79, 124, 125(214), 126(214), 127(214,215), 128, 130(214), 132(214), 133, 135(234), 254, 255(189,190), 303, 343(376), 345
Gorman, J., 383, 384(235), 387(282), 388, 389(235)
Gorniak, H., 313
Gotthammar, B., 102, 286, 338(47)
Gottschalk, A., 351, 360, 379, 385, 386(68)
Goussault, Y., 330
Goutarel, R., 45
Grafen, P., 121
Grams, G. W., 40, 41(165), 43
Grant, S., 290
Grappel, S. F., 202
Graves, D. J., 408
Gray, G. R., 146, 407, 409, 413(25), 433(25), 434(25), 444(41), 445(25)
Gray, G. W., 308
Green, J. R., 323, 324
Green, J. W., 246
Greenblatt, J., 407, 439(28)
Greener, A., 389
Gregoire, R. J., 40, 41(167)
Grehn, M., 363
Greiling, H., 333
Greiner, W., 47
Gress, M. E., 216
Griebel, B., 218
Grieco, P. A., 80
Griffin, B. E., 38, 39
Griffin, T., 410, 411(46), 442(46)
Grimmer, G., 302
Grollman, A. P., 318
Grollman, E. P., 319

Gros, E. G., 31, 221, 239, 338(354), 339
Gross, B., 78, 117
Gross, H., 98, 119
Grossmann, M. K., 373
GrootWassink, J. W., 360, 364(64), 365(64)
Grubb, S. D., 136
Gruezo, F., 314, 316
Günther, A., 280
Guerrera, J., 231, 258(86)
Guha, A. K., 333
Guignard, H., 133
Guilliermond, A., 366
Guiraud, J. P., 401
Guizard, C., 201
Guntz, G., 317
Gupta, B. D., 331
Gussin, A. E. S., 406
Guthrie, R. D., 24, 67, 158, 248

H

Haaland, E., 303
Haber, E., 440
Hackel, R. A., 370, 372, 373(162)
Haddad, A., 320
Haga, M., 26, 103, 222, 225(51), 226(51), 228(67), 229(82), 235(67), 237(82), 244(67), 265(51), 266(67), 267(67,82), 268(82), 269(67,82), 271(67), 272(51, 67,82), 273(51,82), 274(67,82), 276(67), 278(67,82)
Hagopian, A., 320
Haines, A. H., 13, 19, 20(41), 56(1), 72, 113(8), 115(8), 156(8), 225, 226(74), 231(74), 233(74), 284, 291(37), 292(37)
Hajduković, G., 297, 300(88), 341(88), 343(88)
Hakomori, S.-I., 280, 281(4), 292, 308, 309(4), 310, 313, 316(199), 331(172)
Hall, G. E., 59
Hall, R. H., 147, 149(286)
Hallgren, P., 330
Halvorson, H. O., 353, 387, 388(33,34), 380, 382(210,211,241), 383(210), 384(210,211),387, 388, 389, 392(32), 393(32,335,347), 394(338) 395(344), 396
Ham, G. E., 176, 196(82)

Hamer, G. K., 254(192,193), 255, 256(193), 257(192)
Hamilton, J. K., 409
Hanahoe, T. H. P., 204
Hancock, W. S., 415, 416(67), 418(67), 436(67)
Hanessian, S., 64, 96, 97, 98, 99(116), 100(116), 101(132), 102, 103(133), 108, 109(152), 110(152), 117, 280, 286, 287, 337(49)
Hann, R. M., 24, 25, 169, 340(361), 341(361), 344
Hansen, H., 331
Haq, S., 159, 207
Harangi, J., 133
Hardy, F. E., 116
Harnden, M. R., 64
Harold, F. M., 381
Harpaz, N., 309, 441
Harris, G., 381, 383(229)
Harris, J. F., 248
Harris, P. J., 303, 323(130)
Harrison, J. M., 61
Hartgerink, J. W., 97
Hartlief, R., 389
Hasaka, O., 301, 309(117)
Hasamatsu, T., 243, 278(126)
Hasegawa, A., 73, 238
Hasegawa, M., 210
Hasenclever, H. F., 202
Hashimoto, Y., 352
Haskin, M. A., 323
Haskins, W. T., 24
Hassid, W. Z., 318
Hatton, L. R., 16
Haug, A. 280(12), 281
Hautera, P., 383
Haverkamp, J., 249, 254(163)
Haworth, W. N., 215, 223(6), 242, 265(6), 278(123), 362
Hawthorne, D. C., 372, 383, 389
Hay, G. W., 33
Hayes, C. E., 406, 407(8), 444
Hayes, D. H., 36
Hayes, L. J., 422
Haylock, C. R., 117
Haynes, R. H., 387(281), 388
Hayward, L. D., 57, 58
Hearn, M. T. W., 415, 416(67), 418(67), 436(67)

AUTHOR INDEX

Heath, E. C., 295, 316, 327, 341(77), 343(77)
Heathcock, C. H., 138
Hederich, V., 107
Hedgley, E. J., 15
Hegedüs, H., 338(353), 339
Hehre, E. J., 265, 271(207), 351
Heidelberger, M., 311, 437
Heintze, S., 354
Heinze, B., 394
Helferich, B., 44, 47(184), 217, 218, 219, 220(37), 223, 229(37), 231(37), 264(37), 226(37,57), 268(37), 270(37), 274(37)
Hellerqvist, C. G., 249, 250(161), 307
Hemsworth, B. A., 325, 326(264)
Henderson, M., 321, 331(241)
Hendricks, K. B., 299, 330(97), 343(97)
Hengstenberg, W., 63
Henkart, P. A., 437
Héran, N., 35
Herman, A., 392, 394(338)
Herrero, P., 402
Hers, H. G., 332
Hersh, L. S., 418
Hess, B., 351, 391
Hess, G. P., 413, 420(53), 434(53)
Hess, K., 164, 165
Hestrin, S., 354, 378
Heumann, K. E., 165
Hewson, K., 39
Heyns, K., 16, 242, 337(340), 339
Heyraud, A., 254(194), 255, 256(194)
Hickinbottom, W. J., 158
Hickman, J., 433
Hicks, D. R., 80, 94(45)
Hiegel, G. A., 138
Hijioka, Y., 76
Hildesheim, J., 340(368), 345
Hill, A., 63
Hill, R. L., 319, 320, 406, 420, 423, 424(80), 430(80), 441(80), 442
Hines, J. N., 139
Hirao, I., 40
Hirao, N., 154
Hirasaka, Y., 219, 223(40), 246, 264(40), 266(40,55), 272(55), 277(40,55)
Hirschhorn, K., 333
Hirst, E. L., 242, 278(123), 362
Hishikawa, H., 135

Hixon, R. M., 116
Hjertén, S., 413
Ho, T.-L., 50
Hochester, R. M., 246
Hockett, R.C., 293, 340(362), 341(71), 342(362), 344
Hodge, J. E., 30, 217, 218(30), 222, 225(50), 234(50), 243(30), 245, 252(50), 263(30), 265(50), 266(50), 276(30), 278(50)
Hönig, W., 387(284), 388
Hofer, D. C., 255
Hoffmann-Ostenhof, O., 406, 421(4)
Hoffmeyer, L., 108, 113(157)
Hofmann, E., 386, 392, 395(345,384), 396
Hofstad, T., 331
Holbein, B. E., 370, 371
Hollenberg, D. H., 29
Holley, R. A., 371
Holtzman, N. A., 433
Holzer, H., 403
Homberg, K., 117
Homsma, T., 154
Honeyman, J., 19, 55, 56(232,233), 57(232,233,241,243), 58(243,244)
Honma, T., 215, 266(11), 299
Hood, L., 440
Hořejši, V., 340(369), 345, 406, 420(19), 427(12), 428(12)
Hori, T., 301, 309(117)
Horne, D. A., 136
Horton, D., 63, 72, 73, 74(16,17), 75(10, 13), 76(7,16), 77, 79, 94, 99, 140, 142(270), 143(275), 144(275), 146(7), 147(281), 149(277), 154, 217, 225(20), 228(20), 237(20), 238, 240, 250, 263(20), 268(20), 272(20), 273(108), 421, 422, 426(97)
Hoshino, J., 366
Hotta, K., 302
Hough, L., 21, 23(46a), 32, 166, 214, 217, 220, 225, 229(66), 230(45,66), 231(24, 45), 232(24), 233(66), 235(45), 236(45,66), 237(66), 238, 239, 241(45), 246(66), 247(66), 249(24,45, 66,95), 251(24,45,66,95), 252(45,66,- 95), 253(45,66), 258(24,95), 260(24), 264(45,66), 266(45,66), 267(66), 268(45,66), 269(66,95), 270(24,66,95), 272(66), 273(105), 275(45,66,113), 276(45), 278(66), 281, 303, 342(23)

Howard, J., 161
Howard, W. L., 136
Howarth, G. B., 99
Howell, S. F., 444
Hu, A. S. L., 290, 392, 393(347)
Hudson, C. S., 24, 25, 169, 217, 219, 223, 205, 266(58,203), 293, 340(361), 341(71,361), 344, 360, 363(62)
Huebner, C. F., 22
Hünig, S., 98
Huggard, A. J., 29, 36(96)
Hughes, N. A., 16, 17(25), 26, 77
Hughes, R.C., 304
Hullar, T. L., 96, 98, 159, 179(22)
Hultberg, B., 326
Hunedy, F., 166, 168
Hung, Y.-L., 217, 225(20), 228(20), 237(20), 263(20), 268(20), 272(20)
Hunger, A., 302
Hunt, K., 281, 303(25), 342(24), 344(25)
Hurang, K. M., 331
Hurst, D. T., 63
Huseman, E., 164, 245
Hussey, H., 323
Hutson, D. H., 421
Hutten, V., 159, 180(26), 182(26), 184
Huyser, E. S., 150
Hwang, D. S., 385

I

Igarashi, K., 215, 266(11), 299
Iizuka, M., 368
Ikeda, Y., 383
Ikenaka, T., 171, 221, 222(47), 228(47), 264(47), 265(47), 266(47), 267(47), 268(47), 269(47), 270(47), 272(47), 274(47), 276(47)
Iley, D. E., 80, 94(45)
Iliescu, I., 87
Inai, T., 92, 93(86)
Inamdar, A. N., 392, 393(355,356)
Inch, T. D., 27, 61
Ingram, M., 373, 374
Inman, J. K., 417, 436(68,69)
Inouye, Y., 217, 218(29)
Ioannou, P. V., 67
Ionescu, M., 87
Irasawa, J., 299

Ireland, R. E., 147, 148(287), 149(287)
Irisawa, J., 215, 266(11)
Irvine, J. C., 17, 159, 179, 217
Isagulyants, V. I., 150
Isajev, V. I., 375
Isaka, S., 302
Isbell, H. S., 246, 277(137), 420, 426
Iselin, B., 337(341), 339, 340(359), 344
Isemura, M., 221, 222(47), 228(47), 264(47), 265(47), 266(47), 267(47), 268(47), 269(47), 270(47), 272(47), 274(47), 276(47)
Ishak, M. F., 201
Ishido, R., 76
Ishido, Y., 39, 40(164), 76, 80, 88, 92(81), 93(81), 94(43,44)
Issidorides, C. H., 135
Ito, H., 163, 170, 182(98,103,104,107), 183, 184(103), 187(107), 188(98,103, 104,107), 190, 191(103,107), 194(104, 107), 195(98), 196(98), 197(98), 198(98), 199(98), 200(98), 201(98), 207
Ito, K., 158, 173(12), 181, 182(97), 185(97), 206, 210(12), 211(150)
Ivanov, A. I., 390
Iwacha, D. J., 64

J

Jackson, D. C., 334
Jackson, M., 58
Jacobsen, S., 108, 109, 110(153), 111(153), 112(153,154), 113(155,156, 157), 151(156), 152(156)
Jacobson, R. A., 215, 216(12)
Jacquinet, J.-C., 53, 240, 296, 299, 300, 343(86,101), 344(102,105,106)
Järnefelt, J., 290, 313, 315(205)
James, K., 36, 83, 299, 330(97), 343(97)
James, S. P., 338(357), 339
Janáček, K., 380
Janaki, N., 136
Janda, S., 354
Jann, B., 280, 281(8), 301
Jann, K., 280, 281(8), 301
Janson, J.-C., 414
Jansse, P. L., 69
Jarman, M., 39

AUTHOR INDEX

Jarvis, J. A., 64
Jary, J., 29, 30, 226
Jaspers, H. T. A., 381
Jayasankar, N. P., 392
Jeanloz, R. W., 298, 310, 343(90), 426
Jedliński, Z., 152
Jefferis, R., 22, 26(56), 26(56)
Jeffersen, L. M., 337(344), 339
Jeffrey, A. M., 407, 420(27), 426(27), 440(27)
Jeffrey, G. A., 216, 220(14), 225(14), 284
Jenner, M. R., 33, 214, 233
Jennings, D. H., 380
Jennings, H. J., 200, 231, 232(89,90), 233, 254(90), 257(90), 258(90), 269(90)
Jeppesen, L. M., 150, 151(297)
Jewell, J. S., 80, 94(41)
Jiang, K. S., 301
Jirku, V., 395(382), 396
Jocius, I. B., 446
Jodál, I., 128, 129(220), 130(220)
Johnson, B., 366, 391(109)
Johnson, E. A. Z., 298, 406, 415(20), 444(20), 445(20)
Johnson, J. G., 323
Johnson, J. L., 331
Johnson, J. M., 265, 266(203)
Jonas, J., 98
Joncich, M. J., 96
Jones, G., 217
Jones, H. F., 214
Jones, J. K. N., 22, 45, 99, 161, 231, 232(89), 233(91), 247, 281, 342(24)
Joniak, D., 79, 130
Jordaan, A., 136, 147, 149(286)
Jordan, M., 35
Jørgensen, O. S., 368
Josephson, K., 219
Juvvik, P., 27

K

Kabat, E. A., 203, 313, 314, 316, 420, 439, 446
Kaboré, L., 45
Kadentsev, V. I., 261
Kärkkäinen, J., 290
Kaiser, E. M., 136
Kaiser, S., 281, 302(18), 337(18)
Kalanthar, A., 391
Kamai, G. K., 116
Kamerling, J. P., 249, 250(162)
Kamienski, L., 235
Kamiya, Y., 119
Kane, J. A., 424
Kane, T., 177
Kaplan, J. G., 354, 392(36), 393(355,356), 394
Karakawa, W. W., 424, 438
Karasawa, I., 217, 218(29)
Kariyone, K., 50
Karlsson, K.-A., 309
Karrer, P., 235, 286
Karush, F., 407, 437
Katchalslci, E., 202, 311
Kateva, V. A., 116
Kaufman, R. L., 306
Kawana, M., 154
Kaya, T., 366
Kefurt, K., 226
Keglević, D., 283
Keilich, G., 250
Kelen, J., 189
Kelly, P. J., 390
Kemori, Y., 331
Kenne, L., 301
Kennedy, J. P., 174, 189
Kepes, A., 353, 381
Kessel, D., 321, 322, 331(239, 241)
Kevei, F., 386, 387, 388(269)
Kew, O. M., 378
Khamani, D., 373
Khan, N. A., 372, 373(162), 386, 387, 389, 390
Khan, R., 17, 33, 214, 231(1), 233(1), 238(1)
Khare, M. P., 302, 337(346), 339, 341(346)
Khilanani, P., 321, 331(239)
Khorana, H. G., 20, 37, 39, 47(45), 59, 60(269,271), 61(274), 62(279), 413, 420(52), 434(52)
Khuong-Huu, Q., 45, 104, 105(144), 106(144), 285, 342(44)
Kidby, D. K., 366, 367, 368(117), 370, 371
Kidd, J., 223
Kieda, C. M. T., 407, 439(26)
Kilara, A., 396

Killander, J., 408
Kim, K. S., 78, 93
Kim, U., 322
Kim, Y. D., 407
Kim, Y. S., 311
Kindt, T. J., 441
King, J. F., 102
Kirkman, H. N., 317
Kirschenlohr, W., 217
Kirsh, R., 335
Kisaïlus, E. C., 314
Kiss, L., 28
Kit, S., 310
Kitahara, T., 35
Kitano, K., 158, 173(12,13), 210(12,13)
Kitchell, T. J., 399
Kito, E., 364
Klar, J., 164
Klein, R. S., 153
Kleinheidt, E. A., 172
Klemer, A., 30, 117, 139, 140, 141(272), 142, 144(273), 145(278,279), 146(276), 184, 244
Klimov, E. M., 220, 264(44), 271(44), 276(44)
Klobe, R. J., 329
Klöcking, H. P., 22
Kluyver, A. J., 397
Kobata, A., 281, 306, 313, 316(199), 324, 325(257), 327, 330(14,282)
Kobayashi, K. I., 181, 182(97,101,102, 105,105a), 183, 185(97,101), 187(101, 102), 188(101,102), 192(101), 193(101), 194(101), 195(105,105a), 196(102), 197(101), 199(102), 200(102), 211(105,105a)
Koch, H. J., 254(191,192), 255(188), 256(191), 257(191,192)
Kochetkov, N. K., 115, 116(164), 172, 220, 258, 264(44), 271(44), 276(44), 287
Kocourek, J., 340(369), 345, 406, 420(19), 427(12), 429(12)
Koebernick, W., 341(372), 342(372), 345
Koehler, R., 331
Köhler, W., 67
Koell, P., 18
Koeners, H. J., 42
Koeppen, B., 35, 224, 229(61), 252(61), 266(61), 267(61), 271(61)

Kőrösy, F., 422
Köster, R., 53, 54, 55(225,228)
Köttgen, E., 321, 331(240,242)
Kohen, F., 154
Kohn, J., 414, 431(63)
Kohtès, L., 367
Koide, N., 305
Koizumi, K., 219, 220(41,43), 229(41), 235(41), 249, 250(166), 253, 264(41, 43), 266(41)
Kolahi-Zanouzi, M., 391
Koláv̌, C., 300, 344(107)
Kolecki, B., 312, 339(197), 342(197), 343(197)
Koller, F., 406, 421(4)
Kollonitsch, V., 214
Komada, M., 210
Koningsberger, V. V., 382(238),383, 384(238), 389
Kopp, F., 371
Kops, J., 165, 204, 205(147)
Kornberg, H. L., 381
Kornfeld, R., 281, 304(16)
Kornfeld, S., 281, 304(16), 318, 445
Korshak, V. V., 159, 180(27), 182(27)
Kosáková, L., 79, 130
Koscielák, J., 313
Koshland, D. E., 351
Kośiková, B., 79, 130
Kosloskì, C. L., 80
Kotake, M., 133
Kotera, K., 34
Kotowycz, G., 254
Kotyk, A., 380, 384
Kovaleva, N. S., 363, 366(75)
Kovář̌, J., 29
Kowollik, G., 48
Koyama, Y., 158, 173(14), 211(14)
Kraft, D., 406, 421(4)
Kramer, S., 170
Krantz, M. J., 433
Kratochvil, M., 98
Krause, R. M., 437, 440
Kreger-van-Rij, N. J. W., 374, 378
Kreutzer, U., 164
Krishnamahan, M. V. V., 334
Krishnamurthy, S., 137
Krishnamurthy, T. N., 222, 223(52), 237(52), 246(52), 247(52), 258(52), 265(52), 266(52), 273(52), 277(52)

Kristiansen, T., 405, 406(1), 407, 430(1), 444
Kronau, R., 403
Kruckenberg, W., 47
Krüger, J., 391
Krumphanzl, V., 395(382), 396
Krupery, J., 437
Krusius, T., 290, 313
Kryczka, B., 95
Kubo, H., 366
Kubrick, R. L., 407, 439(23)
Kuchynkova, A., 395(382), 396
Kuge, T., 35
Kuhn, L. P., 150
Kuhn, R., 217, 343(375), 344(378,379, 380,381), 345, 360, 366
Kula, M., 387(284), 388
Kulikova, A. K., 395
Kuno, S., 331
Kuo, K. C., 331
Kuo, S. C., 367, 370, 371(146)
Kurita, H., 35
Kurokawa, M., 302
Kurz, G., 351
Kuszmann, J., 28, 41
Kutty, K. M., 333
Kuzuhara, H., 15
Kwiatkowska, S., 58

L

Låås, T., 414
Labadie, G., 333
Lackland, H., 440
Lai, H. L., 389
Laine, R. A., 313, 315(205)
Lampen, J. O., 202, 355, 364, 366, 367(59,90), 368(90,102), 370(102), 371(146)
Lance, D. G., 22
Lancefield, R. C., 437
Landini, M. P., 395, 396(380)
Lange, C., 366, 391(109)
Langen, P., 48
Lapkowski, M., 92
Larm, O., 249, 250(161)
Larner, J., 385, 392
Larsen, B., 280(12), 281, 303, 344(132)

Larsen, J. W., 102, 107(139), 110(139)
Laurent, T. C., 408
Lavallée, P., 64, 287
Lawler, J. V., 22
Lazo, J. S., 331
Lazo, P. S., 375
Lazurkina, T. Yu., 34
Leaback, D. H., 295, 341(77), 343(77)
Leblond, C. P., 320
Lederberg, J., 353
Lederer, E., 63
Lee, E., 285
Lee, E. E., 256(202), 265
Lee, J. B., 115
Lee, R. T., 420, 427(81), 429(81)
Lee, Y. C., 407, 408, 419(35), 420, 421, 427(81), 429(81), 433(31), 436(35)
Leggetter, B. E., 122, 123(199-203), 124(200,202,203), 127(201)
Legler, G., 387, 388
Lehle, L., 368
Lehmann, H., 36
Lehmann, J., 19, 20(41), 21, 22, 23(46), 63
Lehn, J. M., 33
Leibowitz, J., 28, 378, 389
Leitch, G. C., 215, 223(6), 265(6)
Leloir, L. F., 395
Lemieux, R. U., 161, 215, 224, 254, 299, 300(100), 330(97), 335, 343(97), 344(100,104), 420, 423, 433, 437
Lenz, J., 16
Lenz, R. W., 173, 175(73)
Leon, M. A., 408, 419(34), 436(34), 446
Leone, R. E., 43
Lerman, L. S., 412
Lerner, L. M., 18
Letsinger, R. L., 40, 41(165), 43, 48
Levene, P. A., 27, 168, 215
Levine, M. L., 429
Levy, A. A., 421
Levy, G. A., 295, 297(76), 324(76), 325, 337(76), 340(76,81), 341(76)
Lewin, M., 247
Lewis, D., 22, 26, 27(80)
Lewis, D. E., 389
Lewis, F. M., 190
Lewis, M. J., 367
Lewis, T. A., 351
Lezerovich, A., 31

AUTHOR INDEX

Li, Y.-T., 313, 315(205)
Liao, J., 203, 446
Liav, A., (Levy), 296, 298(81), 340(370), 342(370), 343(89), 344(81,89), 345
Libert, H., 182(108), 183, 187(108), 199(108)
Liddle, A. M., 223
Liebowitz, J., 172
Lim, M. T., 242
Lin, J. W. P., 181, 182(94,99), 183, 187(94), 188(94), 197(94), 199(94)
Lindberg, B., 30, 146, 218, 221(32), 223(32), 225(32), 228(32), 229(32), 231(32), 235(32), 249, 250(161), 263(32), 266(32), 267(32), 269(32), 272(32), 280, 290, 301, 303(7,63), 304(7,138), 308, 437
Lindegren, C. C., 385
Lindenberger, W. H., 182(100), 183, 188(100), 196(100)
Lindh, F., 146
Lindner, P., 375, 378(186), 391
Lindqvist, B., 308
Lindseth, H., 17, 33
Link, K. P., 22, 420, 421
Linnenbaum, F. J., 142, 144(273)
Lipke, P. N., 424
Lipkin, D., 61
Lipp, G., 302
Lipták, A., 128, 129(220), 130(220), 131(227), 132, 133
Liras, P., 370
Lis, H., 311, 313(188), 406, 426, 444(106)
Liu, H. H., 421
Llanillo, M. L., 325, 326(269)
Llewellyn, D. R., 351
Lloyd, D., 375
Lloyd, K. O., 281, 303(15), 304(15), 313(15), 314(15)
Lloyd, P. F., 29, 340(365), 345
Lodder, J., 374, 395(387), 396
Lönngren, J., 146, 290, 301, 303(63), 308, 437
Lövgren, T., 383, 387, 388
Loman, P. L., 321, 331(239)
London, R. E., 254, 255(185)
Long, L., Jr., 77, 233, 238
Longchambon, F., 284
Losada, M., 376, 377(198)

Lotan, R., 406, 426, 444(106)
Lotz, W., 387, 388
Louden, L., 364
Luche, J. L., 81
Lüderitz, D., 308
Lüderitz, O., 312, 437
Luescher, E., 412
Luetzow, A. E., 99, 296
Lukes, T. M., 391
Lukowski, H., 30
Lundblad, A., 327, 330(278,279), 332(279)
Lundstein, J., 327, 330(279), 332(279)
Lundt, I., 150, 151(297), 337(337,344), 339

M

McAllan, A., 295, 297(76), 324(76), 337(76), 340(76), 341(76)
McArthur, N., 56
Macbeth, R. A. L., 331
McBride, G., 331
McCarthy, J. F., 24, 248
McCarthy, W. C., 96
McCasland, G. E., 114
McCleary, B. V., 407, 441(29), 443(29)
McCleland, C. W., 99
McCloskey, C. M., 44, 45(180), 50(180)
McCormick, M. H., 60
McElwee, J., 98
McFarlane, R. C., 177, 189, 190(84,117)
McFarren, E. F., 395
McGale, E. H. F., 330
Machell, G., 247
Machinami, T., 243, 278(126)
Maciel, G. E., 255
McInnes, A. G., 62, 63(289)
Mackenzie, G., 18
McKenzie, I. F. C., 334
MacKenzie, S. L., 396
McKeown, G. G., 58
McKibbin, J. M., 280, 281(5), 308(5), 309(5)
McManus, S. P., 102, 107(139), 110(139)
McNally, S., 44
Macovski, E., 86, 87(70,71,72), 88(67–72)
MacPhillamy, H. B., 293, 338(72)
MacQuillan, A. M., 392

MacWilliam, I. C., 385
Madeira-Lopes, A., 402
Maeda, S., 80, 88, 92(81), 93(81), 94(43)
Mage, M., 420
Mage, R. C., 437
Magnus, P. D., 85, 86
Magrath, D. I., 59
Mahadevan, V., 43
Mahoney, R. R., 394, 395(359), 396
Maichuk, D. T., 43
Mainzer, K., 331
Majkova, A. I., 116
Makamoto, T., 301
Malaval, A., 81
Maley, F., 368, 369(133)
Malhotra, O. P., 351, 394(10)
Malkov, Yu. K., 116
Mall, S., 229, 236(84), 242(84), 243(84), 268(84), 270(84), 274(84), 278(84)
Mallory, R. A., 154
Mandels, M., 366
Mandl, I., 355
Manjula, B. N., 446, 447(163)
Manners, D. J., 390
Marans, N. S., 58
March, S. C., 414 , 431
Marchessault, R. H., 197, 215, 216(13), 220(13), 225(13)
Marchin, G. L., 392
Marcus, D. P., 318
Mark, H., 176, 196(81)
Markham, R., 61
Markiewicz, W. T., 68(317), 69
Markin, J. S., 395(381), 396, 397, 403
Marktscheffel, F., 137
Marmur, J., 387, 388(275)
Marnay, A., 325, 326(270)
Marousek, V., 182(107), 183, 187(107), 188(107), 191(107), 194(107)
Marquet, A., 97
Marsh, J. M., 217, 218(26), 429, 430
Marshall, C. R., 57
Marshall, J. J., 290, 408, 419(37), 436(37), 443(37)
Martemyanov, V. S., 118
Martin, O. R., 79, 80
Martinez, E., 23
Martin-Lomas, M., 226
Marvell, E. N., 96

Marx, F., 386
Masaki, M., 35
Maślinska-Solich, J., 152
Masson, M., 59
Mastronardi, I. O., 221
Masura, V., 182(111), 183, 195(111)
Matacic, S. S., 333
Matalon, R., 332
Matile, P., 371, 372, 401(157)
Matsubara, S., 217, 218(27), 263(27,34)
Matsubara, T., 308, 331(172)
Matsuda, I., 327, 330(282)
Matsuda, K., 249, 254, 256(177), 257(177)
Matsuhashi, M., 323
Matsumoto, I., 406, 419(18), 436(18)
Matsumoto, K., 327
Matsunaga, I., 219, 223(40), 264(40), 266(40,55), 272(55), 277(40,55)
Matsuoka, H., 367
Matsusaka, K., 387
Matsushima, R., 92, 93(86)
Matsushima, Y., 171, 221, 222(47), 228(47), 229, 264(47), 265(47), 266(47), 267(47,83), 268(47), 269(47,83), 270(47,83), 272(47,83), 274(47), 276(47)
Matsuura, K., 80, 88, 92(81), 93(81), 94(43,44)
Matsuzaki, K., 158, 173(12,13,14), 181, 182 (96,97,106), 183, 185(96,97), 187(106), 194(106), 197(96), 199(96), 206, 210(12,13), 211(14,150)
Matta, K. L., 35, 298, 335
Matwiyoff, N. A., 254, 255(185)
Maurer, P. H., 439
Mayer, W., 35, 338(348), 339, 342(348), 343(348)
Mayo, F. R., 190
Meada, K., 281, 302(22)
Meerwein, H., 107
Mehltretter, C. L., 169, 421
Meier, R., 355
Melnick, J. L., 310
Melton, L. D., 117, 241, 269(122)
Merdinger, E., 366, 391(109)
Merész, M., 41
Merle, J. P., 197
Merlis, N. M., 159, 180(27), 182(27)
Messmer, A., 337(338), 339

AUTHOR INDEX

Mestres, R., 23
Metzner, E. K., 285
Meyer, J., 371, 372, 401(157)
Meyers, C. Y., 44
Mian, A. J., 159, 179(28), 303
Michaljaničová, D., 384
Michaud, S., 97, 118
Micheel, F., 34, 164, 172, 205, 206
Michelson, A. M., 36, 37(149), 60
Miersch, O., 36
Milandre, J., 441
Milat, M.-L., 296, 343(86)
Miller, E. E., 329
Miller, K. B., 437
Miller, O. N., 326
Miller, P. S., 40, 41(165)
Millin, D. J., 385, 389
Mills, J. A., 14, 15(7a), 19(7a), 21(7a), 23(7a), 25(7a), 72
Minaker, E., 326
Minamoto, K., 37
Mine, K., 35
Minshall, J., 166
Mistele, P., 224
Mitchell, W. O., 202
Mitra, A. K., 245
Mitranic, M. M., 326
Mitura, W., 222, 223(52), 237(52), 246(52), 247(52), 258(52), 265(52), 266(52), 273(52), 277(52)
Mityushova, N. M., 366(114), 367
Miyashita, N., 80
Mizunaga, T., 370
Moffatt, J. G., 39, 48, 60, 61(274), 62, 288
Molodtsov, N. V., 258
Molotkovskii, Yu. G., 34
Mols, O., 108, 113(157)
Mompon, B., 63
Monneret, C., 45, 104, 105(144), 106(144), 107(147), 285, 342(44)
Monsigny, M. L. P., 297, 300(87), 341(87), 343(87), 407, 439(26)
Montague, M. D., 59, 60(270)
Montgomery, J. A., 39
Montgomery, R., 304
Montreuil, J., 327, 330(280,281), 333(281)
Moore, J. A., 281, 282(20), 337(20)
Moore, S., 420
Moreau, C., 81, 82
Morell, A. G., 433
Moreno, F., 369, 370(144), 375, 402

Morfaux, J. N., 380
Morgan, J. W. W., 55, 56(232,233), 57(232,233)
Morgan, W. T. J., 318, 319(226), 345, 409
Morgenlie, S., 15, 338(347), 339
Mori, A., 331
Mori, M., 222, 225(51), 226(51), 228(67), 229(82), 235(67), 237(82), 244(67), 265(51), 266(67), 267(67,82), 268(82), 269(67,82), 270(67), 272(51,67,82), 273(51,82), 274(67,82), 276(67), 278(68,82)
Morrison, J. M., 281, 303(25), 344(25)
Morschel, H., 107
Morse, M. L., 63
Mortimer, R. K., 372, 389
Mortlock, R. P., 327
Mosbach, K., 406
Moscarello, M. A., 326
Mosher, D. F., 305
Moskal, J. R., 318, 322(228)
Moskva, V. V., 116
Moss, R. J., 59, 60(270)
Moulik, S. P., 245
Moulin, G., 402
Mowery, D. F., Jr., 284, 293(35a), 294(35b), 340(362), 342(362), 344
Mowshowitz, D. B., 402
Müller, H., 16
Müther, A., 340(363), 344
Mufti, K. S., 214, 233
Mukaiyama, T., 135
Mukherjee, S., 197
Munasinghe, V. R. N., 89, 90(85), 91(85), 92(85)
Muntz, J. A., 355
Murai, T., 80, 94(43)
Muramatsu, T., 305, 306, 325
Murray, D. H., 15
Muschel, L. H., 313
Mushinski, E., 446, 447(163)
Myrbäck, K., 354, 355, 356(58), 367, 390, 391, 395(383), 396

N

Nagabhushan, T. L., 423
Nagasaki, S., 366, 367
Nagel, J., 38

Nakada, S., 227, 268(79), 271(79), 275(79)
Nakajima, M., 238
Nakajima, T., 368
Nakamura, Y., 389
Nakazaki, N., 39, 40(164)
Nánási, P., 128, 129(220), 130(220), 131(227), 132, 133
Nardin, R., 254, 255(184)
Nashed, M. A., 50
Nassr, M. A. M., 240, 297, 343(86)
Nathenson, S. G., 334
Naumova, T. I., 373
Naves, R. G., 246
Naya, Y., 133
Nayak, U. G., 44
Neal, J. L., 216
Needham, L. L., 48
Needleman, R. B., 387, 388(275), 390
Negoro, H., 360, 364(63), 365(63)
Neher, H. T., 76
Nehmsmann, L. J., 138
Neilson, T., 40, 41(167), 420
Nelson, E. L., 245
Ness, A. T., 25, 340(361), 341(361), 344
Neszmélyi, A., 133, 337(338), 339
Neuberg, C., 355, 386, 392, 395(345)
Neuberger, A., 406
Neufeld, E., 390, 391(314)
Neuman, A., 284
Neumann, N. P., 364, 366, 367(90), 368(90,102), 370(102)
Nevin, R. S., 158, 173(11), 209(11), 210(11)
Newlin, M. R., 244, 276(127)
Newman, M. S., 34, 62, 98
Newth, F. H., 217, 228(23), 237(23), 263(23), 266(23), 268(23), 272(23), 422
Nicholas, S. D., 217, 228(23), 237(23), 263(23), 266(23), 268(23), 272(23)
Nicholson, W. H., 161
Nickerson, T. A., 395, 396(376)
Nielsen, B., 108, 113(155)
Nieswandt, G., 24
Nigam, V. N., 265, 278(209)
Nikaido, H., 312
Nikolaeva, I., 164
Nimmich, W., 146
Nishigaki, M., 327, 330(282)
Nishisawa, Y., 217
Nishiyama, K., 80, 94(44)
Nishizawa, K., 375
Nisizawa, K., 352
Norberg, E., 429
Norberg, T., 292, 330(67), 342(67)
Nordén, N. E., 327, 330(279), 332(279)
Nordin, P., 429
Northcote, D. H., 303, 323(130,131), 324
Norula, J., 229, 236(84), 242(84), 243(84), 268(84), 270(84), 274(84), 278(84)
Notario, V., 392, 393(358)
Novák, E. K., 386, 387, 388
Novogrodsky, A., 311
Nowak, B. E., 122, 135(198)
Nurokoro, N. A., 304, 327, 328(147)

O

O'Brien, E., 167, 285
Obruchnikov, I. V., 172
O'Carra, P., 410, 411(46), 442(46)
Ochoa, A. G., 369, 370(144), 375
O'Colla, P. S., 285
O'Driscoll, K. F., 177, 189, 190(84,117)
Öckerman, P.-A., 326, 327, 330(279)
Ötvös, L., 29
Ogata, M., 305
Ogata, S.-I., 306
Ogata-Arakawa, M., 325
Ogawa, S., 34
Ogilvie, K. K., 41, 64, 65(303,304), 66(303,304,308,309), 67(309)
Ohaku, K. I., 181, 182(96), 185(96), 197(96), 199(96)
Ohannessian, J. O., 284
Ohms, J. J., 440
Ohrui, H., 15, 18, 153
Ohta, M., 35
Oishi, K., 326
Okada, H., 380, 382(210,211), 383(210), 384(210,211,235), 389(235)
Okada, M., 153, 210
Okano, A., 302
Okano, S., 225, 226(69), 229(69), 230(69), 231(69), 233(69), 265(69), 266(69), 267(69), 268(63,69), 270(69), 273(69), 278(69)
Okazaki, R., 318
Okazaki, T., 318
Okugawa, T., 37

Oláh, B., 386, 387, 388(269)
Old, D. C., 327
Oldham, J. W. H., 56, 159, 179
Oldham, M. A., 19
Olsen, K. W., 420, 424(80), 430(80), 441(80)
Olson, N. F., 396
O'Meara, D., 56
Ondetti, M. A., 239
Onishi, H., 374
Onn, T., 301
Onodera, K., 217, 218(29)
Oparaeche, N. N., 89, 90(84,85), 91(85), 92(84,85), 93(84)
Oparin, A. I., 389
Opheim, D. J., 324
Oppenheimer, G., 378, 395(204)
Orchimakov, M. U., 220, 264(44), 271(44), 276(44)
Orsino, J. A., 295, 340(78)
Ørskov, F., 301
Ørskov, I., 301
Örtenblad, B., 390, 391
Orzaez, M., 23
Osawa, T., 406, 419(18), 436(18)
Osborn, M. J., 323
Oshima, Y., 383
Otake, T., 292, 339(70)
Otea, A., 87
Othman, A. A., 95
Otterbach, D. H., 338(358), 339
Ottolenghi, P., 355, 364, 366, 369(54), 371, 372, 375(55)
Oura, E., 380, 389
Ouwehand, J., 389
Overend, W. G., 14, 15(7b), 19, 283, 284(32), 352, 420, 421, 423(76)
Ovodov, Yu. S., 301
Owen, G. R., 37

P

Pacák, J., 115
Painter, T. J., 201, 280(12), 281, 345
Pakhomov, A. M., 164
Palacios, J., 384
Palmer, E. T., 354
Panek, A. D., 390, 391
Pankhurst, R. J., 398
Pankratz, R. A., 22
Pappagianis, D., 374
Parikh, I., 410, 414, 431
Parish, C. R., 334
Parker, K. J., 214
Parra, F., 402
Parrish, F. W., 19, 24(38), 77, 79, 201, 227, 229(78), 233, 238
Parsons, M. A., 59, 60(270)
Pater, R. H., 284, 294(35b)
Pathak, K. D., 136
Patino-Leal, H., 177
Paulsen, H., 76, 78(24), 96, 107(111), 110(111), 134, 300, 337(340,342), 338(342), 339, 341(372), 342(372), 344(107), 345
Paulson, J. C., 320, 442
Paushkin, Ya. M., 150
Pauwels, P. J. S., 254
Payne, R. W., 348, 361(5), 362(5), 374(5), 395(5), 400(5), 402(5)
Pazur, J. H., 217, 218(26), 318, 352, 406, 407(9,17), 409(17), 424, 426, 429(17), 430(17), 433, 437, 438(9,135,143), 439(9,17,23,107), 446(17), 447(17, 107)
Peagram, M. J., 139
Pearson, D. E., 118
Pearson, R. G., 96, 118(105)
Peat, S., 121, 158, 161(15), 215, 217, 218(25)
Pedersen, C., 34, 108, 109, 110(153), 111(153), 112(153,154), 113(155,156, 157), 114, 115, 117, 120, 150, 151(156,297), 152(156), 337(337,344), 339
Pellé, G., 79
Penglis, A. A. E., 50, 51(212)
Pentchev, P. G., 351
Percival, E. E., 280(11), 281, 282(11) 291, 293(65), 301, 303, 341(65,371), 342(65, 371), 343(371), 345
Percival, E. G. V., 341(371), 342(371, 373), 343(371,373), 345, 362
Pereira, M. E. A., 314
Perez, N., 325, 326(269)
Perlin, A. S., 135, 246, 254(192, 193), 255(187,188), 256(186,193), 257(192)
Pernikis, (n. Schneer) R. Ya., 159, 180, 182(27)

AUTHOR INDEX

Perry, A. R., 225
Perry, M. B., 45
Pete, J. -P., 285, 342(44)
Petek, F., 391
Peterson, M., 138
Peterson, S. R., 218
Pétrequin, D., 118
Pfannemüller, B., 245
Pfitzner, K. E., 288
Pfleiderer, W., 67
Phadnis, S. P., 214
Phaff, H. J., 363, 374, 391
Phelps, F. P., 293, 341(71)
Philipson, L., 413
Phillips, A. W., 386, 387, 388
Philo, R., 384
Piasek, A., 313
Pickles, V. A., 75
Pictet, A., 157, 158, 159(2), 179
Pilato, L. A., 122
Pincus, J. H., 437
Pinter, I., 337(338), 339
Piria, R., 392
Piskorsha-Chlebowska, A., 32
Pittet, A. O., 224, 266(65)
Pittman, C. U., Jr., 102, 107(139), 110(139)
Pizer, F. L., 59, 60(266)
Plénat, F., 99
Plessas, N. R., 98, 99(116), 100(116), 101(132), 102, 103(133), 117, 287, 337(49)
Plummer, T. H., 368, 369(133)
Poitau, A., 327, 330(280)
Pol, E. H., 61
Pollard, H. B., 406, 420(22), 421(22), 441(22), 442(22)
Ponomarev, F. G., 124
Ponpipom, M. M., 46, 103, 287
Poráth, J., 405, 406(1), 407, 408(32), 414(32), 415(21), 430(1,32), 431(66), 432(21,66)
Poretz, R. D., 202
Portella, C., 285, 342(44)
Poste, G., 335
Postermak, T., 59, 60
Postma, P. W., 384
Potmesil, R., 340(360), 344
Potter, M., 446, 447(163)

Pougny, J.-R., 296, 297, 300(88), 341(88), 343(86,88)
Pouyssegur, J., 305
Powell, D. A., 308
Powell, D. B., 303
Powell, R. W., 48
Praill, P. G. F., 45
Prather, J., 122
Preckel, R. F., 58
Preston, J., 37
Price, C. C., 49
Pridham, J. B., 246, 375
Prieels, J.-P., 319, 320
Prihar, H. S., 295, 340(80)
Přikrylová, V., 115
Pringsheim, H., 179
Prins, W., 179
Prior, E., 378
Privat de Garilhe, M., 59
Prohaska, R., 319, 322, 333(234)
Prokop, J., 15
Prosser, T. J., 49
Prugh, J. D., 96
Puglisi, P. P., 395(385), 396(380)
Pulkownik, A., 202
Purves, C. B., 57, 360

Q

Quayle, J. R., 352
Quigley, G. J., 215, 216(13), 220(13), 225(13)
Quilliam, M. A., 64, 65(303,304), 66(303, 304)
Quinn, E. J., 207
Quinn, E. V., 159, 179(28)

R

Rabinovitch, P. D., 334
Rabinowitz, Z., 305
Rabinsohn, Y., 158
Rachaman, E. S., 298, 343(90)
Racke, F., 354
Rafestin, M.-E., 297, 300(87), 341(87), 343(87), 441
Rakhamankulov, D. L., 118, 150
Ramaswami, S., 138
Rammler, D. H., 20, 37, 47(45)
Randall, M. H., 46

Raschig, K., 285, 338(42), 340(367), 345
Raschke, W. C., 424
Rathore, B. S., 138
Rauchalles, G., 360
Rauvala, H., 290, 309, 313
Ravindranathan, T., 37
Raymond, A. L., 26
Razumov, A. I., 116
Rebers, P. A., 409
Recondo, E. F., 31
Record, B. R., 362
Redlich, H., 76, 78(24), 337(342), 338(342), 339
Reed, W. P., 312
Reegen, S. L., 174
Rees, B. H., 21, 22, 23(46,56), 26(56)
Rees, C. W., 283
Rees, D. A., 216
Reese, C. B., 37, 38, 39
Reese, E. T., 201, 366
Reeves, R. E., 26, 158
Rege, V. P., 345
Reggiani, M., 225, 249, 250(165), 266(72)
Reglero, A., 324, 325(258)
Reichman, U., 36
Reichstein, T., 16, 280, 281, 282(20), 302(2), 337(20,341,346), 338(352, 353), 339, 340(359), 341(346), 344
Reilly, P. M., 177, 189, 190(84,117)
Reinhard, H., 24, 82
Reinking, K., 205
Reist, E. J., 44, 168, 285
Rembarz, G., 22
Rémy, G., 95
Rendleman, J. A., Jr., 245
Rennie, R. A. C., 15
Repke, D. B., 39
Rerick, M. N., 122, 123, 136(196)
Restelli de Labriola, E., 31
Reuss, H., 34
Reutter, W. G., 321, 331(240,242)
Revallier-Warffemius, J. G., 395
Rexová-Benková, L., 408, 419(36), 436(36), 442(36), 442(36)
Reynolds, R. J. W., 242, 278(123)
Rhee, B., 122, 123(204)
Rheiner, A., 302
Riazi-Farzard, T., 327, 330(281), 333(281)
Richard, B., 22
Richards, E. L., 247

Richards, G. N., 247
Richardson, A. C., 19, 21, 23(46a), 32, 166, 217, 220, 225, 229(66), 230(45, 66), 231(24,45), 232(24,85), 233(66, 85), 235(45), 236(45,66), 237(66,85), 238, 241(45), 246(66), 247(66), 249(24,45,66,95), 251(24,45,66,95), 252(45,66,95), 253(45,66), 258(24,95), 260(24), 264(45,66), 266(45,66), 267(66), 268(45,66), 269(66,95), 270(24,66,95), 272(66), 273(105), 275(45,66), 276(45), 278(66), 285, 290, 291(40), 341(64)
Richter, G. C., 245
Richter, W., 202, 203(138)
Richtmyer, N. K., 360, 363(62)
Rieche, A., 96
Rinaudo, M., 254(194), 255, 256(194)
Rist, C. E., 30
Rittel, W., 302
Ritzmann, G., 153
Rivaille, P., 61
Robbins, P. W., 323
Roberts, C., 372, 376
Roberts, H. R., 395
Roberts, J. D., 254, 257(176), 261
Roberts, R. M., 305
Robertson, G. J., 18
Robertson, J. J., 382(241), 383
Robinson, R., 202
Rodemeyer, G., 139, 140, 141(272), 142, 144(273)
Rodinov, A. V., 172
Rodriguez, L., 371
Rollin, P., 131
Romanic, B. M., 406, 407(17), 409(17), 429(17), 430(17), 439(17), 446(17), 447(17)
Romanovich, A. Yu., 78
Rorive, F., 340(364), 345
Rosato, E. F., 331
Rosato, F., 331
Rosato, F. E., 329
Rose, J. B., 174
Roseman, J. M., 329
Roseman, S., 295, 316, 318, 341(77), 343(77), 381
Rosenberger, P.G., 329
Rosenfeld, L., 248, 408, 419(35), 436(35)
Rosenthal, I., 80, 94(42)

AUTHOR INDEX

Ross, S. D., 189
Rosselet, J. P., 302
Rossi, C., 395, 396(380)
Rossi, J., 395(386), 396
Roth, W. W., 392
Rothstein, A., 355, 372, 380(49)
Rovinski, S., 154
Rovis, L., 316
Rowell, R. M., 30, 224, 266(60)
Roy, N., 245, 246
Royer, R., 81
Rozenfel'd, E. I., 325
Rózsa, J., 395
Rubin, S., 331
Rubin, T., 34
Ruckel, E. R., 181, 182(91,91a), 183(91), 185(91), 188(91), 199(91a)
Ruiz, T., 391
Rush, J., 313, 315(205)
Russetto, M. A. J., 262
Rutherford, J. K., 56
Ruttloff, H., 403
Ryan, K. J., 296(85)

S

Sabel'nikova, M. M., 390
Sachdev, H. S., 43
Sachs, L., 305
Sadana, K. L., 64, 66(308,309), 67(309)
Saegusa, T., 174
Saeki, T., 387, 388(272)
Sakairi, N., 39, 40(164)
Salas, M., 351
Salzburg, H., 76, 78(24)
Samoilenko, E. S., 390
Samuelsson, B. E., 309
Samuelsson, K., 307
Sano, H., 34
San Romáo, M. V., 399
Santer, U. V., 306
Sarasin, V̇., 157, 159(2)
Sarel-Imber, M., 172
Sarid, S., 202
Sarkanen, K., 158, 173(11), 209(11), 210(11)
Sarko, A., 197, 198(125), 201, 215, 216(13), 220(13), 225(13)
Sasaki, T., 37
Sato, H., 331

Sato, T., 76, 366
Saunders, J., 80
Saville, N. M., 44
Sawi, T., 386
Sawyer, W. H., 406
Sayre, R., 223, 266(58)
Scanlin, T. F., 333
Scarr, M. P., 374
Schaafsma, S. E., 99
Schacter, H., 304, 327, 328(147)
Schaefer, H. J., 49
Schaffer, R., 15, 74(18), 75, 246, 277(137), 420
Schamhart, D. H. J., 390
Scheck, H., 395
Scheel, I., 402
Scheer, I., 154
Scheeren, J. W., 99
Scheffold, R., 50
Scheinberg, I. H., 433
Scheiner, O., 406, 421(4)
Schenkel-Brunner, H., 319, 322, 333(234)
Scherrer, R., 364
Schexnayder, D. A., 43
Schifman, A. L., 64, 66(309), 67(309)
Schilling, G., 35
Schilling, W., 367
Schindler, O., 281, 282(19), 302(19), 337(19,346), 338(352), 339, 341(346)
Schlanderer, G., 389
Schlenk, F., 364
Schlubach, H. H., 287, 360, 363
Schlutt, M., 22
Schmadel, H., 34
Schmalz, K., 179
Schmid, H., 286
Schmidt, O. T., 15, 24, 34, 82, 338(348, 349,351), 339, 342(348), 343(348)
Schmidt, U., 121
Schmitz, E., 96
Schneider, J., 421
Schneider, K. F., 301
Schoch, T. J., 217
Schramm, M., 240, 241(120)
Schray, K. J., 352
Schuerch, C., 47, 153, 158, 159, 163, 165, 170, 172, 173(11), 179(28), 180, 181, 182(91,92,93,94,95,98–103,107–111), 183(91,92), 184(92,103), 185(91,92, 101), 186(92), 187(93,94,101,102,107,

108), 188(91,92,98–103,107), 190,
191(103,107), 192(101), 193(101),
194(101,107), 195(98,110,111),
196(98,100,102), 197(94,98,101),
198(98), 199(91a,93,94,95,98,102,108,
109,110), 200(98,102), 201(93,98),
202, 204, 205(147), 207, 208(92),
290(11), 210(11), 223, 244(53),
272(53)
Schütte, H. R., 36
Schultz, G., 35
Schulze, H., 378
Schussler, W., 54
Schwarcz, J. A., 254, 256(186)
Schwesinger, B., 20
Scoffone, E., 35
Seaston, A., 384
Seeley, D. A., 98
Segal, S., 316, 328(215)
Segall, G. H., 57
Seib, P. A., 188, 221
Seidman, M., 421
Sela, B.-A., 441
Sellinger, O. Z., 326
Selinger, Z., 240, 241(120)
Seltzer, M. H., 329
Sen Gupta, K. K., 245
Sengupta, U., 331
Sentandreu, R., 372
Sequiera, J. S., 283
Sergeev, V. A., 159, 180(27), 182(27)
Serrano, R., 384, 401
Seto, S., 249, 254, 256(177), 257(177)
Shadaksharaswamy, M., 430
Shade, W., 96
Shah, R. H., 337(339), 339, 340(339),
341(339)
Shahani, K. M., 396
Shall, S., 360, 366, 368
Shamshurina, N. V., 373
Shaper, J. H., 406, 420, 424(80), 430(80),
441(80)
Shapiro, D., 158, 284
Sharma, N. K., 229, 236(84), 242(84),
243(84), 268(84), 270(84), 274(84),
278(84)
Sharon, N., 239, 293, 296, 298(81), 305,
309, 311, 313(188), 324(150), 340(81,
370), 341(73), 342(73,81,370), 344(81,
89), 345, 406, 426, 441, 444(106)
Shaug, N., 331

Shaw, C. J. G., 56
Shaw, G., 18
Shaw, P., 254
Shechter, H., 19
Sheehan, J. C., 413, 420(53), 434(53)
Shen, L., 319
Shepherd, D. M., 56
Shevchenko, V. P., 34
Shimomura, T., 386, 387, 388(272), 389
Shoda, M., 366
Shostakovskii, M. F., 154
Shneer, R. Ya. (Pernikis), 159, 180(27),
182(27)
Shuey, E. W., 318
Siddiqui, B., 311
Siefert, E., 250
Simmons, D. A. R., 308
Simoni, R. D., 384
Sims, A. P., 397, 398, 399(397), 400(397),
403
Sinaÿ, P., 53, 131, 240, 296, 297, 299,
300(88), 341(88), 343(86,88,101),
344(102,105,106)
Sinclair, H. B., 228, 229(80), 231(80),
237(80), 268(80), 269(80), 274(80)
Sinda, E., 389
Singh, P. P., 80, 338(356), 339
Singh, S., 429
Sinott, M., 337(345), 339
Siro, M. R., 387, 388
Siskin, S. B., 96, 98(108)
Sitadevi, C., 334
Sjöberg, K., 16, 24(18)
Sjöblad, S., 327, 330(279), 332(279)
Slator, A., 378
Slaughter, M., 441
Slavenburg, J. H., 383, 389
Sleeter, R. T., 228, 229(80), 231(80),
237(80), 268(80), 269(80), 274(80)
Slessor, K. N., 117,220, 222(46), 228(46),
229(46), 231(46), 232(90), 235(46),
241(46), 244(46), 246, 254(90),
257(90), 258(90), 263(143), 264(46),
268(46), 269(46,90,122), 272(46),
274(46), 275(46), 276(46), 277(143)
Slomiany, A., 315, 316
Slomiany, B. L., 315, 316
Slooff, W. C., 367
Smets, L. A., 305
Smit, A. L. C., 262
Smith, D. C. C., 116, 161

Smith, D. F., 439
Smith, F., 217, 228(23), 237(23), 263(23), 266(23), 268(23), 272(23), 338(357), 339, 409
Smith, G., 85
Smith, H. C., 333
Smith, J. W., 136
Smith, M., 20, 47(45), 61, 62(279)
Smith, P. J. C., 200, 216, 231, 232(90), 254(90), 257(90), 258(90), 269(90)
Smith, W. L., 367, 369
Smrt, J., 37
Snyder, H. E., 363
Snyder, J., 331
Snyder, W. H., 49
So, L. L., 202
Sobell, M., 19
Sobotka, H., 215
Sohár, P., 28, 41
Sols, A., 351, 355, 363(50,51), 375(50,51), 385(51), 395(51)
Soltes, E. J., 283
Solv'yov, A. A., 261
Somers, K. D., 310
Somers, P. J., 247
Sommer, A., 367
Sondheimer, S. J., 47, 163, 172
Sonobe, T., 292, 339(70)
Šorm, F., 37
Soutome, S., 363, 366(77)
Souza, N. O., 390, 391
Speakman, P. R. H., 16, 17(25)
Speicher, W., 218, 220(37), 229(37), 231(37), 264(37), 266(37), 268(37), 270(37), 274(37)
Spencer, F., 283
Spencer, J. F. T., 303, 343(376), 345
Spencer, R. R., 168
Spencer-Martins, I., 388, 402
Speziali, G. A. G., 389
Spiegelman, S., 387, 388
Spiro, R. G., 304
Spoors, J. W., 121
Spranger, J. W., 332
Springer, G. F., 312, 313(195), 338(350), 339(197), 342(197,350), 343(197)
Springham, D. G., 385
Sproviero, J. O., 239
Srinivasan, V. R., 392
Srivastava, H. C., 80

Staal, G. E. J., 333
Stacey, B. E., 16, 26
Stacey, M., 29
Stahly, E. E., 244, 276(127)
Staněk, J., Jr., 158, 161(19), 162(19), 163, 164(19), 211(19), 226
Stanko, V. I., 136
Stark, J. H., 63
Starkovsky, N. A., 43
Staub, A. M., 312, 437
Staub, A. P. A., 108, 109(152), 110(152)
Steers, E., Jr., 406, 420(22), 421(22), 441(22), 442(22)
Šteffková, J., 29, 30
Stein, E. A., 352
Stein, S. S., 351
Steinbeck, K., 117
Steiner, M. R., 309, 310(179)
Steiner, S., 309, 310(179)
Steinpreis, R., 223, 266(57)
Stelling-Dekker, N. M., 374
Stellner, K., 310
Stening, T. C., 56, 57(241,243), 58(243)
Stephen, A. M., 249, 280(10), 281, 290(10), 303(10)
Sternbach, L. H., 281, 302(18), 337(18)
Sternlicht, H., 351
Stevens, C. L., 338(358), 339
Steyn-Parvé, E. P., 401
Stevens, E. S., 197, 201(123)
Stevens, J. D., 16
Stewart, J. C. M., 38
Stick, R. B., 406
Stick, R. V., 15, 74, 299, 330(97), 343(97)
Stipanovic, A. J., 197, 201(123)
St.-Jacques, M., 216
Stothers, J. B., 170, 254
Stowell, C. P., 407, 433(31)
Strate, G. D., 118
Strecker, G., 327, 330(280,281), 333(281)
Streefkerk, D. G., 249
Strominger, J. L., 318, 323
Strosberg, A. D., 440
Stuart, C. H., 340(356), 345
Stuart, R. S., 254(191), 255, 256(191), 257(191,192)
Stuhlsatz, H. W., 333
Sturgess, J. M., 326
Su, J.-C., 318
Suami, T., 34, 243, 278(126)
Subba, B. C., 136

Sugawara, S., 389
Suggs, J. W., 50
Sugihara, J. M., 13
Sugita, M., 301, 309(117)
Sugiyama, H., 249, 254, 256(177), 257(177)
Suhadolnik, R. J., 239
Sukegawa, M., 219, 223(40), 264(40), 266(40), 277(40)
Sumitomo, H., 153, 182(105), 183, 195(105), 210, 211(105)
Sumner, J. B., 444
Sundaralingam, M., 284
Sundararajan, P. R., 216
Sundberg, L., 407, 415, 431(66), 432(66)
Sung, S.-S. J., 330, 332(300)
Suomalainen, H., 380, 389
Surna, Ya. A., 180
Suryanarayan, K., 334
Sussman, M., 387(276), 388
Sutherland, I. W., 301, 323, 326(248)
Sutherland, J. D., 337(345), 339
Sutoh, K., 408, 419(35), 436(35)
Sutton, D. D., 355
Suzuki, M., 92, 93(86)
Suzuki, S., 289
Svanberg, O., 16, 24(18)
Svensson, S., 280, 290, 303(7,63), 304(7), 307(138), 327, 330(279), 332(279)
Sviridov, A. F., 78
Swan, E. P., 58
Swanny, B., 334
Sykes, E., 321, 331(241)
Symes, K. C., 308
Synge, R. L. M., 56
Szabó, G., 395
Szabó, L., 36, 37(149), 61
Szabó, P., 18, 61
Szarek, W. A., 78, 80, 93, 94(41), 96, 99, 231
Szeja, W., 92
Szulman, A. E., 313

T

Tachikawa, H., 158, 173(12), 181, 182(96), 185(96), 197(96), 199(96), 210(12)
Tacreiter, W., 354, 392(36), 394
Täufel, A., 403

Takahashi, M., 363, 366(77)
Takahashi, T., 383
Takano, M., 26
Takasaki, S., 313
Takemura, S., 119
Takeo, K., 35, 224, 225, 226(69), 228, 229(63,69,81), 230(69), 231(69), 233(69), 265(69), 266(63,69), 267(69), 268(69,81), 269(81), 270(69,81), 271(63), 273(69), 274(81), 278(69)
Takusagawa, F., 215, 216(12)
Tam, S. Y. K., 80, 94(45)
Tamm, C., 281, 282(20), 337(20)
Tanaka, T., 60
Tănăsescu, H., 86, 88(66), 89(66)
Tănăsescu, I., 86, 87(70,71,71), 88, 89(65 -72)
Tanenbaum, S. W., 392, 419, 420
Tanimura, A., 288
Tanner, W., 368, 371
Tanret, C. R., 157
Tanret, G., 390
Taravel, F. R., 254, 255(185), 257(180)
Tarelli, E., 19, 233, 249(95), 251(95), 252(95), 258(95)
Tarentino, A. L., 368, 369(133)
Tarrow, A. B., 439
Tate, M. E., 46
Tatlow, C. E. M., 29
Tatlow, J. C., 25, 29, 36(96)
Tatsumura, T., 331
Tatsuta, K., 227, 268(79), 271(79), 275(79)
Taunton-Rigby, A., 44
Taylor, B., 387(276), 388
Taylor, K. G., 338(358), 339
Taylor, K. J., 216
Tejima, S., 26, 103, 222, 225(51), 226(51), 228(67), 229(82), 235(67), 237(82), 244(67), 265(51), 266(67), 267(67,82), 268(82), 269(67,82), 271(68), 272(51, 67,82), 273(51,82), 274(67,82), 276(67), 278(67,82)
ten Berge, A. M. A., 390
Tener, G. M., 59, 60(271), 61(274)
Terada, O., 395
Terayama, H., 15
Terry, W. D., 437
Terui, G., 383
Terui, K., 181, 182(96), 185(96), 197(96), 199(96)

Thain, E. M., 62
Thaisrivongs, S., 147, 148(287), 149(287)
Theander, O., 15, 30
Thelwall, L. A. W., 238, 273(105)
Theriault, N. Y., 64, 66(308,309), 67(309)
Thiel, I. M. E., 31, 32(111,112), 225, 227(68), 240(68,71), 249, 250(167, 168), 263(68), 266(68), 276(76)
Thiem, J., 106, 107(146), 120
Thomas, E. J., 139
Thomas, G. H. S., 240, 241(117)
Thompson, A., 179, 223, 266(54), 422
Thompson, C. C. 381, 383(229)
Thompson, E. A., 64, 65(303,304), 66(303,304)
Thorpe, S. R., 332
Tibenský, V., 408, 419(36), 436(36), 442(36), 443(36)
Tierney, B., 16
Tikhomirova, A. S., 395
Timell, T. E., 283
Tingle, M. A., 392, 395(344)
Tipson, R. S., 16, 17, 24(19), 227, 229(77), 288, 423
Tipton, C. L., 430
Tkacz, J. S., 202, 370, 371(146)
Toda, K., 366
Todd, A. R., 36, 37(149), 59, 422
Todt, K., 134
Tollens, B., 280, 340(363,364), 344, 345
Tominaga, Y., 406, 407(17), 409(17), 429(17), 430(17), 439(17), 446(17), 447(17)
Tomshich, S. V., 301
Topper, Y. J., 316, 328(215)
Tormey, D. C., 331
Tosi, C., 394
Toubiana, M. J., 63
Toubiana, R., 63
Touster, O., 324
Towers, G. H. N., 392
Tracey, A. S., 117
Tracey, M. V., 366
Tramp, D., 136
Trayer, I. P., 423
Trefts, P. E., 334
Trentesaux-Chauvet, C., 327, 330(280)
Trevelyan, W. E., 390
Trimble, R. B., 368, 369

Tristram, H., 351
Troest, J., 333
Trofimov, B. A., 154
Tronchet, J. M. J., 79
Trovarelli, G., 395(386), 396
Trucco, R. E., 395
Tsai, C.-M., 439
Tsai, J.-H., 296, 343(82)
Tsay, G. C., 289, 324, 330, 332(300)
Tschesche, R., 302
Tsuchiya, T., 227, 268(79), 271(79), 275(79)
Tsujimoto, S., 92
Tu, C.-C., 180
Tucker, L. C. N., 23, 165, 166, 168
Tudos, F., 189
Tulloch, A. P., 63
Tuppy, H., 319, 322, 333(234)
Turino, G. M., 439
Turkova, J., 395(382), 396
Turner, A. F., 61
Turner, B. M., 333
Tuszynski, G. P., 304
Tuzimura, K., 249, 254, 256(177), 256(177)
Tzumori, K., 327

U

Ueda, T., 58
Ugolev, A. M., 366(114), 367
Uhlenbruck, G., 406, 446, 447(164)
Ulbricht, T. L. V., 138
Ulezlo, I. V., 402
Umezawa, H., 281, 302(22)
Umezawa, S., 227, 268(79), 271(79), 275(79)
Unemoto, K., 219, 223(40), 264(40), 266(40), 277(40)
Urbański, T., 58
Uryu, T., 158, 163, 173(12,13,14), 181, 182(93,95,96,97,108), 183, 185(96, 97), 187(93,108), 195, 197(96), 199(93,95,96,108), 201(93), 206, 210(12,13), 211(14,119,150)
Usov, A. I., 115, 116(164), 287
Usui, T., 249, 254, 256(177), 257(177)

Utamura, T., 219, 220(43), 249, 250(166), 253
Utille, J. P., 35
Uwajima, T., 395
Uy, R., 406, 407(14), 410(14), 431(14), 432(14), 442(14), 443(14)
Uzikova, V. N., 150

V

Vaheri, A., 305
Valentin, F., 281, 337(17)
van Beek, W. P., 305
van Boom, J. H., 37, 38, 42, 69
van Dam, B., 395
van Dam-Schermerhorn, L. C., 395
Vandenberg, E. J., 175
van den Bos, T., 389
Van Den Hamer, C. J. A., 433
van de Poll, K. W., 389, 390
Van der Heijden, M. C. M., 333
Van der Laan, L. C. J., 97
van der Stouwe, C., 49
van der Veen, J. M., 249
van der Walt, J. P., 364, 366, 374, 378
van Es, T., 24
van Hoof, F., 332
van Rooij, H., 305
van Stevenick, J., 381, 382(224), 384(224)
van Uden, N., 386, 388, 402
van Wijk, R., 389
Varadarajan, R., 229, 236(84), 242(84), 243(84), 268(84), 270(84), 274(84), 278(84)
Varadarajan, S., 36
Varma, A. J., 170, 171(62), 207(62)
Varma, R. K., 64
Vasseur, E., 395(383), 396
Vazquez, I. M., 31
Veibel, S., 392
Venkataramu, S. D., 118
Venkateswarlu, A., 64
Vercellotti, J. R., 240, 241(117)
Verdegaal, C. H. M., 69
Verhegge, G., 249
Verheyden, J. P. H., 48
Verhoeven, J., 42
Vernon, C. A., 351
Veruovic, B., 158, 182(110), 183, 195(110), 199(110), 223, 244(53), 272(53)

Veselý, V., 340(366), 345
Veyrières, A., 15
Veyssières-Rambaud, S., 134
Viard-Gaudin, C., 401
Vidershaim, G. Ya., 325, 326
Vigerani, A., 225, 249, 250(165), 266(72)
Vignon, M. R., 254(194), 255(184), 256(194), 257(180)
Villa, T. G., 392, 393(358)
Villanueva, J. R., 369, 370(144), 371, 372, 392, 393(358)
Vincendon, M., 254(194), 255, 256(194)
Viñuela, E., 351
Vis, E., 165
Vizsolyi, J. P., 61
Vliegenthart, J. F. G., 249, 250(162), 254(163)
von Euler, H., 354
von Grundherr, G. E., 360
von Hedenström, M., 354
von Schuching, S., 16
Voorhees, V., 420
Votoček, E., 281, 337(17,336), 339, 340(360,366), 344, 345
Vottero, P. J. A., 35
Vretblad, P., 406, 414, 430(62), 431(13)
Vreugdenhil, A. D., 154

W

Waalkes, T. P., 331
Wagenitz, E., 287
Wagner, H., 35
Wågstrom, B., 303
Waheed, A., 368
Waites, G. M. H., 214
Wakabayashi, K., 375
Walden, P., 351
Walker, A., 305
Walker, D. E., 325, 326(271)
Walker, D. L., 80, 94(45)
Walker, G. J., 202
Walker, T. E., 254, 255(185)
Wallack, M. K., 331
Wallenfels, K., 351, 394(10)
Wander, J. D., 72, 76(7), 146(7), 147(281)
Wandrey, C., 387, 388
Wang, J. L., 441
Warborg, E. F., Jr., 306
Ward, R. B., 179
Warner, G. A., 310

AUTHOR INDEX

Warren, C. D., 49, 50(204), 51(207), 52(207), 130
Warren, L., 304, 305, 306
Watanabe, K. A., 29, 36, 308, 309, 331(172)
Watkins, W. M., 313, 318, 319(226), 324(200), 325, 345
Way, R. C., 333
Webber, J. M., 21, 22, 23(46,56), 26(56), 27(77), 225
Weckerle, W., 79, 140, 142(270), 143(275), 144(275), 149(277)
Wedemeyer, K. F., 217
Weetall, H. H., 417, 418(70)
Wehrli, F. W., 254
Weidenhagen, R., 354, 360, 363, 386
Weigel, H., 246
Weil, C. E., 231, 258(86)
Weil, R., 394
Weimann, G., 39
Weiner, I. M., 323
Weisleder, D., 18, 217, 218(30), 243(30), 276(30)
Weiss, E., 280, 302(2)
Wellman, C., 150
Welton-Verstegen, G. W., 401
Wendorff, W. L., 395(388), 396
Wernicke, E., 338(349,351)
West, B. F., 16, 24(19)
West, G. B., 204
Westera, G., 154
Westmore, J. B., 64, 65(303,304), 66(303, 304)
Westphal, O., 295, 301, 308, 312, 340(79), 341(79), 421, 437
Whaley, T. W., 254, 255(185)
Wharton, P. S., 138
Whelan, W. J., 159, 207, 217, 218(25)
Whistler, R. L., 44, 63, 246, 247, 277(136), 278(136)
Whitaker, J. R., 394, 395(359), 396(376)
Whitehead, J. S., 311
Whitham, G. H., 139
Whyte, J. N. C., 303
Wiame, J. M., 398
Wickerham, L. J., 366
Wickner, R. B., 364
Wiegandt, H., 309
Wiggins, L. F., 22, 121, 217, 228(23), 237(23), 263(23), 266(23), 268(23), 272(23)

Wightman, R. H., 27
Wigner, J. H., 57, 58(249)
Wilchek, M., 411, 414, 431(63), 434(47)
Wilcox, C. S., 147, 148(287), 149(287)
Wild, G. M., 429
Wilkes, B. J., 354
Wilkinson, J. F., 316, 327(212)
Wilkinson, S. G., 280(9), 281(8), 303, 307(9,137), 308
Williams, E. H., 99
Williams, J. M., 167, 285, 290, 291(40), 341(64)
Williams, N. J., 367
Williams, N. R., 158
Williams, R. C., Jr., 312
Williamson, P., 338(350), 339, 342(350)
Willstaedt, E., 354
Willstätter, R., 354, 378, 395(204)
Wilson, D. B., 323
Wilson, E. J., Jr., 217
Wilson, K. J., 340(368), 345
Wilson, N. K., 254
Winderman, S., 387(282), 388
Winfield, J. B., 437
Wing, R. E., 44
Winge, Ø., 372, 376
Winkler, K., 403
Winstein, S., 107, 152(151)
Winzler, R. J., 326
Wiseman, A., 367
Witt, I., 403
Wittenburg, E., 22
Wofsy, L., 406, 407(7), 438(7)
Wold, F., 406, 407(14), 410(14), 431(14), 432(14), 442(14), 443(14)
Wolfe, D. E., 329
Wolfe, J. K., 25
Wolfrom, M. L., 58, 74, 179, 217, 219, 220(41), 223, 224, 225(20), 228(20), 229(41), 235(41), 237(20), 240, 241, 244, 263(20), 264(41), 266(41,54,65), 268(20), 272(20), 276(127), 295, 340(78), 422
Wollwage, P. C., 188
Woloszczuk, W., 408, 419(37), 436(37), 443(37)
Wong, C. M., 50
Wong, M. Y. H., 146
Woo, K. B., 331
Wood, C. J., 17, 77
Wood, H. B., Jr., 34, 121

Wood, H. C. S., 420
Wood, J. O., 256(202), 265
Wood, S., 333
Woods, B. M., 303
Woodward, F. N., 282
Worrall, R., 29
Wrede, S., 35
Wright, A., 323
Wright, G., 58
Wright, J. B., 96
Wright, K., 303, 323(131)
Wright, M., 139
Wright, R. S., 60, 61(274)
Wu, Y.-T., 408
Wunderlich, K., 107
Wurster, B., 351

Y

Yagi, H., 395
Yagi, K. J., 246, 277(136), 278(136)
Yamada, K., 80, 94(44), 305
Yamaguchi, H., 172, 208, 301
Yamamoto, K., 331
Yamamoto, S., 366, 367
Yamamoto, T., 368
Yamanaka, K., 327
Yamaoka, N., 249, 254, 256(177), 257(177)
Yamashita, K., 313, 327, 330(282)
Yang, Y., 202
Yariv, J., 340(368), 345
Yarotskii, S. V., 78
Yarrow, D., 348, 361(5), 362(5), 374(5), 395(5), 400(5), 402(5)
Yashphe, J., 395, 396
Yasui, A., 182(105), 183, 195(105), 211(105)
Yates, A. D., 325
Yazawa, H., 50
Yeger, H., 441
Yokoyama, M., 249
Yoshida, T., 335
Yoshikoshi, A., 80

Young, N. M., 408, 419(34), 436(34), 446
Young, R. N., 86
Young, W. W., Jr., 310
Youssefyeh, R. D., 87
Yuen, G. U., 217, 225(20), 228(20), 237(20), 263(20), 268(20), 272(20)
Yunker, M. B., 80, 94(45)
Yurchenco, P. D., 306
Yurkevich, V. V., 363, 366(75), 373

Z

Zachoval, J., 158, 181, 182(92,108), 183(92), 184(92), 185(92), 186(92, 108), 188(92), 199(108), 207, 208(92)
Zajac, W. W., Jr., 122, 123(204), 129, 136
Zakharkin, L. I., 136
Zanlungo, A. B., 114
Zapromctova, O. M., 402
Zarkov, G. E., 172
Zarubinskii, G. M., 88, 95(82)
Zbiral, E., 67
Zehavi, U., 293, 341(73), 342(73)
Zeile, K., 47
Zelnik, R., 281, 282(19), 302(19), 337(19)
Žemlička, J., 37
Zemplén, G., 45, 215, 223
Zhukova, I. G., 389
Zickler, F., 403
Ziderman, I., 247
Zimmermann, F. K., 373, 386, 389, 390, 402
Zinner, H., 22
Ziv, O., 390, 391(314)
Zlotskii, S. S., 118, 150
Zopf, D. A., 407, 420(27), 426(27), 439, 440(27)
Zoutewelle, G., 390
Zsoldos, V., 337(338), 339
Zsolt, J., 386, 387, 388(269)
Zurabyan, S. E., 217, 218(28), 263(28)
Zurr, D., 85
Zydek-Cwick, C. R., 364

SUBJECT INDEX

A

Acetalation, see also Transacetalation
 by 2-acetoxypropene, 76
 agents, 76
 catalysts, 80
Acetals
 acetolysis, 24-26
 bicyclic, hydrogenolysis, 133
 cyclic, acetobrominolysis, 114
  Grignard reagent cleavage, 153
  halogenation, 95-121
  halogenation, reagents causing migration of acetal group, 115-121
  hydrogenolysis, 121-138
  hydrogenolysis of unsaturated, 136-138
  maltose derivatives, 237, 238
  migration, 115
  oxidation by chromium trioxide, 82-85
  oxidation by potassium permanganate, 82
  oxidation by triphenylmethyl fluoroborate, 85, 86
  ozonolysis, 81, 82
  peroxide-induced rearrangement, 150
  photolysis, 86-95
  polymerization, 152
  as protecting groups, and as functional groups, 155, 156
  reaction with bromine, 118
  reaction with N-bromosuccinimide, 96-107
  reaction with N-chlorosuccinimide, 118
  reaction with dibromomethyl methyl ether, 119-121
  reaction with hydrogen fluoride, 117
  reaction with 1,3,5-trichloro-1,3,5-triazine-2,4,6-trione, 118
  reaction with triphenylmethyl fluoroborate, 107-113
  reactivity, of aldoses and aldosides, 71-156
 synthesis, 73
 hydrolysis, selective, 14-24
 isomerization, 26-28
Acetobrominolysis, of cyclic acetals, 114
Acetolysis
 acetal, 24-26
 benzyl ethers, selective, 45
 reagents, 25
Acrolein, acetal, hydrogenolysis and regioselectivity, 137
Adenosine
 3',5'-cyclic phosphate, hydrolysis, 61
 fusion reaction with 2',3',5'-tri-O-acetyladenosine, 37
 —, 5'-O-acetyl-
  fusion reaction with 2',3',5'-tri-O-acetyladenosine, 37
  preparation, 38
 —, N^6-benzoyl-3'-O-(3-benzoylpropanoyl)-2'-deoxy-, preparation, 43
 —, N^6-benzoyl-3',5'-di-O-(chloroacetyl)-, selective deacylation, 43
 —, 3',5'-di-O-acetyl-
  preparation, 40
  reaction with morpholine, 38
 —, 3',5'-di-O-acetyl-2'-deoxy-, selective deacylation, 36
 —, 3',5'-di-O-benzoyl-, preparation, 39
 —, N^6, N^6-dibenzoyl-2',3',5'-tri-O-benzoyl-, selective deacylation, 39
 —, 2',3',5'-tri-O-acetyl-
  fusion reaction with 5'-O-acetyladenosine, 37
  with adenosine, 37
 hydrazinolysis, 40
 selective deacylation, 36
Agarose
 activation by 1,1'-bis(imidazole), 415, 416
  by bisoxirane method, 415
  by cyanogen bromide, 414, 430, 431
  by divinyl sulfone, 415
 affinity chromatography adsorbent, 407, 410, 413-416
Aldehydes, reaction with alditols, 21

SUBJECT INDEX

Alditols
 1-[2-(p-aminophenyl)ethyl]amino-1-deoxy-, as affinity chromatography adsorbents, 420, 426, 427, 431
 reactions with aldehydes, 21
 with ketones, 21
Aldonic acids, preparation, and benzimidazole derivatives, 5
Aldoses
 cyclic acetals, reactivity, 71–156
 2-deoxy-, synthesis, 147
Aldosides, cyclic acetals, reactivity, 71–156
Alkenes, photoaddition to 1,3-dioxolanes, 80, 94
Allofuranose, 1,6-anhydro-β-D-, synthesis, 163
—, 1,2:5,6-di-O-isopropylidene-α-D-, isomerization, 26
Allofuranoside, methyl 6-deoxy- 2,3-O-isopropylidene-β-D-, synthesis, 167
Allofuranosyl bromide, 2,3:5,6-di-O-isopropylidene-β-D-, reaction with base, 147
Allopyranose, 1,6-anhydro-2,3,4-tri-O-benzyl-β-D-, polymerization, 182, 187, 194
—, 4,6-O-isopropylidene-D-, synthesis, 73
—, 1,2,6-tri-O-acetyl-3-chloro-3-deoxy-4-O-(2,3,4,6-tetra-O-acetyl-α-D-glucopyranosyl)-β-D-, synthesis, 233
Allopyranoside, methyl 3-acetamido-2,4-di-O-acetyl-3,6-dideoxy-α-L-, selective deacylation, 30
—, methyl 2-O-acetyl-3,6-dichloro-3,6-dideoxy-4-O-(2,3-di-O-acetyl-4,6-dichloro-4,6-dideoxy-α-D-galactopyranosyl)-β-D-
 mass spectrum, 260
 proton nuclear magnetic resonance spectroscopy, 251
—, methyl 2-O-acetyl-3,6-dichloro-3,6-dideoxy-4-O-(2,3,4-tri-O-acetyl-6-chloro-6-deoxy-α-D-glucopyranosyl)-β-D-, proton nuclear magnetic resonance spectroscopy, 252
—, methyl 2,3-anhydro-4,6-O-benzylidene-α-D-, chlorination and migration of acetal group, 117
—, methyl 2,3-anhydro-6-deoxy-4-O-(2,3,-4-tri-O-acetyl-6-deoxy-α-D-glucopyranosyl)-β-D-, synthesis, 236
—, methyl 2,3-anhydro-4-O-(2,3,4-tri-O-acetyl-6-O-trityl-α-D-glucopyranosyl)-6-O-trityl-β-D-, synthesis, 236
—, methyl 1,3:4,6-di-O-benzylidene-β-D-, reaction with butyllithium, 144
—, methyl 2,3:4,6-di-O-benzylidene-α-D-, reaction with N-bromosuccinimide, 106, 107
—, methyl 3,6-dichloro-4-O-(6-chloro-6-deoxy-α-D-glucopyranosyl)-3,6-dideoxy-β-D-, synthesis, and tetraacetate, 234
—, methyl 3,6-dichloro-3,6-dideoxy-4-O-(4,6-dichloro-4,6-dideoxy-α-D-galactopyranosyl)-β-D-, triacetate, synthesis, 232, 234
—, methyl 2,6-dichloro-2,6-dideoxy-3,4-O-isopropylidene-α-D-, synthesis, 117
—, methyl 4,6-di-O-methyl-2,3-O-(o-nitrobenzylidene)-β-D-, irradiation, 89
—, methyl 2,3,6-tri-O-benzoyl-4-O-(2,3,4,-6-tetra-O-benzoyl-α-D-galactopyranosyl)-β-D-, synthesis, 230, 231
Allose, 1,2:5,6-di-O-isopropylidene-D-, selective hydrolysis, 15
Alluronic acid, DL, crystalline, 4
Allyl group, as protecting group, 49
Altropyranose, 1,6-anhydro-2,3,4-tri-O-benzyl-β-D-, polymerization, 182, 187
Altropyranoside, methyl 3-acetamido-2,4-di-O-acetyl-3,6-dideoxy-α-D-, selective deacylation, 29
—, methyl 2-azido-3,4-O-benzylidene-6-chloro-2,6-dideoxy-α-D-, reaction with N-bromosuccinimide, 102
—, methyl 2-O-benzoyl-4,6-O-benzylidene-3-bromo-3-deoxy-D-, synthesis, 112
—, methyl 2-O-benzoyl-3-bromo-3,6-dideoxy-α-L-, synthesis, 106
—, methyl 4,6-O-benzylidene-2-deoxy-2-iodo-α-D-, isomerization, 27
—, methyl 3,4-O-benzylidene-2,6-dichloro-2,6-dideoxy-α-D-, reaction with N-bromosuccinimide, 102
—, methyl 4,6-O-isopropylidene-α-D-, isomerization, 27

SUBJECT INDEX

Amino sugars, 2-deoxy, hydrogenolysis of acetals, 127
Ammonolysis, selective deacylation, 30–33
Amygdalin, structure, 359
Amylase, macerans, glycoside synthesis, 429
α-Amylase, purification by affinity chromatography, 443
Amylose, 6-O-acetylation, 63
1,5-Anhydroaldofuranoses, synthesis, 164–169
1,6-Anhydroaldofuranoses
  polymerization, 179–197
  synthesis, 161–164
1,6-Anhydroaldopyranans, characterizations and applications, 197–204
1,2-Anhydroaldopyranoses
  polymerization, 207–209
  synthesis, 171, 172
1,3-Anhydroaldopyranoses
  polymerization, 207
  synthesis, 169–171
1,4-Anhydroaldopyranoses
  polymerization, 204–206
  synthesis, 164–169
1,6-Anhydroaldopyranoses
  polymerization, 179–197
  synthesis, 161–164
Anhydroaldoses
  copolymerization, mechanism, 176–179
  nomenclature, 160
  polymerization, mechanism, 173–176
  synthesis, 160–173
Anhydroketoses, synthesis, 160–173
Anhydro sugars
  polymerization, 158–160, 173–212
  synthesis, 157–173
Antiobiotics, aminoglycosidic, synthesis, 95
Antibodies, anti-glycosyl, affinity chromatography, 437–441
Anticoagulant, Dicumarol, discovery, 1
Antigens
  bacterial, immunology, 311, 312
  blood-group, immunology, 313–316
  plant, immunology, 312
Antitumor activity, 1-(2-chloroethyl)-3-β-maltosyl-1-nitrosourea, 243

Arabinitol, 1,1-bis(benzamido)-1,5-dideoxy-, preparation, 31
Arabinofuranose, 3-O-acetyl-1,5-anhydro-2-O-benzoyl-β-L-, synthesis, 172
—, 5-O-benzoyl-2-deoxy-2-fluoro-3-O-formyl-D-, formyl group removal, 36
Arabinopyranose, 1,4-anhydro-2,3-di-O-methyl-α-L-, polymerization, 204
—, 1,2:3,4-di-O-benzylidene-β-D-, reaction with triphenylmethyl borofluoride, 112
Arabinopyranoside, benzyl 2-O-benzyl-3,4-O-benzylidene-β-D-, *endo*- and *exo*-, hydrogenolysis regioselectivity, 131
—, benzyl 3,4-O-benzylidene-β-D-, *endo*- and *exo*-, hydrogenolysis regioselectivity, 131
—, methyl 2-O-benzoyl-3,4-O-benzylidene-β-D-
  reaction with N-bromosuccinimide, 101
  with triphenylmethyl fluoroborate, 110
—, methyl 2-O-benzoyl-3,4-O-benzylidene-β-L-, diastereoisomers, hydrolysis, 19
—, methyl 2,4-di-O-benzyl-3-bromo-3-deoxy-β-D-, synthesis, 111
—, methyl 3,4-O-*endo*-ethylidene-β-L-
  irradiation, 93
  synthesis, 79
—, methyl 3,4-O-isopropylidene-2-O-methyl-β-D-, reaction with butyllithium, 142
Arabinose
  D-, diethyl dithioacetal, irradiation, 94
  L-, 1,2:3,4-diacetal, selective hydrolysis, 16
  dimethyl acetal, synthesis, 8
—, 1,4-anhydro-2,3-di-O-methyl-α-L-, synthesis, 165
—, 2,3:4,5-di-O-benzylidene-D-, diethyl dithioacetal, selective hydrolysis, 22
—, 2,3:4,5-di-O-benzylidene-L-, diethyl dithioacetal, selective hydrolysis, 22
—, 1,2:3,4-di-O-isopropylidene-β-D-, reaction with butyllithium, 142
—, 2,3:4,5-di-O-isopropylidene-D-dialkyl dithioacetals, selective hydrolysis, 22

SUBJECT INDEX

diphenyl dithioacetal, reaction with bases, 147
Arbutin, structure, 358
6-Azacytidine, 2′,3′,5′-tri-O-formyl-, methanolysis, selective, 37
Azaoxabicyclo compounds, hydrogenolysis, 134, 135
6-Azauridine, 2′,3′,5′-tri-O-formyl-, methanolysis, selective, 37

B

Bacteria, serological identification of pathogenic, 437
Benzenesulfonamide, N,N-dibromo-, brominations, 119
Benzoic acid, 2,4,6-trimethyl-, esters, hydrolysis, 34
Benzoylation, selective, of maltose derivatives, 225–227
Benzoylformyl group, in nucleoside chemistry, 41
3-Benzoylpropanoyl group, in nucleoside chemistry, 41
Benzoxonium ions
  reaction with nucleophiles, 113, 151, 152
  rearrangement, 101, 102
Bisoxirane
  affinity chromatography adsorbent supports, 431, 432
  agarose activation for affinity chromatography, 415
Brigl anhydride, synthesis, 161
Bromine, reaction with cyclic acetals, 118
Butanal, reaction with hexitols, 26
2-Butenyl group, as protective group, 51
tert-Butyldimethylsilyl group, O-protecting group, 64, 66, 67
Butyllithium, reaction with 1,3-dioxolanes, 138–148

C

Carbodiimide method, affinity chromatography adsorbent supports, 434, 435

Carbohydrates
  affinity chromatography adsorbents for macromolecular substances, 405–447
  selective chlorination, 231–235
  selective removal of protecting groups, 13–70
Carbon-13 nuclear magnetic resonance spectra
  aldopyranans, 198, 200
  maltose and derivatives, 254–258
Catalysts
  acetalation, 80
  polymerization, 183, 184
    for anhydroaldoses, 174
Cellobiitol, 6-O-benzoyl-, ammonolysis, 32
Cellobiose, structure, 358
—, 6-O-benzoyl-, ammonolysis, 31
α-Cellobiose, 2,3,6,2′,3′,4′,6′-hepta-O-acetyl-, preparation, 30
β-Cellobiose octaacetate, reaction with piperidine, 30
β-Cellobiose octabenzoate, ammonolysis, 31
Cellulose, affinity chromatography adsorbent, 407, 412, 413
Chartreusin, 302
D-fucose, 281
Cheiroside, 302
Chemical ionization mass spectra, maltose derivatives, 261–263
Chlorination, selective, of carbohydrates, 231–235
Chloroacetyl group
  as blocking group, 35
  in nucleoside chemistry, 42, 43
Chromatography
  affinity, adsorbent preparation, 406–409
  adsorbents with carbohydrate ligands for macromolecular substances, 405–447
  applications, 437–443
  of enzymes, 441–443
  of lectins, 444, 445
  techniques, 410–412
Chromium trioxide, cyclic acetal oxidation, 82–85
α-Chymotrypsin, selective deacylation agent, 43

Circular dichroism spectra, aldopyranans, 197, 201
Concanavalin A, affinity chromatography, 444
Conformation, changes during polymerization of anhydro sugars, 192–194
Convolvulin, D-fucose, 281
Copolymerization
  of anhydroaldoses, 176–179
  of 1,6-anhydrosugars, reactivity ratios, 188–194
Coumarin, 4,4'-methylenebis-4-hydroxy-, see Dicumarol
Curacose, occurrence, 281
Cyanogen bromide, agarose activation, for affinity chromatography, 414, 430, 431
Cycloheptaamylose, heptakis(6-azido-6-deoxy)-, hepta(2,3-diacetate), preparation, 33
Cyclohexaamylose, 6'-azido-6'-deoxy-, amylase reaction, 241
—, hexakis(6-azido-6-deoxy)-, preparation, 33
Cyclohexaamylose octadecabenzoate, selective deacylation, 33
Cystic fibrosis, α-L-fucosidase activity, 333, 334
Cytidine, 3',5'-cyclic phosphate, hydrolysis, 61
—, N^4-anisoyl-3'-O-(chloroacetyl)-2'-deoxy-5'-O-(p-methoxytrityl)-, selective deacylation, 43
—, N^4-benzoyl-2',3',5'-tri-O-benzoyl-methanolysis, selective, 37
  selective deacylation, 39, 40
—, 2'-deoxy-, 3'5'-cyclic phosphate, hydrolysis, 61

D

Daunosamine, synthesis, 140, 142
Deacetalation, hydrolysis, 14–24
Deacylation
  enzymic, 33, 34, 36, 43
  regioselective, 40
  selective, 28–44
    in nucleoside derivatives, 36–44
Dealkylation, selective, 44–53
Debenzylation, by acetolysis, 45
Deborination, selective, 53–55

Deboronation, selective, 53–55
Decasaccharide, in fucosidosis urine, structure, 332
4-Decene-3,6-diol, 6-butyl-1,2-dimethoxy-, synthesis, 141
Demethylation, of methylated sugars, 44, 45
Denitration, selective, 55–58
Dephosphonylation, selective, 58–62
Deprotection, see Protecting groups
De(trialkylsilyl)ation, selective, 62–70
Detritylation, selective, 48
Dextrans
  affinity chromatography adsorbent, 419
  synthesis, 196
Dibromomethyl methyl ether, reaction with cyclic acetals, 119–121
Dicumarol
  discovery, 1
  structure, 6
Digipronin, 302
Digitalose
  occurrence, 281
  steroid glycosides, 302
Dihydrocinnamoyl groups, enzymic hydrolysis, 43, 44
D-(+)-Dihydrocoumariloyl group, enzymic hydrolysis, 44
2,4-Dinitrophenylsulfenyl group, as protecting group, in nucleoside chemistry, 43
Diogenine, hydrogenolysis, 122
2,6-Dioxabicyclo[3.1.1]heptanes, synthesis, 169–171
2,7-Dioxabicyclo[2.2.1]heptanes
  polymerization, 210
  synthesis, 164–169
2,7-Dioxabicyclo[4.1.0]heptanes, synthesis, 171, 172
2,8-Dioxabicyclo[3.2.1]octanes, synthesis, 161–164
6,8-Dioxabicyclo[3.2.1]octanes
  hydrogenolysis, 133
  polymerization, 210
  synthesis, 161–164
6,8-Dioxabicyclo[3.2.1]octenes, polymerization, 210
1,3-Dioxolane, 2-(o-nitrophenyl)-, irradiation, 86
1,3-Dioxolanes, photoaddition to alkenes, 80, 94

Disaccharides
ammonolysis of perbenzoylated, 30–32
utilization by yeasts, 347–404
Divinyl sulfone
affinity adsorbent support activation, 432
agarose activation for affinity chromatography, 415

E

Enzymes, affinity chromatography purification, 441–443
Enzymic deacylation, see Deacylation
Erythritol, cis-2,3-O-benzylidene-, isomerization, 27
—, 1,3:2,4-di-O-benzylidene-, selective hydrolysis, 19
Erythrose, 2,4-O-benzylidene-D-, isomerization 26
—, 2,4-O-ethylidene-D-, isomerization, 26
Esculin, structure, 358
Esters, formation by peroxide-induced rearrangement of acetals, 150
1,1-Ethylenedioxacyclohexane, reaction with organolithium reagents, 138
1,1-Ethylenedioxacyclopentane, reaction with organolithium reagents, 138

F

Fibronectin, role in glycoproteins, 305
Formyl group, selective removal, 35, 37
Fructofuranose, 2,6-anhydro-1-O-methyl-D-, synthesis, 172
—, 2,3-anhydro-1,4,6-tri-O-nitro-D-, synthesis, 172
β-D-Fructofuranosidase
genetic analysis, 372
hydrolysis rates for sugars, 365
kinetics, 366
location in yeast cell, 366–371
substrate specificity, 356–361
Fructofuranosides
β-D-, regulation, 371–374
utilization by yeasts, 355–374
Fructopyranose, 1,2:4,5-di-O-isopropylidene-β-D-, selective hydrolysis, 17
—, 1,2:4,5-di-O-isopropylidene-3-O-methyl-β-D-, reaction with butyllithium, 145
—, 2,3:4,5-di-O-isopropylidene-1-O-methyl-β-D-, reaction with butyllithium, 144
—, 2,3:4,5-di-O-isopropylidene-1-O-(tetrahydropyran-2-yl)-β-D-, reaction with lithium diisopropylamide, 146
Fructose, 1,3,5,6-tetra-O-acetyl-4-O-benzoyl-*keto*-D-, synthesis, 83
Fucans, immunological aspects of complex, 311–316
Fucofuranoside, methyl α-L-, preparation, crystalline, 293
—, methyl β-L-, preparation, syrupy, 293
Fucoidans, L-fucose, 301
Fucolipids
biological functions, 309–311
isolation, 315
Fucomannan, 303
Fucopyranose, tetra-O-acetyl-α-D-, preparation, 295
—, tetra-O-acetyl-α-L-, preparation, 295
—, tetra-O-acetyl-β-D-, preparation, 295
—, tetra-O-acetyl-β-L-, preparation, 295
—, 2,3,4-tri-O-benzyl-1-O-(N-methylacetimidyl)-β-L-, preparation, 296
Fucopyranoside, methyl α-D-, preparation, 293
—, methyl α-L-, preparation, 293
—, methyl β-L-, preparation, 293
—, methyl 3,4-O-benzylidene-6-deoxy-β-D-, reaction with N-bromosuccinimide, 102
—, phenyl 2-O-benzyl-3,4-O-benzylidene-β-D-, *endo* and *exo*-, hydrogenolysis regioselectivity, 131
Fucopyranosides, L-, synthesis of complex, 297–299
Fucopyranosyl bromide, 2-O-benzyl-3,4-di-O-(p-nitrobenzoyl)-α-L-, preparation, 296
—, tri-O-acetyl-α-D-, preparation, 295, 296
—, tri-O-acetyl-α-L-, preparation, 295, 296
Fucopyranosyl chloride, tri-O-acetyl-α-L-, preparation, 295–297
Fucose
D- and L-
biological significance, 282

chemistry and biochemistry, 279–345
detection and determination, 289, 290
glycoside formation, by acid equilibration, 293, 294
glycoside formation, by ion-exchange resins, 293
isolation, 282
occurrence, 280, 281, 301–311
relative reactivities of hydroxyl groups, conformational effects, 290, 291
relative reactivities of hydroxyl groups, enhanced reactivity of OH-4, 291, 292
relative reactivities of hydroxyl groups, tritylation, 292
steric and electronic effects, of 5-substituent, 283–285
substituents and protecting groups, 290
D-, biosynthesis and degradation, 336
metabolism, 328, 329
properties of, and derivatives (tables), 337–339
synthesis, 285–288
toxicity, 329
L-, biosynthesis, 316–318
biosynthesis and degradation, 336
degradation, 326
degradation, by bacterial metabolism, 327
degradation, by mammalian metabolism, 327, 328
enzymic transfer to glycoproteins, 318–322
immune-defense system, 334, 335
properties of, and derivatives (tables), 340–345
synthesis, 288, 289
toxicity, 329
Fucose, 2,3-di-O-methyl-D-, see also Labilose
steroid glycosides, 302
—, 2-O-α-L-fucopyranosyl-L-, synthesis, 298
—, 2-O-methyl-L-, occurrence, 281
—, 3-O-methyl-D-, see Digitalose
—, 4-O-methyl-D-, see Curacose
α-D-Fucosidases, 326
α-L-Fucosidases, 324, 325

β-D-Fucosidases, 325, 326
β-L-Fucosidases, 326
Fucosides, L-, urinary, 330
Fucosidosis, 332, 333
Fucosphingolipids, occurrence, 308, 309
L-Fucosyl-agarose, affinity chromatography adsorbent, 444, 445
Fucosyltransferases
L-fucose transfer, 318–322
glycolipid substrates 322–324
in malignancy, 331

G

Galactitol
1,6-bis(diethylborinate) 2,3:4,5-bis(ethylboronate), selective cleavage, 55
hexanitrate, selective denitration, 58
1,6:2,3:4,5-tris(ethylboronate), selective cleavage, 55
Galactofuranose, 1,6-anhydro-α-D-, synthesis, 163
—, 1,6-anhydro-β-D-, synthesis, 169
—, 1,6-anhydro-2,3,5-tri-O-benzyl-β-D-, polymerization, 182
—, 3-O-benzoyl-1,2:5,6-di-O-isopropylidene-α-D-, selective hydrolysis, 15
—, 1,2:5,6-di-O-isopropylidene-α-D-, selective hydrolysis, 15
Galactopyranose, 1,6-anhydro-β-D-, synthesis, 163
—, 1,6-anhydro-2-O-benzyl-3,4-O-isopropylidene-β-D-, polymerization, 195
—, 1,6-anhydro-2-O-(methylsulfonyl)-3,4-O-(o-nitrobenzylidene)-β-D-, photo-rearrangement, 90
—, 1,6-anhydro-2,3,4-tri-O-benzyl-β-D-, polymerization, 182, 187
—, 1,4-anhydro-2,3,6-tri-O-methyl-β-D-polymerization, 204
synthesis, 165
—, 1,2:3,4-di-O-isopropylidene-α-D-, 6-nitrate, denitration, 57
—, 1,2:3,4-di-O-isopropylidene-6-O-[3,4,6-tri-O-acetyl-2-O-(4-oxopentanoyl)-β-D-galactopyranosyl]-α-D-, preparation and hydrazinolysis, 42
—, 4,6-O-isopropylidene-D-, synthesis, 73
—, 1,3,4,6-tetra-O-acetyl-β-D-, synthesis, 35

—, O-[3,4,6-tri-O-acetyl-2-O-(4-oxopentanoyl)-β-D-galactopyranosyl]-(1 → 2)-O-(3,4,6-tri-O-acetyl-β-D-galactopyranosyl)-(1 → 6)-(1,2:3,4-di-O-isopropylidene-α-D-, preparation and hydrazinolysis, 42
—, 2,3,4-tri-O-benzyl-D-, preparation, 50
Galactopyranoside, allyl 6-O-allyl-2,3,4-tri-O-benzyl-α-D-, deallylation, 50
—, allyl 2,3,4-tri-O-benzyl-6-O-(2-butenyl)-α-D-, isomerization and cleavage, 52
—, allyl 2,4,6-tri-O-benzyl-3-O-(2-butenyl)-α-D-, selective cleavage, 51
—, benzyl 6-O-allyl-2-O-benzyl-3,4-O-benzylidene-α-D-, endo- and exo-, hydrogenolysis regioselectivity, 132
—, benzyl 3-O-benzyl-4,6-O-benzylidene-2-O-(2,3,4,6-tetra-O-benzyl-β-D-galactopyranosyl)-β-D-, hydrogenolysis regioselectivity, 130
—, benzyl 2,3-di-O-benzyl-4,6-O-benzylidene-α-D-, hydrogenolysis regioselectivity, 129
—, benzyl 2,3-di-O-benzyl-4,6-O-benzylidene-β-D-, hydrogenolysis regioselectivity, 129
—, benzyl 2,6-di-O-benzyl-3,4-O-benzylidene-β-D, endo- and exo-, hydrogenolysis regioselectivity, 131
—, benzyl 2,3-di-O-methyl-4,6-O-benzylidene-β-D-, hydrogenolysis regioselectivity, 129
—, methyl α-D-, 4,6-phosphate, hydrolysis, 61
—, methyl 6-O-benzoyl-α-D-, synthesis, 29
—, methyl 4,6-O-benzylidene-α-D- hydrogenolysis regioselectivity, 125 reaction with N-bromosuccinimide, 99 synthesis, 78
—, methyl 4,6-O-benzylidene-β-D-, reaction with N-bromosuccinimide, 100
—, methyl 2,3-di-O-benzoyl-4,6-O-benzylidene-α-D-, irradiation, 92
—, methyl tetra-O-benzoyl-α-D-, selective deacylation, 29
—, methyl 2,3,4-tri-O-benzoyl-6-bromo-6-deoxy-β-D-, synthesis, 100
—, o-nitrophenyl β-D-, 6-phosphate, preparation, 63

—, phenyl 2,3-di-O-benzyl-4,6-O-benzylidene-α-D-, hydrogenolysis regioselectivity, 129
—, phenyl 2,3-di-O-benzyl-4,6-O-benzylidene-β-D-, hydrogenolysis regioselectivity, 129
Galactopyranosides
α-D-, utilization by yeasts, 374–378
β-D-, utilization by yeasts, 394–397
Galactose
D-, diethyl dithioacetal, irradiation, 94
utilization by yeasts, oxygen requirement, 397–401
—, 1,2:5,6-di-O-isopropylidene-D-, selective hydrolysis, 15
—, 2-O-α-L-fucopyranosyl-D-, synthesis, 296
—, 1,2,3,4,6-penta-O-benzoyl-D-, ammonolysis, 31
—, 2,3,4,6-tetra-O-benzyl-D-, synthesis, 50
α-D-Galactosidase, 375–378
β-D-Galactosidase, 394–397
purification by affinity chromatography, 442
Galactoside, p-aminophenyl β-D-, synthesis, 421, 422
Galactosiduronic acid, methyl D-, preparation, 3, 5
D-Galactosyltransferase, glycoside synthesis, 430
Galacturonic acid
D-, degradation, 4
diethyl dithioacetal, synthesis, 5
isolation, 3
DL-, crystalline, isolation, 4
D- and L-, preparation of crystalline, 4
Galacturonic acid, tetra-O-acetyl-aldehydo-D-
methyl ester ethyl hemiacetal, isolation, 5
synthesis, 5
Gitostin, 302
Glass, porous, for affinity chromatography adsorbent, 407, 417, 418
Glucan, carbon-13 nuclear magnetic resonance spectra, 200
Glucitol
D-, hexanitrate, selective denitration, 58
reaction with butanal, 26

—, 5-O-acetyl-1,6-di-O-benzoyl-2,4-O-methylene-D-, preparation, 25
—, 6-O-benzoyl-, ammonolysis, 31
—, 2,3-O-benzylidene-D-, isomerization, 27
—, 1,1-bis(benzamido)-1-deoxy-4-O-α-D-glucopyranosyl-D-, synthesis, 227, 239
—, 2,3-O-butylidene-D-, isomerization, 27
—, 2,3-O-butylidene-1-deoxy-D-, isomerization, 27
—, 1,6-di-O-benzoyl-2,4:3,5-di-O-methylene-D-, acetolysis, 25
—, 1,3:2,4-di-O-ethylidene-D-, 5,6-dinitrate, selective denitration, 57
—, 1,3:2,4-di-O-ethylidene-5,6-bis-O-(trifluoroacetyl)-D-, selective deacylation, 29
—, 2,4:5,6-di-O-isopropylidene-D-, selective hydrolysis, 23
—, 1,3:5,6-di-O-isopropylidene-2,4-O-methylene-D-, selective hydrolysis, 23
—, 2,4:3,5-di-O-methylene-D-, acetolysis, selective, 25
—, 2,3-O-ethylidene-D-, isomerization, 27
—, 1,3,5,6-tetra-O-acetyl-2,4-O-benzylidene-D-, oxidation by chromium oxide, 83
—, 1,2,3,4-tetra-O-benzoyl-5,6-O-isopropylidene-D-, oxidation by triphenylmethyl fluoroborate, 86
—, 1,3:2,4:5,6-tri-O-benzylidene-D-, selective hydrolysis, 22
Glucofuranose, 6-O-acetyl-1,2:3,5-di-O-isopropylidene-α-D-, selective hydrolysis, 24
—, 6-O-acetyl-1,2:3,5-di-O-methylene-α-D-
deacetalation, 24
oxidation by potassium permanganate, 82
—, 6-O-acetyl-1,2-O-isopropylidene-α-D-, 3,5-dinitrate, selective denitration, 57
—, 3,6-anhydro-5-O-benzoyl-1,2-O-isopropylidene-α-D-, synthesis, 103, 104
—, 5,6-anhydro-1,2-O-isopropylidene-α-D-, polymerization, 210

—, 5,6-anhydro-1,2-O-isopropylidene-3-O-methyl-α-D-, polymerization, 209
—, 5-O-benzoyl-6-bromo-1,2-O-isopropylidene-α-D-, synthesis, 103, 104
—, 3-O-benzyl-1,2-O-isopropylidene-α-D-, synthesis, 126
—, 3,5-O-benzylidene-1,2-O-isopropylidene-α-D-
hydrogenolysis regioselectivity, 126
isomerization, 27
reaction with N-bromosuccinimide, 103
—, 5,6-O-benzylidene-1,2-O-isopropylidene-α-D-, hydrogenolysis regioselectivity, 125
—, 5,6-O-benzylidene-1,2-O-isopropylidene-3-O-(methylsulfonyl)-α-D-, reaction with triphenylmethyl fluoroborate and tetrabutylammonium bromide, 108, 109
—, 6-bromo-6-deoxy-1,2:3,5-di-O-isopropylidene-α-D-, synthesis, 115, 116
—, 6-chloro-6-deoxy-1,2:3,5-di-O-isopropylidene-α-D-, synthesis, 116, 117
—, 6-deoxy-6-iodo-1,2:3,5-di-O-isopropylidene-α-D-, synthesis, 115, 116
—, 1,2:5,6-di-O-cyclohexylidene-α-D-, hydrogenolysis regioselectivity, 125
—, 1,2:5,6-di-O-isopropylidene-α-D-
hydrogenolysis, over copper chromite, 135
hydrogenolysis regioselectivity, 125
irradiation, 88
migration of 5,6-acetal group, 115, 116
—, 1,2:5,6-di-O-isopropylidene-3-O-(1-propenyl)-α-D-, acid hydrolysis, 49
—, 1,2-O-isopropylidene-α-D-, hydrogenolysis, over copper chromite, 135
—, 1,2,6-tri-O-acetyl-3,5-di-O-methyl-α-D-, selective deacylation, 28
—, 3,5,6-tri-O-acetyl-1,2-O-isopropylidene-α-D-
acetolysis and bromination, 114
reaction with hydrogen fluoride, 117
—, 3,5,6-tri-O-benzoyl-1,2-O-isopropylidene-α-D-, oxidation by triphenylmethyl fluoroborate, 85
Glucomannan, carbon-13 nuclear magnetic resonance spectra, 200
Glucopyranose, 1-O-abscisoyl-β-D-, synthesis, 36

—, 1-O-(p-acetoxybenzoyl)-2,3,4,6-tetra-O-acetyl-α-D-, deacetylation, 34
—, 1-O-(p-acetoxybenzoyl)-2,3,4,6-tetra-O-acetyl-β-D-, deacetylation, 34
—, 2-amino-2-deoxy-4-O-α-D-glucopyranosyl-D-, isolation and synthesis, 240
—, 1,4-anhydro-β-D-, synthesis, 165
—, 1,6-anhydro-β-D-
  acetobrominolysis, 115
  hydrogenolysis, 133
  polymerization, 179
  synthesis, 160, 163
  triesters, polymerization, 183
—, 1,6-anhydro-2,4-di-O-benzyl-3-O-crotyl-β-D-, polymerization, 181–183, 195
—, 1,6-anhydro-2,3-di-O-benzyl-4-O-(2,3,-4,6-tetra-O-benzyl-α-D-glucopyranosyl)-β-D-, polymerization, 195
—, 1,6-anhydro-2,3-di-O-benzyl-4-O-(2,3,-4,6-tetra-O-benzyl-β-D-glucopyranosyl)-β-D-, polymerization, 195
—, 5,6-anhydro-1,2-O-isopropylidene-α-D-, synthesis, 173
—, 5,6-anhydro-1,2-O-isopropylidene-3-O-methyl-α-D-, synthesis, 173
—, 1,2-anhydro-3,4,6-tri-O-benzyl-α-D-, synthesis, 172
—, 1,3-anhydro-2,4,6-tri-O-benzyl-β-D-
  polymerization, 207
  synthesis, 170, 171
—, 1,4-anhydro-2,3,6-tri-O-benzyl-α-D-
  polymerization, 205
  synthesis, 164
—, 1,6-anhydro-2,3-di-O-benzyl-4-O-(2,3,-4,6-tetra-O-benzyl-α-D-glucopyranosyl)-β-D-, synthesis, 223
—, 1,6-anhydro-4-O-α-D-glucopyranosyl-β-D-, synthesis, and hexaacetate, 235
—, 1,6-anhydro-2,3,4-tri-O-benzyl-β-D-, polymerization, 179, 181, 182, 187
—, 1,6-anhydro-2,3,4-tri-O-(p-bromobenzyl)-β-D-, polymerization, 182, 183
—, 1,6-anhydro-2,3,4-tri-O-ethyl-β-D-, polymerization, 181–183
—, 1,4-anhydro-2,3,6-tri-O-methyl-D-, synthesis, 164
—, 1,6-anhydro-2,3,4-tri-O-methyl-β-D-, polymerization, 159, 179–180, 182
—, 1,6-anhydro-2,3,4-tri-O-(p-methylbenzyl)-β-D-, polymerization, 182, 183
—, 1,6-anhydro-2,3,4-tri-O-p-xylyl-β-D-, polymerization, 183
—, 3,6-di-O-acetyl-1,2-anhydro-4-O-(2,3,-4,6-tetra-O-acetyl-α-D-glucopyranosyl)-α-D-, synthesis, 171
—, 2,3:4,6-di-O-isopropylidene-5-thio-α-D-, synthesis, 77
—, O-α-L-fucopyranosyl-(1 → 3)-O-β-D-galactopyranosyl-(1 → 4)-2-acetamido-2-deoxy-D-, preparation, 53
—, penta-O-acetyl-α-D-, reaction with piperidine, 30
—, penta-O-acetyl-β-D-, reaction with piperidine, 30
  selective deacylation, 28
—, 1,3,4,6-tetra-O-acetyl-β-D-, synthesis, 35
—, 1,2,4,6-tetra-O-acetyl-3-O-(chloroacetyl)-β-D-, selective deacylation, 35
—, 1,3,4,6-tetra-O-acetyl-2-O-[3,6-di-O-acetyl-2-deoxy-2-iodo-4-O-(2,3,4,6-tetra-O-acetyl-α-glucopyranosyl)]-α-D-, synthesis, 243
—, 1,2,4,6-tetra-O-acetyl-3-O-formyl-β-D-, selective removal of formyl group, 36
—, 2,3,4,6-tetra-O-acetyl-1-O-(tri-O-acetylgalloyl)-α-D-, deacetylation, 34
—, 2,3,4,6-tetra-O-benzyl-D-, selective acetolysis, 47
—, 2,3,4-tri-O-acetyl-1,6-anhydro-β-D-, polymerization, 179, 180, 183
—, 2,3,6-tri-O-acetyl-1,4-anhydro-α-D-, synthesis, 172
—, 3,4,6-tri-O-acetyl-1,2-anhydro-α-D-
  Brigl's synthesis, 171
  polymerization, 207, 208
—, 1,2,3-tri-O-acetyl-4,6-O-benzylidene-β-D-, irradiation, 92
Glucopyranoside, allyl 3-O-allyl-2,3,6-tri-O-benzyl-D-, synthesis, 170
—, benzyl 2-acetamido-3,4-di-O-acetyl-2-deoxy-β-D-, synthesis, 35
—, benzyl 4,6-O-benzylidene-2-(benzyloxycarbonyl)amino-2-deoxy-α-D-, hydrogenolysis regioselectivity, 127
—, benzyl 2,3-di-O-allyl-4-O-benzyl-β-D-, hydrogenolysis, 130
—, benzyl 2,3-di-O-benzyl-4,6-O-benzylidene-β-D-, hydrogenolysis regioselectivity, 128

—, benzyl 3,4,6-tri-*O*-acetyl-2-amino-2-deoxy-β-D-, selective deacylation, 29
—, ethyl 1-thio-α-D-, structure, 357
—, methyl α-D-
 2,3-bis(diethylborinate) 4,6-ethylboronate, selective cleavage, 55
 2,3-(diphenylcyclodiboronate)-4,6-phenylboronate, selective cleavage, 53
 4,6-phosphate, hydrolysis, 61
 structure, 356
—, methyl β-D-
 4,6-(phenyl phosphate), hydrolysis, 61
 2,3,4,6-tetranitrate, selective denitration, 56, 57
—, methyl 3-acetamido-2,4-di-*O*-acetyl-3,6-dideoxy-α-L-, selective deacylation, 29
—, methyl 6-*O*-acetyl-β-D-, preparation, 63
—, methyl 2-*O*-acetyl-6-deoxy-6-iodo-3-*O*-(methylsulfonyl)-4-*O*-(2,3,4-tri-*O*-acetyl-6-deoxy-6-iodo-α-D-glucopyranosyl)-β-D-, synthesis, 229, 230
—, methyl 2-*O*-acetyl-6-deoxy-6-iodo-3-*O*-(methylsulfonyl)-4-*O*-(2,3,4-tri-*O*-acetyl-6-*O*-*p*-tolylsulfonyl-α-D-glucopyranosyl)-β-D-, synthesis, 229, 230
—, methyl 2-*O*-acetyl-6-deoxy-3-*O*-(methylsulfonyl)-4-*O*-(2,3,4-tri-*O*-acetyl-6-deoxy-α-D-glucopyranosyl)-β-D-, synthesis, 236
—, methyl 3-*O*-acetyl-2-*O*-methyl-β-D-, 4,6-dinitrate, selective denitration, 56
—, methyl 2-*O*-allyl-3,4,6-tri-*O*-benzyl-α-D-, selective acetolysis, 47
—, methyl 4,6-*O*-anisylidene-α-D-, synthesis, 79
—, methyl 6-*O*-benzoyl-α-D-, preparation, 29
—, methyl 3-*O*-benzoyl-4,6-*O*-benzylidene-2-*O*-(trifluoroacetyl)-α-D-, selective deacylation, 36
—, methyl 4-*O*-benzoyl-6-bromo-6-deoxy-α-D-, preparation, 98
—, methyl 3-*O*-benzyl-4,6-*O*-benzylidene-α-D-, hydrogenolysis regioselectivity, 128
—, methyl 4,6-*O*-benzylidene-α-D-2,3-dinitrate, selective denitration, 56, 57
 hydrogenolysis regioselectivity, 125, 130
 reaction with *N*-bromosuccinimide, 97, 98
 synthesis, 78
—, methyl 4,6-*O*-benzylidene-2,3-di-*O*-benzyl-β-D-, hydrogenolysis regioselectivity, 125, 128
—, methyl 4,6-*O*-benzylidene-2,3-di-*O*-(trifluoroacetyl)-α-D-, selective deacylation, 29
—, methyl 6-chloro-6-deoxy-4-*O*-(4,6-dichloro-4,6-dideoxy-α-D-galactopyranosyl)-D-, synthesis, 232
—, methyl 2,3-di-*O*-acetyl-4-*O*-benzoyl-6-deoxy-α-D-, synthesis, 150
—, methyl 2,3-di-*O*-acetyl-4,6-*O*-benzylidene-α-D-, reaction with di-*tert*-butyl peroxide, 150
—, methyl 2,3-di-*O*-allyl-4,6-*O*-benzylidene-α-D-, selective deallylation, 51
—, methyl 2,3-di-*O*-benzoyl-4,6-*O*-benzylidene-α-D-
 irradiation, 92
 peroxide-induced rearrangement, 150
 reaction with di-*tert*-butyl peroxide, 150
—, methyl 2,3-di-*O*-benzyl-4,6-*O*-benzylidene-α-D-, hydrogenolysis regioselectivity, 128
—, methyl 2,3:4,6-di-*O*-isopropylidene-α-D-, synthesis, 77
—, methyl 4,6-*O*-ethylidene-α-D-, 2,3-dinitrate, selective denitration, 56
—, methyl 4,6-*O*-ethylidene-β-D-, 2,3-dinitrate, selective denitration, 56
—, methyl 4,5-*O*-(4-methoxybenzylidene)-2,3-di-*O*-methyl-α-D-, hydrogenolysis regiospecificity, 130
—, methyl 6-*O*-octadecanoyl-β-D-, preparation, 63
—, methyl 4,6-*O*-propylidene-α-D-, 2,3-dinitrate, selective denitration, 56
—, methyl 4,6-*O*-syringylidene-α-D-, synthesis, 79
—, methyl tetra-*O*-acetyl-α-D-, selective deacylation, 28
—, methyl 2,3,4,6-tetra-*O*-acetyl-β-D-
 oxidation by chromium oxide, 83
 selective deacylation, 28
—, methyl 4,6-*O*-(tetraisopropyldisilox-

ane-1,3-diyl)-α-D-, preparation and isomerization, 69
—, methyl 2,3,4,6-tetrakis-O-(trimethylsilyl)-α-D-, selective methanolysis, 62, 63
—, methyl 5-thio-α-D-, 6-phosphate, preparation, 63
—, methyl 2,3,4-tri-O-(trimethylsilyl)-α-D-, preparation, 62
—, methyl 4,6-O-veratrylidene-α-D-, synthesis, 79
—, p-nitrophenyl α-D-, structure, 357
—, o-nitrophenyl tetra-O-acetyl-β-D-, selective deacylation, 28
—, phenyl 2,3-di-O-benzyl-4,6-O-benzylidene-α-D-, hydrogenolysis regioselectivity, 128
α-D-Glucopyranoside hydrolases
  molecular weights, 388
  yeast, substrate specificities, 384–390
Glucopyranosides
  α-D-, uptake by yeasts, 381–391
  utilization by yeasts, Michaelis constants, 382
  β-D-, utilization by yeasts, 391–394
β-D-Glucopyranosiduronic acid, methyl 2,3-di-O-methyl-4-O-(2,3,4,6-tetra-O-methyl-α-D-glucopyranosyl)-
  methyl ester, degradation, 247, 248
  synthesis, 222, 246
Glucopyranosylamine, 2,3,6-tri-O-acetyl-4-O-(2,3,4,6-tetra-O-acetyl-α-D-glucopyranosyl)-β-D-, synthesis, 239
Glucopyranosyl azide, 2,3,6-tri-O-acetyl-4-O-(2,3,4,6-tetra-O-acetyl-α-D-glucopyranosyl)-β-D-, synthesis, 239
Glucopyranosyl bromide, 4-O-benzoyl-2-bromo-2,6-dideoxy-3-O-formyl-α-L-, synthesis, 120
Glucopyranosyl chloride, 3,4,6-tri-O-acetyl-2-O-(trichloroacetyl)-β-D-, selective deacylation, 35
Glucopyranosyl gallate, 2-O-cinnamoyl-β-D-, hydrolysis, 35
Glucopyranosyl 2,4,6-trimethylbenzoate, 2,3,4,6-tetra-O-acetyl-α-D-, deacetylation, 34
—, 2,3,4,6-tetra-O-acetyl-β-D-, deacetylation, 34
Glucose, D-, diethyl dithioacetal, irradiation, 94

—, 1,6-anhydro-2,3,4-tri-O-benzyl-β-D-, acetolysis, 45
—, 6-O-benzoyl-D-, ammonolysis, 31
—, 2-O-benzyl-D-, synthesis, 30
—, 2-O-carbanilino-D-, synthesis, 30
—, 1,2:5,6-di-O-isopropylidene-D-, selective hydrolysis, 15
—, 2,3:5,6-di-O-isopropylidene-4-O-(4,6-O-isopropylidene-α-D-glucopyranosyl)-aldehydo-D-, dimethyl acetal, synthesis, 238
—, 1,2:3,4-di-O-isopropylidene-5-thio-D-, synthesis, 77
—, 4-O-α-D-glucopyranosyl-2,3:5,6-di-O-isopropylidene-aldehydo-D-, dimethyl acetal, synthesis, 238
—, 4-O-(α-D-glucopyranosyluronic acid)-, D-, synthesis, 246
—, 4-O-(α-D-ribo-hexosyl-3-ulose)-D-, by maltose microbial oxidation, 246
—, 2-O-methyl-D-, synthesis, 30
—, 1,2,3,4,6-penta-O-benzoyl-D-, ammonolysis, 31
—, 2,3,4,6-tetra-O-benzyl-D-, synthesis, 50
Glucose aldehydrol, 1(R)-1,3,4,5,6-penta-O-acetyl-1,2-O-isopropylidene-aldehydo-D-, reaction with hydrogen bromide in acetic acid, 114
Glucoseptanose, D-, 1,2:3,4-diacetal, selective hydrolysis, 16
α-D-Glucosidase
  regulation, 388–390
  substrate specificity, 387
β-D-Glucosidases, substrate specificities, 392, 393
D-Glucosyltransferase, glycoside synthesis, 429
Glucuronic acid
  D-, degradation, 4
  L-, crystalline, 4
Glycals, synthesis, 147
Glycans
  biosynthesis, 318
  fucose occurrence, 301–311
Glycerol, 1,2-O-isopropylidene-DL-, isomerization, 27
—, 1,2 (or 2,3)-O-isopropylidene-DL-, hydrogenolysis, stereoselectivity, 124
Glycerol 1-(diethylborinate) 2,3-ethylboronate, selective cleavage, 54

Glycogen, affinity chromatography adsorbent, 419
Glycolipids
  animal, occurrence, 308
  structure determination, 309
  biosynthesis, 310
  fucose-containing, 307–311
Glycolipid substrates, fucosyltransferases, 322–324
Glycopeptides, L-fucose-containing, 306
Glycoproteins
  biosynthesis, 320, 321
  L-fucose-containing, 301, 304–307
  L-fucose enzymic transfer, 318–322
Glycosidases, 304, 324
  glycoside hydrolysis exogenously, 353–378
  yeast, 350
Glycosides
  alkali-sensitive, 6
  allyl, as affinity chromatography adsorbents, 420, 427–429
  aminoalkyl, as affinity chromatography adsorbents, 420, 423, 424, 431
  p-aminophenyl, as affinity chromatography adsorbents, 419–421, 431
    enzymic synthesis, 429, 430
    synthesis, 421–423
  cardiac, D-fucopyranosides, 302
  1-thio, as affinity chromatography adsorbents, 420, 421
  utilization by yeasts, entry into cell, 379–381
    hydrolyzed inside plasmalemma, 378–397
    hydrolyzed outside plasmalemma, 353–378
    oxygen requirement, 397–401
    structure and hydrolysis, 349–353
Glycosphingolipids, biosynthesis, 318
Glycosylamine, N-acyl-, as affinity chromatography adsorbents, 420, 426
Glycosyl 6-aminohexyl diphosphates, as affinity chromatography adsorbents, 424–426
Glycosyltransferases, 318
Gracibioside, 302
Grignard reagents, cyclic acetal cleavage, 153
Guanosine, 3′,5′-cyclic phosphate, hydrolysis, 61

—, N^2-benzoyl-5′-O-benzoyl-2′-O-(4-methoxytetrahydropyran-4-yl)-, preparation, 37
—, N^2-benzoyl-2′,3′,5′-tri-O-benzoyl-, selective deacylation, 39, 40
—, 3′,5′-di-O-acetyl-N^2-benzoyl-, preparation, 40
—, 3′,5′-di-O-acetyl-2′-deoxy-, selective deacylation, 36
—, 2′,3′,5′-tri-O-acetyl-N^2-benzoyl-, selective deacylation, 40
Guaran, affinity chromatography adsorbent, 419
Gulofuranose, 3-O-benzyl-1,2:5,6-di-O-isopropylidene-α-D-, selective hydrolysis, 15
Gulopyranoside, methyl 3-acetamido-2,4-di-O-acetyl-3,6-dideoxy-α-D-, selective deacylation, 30
—, methyl 6-O-benzoyl-3-bromo-3-deoxy-β-D-, synthesis, 100
—, methyl 4-O-benzoyl-3-bromo-3,6-dideoxy-D-, synthesis, 102
Gulopyranosyl chloride, 6-deoxy-2,3-O-isopropylidene-4-O-methyl-β-L-, reaction with base, 147, 148
Gulose, 1,2:5,6-di-O-isopropylidene-D-, selective hydrolysis, 15

H

Halogenation
  cyclic acetal, 95–121
  reagents causing migration of acetal groups, 115–121
Heparin, disaccharide, structure, 240, 241
Heptitol, 2,6-anhydro-1,3-O-p-anisylidene-5,7-O-benzylidene-D-*glycero*-L-*manno*-, selective hydrolysis, 20
—, 2,6-anhydro-1,3:5,7-di-O-benzylidene-D-*glycero*-L-*manno*-, selective hydrolysis, 20
—, 3,5:6,7-di-O-isopropylidene-D-*glycero*-D-*gulo*-, selective hydrolysis, 23
Heptonic acid, 7-amino-7-deoxy-, derivatives, synthesis, 63
Heptono-1,4-lactone, 3,5:6,7-di-O-isopropylidene-D-*glycero*-D-*gulo*-, selective hydrolysis, 23

Heptopyranose,1,4-anhydro-2,3:6,7-di-*O*-isopropylidene-α-D-*glycero*-D-*allo*-, synthesis, 168

Hex-1-enitol, 6-*O*-acetyl-1,5-anhydro-3-azido-1,2,3-trideoxy-4-*O*-(2,3,4,6-tetra-*O*-acetyl-α-D-glucopyranosyl)-D-*ribo*-, synthesis, 242

—, 1,5-anhydro-3-*C*-butyl-1,2,6-trideoxy-L-*arabino*-, synthesis, 142, 143

—, 1,2:4,5-di-*O*-isopropylidene-3-*O*-methyl-D-*arabino*-, synthesis, 145

Hex-5-enofuranose, 5,6-dideoxy-1,2-*O*-isopropylidene-α-D-, irradiation, with 1,3-dioxolane, 80

Hex-2-enopyranoside, methyl 2,3,6-trideoxy-α-L-*erythro*-, synthesis, 142, 143

—, methyl 2,3,6-trideoxy-5-*O*-(2,3,4-tri-*O*-acetyl-6-deoxy-α-D-glucopyranosyl)-β-D-*erythro*-, synthesis, 242, 243

α-L-*threo*-Hex-4-enopyranosiduronic acid, methyl 4-deoxy-2,3-di-*O*-methyl-, methyl ester, synthesis, 248

Hex-2-enose, (*E*)-4,5,6-tri-*O*-acetyl-2,3-dideoxy-*aldehydo*-D-*erythro*-, synthesis, 114

Hex-5-enose, 2,3,4-tri-*O*-acetyl-5,6-dideoxy-D-*xylo*-, diethyl acetal, irradiation, 95

Hex-5-enulopyranose, 5-deoxy-2,3-*O*-isopropylidene-1-*O*-(tetrahydropyran-2-yl)-β-D-*threo*-, synthesis, 145

Hexitol, 1,1-bis(benzamido)-6-*O*-benzoyl-1-deoxy-D-, preparation, 31

—, 2-deoxy-D-*arabino*-, reaction with butanal, 27

—, 2,3,4-trideoxy-DL-*glycero*-, 1-diethylborinate 5,6-ethylboronate, selective cleavage, 54

Hexofuranose, 3-deoxy-1,2:5,6-di-*O*-isopropylidene-α-D-*ribo*-, selective hydrolysis, 15

Hexofuranoside, methyl 5,6-*O*-cyclohexylidene-3-deoxy-2-*C*-methyl-β-D-*ribo*-, acetal cleavage by Grignard reagent, 154

Hexofuranos-3-ulose, 1,2:5,6-di-*O*-isopropylidene-α-D-*ribo*-, acetal cleavage by Grignard reagents, 154

Hexopyranoside, methyl 4,6-*O*-alkylidene-D-, 2,3-dinitrate, selective denitration, 56

—, methyl 4-*O*-benzoyl-6-bromo-2,6-dideoxy-3-*C*-methyl-3-*O*-methyl-α-L-*arabino*-, synthesis, 99

—, methyl 4-*O*-benzoyl-6-bromo-2,6-dideoxy-3-*C*-methyl-3-*O*-methyl-α-D-*xylo*-, synthesis, 99

—, methyl 4,6-*O*-benzylidene-D-, 2,3-dinitrate, selective denitration, 56

—, methyl 4,6-*O*-benzylidene-2,3-di-*O*-methyl-α-D-, reaction with butyllithium, 140

Hexose, 4-deoxy-D-*xylo*-, synthesis, 63

Hexos-2-ulose, 3-deoxy-4,6-*O*-benzylidene-, synthesis, 144

Hexos-3-ulose, 2-deoxy-4,6-*O*-benzylidene-, synthesis, 139, 140

—, 1,2,5,6-tetra-*O*-acetyl-4-*O*-formyl-*keto*-D-*arabino*-, synthesis, 83

5-Hexulosonate, methyl 2,3,4,6-tetra-*O*-acetyl-D-*xylo*-, synthesis, 83

Hexuronic acids
history, 2
synthesis, 4

Hydrazine hydrate, deacylating agent, 39, 40

Hydrazinolysis, regioselective, 40

Hydrogen bromide, acetolysis and bromination of cyclic acetals, 113–115

Hydrogen fluoride, reaction with cyclic acetals, 117

Hydrogenolysis
of bicyclic acetals, 133–135
of cyclic acetals, 121–138
rate of reductive cleavage, 122
reducing systems, 135, 136
regioselectivity, 123, 124
stereoselectivity, 124–135
of cyclic orthoesters, 79
unsaturated cyclic acetals, 136–138

Hydrolysis
acetal, selective, 14–24
enzymic, selective, 43

Hydroxyaminolysis, regioselective, 40

I

Iditol, 1,2:5,6-di-*O*-isopropylidene-D-, 3,4-trithiocarbonate, acetobrominolysis, 114

Idopyranose, 1,4-anhydro-2,3,6-tri-*O*-methyl-L-, synthesis, 165

SUBJECT INDEX

Idopyranosyl bromide, 3,4-di-O-benzoyl-6-bromo-6-deoxy-2-O-formyl-α-D-, synthesis, 120, 121
Idose, 1,2:5,6-di-O-isopropylidene-L-, selective hydrolysis, 15
Imidazole, 1,1'-carbonylbis-, agarose activation for affinity chromatography, 415, 416, 436
Immune-defense system, fucose, 334, 335
Immunology, of complex fucans, 311–316
Inosine, 2',3',5'-tri-O-benzoyl-, selective deacylation, 39
muco-Inositol, phophatidyl-, selective deacetylation in preparation, 34
myo-Inositol
  1(3),2-cyclic phosphate, hydrolysis, 60
  hexaphosphate, selective dephosphonylation, 59
—, α-L-fucopyranosyl-, in urine, 330
—, hexa-O-benzyl-, selective acetolysis, 46
—, 1,3,4,5,6-penta-O-benzyl-, selective acetolysis, 46
—, DL-1,4,5,6-tetra-O-benzyl-, selective acetolysis, 46
—, DL-1,4,5,6-tetra-O-benzyl-3-O-methyl-, selective acetolysis, 46
Inulin
  utilization by yeast, 361–366
  structure, 362
Inulinase, 361–366
Isomaltose, structure, 356
Isomerization, acetal, 26–28

K

Kamaloside, 302
Ketones, reactions with alditols, 21
Koenigs–Knorr synthesis, hexuronic acid, 5

L

Labilose, occurrence, 281
Lactose, structure, 359
β-Lactose, 6-O-benzoyl-, preparation, 32
—, 1,2,6,2',3',4',6'-hepta-O-benzoyl-, ammonolysis, 31, 32

Lactoside, benzyl penta-O-benzyl-4',6'-O-benzylidene-β-, hydrogenolysis regioselectivity, 130
Lactosyl bromide, 2,3,6,2',3',4',6'-hepta-O-acetyl-, reaction with pyridine, 32
Lectins, purification by affinity chromatography, 444, 445
Ledienoside, fucose-containing, 281, 302
Levoglucosan, *see* Glucopyranose, 1,6-anhydro-β-D-
Link, Karl Paul Gerhardt, obituary, 1–12
Lipopolysaccharides, bacterial, L-fucose-containing, 307, 308
Lithium diisopropylamide, reaction with acetals, 145
Lyxofuranose, 1,5-anhydro-2,3-O-isopropylidene-β-D-, synthesis, 168
—, 2,3-O-isopropylidene-D-, synthesis, 74
Lyxofuranoside, methyl 5-O-benzoyl-2,3-O-isopropylidene-α-D-, reaction with dibromomethyl methyl ether, 120
Lyxopyranoside, methyl 4-O-benzoyl-2,3-O-benzylidene-α-D-, reaction with triphenylmethyl fluoroborate, 110–112
—, methyl 2,4-di-O-benzoyl-3-bromo-3-deoxy-α-D-, synthesis, 112

M

Macromolecular substances, affinity chromatography, with carbohydrate-bearing adsorbents, 405–447
Macrophages, α-L-fucosidase activity, 334, 335
Malignancy, fucose levels, 331
Maltal, hexa-O-acetyl-, synthesis, 242
Maltitol, 6-O-benzoyl-, ammonolysis, 31
Maltosamine, N-acetyl-, synthesis, 240, 241
—, N-benzoyl-, synthesis, 227, 239
Maltose, 213–278
  acetates and benzoates, properties, 266, 267
  acetolysis, 248
  alkaline degradation, 247, 248
  anhydro derivaties, properties, 272
  carbon-13 nuclear magnetic resonance spectroscopy of, and derivatives, 254–258

carboxylic derivatives, properties, 277
chemical ionization mass spectra, 261–263
complex-forming reactions, 245
cyclic acetals, properties, 273
deoxy derivatives, properties, 274
derivatives, tables of properties, 263–278
halides, properties, 269, 270
hydrolysis, 248
mass spectrometry of, and derivatives, 258–263
methyl ethers, properties, 265
microbial oxidation, 246
nitrogen-containing derivatives, properties, 275, 276
polymer-forming reactions, 244
proton nuclear magnetic resonance spectroscopy, 249–253
selective-oxidation reactions, 246, 247
structure, 214, 215, 356
sulfonates, properties, 267, 268
sulfur-containing derivatives, properties, 276
synthesis, 215, 216
trityl ethers, properties, 264
—, 6'-amino-6'-deoxy-, synthesis, 241
—, 6'-azido-6'-deoxy-, synthesis, 241
—, 6,6'-dichloro-6,6'-dideoxy-, synthesis, 232
—, 1,2,3,2',3',4'-hexa-O-acetyl-6,6'-di-O-trityl-, synthesis, 219
—, per-O-methyl-, synthesis, 223
—, per-O-methyl-6,6'-di-O-trityl-, synthesis, 223
—, per-O-methyl-6'-O-trityl-, synthesis, 223
—, 6'-thio-, derivatives, synthesis, 244
α-Maltose, structure, 215, 216
—, 1,2,3,2',3',4'-hepta-O-acetyl-6-O-trityl-, synthesis, 219
β-Maltose
monohydrate, structure, 215, 216
reaction with sulfuryl chloride, 231–233
selective benzoylation, 225, 226
selective methanesulfonylation and p-toluenesulfonylation, 227–229
structure, 216
—, 6'-amino-1,6-anhydro-6'-deoxy-, synthesis, 241

—, 1,6-anhydro-2'-O-benzoyl-4',6'-O-benzylidene-, synthesis, 226
—, 1,6-anhydro-4',6'-O-benzylidene-selective benzoylation, 225, 226
synthesis, 237
—, 1,6-anhydro-2,2'-di-O-benzoyl-4',6'-O-benzylidene-, synthesis, 226
—, 1,6-anhydro-2',3'-di-O-benzoyl-4',6'-O-benzylidene-, synthesis, 226
—, 1,6-anhydro-2,3,2',3',4',6'-hexa-O-acetyl-6-thio-, synthesis, 244
—, 1,6-anhydro-per-O-benzyl-, polymerization, 244
—, 1,6-anhydro-2,3,2',3'-tetra-O-benzoyl-4',6'-O-benzylidene-, synthesis, 226
—, 1,6-anhydro-2,2',3'-tri-O-benzoyl-4,6-O-benzylidene-, synthesis, 226
—, 6-O-benzoyl-
ammonolysis, 227
preparation, 31, 32
—, 6-O-β-D-galactosyl-, synthesis, 244
—, 1,2,3,2',3',4',6'-hepta-O-acetyl-, synthesis, 225, 235
—, 1,3,6,2',3',4',6'-hepta-O-acetyl-, synthesis, 224, 225
—, 2,3,6,2',3',4',6'-hepta-O-acetyl-, synthesis, 223
—, 1,2,3,2',3',4',6'-hepta-O-acetyl-6-S-acetyl-6-thio-, synthesis, 244
—, 1,2,3,2',3',4',6'-hepta-O-acetyl-6-O-methyl-, synthesis, 221
—, 1,2,6,2',3',4',6'-hepta-O-acetyl-3-O-(methylsulfonyl)-
desulfonylation, 230
proton nuclear magnetic resonance spectroscopy, 252
synthesis, 229
—, 1,2,3,2',3',4',6'-hepta-O-acetyl-6-O-p-tolylsulfonyl-, synthesis, 229
—, 1,2,3,6,2',3',4'-hepta-O-acetyl-6'-O-trityl-, synthesis, 219
—, 1,2,3,6,2',3',4'-hepta-O-acetyl-6'-O-trityl-α-D-, synthesis, 219
—, 1,2,6,2',3',4',6'-hepta-O-benzoyl-
ammonolysis, 31, 32
synthesis, 225
—, 1,2,3,2',3',4'-hexa-O-acetyl-6,6'-di-O-p-tolylsulfonyl-, synthesis, 227
—, 1,2,3,2',3',4'-hexa-O-acetyl-6,6'-di-O-trityl-, synthesis, 219

—, 1,2,3,2',3',4'-hexa-O-acetyl-6,6'-di-O-trityl-α-D-, synthesis, 219
—, 1,2,6,2',3',6'-hexa-O-benzoyl-, synthesis, 226
—, octa-O-acetyl-
 proton nuclear magnetic resonance spectroscopy, 250, 251
 reaction with phosphorus pentachloride, 35, 224
 with piperidine, 224
 selective deacylation, 28
 synthesis, 215
—, octa-O-benzoyl-
 ammonolysis, 31
 selective debenzoylation, 226, 227, 239
 synthesis, 225
—, 2,3,2',3'-tetra-O-acetyl-1,6-anhydro-4',6'-di-O-(methylsulfonyl)-, synthesis, 229
α-Maltoside, methyl, synthesis, 217, 218
—, phenyl, heptaacetate, synthesis, 218
—, phenyl 3-O-methyl-, synthesis, 222
—, phenyl 6-O-methyl-, synthesis, 221
—, phenyl 6'-O-methyl-, synthesis, 221
β-Maltoside, benzyl, synthesis, 218
—, benzyl 4',6'-O-benzylidene-, selective p-toluenesulfonylation, 228
—, benzyl 6'-trityl-, synthesis, 220
—, p-bromophenyl, synthesis, 218
—, ethyl hepta-O-acetyl-1-thio-, synthesis, 243
—, ethyl 1-thio-6'-O-trityl-, synthesis, 220
—, methyl
 reaction with sulfuryl chloride, 231–233
 selective benzoylation, 225, 226
 selective methanesulfonylation and p-toluenesulfonylation, 227, 228
 structure, 215, 216
 synthesis, 217
—, methyl 3',6'-anhydro-6-deoxy-, synthesis, 237
—, methyl 4,6-O-benzylidene-, synthesis, 237
—, methyl 4',6'-O-benzylidene-
 selective benzoylation, 225, 226
 selective p-toluenesulfonylation, 228
—, methyl 3,6:3',6'-dianhydro-, synthesis, 237

—, methyl 6,6'-dichloro-6,6'-dideoxy-, synthesis, 234
—, methyl 6,6'-di-O-p-tolylsulfonyl-, synthesis, 228
—, methyl 2,4,6,2',3',4',6'-hepta-O-acetyl-3-azido-3-deoxy-, proton nuclear magnetic resonance spectroscopy, 252
—, methyl 2,3,2',3',4',6'-hexa-O-acetyl-, synthesis, 235
—, methyl 2,6,2',3',4',6'-hexa-O-acetyl-
 oxidation to 3-keto derivative, 246, 247
 proton nuclear magnetic resonance spectroscopy, 252
—, methyl 2,6,2',3',4',6'-hexa-O-acetyl-3-azido-3-deoxy-, synthesis, 230
—, methyl 2,6,2',3',4',6'-hexa-O-acetyl-3-bromo-3-deoxy-, synthesis, 236, 237
—, methyl 2,3,6,2',3',6'-hexa-O-acetyl-4'-O-(methylsulfonyl)-, proton nuclear magnetic resonance spectroscopy, 252
—, methyl 2,6,2',3',4',6'-hexa-O-benzoyl-, synthesis, 226
—, methyl 2,3,2',3',4',6'-hexa-O-methyl-, preparation and oxidation, 222
—, methyl 3-O-(methylsulfonyl)-6'-trityl-, synthesis, 220
—, methyl 2,3,2',3',4'-penta-O-acetyl-6-deoxy-6-iodo-6'-O-p-tolylsulfonyl-, synthesis, 229
—, methyl 2,3,2',3',4'-penta-O-acetyl-6,6'-dichloro-6,6'-dideoxy-, mass spectrometry, 258–260
—, methyl 2,3,6,2',3'-penta-O-acetyl-4',6'-di-O-methyl-, mass spectrum, 261
—, methyl 2,3,2',3',4'-penta-O-acetyl-6,6'-di-O-(methylsulfonyl)-, synthesis, 227
—, methyl 2,6,2',3',6'-penta-O-benzoyl-, synthesis, 226
—, methyl 2,3,2',3'-tetra-O-acetyl-4',6'-O-benzylidene-6-O-p-tolylsulfonyl-, synthesis, 228
—, methyl 2,2',3',4'-tetra-O-acetyl-6,6'-di-O-p-tolylsulfonyl-, synthesis, 236
—, methyl 2,2',3',4'-tetra-O-acetyl-3-O-(methylsulfonyl)-6,6'-di-O-trityl-, proton nuclear magnetic resonance spectroscopy, 253

—, methyl 2,6,2',3'-tetra-O-benzoyl-4',6'-O-benzylidene-, synthesis, 226
—, methyl 6-O-p-tolylsulfonyl-, synthesis, 228
—, methyl 6'-O-p-tolylsulfonyl-, synthesis, 228
—, methyl 2,6,2'-tri-O-benzoyl-4',6'-O-benzylidene-, synthesis, 226
—, methyl 2,6,3'-tri-O-benzoyl-4',6'-O-benzylidene-, synthesis, 226
—, phenyl, heptaacetate, synthesis, 218
Maltosides, properties, 263
Maltosyl bromide, 2,3,6,2',3',4',6'-hepta-O-acetyl-, reaction with pyridine, 32
Maltosyl chloride, 6,6'-dichloro-6,6'-dideoxy-, penta(chlorosulfate), synthesis, 232
—, 3,6,2',3',4',6'-hexa-O-acetyl-2-O-(trichloroacetyl)-β-, structure, 224
Maltosyl halides, properties, 271
Maltotriose, structure, 357
β-Maltotriose hendecaacetate, reaction with phosphorus pentachloride, 35
Maltouronic acid, synthesis, 246
Mannan, carbon-13 nuclear magnetic resonance spectra, 200
Mannitol
  D-, hexanitrate, selective denitration, 57, 58
  1,2:5,6-tetrakis(diethylborinate) 3,4-ethylboronate, selective cleavage, 54, 55
  1,2:3,4:5,6-tris(ethylboronate), selective cleavage, 54
—, 1,4-anhydro-3,5-O-benzylidene-D-, isomerization, 26, 27
—, 6-O-(tert-butyldimethylsilyl)-1-O-(tert-butyldiphenylsilyl)-2,3:4,5-di-O-methylene-D-, selective hydrolysis, 64
—, 1-O-(tert-butyldiphenylsilyl)-2,3:4,5-di-O-methylene-D-, preparation, 64
—, 1,4:3,6-dianhydro-2,5-di-O-benzyl-D-, acetolysis, 45
—, 1,2:4,5-di-O-isopropylidene-D-, selective hydrolysis, 23
—, 1,2:5,6-di-O-isopropylidene-D-, 3,4-trithiocarbonate, acetobrominolysis, 114
—, 1,2,5,6-tetra-O-acetyl-3,4-O-methylene-D-, oxidation by chromium oxide, 83, 84

—, 1,2,5,6-tetra-O-benzoyl-3,4-O-isopropylidene-D-, oxidation by triphenylmethyl fluoroborate, 85
—, 1,3:2,5:4,6-tri-O-ethylidene-D-, selective hydrolysis, 22
—, 1,2:3,4:5,6-tri-O-isopropylidene-D-, selective hydrolysis, 22
—, 1,3:2,5:4,6-tri-O-methylene-D-acetolysis, selective, 25
selective hydrolysis, 22
Mannofuranose, 1,6-anhydro-β-D-, synthesis, 163
—, 2,3-O-cyclohexylidene-α-D-, synthesis, 76
—, 2,3:4,6-di-O-cyclohexylidene-α-D-, synthesis, 76
—, 2,3:5,6-di-O-isopropylidene-α-D-, acetobrominolysis, 114, 115
—, 2,3-O-isopropylidene-D-, acetobrominolysis, 115
Mannofuranoside, methyl 2,3-O-benzylidene-5,6-di-O-methyl-α-D-, hydrogenolysis, 127
—, methyl 2,6-di-O-benzyl-α-D-, synthesis, 127
—, methyl 2,3:5,6-di-O-benzylidene-α-D-, hydrogenolysis regioselectivity, 127
Mannofuranosyl bromide, 5-O-acetyl-6-bromo-6-deoxy-2,3-O-isopropylidene-α-D-, synthesis, 115
—, 2,3:5,6-di-O-isopropylidene-α-D-, reaction with base, 147
Mannopyranose, 1,6-anhydro-2,3-O-benzylidene-β-D-, reaction with N-bromosuccinimide, 107
—, 1,4-anhydro-6-deoxy-2,3-O-isopropylidene-α-L-, synthesis, 166
—, 1,2-anhydro-3,4,6-tri-O-benzyl-β-D-polymerization, 208, 209
synthesis, 171, 172
—, 1,3-anhydro-2,4,6-tri-O-benzyl-β-D-polymerization, 207
synthesis, 170
—, 1,6-anhydro-2,3,4-tri-O-benzyl-β-D-polymerization, 182, 187
synthesis, 163
—, 2,3:4,6-di-O-isopropylidene-D-, synthesis, 73
—, 4,6-O-isopropylidene-D-, synthesis, 73
—, 3,4,6-tri-O-benzyl-1,2-O-(1-methox-

SUBJECT INDEX

yethylidene)-β-D-, selective acetolysis, 46
Mannopyranoside, benzyl 2,3-di-O-benzyl-4,6-O-benzylidene-α-D-, hydrogenolysis regioselectivity, 128
—, benzyl exo-2,3:4,6-di-O-benzylidene-α-D-, hydrogenolysis and regioselectivity, 132
—, methyl 4-O-acetyl-6-deoxy-2,3-O-(o-nitrobenzylidene)-α-L-, irradiation, 89
—, methyl 4,6-O-benzylidene-α-D-, hydrogenolysis regioselectivity, 125
—, methyl 2,3-di-O-acetyl-4,6-O-benzylidene-α-D-, reaction with N-bromosuccinimide, 99
—, methyl 4,6-di-O-benzoyl-2,3-O-isopropylidene-α-D-, reaction with dibromomethyl methyl ether, 120
—, methyl 2,3:4,6-di-O-benzylidene-α-D-
  hydrogenolysis regioselectivity, 126, 133
  irradiation, 92
  reaction with N-bromosuccinimide, 103-106
    with butyllithium, 139, 140
    with triphenylmethyl fluoroborate, 112
  selective hydrolysis, 18
  synthesis, 78
—, methyl 2,3:4,6-di-O-isopropylidene-D-, selective hydrolysis, 23, 24
—, methyl 2,3:4,6-di-O-isopropylidene-α-D-, selective hydrolysis, 18, 19
—, methyl 2,3:4,6-di-O-(o-nitrobenzylidene)-α-D-, irradiation, 91
—, methyl 4,6-O-ethylidene-α-D-, 2,3-dinitrate, selective denitration, 56
—, methyl 2,3-O-isopropylidene-α-D-, synthesis, 79
—, methyl 4,6-O-isopropylidene-α-D-, synthesis, 79
Mannuronic acid, D-, and L-, isolation of crystalline anomers, 3, 4
Mass spectrometry, maltose and derivatives, 258-263
Melezitose, structure, 357
Melibiose
  structure, 359
  utilization by yeasts, 374-378
Metabolism

D-fucose, 328, 329
L-fucose, 327, 328
Methanesulfonylation, selective, of maltose and derivatives, 227
Methanesulfonyl chloride–N,N-dimethylformamide reaction, 233-235
1-Methylallyl group, as protecting group, 52
2-Methylallyl group, as protecting group, 52
Methylation, selective, of maltose derivatives, 220-223
Microanalysis, carbohydrate, 2
Monosaccharides, ammonolysis of perbenzoylated, 30, 31
Myeloma proteins, purification by affinity chromatography, 446, 447

N

(1-Naphthyldiphenylmethyl) group, reductive cleavage, 48
Neogitostin, 302
3-Nonene-2,5-diol, 5-butyl-1-methoxy-, synthesis, 141
3-Nonene-1,2,5-triol, 5-butyl-, synthesis, 142
Nucleophiles, reaction with benzoxonium ions, 113, 151, 152
Nucleophilic-displacement reactions, of sulfonates, 229-231
Nucleoside, 2′,3′-O-benzylidene-, reaction with N-bromosuccinimide, 103
Nucleosides
  regioselective ring cleavage of tetraisopropyldisiloxane-1,3-diyl derivatives, 68
  selective deacylation, 36-44
  synthesis, 153
Nucleotides, selective dephosphonylation, 59

O

Obituary, Karl Paul Gerhardt Link, 1-12
Odorotrioside G, 302
Oligodeoxyribonucleotides, synthesis, 64
Oligo-(1 → 6)-D-glucosidase
  regulation, 388-390
  substrate specificity, 387

Oligonucleotides, synthesis, by selective deacylation, 36
Oligoribonucleotides, synthesis, 64, 66
Oligosaccharides
  reductive-amination, affinity adsorbent preparation, 433, 434
  synthesis, 4-oxopentanoyl group hydrazinolysis, 42
  urinary, 330, 331
Ontogeny, fucolipid role, 309
Organolithium compounds, reaction with dioxolanes, 138
Orthoesters
  hydrogenolysis, 127
  hydrogenolysis of cyclic, cyclic acetal synthesis, 79
4-Oxopentanoyl group, as protecting group, 42
Oxygen, requirement for utilization of glycosides and galactose, 397–401
Ozonolysis, cyclic acetal, 81, 82

P

Panstroside, D-fucose-containing, 302
Pectin
  affinity chromatography adsorbent, 419
  methanolyzate, structure, 5
Pent-1-enitol, 1,2-dideoxy-4,5-O-isopropylidene-D-*erythro*-, synthesis, 147
Pent-4-*exo*-enopyranose, 4-deoxy-1,2-O-isopropylidene-β-D-*threo*-, synthesis, 142
Pentitol, 1,1-bis(benzamido)-1-deoxymono-O-benzoyl-, preparation, 31
2-Pentulofuranose, 1,2:3,4-di-O-isopropylidene-β-D-, selective hydrolysis, 16, 17
Permeases, glycoside entry into yeast, 379–381
2-(*o*-Phenylenedioxy)acetyl group, enzymic hydrolysis, 44
Phenyllithium, reaction with dioxolanes, 138
Photolysis, cyclic acetals, 86–95
Piperidine, 1-(3,6,2′,3′,4′,6′-hexa-O-acetyl-β-cellobiosyl)-, preparation, 30
—, 1-(3,4,6-tri-O-acetyl-D-glucopyranosyl)-, preparation, 30
Plasmalemma
  glycoside passage, 349, 352
  glycosides, hydrolyzed inside, 402–404
  hydrolyzed outside, 353–378, 401, 402
  utilization of glycosides hydrolyzed inside, 378–397
Poly(acrylamide) beads, affinity chromatography adsorbent, 407, 416, 417, 436
Polymerization, *see also* Copolymerization
  of anhydroaldoses, mechanism, 173–176
  of anhydro sugars, 158–160, 173–212
  of cyclic acetals, 152
Polysaccharides
  bacterial, L-fucose, 301
  D-fucopyranosides, 302
  stereoregular, synthesis, 158, 199
Polystyrene resins, in acetal synthesis, 79
Potassium permanganate, cyclic acetal oxidation, 82
Propene, 2-acetoxy-, in acetal synthesis, 76
1-Propenyl group, as protecting group, 50
Protecting groups
  cyclic acetals, 72, 155
  selective removal, 13–70
Proteins, *see* Myeloma proteins
Proton nuclear magnetic resonance spectra
  aldopyranans, 198
  maltose and derivatives, 249–253
Pyran, 2-alkoxytetrahydro-, photolysis, 95
Pyrazole, 3(5)-(1-deoxy-1,2:4,5-di-O-isopropylidene-D-*manno*-pentitol-1-yl)-, isomerization, 27, 28
Pyridinium bromide, 3,6,2′,3′,4′,6′-hexa-O-acetylmaltosyl-, preparation, 32
—, 3,6,3′,4′,6′-penta-O-acetyllactosyl-, preparation, 32
—, 3,3′,4′-tri-O-acetyllactosyl-, preparation, 32

R

Raffinose
  structure, 359

utilization by yeasts, 375–378
Reactivity, of cyclic acetals of aldoses and aldosides, 71–156
Rearrangement, peroxide-induced, of acetals to esters, 150
Regioselectivity, hydrogenolysis, 123, 124
Rhamnitol, 1,2:3,4-di-O-isopropylidene-L-, selective hydrolysis, 22
Rhamnofuranose, 1,5-di-O-acetyl-L-, preparation, 55
Rhamnofuranoside, methyl 2,3-O-isopropylidene-α-L-, synthesis, 166
—, methyl 2,3-O-isopropylidene-β-L-, synthesis, 167
Rhamnopyranose, L-, 1,2:3,5-bis(ethylboronate), selective cleavage, 55
Rhamnopyranoside, benzyl 4-O-allyl-2,3-O-benzylidene-L-, exo-, hydrogenolysis regioselectivity, 132
—, benzyl 4-O-benzyl-2,3-O-benzylidene-L-, $endo$- and exo-, hydrogenolysis regioselectivity, 132
—, benzyl 2,3-O-benzylidene-L-, $endo$- and exo-, hydrogenolysis regioselectivity, 132
—, benzyl exo-2,3-O-benzylidene-4-O-(2,3,4,6-tetra-O-benzyl-α-D-galactopyranosyl)-α-L-, synthesis, 133
—, methyl 4-O-benzoyl-2,3-isopropylidene-α-L-, reaction with dibromomethyl methyl ether, 120
—, methyl 2,3-O-benzylidene-α-L- reaction with butyllithium, 142, 143 synthesis, 78
—, methyl 2,3-O-benzylidene-5-O-acetyl-α-L-, reaction with N-bromosuccinimide, 107
—, methyl 4-O-methyl-2,3-di-O-(trifluoroacetyl)-α-L-, selective alcoholysis, 29
Rhamnopyranosyl chloride, 2,3-O-isopropylidene-4-O-methyl-α-L-, reaction with base, 147
Ribitol, 5-O-benzoyl-4-C-(benzoyloxymethyl)-1,2:3,4-di-O-isopropylidene-L-, selective hydrolysis, 23
Riboflavine 5'-(alkyl phosphate), preparation, 60
Ribofuranose, 1,5-anhydro-β-D-, synthesis, 165

—, 1,5-anhydro-2,3-O-benzylidene-β-D- polymerization, 206 synthesis, 165
—, 3,5-di-O-benzoyl-1,2-O-(diphenylmethylidene)-α-D-, reaction with 2,4-bis(trimethylsilyl)oxypyrimidine and stannic chloride, 153
—, 1,5-di-O-trityl-D-, selective cleavage, 47
—, 2,3-O-isopropylidene-D-, synthesis, 75
—, 1,3(2),5-tri-O-trityl-D-, selective cleavage, 47
—, 1,3(2),5-tri-O-trityl-2(3)-O-acetyl-D-, selective cleavage, 47, 48
Ribofuranoside, methyl 5-O-benzoyl-2,3-O-isopropylidene-β-D-, reaction with dibromomethyl methyl ether, 120
—, methyl 2,3-O-benzylidene-β-D-, irradiation, 88
—, methyl 2,3-O-benzylidene-5-O-methyl-β-D-, reaction with N-bromosuccinimide, 101
Ribofuranosyl cyanide, 5-O-benzoyl-β-D-, preparation, 39
—, 2,3,5-tri-O-benzoyl-β-D-, selective debenzoylation, 39
Ribonucleoside, 3'-O-acyl-2'-O-(4-methoxytetrahydropyran-4-yl)-, preparation, 38
Ribopyranose, 1,2:3,4-di-O-isopropylidene-5-thio-α-D-, selective hydrolysis, 17
—, 3,4-O-isopropylidene-D-, synthesis, 75
Ribopyranoside, methyl 5-O-benzoyl-2,3-O-benzylidene-β-D- reaction with tetraethylammonium p-toluenesulfonate, 152 with triphenylmethyl fluoroborate, 108, 152
—, methyl 3,4-O-(R)-benzylidene-β-D-, isomerization, 26
Ribose
D-, 1,2:3,4-diacetal, selective hydrolysis, 16
diethyl dithioacetal, irradiation, 94
—, 2,4:3,5-di-O-benzylidene-D-, ethylene dithioacetal, selective hydrolysis, 22
—, 1,2:3,4-di-O-isopropylidene-5-thio-D-, synthesis, 77
Rodenticide, Warfarin, discovery, 1

S

Saccharomyces cerevisiae, sugar utilization, 348
Salicin, structure, 358
Starch, affinity chromatography adsorbent, 419
Starch Round Table, 9
Stereoselectivity, hydrogenolysis, 124–135
Strebloside, 302
Streptamine, preparation, selective deacetylation, 34
Strospeside, 302
Succinimide, N-bromo-, cyclic acetal bromination, 96–107
Succinimide, N-chloro-, reaction with cyclic acetals, 118
Sucrose
  structure, 356
  utilization by yeasts, delayed, 374
—, 2,1′:4,6-di-O-isopropylidene-, tetraacetate, ammonolysis, 33
—, 2,3,4,1′,3′,4′-hexa-O-benzoyl-6,6′-di-O-(tert-butyldimethylsilyl)-, selective cleavage, 67
—, octa-O-acetyl-, selective deacylation, 28
—, octa-O-benzoyl-, selective deacylation, 32
—, tetra-O-acetyl-2,1′:4,6-di-O-isopropylidene-, selective hydrolysis, 17, 18
Sugars, see also Amino sugars; Anhydro sugars
  utilization by yeasts, 347–404
Sulfonates, nucleophilic-displacement reactions, 229–231
Sulfuryl chloride reaction, maltose derivatives, 231–233
Sweet-clover disease, cattle, 2, 5, 6

T

Talitol, 1,3:2,4:5,6-tri-O-methylene-D-, acetolysis, selective, 24
Talofuranose, 1,6-anhydro-α-D-, synthesis, 163
Talofuranoside, methyl 6-deoxy-2,3-O-isopropylidene-α-L-, synthesis, 166
Talopyranose, 1,4-anhydro-6-deoxy-2,3-O-isopropylidene-β-L-, synthesis, 165, 166
—, 1,6-anhydro-3,4-O-isopropylidene-β-D-, hydrolysis and isomerization, 26
—, 4,6-O-isopropylidene-D-, synthesis, 73
Tetradeca(ribonucleotide), synthesis, 42
Tetritol, 2-deoxy-DL-glycero-, 4-diethylborinate 1,3-ethylboronate, selective cleavage, 54
—, 1,3:2,4-di-O-benzylidene-, selective hydrolysis, 20
Threitol, 1,3:2,4-di-O-benzylidene-L-, selective hydrolysis, 19
Threonine, β-D-glucopyranosyl-(1 → 3)-α-L-fucopyranosyl-(1 → 3)-L-, isolation and synthesis, 330
Thymidine
  3′,5′-cyclic phosphate, hydrolysis, 61
  3′- and 5′-(trialkylsilyl) ethers, acid hydrolysis, 65
—, 3′-O-acetyl-, preparation, 41
—, 5′-O-acetyl-
  hydrolysis, 39
  preparation, 41, 43
—, 3′-O-acetyl-5′-O-(chloroacetyl)-, selective deacylation, 43
—, 5′-O-acetyl-3′-O-(2,4-dinitrophenylsulfenyl)-, selective deacylation, 43
—, 3′-O-acetyl-5′-O-(4-methoxycrotonoyl)-, selective deacylation, 39
—, 3′-O-acetyl-5′-O-(2,2,2-tribromoethoxycarbonyl)-, selective deacylation, 41
—, 5′-O-acetyl-3′-O-(2,2,2-tribromoethoxycarbonyl)-, selective deacylation, 41
—, 3′-O-(p-anisyldiphenylmethyl)-5′-O-(1-naphthyldiphenylmethyl)-, selective cleavage, 48
—, 5′-(3-benzoylpropanoyl)-, preparation, 41
—, 3′,5′-di-O-acetyl-, selective deacylation, 36
—, 5′-O-pivaloyl-, hydrolysis, 39
Thymidine 3′-phosphate, 5′-O-(2,4,6-trimethylbenzoyl)-, hydrolysis, 39
p-Toluenesulfonylation, selective, of maltose and derivatives, 227–229

SUBJECT INDEX

Transacetalation, 78
α,α-Trehalose
  structure, 353, 358
  utilization by yeasts, 390, 391
—, 6,6'-di-O-hexadecanoyl-, preparation, 63
—, 2,3,2',3'-tetra-O-benzoyl-4,6:4',6'-di-O-benzylidene-, selective methanolysis, 19
Trialkylsilyl group, selective removal, 62–70
1,3,5-Triazine-2,4,6-trione, 1,3,5-trichloro-, reaction with cyclic acetals, 118
2,2,2-Tribromoethoxycarbonyl group, non-hydrolytic removal, 41
Trifluoroacetyl group, as blocking group, 36
3,7,9-Trioxabicyclo[4.2.1]nonanes, hydrogenolysis, 134
3,6,8-Tioxabicyclo[3.2.1]octanes, hydrogenolysis, 134
Triphenylmethyl fluoroborate
  acetal oxidation, 85, 86
  cyclic acetal oxidation and halogenation, 107–113, 152
Trisaccharides, synthesis, 53
Tritylation, selective, of maltose and derivatives, 219, 220
Tritylone group, as protecting group, 48, 49
Tumorigenesis
  fucolipid role, 309
  fucosyltransferase levels, 331
  glycopeptides, 305
Turanose, structure, 357

U

Uracil, 1-(3-deoxy-3-iodo-5-O-trityl-β-D-xylofuranosyl)-, preparation, 48
Urea, 1-(2-chloroethyl)-3-(hepta-O-acetyl-β-maltosyl)-, synthesis, 243
—, 1-(2-chloroethyl)-3-(hepta-O-acetyl-β-maltosyl)-1-nitroso-, synthesis and antitumor activity, 243
—, 1-(2-chloroethyl)-3-β-maltosyl-, synthesis, 243
—, 1-(2-chloroethyl)-3-β-maltosyl-1-nitroso-, synthesis and antitumor activity, 243
Uridine, 3',5'-cyclic phosphate, hydrolysis, 61
—, 5'-O-acetyl-, deacylation rate, 38
—, 3'-O-acetyl-5'-O-pivaloyl-2'-O-(tetrahydropyran-4-yl)-, selective deacylation, 39
—, 2'-O-acetyl-3',5'-O-(tetraisopropyldisiloxane-1,3-diyl)-, selective cleavage, 68
—, 5'-O-acyl-, ammonolysis half-times, 38
—, 5'-O-(p-anisyldiphenylmethyl)-, hydrolysis rate, 47
—, 5'-O-(chloroacetyl)-, deacylation rate, 38
—, 2'-deoxy-3',5'-di-O-(dihydrocinnamoyl)-, enzymic hydrolysis, 43
—, 2',3'-di-O-acetyl-, synthesis, 37
—, 3',5'-di-O-acetyl-2'-deoxy-2'-halo-, regioselective deacylation, 41
—, 3',5'-di-O-benzoyl-, synthesis, 153
—, 2',5'-di-O-trityl-, selective detritylation, 48
—, 3',5'-di-O-trityl-, selective detritylation, 48
—, 5'-O-formyl-, deacylation rate, 38
—, 5'-O-(methoxyacetyl)-, deacylation rate, 38
—, 5'-O-(phenoxyacetyl)-, deacylation rate, 38
—, 2',3'-O-(tetraisopropyldisiloxane-1,3-diyl)-, preparation, 69
—, 2',3',5'-tri-O-acetyl-, selective deacylation, 36
—, 5'-O-(tri-p-anisylmethyl)-, hydrolysis rate, 47
—, 2',3',5'-tri-O-benzoyl-, selective deacylation, 39
—, 2',3',5'-tri-O-formyl-, methanolysis, selective, 37
—, 5'-O-trityl-, hydrolysis rate, 47
Uronic acids, assay, 3

V

Vinyl sulfone, *see* Divinyl sulfone

W

Warfarin, discovery, 1

X

Xylitol, DL-, 4-diethylborinate 1,2:3,5-bis(ethylboronate), selective cleavage, 54
Xylofuranose, 3,5-anhydro-1,2-O-isopropylidene-α-D-
  polymerization, 211
  synthesis, 173
—, 3,5-O-benzylidene-1,2-O-isopropylidene-α-D-, irradiation, 88
—, 2,3-di-O-acetyl-1,5-anhydro-β-D-, synthesis, 172
—, 2,3-di-O-acetyl-5-O-(chloroacetyl)-D-, selective deacylation, 35
—, 2,3-di-O-benzoyl-1,5-di-O-trityl-D-, selective cleavage, 47
—, 1,2:3,5-di-O-cyclohexylidene-α-D-, acetal cleavage by Grignard reagent, 155
—, 1,2:3,5-di-O-isopropylidene-α-D-, irradiation, 88
—, 1,2:3,5-di-O-methylene-α-D-
  deacetalation, 24
  oxidation by potassium permanganate, 82
—, 1,2-O-isopropylidene-α-D-, 3,5-cyclic phosphate, deacetalation and hydrolysis, 62
Xylofuranoside, methyl 2-bromo-2-deoxy-3-O-formyl-5-O-benzoyl-β-D-, synthesis, 120

—, methyl 2,5-di-O-benzoyl-3-bromo-3-deoxy-D-, synthesis, 109
Xylopyranose, 3-O-benzoyl-1,2-O-benzylidene-4-bromo-4-deoxy-α-L-, synthesis, 112
—, 1,2:3,4-di-O-isopropylidene-5-thio-α-D-, selective hydrolysis, 17
Xylopyranoside, methyl β-D-, 3-diethylborinate 2,4-ethylboronate, selective cleavage, 55
—, methyl 2,3-di-O-benzoyl-4-bromo-4-deoxy-α-L-, synthesis, 111
Xylose
  D-, 1,2:3,5-diacetal, selective hydrolysis, 16
  diethyl dithioacetal, irradiation, 94
  L-, 1,2:3,5-diacetal, selective hydrolysis, 16
—, 2,4:3,5-di-O-benzylidene-D-, diethyl dithioacetal, selective hydrolysis, 22
—, 2,3:4,5-di-O-cyclohexylidene-D-, ethylene acetal, synthesis, 76
—, 2,3:4,5-di-O-isopropylidene-D-, dialkyl dithioacetals, selective hydrolysis, 22
—, 2,4:3,5-di-O-isopropylidene-D-, diethyl dithioacetal, selective hydrolysis, 24

Y

Yeast cell, see Plasmalemma
Yeast cell-wall, diagram, 368
Yeasts, sugar utilization, 347–404

RAYMOND H. FOGLER LIBRARY

DATE DUE